Tenth Edition

Electronics
Principles & Applications

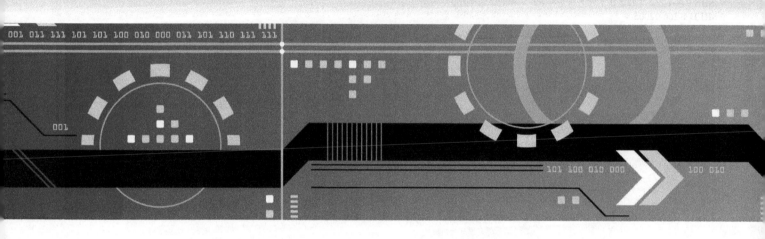

Charles A. Schuler

McGraw Hill

ELECTRONICS

Published by McGraw Hill LLC, 1325 Avenue of the Americas, New York, NY 10019. Copyright ©2024 by
McGraw Hill LLC. All rights reserved. Printed in the United States of America. No part of this publication may
be reproduced or distributed in any form or by any means, or stored in a database or retrieval system, without
the prior written consent of McGraw Hill LLC, including, but not limited to, in any network or other electronic
storage or transmission, or broadcast for distance learning.

Some ancillaries, including electronic and print components, may not be available to customers outside the
United States.

This book is printed on acid-free paper.

2 3 4 5 6 7 8 9 CPI 28 27 26 25 24

ISBN 978-1-266-22005-0
MHID 1-266-22005-4

Cover Image: *Image Source/Perry Mastrovito*

mheducation.com/highered

Contents

Preface

Electronics: Principles and Applications, 10e, introduces analog devices, circuits, and systems. It also presents various digital techniques that are now commonly used in what was once considered the sole domain of analog electronics. It is intended for students who have a basic understanding of Ohm's law; Kirchhoff's laws; power; schematic diagrams; and basic components such as resistors, capacitors, and inductors. The digital material is self-contained and will not pose a problem for those students who have not completed a course in digital electronics. The only mathematics prerequisite is a command of basic algebra.

The major objective of this text is to provide entry-level knowledge and skills for a wide range of occupations in electricity and electronics. Its purpose is to assist in the education and preparation of technicians who can effectively diagnose, repair, verify, install, and upgrade electronic circuits and systems. It also provides a solid and practical foundation in analog electronic concepts, device theory, and modern digital solutions for those who may need or want to go on to more advanced study.

The tenth edition, like the earlier ones, combines theory and applications in a logical, evenly paced sequence. It is important that a student's first exposure to electronic devices and circuits be based on a smooth integration of theory and practice. This approach helps the student develop an understanding of how devices such as diodes, transistors, and integrated circuits function and how they are used in practice. Then the understanding of these functions can be applied to the solution of practical problems such as performance analysis and troubleshooting.

This is an extremely practical text. The devices, circuits, and applications are typical of those used in all phases of electronics. Reference is made to common aids such as Internet sites, component identification systems, and substitution guides, and real-world troubleshooting techniques are applied whenever appropriate. The information, theory, and calculations presented are the same as those used by practicing technicians. The formulas presented are immediately applied in examples that make sense and relate to the kinds of calculations actually made by technical workers.

The 16 chapters progress from an introduction to the broad field of electronics through solid-state theory, transistors, and the concepts of gain, amplifiers, oscillators, electronic communications and data transfer, integrated circuits, control circuitry,

regulated power supplies, and digital signal processing. As an example of the practicality of the text, an entire chapter is devoted to troubleshooting circuits and systems. In other chapters, entire sections are devoted to this vital topic. Since the last edition, the electronics industry has continued its march toward more digital and mixed-signal applications to replace what used to be purely analog functions. The distinction between analog and digital continues to blur. This is the only text of its kind that addresses this issue.

New to this Edition

This edition introduces significant new material in the areas of software updates, integrated circuit communication protocols and power factor correction. Software updates are becoming more of an issue as devices rely more on computers and microcontrollers. It is possible for an update to go awry and render a device inoperable. Mixed signal circuitry is expanded in this edition. The division between the analog and digital worlds has become blurred into non-existence. Chip to chip communication protocols have been added for a better understanding of modern circuitry. Finally, power factor correction has been added. This technology is now becoming commonplace as manufacturers come into compliance with newer standards such as Energy Star Plus put forth by the U.S. Environmental Protection Agency.

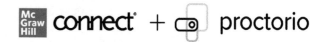

This edition is available online with Connect, McGraw Hill Education's integrated assignment and assessment platform. Connect also offers SmartBook for the new edition, which is the first adaptive reading experience proven to improve grades and help students study more effectively. All of the title's website and ancillary content is also available through Connect, including extra review questions, links to industry sites, chapter study overviews, assignments, the Instructor's Manual, and a MultiSim Primer, all for students. The following is a list of features that can be found through Connect.

Instructor Resources

MultiSim Primer (by Patrick Hoppe of Gateway Technical College), which provides a tutorial for new users of the software

Instructor's Manual

PowerPoint presentations for classroom use

Electronic test bank questions for each chapter

Parts and equipment lists

Learning Outcomes

Answers to textbook questions:
- Chapter review questions
- Critical thinking questions

Answers and data for lab experiments and assignments

Projects

Instrumentation lab experiments in .pdf format

Soldering (.pdf file)

Circuit simulation files (EWB 5 and MultiSim versions 6, 7, 8, 11, and 14)

Experiments Manual

A correlated Experiments Manual provides a wide array of hands-on labwork, problems, and circuit simulations. MultiSim files are provided for both the simulation activities and the hands-on activities. These files are located on the Student Side of the resources in Connect.

About the Author

Charles A. Schuler received his Ed.D. from Texas A&M University in 1966, where he was an N.D.E.A. fellow. He has published many articles and seven textbooks on electricity and electronics, almost as many laboratory manuals, and another book that deals with ISO 9000. He taught electronics technology and electrical engineering technology at California University of Pennsylvania for 30 years. He is currently a full-time writer, as he continues his passion to make the difficult easy to understand.

Instructors
Student Success Starts with You

Tools to enhance your unique voice

Want to build your own course? No problem. Prefer to use an OLC-aligned, prebuilt course? Easy. Want to make changes throughout the semester? Sure. And you'll save time with Connect's auto-grading, too.

65%
Less Time Grading

Laptop: Getty Images; Woman/dog: George Doyle/Getty Images

A unique path for each student

In Connect, instructors can assign an adaptive reading experience with SmartBook® 2.0. Rooted in advanced learning science principles, SmartBook 2.0 delivers each student a personalized experience, focusing students on their learning gaps, ensuring that the time they spend studying is time well-spent.
mheducation.com/highered/connect/smartbook

Affordable solutions, added value

Make technology work for you with LMS integration for single sign-on access, mobile access to the digital textbook, and reports to quickly show you how each of your students is doing. And with our Inclusive Access program, you can provide all these tools at the lowest available market price to your students. Ask your McGraw Hill representative for more information.

Solutions for your challenges

A product isn't a solution. Real solutions are affordable, reliable, and come with training and ongoing support when you need it and how you want it. Visit **supportateverystep.com** for videos and resources both you and your students can use throughout the term.

Students
Get Learning that Fits You

Effective tools for efficient studying

Connect is designed to help you be more productive with simple, flexible, intuitive tools that maximize your study time and meet your individual learning needs. Get learning that works for you with Connect.

Study anytime, anywhere

Download the free ReadAnywhere® app and access your online eBook, SmartBook® 2.0, or Adaptive Learning Assignments when it's convenient, even if you're offline. And since the app automatically syncs with your Connect account, all of your work is available every time you open it. Find out more at **mheducation.com/readanywhere**

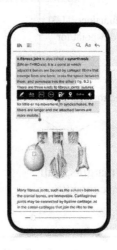

"I really liked this app—it made it easy to study when you don't have your textbook in front of you."

- Jordan Cunningham, Eastern Washington University

iPhone: Getty Images

Everything you need in one place

Your Connect course has everything you need—whether reading your digital eBook or completing assignments for class, Connect makes it easy to get your work done.

Learning for everyone

McGraw Hill works directly with Accessibility Services Departments and faculty to meet the learning needs of all students. Please contact your Accessibility Services Office and ask them to email accessibility@mheducation.com, or visit **mheducation.com/about/accessibility** for more information.

Walkthrough

Electronics: Principles and Applications takes a concise and practical approach to this fascinating subject. The textbook's easy-to-read style, color illustrations, and basic math level make it ideal for students who want to learn the essentials of modern electronics and apply them to real job-related situations.

Learning Outcomes

This chapter will help you to:

1-1 *Identify* some major events in the history of electronics. [1-1]

1-2 *Classify* circuit operation as digital or analog. [1-2]

1-3 *Name* major analog circuit functions. [1-3]

1-4 *Begin* developing a system viewpoint for troubleshooting. [1-3]

1-5 *Analyze* circuits with both dc and ac sources. [1-4]

1-6 *List* the current trends in electronics. [1-5]

Each chapter starts with *Learning Outcomes* that give the reader an idea of what to expect in the following pages, and what he or she should be able to accomplish by the end of the chapter. These outcomes are distinctly linked to the chapter subsections.

Key Terms, noted in the margins, call the reader's attention to key concepts.

1-2   Digital or Analog

Today, electronics is such a huge field that it is often necessary to divide it into smaller subfields. You will hear terms such as medical electronics, instrumentation electronics, automotive electronics, avionics, consumer electronics, industrial electronics, and others. One way that electronics can be divided is into digital or analog.

Digital electronic device

A *digital electronic device* or circuit will recognize or produce an output of only several limited states. For example, most digital circuits will respond to only two input conditions: low or high. *Digital circuits* may also be called *binary* since they are based on a number system with only two digits: 0 and 1.

Digital circuit

Analog circuit

Linear circuit

An *analog circuit* can respond to or produce an output for an infinite number of states. An analog input or output might vary between 0 and 10 volts (V). Its actual value could be 1.5, 3.8, or even 5.2 V. In th___ *infinite* ___

EXAMPLE I-I

An audio compact disk (CD) uses 16 bits to represent each sample of the signal. How many steps or volume levels are possible? Use the appropriate power of 2:

$$2^{16} = 65,536$$

This is easy to solve using a calculator with an x^y key. Press 2, then x^y, and then 16 followed by the = key.

Numerous solved *Example* problems throughout the chapters demonstrate the use of formulas and the methods used to analyze electronic circuits.

HISTORY OF ELECTRONICS

Niels Bohr and the Atom

Scientists change the future by improving on the ideas of others. Niels Bohr proposed a model of atomic structure in 1913 that applied energy levels (quantum mechanics) to the Rutherford model of the atom. Bohr also used some of the work of Max Planck.

Everett Collection Historical/Alamy Stock Photo

You May Recall

Chokes are so named because they "choke off" high-frequency current flow.

History of Electronics, *You May Recall*, and *About Electronics* add historical depth to the topics and highlight new and interesting technologies or facts.

ABOUT ELECTRONICS

Materials Used for Dopants, Semiconductors, and Microwave Devices

- Gallium arsenide (GaAs) works better than silicon in microwave devices because it allows faster movement of electrons.
- Materials other than boron and arsenic are used as dopants.
- It is theoretically possible to make semiconductor devices from crystalline carbon.
- Crystal radio receivers were an early application of semiconductors.

All critical facts and principles are reviewed in the **Summary and Review** section at the end of each chapter.

Chapter 1 Summary and Review

Summary

1. Electronics is a relatively young field. Its history began in the 20th century.
2. Electronic circuits can be classified as digital or analog.
3. The number of states or voltage levels is limited in a digital circuit (usually to two).
4. An analog circuit has an infinite number of voltage levels.
5. In a linear circuit, the output signal is a replica of the input.
6. All linear circuits are analog, but not all analog circuits are linear. Some analog circuits distort

10. The number of output levels from a D/A converter is equal to 2 raised to the power of the number of bits used.
11. Digital signal processing uses computers to enhance signals.
12. Block diagrams give an overview of electronic system operation.
13. Schematic diagrams show individual part wiring and are usually required for component-level troubleshooting.
14. Troubleshooting begins at the system level.
15. Alternating current and direct current signals are

All of the important chapter formulas are summarized at the end of each chapter in **Related Formulas**. **Chapter Review Questions** are found at the end of each chapter; and separate, more challenging **Chapter Review Problems** sections are available in appropriate chapters.

Related Formulas

Number of levels in a binary system: levels $= 2^n$

Inductive reactance: $X_L = 2\pi f L$

Capacitive reactance: $X_C = \dfrac{1}{2\pi f C}$

Chapter Review Questions

Determine whether each statement is true or false.

1-1. Most digital circuits can output only two states, high and low. (1-2)
1-2. Digital circuit outputs are usually sine waves. (1-2)
1-3. The output of a linear circuit is an exact replica of the input. (1-2)
1-4. Linear circuits are classified as analog. (1-2)
1-5. All analog circuits are linear. (1-2)

1-6. The output of a 4-bit D/A converter can produce 128 different voltage levels. (1-2)
1-7. An attenuator is an electronic circuit used to make signals stronger. (1-3)
1-8. Block diagrams are best for component-level troubleshooting. (1-3)
1-9. In Fig. 1-8, if the signal at point 4 is faulty, then the signal at point 3 must also be faulty. (1-3)
1-10. Refer to Fig. 1-8. The power supply should be checked first. (1-3)

Finally, each chapter ends with *Critical Thinking Questions* and *Answers to Self-Tests*.

Answers to Self-Tests

1. T	7. T	13. F	19. capacitors
2. T	8. F	14. F	20. bypass
3. F	9. T	15. T	21. coupling (dc block)
4. T	10. F	16. F	22. F
5. F	11. F	17. −7.5 V	23. T
6. T	12. T	18. 12.5 V, 0 V	24. F

Critical Thinking Questions

1-1. Functions now accomplished by using electronics may be accomplished in different ways in the future. Can you think of any examples?

1-2. Can you describe a simple system that uses only two wires but will selectively signal two different people?

1-3. What could go wrong with capacitor C_2 in Fig. 1-10, and how would the fault affect the waveform at Node D?

1-4. What could go wrong with capacitor C_2 in Fig. 1-13, and how would the fault affect the waveform at Node D?

Acknowledgments

Where does one begin? This book was started with a research project. It was tested both in industry and education. The testing led to many refinements that produced a truly effective textbook. Many people contributed to that effort . . . both in education and in industry. Their dedication and diligence helped launch what has become a remarkably long-lasting and practical work. Then, there are all those instructors and students who have given sage and thoughtful advice over the years. And there are those gifted and hardworking folks at McGraw Hill. Finally, there is my family, who indulge my passion and encourage my efforts.

Safety

Electric and electronic circuits can be dangerous. Safe practices are necessary to prevent electrical shock, fires, explosions, mechanical damage, and injuries resulting from the improper use of tools.

Perhaps the greatest hazard is electrical shock. A current through the human body in excess of 10 milliamperes can paralyze the victim and make it impossible to let go of a "live" conductor or component. Ten milliamperes is a rather small amount of current flow: It is only *ten one-thousandths* of an ampere. An ordinary flashlight can provide more than 100 times that amount of current!

Flashlight cells and batteries are safe to handle because the resistance of human skin is normally high enough to keep the current flow very small. For example, touching an ordinary 1.5-V cell produces a current flow in the microampere range (a microampere is one one-millionth of an ampere). This amount of current is too small to be noticed.

High voltage, on the other hand, can force enough current through the skin to produce a shock. If the current approaches 100 milliamperes or more, the shock can be fatal. Thus, the danger of shock increases with voltage. Those who work with high voltage must be properly trained and equipped.

When human skin is moist or cut, its resistance to the flow of electricity can drop drastically. When this happens, even moderate voltages may cause a serious shock. Experienced technicians know this, and they also know that so-called low-voltage equipment may have a high-voltage section or two. In other words, they do not practice two methods of working with circuits: one for high voltage and one for low voltage. They follow safe procedures at all times. A 5-V power supply could have internal circuitry that operates at 400 V! They do not assume protective devices are working. They do not assume a circuit is off even though the switch is in the OFF position. They know the switch could be defective.

Even a low-voltage, high-current-capacity system like an automotive electrical system can be quite dangerous. Short-circuiting such a system with a ring or metal watchband can cause very severe burns—especially when the ring or band welds to the points being shorted.

As your knowledge and experience grow, you will learn many specific safe procedures for dealing with electricity and electronics. In the meantime,

1. Always follow procedures.
2. Use service manuals as often as possible. They often contain specific safety information. Read, and comply with, all appropriate material safety data sheets.
3. Investigate before you act.
4. When in doubt, *do not act*. Ask your instructor or supervisor.

General Safety Rules for Electricity and Electronics

Safe practices will protect you and your fellow workers. Study the following rules. Discuss them with others, and ask your instructor about any you do not understand.

1. Do not work when you are tired or taking medicine that makes you drowsy.
2. Do not work in poor light.
3. Do not work in damp areas or with wet shoes or clothing.
4. Use approved tools, equipment, and protective devices.
5. Avoid wearing rings, bracelets, and similar metal items when working around exposed electric circuits.
6. Never assume that a circuit is off. Double-check it with an instrument that you are sure is operational.
7. Some situations require a "buddy system" to guarantee that power will not be turned on while a technician is still working on a circuit.
8. Never tamper with or try to override safety devices such as an interlock (a type of switch that automatically removes power when a door is opened or a panel removed).
9. Keep tools and test equipment clean and in good working condition. Replace insulated probes and leads at the first sign of deterioration.
10. Some devices, such as capacitors, can store a *lethal* charge. They may store this charge for long periods

of time. You must be certain these devices are discharged before working around them.

11. Do not remove grounds, and do not use adaptors that defeat the equipment ground.

12. Use only an approved fire extinguisher for electrical and electronic equipment. Water can conduct electricity and may severely damage equipment. Carbon dioxide (CO_2) or halogenated-type extinguishers are usually preferred. Foam-type extinguishers may also be desired in *some* cases. Commercial fire extinguishers are rated for the type of fires for which they are effective. Use only those rated for the proper working conditions.

13. Follow directions when using solvents and other chemicals. They may be toxic or flammable, or they may damage certain materials such as plastics. Always read and follow the appropriate material safety data sheets.

14. A few materials used in electronic equipment are toxic. Examples include tantalum capacitors and beryllium oxide transistor cases. These devices should not be crushed or abraded, and you should wash your hands thoroughly after handling them. Other materials (such as heat shrink tubing) may produce irritating fumes if overheated. Always read and follow the appropriate material safety data sheets.

15. Certain circuit components affect the safe performance of equipment and systems. Use only exact or approved replacement parts.

16. Use protective clothing and safety glasses when handling high-vacuum devices such as picture tubes and cathode-ray tubes.

17. Don't work on equipment before you know proper procedures and are aware of any potential safety hazards.

18. Many accidents have been caused by people rushing and cutting corners. Take the time required to protect yourself and others. Running, horseplay, and practical jokes are strictly forbidden in shops and laboratories.

19. Never look directly into light-emitting diodes or fiber-optic cables. Some light sources, although invisible, can cause serious eye damage.

20. Lithium batteries can explode and start fires. They must be used only as intended and only with approved chargers. Lead-acid batteries produce hydrogen gas, which can explode. They too must be used and charged properly.

Circuits and equipment must be treated with respect. Learn how they work and the proper way of working on them. Always practice safety: your health and life depend on it.

Electronics workers use specialized safety knowledge.

Yamato1987/IStock/Getty Images; suphakit73/Shutterstock

Introduction

Learning Outcomes

This chapter will help you to:

1-1 *Identify* some major events in the history of electronics. [1-1]

1-2 *Classify* circuit operation as digital or analog. [1-2]

1-3 *Name* major analog circuit functions. [1-3]

1-4 *Begin* developing a system viewpoint for troubleshooting. [1-3]

1-5 *Analyze* circuits with both dc and ac sources. [1-4]

1-6 *List* the current trends in electronics. [1-5]

Looking back at electronics in the 20th century shows the inception of major discoveries such as radio, TV, cell phones, computers, medical electronics, and others that changed the path of mankind. The 21st century is seeing even more amazing breakthroughs. There will be ongoing rapid growth and solutions for an imperiled planet. As a career, it is a solid choice. This book was written to help move you along your path. This chapter introduces you to the broad field of electronics and helps you sort things out by building on what you already know about electric circuits and learning how basic electronic functions fit together to make solutions, products, and systems.

1-1 A Brief History

It is hard to place an exact date on the beginning of electronics. The year 1899 is one possibility. During that year, J. J. Thomson, at the University of Cambridge in England, discovered the electron. Two important developments at the beginning of the 20th century made people interested in electronics. The first was in 1901, when Guglielmo Marconi sent a message across the Atlantic Ocean using *wireless* telegraphy. Today we call wireless communication *radio*. The second development came in 1906, when Lee De Forest invented the audion vacuum tube. The term *audion* related to its first use, to make sounds ("audio") louder. It was not long before the wireless inventors used the *vacuum tube* to improve their equipment.

Audion

Vacuum tube

Another development in 1906 is worth mentioning. Greenleaf W. Pickard used the first crystal radio detector. This great improvement helped make radio and electronics more popular. It also suggested the use of *semiconductors* (crystals) as materials with future promise for the new field of radio and electronics.

Semiconductor

Commercial radio was born in Pittsburgh, Pennsylvania, at station KDKA in 1920. This development marked the beginning of a new era, with electronic devices appearing in the average home. By 1937 more than half the homes in the United States had a radio. Commercial television began around 1946. In 1947 several hundred thousand home radio receivers were manufactured and sold. Complex television receivers and complicated electronic devices made technicians wish for something better than vacuum tubes.

The first vacuum tube computer project was funded by the U.S. government, and the research began in 1943. Three years later, the ENIAC was formally dedicated at the Moore School of Electrical Engineering of the University of Pennsylvania on February 15, 1946. It was the world's first electronic digital computer:

- Size: 30 ft × 50 ft
- Weight: 30 tons
- Vacuum tubes: 17,468

- Resistors: 70,000
- Capacitors: 10,000
- Relays: 1,500
- Switches: 6,000
- Power: 150,000 W
- Cost: $486,000 (about $8 million today)
- Reliability: 7 minutes mean time between failures (MTBF)

A group of students at the Moore School participated in the fiftieth-year anniversary celebration of the ENIAC by developing an equivalent complementary metal oxide semiconductor (CMOS) chip:

- Size: 7.44 mm × 5.29 mm
- Package: 132 pin grid array (PGA)
- Transistors: 174,569
- Cost: several dollars (estimated, per unit, if put into production)
- Power: approximately 1 W
- Reliability: 50 years (estimated)

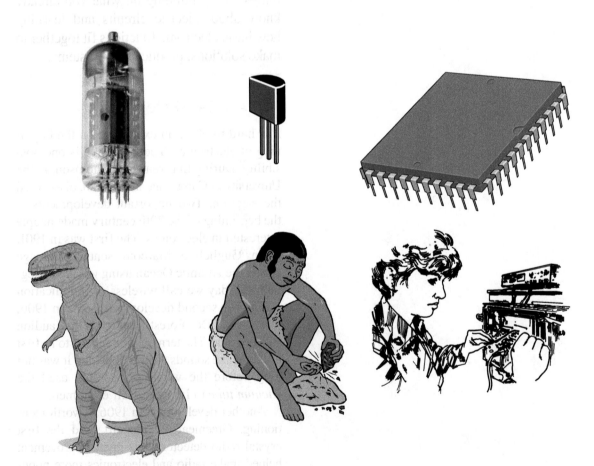

The vacuum tube, the transistor, and then the integrated circuit. The evolution of electronics can be compared with the evolution of life.

(top left): begemot_30/123RF

Scientists had known for a long time that many of the jobs done by vacuum tubes could be done more efficiently by semiconducting crystals, but they could not make crystals pure enough to do the job. The breakthrough came in 1947. Three scientists working with Bell Laboratories made the first working transistor. This was such a major contribution to science and technology that the three men—John Bardeen, Walter H. Brattain, and William B. Shockley—were awarded the Nobel Prize.

Around the same time (1948) Claude Shannon, also then at Bell Laboratories, published a paper on communicating in binary code. His work formed the basis for the digital communications revolution, from cell phones to the Internet. Shannon was also the first to apply Boolean algebra to telephone switching networks when he worked at the Massachusetts Institute of Technology in 1940. Shannon's work forms much of the basis for what we now enjoy in both telecommunications and computing.

Improvements in transistors came rapidly, and now they have all but completely replaced the vacuum tube. *Solid state* has become a household term. Many people believe that the transistor is one of the greatest developments ever.

Solid-state circuits were small, efficient, and more reliable. But the scientists and engineers still were not satisfied. Work done by Jack Kilby of Texas Instruments led to the development of the *integrated circuit* in 1958. Robert Noyce, working at Fairchild, developed a similar project. The two men shared a Nobel Prize in Physics for inventing the integrated circuit.

Integrated circuits are complex combinations of several kinds of devices on a common base, called a *substrate,* or in a tiny piece of silicon. They offer low cost, high performance, good efficiency, small size, and better reliability than an equivalent circuit built from separate parts.

The complexity of some integrated circuits allows a single chip of silicon only 0.64 centimeter (cm) [0.25 inch (in.)] square to replace huge pieces of equipment. Although the chip can hold thousands of transistors, it still has diodes, resistors, and capacitors too!

In 1971 Intel Corporation in California announced one of the most sophisticated of all integrated circuits—the microprocessor. A *microprocessor* is most of the circuitry of a computer reduced to a single integrated circuit. Microprocessors, some containing the equivalent of billions of transistors, have provided billions of dollars worth of growth for the electronics industry and have opened up entire new areas of applications.

The Intel 4004 contained 2,300 transistors, and today a Xeon processor has more than 6 billion. The 4004 had features as small as 10 micrometers (μm), and today the feature size is shrinking toward 10 nanometers (nm).

In 1977 the cellular telephone system entered its testing phase. Since then, the system has experienced immense growth. Its overwhelming success has fostered the development of new technology, such as digital communications and linear integrated circuits for communications.

In 1982, Texas Instruments offered a single chip digital signal processor (DSP). This made it practical to apply DSP to many new product designs. The growth has continued ever since, and DSP is now one of the most rapidly expanding segments of the semiconductor industry.

The integrated circuit is producing an electronics explosion. Now electronics is being applied in more ways than ever before. At one time radio was almost its only application. Today electronics makes a major contribution to our society and to every field of human endeavor. It affects us in ways we may not be aware of. We are living in the electronic age.

Microprocessor

Solid state

Integrated circuit

Substrate

Self-Test

Determine whether each statement is true or false.

1. Electronics is a young technology that began in the 20th century.
2. The early histories of radio and electronics are closely linked.
3. Transistors were invented before vacuum tubes.
4. A modern integrated circuit can contain thousands of transistors.
5. A microprocessor is a small circuit used to replace radio receivers.

1-2 Digital or Analog

Digital electronic device

Digital circuit

Analog circuit

Linear circuit

Today, electronics is such a huge field that it is often necessary to divide it into smaller subfields. You will hear terms such as medical electronics, instrumentation electronics, automotive electronics, avionics, consumer electronics, industrial electronics, and others. One way that electronics can be divided is into digital or analog.

A *digital electronic device* or circuit will recognize or produce an output of only several limited states. For example, most digital circuits will respond to only two input conditions: low or high. *Digital circuits* may also be called *binary* since they are based on a number system with only two digits: 0 and 1.

An *analog circuit* can respond to or produce an output for an infinite number of states. An analog input or output might vary between 0 and 10 volts (V). Its actual value could be 1.5, 2.8, or even 7.653 V. In theory, an *infinite* number of voltages are possible. On the other hand, one type of digital circuit recognizes inputs ranging from 0 to 0.4 V as low (binary 0) and those ranging from 2.0 to 5 V as high (binary 1). This type of digital circuit does not respond any differently for an input of 2 V than it does for one at 4 V. Both of these voltages are in the high range. Input voltages between 0.4 and 2.0 V are not allowed in digital systems because they cause an output that is unpredictable.

For a long time, almost all electronic devices and circuits operated in the analog fashion. This seemed to be the most obvious way to do a particular job. After all, most of the things that we measure are analog in nature. Your height, weight, and the speed at which you travel in a car are all analog quantities. Your voice is analog. It contains an infinite number of levels and frequencies. So, if you wanted a circuit to amplify your voice, you would probably think of using an analog circuit.

Telephone switching and computer circuits forced engineers to explore digital electronics. They needed circuits and devices to make logical decisions based on certain input conditions. They needed highly reliable circuits that would always operate the same way. By limiting the number of conditions or states in which the circuits must operate, they could be made more reliable. An infinite number of states—the analog circuit—were not what they needed.

Figure 1-1 gives examples of circuit behavior to help you identify digital or analog operation. The signal going into the circuit is on the left, and the signal coming out is on the right. For now, think of a signal as some electrical quantity, such as voltage, that changes with time. The circuit marked *A* is an example of a digital device. Digital waveforms are rectangular. The output signal is a rectangular wave; the input signal is not exactly a rectangular wave. Rectangular waves have only two voltage levels and are very common in digital devices.

Circuit *B* in Fig. 1-1 is an analog device. The input and the output are sine waves. The output is larger than the input, and it has been shifted above the zero axis. The most important feature is that the output signal is a combination of an infinite number of voltages. In a *linear circuit,* the output is an exact replica of the input. Though circuit *B* is linear, not all analog circuits are linear. For example, a certain audio amplifier could have a distorted sound. This amplifier would still be in the analog category, but it would be nonlinear.

Circuits *C* through *F* are all digital. Note that the outputs are all *rectangular* waves (two levels of voltage). Circuit *F* deserves special attention. Its input is a rectangular wave. This could be an analog circuit responding to only two voltage levels except that something has happened to the signal, which did not occur in any of the other examples. The output frequency is different from the input frequency. Digital circuits that accomplish this are called *counters,* or *dividers.*

It is now common to convert analog signals to a digital format that can be stored in computer memory, on magnetic or optical disks, or on magnetic tape. Digital storage has advantages. Everyone who has heard music played from a digital disk knows that it is usually noise free. Digital recordings do not deteriorate with use as analog recordings do.

Another advantage of converting analog signals to digital is that computers can then be used to enhance the signals. Computers are digital machines. They are powerful, high-speed number crunchers. A computer can do various things to signals such as eliminate noise and distortion, correct for frequency and phase errors, and identify signal patterns. This area of electronics is known as digital signal

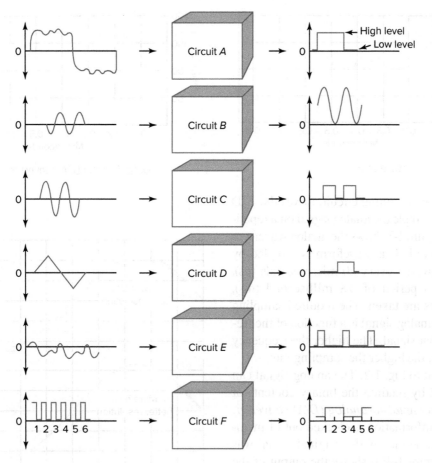

Fig. I-I A comparison of digital and analog circuits.

processing (DSP). DSP is used in medical electronics to enhance scanned images of the human body, in audio to remove noise from old recordings, and in many other ways. DSP is covered in Chap. 16.

Figure 1-2 shows a system that converts an analog signal to digital and then back to analog. An *analog-to-digital (A/D) converter* is a circuit that produces a binary (only 0s and 1s) output. Note that the numbers stored in memory are

A/D converter

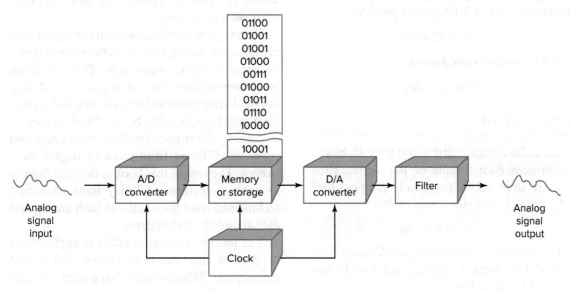

Fig. I-2 An analog-to-digital-to-analog system.

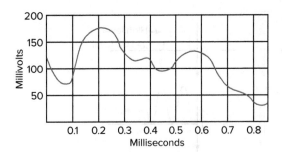

Fig. I-3 An analog waveform.

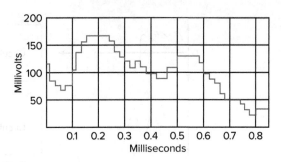

Fig. I-4(*a*) Output of the D/A converter.

binary. A clock (a timing circuit) drives the A/D converter to sample the analog signal on a repetitive basis. Figure 1-3 shows the analog waveform in greater detail. This waveform is sampled by the A/D converter every 20 microseconds (μs). Thus, over a period of 0.8 millisecond (ms), forty samples are taken. The required sampling rate for any analog signal is a function of the frequency of that signal. The higher the frequency of the signal, the higher the sampling rate.

Refer back to Fig. 1-2. The analog signal can be recreated by sending the binary contents of **D/A converter** memory to a *digital-to-analog (D/A) converter*. The binary information is clocked out of memory at the same rate as the original signal was sampled. Figure 1-4(*a*) shows the output of the D/A converter. It can be seen that the waveform is not exactly the same as the original analog signal. It is a series of discrete steps. However, by using more steps, a much closer representation of the original signal can be achieved, as shown in Fig. 1-4(*b*). Step size is determined by the number of binary digits (bits) used. The number of steps is found by raising 2 to the power of the number of bits. A 5-bit system provides

$$2^5 = 32 \text{ steps}$$

An 8-bit system would provide

$$2^8 = 256 \text{ steps}$$

EXAMPLE I-I

An audio compact disk (CD) uses 16 bits to represent each sample of the signal. How many steps or volume levels are possible? Use the appropriate power of 2:

$$2^{16} = 65,536$$

This is easy to solve using a calculator with an x^y key. Press 2, then x^y, and then 16 followed by the = key.

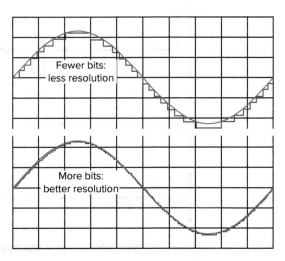

Fewer bits: less resolution

More bits: better resolution

Fig. I-4(*b*) More bits gives a better signal representation.

Actually, the filter shown in Fig. 1-2 smooths the steps, and the resulting analog output signal would be quite acceptable for many applications such as speech.

If enough bits and an adequate sampling rate are used, an analog signal can be converted into an accurate digital equivalent. The signal can be converted back into analog form and may not be distinguishable from the original signal. Or it may be noticeably better if DSP is used.

Analog electronics involves techniques and concepts different from those of digital electronics. The rest of this book is devoted mainly to analog electronics. Today most electronic technicians must have skills in both analog and digital circuits and systems.

The term *mixed signal* refers to applications or devices that use both analog and digital techniques. Mixed-signal integrated circuits are covered in Chap. 13.

Determine whether each statement is true or false.

6. Electronic circuits can be divided into two categories, digital or analog.
7. An analog circuit can produce an infinite number of output conditions.
8. An analog circuit recognizes only two possible input conditions.

9. Rectangular waves are common in digital systems.
10. D/A converters are used to convert analog signals to their digital equivalents.
11. The output of a 2-bit D/A converter can produce eight different voltage levels.

I-3   Analog Functions

This section presents an overview of some functions that analog electronic circuits can provide. Complex electronic systems can be broken down into a collection of individual functions. An ability to recognize individual functions, how they interact, and how each contributes to system operation will make system analysis and troubleshooting easier.

Analog circuits perform certain operations. These operations are usually performed on *signals*. Signals are electrical quantities, such as voltages or currents, that have some merit or use. For example, a microphone converts a human voice into a small voltage whose frequency and level change with time. This small voltage is called an *audio signal*.

Analog electronic circuits are often named after the function or operation they provide. *Amplification* is the process of making a signal larger or stronger, and circuits that do this are called *amplifiers*. Here is a list of the major types of analog electronic circuits.

1. *Adders:* Circuits that add signals together. *Subtractors,* also called *difference amplifiers,* are also available.
2. *Amplifiers:* Circuits that increase signal voltage, current, or power.
3. *Attenuators:* Circuits that decrease signal levels.
4. *Clippers:* Devices that prevent signals from exceeding a fixed amplitude limit or limits.
5. *Comparators:* Devices that compare signal voltage to a reference voltage. Some have one threshold voltage, and others have two.

6. *Controllers:* Devices that regulate signals and load devices. For example, a controller might be used to set and hold the speed of a motor.
7. *Converters:* Devices that change a signal from one form to another (e.g., voltage-to-frequency and frequency-to-voltage converters).
8. *Differentiators:* Circuits that respond to rapidly changing events. They may also be called *high-pass filters.*
9. *Demultiplexer:* A device that routes one circuit or device into many or one output path into several.
10. *Detectors:* Devices that remove or recover information from a signal (a radio detector removes voice or music from a radio signal). They are also called *demodulators.*
11. *Dividers:* Devices that arithmetically divide a signal.
12. *Filters:* Devices that remove unwanted frequencies from a signal by allowing only those that are desired to pass through.
13. *Integrator:* A circuit that sums over some time interval.
14. *Inverters:* Devices that convert direct current (dc) to alternating current (ac).
15. *Mixers:* Another name for adders; also, nonlinear circuits that produce the sum and difference frequencies of two input signals.
16. *Modulators:* Devices that allow one signal to control another's amplitude, frequency, or phase.
17. *Multiplexer:* A device that routes many circuits or devices into one; several signal

Signals

sources are combined or selected for
one output.

18. *Multipliers:* Devices that perform arithmetic
multiplication of some signal characteristic.
There are frequency and amplitude
multipliers.

19. *Oscillators:* Devices that convert dc to ac.

20. *Rectifiers:* Devices that change ac to dc.

21. *Regulators:* Circuits that hold some value,
such as voltage or current, constant.

22. *Sensors:* Circuits that convert some
physical characteristic into a voltage
or current.

23. *Source:* The origin of a type of energy—
voltage, current, or power.

24. *Switches:* Devices that turn signals on
or off or change the signal path in an
electronic system.

25. *Timers:* Devices that control or measure
time.

26. *Trigger:* A circuit that activates at some
circuit value and usually produces an
output pulse.

A *schematic diagram* shows all the individual parts of a circuit and how they are
interconnected. Schematics use standard
symbols to represent circuit components.
A *block diagram* shows all the individual
functions of a system and how the signals
flow through the system. Schematic diagrams
are usually required for what is known as

Technician inspecting a circuit board.
John A. Rizzo/Getty Images

component-level troubleshooting. A component is a single part, such as a resistor, capacitor, or an integrated circuit. Component-level
repair requires the technician to isolate and
replace individual parts that are defective.

System-level repair often requires only a
block diagram or a knowledge of the block diagram. The technician observes symptoms and
makes measurements to determine which function or functions are improper. Then an entire
module, panel, or circuit board is replaced.
Component-level troubleshooting usually takes
longer than system-level does. Since time is
money, it may be economical to replace entire
modules or circuit boards.

Troubleshooting begins at the system level.
Using a knowledge of circuit functions and the
block diagram, observation of the symptoms,
and measurements, the technician isolates the
difficulty to one or more circuit functions. If
replacement boards or modules are on hand, one
or more functions can be replaced. However,
if component-level troubleshooting is required,
the technician continues the isolation process to
the component level, often by using a voltmeter
and an oscilloscope.

Figure 1-5 shows one block of a block diagram for you to see the process. Troubleshooting is often a series of simple yes or no decisions.
For example, is the output signal shown in
Fig. 1-5 normal? If so, there is no need to troubleshoot that circuit function. If it is not normal,
four possibilities exist: (1) a power supply problem, (2) an input signal problem, (3) defective
block (function), or (4) some combination of
these three items. Voltmeters and/or oscilloscopes are generally used to verify the power
supply and the input signal to a block. If the
supply and input signals are normal, then the
block can be replaced or component-level
troubleshooting on that circuit function can
begin. The following chapters in this book
detail how electronic circuits work and cover
component-level troubleshooting.

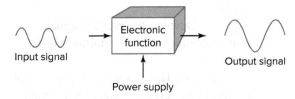

Fig. I-5 One block of a block diagram.

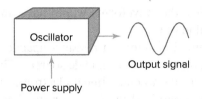

Fig. I-6 A block with only a power supply input.

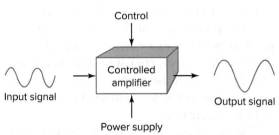

Fig. I-7 Amplifier with a control input.

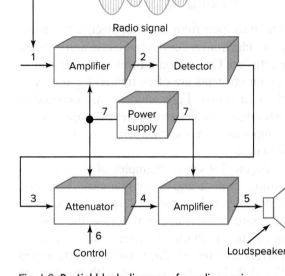

Fig. I-8 Partial block diagram of a radio receiver.

Figure 1-6 shows a block with only one input (power) and one output. Assuming the output signal is missing or incorrect, the possibilities are: (1) the power supply is defective, (2) the oscillator is defective, or (3) both are defective.

Figure 1-7 shows an amplifier that is controlled by a separate input. If its output signal is not correct, the possible causes are: (1) the power supply is defective, (2) the input signal is defective, (3) the control input is faulty, (4) the amplifier has malfunctioned, or (5) some combination of these four items.

Figure 1-8 illustrates a partial block diagram for a radio receiver. It shows how signals flow through the system. A radio signal is amplified, detected, attenuated, amplified again, and then sent to a loudspeaker to produce sound. Knowing how the signal moves from block to block enables a technician to work efficiently. For

example, if the signal is missing or weak at point 5, the problem could be caused by a bad signal at point 1, or any of the blocks shown might be defective. The power supply should be checked first, since it affects most of the circuit functions shown. If it checks out good, then the signal can be verified at point 1, then point 2, and so on. A defective stage will quickly be located by this orderly process. If the signal is normal at point 3 but not at point 4, then the attenuator block and/or its control input is bad.

Much of this book is devoted to the circuit details needed for component-level troubleshooting. However, you should remember that troubleshooting begins at the system level. Always keep a clear picture in your mind of what the individual circuit function is and how that function can be combined with other functions to accomplish system operation.

Self-Test

Determine whether each statement is true or false.

12. Amplifiers make signals larger.
13. If a signal into an amplifier is normal but the output is not, then the amplifier has to be defective.

14. Component-level troubleshooting requires only a block diagram.
15. A schematic diagram shows how individual parts of a circuit are connected.
16. The first step in troubleshooting is to check individual components for shorts.

1-4 Circuits with Both DC and AC

The transition from the first electricity course to an electronics course can cause some initial confusion. One reason for this is that dc and ac circuit concepts are often treated separately in the first course. Later, students are exposed to electronic circuits that have both dc and ac components. This section will make the transition easier.

Figure 1-9 shows examples of circuits containing both dc and ac components. A battery, a dc source, is connected in series with an ac source. The waveform across the resistor shows that both direct current and alternating current are present. The waveform at the top in Fig. 1-9 shows a sine wave with an average value that is positive. The waveform below this shows a sine wave with a negative average value. The average value in both waveforms is called the *dc component of the waveform,* and it is equal to the battery voltage. Without the

batteries, the waveforms would have an average value of 0 V.

Figure 1-10 shows a resistor-capacitor (*RC*) circuit that has both ac and dc sources. This circuit is similar to many linear electronic circuits that are energized by dc power supplies, such as batteries, and that often process ac signals. Thus, the waveforms in linear electronic circuits often show both ac and dc components.

Figure 1-11 shows the waveforms that occur at the various nodes in Fig. 1-10. A node is a point at which two or more circuit elements (resistors, inductors, etc.) are connected. These two figures will help you understand some important ideas that you will need in your study of linear electronics.

The waveform for Node A, in Fig. 1-11, shows *pure direct current.* The word "pure" is used because there is no ac component. This is the waveform expected from a dc source such as a battery. Since Node A in Fig. 1-10 is the positive terminal of the battery, the dc waveform is no surprise.

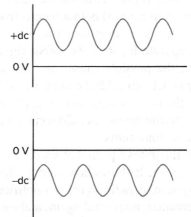

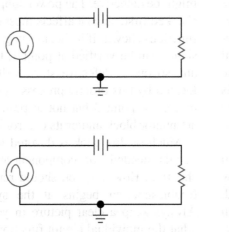

Fig. 1-9 Circuits with dc and ac sources.

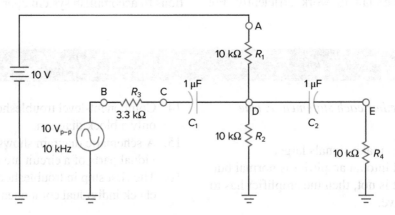

Fig. 1-10 An *RC* circuit with two sources.

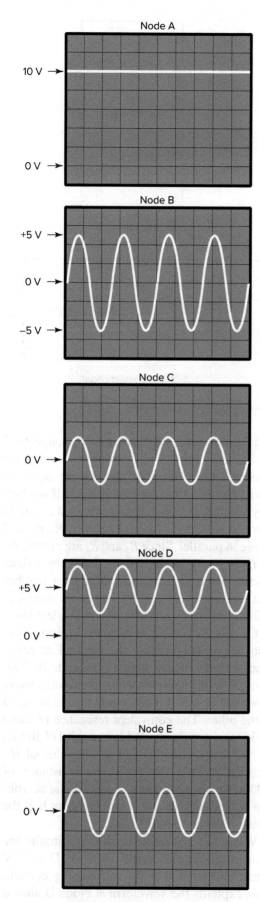

Fig. I-II Waveforms for Fig. I-IO.

Node B, in Fig. 1-11, shows *pure alternating current* (there is no dc component). Node B is the ac source terminal in Fig. 1-10, so this waveform is what one would expect it to be.

The other waveforms in Fig. 1-11 require more thought. Starting with Node C, we see a pure ac waveform with about half the amplitude of the ac source. The loss in amplitude is caused by the voltage drop across R_3, discussed later. Node D shows an ac waveform with a 5 V dc component. This dc component is established by R_1 and R_2 in Fig. 1-10, which act as a voltage divider for the 10 V dc battery. Finally, Node E in Fig. 1-11 shows a pure ac waveform. The dc component has been removed by C_2 in Fig. 1-10. A dc component is present at Node D but is missing at Node E because *capacitors block or remove the dc component of signals or waveforms.*

Capacitors block dc component

You May Recall

. . . that capacitors have infinite reactance (opposition) for direct current and act as open circuits.

The formula for capacitive reactance is

$$X_C = \frac{1}{2\pi fC}$$

As the frequency (f) approaches direct current (0 Hz), the reactance approaches infinity. In capacitors, the relationship between frequency and reactance is *inverse*. As one goes down, the other goes up.

EXAMPLE I-2

Determine the reactance of the capacitors in Fig. 1-10 at a frequency of 10 kHz and compare this reactance with the size of the resistors:

$$X_C = \frac{1}{2\pi fC}$$

$$= \frac{1}{6.28 \times 10 \times 10^3 \times 1 \times 10^{-6}}$$

$$= 15.9\ \Omega$$

The reactance 15.9 Ω is low. In fact, we can consider the capacitors to be short circuits at 10 kHz because the resistors in Fig. 1-10 are 10 kΩ, which is much larger.

Let's summarize two points: (1) the capacitors are open circuits for direct current, and (2) the capacitors are short circuits for ac signals when the signal frequency is relatively high. These two concepts are applied over and over again in analog electronic circuits. Please try to remember them.

What happens at other frequencies? At higher frequencies, the capacitive reactance is even lower, so the capacitors can still be viewed as shorts. At lower frequencies, the capacitors show more reactance, and the short-circuit viewpoint may no longer be correct. As long as the reactance is less than one-tenth of the effective resistance, the short-circuit viewpoint is generally good enough.

EXAMPLE I-3

Determine the reactance of the capacitors in Fig. 1-10 at a frequency of 100 Hz. Will the short-circuit viewpoint be appropriate at this frequency?

$$X_C = \frac{1}{2\pi f C}$$

$$= \frac{1}{6.28 \times 100 \times 1 \times 10^{-6}}$$

$$= 1.59 \text{ k}\Omega$$

This reactance is in the 1,000-Ω range, so the capacitors *cannot* be viewed as short circuits at this frequency.

Figure 1-12 illustrates the equivalent circuits for Fig. 1-10. The dc equivalent circuit shows the battery, R_1, and R_2. Where did the other resistors

Superposition theorem

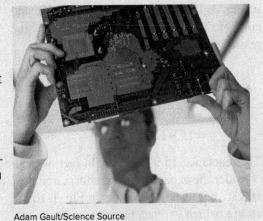
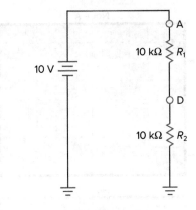

DC equivalent circuit

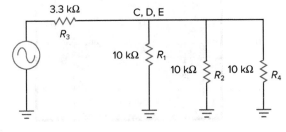

AC equivalent circuit

Fig. I-12 Equivalent circuits for Fig. I-10.

and the ac source go? They are "disconnected" by the capacitors, which are open circuits for direct current. Since R_1 and R_2 are equal in value, the dc voltage at Node D is half the battery voltage, or 5 V. The ac equivalent circuit is more complicated. Note that resistors R_1, R_2, and R_4 are in parallel. Since R_2 and R_4 are connected by C_2 in Fig. 1-10, they can be joined by a short circuit in the ac equivalent circuit. Remember that the capacitors can be viewed as short circuits for signals at 10 kHz. An equivalent short at C_2 puts R_2 and R_4 in parallel. Resistor R_1 is also in parallel because the internal ac resistance of a dc voltage source is taken to be 0 Ω. Thus, R_1 in the ac equivalent circuit is effectively grounded at one end and connected to Node D at the other. The equivalent resistance of three 10-kΩ resistors in parallel is one-third of 10 kΩ, or 3.33 kΩ—almost equal to the value of R_3. Resistor R_3 and the equivalent resistance of 3.33 kΩ form a voltage divider. So, the ac voltage at Nodes C, D, and E will be about half the value of the ac source, or 5 V$_{p-p}$.

When the dc and ac equivalent circuits are taken together, the result at Node D is 5 V direct current and 5 V$_{p-p}$ alternating current. This explains the waveform at Node D shown in Fig. 1-11. The *superposition theorem*, which

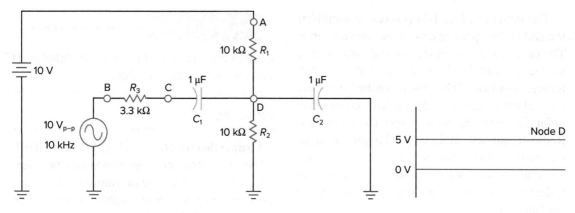

Fig. I-13 The concept of bypassing.

you may have studied, provides the explanation for the combining effect.

There is another very important concept used in electronic circuits, called *bypassing*. Look at Fig. 1-13 and note the C_2 is grounded at its right end. This effectively shorts Node D as far as the ac signal is concerned. The waveform shows that Node D has only 5 V dc, since the ac signal has been *bypassed*. Bypassing is used at nodes in circuits in which the ac signal must be eliminated.

Capacitors are used in many ways. Capacitor C_2 in Fig. 1-10 is often called a *coupling capacitor*. This name serves well since its function is to couple the ac signal from Node D to Node E. However, while it couples the ac signal, it *blocks* the dc component. So, it may also be called a

blocking capacitor. Capacitor C_2 in Fig. 1-13 serves a different function. It eliminates the ac signal at Node D and is called a *bypass capacitor*.

Figure 1-14 shows a clever application of the ideas presented here. Suppose there is a problem with weak signals from a television station. An amplifier can be used to boost a weak signal. The best place for one is at the antenna, but the antenna is often on the roof. The amplifier needs power, so one solution would be to run power wires to the roof along with a separate cable for the television signal. The one coaxial cable can serve both needs (power and signal). The inductor and capacitor are often packaged, along with RF connectors, as a unit called a bias-T. The idea can be called power over coax.

Blocking capacitor

Bypass capacitor

Bypassing

Coupling capacitor

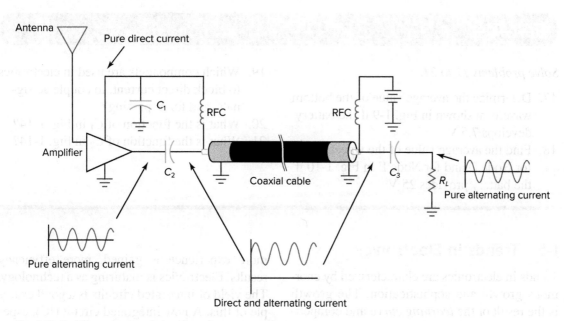

Fig. I-14 Sending power and signal on the same cable.

The battery in Fig. 1-14 powers an amplifier located at the opposite end of the coaxial cable. The outer conductor of the coaxial cable serves as the ground for both the battery and the remote amplifier. The inner conductor of the coaxial cable serves as the positive connection point for both the battery and the amplifier. Radio-frequency chokes (RFCs) are used to isolate the signal from the power circuit. RFCs are coils wound with copper wire. They are inductors and have more reactance for higher frequencies.

You May Recall

. . . that inductive reactance increases with frequency:

$$X_L = 2\pi fL$$

Frequency and reactance are *directly* related in an inductor. As one increases, so does the other.

At direct current ($f = 0$ Hz), the inductive reactance is zero. The dc power passes through the chokes with no loss. As frequency increases, so does the inductive reactance. In Fig. 1-14 the inductive reactance of the choke on the right side of the figure prevents the battery from shorting the high-frequency signal to ground. The inductive reactance of the choke on the left side of Fig. 1-14 keeps the ac signal out of the power wiring to the amplifier.

You May Recall

Chokes are so named because they "choke off" high-frequency current flow.

EXAMPLE 1-4

Assume that the RFCs in Fig. 1-14 are 10 µH. The lowest-frequency television channel starts at 54 MHz. Determine the minimum inductive reactance for television signals. Compare the minimum choke reactance with the impedance of the coaxial cable, which is 72 V.

$$X_L = 2\pi fL = 6.28 \times 54 \times 10^6 \times 10 \times 10^{-6}$$
$$= 3.39 \text{ k}\Omega$$

The reactance of the chokes is almost 50 times the cable impedance. This means the chokes effectively isolate the cable signal from the battery and from the power circuit of the amplifier.

Capacitors C_2 and C_3 in Fig. 1-14 are coupling capacitors. They couple the ac signal into and out of the coaxial cable. These capacitors act as short circuits at the signal frequency, and they are open circuits for the dc signal from the battery. Capacitor C_1 is a bypass capacitor. It ensures that the amplifier is powered by pure direct current. Resistor R_L in Fig. 1-14 is the load for the ac signal. It represents the television receiver.

Self-Test

Solve problems 17 to 21.

17. Determine the average value of the bottom waveform shown in Fig. 1-9 if the battery develops 7.5 V.
18. Find the average value of the waveform for Node D and for Node E in Fig. 1-10 if the battery provides 25 V.
19. Which components are used in electronics to block direct current, to couple ac signals, and for bypassing?
20. What is the function of C_1 in Fig. 1-14?
21. What is the function of C_2 in Fig. 1-14?

1-5 Trends in Electronics

Trends in electronics are characterized by enormous growth and sophistication. The growth is the result of the *learning curve* and competition. The learning curve simply means that as more experience is gained, more efficiency results. Electronics is maturing as a technology. The yield of integrated circuits is a good example of this. A new integrated circuit (IC), especially a sophisticated one, may yield less than

Learning curve

10 percent. Nine out of ten do not pass the test and are thrown away, making the price of a new device very high. Later, after much is learned about making that part, the yield goes up to 90 percent. The price drops drastically, and many new applications are found for it because of the lower price. Although the new parts are complex and sophisticated, the usual result is a product that is easier to use. In fact, "user-friendly" is a term used to describe sophisticated products.

The IC is the key to most electronic trends. These marvels of *microminiaturization* keep expanding in performance and usually decrease the cost of products. They also require less energy and offer high reliability. One of the most popular ICs, the microprocessor, has created many new products. Microprocessor chips are now fast and inexpensive, encouraging rapid growth.

Along with ICs, *surface-mount technology* (SMT) also helps to expand electronics applications. SMT is an alternative to insertion technology for the fabrication of circuit boards. With insertion technology, device leads pass through holes in the circuit board. The insides

of the holes are usually plated with metal to electrically connect the various board layers. Circuit boards designed for insertion technology have more plated-through holes, are larger, and cost more.

The devices intended for SMT have a different appearance. As Fig. 1-15 shows, the device packages have very short leads or just end terminals. These packages are designed to be soldered onto the surface of printed circuit boards. The short leads save material and reduce the

Microminiaturization

Surface-mount technology (SMT)

Fig. I-I5 Circuit board with surface mount parts.
Kenishirotie/Shutterstock

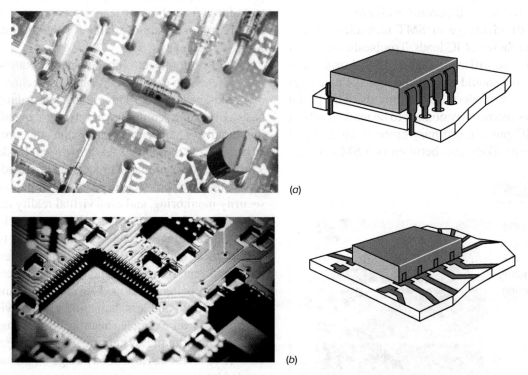

(a)

(b)

A comparison of conventional-mount and surface-mount technologies. (*a*) The photo and the drawing show conventional component mounting. (*b*) Photo and drawing of a surface-mount technology (SMT) circuit board.

(top left): ancher/123RF; (bottom left): Montypeter/Shutterstock

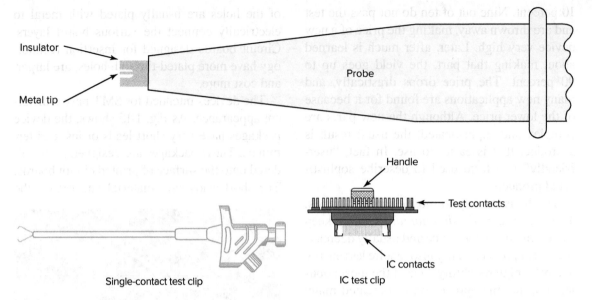

Fig. 1-16 Tools for SMT measurements.

stray effects associated with the longer leads used in insertion technology. SMT provides better electrical performance, especially in high-frequency applications.

Two other advantages of SMT are lower circuit assembly cost, since it is easier to automate, and a lower profile. Since more boards can be packed into a given volume, smaller, less expensive products will become available.

A disadvantage of SMT technology is the close spacing of IC leads. Troubleshooting and repair are difficult. Figure 1-16 shows some tools that should be on hand to make measurements on modern circuit boards. The probe allows momentary contact to be made safely at one IC pin. An ordinary probe is uninsulated and will likely slip between two SMT device leads. When this happens, the two leads will be shorted together, and damage could result. The single contact test clip in Fig. 1-16 is preferred for making connections that will be used for more than one measurement. The IC test clip in Fig. 1-16 is the best tool for SMT IC measurements. It clips onto an SMT IC and provides larger and widely spaced test contacts for safe probing or test-clip connections. Different models are available for the various SMT IC packages.

The uses for electronic devices, products, and systems are expanding. Computer technology finds new applications almost on a daily basis. Electronic communications are expanding rapidly. Thanks to compression and processing breakthroughs, the growth is brisk. Three-dimensional image processing is providing systems for product inspection, automated security monitoring, and even virtual reality for education and entertainment. Computer technology is merging with telecommunications to provide new methods of information transfer, education, entertainment, and shopping. New sensors are being developed to make systems energy efficient and less damaging to the environment. As an example, heating, ventilating, and air-conditioning systems will use oxygen sensors to direct airflow in buildings on an as-needed basis.

The information age is merging databases to reduce errors and improve safety and efficiency. A patient is more likely to get the tests she or he

ABOUT ELECTRONICS

It is possible to probe some surface-mount integrated circuits safely. Probing surface mount devices requires great care to avoid shorting device pins together.

Janka Dharmasena/iStock/Getty Images

needs, the correct medications, the correct procedures, and all in a timely fashion. Health care professionals have instant access to medical history, test results, notes, and comments from other professionals. And the patient wrist tag might have an embedded radio-frequency (RF) chip. Medical imaging continues to improve to hasten the diagnostic procedure, increase accuracy, and

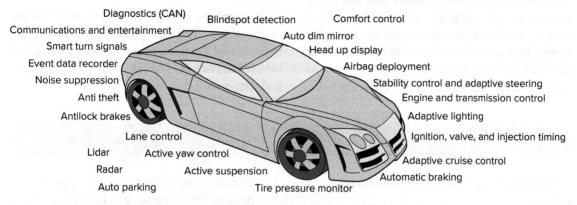

Diagnostics (CAN)
Communications and entertainment
Smart turn signals
Event data recorder
Noise suppression
Anti theft
Antilock brakes
Lane control
Lidar
Radar
Auto parking
Blindspot detection
Active yaw control
Active suspension
Tire pressure monitor
Comfort control
Auto dim mirror
Head up display
Airbag deployment
Stability control and adaptive steering
Engine and transmission control
Adaptive lighting
Ignition, valve, and injection timing
Adaptive cruise control
Automatic braking

On the way to driverless automobiles, many new electronic systems are added all the time.

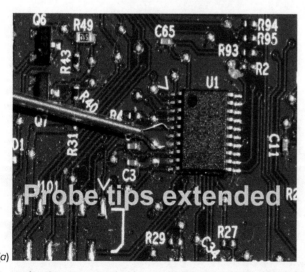

A large array of photovoltaic panels.
Stocktrek Images/Getty Images

(a)

(b)

The probe tips are extended by squeezing the release. The probe tips are moved around the desired pin. When released, the tips retract into the probe body and snug up to the pin.

(both image): Charles A Schuler

eliminate the need for some invasive procedures or more costly or dangerous tests.

Homes and other structures are becoming more energy efficient thanks to sophisticated but affordable control systems and improved appliances and lighting. Renewable sources such as photovoltaic arrays can feed surplus energy into the grid; this would not be safe or practical without electronic devices such as inverters, controllers, and smart converters.

The automobile is becoming an array of various electronic systems. Almost half the cost of a new car is in the electronics and that is not limited to just electric and hybrid vehicles.

The outlook is bright for those with careers in electronics. The new products, the new applications, and the tremendous growth mean good jobs for the future. The jobs will be challenging and marked by constant change.

Self-Test

Determine whether each statement is true or false.

22. Integrated circuits will be used less in the future.

23. The learning curve makes electronic devices less expensive as time goes on.

24. In the future, more circuits will be fabricated using insertion technology and fewer with SMT.

Chapter 1 Summary and Review

Summary

1. Electronics is a relatively young field. Its history began in the 20th century.
2. Electronic circuits can be classified as digital or analog.
3. The number of states or voltage levels is limited in a digital circuit (usually to two).
4. An analog circuit has an infinite number of voltage levels.
5. In a linear circuit, the output signal is a replica of the input.
6. All linear circuits are analog, but not all analog circuits are linear. Some analog circuits distort signals.
7. Analog signals can be converted to a digital format with an A/D converter.
8. Digital-to-analog converters are used to produce a simulated analog output from a digital system.
9. The quality of a digital representation of an analog signal is determined by the sampling rate and the number of bits used.
10. The number of output levels from a D/A converter is equal to 2 raised to the power of the number of bits used.
11. Digital signal processing uses computers to enhance signals.
12. Block diagrams give an overview of electronic system operation.
13. Schematic diagrams show individual part wiring and are usually required for component-level troubleshooting.
14. Troubleshooting begins at the system level.
15. Alternating current and direct current signals are often combined in electronic circuits.
16. Capacitors can be used to couple ac signals, to block direct current, or to bypass alternating current.
17. SMT is replacing insertion technology.

Related Formulas

Number of levels in a binary system: levels $= 2^n$

Capacitive reactance: $X_C = \dfrac{1}{2\pi fC}$

Inductive reactance: $X_L = 2\pi fL$

Chapter Review Questions

Determine whether each statement is true or false.

1-1. Most digital circuits can output only two states, high and low. (1-2)

1-2. Digital circuit outputs are usually sine waves. (1-2)

1-3. The output of a linear circuit is an exact replica of the input. (1-2)

1-4. Linear circuits are classified as analog. (1-2)

1-5. All analog circuits are linear. (1-2)

1-6. The output of a 4-bit D/A converter can produce 128 different voltage levels. (1-2)

1-7. An attenuator is an electronic circuit used to make signals stronger. (1-3)

1-8. Block diagrams are best for component-level troubleshooting. (1-3)

1-9. In Fig. 1-8, if the signal at point 4 is faulty, then the signal at point 3 must also be faulty. (1-3)

1-10. Refer to Fig. 1-8. The power supply should be checked first. (1-3)

Chapter Review Questions...continued

1-11. Refer to Fig. 1-10. Capacitor C_2 would be called a bypass capacitor. (1-4)

1-12. Node C in Fig. 1-10 has no dc component since C_1 blocks direct current. (1-4)

1-13. In Fig. 1-11, Node D is the only waveform with dc and ac components. (1-4)

1-14. Refer to Fig. 1-14. The reactance of the coils is high for dc signals. (1-4)

Critical Thinking Questions

1-1. Functions now accomplished by using electronics may be accomplished in different ways in the future. Can you think of any examples?

1-2. Can you describe a simple system that uses only two wires but will selectively signal two different people?

1-3. What could go wrong with capacitor C_2 in Fig. 1-10, and how would the fault affect the waveform at Node D?

1-4. What could go wrong with capacitor C_2 in Fig. 1-13, and how would the fault affect the waveform at Node D?

Answers to Self-Tests

1. T
2. T
3. F
4. T
5. F
6. T
7. T
8. F
9. T
10. F
11. F
12. T
13. F
14. F
15. T
16. F
17. −7.5 V
18. 12.5 V, 0 V
19. capacitors
20. bypass
21. coupling (dc block)
22. F
23. T
24. F

Semiconductors

Learning Outcomes

This chapter will help you to:

2-1 *Identify* some common electronic materials as conductors or semiconductors. [2-1]

2-2 *Predict* the effect of temperature on conductors. [2-1]

2-3 *Predict* the effect of temperature on semiconductors. [2-2]

2-4 *Show* the directions of electron and hole currents in semiconductors. [2-3, 2-4]

2-5 *Identify* the majority and minority carriers in N-type semiconductors. [2-5]

2-6 *Identify* the majority and minority carriers in P-type semiconductors. [2-5]

2-7 *Explain* the term band gap. [2-7]

Electronic circuits used to be based on the flow of electrons in devices called vacuum tubes. Today, almost all electronic circuits are based on current flow in semiconductors. The term "solid state" means that semiconducting crystals are being used to get the job done. The mechanics of current flow in semiconductors is different from that in conductors. Some current carriers are not electrons. High temperatures create additional carriers in semiconductors. These are important differences between semiconductors and conductors. The transistor is considered to be one of the most important developments of all time. It is a semiconductor device. Diodes and integrated circuits are also semiconductors. This chapter covers the basic properties of semiconductors.

2-1 Conductors and Insulators

All materials are made from atoms. At the center of any atom is a small, dense core called the *nucleus*. Figure 2-1(*a*) shows that the nucleus of a copper atom is made up of positive (+) particles called *protons* and neutral (N) particles called *neutrons*. Around the nucleus are orbiting *electrons* that are negative (−) particles. Copper, like all atoms, has an equal number of protons and electrons. Thus, the net atomic charge is zero.

In electronics, the main interest is in the orbit that is farthest away from the nucleus. It is called the *valence orbit*. In the case of copper, there is only one valence electron. A *copper atom* can be simplified as shown in Fig. 2-1(*b*). Here, the nucleus and the first three orbits are combined into a net positive (+) charge of one. This is balanced by the single valence electron.

Nucleus

Proton

Neutron

Electron

Valence orbit

Copper atom

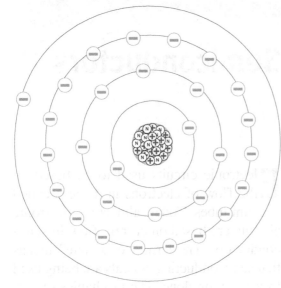

(a) Bohr model of the copper atom (not to scale)

Low resistance

(b) Simplified model

Fig. 2-1 Atomic copper.

Conductors

Positive
temperature
coefficient

Conductors form the fundamental paths for electronic circuits. Figure 2-2 shows how a copper wire supports the flow of electrons. A

ABOUT ELECTRONICS

Superconductivity occurs at extremely low temperatures. MRI machines used in medicine use liquid hydrogen to achieve −442°F.

copper atom contains a positively charged nucleus and negatively charged electrons that orbit around the nucleus. Figure 2-2 is simplified to show only the outermost orbiting electron, the *valence electron*. The valence electron is very important since it acts as the *current carrier*.

Even a very small copper wire contains billions of atoms, each with one valence electron. These electrons are only weakly attracted to the nuclei of the atoms. They are very easy to move. If an *electromotive force (a voltage)* is applied across the wire, the valence electrons will respond and begin drifting toward the positive end of the source voltage. Since there are so many valence electrons and since they are so easy to move, we can expect tremendous numbers of electrons to be set in motion by even a small voltage. Thus, copper is an excellent electric conductor. It has very *low resistance*.

Heating a copper wire will change its resistance. As the wire becomes warmer, the valence electrons become more active. They move farther away from their nuclei, and they move more rapidly. This activity increases the chance for collisions as current-carrying electrons drift toward the positive end of the wire. These collisions absorb energy and increase the resistance to current flow. The resistance of the wire increases as it is heated.

All conductors show this effect. As they become hotter, they conduct less efficiently, and their resistance increases. Such materials are said to have a *positive temperature coefficient*. This simply means that the relationship between temperature and resistance is positive—that is, they increase together.

Valence electrons

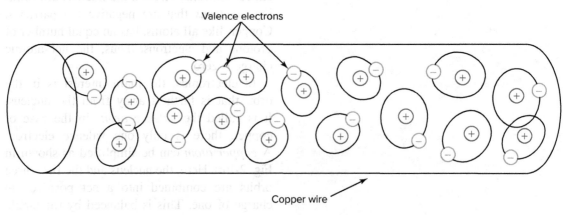

Copper wire

Fig. 2-2 The structure of a copper conductor.

Copper is the most widely applied conductor in electronics. Most of the wire used in electronics is made from copper. *Printed circuits* use copper foil to act as circuit conductors. Copper is a good conductor, and it is easy to solder. This makes it very popular.

Aluminum is a good conductor, but not as good as copper. It is used more in power transformers and transmission lines than it is in electronics. Aluminum is less expensive than copper, but it is difficult to solder and tends to corrode rapidly when brought into contact with other metals.

Silver is the best conductor because it has the least resistance. It is also easy to solder. The high cost of silver makes it less widely applied than copper. However, silver-plated conductors are sometimes used in critical electronic circuits to minimize resistance.

Gold is a good conductor. It is very stable and does not corrode as badly as copper and silver. Some sliding and moving electronic contacts are gold-plated. This makes the contacts very reliable.

The opposite of a conductor is called an *insulator*. In an insulator, the valence electrons are tightly bound to their parent atoms. They are not free to move, so little or no current flows when a voltage is applied. Practically all insulators used in electronics are based on compounds. A *compound* is a combination of two or more different kinds of atoms. Some of the widely applied insulating materials include rubber, plastic, Mylar, ceramic, Teflon, and polystyrene.

Whether a material will insulate depends on how the atoms are arranged. Carbon is such a material. Figure 2-3(*a*) shows carbon arranged in the diamond structure. With this crystal or diamond structure, the valence electrons cannot move to serve as current carriers. Diamonds are insulators. Figure 2-3(*b*) shows carbon arranged in the graphite structure. Here, the valence electrons are free to move when a voltage is applied. It may seem odd that both diamonds and graphite are made from carbon. One insulates, and the other does not. It is simply a matter of whether the valence electrons are locked into the structure. Carbon in graphite form is used to make resistors and electrodes. So far, the diamond structure of carbon has not been widely used to make electrical or electronic devices.

Printed circuit

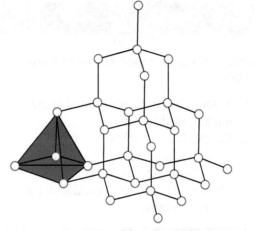

(*a*) Diamond

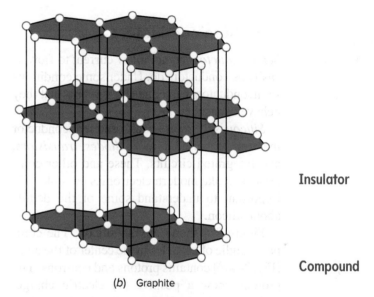

(*b*) Graphite

Insulator

Compound

Fig. 2-3 Structures of diamond and graphite.

ABOUT ELECTRONICS

Materials Used for Dopants, Semiconductors, and Microwave Devices

- Gallium arsenide (GaAs) works better than silicon in microwave devices because it allows faster movement of electrons.
- Materials other than boron and arsenic are used as dopants.
- It is theoretically possible to make semiconductor devices from crystalline carbon.
- Crystal radio receivers were an early application of semiconductors.

Determine whether each statement is true or false.

1. Valence electrons are located in the nucleus of the atom.
2. Copper has one valence electron.
3. In conductors, the valence electrons are strongly attracted to the nucleus.
4. The current carriers in conductors are the valence electrons.
5. Cooling a conductor will decrease its resistance.
6. Silver is not often used in electronic circuits because of its high resistance.
7. Aluminum is not used as much as copper in electronic circuits because it is difficult to solder.

2-2   Semiconductors

Semiconductor

Semiconductors do not allow current to flow as easily as conductors do. Under some conditions semiconductors can conduct so poorly that they behave as insulators.

Diode

Transistor

Silicon is the most widely used semiconductor material. It is used to make *diodes, transistors,* and integrated circuits. These and other components make modern electronics possible. It is important to understand some of the details about silicon.

Figure 2-4 shows atomic silicon. The compact bundle of particles in the center of the atom [Fig. 2-4(*a*)] contains protons and neutrons. The protons show a positive (+) electric charge, and the neutrons show no electric charge (N). Negatively charged electrons travel around the nucleus in orbits. The first orbit has two electrons. The second orbit has eight electrons. The last, or outermost, orbit has four electrons.

Because we are interested mainly in the valence orbit, it is possible to simplify the drawing of the silicon atom. Figure 2-4(*b*) shows only the nucleus and the valence orbit. The valence orbit is the most important feature.

Active material

Materials with four valence electrons are not stable. They tend to combine chemically with other materials. They can be called *active materials.* This activity can lead them to a more

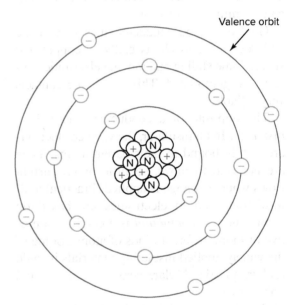

(*a*)   The structure of a silicon atom

Valence orbit

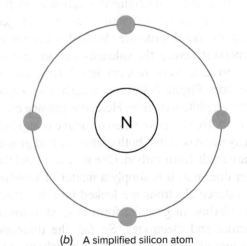

(*b*)   A simplified silicon atom

Fig. 2-4 Atomic silicon.

stable state. A law of nature makes certain materials tend to form combinations that will make eight electrons available in the valence orbit. Eight is an important number because it gives stability.

One possibility is for silicon to combine with oxygen. A single silicon atom can join, or link, with two oxygen atoms to form *silicon dioxide* (SiO$_2$). This linkage is called an *ionic bond*. The new structure, SiO$_2$, is much more stable than either silicon or oxygen. It is interesting to consider that chemical, mechanical, and electrical properties often run parallel. Silicon dioxide is stable chemically. It does not react easily with other materials. It is also stable mechanically. It is a hard, glasslike material. Finally, it is stable electrically. It does not conduct; in fact, it is used as an *insulator* in integrated circuits and other solid-state devices. SiO$_2$ insulates because all of the valence electrons are tightly locked into the ionic bonds. They are not easy to move and therefore do not support the flow of current.

Sometimes oxygen or another material is not available for silicon to combine with. The silicon still wants the stability given by eight valence electrons. If the conditions are right, silicon atoms will arrange to share valence electrons. This process of sharing is called *covalent bonding*. The structure that results is called a *crystal*. Figure 2-5 is a symbolic diagram of a crystal of pure silicon. The dots represent valence electrons.

Count the valence electrons around the nucleus of one of the atoms shown in Fig. 2-5(*a*). Select one of the internal nuclei as represented by a circled N. You will count eight electrons. Thus, the silicon crystal is very stable. Figure 2.5(*b*) shows a three-dimensional representation of a portion of a silicon crystal.

At room temperature, pure silicon is a very poor conductor. If a moderate voltage is applied across the crystal, very little current will flow. The valence electrons that normally would support current flow are all tightly locked up in covalent bonds.

Pure silicon crystals behave like *insulators*. Yet silicon itself is classified as a semiconductor. Pure silicon is sometimes called *intrinsic silicon*. Intrinsic silicon contains very few free electrons to support the flow of current and therefore acts as an insulator.

Crystalline silicon can be made to semiconduct. One way to improve its conduction is to

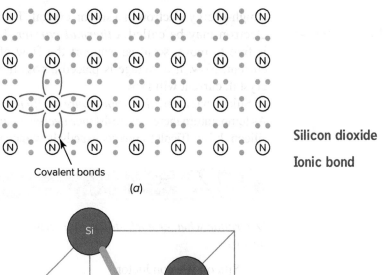

Silicon dioxide

Ionic bond

Covalent bonds

(a)

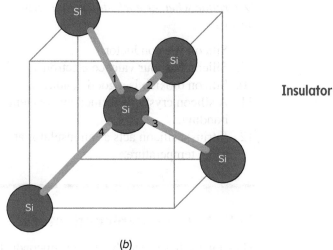

(b)

Fig. 2-5 A crystal of pure silicon.

Insulator

Covalent bonding

Crystal

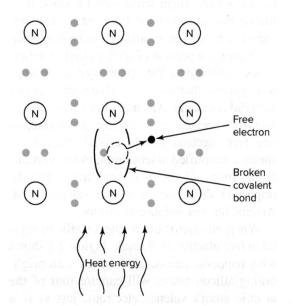

Free electron

Broken covalent bond

Heat energy

Fig. 2-6 Thermal carrier production.

Intrinsic silicon

heat it. Heat is a form of energy. A valence electron can absorb some of this energy and move to a higher orbit level. The high-energy electron has *broken* its covalent bond. Figure 2-6 shows

Thermal carrier

a high-energy electron in a silicon crystal. This electron may be called a *thermal carrier*. It is free to move, so it can support the flow of current. Now, if a voltage is placed across the crystal, current will flow.

Silicon has a *negative temperature coefficient*. As temperature increases, resistance *decreases* in silicon. It is difficult to predict exactly how much

the resistance will change in a given case. One rule of thumb is that the resistance will be cut in half for every 6°C rise in temperature.

The semiconductor material germanium was once used to make diodes and transistors but has become less popular. Silicon has better thermal characteristics and can operate at higher temperatures.

Self-Test

Determine whether each statement is true or false.

8. Silicon is a conductor.
9. Silicon has four valence electrons.
10. Silicon dioxide is a good conductor.
11. A silicon crystal is formed by covalent bonding.
12. Intrinsic silicon acts as an insulator at room temperature.

13. Heating semiconductor silicon will decrease its resistance.
14. An electron that is freed from its covalent bond by heat is called a thermal carrier.
15. Silicon has a negative temperature coefficient.
16. Semiconductor germanium finds wider application than silicon.
17. Germanium devices can safely operate at higher temperatures than silicon devices.

2-3 N-Type Semiconductors

Thus far we have seen that pure semiconductor crystals are very poor conductors. High temperatures can make them semiconduct because thermal carriers are produced. For most applications, there is a better way to make them semiconduct.

Doping

Doping is a process of adding other materials called *impurities* to the silicon crystal to change its electrical characteristics. One such impurity

Arsenic

material is *arsenic*. Arsenic is known as a *donor impurity* because each arsenic atom donates one free electron to the crystal. Figure 2-7 shows a simplified arsenic atom. Arsenic is different from silicon in several ways, but the important difference is in the valence orbit. Arsenic has *five* valence electrons.

When an arsenic atom enters a silicon crystal, a free electron will result. Figure 2-8 shows what happens. The covalent bonds with neighboring silicon atoms will capture four of the arsenic atom's valence electrons, just as if it were another silicon atom. This tightly locks the arsenic atom into the crystal. The fifth valence electron cannot form a bond. It is a *free* electron as far as the crystal is concerned. This makes the electron very easy to move. It can serve as a current carrier. Silicon with some

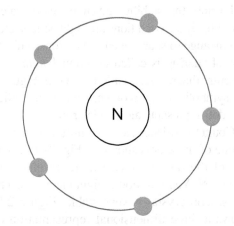

Fig. 2-7 A simplified arsenic atom.

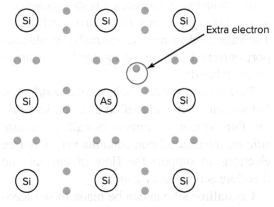

Fig. 2-8 N-type silicon.

arsenic atoms will semiconduct even at room temperature.

Doping lowers the resistance of the silicon crystal. When donor impurities with five valence electrons are added, free electrons are produced. Since electrons have a negative charge, we say that an *N-type semiconductor material* results.

Self-Test

Supply the missing word in each statement.

18. Arsenic is a _____ impurity.
19. Arsenic has _____ valence electrons.
20. When silicon is doped with arsenic, each arsenic atom will give the crystal one free _____.
21. Free electrons in a silicon crystal will serve as current _____.
22. When silicon is doped, its resistance _____.

2-4 P-Type Semiconductors

Doping can involve the use of other kinds of impurity materials. Figure 2-9 shows a simplified *boron atom*. Note that boron has only three valence electrons. If a boron atom enters the silicon crystal, another type of current carrier will result.

Figure 2-10 shows that one of the covalent bonds with neighboring silicon atoms cannot be formed. This produces a *hole*, or *missing electron*. The hole is assigned a *positive charge* since it is capable of attracting, or being filled by, an electron.

Boron is known as an *acceptor impurity*. Each boron atom in the crystal will create a hole that is capable of accepting an electron.

Holes serve as current carriers. In a conductor or an N-type semiconductor, the carriers are

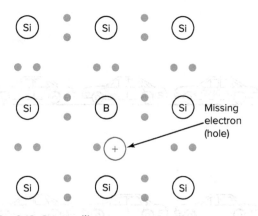

Fig. 2-10 P-type silicon.

electrons. The free electrons are set in motion by an applied voltage, and they drift toward the positive terminal. But in a P-type semiconductor, the holes move toward the negative terminal of the voltage source. Hole current is equal to electron current but *opposite* in direction. Figure 2-11 illustrates the difference between N-type and *P-type semiconductor materials*. In Fig. 2-11(*a*) the carriers are electrons, and they drift toward the positive end of the voltage source. In Fig. 2-11(*b*) the carriers are holes, and they drift toward the negative end of the voltage source.

Figure 2-12 shows a simple analogy for hole current. Assume that a line of cars is stopped for a red light, but there is space for the first car to move up one position. The driver of that car takes the opportunity to do so, and this makes a space for directly behind it. The

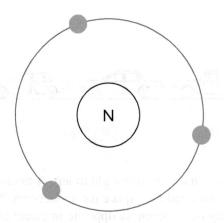

Fig. 2-9 A simplified boron atom.

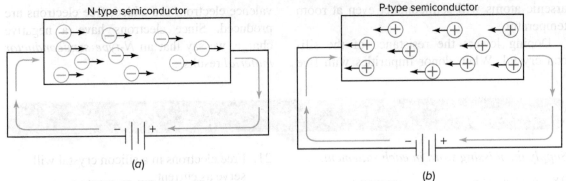

Fig. 2-11 Conduction in N- and P-type silicon.

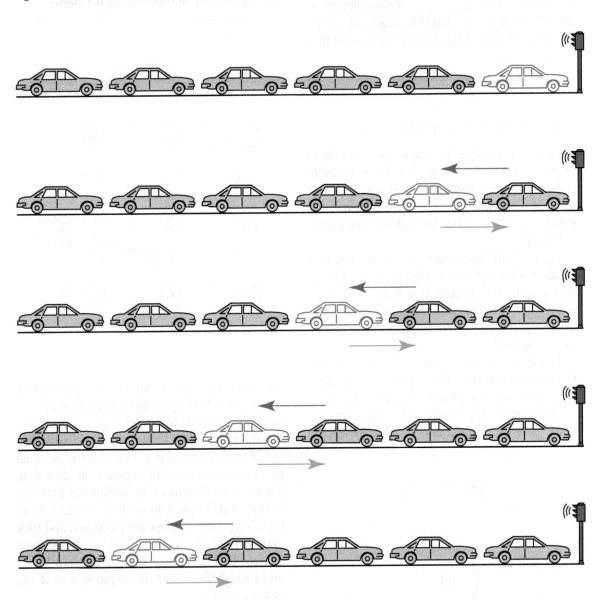

Fig. 2-12 Hole current analogy.

driver of the second car also moves up one position. This continues with the third car, the fourth car, and so on down the line. The cars are moving from left to right. Note that the space is moving from right to left. A hole may be considered as a space for an electron. This is why hole current is opposite in direction to electron current.

Supply the missing word in each statement.

23. Boron is an _____ impurity.
24. Boron has _____ valence electrons.
25. Electrons are assigned a negative charge, and holes are assigned a _____ charge.

26. Doping a semiconductor crystal with boron will produce current carriers called _____.

27. Electrons will drift toward the positive end of the energy source, and holes will drift toward the _____ end.

2-5 Majority and Minority Carriers

When N- and P-type semiconductor materials are made, the doping levels can be as small as 1 part per million or 1 part per billion. Only a tiny trace of impurity materials having five or three valence electrons enters the crystal. It is not possible to make the silicon crystal absolutely pure. Thus, it is easy to imagine that an occasional atom with three valence electrons might be present in an N-type semiconductor. An unwanted hole will exist in the crystal. This hole is called a *minority carrier*. The free electrons are the *majority carriers*.

In a P-type semiconductor, one expects holes to be the carriers. They are in the majority. A few free electrons might also be present. They will be the minority carriers in this case.

The majority carriers will be electrons for N-type material and holes for P-type material. Minority carriers will be holes for N-type material and electrons for P-type material.

Today very high-grade silicon can be manufactured. This high-grade material has very few unwanted impurities. Although this keeps the number of minority carriers to a minimum, their numbers are increased by high temperatures. This can be quite a problem in electronic circuits. To understand how heat produces minority carriers, refer to Fig. 2-6. As additional heat energy enters the crystal, more and more electrons will gain enough energy to break their bonds. Each broken bond produces both a free electron and a hole. Heat produces carriers in *pairs*. If the crystal was manufactured to be N-type material, then every thermal hole becomes a minority carrier and the thermal electrons join the other majority carriers. If the crystal was made as P-type material, then the thermal holes join the majority carriers and the thermal electrons become minority carriers.

Carrier production by heat decreases the crystal's resistance. The heat also produces minority carriers. Heat and the resulting minority carriers can have an adverse effect on the way semiconductor devices work.

This chapter has focused on silicon because most semiconductors are made from it. However, other materials called *compound semiconductors* are becoming important. Some of these are discussed in the next section.

Compound semiconductors

Minority carrier

Majority carrier

HISTORY OF ELECTRONICS

Niels Bohr and the Atom

Scientists change the future by improving on the ideas of others. Niels Bohr proposed a model of atomic structure in 1913 that applied energy levels (quantum mechanics) to the Rutherford model of the atom. Bohr also used some of the work of Max Planck.

Everett Collection Historical/Alamy Stock Photo

Determine whether each statement is true or false.

28. In the making of N-type semiconductor material, a typical doping level is about 10 arsenic atoms for every 90 silicon atoms.
29. A free electron in a P-type crystal is called a majority carrier.
30. A hole in an N-type crystal is called a minority carrier.

31. As P-type semiconductor material is heated, one can expect the number of minority carriers to increase.
32. As P-type semiconductor material is heated, the number of majority carriers decreases.
33. Heat increases the number of minority and majority carriers in semiconductors.

2-6   Other Materials

Silicon is a key semiconductor material and will remain so for some time. However, there are newer materials to address the need for:

1. Higher power—ability to conduct large currents at higher voltages
2. Faster operation—ability to work at high frequencies and perform rapid calculations
3. Improved efficiency—less conversion of energy to waste heat

Table 2-1 lists some semiconductor materials.

Bandgaps are discussed in the next section. Breakdown is the ability to safely withstand high voltage and is an important characteristic as semiconductors are being applied in areas such as electric vehicles, electrical energy conversion and management. Note that diamond semiconductors offer the promise of high voltage operation. Also, due to superior thermal conductivity, diamond devices could offer unheard of power levels. Carrier mobility predicts how a material will serve at high frequencies and at high switching speeds. Note the high mobility for gallium arsenide—it explains why some microwave transistors use that material. There are now an expanding number of gallium nitride and silicon carbide devices. As seen in Table 2-1, their superior (compared to silicon) breakdown and thermal conductivity make them attractive for high power devices.

Organic semiconductors are finding more applications. These devices use semiconducting and sometimes conducting materials that are made of molecules containing carbon, mostly in combination with hydrogen and oxygen. Slower than silicon, but more flexible and potentially much cheaper, organic electronics has already produced circuits with hundreds of transistors printed on plastic, experimental sensors and memories, and displays that bend like paper. Organic displays might compete with liquid crystal displays, as they are brighter and faster and don't suffer from a limited viewing angle.

ABOUT ELECTRONICS

Silicon carbide devices can safely handle thousands of volts.

Table 2-I   Semiconductor Materials

Material	Availability	Bandgap (eV)	Breakdown (MV/cm)	Carrier Mobility $(cm^2/(V \cdot s))$	Thermal Conductivity $(W/m \cdot K)$
Silicon (S)	Now	1.12	0.3	1,500	148
Gallium Arsenide (GaAs)	Now	1.42	0.4	10,000	55
Gallium Nitride (GaN)	Now	3.3	3.3	1,000–2,000	250
Silicon Carbide (SiC)	Now	3.5	3.5	500–950	250–400
Diamond (C)	Future?	5.47	10	300	1,000–2,000

2-7 Band Gaps

In a semiconductor, such as silicon, the energy difference between the top of the valence band and the bottom of the conduction band is called the *band gap*. Or it is the amount of energy, in electron volts (eV), required to free a valence electron from its orbit and boost it to the conduction level.

$$1 \text{ eV} = 1.602 \times 10^{-19} \text{ joules}$$

The joule is the SI unit of work or energy and amounts to a force of 1 newton applied over a distance of 1 meter, or to a current of 1 ampere through a 1-ohm resistor for 1 second. The band gap for silicon is 1.1 eV, and for gallium arsenide, it's 1.43 eV.

As Fig. 2-13 shows, there is no energy gap between the valence band and the conduction band in a conductor. In fact, the bands overlap as shown in red. An insulator has a large band gap. This means that it is very difficult to move a valence electron into the conduction band. However, it can be done. This is why insulators can break down and conduct if subjected to very high voltages. Now, look at the graph for intrinsic silicon. The band gap is smaller than that of an insulator, but it's still too large for most applications. Finally, look at doped silicon. The electrons provided by the dopant material (green) fall just below the conduction band. The band gap is small for doped semiconductors. This is important for the operation of devices such as diodes and solar cells, both of which are explained in the next chapter.

In the case of a solar cell, to free an electron, the energy of a photon (a light particle or a quantum unit of light energy) must be at least as great as the band gap energy. Photons with more energy than the band gap energy will expend the extra energy as heat. So it's important for a solar cell to be optimized through slight modifications to the silicon's molecular structure. A key to obtaining an efficient solar cell is to convert as much sunlight as possible into electricity.

The photon energy of light varies according to the different wavelengths of the light. The entire spectrum of sunlight, from infrared to ultraviolet, covers a range from about 0.5 eV to about 2.9 eV. For example, red light has an energy of about 1.7 eV, and blue light has an energy of about 2.7 eV. Most solar cells cannot use about 55 percent of the energy of sunlight, because this energy is either below the band gap of the material or is excessive. There is currently intense interest in finding new semiconductor materials to improve the efficiency and lower the cost of solar cells. It is possible to stack cells that have different band gaps to increase efficiency.

ABOUT ELECTRONICS

Diamond might someday make extremely high-voltage/high-power devices possible. Diamond has a band gap of 5.5 eV and excellent heat conductivity.

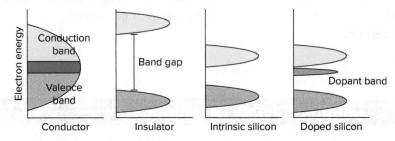

Fig. 2-I3 Energy band diagrams.

Self-Test

Determine whether each statement is true or false.

34. The band gap of materials is measured in volts.
35. The band gap for copper or silver is zero.
36. The electron volt is a unit of work or energy.
37. If a photon has more energy than the band gap of a solar cell, it cannot boost an electron into the conduction band.
38. Doping semiconductors increases their band gaps.

Chapter 2 Summary and Review

Summary

1. Good conductors, such as copper, contain a large number of current carriers.
2. In a conductor, the valence electrons are weakly attracted to the nuclei of the atoms.
3. Heating a conductor will increase its resistance. This response is called a positive temperature coefficient.
4. Silicon atoms have four valence electrons. They can form covalent bonds that result in a stable crystal structure.
5. Heat energy can break covalent bonds, making free electrons available to conduct current. This gives silicon and other semiconductor materials a negative temperature coefficient.
6. Germanium semiconductor devices are no longer widely used due to temperature limitations.
7. The process of adding impurities to a semiconductor crystal is called doping.
8. Doping a semiconductor crystal changes its electrical characteristics.
9. Donor impurities have five valence electrons and produce free electrons in the crystal. This forms N-type semiconductor material.
10. Free electrons serve as current carriers.
11. Acceptor impurities have three valence electrons and produce holes in the crystal.
12. Holes in semiconductor materials serve as current carriers.
13. Hole current is opposite in direction to electron current.
14. Semiconductors with free holes are classified as P-type materials.
15. Impurities with five valence electrons produce N-type semiconductors.
16. Impurities with three valence electrons produce P-type semiconductors.
17. Holes drift toward the negative end of a voltage source.
18. Electrons are majority carriers for N-type material. Holes are majority carriers for P-type material.
19. Holes are minority carriers for N-type material. Electrons are minority carriers for P-type material.
20. The number of minority carriers increases with temperature.
21. To move a valence electron to the conduction band, an amount of energy equal to or greater than the band gap must be applied.

Chapter Review Questions

Determine whether each statement is true or false.

2-1. The current carriers in conductors such as copper are holes and electrons. (2-1)

2-2. It is easy to move the valence electrons in conductors. (2-1)

2-3. A positive temperature coefficient means the resistance goes up as temperature goes down. (2-1)

2-4. Conductors have a positive temperature coefficient. (2-1)

2-5. Silicon does not semiconduct unless it is doped or heated. (2-2)

2-6. Silicon has five valence electrons. (2-2)

2-7. A silicon crystal is built by ionic bonding. (2-2)

2-8. Materials with eight valence electrons tend to be unstable. (2-2)

2-9. Semiconductors have a negative temperature coefficient. (2-2)

2-10. Silicon has better high temperature characteristics than germanium. (2-2)

2-11. When a semiconductor is doped with arsenic, free electrons are placed in the crystal. (2-3)

2-12. N-type material has free electrons available to support current flow. (2-3)

2-13. Doping a crystal increases its resistance. (2-3)

2-14. Doping with boron produces free electrons in the crystal. (2-4)

2-15. Hole current is opposite in direction to electron current. (2-4)

2-16. Holes are current carriers and are assigned a positive charge. (2-4)

2-17. If a P-type semiconductor shows a few free electrons, the electrons are called minority carriers. (2-5)

2-18. If an N-type semiconductor shows a few free holes, the holes are called minority carriers. (2-5)

Critical Thinking Questions

2-1. Suppose that you could perfect a method of inexpensively making ultrapure carbon crystals and then doping them. How could these crystals be used in electronics? (*Hint:* Diamonds are noted for their extreme hardness and ability to withstand high temperatures.)

2-2. Some semiconductors, such as gallium arsenide, show better carrier mobility than silicon. That is, the carriers move faster in the crystal. What kinds of devices could benefit from this?

2-3. Semiconductors respond to temperature by showing decreased resistance leading to problems in many, but not all, electronic products. Can you think of an application where their temperature sensitivity is desired?

2-4. You have learned that conductors and semiconductors have opposite temperature coefficients. How could you use this knowledge to design a circuit that remains stable over a wide temperature range?

Answers to Self-Tests

1. F	11. T	21. carriers	30. T
2. T	12. T	22. decreases	31. T
3. F	13. T	23. acceptor	32. F
4. T	14. T	24. three	33. T
5. T	15. T	25. positive	34. F
6. F	16. F	26. holes	35. T
7. T	17. F	27. negative	36. T
8. F	18. donor	28. F	37. F
9. T	19. five	29. F	38. F
10. F	20. electron		

Diodes

Learning Outcomes

This chapter will help you to:

3-1 *Predict* the conductivity of diodes under the conditions of forward and reverse bias. [3-1]

3-2 *Interpret* volt-ampere characteristic curves for diodes. [3-2]

3-3 *Identify* the cathode and anode leads of some diodes by visual inspection. [3-3]

3-4 *Identify* the cathode and anode leads of diodes by ohmmeter testing. [3-3]

3-5 *Identify* diode schematic symbols. [3-3]

3-6 *List* several diode types and applications. [3-4]

3-7 *Describe* the structure and characteristics of photovoltaic devices. [3-5]

This chapter introduces the most basic semiconductor device, the diode. Diodes are very important in electronic circuits. Everyone working in electronics must be familiar with them. Your study of diodes will enable you to predict when they will be on and when they will be off. You will be able to read their characteristic curves and identify their symbols and their terminals. This chapter also introduces several important types of diodes and some of the many applications for them.

3-1 The PN Junction

A basic use for P- and N-type semiconductor materials is in *diodes*. Figure 3-1 shows a representation of a *PN-junction diode*. Notice that it contains a P-type region with free holes and an N-type region with free electrons. The diode structure is continuous from one end to the other. It is one complete crystal of silicon. Other types, called barrier diodes, are covered later in this chapter.

The junction shown in Fig. 3-1 is the boundary, or dividing line, that marks the end of one section and the beginning of the other. It does not represent a mechanical joint. In other words, the *junction* of a diode is that part of the crystal where the P-type material ends and the N-type material begins.

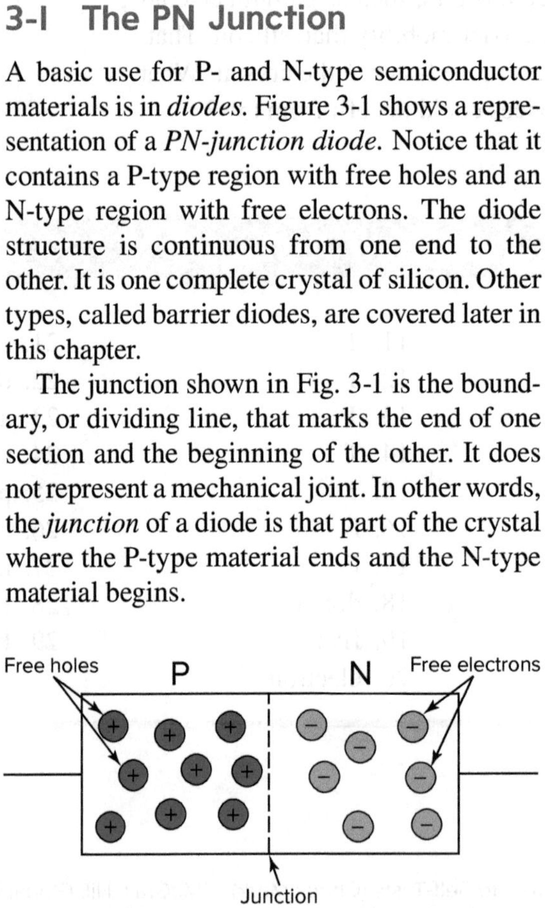

Diode

PN-junction diode

Fig. 3-1 The structure of a junction diode.

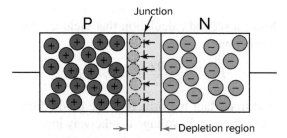

Fig. 3-2 The diode depletion region.

Because the diode is a continuous crystal, free electrons can move across the junction. When a diode is manufactured, some of the free electrons cross the junction to fill some of the holes. Figure 3-2 shows this effect. The result is that a *depletion region* is formed. The electrons that have filled holes are effectively captured (shown in gray) and are no longer available to support current flow. With the electrons gone and the holes filled, no free carriers are left. The region around the junction has become *depleted* (shown in yellow).

The depletion region will not continue to grow for very long. An electric potential, or force, forms along with the depletion region and prevents all the electrons from crossing over and filling all the holes in the P-type material.

Figure 3-3 shows why this potential is formed. Any time an atom loses an electron, it becomes unbalanced. It now has more protons in its nucleus than it has electrons in orbit. This gives it an overall positive charge. It is called a *positive ion*. In the same way, if an atom gains an extra electron, it shows an overall negative charge and is called a *negative ion*. When one of the free electrons in the N-type material leaves its parent atom, that atom becomes a positive ion. When the electron joins another

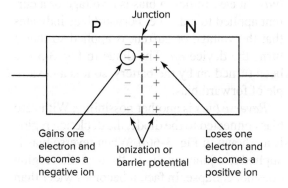

Fig. 3-3 Formation of the barrier potential.

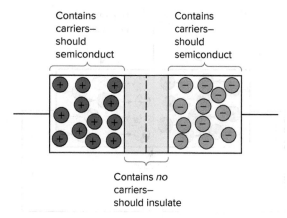

Fig. 3-4 Depletion region as an insulator.

atom on the P-type side, that atom becomes a negative ion. The ions form a charge that prevents any more electrons from crossing the junction.

So when a diode is manufactured, some of the electrons cross the junction to fill some of the holes. The action soon stops because a negative charge forms on the P-type side to repel any other electrons that might try to cross over. This negative charge is called the *ionization potential*, or the *barrier potential*. "Barrier" is a good name since it does stop additional electrons from crossing the junction.

Now that we know what happens when a PN junction is formed, we can investigate how it will behave electrically. Figure 3-4 shows a summary of the situation. There are two regions with free carriers. Since there are carriers, we can expect these regions to *semiconduct*. But right in the middle there is a region with no carriers. When there are no carriers, we can expect it to *insulate*.

Any device having an insulator in the middle will not conduct. So we can assume that PN-junction diodes are insulators. However, a depletion region is not the same as a fixed insulator. It was formed in the first place by electrons moving and filling holes. An external voltage can *eliminate* the depletion region.

In Fig. 3-5, a PN-junction diode is connected to an external battery in such a way that the depletion region is eliminated. The positive terminal of the battery repels the holes on the P-type side and pushes them toward the junction. The negative terminal of the battery repels the electrons and pushes them toward the junction. This *collapses* (eliminates) the depletion region.

Depletion region

Barrier potential

Positive ion

Negative ion

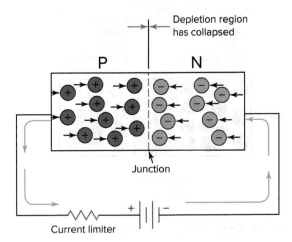

Depletion region has collapsed

P   N

Junction

Current limiter   + ||| −

Fig. 3-5 Forward bias.

With the depletion region collapsed, the diode can semiconduct. Figure 3-5 shows electron current leaving the negative side of the battery, flowing through the diode, through the current limiter (a resistor), and returning to the positive side of the battery. The current-limiting resistor is needed in some cases to keep the current flow at a safe level. Diodes can be destroyed by excess current. Ohm's law can be used to find current in diode circuits. For example, if the battery in Fig. 3-5 is 6 V and the resistor is 1 kilohm (kΩ),

$$I = \frac{V}{R} = \frac{6 \text{ V}}{1 \text{ k}\Omega} = 6 \text{ milliamperes (mA)}$$

The above calculation ignores the diode's resistance and voltage drop. It is only an *approximation* of the circuit current. If we know the drop across the diode, it is possible to accurately predict the current. The diode drop is simply subtracted from the supply voltage:

$$I = \frac{6 \text{ V} - 0.6 \text{ V}}{1 \text{ k}\Omega} = 5.4 \text{ mA}$$

Forward bias

A typical silicon diode drops about 0.6 V when it is conducting. This is still an approximation, but it is more accurate than our first attempt.

EXAMPLE 3-1

Reverse bias

Calculate the current in Fig. 3-5 for a 1-V battery and a 1-kΩ resistor. Determine the importance of correcting for the diode voltage drop. First, calculate the current without correcting for the diode drop:

$$I = \frac{1 \text{ V}}{1 \text{ k}\Omega} = 1 \text{ mA}$$

Make a second calculation that includes the correction:

$$I = \frac{1 \text{ V} - 0.6 \text{ V}}{1 \text{ k}\Omega} = 0.4 \text{ mA}$$

It is important to correct for the diode drop when the supply voltage is relatively low.

EXAMPLE 3-2

Schottky diodes drop about 0.3 V when conducting. These diodes are explained in Sec. 3-4. Calculate the current in Fig. 3-5 for a Schottky diode, a 1-V battery, and a 1-kV resistor.

$$I = \frac{1 \text{ V} - 0.3 \text{ V}}{1 \text{ k}\Omega} = 0.7 \text{ mA}$$

The small voltage drop of Schottky diodes makes a significant difference in low-voltage circuits.

EXAMPLE 3-3

Calculate the current in Fig. 3-5 for a 100-V battery and a 1-kΩ resistor. Determine the importance of correcting for the voltage drop of a silicon diode.

$$I = \frac{100 \text{ V}}{1 \text{ k}\Omega} = 100 \text{ mA}$$

$$I = \frac{100 \text{ V} - 0.6 \text{ V}}{1 \text{ k}\Omega} = 99.4 \text{ mA}$$

It is not as important to correct for the diode drop when the supply voltage is relatively high.

The condition of Fig. 3-5 is called *forward bias*. In electronics, a bias is a voltage or a current applied to a device. Forward bias indicates that the voltage or current is applied so that it turns the device *on*. The diode in Fig. 3-5 has been turned on by the battery, so it is an example of forward bias.

Reverse bias is another possibility. With zero bias connected to the diode, the depletion region is as shown in Fig. 3-6(a). When reverse bias is applied to a junction diode, the depletion region does not collapse. In fact, it becomes wider than it was. Figure 3-6(b) shows a diode with reverse

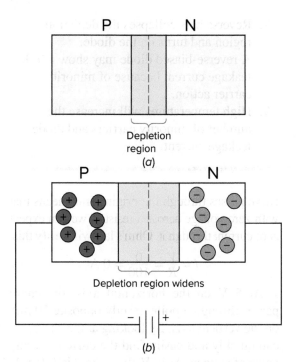

Fig. 3-6 The effect of reverse bias on the depletion region.

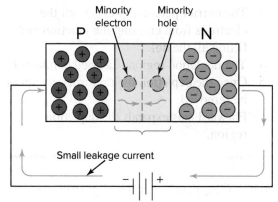

Fig. 3-7 Leakage current due to minority carriers.

bias applied. The positive side of the battery is applied to the N-type material. This attracts the free electrons away from the junction. The negative side of the battery attracts the holes in the P-type material away the junction. This makes the depletion region wider than it was when no voltage was applied. The transition from conducting to not conducting is not instantaneous. The reverse recovery time of a diode is important in high-frequency and fast-switching applications.

Because reverse bias widens the depletion region, it can be expected that no current flow will result. The depletion region is an insulator, and it will block the flow of current. Actually, a small current will flow because of *minority carriers*. Figure 3-7 shows why this happens. The P-type material has a few minority electrons. These are pushed to the junction by the repulsion of the negative side of the battery. The

N-type material has a few minority holes. These are also pushed toward the junction. Reverse bias forces the minority carriers together, and a small *leakage current* results. Diodes are not perfect, but modern silicon diodes usually show a leakage current so small that it cannot be measured with ordinary meters. At room temperature, there are only a few minority carriers in silicon, so the reverse leakage can be ignored.

In summary, the PN-junction diode will conduct readily in one direction and very little in the other. The direction of easy conduction is from the N-type material to the P-type material. If a voltage is applied across the diode to move the current in this direction, it is called forward bias. The diode is very useful because it can steer current in a given direction. It can also be used as a switch and a means of changing alternating current to direct current. Other diodes perform many special jobs in electric and electronic circuits.

Leakage current

ABOUT ELECTRONICS

Diodes Provide Protection from Reverse Polarity A diode can provide reverse polarity protection. One approach is to use a series protection diode, and the other is to use a shunt protection diode that causes a fuse to blow when polarity is reversed.

Self-Test

Determine whether each statement is true or false.

1. A junction diode is doped with both P- and N-type impurities.

2. The depletion region is formed by electrons crossing over the P-type side of the junction to fill holes on the N-type side of the junction.

3. The barrier potential prevents all the electrons from crossing the junction and filling all the holes.
4. The depletion region is a good conductor.
5. Once the depletion region forms, it cannot be removed.
6. Forward bias expands the depletion region.

7. Reverse bias collapses the depletion region and turns on the diode.
8. A reverse-biased diode may show a little leakage current because of minority-carrier action.
9. High temperatures will increase the number of minority carriers and diode leakage current.

3-2 Characteristic Curves of Diodes

Diodes conduct well in one direction but not in the other. This is the fundamental property of diodes. They have other characteristics too, and some of these must be understood in order to have a working knowledge of electronic circuits.

Characteristics of electronic devices can be shown in several ways. One way is to list the amount of current flow for each of several values of voltage. These values could be presented in a table. A better way to do it is to show the values on a graph. Graphs are easier to use than tables of data.

One of the most frequently used graphs in electronics is the volt-ampere characteristic curve. Units of voltage make up the horizontal axis, and units of current make up the vertical axis. Figure 3-8 shows a volt-ampere characteristic curve for a 100-Ω resistor. The origin is the point where the two axes cross. This point indicates zero voltage and zero current. Note that the resistor

curve passes through the origin. This means that with zero voltage across a resistor, we can expect zero current through it. Ohm's law will verify this:

$$I = \frac{V}{R} = \frac{0}{100} = 0 \text{ A}$$

At 5 V on the horizontal axis, the curve passes through a point exactly opposite 50 mA on the vertical axis. By looking at the curve, we can quickly and easily find the current for any value of voltage. At 10 V, the current is 100 mA. We can check this using Ohm's law:

$$I = \frac{V}{R} = \frac{10}{100} = 0.1 \text{ A} = 100 \text{ mA}$$

Moving to the left of the origin in Fig. 3-8, we can obtain current levels for values of reverse voltage. Reverse voltage is indicated by V_R, and V_F indicates the forward voltage. At −5 V, the current through the resistor will be −50 mA. The minus signs indicate that when the voltage across a resistor is reversed in polarity, the resistor current will reverse (change direction). Forward current is indicated by I_F, and I_R indicates reverse current.

The characteristic curve for a resistor is a straight line. For this reason, it is said to be a linear device. Resistor curves are not necessary. With Ohm's law to help us, we can easily obtain any data point without a graph.

EXAMPLE 3-4

How would Fig. 3-8 appear for a 50-Ω resistor?

$$I = \frac{V}{R} = \frac{10 \text{ V}}{50 \text{ Ω}} = 200 \text{ mA}$$

The curve would be a straight line passing through the origin and through data points at ±10 V and ±200 mA. Thus, the 50-Ω curve would be steeper (have more slope) than the 100-Ω curve.

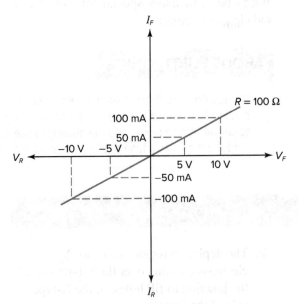

Fig. 3-8 A volt-ampere characteristic curve for a resistor.

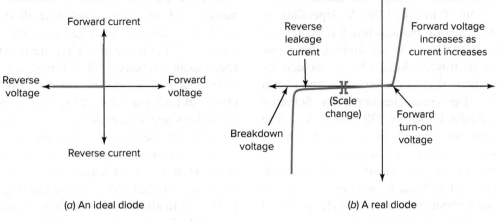

(a) An ideal diode   (b) A real diode

Fig. 3-9 Diode volt-ampere characteristic curves.

Diodes are more complicated than resistors. Their volt-ampere characteristic curves give more information than can be provided with a simple equation. Figure 3-9 shows volt-ampere curves for both an ideal diode and a real diode. These curves are not linear like the one shown in Fig. 3-8. Ideal diodes do not exist, but real diodes can come close to being ideal in some situations. It was already mentioned that the forward voltage drop can be ignored in high-voltage circuits. Thus, the ideal diode volt-ampere curve shows zero forward voltage. Also, an ideal diode has no leakage current and never conducts at all when subjected to reverse voltage, no matter how much.

The real diode shown in Fig. 3-9 has some forward voltage drop and a small amount of leakage current, and perhaps most important, it has a limit called the *breakdown voltage*. This breakdown usually occurs at hundreds of volts, so the scale of the horizontal axis is much larger to the left of the origin. The scale for the left side is perhaps from 0 to 1,000 V and from 0 to 2 V on the right side. The forward turn-on voltage is about 0.65 V for a silicon diode. This occurs with a small value of forward current, perhaps 1 mA. With larger values of forward current, the forward voltage increases, perhaps to 1 V at 1 A. The reverse leakage current is often less than 1 mA, and so the reverse current axis is often calibrated in much smaller units of current.

A comparison of the characteristic curves for a silicon diode and a germanium diode is shown in Fig. 3-10. Barrier diodes, which are discussed in this section, also have a lower

turn-on voltage. Barrier diodes are becoming more widely applied. Their lower forward bias voltage and fast turn-off are major advantages. Germanium diodes are not used in new designs and suppliers can be hard to find.

Figure 3-10 also shows how silicon and germanium (or barrier) diodes compare under conditions of reverse bias. At reasonable levels of V_R, the leakage current of the silicon diode is very low. The germanium diode shows much more leakage. However, if a certain critical value of V_R is reached, the silicon diode will show a rapid increase in reverse current. This is shown as the *reverse breakdown point*. It is also referred to as the *avalanche voltage*. Avalanche breakdown occurs when carriers accelerate and gain enough energy to collide with valence electrons and knock them loose. This causes an "avalanche" of carriers, and the reverse current flow increases tremendously.

Reverse breakdown point

Avalanche voltage

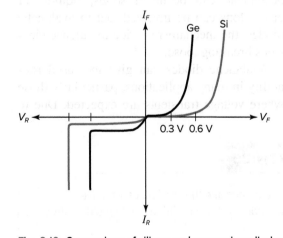

Fig. 3-10 Comparison of silicon and germanium diodes.

The avalanche voltage for silicon diodes ranges from 50 to over 1,000 V, depending on how the diode was manufactured. If the reverse current at avalanche is not limited, the diode will be destroyed. Avalanche is avoided by using a diode that can safely withstand circuit voltages. The avalanche rating for Schottky barrier diodes is usually 200 volts or less. The avalanche rating for silicon carbide barrier diodes can be well over 1,000 volts.

Some diodes are manufactured to break down, or avalanche, at a specified voltage and to do so without harm to the diode, provided that the energy is limited. Ordinary diodes are often destroyed by reverse breakdown. The reverse current tends to be concentrated in one spot, which causes heat and damage. Avalanche diodes can be used to safely absorb high-voltage transients and, by doing so, protect the rest of the circuit or another piece of equipment from damage. There are several categories of devices that can absorb transient voltages, but the avalanche types are the fastest acting (rated in picoseconds) and are preferred for some applications.

Zener diodes, covered in the next section, are also manufactured to break down at a specified voltage. However, the voltages are usually less, and the actual breakdown mechanism is different. *Avalanche* implies what the term refers to; for example, on a steep hillside one rock can break loose and strike other rocks and result in a shower of rocks flowing down the hill. In an avalanche diode, a valence electron, subject to the field of the reverse voltage, can break loose and strike other valence electrons, leading to a large increase of reverse current. Avalanche can occur in solids, liquids, or gases. Ions can be involved, but in avalanche diodes, the mechanism is due to valence electrons breaking loose.

Avalanche diodes can give increased reliability in many applications, particularly those where voltage transients are expected. Due to their high speed and ability to withstand large numbers of transients, avalanche diodes are used to protect circuits against surges, lightning, and other transients. They are faster than metal oxide variances (MOVs), zeners, and gas tubes. Avalanche diodes are the diodes of choice in high-voltage circuits, such as voltage multipliers and where diodes are connected in series to achieve high-voltage operation.

Inductive loads often generate voltage transients when the circuit is interrupted. Diodes are often used to control these transients (Fig. 3-27, p. 49) and to allow current to flow so as to discharge the inductor. These are often called freewheeling diodes and are discussed in more detail in later chapters. Avalanche diodes are often preferred for free-wheeling applications.

Figure 3-11 shows how volt-ampere characteristic curves can be used to indicate the effects of temperature on diodes. The temperatures are in degrees Celsius (°C). Electronic circuits may have to work over a range of temperatures from −50° to +100°C. At the low end mercury will freeze; at the high end water will boil. The range for military-grade electronic circuitry is −55° to +125°C. For circuits to operate in such a wide temperature range, extreme care must be taken in the selection of materials, the manufacturing processes used, and the handling and testing of the finished product. This is why military-grade devices are more expensive than industrial- and commercial-grade devices.

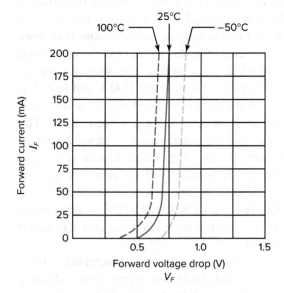

Fig. 3-11 Characteristic curves showing the effect of temperature on a typical silicon diode.

By examining the curves in Fig. 3-11, you can conclude that silicon conducts better at elevated temperatures. Since the forward voltage drop V_F decreases as temperature goes up, its resistance must be going down. This agrees with silicon's negative temperature coefficient. Figure 3-11 also shows that diodes can be used as temperature sensors.

Self-Test

Supply the missing word in each statement.

10. The characteristic curve for a linear device is shaped like a _____.
11. A volt-ampere characteristic curve for a resistor is shaped like a _____.
12. A volt-ampere characteristic curve for a 1,000-Ω resistor will, at 10 V on the horizontal axis, pass through a point opposite _____ on the vertical axis.
13. The volt-ampere characteristic curve for an open circuit (∞ Ω) will be a straight line on the _____ axis.
14. The volt-ampere characteristic curve for a short circuit (0 Ω) will be a straight line on the _____ axis.
15. Resistors are linear devices. Diodes are _____ devices.
16. A silicon diode does not begin conducting until _____ V of forward bias is applied.
17. Diode avalanche, or reverse breakdown, is caused by excess reverse _____.

3-3 Diode Lead Identification

Diodes have *polarity*. Components such as resistors can be wired either way into a circuit, but diodes must be installed properly. Connecting a diode backward can destroy it and may also damage many other parts of a circuit. A technician must always be absolutely sure that the diodes are correctly connected.

Technicians often refer to schematic diagrams when checking diode polarity. Figure 3-12 shows the *schematic symbol* for a diode. The P-type material makes up the *anode* of the diode. The word "anode" is used to identify the terminal that attracts electrons. The N-type material makes up the *cathode* of the diode.

The word "cathode" refers to the terminal that gives off, or emits, electrons. Note that the forward electron current moves from the cathode to the anode (against the arrowhead).

Diodes are available in many package styles. Some examples are shown in Fig. 3-13. Manufacturers use plastic, glass, metal, ceramic, or a combination of these to package diodes. There are quite a few sizes and shapes available. Generally, the larger devices have higher current ratings. The diode package is often marked to denote the *cathode lead*. This can be done with one or more bands near the cathode lead. An example of this method is shown on the DO-41 package in Fig. 3-13. Some older package styles used a bevel or a plus sign (+) to denote the cathode lead.

Other packages use various schemes for lead identification. A few use an imprint of the diode symbol. This method can be used with the 194-05 package in Fig. 3-13, although the illustration does not show it. The TO-220AC style has both a cathode lead and a metal tab, which also

Polarity

Schematic symbol

Cathode lead

Anode

Cathode

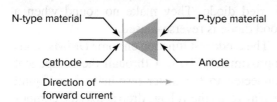

Fig. 3-12 Diode schematic symbol.

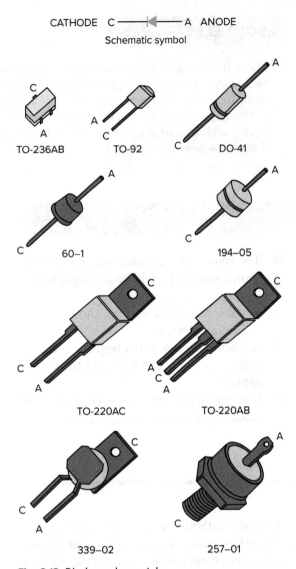

CATHODE C ——►|—— A ANODE
Schematic symbol

TO-236AB   TO-92   DO-41

60–1   194–05

TO-220AC   TO-220AB

339–02   257–01

Fig. 3-13 Diode package styles.

serves as a cathode contact. Either the lead or the tab can be used to connect the diode to the rest of the circuit. The TO-220AB case shows two anode leads. This is a different situation because there are two diodes inside the package. The anodes of the two diodes are available as separate terminals, but the cathodes are connected internally.

Manufacturers can offer diodes in both a normal polarity version and a reverse polarity version. For example, the threaded stud end of the 257-01 package in Fig. 3-13 is the anode in the reverse polarity version. The part number is followed by an "R" to denote the reverse polarity version. However, the part number is rarely marked on the device. Another problem is that manufacturers use the same package to house different devices. Both the TO-236AB package

and the TO-220AB package shown in Fig. 3-13 can also be used for transistors. In other words, a casual inspection of an electronic circuit will not always allow you to positively identify components and their leads. You should use schematics or other service literature to be certain.

It is easy to check a diode and identify the leads using a volt-ohm-milliammeter (VOM), or a digital multimeter (DMM). This check uses the ohmmeter function of the meter or, in the case of some DMMs, the diode test function. The ohmmeter is connected across the diode and the resistance is noted as in Fig. 3-14(a). When the diode is on or forward biased a relatively low reading on the ohm's scale will occur or the forward voltage drop will be displayed in the case of a DMM. Then the ohmmeter leads are reversed as shown in Fig. 3-14(b). The resistance should change drastically—usually to infinity, or the DMM should indicate over range or overload (OL). In Fig. 3-14 we conclude that the diode is good and that the cathode lead is at the left. When the positive lead of the ohmmeter was on the right lead, the diode was turned on. Forward current is from cathode to anode. Making the anode positive is necessary if the anode is going to attract electrons. Remember, in order to turn on the diode, the anode must be positive with respect to the cathode.

Diode testing is usually straight forward, but there are a few qualifiers to consider. An older meter might have reverse polarity on resistance ranges. Another meter might not apply enough voltage to turn on a diode. Yet another meter could have a low ohms function that will show a good diode to be open circuit. You must know the characteristics and limitations of your test equipment.

Modern DMMs have an ohms range and a diode range. The diode range is usually marked with the diode schematic symbol. Use the diode range when testing diodes.

Some DMMs have an audible output on the diode range. They beep once when a good diode is forward-biased and beep continuously for a shorted diode. They make no sound when a good diode is reverse-biased.

The diode test function on some DMMs sends approximately 0.6 mA through the component connected to the meter terminals. The digital display reads the voltage drop across the component. A normal, forward-biased junction will

(a) The diode is on (forward bias)

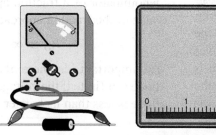

(b) The diode is off (reverse bias)

Fig. 3-14 Diode testing and lead identification.

read somewhere between 0.250 and 0.700 using this type of meter. A reverse-biased junction will cause the meter display to indicate overrange.

Table 3-1 shows some typical readings obtained using a DMM on its ohms function and on its diode function to test various diode types. In every case, the diode was normal and was forward-biased by the meter. Notice that as the current capacity (size) of the silicon diodes increases, the diode's forward resistance decreases when using the ohms function, and the voltage drop across the diode is smaller when using the diode function. Also notice that the Schottky and germanium diodes show the

lowest resistances and voltage drops. Schottky diodes are explained in the next section.

Diodes are nonlinear devices. They will not show the same resistance when operated at different levels of forward bias. For example, a silicon diode might show 500 Ω of forward resistance when measured on a 2-kΩ range and 5 kΩ of forward resistance when measured on a 20-kΩ range. This is to be expected since the ohmmeter operates the diode at different points on its characteristic curve when different ranges are selected. Figure 3-15 illustrates this idea.

Table 3-1	Typical Results of Diode Testing with a Digital Multimeter (DMM)

	Results	
Device Tested	Ohms Function (kΩ)	Diode Function
Small silicon diode	19	0.571
1-A silicon diode	17	0.525
5-A silicon diode	14	0.439
100-A silicon diode	8.5	0.394
Schottky (barrier) diode	7	0.199
Small germanium diode	3	0.277

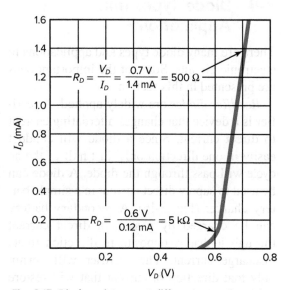

Fig. 3-15 Diode resistance at different operating points.

Ohm's law is used to calculate diode resistance at two different operating points on the *characteristic curve*. At the upper operating point the diode's resistance is 500 Ω, and it is 5 kΩ at the lower operating point.

Beginners may be confused by diode polarity. There is a good reason, too. One of the older ways to mark the cathode lead was to use a plus (+) symbol (this is no longer done by diode manufacturers). Yet, we have said that the diode is turned on when its *anode lead* is made positive. This seems to be a contradiction. However, the reason the plus sign was used to indicate the cathode lead is related to how the diode behaves in a *rectifier circuit*. In a rectifier circuit, it is the cathode lead that is in contact with the positive end of the load. So, the plus sign was used to help technicians find the load polarity.

Rectifier circuits are covered briefly in the next section and in detail in Chap. 4.

EXAMPLE 3-5

Find R_D for Fig. 3-15 when $V_D = 0.2$ V. If we attempt to use Ohm's law,

$$R_D = \frac{V_D}{I_D} = \frac{0.2 \text{ V}}{0} = \text{undefined}$$

Division by 0 is undefined. However, as the denominator of a fraction approaches 0, the value of the fraction approaches infinity:

$$R_D \Rightarrow \infty$$

The important idea here: the resistance of a diode is infinite if the voltage drop across the diode is less than its barrier potential.

Self-Test

Supply the missing word in each statement.

18. Assume that a diode is forward-biased. The diode lead that is connected to the negative side of the source is called the _____.

19. The diode lead near the band or bevel on the package is the _____ lead.

20. A plus (+) sign on an older diode indicates the _____ lead.

21. An ohmmeter is connected across a diode. A low resistance is shown. The leads are reversed. A low resistance is still shown. The diode is _____.

22. When the positive lead from an ohmmeter is applied to the anode lead of a diode, the diode is turned _____.

23. Diodes show different values of forward resistance on different ohmmeter ranges because they are _____.

3-4 Diode Types and Applications

There are many diode types and applications in electronic circuits. Some of the important ones are presented in this section.

Rectifier diodes are widely applied. A rectifier is a device that changes alternating current to direct current. Since a diode will conduct easily in one direction only, just half of the ac cycle will pass through the diode. A diode can be used to supply direct current in a simple battery charger (Fig. 3-16). A secondary battery can be charged by passing a direct current through it that is opposite in direction to its discharge current. The rectifier will permit only that direction of current that will restore (recharge) the battery.

Notice in Fig. 3-16 that the diode is connected so the current flow during charging is opposite to the current flow during discharging. The cathode of the diode *must* be connected to the positive terminal of the battery. A mistake in this connection would discharge the battery or damage the diode. It is very important to connect diodes correctly.

D_1 and D_2 in Fig. 3-17 are called *steering diodes*. Normally, the main supply energizes the load with D_1 on and D_2 off. With D_1 on, the voltage at the cathodes is about +12.3 V, which means D_2 is reverse-biased (its cathode is more positive than its anode). If the main supply fails, D_2 is forward-biased and the backup supply provides the load current through D_2.

The load voltage is now less than it is with normal operation: about 11.3 V instead of 12.3 V.

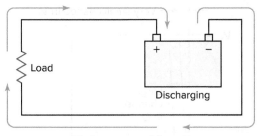

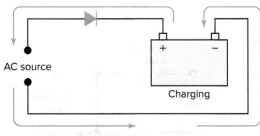

Discharge current flows in this direction

Diode will allow current to flow in this direction only

Fig. 3-16 Battery charging with a diode.

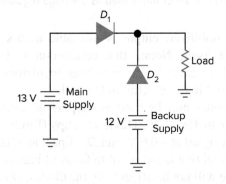

Fig. 3-17 Steering diodes.

Note that a shorted main supply will have no effect on the load voltage because D_1 is reverse-biased.

An ideal rectifier would turn off at the instant it is reverse-biased. PN-junction diodes cannot turn off instantaneously. There are quite a few holes and electrons around the junction when a diode is conducting. Applying reverse bias will not immediately turn the diode off since it takes time to sweep these carriers away from the junction and establish a depletion region. This effect is not a problem when rectifying low frequencies such as 60 Hz. However, it is a factor in high-frequency circuits. Barrier diodes do not have to clear carriers from a PN junction when turning off which allows them to rectify frequencies much higher than 60 Hz.

Some metal-to-semiconductor interfaces will also rectify. This type of interface is called a *barrier. Schottky diodes* (or *barrier diodes*) use an N-type chip of silicon bonded to platinum. This semiconductor-to-metal barrier provides diode action and turns off much more quickly than a PN junction. Figure 3-18 shows the schematic symbol for a Schottky diode.

Schottky diode

When a Schottky diode is forward-biased, electrons in the N-type cathode must gain energy to cross the barrier to the metal anode. The term *hot-carrier diode* is sometimes used because of this fact. Once the "hot carriers" reach the metal, they join the great number of free electrons there and quickly give up their extra energy. When reverse bias is applied, the diode stops conducting almost immediately since a depletion region does not have to be established to block current flow. A PN junction diode might have a reverse recovery time measured in microseconds while a Schottky barrier diode is measured in nanoseconds. The electrons cannot cross back over the barrier because they have lost the extra energy required to do so. However, if more than about 50 V of reverse bias is applied, the electrons will gain the required energy, and the barrier will break over and conduct. This prevents some barrier-type devices from being used in high-voltage circuits. Another barrier diode using silicon carbide (SiC) is noted for its high reverse breakdown capability. Some SiC diodes can operate safely well beyond 1,000 V. Schottky diodes require only about 0.3 V of forward bias to establish forward current. They are well suited for high-frequency, low-voltage applications. They are commonly used in switch mode power supplies, which are covered in Chap. 15.

Hot-carrier diode

A diode can be used to hold a voltage constant. This is called *voltage regulation*. A special type called a *zener diode* is used as a voltage regulator.

Voltage regulation

Zener diode

Fig. 3-18 Schottky diode schematic symbol.

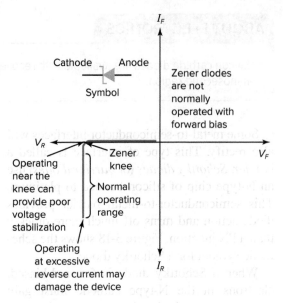

Fig. 3-19 Characteristic curve and symbol of a zener diode.

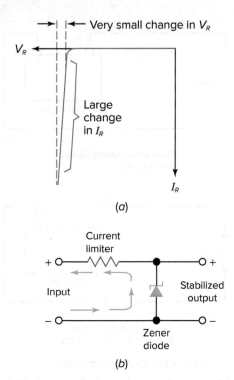

(a)

(b)

Fig. 3-20 A zener diode used as a voltage regulator.

The characteristic curve and symbol for a zener diode are shown in Fig. 3-19. The symbol is similar to that of a rectifier diode except that the cathode is drawn as a bent line representing the letter Z. Zener diodes are manufactured to regulate voltages from 3.3 to 200 V. As an example, the 1N4733 is a popular 5.1-V zener.

The important difference between zener diodes and rectifier diodes is in how they are used in electronic circuits. As long as zeners are operated within their normal range, their voltage drop will equal their rated voltage plus or minus a small error voltage. They are operated *backward* compared with a rectifier diode. In a rectifier, the normal current is from cathode to anode. Zeners are operated in reverse breakover and conduct from anode to cathode.

A change in zener diode current will cause only a small change in the zener voltage. This can be seen clearly in Fig. 3-20(*a*). Within the normal operating range, the zener voltage is reasonably stable.

Figure 3-20(*b*) shows how a zener diode can be used to stabilize a voltage. A current-limiting resistor is included to prevent the zener diode from conducting too much and overheating.

The stabilized output is available across the diode itself. Notice that conduction is from anode to cathode. Zener voltage regulators are covered in more detail in Chap. 4.

Diodes may be used as *clippers* or *limiters*. Refer to Fig. 3-20. Diode D_1 clips (limits) the input signal at -0.6 V, and D_2 clips it at $+0.6$ V. A signal that is too small to forward-bias either diode will not be affected by the diodes. Diodes have a very high resistance when they are off. However, a large signal will turn the diodes on, and they will conduct. When this happens, the excess signal voltage is dropped across R_1. Therefore, the total output swing is limited to 1.2 V peak-to-peak. This kind of limiting action may be used if a signal gets too large. For example, clippers can be used to keep audio signals from exceeding some loudness limit.

Figure 3-21 shows that the input signal is a sine wave, but the output signal is more like a square wave. Sometimes a clipping circuit is used to change the shape of a signal. A third way that clippers can be used is to remove noise pulses riding on a signal. If the noise pulses exceed the clipping points, they will be clipped off or limited. The resulting signal is more noise-free than the original.

Diode D_2 clips the positive part of the signal in Fig. 3-21. As the signal voltage begins

Clippers or limiter

ABOUT ELECTRONICS

Barrier diodes are used in high-frequency rectifier circuits and radio frequency detectors.

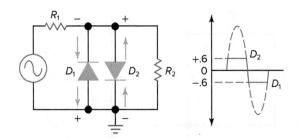

Fig. 3-21 Diode clipper.

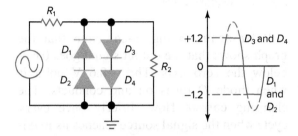

Fig. 3-22 Clipping at a higher threshold.

increasing from 0 V, nothing happens at first. Then, when the signal voltage reaches 0.6 V, D_2 turns on and begins to conduct. Now its resistance is much less than the resistance of R_1. Resistor R_1 drops the signal source voltage that exceeds 0.6 V. Later the negative alternation begins. As the signal first goes negative, nothing happens. When it reaches −0.6 V, D_1 turns on. As D_1 conducts, R_1 drops the signal voltage in excess of −0.6 V. The total output swing is the difference between +0.6 and −0.6 V, or 1.2 V peak-to-peak. Germanium diodes would turn on at 0.3 V and produce a total swing of 0.6 V peak-to-peak if used in a clipper circuit.

The clipping points can be changed to a higher voltage by using series diodes. Examine Fig. 3-22. It will require 0.6 V + 0.6 V, or 1.2 V, to turn on D_3 and D_4. Notice that the positive clipping point is now shown on the graph at +1.2 V. In a similar fashion, D_1 and D_2 will turn on when the signal swings to −1.2 V. The output signal in Fig. 3-21 has been limited to a total swing of 2.4 V peak-to-peak. Higher clipping voltages can be obtained by using zener diodes, as shown in Fig. 3-23(a). Assume that D_2 and D_4 are 4.7-V zeners. The positive-going signal will be clipped at +5.3 V since it takes 4.7 V to turn on D_4 and another +0.6 V to turn on D_3. Diodes D_1 and D_2 clip the negative alternation at −5.3 V. The total peak-to-peak output signal in Fig. 3-23(a) is limited to 10.6 V.

When a zener diode is *forward-biased*, it drops a bit more than a rectifier diode (about

0.7 V). Therefore, the circuit in Fig. 3-23(a) can be simplified by using two zeners back to back, as shown in Fig. 3-23(b). If the current is flowing up, then the bottom zener will drop 0.7 V, and the top zener will drop its rated voltage. When the current is flowing down, the top zener will drop 0.7 V, and the bottom zener will drop its rated voltage. For example, if the circuit uses two 1N4733s (5.1-V devices), the total output swing will be limited to 5.1 + 0.7 = 5.8 V peak voltage, or 11.6 V peak-to-peak.

Diodes may also be used as *clamps* or *dc restorers*. (Refer to Fig. 3-24.) The signal source

Clamps or dc restorer

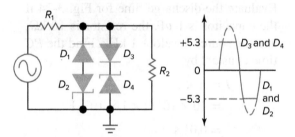

Fig. 3-23(a) Using zener diodes to set a higher clipping threshold.

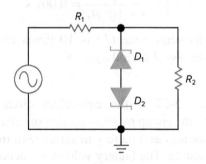

Fig. 3-23(b) A simplified high-threshold clipper.

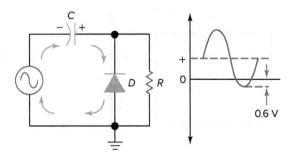

Fig. 3-24 Positive clamp.

generates an ac waveform. The graph shows that the output signal that appears across the resistor is not ordinary alternating current. It does not have an average value of 0 V. It averages to some positive voltage. Such signals are common in electronic circuits and are said to have both an ac component and a dc component. Where does the dc component come from? The diode creates it by charging the capacitor. Note that diode D in Fig. 3-24 will allow a charging current to flow into the left side of capacitor C. This current places extra electrons on the left side of the capacitor, and a negative charge results. Electrons flow off the right plate of the capacitor and make it positive. If the discharge time of the circuit ($T = R \times C$) is long compared with the period of the signal, the capacitor will maintain a steady charge from cycle to cycle.

EXAMPLE 3-6

Evaluate the discharge time for Fig. 3-24 if the capacitor is 1 μF, the resistor is 10 kΩ, and the source develops 1 kHz. Find the RC time constant by

$$T = R \times C$$
$$= 10 \times 10^3 \, \Omega \times 1 \times 10^{-6} \, F$$
$$= 0.01 \, s$$

Find the period of the signal:

$$t = \frac{1}{f} = \frac{1}{1 \times 10^3 \, Hz} = 0.001 \, s$$

The discharge time (T) is 10 times larger than the signal period (t).

Figure 3-25 is the equivalent circuit. It explains the clamp by showing that the charged capacitor acts as a battery in series with the ac signal source. The battery voltage V_{dc} accounts for the upward shift shown in the graph.

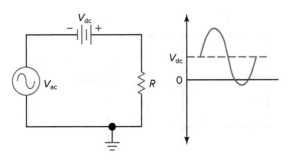

Fig. 3-25 Clamp equivalent circuit.

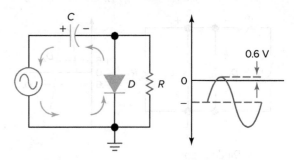

Fig. 3-26 Negative clamp.

Refer again to Fig. 3-24. Note that the graph shows that the output signal goes 0.6 V below the zero axis. This −0.6 V point is when diode D turns on and conducts. The charging current flows briefly once every cycle when the signal source reaches its maximum negative voltage.

Figure 3-26 shows what happens if the diode is reversed. The charging current is reversed, and the capacitor develops a negative voltage on its right plate. Notice that the graph shows that the output signal has a negative dc component. This circuit is called a *negative clamp*.

Clamping sometimes happens when we do not want it. For example, a signal generator is often used for circuit testing. Some signal generators use a coupling capacitor between their output circuitry and their output jack. If you connect such a generator to an unbalanced diode load that allows a charge to build up on the built-in coupling capacitor, confusing results may occur. The resulting dc charge will act in series with the ac signal and may change the way the test circuit works. A dc voltmeter or a dc-coupled oscilloscope can be connected from ground to the output jack to verify that clamping is occurring.

Figure 3-27 shows how diodes are sometimes used to prevent arcing and component damage. When the current is suddenly interrupted in a coil, a large counterelectromotive force (CEMF) is generated across the coil. This high voltage can cause arcing and can also destroy sensitive devices, such as integrated circuits and transistors. Note that in Fig. 3-27(a), there is an arc when the switch in series with the relay coil opens. In Fig. 3-27(b), there is a protection diode across the coil. This diode is forward-biased by the CEMF. The diode safely discharges the coil and prevents arcing or damage.

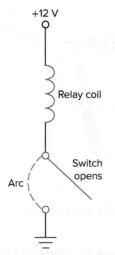

(a) Inductive kick causes an arc when switch opens

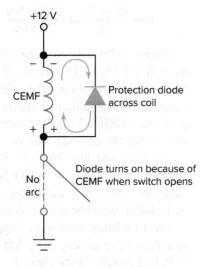

(b) No arc

Fig. 3-27 Using a diode to stop "inductive kick."

Another important diode type is the *light-emitting diode,* or LED. Its schematic symbol is shown in Fig. 3-28(a). Figure 3-28(b) shows that as the electrons of the LED cross the junction, they combine with holes. This changes their status from one energy level to a lower energy level. The extra energy they had as free electrons must be released. Silicon diodes give off this extra energy as heat. *Gallium arsenide diodes* release some of the energy as heat and some as infrared light. This type of diode is called an infrared-emitting diode (IRED). Infrared light is not visible to the human eye. By doping gallium arsenide with various materials, manufacturers can produce diodes with visible outputs of red, green, or yellow light.

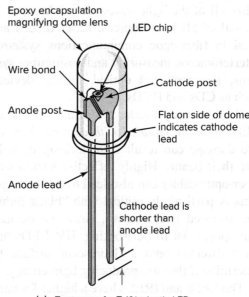

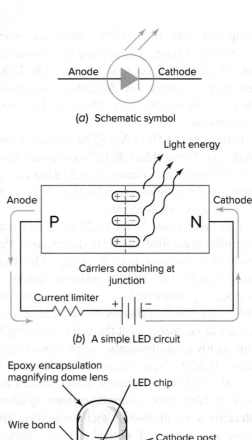

(a) Schematic symbol

(b) A simple LED circuit

(c) Features of a T-1¾ plastic LED

Fig. 3-28 Light-emitting diode.

More recently, blue LEDs have become more efficient and less expensive to manufacture. A white LED is a blue LED that's surrounded by a phosphorescent dye that glows white when it is struck by blue light. This is a similar process to that in fluorescent lamps where the coating glows white when it is irradiated by the ultraviolet light generated inside the tube. White LEDs are now replacing incandescent lamps in some applications. They are more efficient, don't produce as much unwanted infrared, and have an operating life of 100,000 hours

Light-emitting diode

compared with only 8,000 hours for many incandescent types. LEDs have now exceeded the efficiency of compact fluorescent lights and don't contain any hazardous materials. They are becoming very attractive for many applications.

Ultraviolet LEDs (UV LEDs) are now being produced. These "black light" sources are finding applications in currency validation equipment, medical and biological detectors, security systems, and leak detectors.

The laser diode is an LED or IRED with carefully controlled physical dimensions that produce a resonant optical cavity. The light energy builds up as the resonant cavity is pumped by semiconductor photon emission. The cavity acts as a sharply tuned filter, and all of the output energy is at the same wavelength. This yields monochromatic (single-color) light. Also, all of the light waves are in phase, as is typical of all laser sources. Laser diodes are used in fiber-optic communications systems, interferometric measuring and positioning systems, scanners, and optical storage devices such as CDs and DVDs.

High-intensity LEDs, UV LEDs, and laser LEDs *must be handled with caution.* Serious eye damage can result from looking directly into their beams. Highly reflective surfaces or fiber-optic cables can also lead to eye damage. This is particularly critical with "black light" and infrared laser sources, since the devices can appear not to be working. UV LEDs are often directed onto a fluorescent surface to determine if they are producing light energy.

The LEDs and IREDs have a higher forward voltage drop than do silicon diodes. This drop varies from 1.5 to 2.5 V depending on diode current, the diode type, and its color. If the manufacturer's data is not available, 2 V is a good starting point. Assume that the diode circuit in Fig. 3-28(b) is being designed for an LED current of 20 mA and that the supply (battery) produces 5 V. Ohm's law is used to find the value of the current-limiter resistor. The

Seven-segment display

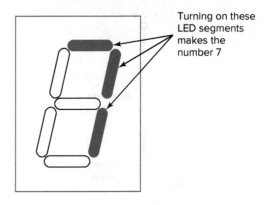

Fig. 3-29 An LED numeric display.

diode drop must be subtracted from the supply to find the voltage across the resistor:

$$R = \frac{V_S - V_D}{I_D} = \frac{5\text{ V} - 2\text{ V}}{20\text{ mA}} = 150\ \Omega$$

Figure 3-28(c) shows the physical appearance of a T-1¾ LED package. The T-1¾ package is 5 millimeters (mm) in diameter and is a common size. Another common size is the T-1 package, which is 3 mm in diameter. The figure shows that the cathode lead is shorter than the anode lead and also that the flat side of the dome can be used to identify the cathode lead. As with other diode types, LEDs *must* be installed with the correct polarity.

Light-emitting diodes are rugged and small, and they have a very long life. They can be switched rapidly since there is no thermal lag caused by gradual cooling or heating in a filament. They lend themselves to certain photochemical fabrication methods and can be made in various shapes and patterns. They are much more flexible than incandescent lamps. Light-emitting diodes may be used as numeric displays to indicate the numerals 0 through 9. A typical *seven-segment display* is shown in Fig. 3-29. By selecting the correct segments, the desired number is displayed.

EXAMPLE 3-7

Select a current-limiting resistor for an automotive circuit in which the diode current needs to be 15 mA. Such circuits use 12 V, and we can assume a 2-V diode drop:

$$R = \frac{12\text{ V} - 2\text{ V}}{15\text{ mA}} = 667\ \Omega$$

The power dissipation in a current-limiting resistor can also be important:

$$P = I^2R = (15 \text{ mA})^2 \times 667 \ \Omega = 150 \text{ mW}$$

For better reliability, power dissipation is normally doubled. Since 300 mW is more than $\frac{1}{4}$ W, a $\frac{1}{2}$-W resistor would be a good choice.

Photodiodes are silicon devices sensitive to light input. They are normally operated in reverse bias. When light energy enters the depletion region, pairs of holes and electrons are generated and support the flow of current. Thus, a photodiode shows a very high reverse resistance with no light input and less reverse resistance with light input. Figure 3-30 shows an optocoupler circuit. An *optocoupler* is a package containing an LED or IRED and a photodiode or phototransistor. When S_1 is open, the LED is off and no light enters the photodiode. The resistance of the photodiode is high, and the output signal will be high. When S_1 is closed, the LED is on. Light enters the photodiode so its resistance drops, and the output signal drops to a lower level because of the voltage drop across R_2. Optocouplers are used to electrically isolate one circuit from another. They are also called *optoisolators*. The only thing connecting the input circuit to the output circuit in Fig. 3-30 is light, so they are electrically isolated from each other.

Light-emitting diodes and photodiodes are often used in conjunction with fiber-optic cable for the purpose of data transmission. Compared with wire, fiber-optic cable is more expensive but has several advantages:

1. Elimination of electrical and magnetic field interference
2. Greater data capacity for long runs
3. Data security
4. Safe in explosive environments
5. Smaller and lighter

Both LEDs and laser diodes can be pulsed rapidly to allow high-speed data transmission. At the other end, a light detector is needed to change the light back into electrical pulses. Photodiodes are used to accomplish this. Figure 3-31 shows that light from a diode enters one end of a cable and leaves the other end where it strikes another diode.

Figure 3-31 also shows the construction and types of *fiber-optic cables*. These cables are light pipes. The principle of operation is total internal reflection of light. When light strikes a transparent surface, it is divided into a reflected beam and a refracted (bent) beam. If a ray of light strikes at some angle less than a so-called critical angle, all the light is reflected. If a core material is cladded with a different material having a smaller refractive index, total reflection is achieved for those rays that strike the cladding at shallow angles. Most light cables use various blends of silica glass for the core and for the cladding.

The step-index multimode fiber shown in Fig. 3-31 uses a relatively large core. Thus, some of the light rays that make up a light pulse may travel a direct route, whereas others zig and zag as they bounce off the cladding. Different rays arrive at the detector diode at different times, depending on different path lengths. The output pulse is spread in time. Look closely at

Photodiode

Fiber-optic cables

Optocoupler

Optoisolator

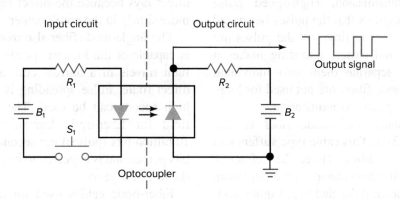

Fig. 3-30 An optocoupler circuit.

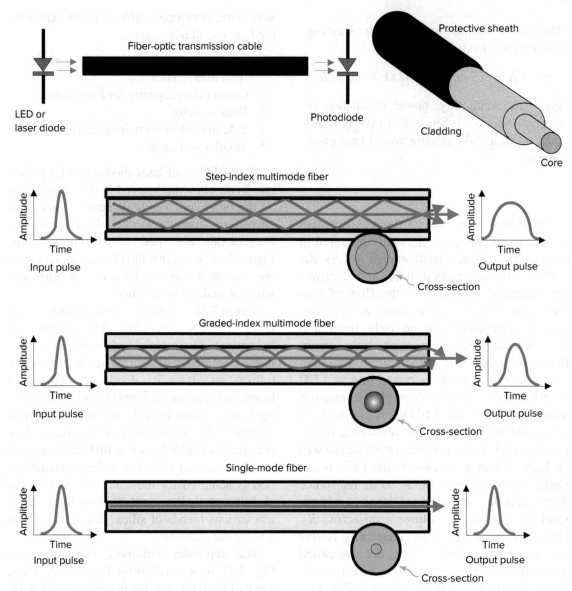

Fig. 3-31 Fiber-optic cables.

the relationship between input and output pulses in Fig. 3-31. You can see that the pulse spreading in the multimode cable does not allow high-speed transmission. High-speed pulse transmission requires that the pulses be spaced very close together in time. As the pulses are spaced closer and closer, spreading makes it impossible to separate them into individual pulses. Multimode fibers are not used for long-distance, high-speed communication.

A graded-index multimode fiber is also shown in Fig. 3-31. This cable type suffers less output pulse spreading. Here, the refractive index of a smaller core changes gradually from the center out toward the cladding. Light traveling down the core curves rather than zigzags;

this is due to the gradual change in the index. Also, the bent (curved) rays wind up arriving at the diode detector at about the same time as the direct rays because the direct rays must travel more slowly in the core's center.

The single-mode fiber also shown in Fig. 3-31 is capable of the highest speeds. Note that the light travels in a narrow core and only by a direct route. Pulse spreading is minimal, and high speeds can be used. The current speed limit for fiber-optic transmission is about 10 billion bits (pulses) per second. One trillion bits per second is expected to be reached within the next few years.

Fiber-optic cables used for data transmission typically carry light signals at levels of

100 microwatts (µW) or less. Eye damage is not possible at these levels. However, other applications may use much higher power levels. Never look into the end of a fiber-optic cable unless the power level has been verified as absolutely safe. Also, remember that some systems use infrared light. What you can't see can hurt you.

The *varicap* or *varactor* diode is a solid-state replacement for the variable capacitor. Much of the tuning and adjusting of electronic circuits involves changing capacitance. Variable capacitors are often large, delicate, and expensive parts. If the capacitor must be adjusted from the front panel of the equipment, a metal shaft or a complicated mechanical connection must be used. This causes some design problems. The varicap diode can be controlled by voltage. No control shaft or mechanical linkage is needed. The varicap diodes are small, rugged, and inexpensive. They are used instead of variable capacitors in modern electronic equipment.

The capacitor effect of a PN junction is shown in Fig. 3-32. A capacitor consists of two conducting plates separated by a dielectric material or insulator. Its capacitance depends on the area of the plates as well as on their separation. A reverse-biased diode has a similar electrical format. The P-type material semiconducts and forms one plate. The N-type material also semiconducts and forms the other plate. The depletion region is an insulator and forms the dielectric. By adjusting the reverse bias, the width of the depletion region, that is, the dielectric, is changed; and this changes the capacity

of the diode. With a high reverse bias, the diode capacitance will be low because the depletion region widens. This is the same effect as moving the plates of a variable capacitor farther apart. With little reverse bias, the depletion region is narrow. This makes the diode capacitance increase.

Figure 3-33 shows the capacitance in picofarads (pF) versus reverse bias for a varicap *tuning diode*. Capacitance decreases as reverse bias increases. The varicap diode can be used in a simple *LC* tuning circuit, as shown in Fig. 3-34. The tuned circuit is formed by an inductor (L) and two capacitors. The top capacitor C_2 is usually much higher in value than the bottom varicap diode capacitor C_1. This makes the resonant frequency of the tuned circuit mainly dependent on the inductor and the varicap capacitor.

Varicap or Varactor

Tuning diode

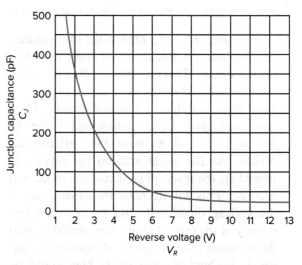

Fig. 3-33 Junction capacitance versus reverse voltage characteristic curve of a varicap diode.

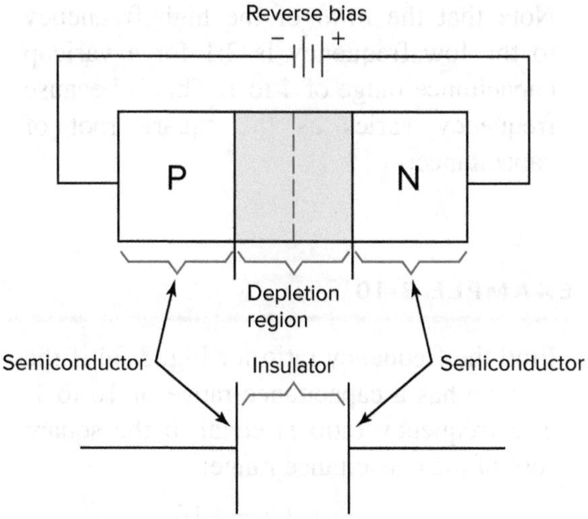

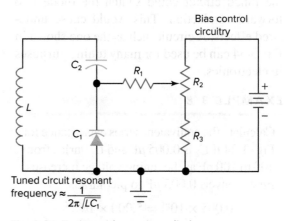

Fig. 3-32 Diode capacitance effect.

Fig. 3-34 Tuning with a varicap diode.

The series capacitance tunes the inductor in Fig. 3-34. This capacitance is determined by the bias control circuitry, so adjusting R_2 will change the resonant frequency of the LC tuning circuit.

R_1 in Fig. 3-34 is a high value of resistance and isolates the tuned circuit from the bias-control circuit. This prevents the Q of the tuned circuit, that is, the sharpness of the resonance, from being lowered by resistive loading. High resistance gives light loading and better Q. Resistors R_2 and R_3 form the variable-bias divider. As the wiper arm on the resistor is moved up, the reverse bias across the diode will increase. This will decrease the capacitance of the varicap diode and raise the resonant frequency of the tuned circuit. You should inspect the resonant frequency formula and verify this trend. Without R_3, the diode bias could be reduced to zero. In a varicap tuning diode, zero bias is not usually acceptable. An ac signal in the tuned circuit could switch the diode into forward conduction. This would cause undesired effects. A circuit such as the one shown in Fig. 3-34 can be used for many tuning purposes in electronics.

EXAMPLE 3-8

Calculate the equivalent series capacitance for Fig. 3-34 if C_2 is 0.005 μF and C_1 varies from 400 to 100 pF as the tuning voltage increases. First, convert 0.005 μF to picofarads:

$$0.005 \times 10^{-6} = 5{,}000 \times 10^{-12}$$

Next, determine the series capacitance for

$$C_1 = 400 \text{ pF}:$$

$$C_S = \frac{400 \times 5{,}000}{400 + 5{,}000} = 370 \text{ pF}$$

Then, determine the series capacitance for

$$C_1 = 100 \text{ pF}:$$

$$C_S = \frac{100 \times 5{,}000}{100 + 5{,}000} = 98 \text{ pF}$$

In both cases, the series capacitance is close to the value of C_1 alone.

EXAMPLE 3-9

Find the frequency range for Fig. 3-34 for a varicap range of 100 to 400 pF if the coil is 1 μH. Assume that C_2 is large enough so that its value will not have a significant effect. Find the high frequency:

$$f_h = \frac{1}{6.28 \times \sqrt{100 \times 10^{-12} \times 1 \times 10^{-6}}}$$

$$= 15.9 \text{ MHz}$$

Find the low frequency:

$$f_l = \frac{1}{6.28 \times \sqrt{400 \times 10^{-12} \times 1 \times 10^{-6}}}$$

$$= 7.96 \text{ MHz}$$

Subtract to find the frequency range:

$$f_{\text{range}} = f_h - f_l = 15.9 \text{ MHz} - 7.96 \text{ MHz}$$

$$= 7.94 \text{ MHz}$$

Note that the *ratio* of the high frequency to the low frequency is 2:1 for a varicap capacitance range of 4 to 1. This is because frequency varies as the square root of capacitance.

EXAMPLE 3-10

Find the frequency ratio for Fig. 3-34 if the varicap has a capacitance range of 10 to 1. The frequency ratio is equal to the square root of the capacitance range:

$$f_{\text{ratio}} = \sqrt{10} = 3.16$$

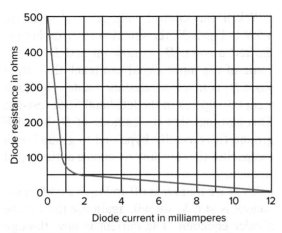

Fig. 3-35 PIN diode resistance versus current.

Some diodes are built with an *intrinsic* layer between the P and the N regions. These are called *PIN diodes,* where the "I" denotes the intrinsic layer between the P material and the N material. The intrinsic layer is pure silicon (not doped). When a PIN diode is forward-biased, carriers are injected into the intrinsic region. Then, when the diode is reverse-biased, it takes a relatively long time to sweep these carriers out of the intrinsic region. This makes PIN diodes useless as high-frequency rectifiers.

The value of PIN diodes is that they can act as variable resistors for RF currents. Figure 3-35

shows how the resistance of a typical PIN diode varies with the direct current flowing through it. As the direct current increases, the diode's resistance drops.

PIN diodes are also used for RF switching. They can be used to replace relays for faster, quieter, and more reliable operation. A typical situation that occurs in two-way radios is shown in Fig. 3-36. The transmitter and receiver share an antenna. The receiver must be isolated from the antenna when the transmitter is on or it may be damaged. This is accomplished by applying a positive voltage to the bias terminal in Fig. 3-36, which turns on both PIN diodes. Direct current will flow from ground, through D_2, through the coil, through D_1, and through the radio-frequency choke (RFC) into the bias terminal. Both diodes will have a low resistance, and the radio signal from the transmitter will pass through D_1 and on to the antenna with little loss. D_2 also has low resistance when transmitting, and this prevents any significant RF voltage from appearing across the receiver input. The bias voltage is removed when receiving, and both diodes will then show a high resistance. The antenna is effectively disconnected from the transmitter by D_1.

In addition to switching, PIN diodes can also provide *attenuation* of RF signals. Figure 3-37

PIN diode

Attenuation

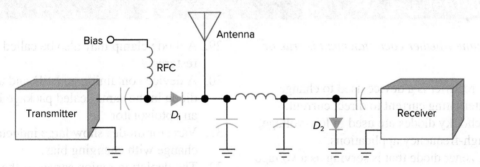

Fig. 3-36 PIN diode transmit-receive switching.

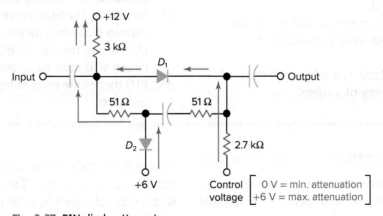

Fig. 3-37 PIN diode attenuator.

shows a PIN diode attenuator circuit. When the control point is at 0 V, signals pass through from input to output with little loss. This is because D_1 is forward-biased and in a low-resistance state. D_2 is now reverse-biased, and it has almost no effect on the signal. The bias conditions can be determined by solving for the dc voltage drop across the 3,000-Ω resistor. With the control point at 0 V, there is 12 V across the series circuit containing the 3,000-Ω resistor, D_1, and the 2,700-Ω resistor. Follow the blue arrows. The diode resistance is small enough to be ignored. The drop across the 3,000-Ω resistor can be found with the voltage divider equation:

$$V = \frac{3,000}{3,000 + 2,700} \times 12 \text{ V} = 6.32 \text{ V}$$

The voltage at the top end of the left-hand 51-Ω resistor is found by subtracting the 6.32-V drop from the 12-V supply:

$$V = 12 \text{ V} - 6.32 \text{ V} = 5.68 \text{ V}$$

Thus, the cathode of D_2 is at +6 V, and the anode connects through a 51-Ω resistor to a voltage of 5.68 V. With the cathode more positive than the anode, D_2 is reverse-biased and has a very high resistance.

When the control voltage is changed to +6 V in Fig. 3-37, the situation reverses. Follow the red arrows. D_2 is now on and D_1 is off. Little of the input signal can reach the output since D_2 is in a low-resistance state. The input signal dissipates in the left-hand 51-Ω resistor. This assumes that the cathode of D_2 is at RF ground (it is usually bypassed to ground with a capacitor that has low reactance at the signal frequency).

To prove that D_1 is off in Fig. 3-37 when the control is at 6 V, we will again use the voltage divider equation. The current is now through D_2, the 51-Ω resistor, and the 3,000-Ω resistor (look at the red arrows). The drop across the 3,000-Ω resistor is found by

$$V = \frac{3,000}{3,000 + 51} \times (12 \text{ V} - 6 \text{ V}) = 5.9 \text{ V}$$

The voltage at the anode end of D_1 is found by subtracting the drop from the 12-V supply:

$$V = 12 \text{ V} - 5.9 \text{ V} = 6.1 \text{ V}$$

Thus, the anode end of D_1 is only 0.1 V positive with respect to the cathode end. This is not enough to forward-bias it, so D_1 is off and in a high-resistance state.

Self-Test

Determine whether each statement is true or false.

24. A rectifier is a device used to change alternating current to direct current.
25. Schottky diodes are used in low-voltage, high-frequency applications.
26. A zener diode that is serving as a voltage regulator has electron flow from its anode to its cathode.
27. A normally operating rectifier diode will conduct from its anode to its cathode.
28. A diode clamp is used to limit the peak-to-peak swing of a signal.
29. A diode clamp may also be called a dc restorer.
30. A device containing an LED and a photodiode in the same sealed package is called an optoisolator.
31. Varactor diodes show large inductance change with changing bias.
32. The depletion region serves as the dielectric in a varicap diode capacitor.
33. Increasing the bias (reverse) across a varicap diode will increase its capacitance.
34. Decreasing the capacitance in a tuned circuit will raise its resonant frequency.
35. PIN diodes are used as high-frequency rectifiers.

3-5 Photovoltaic Energy Sources

There is a lot of interest in renewable energy sources. Sunlight is considered a renewable source since it cannot be depleted by using it.

Photovoltaic (PV) devices directly convert sunlight into electric energy. They are often called solar cells, solar panels, solar modules, or solar arrays. Over 95 percent of all PV solar cells

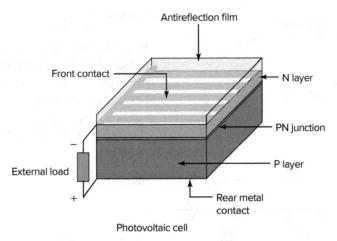

Photovoltaic cell

Fig. 3-38 Silicon PV cell construction.

produced are composed of silicon. To make a PV solar cell, the silicon is doped, and a PN junction is formed in much the same fashion as for the diodes already discussed in this chapter. One difference is that PV cells are designed so that light can enter through the front, or top, as shown in Fig. 3-38. There is an antireflection film, which is transparent. Also, the front contact is in the form of a grid so that light can pass through and reach the semiconductor layers below. The rear metal contact is solid. Thus light cannot enter from the bottom side. The rear metal contact provides the return path for electrons that have traveled through the load, and it also physically supports the semiconductor layers.

Light is made up of energetic particles called photons. A photon with enough energy (equal to or greater than the band gap) can dislodge a valence electron and make it available for conduction. Albert Einstein was the first person to correctly describe photoelectric emission, for which he was awarded a Nobel Prize. If a photon enters the P layer shown in Fig. 3-38 and knocks loose an electron, the liberated electron will be swept across the junction by the barrier potential and enter the N layer. If an external load is connected, the electron will be collected by the front contact, travel through the load, and reenter the P layer via the rear meal contact. You might want to review Fig. 3-3 and verify that the *barrier potential* will indeed attract liberated electrons in the P layer.

In a PV cell, photons must reach into the P layer to be useful. However, they should not penetrate too deeply into the P layer, because

there they would likely combine with holes and thus be lost. The PV cell structure is carefully designed and crafted to absorb as many photons as possible and to keep the liberated electrons from recombining with holes. Ideally, the electrons are freed as close to the junction as possible.

Solar cells produce the most load power when the load is of the correct value. They produce the most voltage (V_{OC}) when unloaded (open circuit) and the most current (I_{SC}) when shorted as shown in Fig. 3-39. Note that both of these conditions produce *zero load power* (the bottom curve). V_{OC} is typically 0.5 V, and V_{MP} (the voltage at maximum power) is typically 0.45 V. Figure 3-39 shows that the maximum power point (P_{max}) occurs at less than short-circuit current and less than maximum output

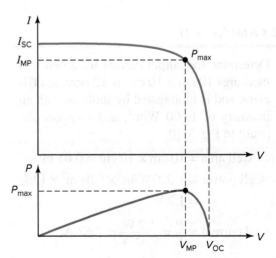

Fig. 3-39 Current and power characteristics of a solar cell.

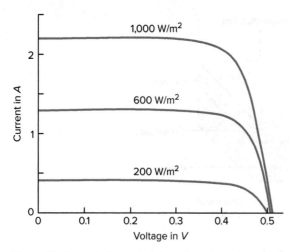

Fig. 3-40 Current characteristics of a solar cell.

voltage. It occurs at only one value of load resistance for a given amount of cell illumination:

$$R_{L(\text{Ideal})} = \frac{V_{\text{MP}}}{I_{\text{MP}}}$$

The available output is a function of the brightness of the sunlight, as shown in Fig. 3-40. The maximum intensity of sunlight at the earth's surface is 1,000 W/m² (watts per square meter) with an average wavelength of 550 nm, which happens to be in the green part of the color spectrum. Human vision is most sensitive there.

At solar noon, on a clear March or September equinox day, the solar radiation at the equator is about 1,000 W/m². Obviously, the brightness varies with the time of day, latitude, atmospheric conditions, time of year, and so on and is almost always less than 1,000 W/m².

EXAMPLE 3-11

Determine the output current for a cell that measures 10 cm × 10 cm, is 12 percent efficient, and is illuminated by sunlight with an intensity of 1,000 W/m², and compare the result to Fig. 3-40.

Cell area = 10 cm × 10 cm = 0.01 m²

Cell power = 1,000 W/m² × 0.01 m² × 12%

              = 1.2 W

Cell current = $\frac{P}{V} = \frac{1.2 \text{ W}}{0.45 \text{ V}} = 2.67$ A

This result agrees with Fig. 3-40.

EXAMPLE 3-12

Find the best load resistance (most load power) for Fig. 3-40 for an illumination intensity of 1,000 W/m². Is one value of load best for all levels of cell illumination?

$$R = \frac{V}{I} = \frac{0.45 \text{ V}}{2.22 \text{ A}} = 0.203 \text{ ohms}$$

No, the ideal load resistance varies with light intensity.

To be useful, PV cells must be combined in modules or arrays of modules, as shown in Fig. 3-41. Series connections provide more voltage and parallel connections more current. The interconnected solar cells are often embedded in transparent ethyl-vinyl-acetate, supported by an aluminum frame and covered with glass on the front side.

The typical power ratings of a solar module are between 10 and 100 peak watts. The characteristic data refer to the standard test conditions of 1,000 W/m² solar radiation at a cell temperature of 25°C. Higher temperatures

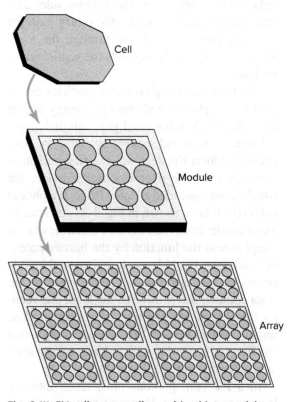

Fig. 3-41 PV cells are usually combined into modules and sometimes arrays.

cause the power output to drop. Luckily, higher temperatures usually correspond to more sunlight, so this effect tends to make the performance more uniform over a range of solar brightness.

There are three cell types according to the type of crystal structure: monocrystalline, polycrystalline, and amorphous. To produce a monocrystalline silicon cell, extremely pure semiconducting material is required. Monocrystalline ingots are extracted from molten silicon and then sawed into thin wafers. This is a tedious process and thus the most expensive. Silicon wafer production is discussed in Chap. 13.

Polycrystalline cells cost less to make. Here, liquid silicon is poured into blocks that are subsequently sawed into plates. During solidification of the material, crystal structures of varying sizes are formed, at whose borders defects (flaws) emerge. The flaws result in decreased cell efficiency.

Amorphous cells are made by depositing a very thin film of silicon on glass or another substrate material. These are sometimes called thin-film cells. The deposited layer thickness amounts to less than 1 μm, so the production costs are lower due to the lower material costs. Unfortunately, the efficiency of amorphous cells is much lower than that of the other two cell types. Because of this, they are primarily used in low-power applications and sometimes where flexibility is required (being very thin, they can withstand more flexing when deposited on a flexible substrate).

Typical efficiencies for the three types are:

Monocrystalline: 14 to 17 percent

Polycrystalline: 13 to 15 percent

Amorphous: 5 to 7 percent

An increase in PV efficiency is probably going to be required before PV arrays start appearing on lots of rooftops. The limiting factors include the following:

- Some wavelengths of light are not absorbed or converted.
- Excess photon energy is converted into heat rather than current flow.
- Electrical resistance losses occur in the crystal, contacts, and cables.
- Reflection losses occur off the face.

- Surface defects prevent photon penetration.
- Crystal flaws and material impurities detract from performance.

The theoretical maximum efficiency for silicon PV devices is about 29 percent. This will likely never be achieved. Researchers are looking at other materials such as gallium arsenide and other technologies such as using more than one PN junction per cell to improve efficiency and provide more power. In a single-junction PV cell, only those photons whose energy is equal to or greater than the band gap of the cell material can free an electron. In other words, the photovoltaic response of single-junction cells is limited to the portion of the sun's spectrum whose energy is above the band gap of the absorbing material. Lower-energy photons are wasted.

One way to get around this limitation is to use cells with more than one band gap and more than one junction to generate current. These are referred to as multijunction cells (or cascade or tandem cells). Multijunction devices can achieve a higher total conversion efficiency because they can convert more of the spectrum of sunlight to electricity. Efficiencies as high as 40 percent have been reached, but the costs are still too high for almost all commercial applications. A solar panel for an earth-orbiting satellite can be very costly because there is little or no competition from other energy sources.

PV troubleshooting involves visual inspection and some basic knowledge and sometimes ordinary test equipment. For example, output voltage is measurable with an ordinary multimeter. *Caution:* Some solar arrays generate potentially lethal voltages. Current flow is always a problem when troubleshooting. People who work on PV energy systems should own or have access to clamp-on ammeters that work at direct current.

Fluke 80i IIOS ac/dc current probe.
Courtesy of Fluke Corporation

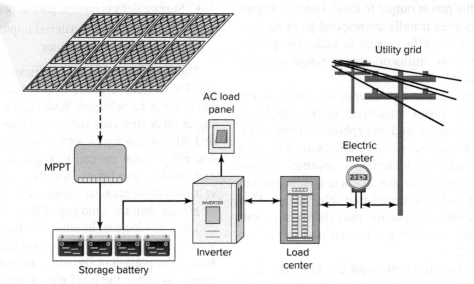

Fig. 3-42 PV system.

Small (low-wattage) PV systems might connect directly to storage batteries. Large PV systems cannot be connected directly. They must be connected via a power conditioner, controller, or inverter (see Fig. 3-42), and often overcurrent protection devices are part of the system. As mentioned earlier in this section, the ideal value of load varies with the light hitting the panel. Just as an automobile needs a transmission to match the engine to the road conditions and vehicle speed, PV systems need maximum power point tracking (MPPT) systems (or similar devices) to maintain good

performance over a range of light, load, and temperature conditions. MPPTs are dc-to-dc converters specially designed to match solar PV arrays to storage batteries. Troubleshooting converters and inverters is not covered here, but later parts of this book (e.g., Chap. 15) deal with them.

Finally, some commonsense items are worth mentioning. If PV system performance is dropping off, it might be time to wash off the built-up dust and dirt. Also, is there a problem caused by partial blockage of the sun's rays? Have you taken the ambient conditions into account (e.g., overcast skies)?

Supply the missing word in each statement.

36. Energy sources that cannot be depleted are said to be _____.

37. A photon entering a PV cell might move an electron from the _____ band to the conduction band.

38. An electron on the P side of the junction in a PV cell that has moved into the conduction band will be swept into the N side by the _____ _____.

39. The liberated electrons in a PV cell can be lost to the load circuit if they are consumed by recombination with _____.

40. The maximum power produced by a PV cell is _____ _____ than $I_{SC} \times V_{OC}$.

41. With more sunlight, more power and more _____ are available from a PV cell.

42. PV cells sawn from silicon ingots are said to be _____.

43. A PV module is a combination of PV _____.

44. PV cells are wired in series to produce more power and _____.

45. Amorphous cells have the _____ cost and the _____ efficiency.

46. An MPPT is a(n) _____ converter.

47. An inverter is a(n) _____ converter.

Chapter 3 Summary and Review

Summary

1. One of the most basic and useful electronic components is the PN-junction diode.
2. When the diode is formed, a depletion region appears that acts as an insulator.
3. Forward bias forces the majority carriers to the junction and collapses the depletion region. The diode conducts. (Technically speaking, it semiconducts.)
4. Reverse bias widens the depletion region. The diode does not conduct.
5. Reverse bias forces the minority carriers to the junction. This causes a small leakage current to flow. It can usually be ignored.
6. Volt-ampere characteristic curves are used very often to describe the behavior of electronic devices.
7. The volt-ampere characteristic curve of a resistor is linear (a straight line).
8. The volt-ampere characteristic curve of a diode is nonlinear.
9. It takes about 0.3 V of forward bias to turn on a germanium diode, about 0.6 V for a silicon rectifier, and about 2 V for an LED.
10. A silicon diode will avalanche at some high value of reverse voltage.
11. Diode leads are identified as the cathode lead and the anode lead.
12. The anode must be made positive with respect to the cathode to make a diode conduct.
13. Manufacturers mark the cathode lead with a band, bevel, flange, or plus (+) sign.
14. If there is doubt, the ohmmeter test can identify the cathode lead. It will be connected to the negative terminal. A low resistance reading indicates that the negative terminal of the ohmmeter is connected to the cathode.
15. Caution should be used when applying the ohmmeter test. Some ohmmeters have reversed polarity. The voltage of some ohmmeters is too low to turn on a PN-junction diode. Some ohmmeters' voltages are too high and may damage delicate PN junctions.
16. A diode used to change alternating current to direct current is called a rectifier diode.

17. Schottky diodes do not have a depletion region and turn off much faster than silicon diodes.
18. A diode used to stabilize or regulate voltage is the zener diode.
19. Zener diodes conduct from anode to cathode when they are working as regulators. This is just the opposite of the way rectifier diodes conduct.
20. A diode clipper or limiter can be used to stabilize the peak-to-peak amplitude of a signal. It may also be used to change the shape of a signal or reduce its noise content.
21. Clamps or dc restorers add a dc component to an ac signal.
22. Light-emitting diodes are used as indicators and transmitters and in optoisolators.
23. Varicap diodes are solid-state variable capacitors. They are operated under conditions of reverse bias.
24. Varicap diodes show minimum capacitance at maximum bias. They show maximum capacitance at minimum bias.
25. PIN diodes are used to switch radio-frequency signals and also to attenuate them.
26. This chapter has presented quite a few diode types. Figure 3-43 will help you remember their names and symbols.

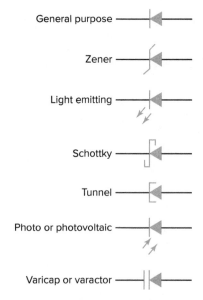

Fig. 3-43 A review of diode types and symbols.

Related Formulas

Diode forward current: $I_F = \dfrac{V_S - 0.6}{R}$ or $\dfrac{V_S - V_D}{R}$

RC time constant: $T = RC$

Resonant frequency: $f_R = \dfrac{1}{2\pi\sqrt{LC}}$

Series capacitance: $C_S = \dfrac{C_1 C_2}{C_1 + C_2}$

Chapter Review Questions

Determine whether each statement is true or false.

3-1. A PN-junction diode is made by mechanically joining a P-type crystal to an N-type crystal. (3-1)

3-2. The depletion region forms only on the P-type side of the PN junction in a solid-state diode. (3-1)

3-3. The barrier potential prevents all the electrons on the N-type side from crossing the junction to fill all the holes in the P-type side. (3-1)

3-4. The depletion region acts as an insulator. (3-1)

3-5. Forward bias tends to collapse the depletion region. (3-1)

3-6. Reverse bias drives the majority carriers toward the junction. (3-1)

3-7. It takes 0.6 V of forward bias to collapse the depletion region and turn on a silicon solid-state diode. (3-1)

3-8. A diode has a linear volt-ampere characteristic curve. (3-2)

3-9. Excessive reverse bias across a rectifier diode may cause avalanche and damage it. (3-2)

3-10. Silicon is a better conductor than germanium. (3-2)

3-11. Less voltage is required to turn on a germanium diode than to turn on a silicon diode. (3-2)

3-12. The behavior of electronic devices such as diodes changes with temperature. (3-2)

3-13. The Celsius temperature scale is used in electronics. (3-2)

3-14. Leakage current in a diode is from the cathode to the anode. (3-2)

3-15. Forward current in a diode is from the cathode to the anode. (3-2)

3-16. Diode manufacturers usually mark the package in some way so as to identify the cathode lead. (3-3)

3-17. Making the diode anode negative with respect to the cathode will turn on the diode. (3-3)

3-18. It is possible to test most diodes with an ohm-meter and identify the cathode lead. (3-3)

3-19. Rectifier diodes are used in the same way as zener diodes. (3-4)

3-20. Zener diodes are normally operated with the cathode positive with respect to the anode. (3-4)

3-21. Two germanium diodes are connected as shown in Fig. 3-20. With a 10-V peak-to-peak input signal, the signal across R_2 would be 0.6 V peak-to-peak. (3-4)

3-22. The function of D in Fig. 3-24 is to limit the output signal swing to no more than 0.6 V peak-to-peak. (3-4)

3-23. Light-emitting diodes emit light by heating a tiny filament red hot. (3-4)

3-24. The capacitance of a varicap diode is determined by the reverse bias across it. (3-4)

3-25. Germanium diodes cost less and are therefore more popular than silicon diodes in modern circuitry. (3-4)

3-26. Diode clippers are also called clamps. (3-4)

3-27. As the wiper arm of R_2 in Fig. 3-34 is moved up, f_r will increase. (3-4)

Chapter Review Problems

3-1. Refer to Fig. 3-5. The diode is silicon, the battery is 3 V, and the current-limiter resistor is 150 OHMS. Find the current flow in the circuit. (Hint: Don't forget to subtract the diode's forward voltage drop.) (3-1)

3-2. Refer to Fig. 3-11. Calculate the forward resistance of the diode at a temperature of 25°C and a forward current of 25 mA. (3-2)

3-3. Refer again to Fig. 3-11. Calculate the forward resistance of the diode at a temperature of 25°C and a forward current of 200 mA. (3-2)

3-4. Refer to Fig. 3-23. Both resistors are 10 kV, both zeners are rated at 3.9 V, and the input signal is 2 V peak-to-peak. Calculate the output signal. (Hint: Don't forget the voltage divider action of R_1 and R_2.) (3-4)

3-5. Find the output signal for Fig. 3-23 for the same conditions as given in Prob. 3-4 but with an input signal of 20 V peak-to-peak. (3-4)

3-6. What value of current-limiter resistor should be used in an LED circuit powered by 8 V if the desired LED current is 15 mA? You may assume an LED forward drop of 2 V. (3-4)

Critical Thinking Questions

3-1. A nearly ideal diode would have, among other characteristics, a very small barrier potential (say a millivolt or so). What would be the advantage of such a tiny barrier potential?

3-2. Can you think of a way to use a diode to measure temperature?

3-3. High-power diodes can get very hot, and heat is a major factor in the failure of electronic devices. Does anything in this chapter suggest a possible solution?

3-4. Infrared remote control units are very popular in products such as television receivers and DVD

players. Can you describe a simple circuit, to be used in conjunction with an oscilloscope, that could help in diagnosing problems with remote control units?

3-5. Can you think of a reason why optocouplers are often used in medical electronics?

3-6. Why is the PIN diode transmit-receive circuit shown in Fig. 3-36 not useful for cellular telephones?

3-7. Can you identify two effects of adding a series rectifier to a string of decorative lights?

Answers to Self-Tests

1. T	13. horizontal	25. T	37. valence
2. F	14. vertical	26. T	38. barrier potential
3. T	15. nonlinear	27. F	39. holes
4. F	16. 0.6	28. F	40. less than
5. F	17. bias (voltage)	29. T	41. current
6. F	18. cathode	30. T	42. monocrystalline
7. F	19. cathode	31. F	43. cells
8. T	20. cathode	32. T	44. voltage
9. T	21. shorted	33. F	45. lowest, lowest
10. straight line	22. on	34. T	46. dc-to-dc
11. straight line	23. nonlinear	35. F	47. dc-to-ac
12. 10 mA (0.01 A)	24. T	36. renewable	

Power Supplies

Learning Outcomes

This chapter will help you to:

4-1 *View* power supplies as systems. [4-1]

4-2 *Identify* and explain common rectifier circuits. [4-2, 4-3]

4-3 *Predict* and measure dc output voltage for unfiltered and filtered power supplies. [4-4, 4-5]

4-4 *Explain* how voltage multipliers work. [4-6]

4-5 *Measure* and calculate ripple and voltage regulation. [4-7]

4-6 *Explain* and make basic calculations for zener voltage regulators. [4-8]

4-7 *Troubleshoot* power supplies. [4-9]

4-8 *Select* replacement parts. [4-10]

Electronic circuits need energy to work. In most cases, this energy is provided by a circuit called the power supply. A power supply failure will affect all of the other circuits. The supply is a key part of any electronic system. Power supplies use rectifier diodes to convert alternating current to direct current. They may also use zener diodes as voltage regulators. This chapter covers the circuits that use diodes in these ways. It also discusses component-level troubleshooting. Knowing what each part of a circuit does and how the circuit functions allows technicians to find faulty components.

4-1 The Power-Supply System

Most of today's power supplies are hybrids; they are a combination of linear and digital circuits. This chapter covers the linear portion. That's the part usually connected directly to the ac line or via a 60-Hz power transformer. It uses diodes and filter capacitors, usually electrolytics, to convert ac to dc. The linear portion is often followed by a digital section called a *switcher* or a *switch-mode supply*. Chapter 15 covers the rest of what is needed to understand modern hybrid power supplies.

The power supply changes the available electric energy (usually ac) to the form required by the various circuits within the system (usually dc). One of the early steps in the troubleshooting of any electronic system is to check the supply voltages at various stages in the circuitry.

Power supplies range from simple to complex, depending on the requirements of the system. A simple power supply may be required to furnish 12 V dc. A more complicated power supply may provide several voltages, some positive and some negative with respect to the chassis ground. A supply that provides voltages at

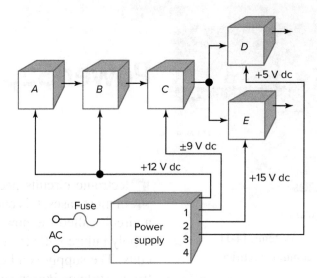

Fig. 4-1 Block diagram of an electronic system.

Bipolar supply

both polarities is called a *bipolar supply*. Some power supplies may have a wide output voltage tolerance. The output may vary ±20 percent. Another power supply may have to keep its output voltage within ±0.01 percent. Obviously, a strict tolerance complicates the design of the supply. Such supplies are covered in Chap. 15.

Figure 4-1 shows a block diagram for an electronic system. The power supply is a key part of the system since it energizes the other circuits. If a problem develops in the power supply, the fuse might "blow" (open). In that case, none of the voltages could be supplied to the other circuits. Another type of problem might involve the loss of only one of the outputs of the power supply. Suppose the +12-V dc output drops to zero because of a component failure in

the power supply. Circuits *A* and *B* would no longer work.

The second output of the power supply shown in Fig. 4-1 develops both positive and negative dc voltages with respect to the common point (usually the metal chassis). This output could fail, too. It is also possible that only the negative output could fail. In either case, circuit *C* would not work normally under such conditions.

Troubleshooting electronic systems can be made much easier with block diagrams. If the symptoms indicate the failure of one of the blocks, then the technician will devote special attention to that part of the circuit. Since the power supply energizes most or all of the other blocks, it is one of the first things to check when troubleshooting.

Self-Test

Supply the missing word in each statement.

1. Power supplies will usually change alternating current to _____.
2. Power-supply voltages are usually specified by using the chassis _____ as a reference.

3. Drawings such as Fig. 4-1 are called _____ diagrams.
4. On a block diagram, the circuit that energizes most or all of the other blocks is called the _____.

Rectification

4-2 Rectification

Most electronic circuits need direct current. Alternating current is supplied by the power companies. The purpose of the power supply is to

change alternating current to direct current by *rectification*. Alternating current flows in both directions, and direct current flows in only one direction. Since diodes conduct in only one direction, they serve as rectifiers.

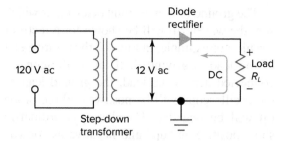

Fig. 4-2 A simple dc power supply.

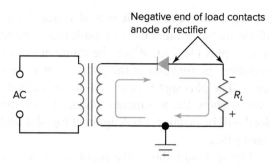

Fig. 4-3 Establishing the polarity in a rectifier circuit.

The ac supply available at ordinary wall outlets is 120 V, 60 hertz (Hz). Electronic circuits often require lower voltages. Transformers can be used to step down the voltage to the level needed. Figure 4-2 shows a simple power supply using a step-down transformer and a diode rectifier.

The load for the power supply in Fig. 4-2 could be an electronic circuit, a battery being charged, or some other device. In this chapter, the loads will be shown as resistors designated R_L.

The transformer in Fig. 4-2 has a voltage ratio of 10:1. With 120 V across the primary, 12 V ac is developed across the secondary. If it were not for the diode, there would be 12 V ac across the load resistor. The diode allows current flow only from its cathode to its anode. The diode is in series with the load. Current is the same everywhere in a series circuit, so the diode current and the load current are the same. Since the load current is flowing in only one direction, it is *direct current*. When direct current flows through a load, a dc voltage appears across the load.

EXAMPLE 4-1

What will the secondary voltage be for Fig. 4-2 if the transformer ratio is 2:1? A voltage ratio of 2:1 means that we can divide the primary voltage by 2 to find the answer:

$$V_{secondary} = \frac{V_{primary}}{2} = \frac{120 \text{ V}}{2} = 60 \text{ V}$$

Note the polarity across the load in Fig. 4-2. Electrons move from negative to positive through a load. The *positive* end of the load is connected to the *cathode* end of the rectifier. In all rectifier circuits, the positive end of the load will be that end that contacts the cathode of the rectifier. It can also be stated that the *negative*

end of the load will be in contact with the *anode* of the rectifier. Figure 4-3 illustrates this point. Compare Figs. 4-2 and 4-3. Note that the diode polarity determines the load polarity.

In Chap. 3 it was stated that to forward-bias a diode, the anode must be made positive with

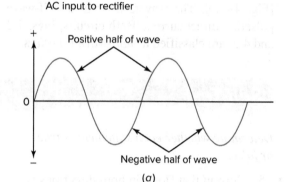

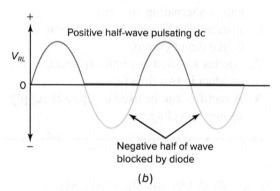

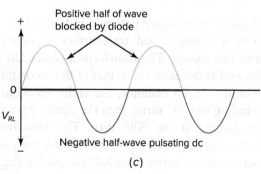

Fig. 4-4 Rectifier circuit waveforms.

respect to the cathode. It was also noted that diode manufacturers used to mark the cathode with a plus (+) sign. When the diode acts as a rectifier, the function of the plus sign becomes clear. The plus sign was placed on the cathode end to show the technician which end of the load will be positive. Look again at Fig. 4-2 and verify this.

Figure 4-4(a) shows the input waveform to the rectifier circuits of Figs. 4-2 and 4-3. Two complete cycles are shown. In Fig. 4-4(b), the waveform that appears across the load resistor of Fig. 4-2 is shown. The negative half of the cycle is missing since the diode blocks it. This waveform is called *half-wave pulsating direct current*. It represents only the positive half of the ac input to the rectifier.

In Fig. 4-3 the diode has been reversed. This causes the positive half of the cycle to be blocked [Fig. 4-4(c)]. The waveform is also half-wave pulsating direct current. Both circuits, Figs. 4-2 and 4-3, are classified as *half-wave rectifiers*.

Half-wave pulsating direct current

Half-wave rectifiers

The ground reference point determines which way the waveform will be shown for a rectifier circuit. For example, in Fig. 4-3 the positive end of the load is grounded. If an oscilloscope is connected across the load, the ground lead of the oscilloscope will be positive and the probe tip will be negative. Oscilloscopes ordinarily show positive as "up" and negative as "down" on the screen. The actual waveform will appear as that shown in Fig. 4-4(c). Waveforms can appear up or down depending on circuit polarity, instrument polarity, and the connection between the instrument and the circuit.

Half-wave rectifiers are usually limited to low-power applications. They take useful output from the ac source for only half the input cycle. They are not supplying any load current half the time. This limits the amount of electric energy they can deliver over a given period of time. High power means delivering large amounts of energy in a given time. A half-wave rectifier is a poor choice in high-power applications.

Self-Test

Determine whether each statement is true or false.

5. Current that flows in both directions is called alternating current.
6. Current that flows in one direction is called direct current.
7. Diodes are used as rectifiers because they conduct in two directions.
8. A rectifier can be used in a power supply to step up voltage.

9. In a rectifier circuit, the positive end of the load will be connected to the cathode of the rectifier.
10. The waveform across the load in a half-wave rectifier circuit is called half-wave pulsating direct current.
11. A half-wave rectifier supplies load current only 50 percent of the time.
12. Half-wave rectifiers are usually used in high-power applications.

4-3 Full-Wave Rectification

Full-wave rectifier

A *full-wave rectifier* is shown in Fig. 4-5(a). It uses a center-tapped transformer secondary and two diodes. The transformer center tap is located at the electrical center of the secondary winding. If, for example, the entire secondary winding has 100 turns, then the center tap will be located at the 50th turn. The waveform across the load in Fig. 4-5(a) is *full-wave pulsating direct current* with half the peak voltage of the secondary because of the center tap. Both

Full-wave pulsating direct current

alternations of the ac input are used to energize the load. Thus, a full-wave rectifier can deliver twice the power of a half-wave rectifier.

The ac input cycle is divided into two parts: a *positive alternation* and a *negative alternation*. The positive alternation is shown in Fig. 4-5(b). The induced polarity at the secondary is such that D_1 is turned on. Electrons leave the center tap and flow through the load, through D_1, and back into the top of the secondary. Note that the positive end of the load resistor is in contact with the cathode of D_1.

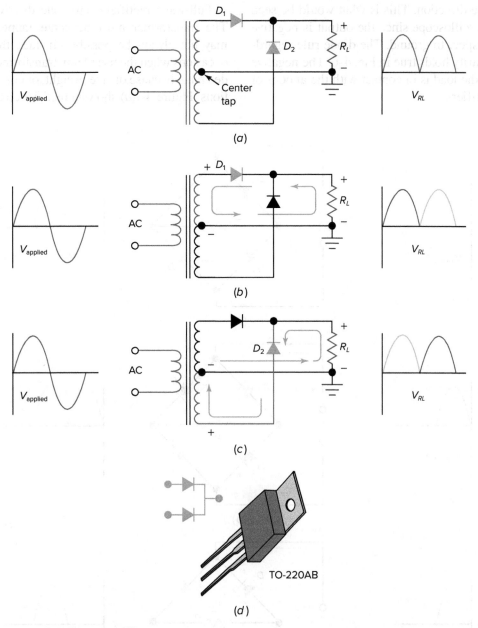

(a)

(b)

(c)

(d)
TO-220AB

Fig. 4-5 Full-wave rectifiers.

On the negative alternation, the polarity across the secondary is reversed. This is shown in Fig. 4-5(c). Electrons leave the center tap and flow through the load, through D_2, and back into the bottom of the secondary. The load current is the same for both alternations: it flows up through the resistor. Since the direction never changes, the load current is *direct current*.

Full-wave rectifiers can be constructed using two separate diodes or by using a package that contains two diodes. An example is shown in Fig. 4-5(d).

Figure 4-6 shows a full-wave rectifier with the diodes reversed. This reverses the polarity

across the load resistor. Note that the output waveform shows both alternations going in a

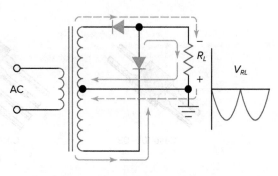

Fig. 4-6 Reversing the rectifier diodes.

negative direction. This is what would be seen on an oscilloscope since the output is negative with respect to ground. The diode rule regarding polarity holds true in Fig. 4-6. The negative end of the load is in contact with the anodes of the rectifiers.

Full-wave rectifiers have one disadvantage: The transformer must be center-tapped. This may not always be possible. In fact, there are occasions when the use of any transformer is not desirable because of size, weight, or cost restrictions. Figure 4-7(a) shows a rectifier circuit that

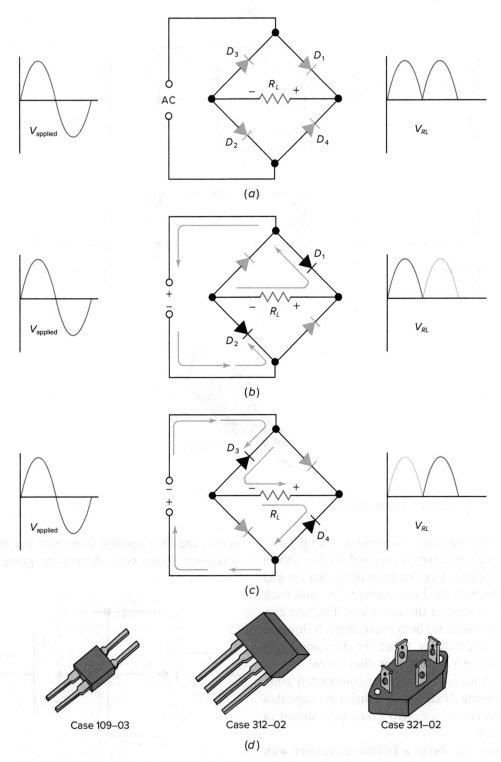

Fig. 4-7 Bridge rectifier circuit and case styles.

gives full-wave performance without the transformer. It is called a *bridge rectifier*. It uses four diodes to give full-wave rectification.

Figure 4-7(*b*) traces the circuit action for the positive alternation of the ac input. The current moves through D_2, through the load, through D_1, and back to the source. The negative alternation is shown in Fig. 4-7(*c*). The current is always moving from left to right through the load. Again, the positive end of the load is in contact with the rectifier cathodes. This circuit could be arranged for either ground polarity simply by choosing the left or the right end of the load as the common point.

A bridge rectifier requires four separate diodes, or a special rectifier package that contains four diodes connected in the bridge configuration. Figure 4-7(*d*) shows three examples of packaged bridge rectifiers.

Bridge rectifier

Self-Test

Supply the missing word in each statement.

13. A transformer secondary is center-tapped. If 50 V is developed across the entire secondary, the voltage from either end to the center tap will be _____.
14. A half-wave rectifier uses _____ diode(s).
15. A full-wave rectifier using a center-tapped transformer requires _____ diodes.
16. Each cycle of the ac input has two _____.
17. In rectifier circuits, the load current never changes _____.
18. A bridge rectifier eliminates the need for a _____.
19. A bridge rectifier requires _____ diodes.

4-4 Conversion of RMS Values to Average Values

There is a significant difference between *pure direct current* and *pulsating direct current* (rectified alternating current). Meter readings taken in rectifier circuits can be confusing if you do not understand the difference. Figure 4-8 compares a pure dc waveform with a pulsating dc waveform. A meter used to make measurements in a pure dc circuit will respond to the steady value of the direct current. In the case of pulsating dc, the meter will try to follow the pulsating waveform. At one instant in time, the meter tries to read zero. At another instant in time, the meter tries to read the peak value. Meter movements cannot react to the rapid changes because of the *damping* in their mechanism. Damping in a meter limits the speed with which the pointer can change position. The meter settles on the *average value* of the waveform.

Digital meters do not have damping, but they produce the same results. The display is not updated often enough to follow the pulsating dc waveform. If it did rapidly follow the waveform, the display would be a useless blur of constantly changing numbers. For this reason, digital meters also display the average value of a pulsating dc waveform.

Alternating current supply voltages are typically specified by their *root-mean-square (rms) values*. It would be convenient to have a way of converting rms values to average values when working with rectifier circuits.

RMS value

> ### You May Recall
>
> . . . that sinusoidal alternating current can be measured in several ways and that it is possible to convert from one to another.

Figure 4-9 shows some measurements and conversion factors. If you have access to a calculator, it might be easier to calculate the conversion factors than to remember them:

$0.707 = \frac{1}{\sqrt{2}}$ (to go from peak to rms values)

$0.637 = \frac{2}{\pi}$ (to go from peak to average values)

As another aid, remember that rms means *root-mean-square*, and you will know which one to use when converting peak to rms.

Damping

Average Value

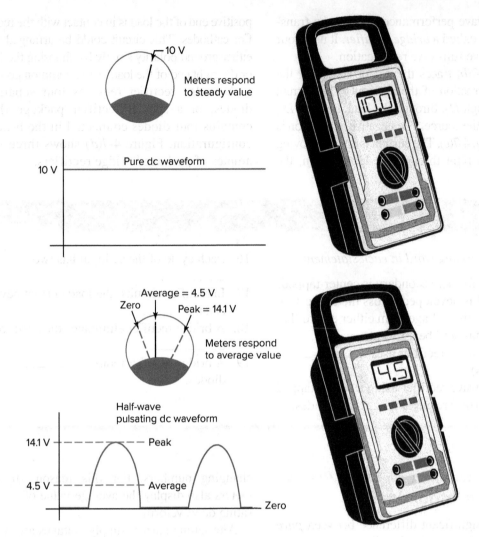

Fig. 4-8 Meters measure steady or average values.

Pure dc waveform

10 V

10 V

Meters respond to steady value

Average = 4.5 V
Zero
Peak = 14.1 V

Meters respond to average value

Half-wave pulsating dc waveform

14.1 V — — — — — Peak

4.5 V — — — — — Average

Zero

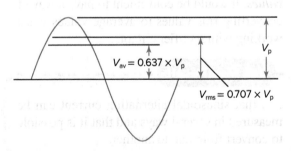

Fig. 4-9 Measuring sinusoidal ac.

V_p

$V_{av} = 0.637 \times V_p$

$V_{rms} = 0.707 \times V_p$

EXAMPLE 4-2

Suppose the peak-to-peak value of the sine wave shown in Fig. 4-9 is 340 V. Find the average and the rms values. First, divide by 2 to find the peak value:

$$V_p = \frac{V_{p-p}}{2} = \frac{340 \text{ V}}{2} = 170 \text{ V}$$

$$V_{av} = V_p \times 0.637 = 170 \text{ V} \times 0.637 = 108 \text{ V}$$

$$V_{rms} = V_p \times 0.707 = 170 \text{ V} \times 0.707 = 120 \text{ V}$$

Algebra can be used to relate rms values to average values:

$$V_{av} = 0.637 \times V_p$$

$$V_{rms} = 0.707 \times V_p$$

Rearranging the second equation gives

$$V_p = \frac{V_{rms}}{0.707}$$

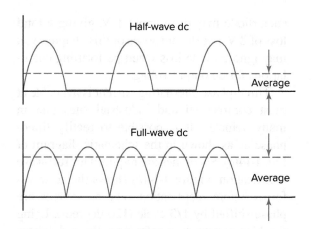

Fig. 4-10 Comparing half-wave and full-wave direct current.

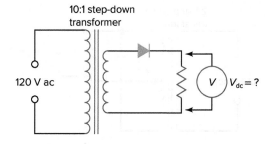

Fig. 4-11 Calculating dc voltage in a half-wave rectifier circuit.

Substituting the right-hand side into the first equation gives

$$V_{av} = 0.637 \times \frac{V_{rms}}{0.707} = 0.9 \times V_{rms}$$

Thus, the *average* value of a rectified sine wave is 0.9, or 90 percent, of the rms value. This means that a dc voltmeter connected across the output of a rectifier should indicate 90 percent of the rms voltage input to the rectifier. This would be true for the entire waveform. But, as Fig. 4-10 shows, the average value must be less for half the waveform. Half-wave pulsating direct current has half of the average value compared with full-wave pulsating direct current. So, for a half-wave rectifier, the average value of the waveform is $0.9/2 = 0.45$, or 45 percent, of the rms value.

EXAMPLE 4-3

What should the dc voltmeter shown in Fig. 4-11 read? Taking the step-down action of the transformer into account first,

$$V_{secondary} = \frac{120}{10} = 12 \text{ V}$$

Next, note that Fig. 4-11 shows a half-wave rectifier. The appropriate conversion factor is 0.45:

$$V_{av} = V_{rms} \times 0.45$$
$$= 12 \times 0.45$$
$$= 5.4 \text{ V}$$

The meter should read 5.4 V.

If the circuit in Fig. 4-11 were constructed, how close could we expect the actual reading to be? The actual reading would be influenced by several factors: (1) the actual line voltage,

(2) transformer winding tolerance, (3) meter accuracy, (4) rectifier loss, and (5) transformer losses. The actual line voltage and the actual transformer secondary voltage can be accounted for by accurate measurements. The meter accuracy can be high with a quality meter that has been checked against a standard. The rectifier loss is caused by the 0.6-V forward drop needed for conduction in a silicon diode. At high current levels, the drop will be greater. For example, if the rectifier current is several amperes, the diode loss will be close to 1 V. Transformer losses also increase at high current levels. Thus, the actual readings can be expected to be a little on the low side, especially at high-load current levels.

EXAMPLE 4-4

What should the dc voltmeter shown in Fig. 4-12 read? Taking the step-down action of the transformer into account first,

$$V_{secondary} = \frac{120}{2} = 60 \text{ V}$$

Since Fig. 4-12 shows a full-wave circuit, the appropriate conversion factor is 0.9. However, we *must* take into account that only *half* of the secondary is conducting at any given time. Please review Fig. 4-5 if you are confused.

$$V_{av} = \frac{V_{secondary}}{2} \times 0.9$$
$$= \frac{60}{2} \times 0.9 = 27 \text{ V}$$

The meter should read 27 V. If the load demands a high current, then the actual voltage will be less. What would happen if one of the diodes should "burn out" (open)? This would change the circuit from *full-wave* to *half-wave*. The dc voltmeter could then be expected to read

$$V_{av} = V_{rms} \times 0.45$$
$$= 30 \times 0.45$$
$$= 13.5 \text{ V}$$

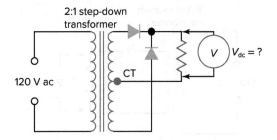

Fig. 4-12 Calculating dc voltage in a full-wave rectifier circuit.

EXAMPLE 4-5

Calculate the average dc voltage for Fig. 4-12, assuming that a bridge rectifier will be used with the entire secondary (in other words, the center tap will not be connected).

$$V_{av} = V_{secondary} \times 0.9 = 60 \text{ V} \times 0.9 = 54 \text{ V}$$

EXAMPLE 4-6

What should the dc voltmeter shown in Fig. 4-13 read? Taking the step-down action of the transformer into account first,

$$V_{secondary} = \frac{120}{4} = 30 \text{ V}$$

Since Fig. 4-13 shows a full-wave bridge circuit, the appropriate conversion factor is 0.9:

$$V_{av} = 30 \times 0.9 = 27 \text{ V}$$

The diode loss in a bridge rectifier is *twice* that of the other circuits. A review of Fig. 4-7 will show that *two* diodes are always conducting in *series*. The 0.6-V drop will be doubled to 1.2 V. In low-voltage rectifier circuits, this can be significant. If the current demand is high,

each diode may drop about 1 V, giving a total loss of 2 V. For the purposes of this chapter, you may ignore diode loss when performing calculations for dc output voltage.

Three-phase alternating current is available at most commercial and industrial sites and in many vehicles. It is possible to rectify three-phase ac as shown in the schematic diagram of Fig. 4-14(*a*). Six diodes are required for full-wave rectification. Figure 4-14(*b*) shows the ac waveforms before rectification. The ac sources are phase shifted by 1/3 cycle (120 degrees). Using the blue source as a reference, the red source phase leads by 120 degrees, and the green source phase lags by 120 degrees. Figure 4-14(*c*) shows that the negative alternations are "folded up" as they are in single-phase, full-wave circuits. Figure 4-14(*d*) shows the waveform across the load resistor. Notice that the load voltage stays fairly constant near the peak value of the three-phase source. This is an advantage of three-phase power. Unfortunately, three-phase power is not available in homes and some other locations. Filters are usually required in single-phase rectifiers to smooth the load current and voltage. Filters are discussed in the next section of this chapter.

If you compare the waveforms in Fig. 4-14 with those in Fig. 4-10, it should be apparent that the average dc value is higher for three-phase rectifiers. For full-wave three-phase circuits, the average value is $V_{rms} \times 1.35$.

EXAMPLE 4-7

The ac source shown in Fig. 4-14(*a*) is 208 V. What will a dc voltmeter read if it is connected across R_L?

$$V_{av} = V_{rms} \times 1.35 = 208 \times 1.35 = 281 \text{ V}$$

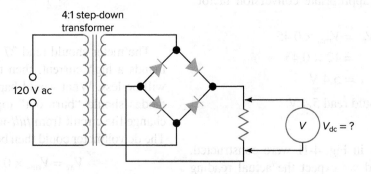

Fig. 4-13 Calculating dc voltage in a bridge rectifier circuit.

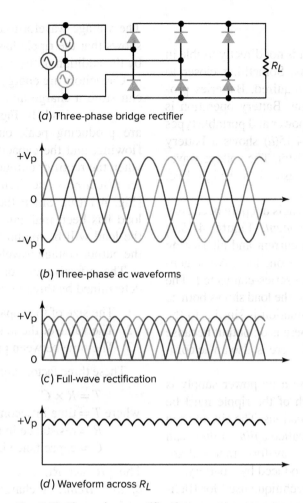

(a) Three-phase bridge rectifier

+V_p

0

−V_p

(b) Three-phase ac waveforms

+V_p

0

(c) Full-wave rectification

+V_p

(d) Waveform across R_L

Fig. 4-14 Three-phase rectification.

Self-Test

Supply the missing word or number in each statement.

20. A transformer has five times as many primary turns as secondary turns. If 120 V ac is across the primary, the secondary voltage should be _____.

21. Suppose the transformer in question 20 is center-tapped and connected to a full-wave rectifier. The average dc voltage across the load should be _____.

22. The average dc load voltage for the data in question 21 will change to _____ if one of the rectifiers burns out (opens).

23. The ac input to a half-wave rectifier is 32 V. A dc voltmeter connected across the load should read _____.

24. The ac input to a bridge rectifier is 20 V. A dc voltmeter connected across the load should read _____.

25. In rectifier circuits, one can expect the output voltage to drop as load current _____.

26. Rectifier loss is more significant in _____voltage rectifier circuits.

27. If each diode in a high-current bridge rectifier drops 1 V, then the total rectifier loss is _____.

4-5 Filters

Pulsating direct current is not directly usable in most electronic circuits. Something closer to pure direct current is required. Batteries produce pure direct current. Battery operation is usually limited to low-power and portable types of equipment. Figure 4-15(a) shows a battery connected to a load resistor. The voltage waveform across the load resistor is a straight line. There are no pulsations.

Pulsating direct current is not pure because it contains an *ac component*. Figure 4-15(b) shows how both direct current and alternating current can appear across one load. An ac generator and a battery are series-connected. The voltage waveform across the load shows both ac and dc content. This situation is similar to the output of a rectifier. There is dc output because of the rectification, and there is also an ac component (the pulsations).

The ac component in a dc power supply is called the *ripple*. Much of the ripple must be removed for most applications. The circuit used to remove the ripple is called a *filter*. Filters can produce a very smooth waveform that will approach the waveform produced by a battery.

The most common technique used for filtering is a capacitor connected across the output. Figure 4-16 shows a simple *capacitive filter* that has been added to a full-wave rectifier circuit.

The voltage waveform across the load resistor shows that the ripple has been greatly reduced by the addition of the capacitor.

Capacitors are energy storage devices. They can store a charge and then later deliver that charge to a load. In Fig. 4-17(a) the rectifiers are producing peak output, load current is flowing, and the capacitor is *charging*. Later, when the rectifier output drops off, the capacitor *discharges* and furnishes the load current [Fig. 4-17(b)]. Since the current through the load has been maintained, the voltage across the load will be maintained also. This is why the output voltage waveform shows less ripple.

The effectiveness of a capacitive filter is determined by three factors:

1. The size of the capacitor
2. The value of the load
3. The time between pulsations

These three factors are related by the formula

$$T = R \times C$$

where T = time in seconds (s)
R = resistance in ohms (Ω)
C = capacitance in farads (F)

The product RC is called the *time constant of the circuit*. A charged capacitor will lose 63.2 percent of its voltage in T seconds. It takes approximately $5 \times T$ seconds to completely discharge the capacitor.

AC component

Ripple

Filter

Capacitive filter

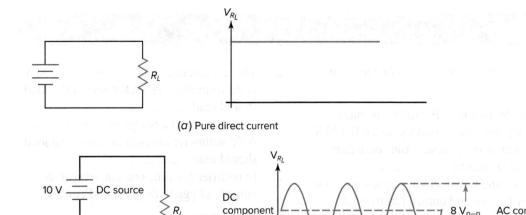

(a) Pure direct current

(b) Alternating current with a dc component

Fig. 4-I5 Dc and ac waveforms.

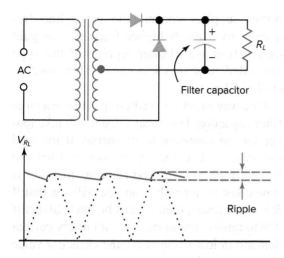

Fig. 4-16 A full-wave rectifier with a capacitive filter.

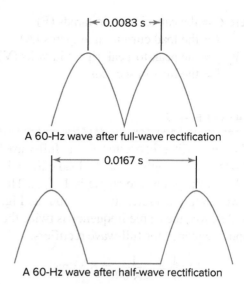

Fig. 4-18 A rectified 60-Hz wave.

To be effective, a filter capacitor should be only *slightly* discharged between peaks. This will mean a small voltage change across the load and, thus, little ripple. The time constant will have to be long when compared with the time between peaks. This makes it interesting to compare half-wave and full-wave filtering. The time between peaks for full-wave and half-wave rectifiers is shown in Fig. 4-18. Obviously, in a half-wave circuit, the capacitor has twice the time to discharge, and the ripple will be greater. Full-wave rectifiers are desirable when most of the ripple must be removed. This is

because it is easier to filter a wave whose peaks are closer together. Looking at it another way, it will take a capacitor twice the size to adequately filter a half-wave rectifier, if all other factors are equal.

EXAMPLE 4-8

Estimate the relative effectiveness of 100-μF and 1,000-μF capacitors when they will be used to filter a 60-Hz half-wave rectifier loaded by 100 Ω. First, find both time constants:

$$T = R \times C$$
$$T_1 = 100\ \Omega \times 100\ \mu\text{F} = 0.01\ \text{s}$$
$$T_2 = 100\ \Omega \times 1{,}000\ \mu\text{F} = 0.1\ \text{s}$$

If we look at Fig. 4-18, we see that the discharge time for 60-Hz half-wave circuits is in the vicinity of 0.01 s. This means that the smaller filter will discharge for about one time constant, losing about 60 percent, creating a significant amount of ripple. The larger capacitor has a 0.1-s time constant, which is long compared with the discharge time. The 1,000-μF capacitor will be a much more effective filter (a lot less ripple).

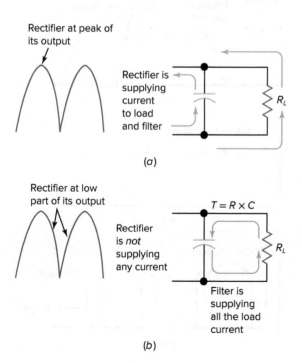

Fig. 4-17 Filter capacitor action.

The choice of a filter capacitor can be based on the following equation:

$$C = \frac{I}{V_{\text{p-p}}} \times T$$

where C = the capacitance in farads (F)
I = the load current in amperes (A)
V_{p-p} = the peak-to-peak ripple in volts (V)
T = the time in seconds (s)

EXAMPLE 4-9

Choose a filter capacitor for a full-wave, 60-Hz power supply when the load current is 5 A and the allowable ripple is 1 V_{p-p}. The power supply operates at 60 Hz, but as Fig. 4-18 shows, the ripple frequency is twice the input frequency for full-wave rectifiers:

$$T = \frac{1}{f} = \frac{1}{2 \times 60} = 8.33 \text{ ms}$$

This agrees with Fig. 4-18. Find the filter size next:

$$C = \frac{I}{V_{p-p}} \times T = \frac{5}{1} \times 8.33 \times 10^{-3}$$
$$= 41.7 \text{ mF}$$
$$= 41,700 \text{ } \mu\text{F}$$

The size of filter capacitors is often expressed in microfarads.

EXAMPLE 4-10

Choose a filter capacitor for a full-wave, 100-kHz power supply when the load current is 5 A and the allowable ripple is 1 V_{p-p}. Compare the capacitor with that found in the previous example.

$$T = \frac{1}{2 \times 100 \times 10^3} = 5 \text{ } \mu\text{s}$$
$$C = \frac{5}{1} \times 5 \times 10^{-6} = 25 \text{ } \mu\text{F}$$

The size of the capacitor is much smaller when compared with the previous example.

Figure 4-18 is based on the 60-Hz power-line frequency. If a much higher frequency were used, the job of the filter could be even easier. For example, if the frequency were 1 kilohertz (kHz), the time between peaks in a full-wave rectifier output would be only 0.0005 s. In this short period of time, the filter capacitor would be only slightly discharged. Another interesting point about high frequencies is that transformers can be made much smaller. Some power supplies convert the power-line frequency to a much higher frequency to gain these advantages. Power supplies of this type are called *switchmode supplies*. They are covered in Chap. 15.

One way to get good filtering is to use a large filter capacitor. This means that it will take longer for the capacitor to discharge. If the load resistance is low, the capacitance will have to be very high to give good filtering. Inspect the time constant formula, and you will see that if R is made lower, then C must be made higher if T is to remain the same. So, with heavy current demand (a low value of R), the capacitor value must be quite high.

Electrolytic capacitors are available with very high values of capacitance. However, a very high value in a capacitive-input filter can cause problems. Figure 4-19 shows waveforms that might be found in a capacitively filtered power supply. The unfiltered waveform is shown in Fig. 4-19(*a*). In Fig. 4-19(*b*) the

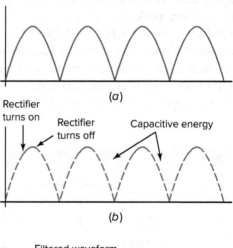

(*a*)

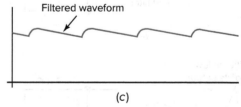

(*b*)

(*c*)

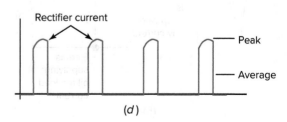

(*d*)

Fig. 4-19 Capacitor filter circuit waveforms.

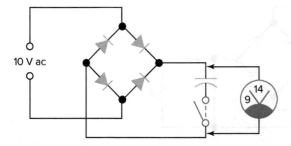

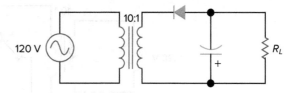

Fig. 4-21 A filtered half-wave rectifier circuit.

Fig. 4-20 Calculating dc output voltage with a capacitive filter.

capacitor supplies energy between peaks. Note that the rectifiers do not conduct until their peak output exceeds the capacitive voltage. The rectifier turns off when the peak output ends. The rectifiers conduct *for only a short time.* Figure 4-19(*d*) shows the rectifier current waveform. Notice the high peak-to-average ratio. This causes ac line distortion which is discussed later in this section.

In some power supplies, the peak-to-average current ratio in the rectifiers may exceed 100:1. This causes the rms rectifier current to be greater than eight times the current delivered to the load. The rms current determines the actual heating effect in the rectifiers. This is why diodes may be rated at 10 A when the power supply is designed to deliver only 2 A.

The dc output voltage of a filtered power supply is *higher* than the output of a nonfiltered supply. Figure 4-20 shows a bridge rectifier circuit with a switchable filter capacitor. Before the switch is closed, the meter will read the average value of the waveform:

$$V_{av} = 0.9 \times V_{rms}$$
$$= 0.9 \times 10$$
$$= 9 \text{ V}$$

After the switch is closed, the capacitor charges to the *peak* value of the waveform:

$$V_p = 1.414 \times V_{rms}$$
$$= 1.414 \times 10$$
$$= 14.14 \text{ V}$$

This represents a significant change in output voltage. However, as the supply is loaded, the capacitor will not be able to maintain the peak voltage, and the output voltage will drop. The more heavily it is loaded (the more current there is), the lower the output voltage will be. Therefore, you can assume that the dc output

voltage in a capacitively filtered supply is equal to the peak value of the ac input *when the supply is lightly loaded,* or not loaded at all as in Fig. 4-20.

Figure 4-21 shows a filtered half-wave rectifier circuit. What is the procedure for predicting the voltage across R_L? When filters are used, do not use the 0.9 or the 0.45 conversion constants. Remember, the filter charges to the *peak* value of the input.

Referring to Fig. 4-21, the input is 120 V ac and is stepped down by the transformer:

$$\frac{120 \text{ V rms}}{10} = 12 \text{ V rms}$$

The peak value is found next:

$$V_p = 1.414 \times 12 \text{ V}$$
$$= 16.97 \text{ V}$$

Assuming a light load, the dc voltage across the load resistor in Fig. 4-21 is nearly 17 V. If the filter capacitor were *open*, the dc output voltage would drop quite a bit. Its average value would be

$$V_{av} = 0.45 \times 12 \text{ V}$$
$$= 5.4 \text{ V}$$

Open

So, a good capacitor in Fig. 4-21 makes the output nearly 17 V, and an open capacitor means that the output will be only 5.4 V. Understanding this can be quite important when troubleshooting power supplies.

Figure 4-22 shows the same transformer and input, but the half-wave rectifier has been replaced with a bridge rectifier. Since the circuit is filtered, the dc output will again be equal to the peak value, or 16.97 V. If the capacitor opens in this circuit, the output will be

$$V_{av} = 0.9 \times 12 \text{ V}$$
$$= 10.8 \text{ V}$$

Obviously, the failure (open type) of a filter capacitor in full-wave circuits will have a less drastic effect on the dc output voltage than it does in half-wave circuits.

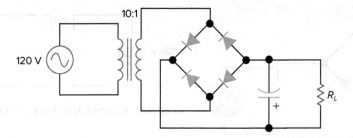

120 V

10:1

R_L

Fig. 4-22 A filtered bridge rectifier circuit.

The fact that a filter capacitor charges to the *peak* value of the ac waveform is important. Filter capacitors must be rated for this higher voltage. Another important point is capacitor *polarity*. If you check Figs. 4-21 and 4-22, you will notice that the + lead is at the bottom. Verify that this is correct by checking the rectifier connections. Most filter capacitors are of the electrolytic type. These can *explode* if connected backward. This includes many tantalum capacitors.

EXAMPLE 4-11

What voltage rating will be required for the filter capacitor in Fig. 4-22 if the transformer ratio is 1:1? The secondary voltage will be equal to the primary voltage, so the capacitor will charge to the peak value of the ac line:

$$V_p = 1.414 \times V_{rms} = 1.414 \times 120 = 170 \text{ V}$$

The capacitor will charge to 170 V. A margin of safety is required, so a capacitor rated at 250 V or more would likely be used in this case.

Capacitor filters are common and do a good job. However, they cause *harmonic distortion* on the ac power line. In much of North America the ac line frequency is sinusoidal 60 Hz. Sine waves have only one frequency. Ideally, the ac power line will be pure 60 Hz. In practice, other frequencies will be present. Harmonic line

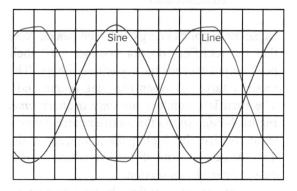

Sine   Line

Fig. 4-23 Line harmonic distortion caused by a capacitor input filter.

distortion results when loads, like capacitor-filtered supplies, give rise to other frequencies. These harmonics occur at integer multiples. For 60 Hz, they are 120 Hz, 180 Hz, 240 Hz, 300 Hz, and so on. On an oscilloscope the distortion looks like peak clipping as shown in Fig. 4-23 in red. *Never connect a bench oscilloscope directly to the ac line.* See Safe Measurements in the next section. The line distortion is caused by the capacitor current loading the ac line more during the ac line peaks as was shown in Fig. 4-19(d).

There is an international standard for power line harmonics (from the 2nd up to the 40th). It is IEC 61000-3-2 (International Electrotechnical Commission). In the United States, the standards are called Energy Star and are more concerned with power supply efficiency. Power factor correction circuits that also control harmonic distortion are covered in Chap. 14.

Self-Test

Supply the missing word or number in each statement.

28. Pure dc contains no _____.
29. Rectifiers provide _____ dc.
30. Power supplies use filters to reduce _____.

31. Capacitors are useful in filter circuits because they store electric _____.
32. In a power supply with a capacitor filter, the effectiveness of the filter is determined by the size of a capacitor, the ac frequency, and the _____.

33. Half-wave rectifiers are more difficult to filter because the filter has more time to _____.

34. Heating effect is determined by the _____ value of a current.

35. In a filtered power supply, the dc output voltage can be as high as _____ times the rms input voltage.

36. The conversion factor that is useful when predicting the dc output voltage of a filtered supply is _____.

37. The conversion factors of 0.45 and 0.90 are useful for predicting the dc output of _____ supplies.

38. A filter capacitor's voltage rating must be greater than the _____ value of the pulsating waveform.

39. The dc output from a lightly loaded supply using a bridge rectifier with 15-V ac input and a filter capacitor at the output will be _____.

4-6 Voltage Multipliers

The typical general-purpose line voltage in this country is about 115 to 120 V ac. Usually, solid-state circuits require lower voltage for operation. Sometimes, higher voltages are required. One way to obtain a higher voltage is to use a step-up transformer. Unfortunately, transformers are expensive devices. They are also relatively large and heavy. For these reasons, designers may not want to use them to obtain high voltages.

Voltage multipliers can be used to produce higher voltages and eliminate the need for a transformer. Figure 4-24(*a*) shows the diagram for a full-wave voltage doubler. This circuit can produce an output voltage as high as 2.8 times the rms input voltage. The output will be a dc voltage with some ripple.

Figure 4-24(*b*) shows the operation of the full-wave doubler. It shows how C_1 charges through D_1 when the ac line is on its positive alternation. Capacitor C_1 can be expected to charge to the peak value of the ac line. Assuming the input voltage is 120 V, we have

$$V_p = 1.414 \times V_{rms}$$
$$= 1.414 \times 120 \text{ V}$$
$$= 169.68 \text{ V}$$

Voltage multipliers

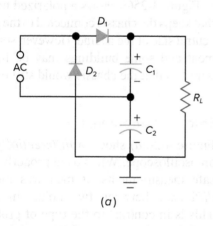

(a)

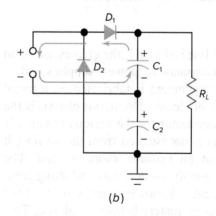

(b)

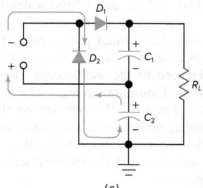

(c)

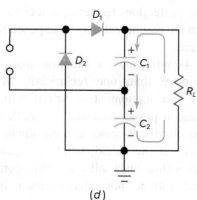

(d)

Fig. 4-24 Full-wave voltage doubler.

On the negative alternation of ac line voltage in Fig. 4-24(c), C_2 charges through D_2 to the peak value of 169.68 V. Now both C_1 and C_2 are charged. In Fig. 4-24(d) we can see that C_1 and C_2 are in series. Their polarities are series-aiding, and they will produce double the peak line voltage across the load:

$$V_{RL} = V_{C_1} + V_{C_2}$$
$$= 169.68 \text{ V} + 169.68 \text{ V}$$
$$= 339 \text{ V}$$

Voltage doublers can come close to producing three times the line voltage. As they are loaded, their output voltage tends to drop rapidly. Thus, a voltage doubler energized by a 120-V ac line might produce a voltage near 240 V dc when delivering current to a load. A voltage multiplier is a poor choice when stable output voltages are required.

EXAMPLE 4-12

Find the voltage across R_L for Fig. 4-24 assuming a light load and an ac input of 230 V. Rather than calculate each capacitor voltage, it is easier to use double the 1.414 factor:

$$V_{RL} = 2.82 \times V_{\text{rms}} = 2.82 \times 230 \text{ V} = 649 \text{ V}$$

Lack of line isolation is the greatest problem with transformerless power supplies. Most electronic equipment is fabricated on a metal framework or chassis. Often, this chassis is the common conductor for the various circuits. If the chassis is not isolated from the ac line, it can present an extreme *shock hazard*. The chassis is usually inside a nonconducting cabinet. The control knobs and shafts are made of nonconducting materials such as plastic. This gives some protection. However, a technician working on the equipment may be exposed to a shock hazard.

Figure 4-25(a) shows a situation that has surprised more than one technician. Most bench-type test equipment is wired with a three-conductor power cord that automatically grounds its chassis, its case, and the shield on the test lead. If the shield, which is usually terminated with a black alligator clip, comes into contact with a "hot" chassis, there is a

ground loop or short circuit across the ac line. Trace the short circuit in Fig. 4-25(a). The path is from the hot wire, through the polarized outlet, through the power cord, through the metal chassis, through the alligator clip lead of the test equipment, and through the power cord of the test equipment to ground. Since ground and the grounded neutral wire are tied together in the breaker panel, this traced path is a short circuit across the ac line. Thus, connecting test equipment to "hot" equipment can open (trip) circuit breakers, blow fuses, damage test leads, and damage circuits. Worse than this, a technician's body may become a part of the ground loop, and a serious electric shock can result. Working on equipment that is not isolated from the ac line is dangerous.

Figure 4-25(b) shows how an *isolation transformer* can be used to solve the hot-chassis problem. The transformer is plugged into the polarized outlet, and the chassis is energized from the secondary. There is very high electrical resistance from the primary of the transformer to the secondary. Now a fault current cannot flow from the hot wire to the metal chassis. The chassis has been *isolated* from the ac line.

Figure 4-25(c) shows a polarized power plug that keeps the chassis connected to the grounded neutral side of the ac line. However, some equipment and some buildings may be improperly wired so that the chassis would still be hot.

Safe Measurements

Figure 4-26(a) shows a *differential probe* for an oscilloscope. When used properly, it allows safe measurements. It measures the voltage *difference* between two points in a circuit. This is in contrast to the type of probe shown in Figs. 4-25(a) and 4-25(b) which measures the difference between a single point and *ground*. Differential probes convert the difference between two signals into one voltage referenced to the oscilloscope ground. The red waveform shown in Fig. 4-23 was obtained with a bench oscilloscope connected to a differential probe which was connected to the 120-V ac power line. Thus, there was no *ground loop* and the measurement was safe to perform.

Isolation transformer

Shock hazard

Ground loop

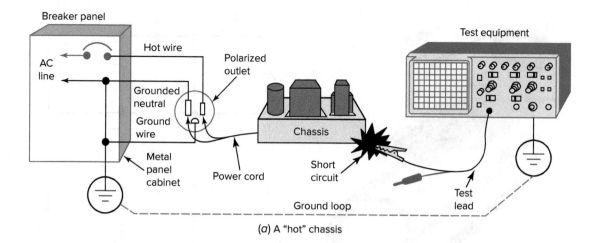

(a) A "hot" chassis

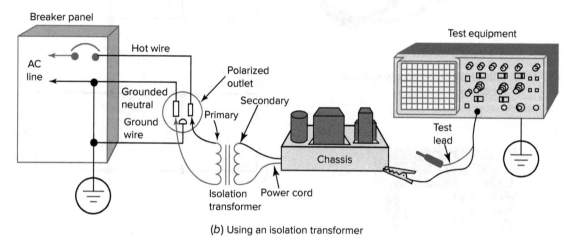

(b) Using an isolation transformer

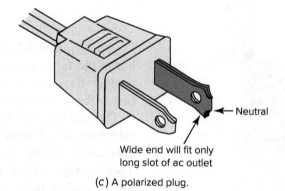

Neutral

Wide end will fit only
long slot of ac outlet

(c) A polarized plug.

Fig. 4-25 The "hot" chassis problem and two solutions.

Differential probes vary in terms of their safe voltage limits. Before using one, refer to the manufacturer's specifications and/or operator's manual.

Figure 4-26(b) shows another approach to safe measurements. It is a battery-powered oscilloscope. It is used to safely make what are called *floating measurements*.

Floating measurements are made with the common connection of the test equipment at some potential other than ground. The maximum ratings for this ScopeMeter are 600 Vrms floating, and to each instrument input. The manual must be consulted for *important safety information*. As an example, the two common inputs for this meter *must be at the same potential*.

**Floating
measurements**

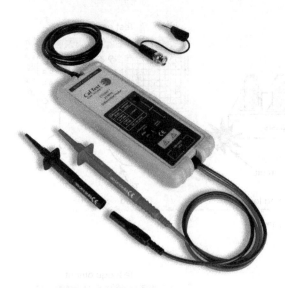

Fig. 4-26(a) Differential probe accessory for an oscilloscope.

Cal Test Electronics

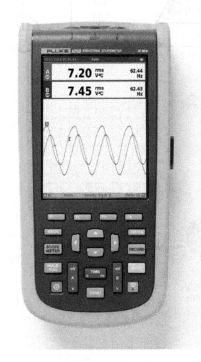

Fig. 4-26(b) Fluke ScopeMeter.

Reproduced with Permission, Fluke Corporation

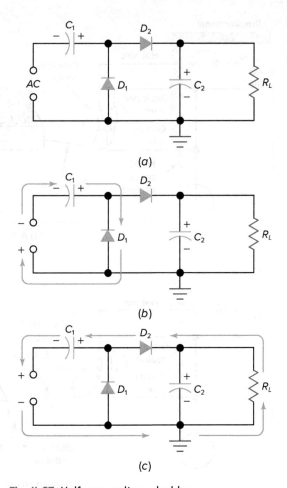

Fig. 4-27 Half-wave voltage doubler.

Half-wave voltage doubler

This is different than some battery-operated oscilloscopes.

The *half-wave voltage doubler* shown in Fig. 4-27(a) offers some improvement in safety over the full-wave voltage doubler. Compare Figs. 4-24(a) and 4-27(a). The chassis is always hot in the full-wave doubler. In the half-wave doubler, the chassis is hot only if the connection to the ac outlet is wrong.

The half-wave doubler works a little differently from the full-wave doubler. On the negative alternation, C_1 will be charged [Fig. 4-27(b)]. Then in Fig. 4-27(c), C_1 adds in series with the ac line's positive alternation, and C_2 will be charged to twice the peak line voltage. Load resistance R_L is in parallel with C_2 and will see a peak voltage of about 340 V with a line voltage of 120 V ac. The key differences are the capacitor voltage ratings and the ripple frequency across the load. Full-wave doublers use two identical capacitors. Each would have to be rated at least equal to the peak line voltage. Half-wave doublers require the load capacitor to be rated at least equal to twice the peak line voltage. The ripple frequency in a full-wave

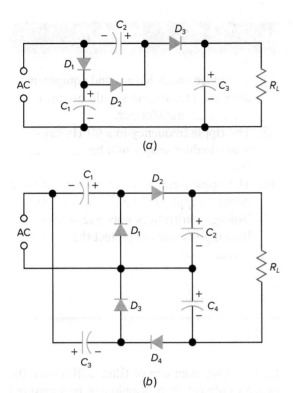

(a)

(b)

Fig. 4-28 Voltage multipliers.

doubler will be twice the line frequency. Half-wave doublers will show a ripple frequency equal to the line frequency.

It is possible to build voltage multipliers that triple, quadruple, and multiply even more. Figure 4-28(a) shows a voltage tripler. On the first positive alternation, C_1 is charged through D_1. On the next alternation, C_2 is charged to twice the peak line voltage through D_2 and C_1. Finally, C_3 is charged to three times the peak line voltage through D_3 and C_2 on the next positive alternation. With 120-V ac input, the load would see a peak voltage of 509 V. A voltage quadrupler is shown in Fig. 4-28(b). This circuit is actually two half-wave doublers connected back to back and sharing a common input. The voltages across C_2 and C_4 will combine to produce four times the peak ac line voltage. Assuming a line input of 120 V, R_L would see a peak voltage of 679 V.

Figure 4-29 shows a Cockroft–Walton voltage multiplier. It can be used to energize low current loads such as photomultiplier tubes. These are used in photon (light) and radiation detectors. It has poor voltage regulation and is not used when moderate- or heavy-load currents are required. The multiplier shown here has four stages (N = 4) and *each stage* can provide as much as *double* the peak input voltage. Assuming a 120 V rms ac source, the load voltage can be as high as:

$$V_{LOAD} = V_{RMS} \times 1.414 \times 2\,N$$
$$= 120\,V \times 1.414 \times 2 \times 4$$
$$= 1{,}357\,V\ dc$$

As do all voltage multipliers, this circuit exhibits poor voltage regulation. Using larger capacitors helps, but those can cause unacceptable current surges when the supply is first energized.

Figure 4-29 shows a surge resistor in series with the ac source. A surge resistor causes some loss of output voltage. Another approach is to use a Negative Temperature Coefficient (NTC) surge limiter instead of an ordinary resistor. This device might be marked "NTC." When an NTC surge limiter is cold, its resistance is high. This limits the turn-on surge current associated with capacitors. As it warms up, its resistance drops allowing a more normal (higher) output voltage. The symbol for an NTC surge resistor shows a line with a bend. This is one method of denoting that a device is not linear.

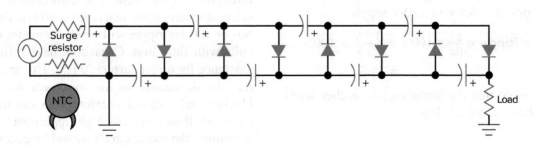

Fig. 4-29 Cockroft-Walton voltage multiplier.

Supply the missing word in each statement.

40. Connecting grounded test equipment to a hot chassis will result in a ground _____.

41. Voltage doublers may be used to obtain higher voltages and eliminate the need for a(n) _____.

42. A lightly loaded voltage doubler will give a dc output voltage that is _____ times the rms input.

43. The output of voltage multipliers tends to _____ quite a bit when the load is increased.

44. To reduce shock hazard and equipment damage, a technician should use a(n) _____ transformer.

45. The ripple frequency in a 60-Hz half-wave doubler supply will be _____ Hz.

46. The ripple frequency in a 60-Hz full-wave doubler supply will be _____ Hz.

47. Voltage multipliers may use surge-limiting resistors to protect the _____.

4-7 Ripple and Regulation

Percentage of ripple

A power-supply filter reduces ripple to a low level. The actual effectiveness of the filter can be checked with a measurement and then a simple calculation. The formula for calculating the *percentage of ripple* is

$$\text{Ripple} = \frac{ac}{dc} \times 100\%$$

where ac is the rms value.

For example, assume the ac ripple remaining after filtering is measured and found to be 1 V in a 20-V dc power supply. The percentage of ripple is

$$\text{Ripple} = \frac{ac}{dc} \times 100\%$$
$$= \frac{1}{20} \times 100\%$$
$$= 5\%$$

EXAMPLE 4-13

Coupling capacitor

Find the percentage of ac ripple if the ac content is 0.5 V in a 20-V supply.

$$\text{Ripple} = \frac{ac}{dc} \times 100\% = \frac{0.5\,V}{20\,V} \times 100\%$$
$$= 2.5\%$$

Notice that the percentage is smaller when the ac content is less.

Ripple should be measured only when the supply is delivering its *full* rated output. At zero load current, even a poor filter will reduce the ripple to almost zero. Ripple can be measured with an oscilloscope or a voltmeter. The oscilloscope will easily give the peak-to-peak value of the ac ripple. Many meters will indicate the approximate value of the rms ripple content. It will not be exact since the ripple waveform is *nonsinusoidal*. In a capacitive filter, the ripple is similar to a sawtooth waveform. This causes an error with most meters since they are calibrated to indicate rms values for sine waves. There are meters that will read the true rms value of nonsinusoidal alternating current. True rms meters are becoming more popular as prices drop.

To measure the ac ripple riding on a dc waveform, an older meter may have to be switched to a special function, or one of the test leads may have to be moved to a special jack. The special function or jack may be labeled *output*. The output jack is connected to the meter circuitry through a *coupling capacitor*. This capacitor is selected to have a low reactance at 60 Hz. Thus, 60- or 120-Hz ripple will reach the meter circuits with little loss. Capacitors have infinite reactance for direct current (0 Hz). This means that the dc content of the waveform will be blocked and will not interfere with the measurement. If an unusually high ripple content is measured, the meter circuit should be checked to be certain the dc component is not upsetting the reading.

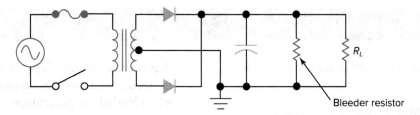

Fig. 4-30 Power supply with a bleeder resistor.

The *regulation* of a power supply is its ability to hold the output steady under conditions of changing input or changing load. As power supplies are loaded, the output voltage tends to drop to a lower value. The quality of the voltage regulation can be checked with two measurements and then a simple calculation. The formula for calculating the *percentage of voltage regulation* is

$$\text{Regulation} = \frac{\Delta V}{V_{FL}} \times 100\%$$

where ΔV = voltage change from no load to full load

V_{FL} = output voltage at full load

For example, a power supply is checked with a dc voltmeter and shows an output of 14 V when no (0) load current is supplied. When the power supply is loaded to its rated maximum current, the meter reading drops to 12 V. The percentage of voltage regulation is

$$\text{Regulation} = \frac{\Delta V}{V_{FL}} \times 100\%$$

$$= \frac{2\text{ V}}{12\text{ V}} \times 100\%$$

$$= 16.7\%$$

EXAMPLE 4-14

Find the percentage of voltage regulation when the output drops from 14.5 V to 14.0 V as the supply is loaded.

$$\text{Regulation} = \frac{\Delta V}{V_{FL}} \times 100\%$$

$$= \frac{0.5\text{ V}}{14\text{ V}} \times 100\% = 3.57\%$$

Notice that the percentage is smaller when the voltage change is less.

The output voltage of some power supplies can increase quite a bit when there is a *no-load condition*. The no-load condition can be avoided by connecting a fixed load called a *bleeder* to the output of a power supply. Figure 4-30 shows the use of a *bleeder resistor*. If R_L is disconnected, the bleeder will continue to load the output of the supply. Thus, some minimum current will always flow. This fixed load can reduce the fluctuations in output voltage with changes in R_L. So one function of a bleeder is to improve supply regulation.

Bleeder resistors perform another important function. They drain the filter capacitors after the power is turned off. Some filter capacitors can store a charge for months. Charged capacitors can present a shock hazard. It is *not* safe to assume that the capacitors have been drained even if there is a bleeder resistor across them. The bleeder could be open. Technicians who work on high-voltage supplies use a shorting rod or a shorting stick to be certain that all the filters are drained before working on the equipment. High-energy capacitors can discharge violently, so it is important that the shorting rod contain a high-wattage resistor of about 100 Ω to keep the discharge current reasonable. Figure 4-31 shows such a device.

Bleeder resistor

Percentage of voltage regulation

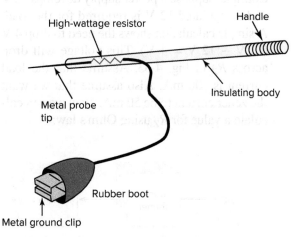

Fig. 4-31 A shorting rod.

Supply the missing word or number in each statement.

48. As the load current increases, the ac ripple tends to _____.
49. As the load current increases, the dc output voltage tends to _____.
50. A power supply develops 13 V dc with 1-V ac ripple. Its percentage of ripple is _____.

51. A power supply develops 28 V under no-load conditions and drops to 24 V when loaded. Its percentage of regulation is _____.
52. A bleeder resistor may improve supply regulation and help to ensure that the capacitors are _____ after the supply is turned off.

4-8 Zener Regulators

Regulator

Zener shunt regulator

Zener diode

Power-supply output voltage tends to change as the load on the power supply changes. The output also tends to change as the ac input voltage changes. This can cause some electronic circuits to operate improperly. When a stable voltage is required, the power supply must be *regulated*. The block diagram of a power supply (Fig. 4-32) shows where the regulator is often located in the system.

   Regulators can be elaborate circuits using integrated circuits and transistors. Such circuits are covered in Chap. 15. For some applications, however, a simple *zener shunt regulator* does the job (Fig. 4-33). The regulator is a *zener diode*, and it is connected in shunt (parallel) with the load. If the voltage across the diode is constant, then the load voltage must also be constant.

   The design of a shunt regulator using a zener diode is based on a few simple calculations. For example, suppose a power supply develops 16 V and a regulated 12 V is required for the load. A simple calculation shows the need to drop 4 V (16 V − 12 V = 4 V). This voltage will drop across R_Z in Fig. 4-33. Assume that the load current is 100 mA. Also assume that we want the zener current to be 50 mA. Now we can calculate a value for R_Z using Ohm's law:

$$R_Z = \frac{V}{I_{\text{total}}}$$

$$= \frac{4\text{ V}}{0.100\text{ A} + 0.050\text{ A}}$$

$$= 26.67\ \Omega$$

The nearest standard value of a resistor is 27 Ω, which is very close to the calculated value. The power dissipation in the resistor can be calculated:

$$P = V \times I$$

$$= 4\text{ V} \times 0.150\text{ A}$$

$$= 0.6\text{ watt (W)}$$

We can use a 1-W resistor, although a 2-W resistor may be required for better reliability. Next, the power dissipation in the diode is

$$P = V \times I$$

$$= 12\text{ V} \times 0.050\text{ A}$$

$$= 0.6\text{ W}$$

A 1-W zener diode may be adequate. However, if the load is disconnected, the zener has to dissipate quite a bit more power. All the current (150 mA) flows through the diode. The diode dissipation increases to

$$P = V \times I$$

$$= 12\text{ V} \times 0.150\text{ A}$$

$$= 1.8\text{ W}$$

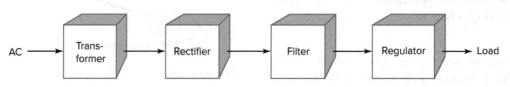

Fig. 4-32 Location of a regulator in a power supply.

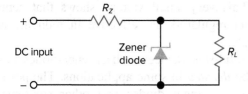

Fig. 4-33 Zener-diode shunt regulator.

Obviously, the zener must be capable of handling more power if there is a possibility of the load being disconnected.

EXAMPLE 4-15

Determine R_Z for a 12-V zener regulator with a dc input of 20 V, a load current of 65 mA, and a zener current of 20 mA.

$$R_Z = \frac{V_{R_Z}}{I_{total}} = \frac{20\text{ V} - 12\text{ V}}{65\text{ mA} + 20\text{ mA}} = 94.1\text{ }\Omega$$

Use 91 Ω, which is the closest standard value. Find the dissipation in R_Z next:

$$P_{R_Z} = V \times I = 8\text{ V} \times 85\text{ mA} = 0.68\text{ W}$$

Use a 2-W resistor for good reliability.

EXAMPLE 4-16

Find the zener diode dissipation for Example 4-14 if the load is disconnected from the regulator.

$$P = V \times I = 12\text{ V} \times 85\text{ mA} = 1.02\text{ W}$$

Use a 2-W zener for good reliability.

Another possibility is that the load might demand more current. Suppose that the load current in Fig. 4-33 increases to 200 mA. Resistor R_Z would drop

$$V = I \times R$$
$$= 0.200\text{ A} \times 27\text{ }\Omega$$
$$= 5.4\text{ V}$$

This would cause a decrease in voltage across the load:

$$V_{load} = V_{supply} - V_{R_Z}$$
$$= 16\text{ V} - 5.4\text{ V}$$
$$= 10.6\text{ V}$$

The regulator is no longer working. Shunt regulators work only up to the point at which

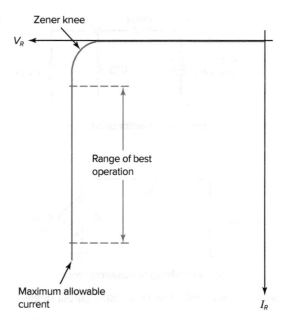

Fig. 4-34 Characteristic curve of a zener diode.

the zener stops conducting. The zener current should not be allowed to approach zero. As shown in Fig. 4-34, the region of the characteristic curve near the zener knee shows poor regulation.

Zener regulators *reduce* ac ripple. This is because zener diodes have a low impedance when biased properly. For example, one manufacturer of the 1N4733 zener diode rates its typical dynamic impedance (Z_Z) at 5 Ω when it is biased at 10 mA. Let's determine what this characteristic can mean in terms of ripple performance.

Figure 4-35(a) shows a regulator circuit based on the 1N4733 zener. This diode regulates at 5.1 V. This will establish a 5.1-V drop across the 470-Ω load resistor and a load current of

$$I = \frac{V}{R}$$
$$= \frac{5.1\text{ V}}{470\text{ }\Omega}$$
$$= 10.9\text{ mA}$$

If we assume a zener current of 10 mA, then the total current through the series resistor is

$$I_T = 10.9\text{ mA} + 10\text{ mA}$$
$$= 20.9\text{ mA}$$

The series resistor drops the difference between the supply voltage and the load voltage:

$$V_{R_Z} = 10\text{ V} - 5.1\text{ V}$$
$$= 4.9\text{ V}$$

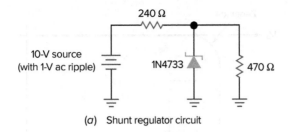

(a) Shunt regulator circuit

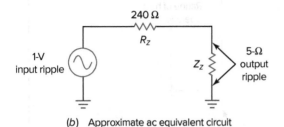

(b) Approximate ac equivalent circuit

Fig. 4-35 Determining zener regulator output ripple.

Ohm's law gives us the value for the series resistor:

$$R_Z = \frac{4.9 \text{ V}}{20.9 \text{ mA}}$$

$$= 234 \text{ }\Omega$$

The closest standard value is 240 Ω, and this is shown in Fig. 4-35(*a*).

Figure 4-35(*b*) shows the approximate *ac equivalent circuit* for the regulator. The 470-Ω load resistor has been ignored because its resistance is much greater than the zener impedance. The 1 V of ac ripple will be divided by R_Z and Z_Z. The voltage divider equation will predict the ac ripple at the output of the regulator:

$$\text{Ripple} = \frac{5 \text{ }\Omega}{240 \text{ }\Omega + 5 \text{ }\Omega} \times 1 \text{ V}$$

$$= 20.4 \text{ mV}$$

This very small voltage shows that zener shunt regulators are effective in reducing ac ripple.

Solid-state devices such as zener diodes have to be *derated* in some applications. The power rating of zener diodes and other solid-state devices must be *decreased* as the device temperature goes up. The temperature inside the cabinet of an electronic system might increase from 25° to 50°C after hours of continuous operation. This increase in temperature decreases the safe dissipation levels of the devices in the cabinet. Figure 4-36 shows a typical power derating curve for a zener diode.

Cabinet temperatures are only part of the problem. A zener diode that is dissipating a watt or so will be self-heating. So depending on dissipation levels and the environment, components like zeners may have to be *derated* for reliable operation.

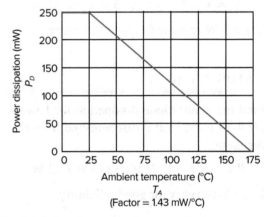

Fig. 4-36 Derating curve of a zener diode.

Self-Test

Supply the missing word or number in each statement.

53. A zener diode shunt regulator uses the zener connected in _____ with the load.
54. A power supply develops 8 V. Regulated 5 V is required at a load current of 500 mA. A zener diode shunt regulator will be used. The zener current should be 200 mA. The value of R_Z should be _____.

55. The dissipation in R_Z in question 54 is _____.

56. The zener dissipation in question 54 is _____.

57. If the load current were interrupted in question 54, the zener would dissipate _____.

58. A zener shunt regulator can provide voltage regulation and reduce _____.

4-9 Troubleshooting

One of the major skills of an electronic technician is *troubleshooting*. The process involves the following steps:

1. Observing the symptoms
2. Analyzing the possible causes
3. Limiting the possibilities by tests and measurements

Good troubleshooting is an orderly process. To help keep things in order, remember the word *"GOAL."* GOAL stands for *G*ood, *O*bserve, *A*nalyze, and *L*imit.

Electronic equipment that is broken usually shows very definite *symptoms*. These are extremely important. Technicians should try to note all the symptoms before proceeding. This demands a knowledge of the equipment. You must know what the normal performance of a piece of equipment is in order to be able to identify what is abnormal. It is often necessary to make some adjustments or run some checks to be sure that the symptoms are clearly identified. For example, if a radio receiver has a hum or whistle on one station, several other stations should be tuned in to determine whether the symptom persists. These kinds of adjustments and checks will help the technician to properly observe the symptoms.

Analyzing possible causes comes after the symptoms are identified. This part of the process involves a general knowledge of the block diagram of the equipment. Certain symptoms are closely tied to certain blocks on the diagram. Experienced technicians "think" the block diagram. They do not need a drawing in front of them. Their experience tells them how the major sections of the circuit work, how signals flow from stage to stage, and what happens when one section is not working properly. For example, suppose a technician is troubleshooting a radio receiver. There is only one major symptom: No sound of any kind is coming from the speaker. Experience and knowledge of the block diagram will tell the technician that two major parts of the circuit can cause this symptom: the power supply and/or the audio output section.

After the possibilities are established, it is time to limit them by tests and measurements. A few voltmeter checks generally will tell the technician if the power-supply voltages are correct. If they are not, then the technician must further limit the possibilities by making more checks. Circuit failure is usually limited to one component. Of course, the failure of one component may damage several others because of the way they interact. A resistor that has burned black is almost always a sure sign that another part has shorted.

Power-supply troubleshooting follows the general process. The symptoms that can be observed are

1. No output voltage
2. Low output voltage
3. Excessive ripple voltage
4. High output voltage

Note that the symptoms are all limited to voltages. This is the way technicians work. Voltages are easy to measure. Current analysis is rarely used because it is necessary to break into the circuit and insert an ammeter. It is also worth mentioning that two of the power-supply symptoms might appear at the same time: low output voltage and excessive ripple voltage.

Once the symptoms are clearly identified, it is time to analyze possible causes. For *no output voltage*, the possibilities include

1. Open fuse or circuit breaker
2. Defective switch, line cord, or outlet
3. Defective transformer
4. Open surge-limiting resistor or NTC resistor
5. Open diode or diodes (rare)
6. Open doubler capacitor

The last step is to limit the list of possibilities. This step is accomplished by making some measurements. Figure 4-37 is the schematic diagram for a half-wave doubler power supply. The technician can make ac voltage measurements as shown at A, B, C, and D to find the cause of no output voltage. For example, suppose the measurement at A is 120 V alternating current but 0 V at B. This indicates a blown fuse. Suppose A and B show line voltage and C shows zero. This would indicate an open surge-limiting resistor. If measurements A, B, and C are 120 V alternating current and if measurement D is zero, then capacitor C_1 is open.

Another failure point that shows up in some equipment is a PPTC (Polymeric Positive

Troubleshooting

Power-supply troubleshooting

GOAL

Symptom

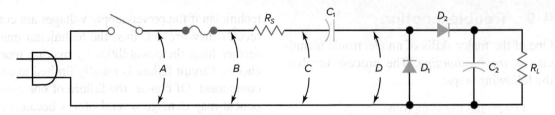

Fig. 4-37 A half-wave doubler schematic.

Temperature Coefficient) device. Figure 4-38 shows the schematic symbol and the physical appearance of such a device. As the symbol suggests, the device is a nonlinear element. As temperature increases resistance increases. At some higher than normal temperature, the device "opens" protecting the equipment from further damage. The device actually switches to a high resistance state. If the fault clears, the PPTC will cool and return to its normal or "on" state. So, PPTCs can be viewed as self-healing fuses. They are the solid-state equivalent of older bimetal switches that warped with current and temperature, breaking the circuit. These also reset when they cooled.

Some defects show the need for more checking. Again referring to Fig. 4-37, if the surge-limiting resistor R_S is open, it may be because another component is defective. Simply replacing R_S may result in the new part burning out. It is a good idea to check the diodes and the

capacitors when a surge limiter opens or a fuse blows. One of the capacitors or diodes could be shorted.

Solid-state rectifier diodes usually do not open (show a very high resistance in both directions). There are exceptions, of course. Their typical failure mode is to short-circuit. Diodes can be checked with an ohmmeter, but this requires disconnecting at least one side of the diode. Sometimes it is possible to obtain a rough check with the diode still in the circuit. *Always remove power* before making ohmmeter tests, and make sure the filter capacitors are *discharged*.

Figure 4-39 shows the schematic for a full-wave power supply. An ohmmeter test (with the supply off) across the diodes will show a low resistance when the diode is forward-biased and a higher resistance roughly equal to the total load resistance when the diode is reverse-biased. This will prove that the diode is not shorted, but it may have excessive leakage.

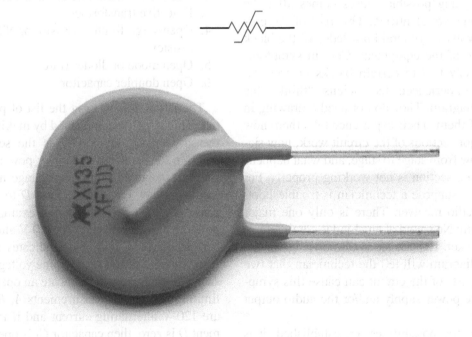

Fig. 4-38 PPTC schematic symbol and physical appearance.

Courtesy of Littelfuse

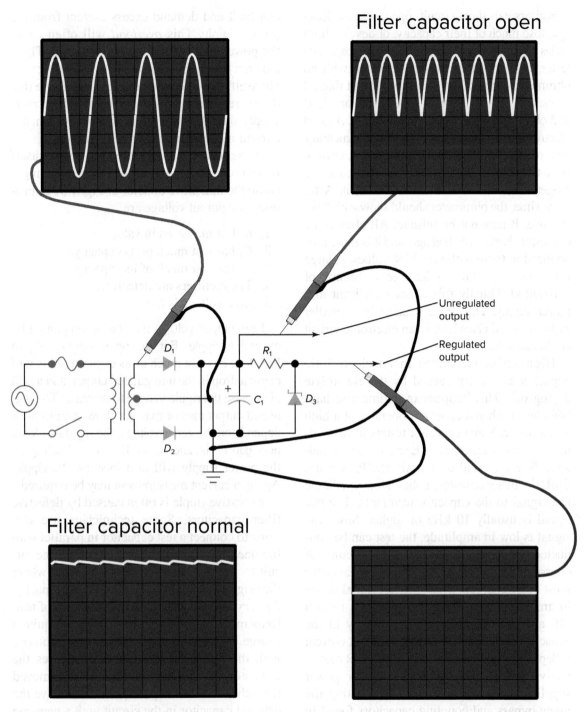

Fig. 4-39 Full-wave supply and waveforms.

The sure method is to disconnect one end of the diode from the circuit. Bridge rectifier diodes can also be checked in circuit with similar results and limitations.

Figure 4-39 also shows some possible waveforms. Many technicians prefer troubleshooting with an oscilloscope. The ac waveform at the transformer secondary proves that the supply is plugged into an ac source, that it is turned on,

and that the fuse is not blown. The full-wave pulsating dc waveform at the unregulated output is *not normal*: It indicates that C_1 is open. With a normal filter capacitor, some ripple might be seen, as also shown in Fig. 4-39. The waveform at the regulated output shows no ripple and a lower dc voltage, which is to be expected.

Many of the filter capacitors used in modern power supplies are of the *electrolytic* type. These

Electrolytic capacitor

capacitors can short-circuit, open, develop leakage, lose much of their capacity, or develop high series resistance. They can be tested on a capacitor tester, or a rough check can be made with an ohmmeter. Be *sure* the supply is off and that all capacitors are discharged. Disconnect one lead and observe polarity when testing them. A good electrolytic capacitor will show a momentary low resistance as it draws a charging current from the ohmmeter. The larger the capacitor, the longer the low resistance will be shown. After some time, the ohmmeter should show a high resistance. It may not be infinite. All electrolytic capacitors have some leakage, and it is more pronounced in those with very high values. A large capacitor may show a leakage resistance of 100,000 Ω. Usually this is not significant in a power supply. This same leakage in a smaller capacitor used elsewhere in an electronic circuit could cause trouble.

High series resistance in an electrolytic capacitor is usually caused by the electrolyte drying out. This happens over time and happens faster when a capacitor operates at a high temperature. Some capacitor testers have an effective series resistance (ESR) mode. Stand-alone ESR meters also are available. To measure ESR, the meter or tester applies a low-amplitude test signal to the capacitor terminals. The test signal is usually 10 kHz or higher. Since the signal is low in amplitude, the test can be conducted while the capacitor is in circuit. Of course, the circuit must be off and the capacitor must be discharged. Since the test signal is low in amplitude, semiconductor junctions remain off, and the test current flows mostly in the capacitor under test. The amount of test current is dependent on the capacitor's ESR. ESR meters are valuable troubleshooting tools for power supplies. They are also useful for testing the many bypass and coupling capacitors found in modern electronic circuits.

Low output voltage

The symptom of *low output voltage* in a power supply can be caused by

1. Excessive load current (overload)
2. Low input (line) voltage
3. Defective surge-limiting resistor
4. Defective filter capacitors
5. Defective rectifiers

Power supplies are often one part of an electronic system. Some other part of the system can fault and demand excess current from the power supply. This *overload* will often cause the power-supply output voltage to drop. There may not be anything wrong with the power supply itself. It is a good idea to first make sure that the current demand is normal when the power-supply output is low. This is one case when a current measurement may be required.

If the load is normal, then the supply itself must be checked. Some of the defects that might cause the half-wave doubler of Fig. 4-37 to produce low output voltage are

1. R_S has increased in value.
2. C_1 has lost much of its capacity.
3. C_2 has lost much of its capacity.
4. The rectifiers are defective.
5. Line voltage is low.

Low output voltage may be accompanied by excessive ripple. For example, suppose C_1 in Fig. 4-39 has lost much of its capacity. This will cause a drop in the unregulated output, and it will also cause the ripple voltage to increase. The regulated output may or may not show symptoms. It depends on the zener voltage, the regulated load, how bad C_1 is, and so on. Excessive loading on the power supply will also increase the ripple. Again, a current measurement may be required.

Excessive ripple is often caused by defective filter capacitors. Some technicians use clip leads to connect a test capacitor in parallel with the one they suspect. This will restore the circuit to normal operation in those cases where the original capacitor is open or low in capacity. *Be very careful* when making this kind of test. Remember, the power supply can store quite a charge. Be *sure* to observe the correct polarity with the test capacitor. If the test proves the capacitor is defective, it should be removed from the circuit. It is poor practice to leave the original capacitor in the circuit with a new one soldered across it.

The last power-supply symptom is high output voltage. Usually this is caused by low load current (underload). The trouble is not in the power supply but somewhere else in the circuit. It may be that a bleeder resistor is open. This decreases the load on the power supply, and the output voltage goes up. High output in a regulated power supply would indicate a defect in the regulator. Regulator troubleshooting is covered in Chap. 15.

Determine whether each statement is true or false.

59. A skilled troubleshooter uses a random trial-and-error technique to find circuit faults.
60. In troubleshooting, it is often possible to limit the problem to one area of the block diagram by observing the symptoms.
61. A resistor that is burned black may indicate that another component in the circuit has failed.
62. A supply that is overloaded will often show low output voltage.
63. Refer to Fig. 4-29. Resistor R_S burns out (opens). The symptom will be zero dc output voltage.
64. Refer to Fig. 4-37. The fuse blows repeatedly. Rectifier D_1 is probably open.
65. Refer to Fig. 4-37. The output voltage is low. Capacitor C_2 could be defective.
66. Refer to Fig. 4-39. The zener diode burns out. Regulated output voltage will be high.
67. Refer to Fig. 4-39. The zener diode is shorted. Both outputs will be zero.
68. Refer to Fig. 4-39. R_1 is open. Both outputs will be zero.

4-10 Replacement Parts

After the defective parts are located, it is time to choose replacement parts. *Exact replacements* are the safest choice. If exact replacements are not available, it may be possible to make substitutions. A substitution should have ratings at least equal to those of the original. It would never do to replace a 2-W resistor with a 1-W resistor. The replacement resistor would probably fail in a short time. It may *not* be a good idea to replace a resistor with one having a higher power rating. In some circuits, the resistor may protect another more expensive part by increasing in value under overload conditions. Also, a fire hazard can result in some circuits if a carbon-composition resistor is substituted for a film resistor. It is easy to see why exact replacements are the safest.

Rectifier diodes have several important ratings. They are rated for average current and for surge current. The current peaks can be much higher than the average current with capacitive filters. However, the current peaks caused by filter capacitors are *repetitive*. Therefore, the average current rating of a rectifier diode is often *greater* than the actual circuit load current. Table 4-1 lists some of the maximum ratings for several common rectifiers. Never make assumptions. For example, a 1N4009 is an ultra-high-speed switching

Exact replacement

Table 4-1	Common Rectifier Diode Ratings		
Device	Peak Inverse Voltage in V	Average Rectified Output Current in A (Resistive Load)	Nonrepetitive Peak Surge Current in A (1 Cycle)
1N4001	50	1	30
1N4002	100	1	30
1N4003	200	1	30
1N4004	400	1	30
1N4005	600	1	30
1N4006	800	1	30
1N4007	1,000	1	30
1N5400	50	3	200
1N5401	100	3	200
1N5402	200	3	200
1N5404	400	3	200
1N5406	600	3	200
1N5407	800	3	200
1N5408	1,000	3	200

ABOUT ELECTRONICS

The 1N4009 has a reverse recovery time of 2 ns, while ordinary rectifiers recover in 30 μs!

Schematic	Name	PIV per diode	PIV per diode with capacitive filter	Diode current
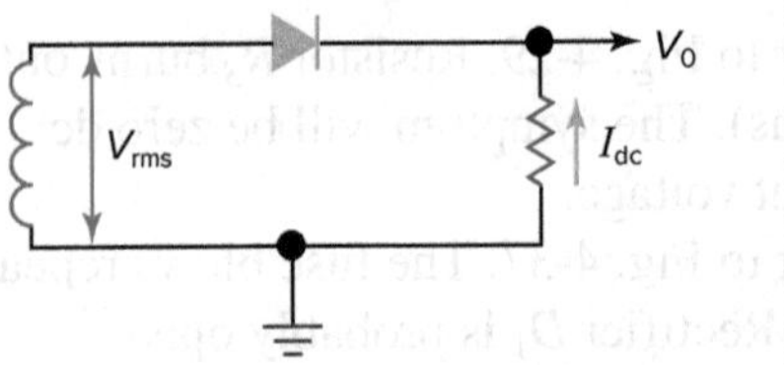	Half-wave	1.41 V_{rms}	2.82 V_{rms}	I_{dc}
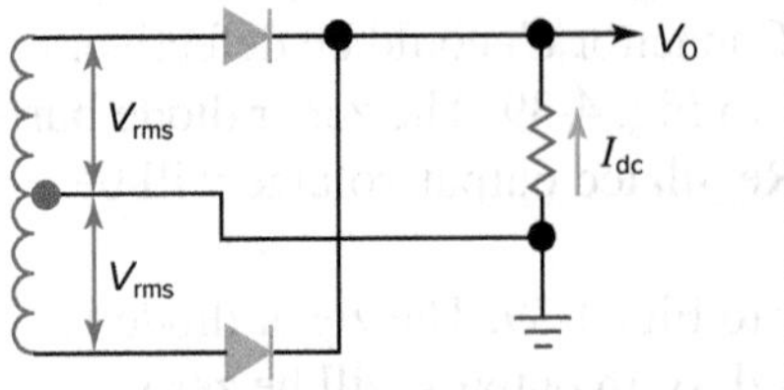	Full-wave	2.82 V_{rms}	2.82 V_{rms}	0.5 I_{dc}
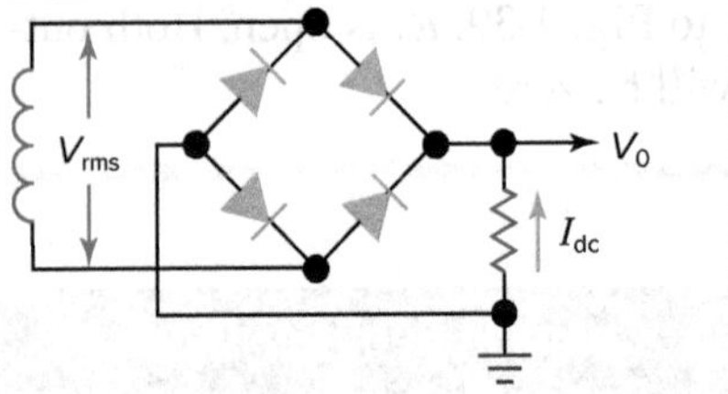	Bridge (full-wave)	1.41 V_{rms}	1.41 V_{rms}	0.5 I_{dc}

Fig. 4-40 Diode ratings for various supply circuits.

diode. One might assume that it is a member of the 1N4001–1N4007 family, but it definitely is *not*.

The maximum reverse-bias voltage that the diode can withstand is another important rating. In a half-wave power supply with a capacitive-input filter or in a full-wave power supply with a center-tapped transformer, the diodes are subjected to a reverse voltage equal to two times the peak value of the ac input. This is because the charged capacitor adds in series with the input when the diodes are off. Thus, the rectifier diodes must block twice the peak input. Figure 4-40 shows the diode ratings for various power-supply circuits.

Electrolytic capacitors are rated for a dc working voltage (dcWV or VdcW or WVdc). This voltage must not be exceeded. Filter capacitors charge to the peak value of the rectified wave. Such a capacitor's dcWV rating should be greater than the peak voltage value.

The capacity of the electrolytic filters is also very important. Substituting a lower value may result in low output voltage and excessive ripple. Substituting a much higher value may cause he rectifiers to run hot and be damaged. A value close to the original is the best choice.

Transformers may also have to be replaced. The replacements should have the same voltage ratings, the same current ratings, and the same taps.

Sometimes, the physical characteristics of the parts are just as important as the electrical characteristics. A replacement transformer may be too large to fit in the same place on the chassis, or the mounting bolt pattern may be different. A replacement filter capacitor may not fit in the space taken by the old one. The stud on a power rectifier may be too large for the hole in the heat sink. It pays to check into the mechanical details when choosing replacement parts.

Substitution guide

Technicians use *substitution guides* to help them choose replacement parts. These are especially helpful for finding replacements for

ABOUT ELECTRONICS

Part names are often another puzzle. PPTCs can be named PolySwitch, Semifuse, Polyfuse, or Multifuse.

solid-state devices. The guides list many device numbers and the numbers for the replacement parts. The guides often include some of the ratings and physical characteristics for the replacement parts. Even though the guides are generally very good, at times the recommended part will not work properly. Some circuits are critical, and the recommended replacement part may be just different enough to cause trouble. There may also be some physical differences between the original and the replacement recommended by the guide.

Solid-state devices have two types of part numbers: registered and nonregistered. There are three major groups of registered devices: *JEDEC*, *PRO-ELECTRON*, and *JIS*.

The JEDEC Solid State Technology Association, formerly known as the Joint Electron Devices Engineering Council (JEDEC), is an independent semiconductor engineering trade organization and standardization body. JEDEC is accredited by the American National Standards Institute (ANSI). It is associated with the Electronic Industries Alliance (EIA), a trade association that represents all areas of the electronics industry in the United States. JEDEC has more than 300 members, including some of the world's largest computer companies. It was founded in 1958 as a joint activity between the EIA and the National Electrical Manufacturers Association (NEMA) to develop standards for semiconductor devices. NEMA dropped out in 1979. In fall 1999, JEDEC became a separate trade association under the current name and maintains an EIA alliance. Earlier in the 20th century, the organization was known as JETEC, the Joint Electron Tube Engineering Council, and was responsible for assigning and coordinating type numbers for vacuum tubes. The type 6L6 vacuum tube, still found in some electric guitar amplifiers, has a type number that was assigned by JETEC.

The early work of JEDEC began as a part-numbering system for devices that became popular in the 1960s. For example, the 1N4001 rectifier diode and the 2N2222 transistor part numbers came from JEDEC. These part numbers are still popular today. JEDEC later developed a numbering system for integrated circuits, but this system did not gain wide acceptance. JEDEC has issued widely used standards for device interfaces, such as the JEDEC memory standards for computer memory, including the DDR SDRAM (double data rate, synchronous dynamic random-access memory) standards. Let's look at how the system works for solid-state devices:

```
1N4001A
 ¦ ¦ ¦  ` - - - Variant (A implies enhanced
 ¦ ¦ ¦              or improved characteristics)
 ¦ ¦ ` - - Serial number (2, 3, or 4 digits)
 ¦ ` - - - - Always N
1 = diode
2 = transistor
3 = MOS field-effect transistor or SCR
4 = optocoupler
6 = optocoupler
```

A 1N4001A is a diode, a 2N5179 is a transistor, a 3N211 is a metallic oxide semiconductor (MOS) field-effect transistor, a 3N84 is a silicon-controlled rectifier or switch (SCR), and a 4N37 is an optocoupler. The first number refers to the number junctions. Diodes have one junction.

JEDEC has also developed a number of popular package drawings for semiconductors such as TO-3, TO-5, and so on. JEP95, the JEDEC registered and standard outlines for solid-state devices, is a compilation of some 3,000 pages of outline drawings for microelectronic packages, including transistors, diodes, dual-inline packages (DIPS), chip carriers, sockets, and package interface ball grid array (BGA) outlines in both inch and metric versions. There are over 500 registrations in JEP95. Examples of JEDEC registered cases or packages are DO-4, TO-9, and TO-92. Also, some packages are known by their acronyms in addition to their JEDEC designations:

SOT (small-outline
    transistor)...................... JEDEC TO-243
DIP.................................... JEDEC MS-001
SOIC (small-outline integrated
    circuit)........................... JEDEC MS-012

Pro Electron is the European type designation and registration system for active components (such as semiconductors, liquid crystal displays, sensor devices, and vacuum tubes). Pro Electron was established in 1966 in Brussels, Belgium. In 1983 it merged with the European

JEDEC

PRO-ELECTRON

JIS

BC549C

Variant (A, B, C for transistors implies low, medium, or high gain)
Serial number (3 or 4 digits or letter and 2 digits)
Device type: A = diode, signal, low power

A = germanium
B = silicon

B = diode, variable capacitance
C = transistor, low power, low frequency
D = transistor, audio frequency power
E = diode, tunnel
F = transistor, high frequency, low power
G = miscellaneous devices
H = diode, magnetic field sensing
K = hall effect device
L = transistor, high frequency, high power
N = optocoupler
P = transistor, light-sensing
Q = light emitter
R = switching device, low power
S = transistor, low power, switching
T = triac or thyristor
U = transistor, switching power
W = surface acoustic wave device
X = diode, frequency multiplier
Y = diode, rectifier
Z = diode, zener

Electronic Component Manufacturers Association (EECA) and since then has operated as an agency of the EECA. Pro Electron supports the unambiguous identification of electronic parts, even when made by different manufacturers. Manufacturers can register new devices with the agency and receive new type designators for them. As an example of how it works, a BC549C device is a high-gain, low-power, silicon audio transistor; an AD162 is a germanium power transistor; and a BY133 is a silicon rectifier diode.

Pro Electron naming for transistors and zener diodes has been widely applied by semiconductor manufacturers around the world. As an example, an Internet search for BS170 reveals many manufacturers for that device. The Pro Electron naming convention for integrated circuits did not take hold, even in Europe. The first letters for European active devices are

A Germanium (or any semiconductor with junctions with a band gap of 0.6 to 1.0 eV)

B Silicon (or any semiconductor with a band gap of 1.0 to 1.3 eV)

C Semiconductors like gallium arsenide with a band gap of 1.3 eV or more

D Semiconductors with a band gap of less than 0.6 eV (infrequently used; most European devices starting with D are 1.4-V filament tubes named under the older Mullard-Philips tube designation)

E Tubes with a 6.3-V filament

P Tubes for a 300-mA series filament supply

R Devices without junctions, such as photoconductive cells

S Solitary digital integrated circuits

T Linear integrated circuits

U Tubes for a 100-mA series heater supply (or mixed signal integrated circuits)

The JIS (Japanese Industrial Standard) uses the following format:

digit, two letters, serial number, [suffix]

Digit: The number of junctions, as in the JEDEC code.

Letters: The letters indicate the intended application for the device according to the following designations:

SA PNP HF transistor

SB PNP AF transistor

SC NPN HF transistor

SD NPN AF transistor

SE Diodes

SF Thyristors

SG Gunn devices

SH UJT

SJ P-channel FET/MOSFET

SK N-channel FET/MOSFET

SM Triac

SQ LED

SR Rectifier

SS Signal diodes

ST Diodes

SV Varicaps

SZ Zener diodes

Serial number: The serial number ranges from 10 to 9999.

Suffix: The (optional) suffix indicates that the type is approved for use by various Japanese organizations.

Some JIS device number examples are 2SA1187, 2SB646, and 2SC733. Since the code for transistors always begins with 2S, it is sometimes omitted. For example, a 2SC733 might be marked C733.

Major manufacturers often use proprietary (nonstandard) numbering systems with their own prefix codes. The following prefixes are examples:

IRF International Rectifier

MJ Motorola power, metal case

MJE Motorola power, plastic case

MPF Motorola field-effect transistor

MPS Motorola low power, plastic case

MRF Motorola HF, VHF, and microwave transistor

RCA RCA

RCS RCS

TIP Texas Instruments power transistor (plastic case)

TIPL Texas Instruments planar power transistor

TIS Texas Instruments small-signal transistor (plastic case)

ZT Ferranti

ZTX Ferranti

A few examples of nonstandard part numbers are IRF510, MJE3055, MPF102, TIP32A, and ZTX302. Unfortunately, part numbers with manufacturers' prefixes have become unreliable indicators of which company made the device. Also, companies change hands. The Motorola semiconductor division is no longer in operation, but its business continues with ON Semiconductor and Freescale (separate corporations).

Some proprietary naming schemes adopt portions of other naming schemes; for example, a PN2222A is a 2N2222A but in a plastic case.

Equipment manufacturers buying large numbers of parts sometimes have them supplied with *house numbers*. These are proprietary in that they usually cannot be cross-referenced. Even if the manufacturer of the device with the house number is known, if contacted it will not be able to supply information about devices it supplied with house numbers. House numbers serve the purpose of limiting service to authorized repairers, and they also help ensure that the integrity and reliability of a product are not compromised by inferior parts.

With so many independent naming schemes, and the abbreviation of part numbers when printed on the devices, ambiguity sometimes occurs. For example, two different devices may

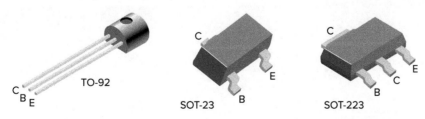

Fig. 4-41 Surface-mount devices can use a variation of the older part number.

be marked "J176" (one the J176 low-power junction FET, the other the higher-powered MOSFET 2SJ176).

As older devices are given surface-mount counterparts, they tend to be assigned many different part numbers because manufacturers have their own systems to cope with the variety of connection arrangements and options for dual or matched devices in one pack. Even when the original device (such as a 2N3904) was assigned by a standards authority and has become well known by engineers and technicians over the years, the new versions are far from standardized in their naming. As shown in Fig. 4-41, the 2N3904 uses the TO-92 package, the MMBT3904 comes in the SOT-23 case (surface mount), and the PZT3904 is in the SOT-223 package (also surface mount).

Self-Test

Determine whether each statement is true or false.

69. It may not be good practice to replace a 1-W resistor with a 2-W resistor.
70. It may not be safe to replace a film resistor with a carbon-composition resistor.
71. It may not be good practice to replace a 1,000-μF filter capacitor with a 2,000-μF capacitor.

72. A transistor is marked 2N3904. This is a house number.
73. The safest replacement part is the exact replacement.
74. The 1N914 is an example of a JEDEC registered part.

Chapter 4 Summary and Review

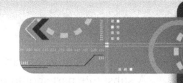

Summary

1. The power supply provides the various voltages for the circuits in an electronic system.
2. Bipolar power supplies develop both polarities with respect to the chassis ground.
3. Diagrams that show the major sections of electronic systems and how they are related are called block diagrams.
4. Power supplies usually change voltage levels and change alternating current to direct current.
5. In a diode rectifier circuit, the positive end of the load will be in contact with the cathode of the rectifier. The negative end of the load will be in contact with the anode of the rectifier.
6. A single diode forms a half-wave rectifier.
7. Half-wave rectification is generally limited to low-power applications.
8. A full-wave rectifier utilizes both alternations of the ac input.
9. One way to achieve full-wave rectification is to use a center-tapped transformer secondary and two diodes.
10. It is possible to achieve full-wave rectification without a center-tapped transformer by using four diodes in a bridge circuit.
11. A dc voltmeter or a dc ammeter will read the average value of a pulsating waveform.
12. The average value of half-wave, pulsating direct current is 45 percent of the rms value.
13. The average value of full-wave, pulsating direct current is 90 percent of the rms value.
14. Pulsating direct current contains an ac component called ripple.
15. Ripple can be reduced in a power supply by adding filter circuits after the rectifiers.
16. Filters for 60-Hz supplies are usually capacitive.
17. NTC surge limiters are used in some power supplies.
18. Capacitive filters cause a heating effect in the rectifiers that requires them to have ratings greater than the dc load current.
19. The factors for predicting dc output voltage are 0.45 for half-wave, 0.90 for full-wave, and 1.414 for any supply with a capacitive filter.
20. Full-wave rectifiers are easier to filter than half-wave rectifiers.
21. Line-operated equipment should always be operated with an isolation transformer to protect the technician and the equipment being serviced.
22. A surge-limiting resistor may be included in power supplies to protect the rectifiers from damaging current peaks.
23. Ripple should be measured when the power supply is delivering its rated full-load current.
24. Ripple is usually nonsinusoidal.
25. The percent regulation is a comparison of the no-load voltage and the full-load voltage.
26. Bleeder resistors can improve voltage regulation and drain the filter capacitors when the power supply is off.
27. A voltage regulator can be added to a power supply to keep the output voltage constant.
28. Zener diodes are useful as shunt regulators.
29. Limiting the possible causes to one or two defects usually involves making tests with meters and other equipment. The schematic diagram is very helpful in this phase of the troubleshooting process.
30. Defects may come in groups. One part shorting out could damage several others.
31. In troubleshooting power supplies, no output voltage is usually caused by open components.
32. Open components can be isolated by voltage measurements or resistance checks with the circuit turned off and the filters drained.
33. Electrolytic capacitors can short-circuit, develop excess leakage, open, or lose much of their capacity.
34. Power-supply voltages are affected by load current.
35. Excessive ripple is usually caused by defective filter capacitors.

36. Maximum ratings of parts must never be exceeded. A substitute part should be at least equal to the original.

37. Substitution guides are very helpful in choosing replacement parts.

Related Formulas

Transformer action (step-down):

$$V_{secondary} = \frac{V_{primary}}{\text{turns ratio}}$$

Transformer action (step-up):

$$V_{secondary} = V_{primary} \cdot \text{turns ratio}$$

Sine wave conversions:

$$V_{rms} = 0.707 \cdot V_p$$
$$V_p = 1.414 \cdot V_{rms}$$
$$V_{av} = 0.9 \cdot V_{rms}\text{(full-wave)}$$
$$V_{av} = 0.45 \cdot V_{rms}\text{(half-wave)}$$

RC time constant: $T = RC$

Filter capacitor size: $C = \dfrac{I}{V_{p-p}} \times T$

Period: $T = \dfrac{1}{f}$

Regulation: $\text{Regulation} = \dfrac{\Delta V}{V_{FL}} \cdot 100\%$

Ripple: $\text{Ripple} = \dfrac{ac}{dc} \cdot 100\%$

Zener resistor: $R_Z = \dfrac{V_{supply} - V_{zener}}{I_{total}}$

$$R_Z = \frac{V_{supply} - V_{zener}}{I_{total} + I_{load}}$$

Power: $P = V \cdot I$

Chapter Review Questions

Determine whether each statement is true or false.

4-1. A schematic shows only the major sections of an electronic system in block form. (4-1)

4-2. In troubleshooting, one of the first checks that should be made is power-supply voltages. (4-1)

4-3. Rectification is the same as filtering. (4-2)

4-4. Diodes make good rectifiers. (4-2)

4-5. A transformer has 120 V alternating current across its primary and 40 V ac across its secondary. It is a step-down transformer. (4-2)

4-6. The positive end of the load will be in contact with the cathode of the rectifier. (4-2)

4-7. A single diode can give full-wave rectification. (4-2)

4-8. Half-wave rectifiers are limited to low-power applications. (4-2)

4-9. A full-wave rectifier uses two diodes and a center-tapped transformer. (4-3)

4-10. A bridge rectifier can provide full-wave rectification without a center-tapped transformer. (4-3)

4-11. A bridge rectifier uses three diodes. (4-3)

4-12. The average value of a sine wave is 0.637 times its rms value. (4-4)

4-13. With pulsating direct current, a dc voltmeter will read the rms value of the waveform. (4-4)

4-14. The ac input to a half-wave rectifier is 20 V. A dc voltmeter connected across the load should read 10 V. (4-4)

4-15. Increasing the load current taken from a power supply will tend to make the output voltage drop. (4-4)

4-16. Diode losses can always be ignored when they are used as rectifiers. (4-4)

4-17. With light loads, power supply filter capacitors hold the dc output near the peak value of the input. (4-5)

4-18. A filter capacitor loses much of its capacity. The symptoms could be excess ripple and low output voltage. (4-5)

4-19. Capacitive filters increase the heating effect in the rectifiers. (4-5)

4-20. Ac power line harmonics increase energy efficiency. (4-5)

4-21. The conversion factors 0.45 and 0.90 are not used to predict the dc output voltage of filtered power supplies. (4-5)

4-22. Pure direct current means that no ac ripple is present. (4-5)

4-23. A lightly loaded voltage doubler may give a dc output voltage nearly 4 times the ac input voltage. (4-6)

4-24. An isolation transformer eliminates all shock hazards for an electronics technician. (4-6)

4-25. The ripple frequency for a half-wave doubler will be twice the ac line frequency. (4-6)

4-26. A 5-V dc power supply shows 0.2 V of ac ripple. The ripple percentage is 4. (4-7)

4-27. From no load to full load, the output of a supply drops from 5.2 to 4.8 V. The regulation is 7.69 percent. (4-7)

4-28. Alternating current ripple can be measured with a dc voltmeter. (4-7)

4-29. It is necessary to load a power supply to measure its ripple and regulation. (4-7)

4-30. The main function of a bleeder resistor is to protect the rectifiers from surges of current. (4-7)

4-31. A zener diode shunt regulator is generally used to filter out ac ripple. (4-8)

4-32. The dissipation in a shunt regulator goes up as the load current goes down. (4-8)

4-33. A power supply blows fuses. The trouble could be a shorted filter capacitor. (4-9)

4-34. A power supply develops too much output voltage. The problem might be high load current. (4-9)

4-35. A burned-out surge resistor is found in a voltage doubler circuit. It might be a good idea to check the diodes and filter capacitors before replacing the resistor. (4-9)

4-36. A shorted capacitor can be found with an ohmmeter check. (4-9)

4-37. A shorted diode can be found with an ohmmeter check. (4-9)

4-38. There is no way to locate data on parts using house numbers. (4-10)

4-39. The EIA is a European association of electronics manufacturers. (4-10)

Chapter Review Problems

4-1. Refer to Fig. 4-3. The ac line is 120 V, and the transformer is 3:1 step-down. What would a dc voltmeter read if connected across R_L? (4-2)

4-2. Refer to Fig. 4-5. The ac line is 120 V, and the primary turns equal the secondary turns. What would a dc voltmeter read if connected across R_L? (4-3)

4-3. Refer to Fig. 4-7. The ac input is 120 V. What would a voltmeter read if connected across the load resistor? (4-4)

4-4. Refer to Fig. 4-16. The ac input is 120 V, and the primary turns equal the secondary turns. What would a dc voltmeter read if connected across R_L? (4-5)

4-5. Refer to Fig. 4-16. Assume a light load and a source voltage of 240 V ac. What would a dc voltmeter read if connected across R_L? (4-5)

4-6. Refer to Fig. 4-27. Assume a light load and an ac source of 240 V. What is the dc voltage across R_L? (4-6)

4-7. Refer to Fig. 4-33. The dc input is 24 V, and the zener is rated at 9.1 V. Assume a zener current of 100 mA and a load current of 50 mA. Calculate the value for R_Z. (4-8)

4-8. What is the dissipation in R_Z in problem 4-7? (4-8)

4-9. What is the dissipation in the zener diode in problem 4-7? (4-8)

4-10. What is the zener dissipation in problem 4-7 if R_L burns out (opens)? (4-8)

4-11. A power supply output drops from 14 to 12.5 V dc when it is loaded. Find its regulation. (4-7)

4-12. The output of the supply in problem 4-11 shows 500 mV ac ripple when it is loaded. Find its ripple percentage. (4-7)

Critical Thinking Questions

4-1. Referring to Fig. 4-1, we see that stage *A* and stage *B* are both energized by the +12 V dc output of the power supply. Is it likely that stage *A* would have a power supply problem that stage *B* would not have?

4-2. Diode manufacturers package two diodes in one case for use in full-wave rectifier circuits. These packages have a metal tab that contacts both cathodes. They also offer a *reverse polarity* version in which the tab contacts both anodes. Why are reverse polarity versions offered?

4-3. Is there ever a situation when there is ac ripple in a circuit that is powered by a battery?

4-4. If an isolation transformer has a short circuit from its primary winding to its secondary winding, will it still work?

4-5. How would you check an isolation transformer to make sure that it does not have a problem such as the one mentioned in question 4-4?

4-6. A friend asks you to help troubleshoot an electronic gadget that she built. You agree, and when you look at the components, you notice a capacitor with a pronounced bulge. What would you do?

4-7. A story in the newspaper relates an incident when a ham radio operator was electrocuted in his basement *during a prolonged power outage.* Does the story make any sense?

Answers to Self-Tests

1. direct current
2. ground (common)
3. block
4. power supply
5. T
6. T
7. F
8. F
9. T
10. T
11. T
12. F
13. 25 V
14. 1
15. 2
16. alternations
17. direction
18. center-tapped transformer
19. 4
20. 24 V ac
21. 10.8 V dc
22. 5.4 V dc
23. 14.4 V dc
24. 18 V dc
25. increases
26. low
27. 2 V
28. ripple (ac)
29. pulsating
30. ripple
31. energy
32. load resistance (current)
33. discharge
34. rms
35. 1.414
36. 1.414
37. unfiltered
38. peak
39. 21.2 V dc
40. loop
41. transformer
42. 2.82
43. decrease
44. isolation
45. 60 Hz
46. 120 Hz
47. diodes (rectifiers)
48. increase
49. decrease
50. 7.69 percent
51. 16.7 percent
52. discharged
53. parallel (shunt)
54. 4.29 Ω
55. 2.1 W
56. 1 W
57. 3.5 W
58. ripple
59. F
60. T
61. T
62. T
63. T
64. F
65. T
66. T
67. F
68. F
69. T
70. T
71. T
72. F
73. T
74. T

Transistors

Learning Outcomes

This chapter will help you to:

5-1 *Define* amplification and power gain. [5-1]

5-2 *Identify* transistor schematic symbols and leads. [5-2, 5-6]

5-3 *Calculate* current gain. [5-2]

5-4 *Identify* power-supply polarity and current flow for NPN and PNP transistors. [5-2]

5-5 *Interpret* characteristic curves and determine current gain and power dissipation. [5-3]

5-6 *Discuss* package styles and data sheets. [5-4]

5-7 *Test* bipolar junction transistors using an ohmmeter. [5-5]

5-8 *Explain* junction field-effect transistors and metal oxide semiconductor field-effect transistors and identify their schematic symbols. [5-6]

5-9 *Compare* the types of transistors. [5-6]

5-10 *Compare* small signal transistors to power transistors. [5-7]

5-11 *Discuss* power transistor ratings and SOA. [5-7]

5-12 *Use* thermal data to find maximum safe power. [5-7]

5-13 *Explain* transistor switching and hard saturation. [5-8]

5-14 *Explain* analog switching. [5-8]

This chapter introduces the transistor. Transistors are solid-state devices similar in some ways to the diodes you have studied. Transistors are more complex and can be used in many more ways. The most important feature of transistors is their ability to amplify signals and act as switches. Amplification can make a weak signal strong enough to be useful in an electronic application. For example, an audio amplifier can be used to supply a strong signal to a loudspeaker.

5-1 Amplification

Amplification is one of the most basic ideas in electronics. Amplifiers make sounds louder and signal levels greater and, in general, provide a function called *gain*. Figure 5-1 shows the general function of an *amplifier*. Note that the

Amplifier

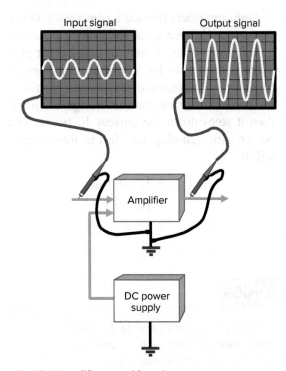

Fig. 5-1 Amplifiers provide gain.

amplifier must be provided two things: *dc power* and the *input signal*. The signal is the electrical quantity that is too small in its present form to be usable. With gain, it becomes usable. As shown in Fig. 5-1, the output signal is greater because of the gain provided by the amplifier.

Gain can be measured in several ways. If an oscilloscope is used to measure the amplifier input signal voltage and the output signal voltage, then the *voltage gain* can be determined. A certain amplifier may provide an output voltage that is 10 times greater than the input voltage. The voltage gain of the amplifier is 10. If an ammeter is used to measure amplifier input and output currents, then the *current gain* can be obtained. With a 0.1-A input signal, an amplifier might produce a 0.5-A output signal for a current gain of 5. If the voltage gain and the current gain are both known, then the *power gain* can be established. An amplifier that produces a voltage gain of 10 and a current gain of 5 will give the following power gain:

$$P = V \times I$$

or

$$P_{\text{gain}} = V_{\text{gain}} \times I_{\text{gain}}$$
$$= 10 \times 5$$
$$= 50$$

Only amplifiers provide a power gain. Other devices might give a voltage gain or a current gain, but not both. A step-up transformer provides voltage gain but is *not* an amplifier. A transformer does not provide any power gain. If the transformer steps up the voltage 10 times, then it steps down the current 10 times. The power gain, ignoring loss in the transformer, will be

$$P_{\text{gain}} = V_{\text{gain}} \times I_{\text{gain}}$$
$$= 10 \times 0.1$$
$$= 1$$

Voltage gain

Current gain

Power gain

Small-signal amplifier

Power amplifier

ABOUT ELECTRONICS

Transistors can be compared to valves because they control flow. They have almost completely replaced vacuum tubes, which used to be called valves in some countries.

EXAMPLE 5-1

Calculate the power gain of an amplifier that has a voltage gain of 0.5 and a current gain of 100.

$$P_{\text{gain}} = V_{\text{gain}} \times I_{\text{gain}} = 0.5 \times 100 = 50$$

Note that an amplifier can show a voltage loss and still have a significant power gain. Likewise, another amplifier might have a current loss and still have a power gain.

A step-down transformer provides a current gain. It cannot be considered an amplifier. The current gain is offset by a voltage loss, and thus, there is no power gain.

Even though power gain seems to be the important idea, some amplifiers are classified as *voltage amplifiers*. In some circuits, only the voltage gain is mentioned. This is especially true in amplifiers designed to handle *small signals*. You will run across many voltage amplifiers or *small-signal amplifiers* in electronic systems. You should remember that they provide power gain, too.

The term *power amplifier* is generally used to refer to amplifiers that develop a *large signal*. Power amplifiers use power transistors, which are covered in Sec. 5-7. A signal can be large in terms of its voltage level, its current level, or both. In the electronic system in Fig. 5-2, the speaker requires several watts for good volume. The signal from the Bluetooth receiver is in the milliwatt (mW) region. A total power gain of hundreds is needed. However, only the final large-signal amplifier is called a power amplifier.

In electronics, *gain* is not expressed in volts, amperes, or watts. If voltage gain is being discussed, it will be a pure number. Gain is the ratio of some output to some input. The letter A is often used as the symbol for gain or amplification. For voltage gain, it is

$$A_V = \frac{V_{\text{out}}}{V_{\text{in}}}$$

EXAMPLE 5-2

Calculate the voltage gain of an amplifier if it has an input signal of 15 mV and an output signal of 1 V.

$$A_V = \frac{V_{\text{out}}}{V_{\text{in}}} = \frac{1 \text{ V}}{15 \times 10^{-3} \text{ V}} = 66.7$$

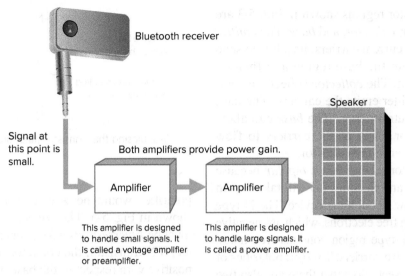

Fig. 5-2 Small-signal and large-signal amplifiers.

The units cancel. So if an amplifier outputs 10 V for 1 V of input, its voltage gain equals 10. It does not equal 10 V. Gain is often expressed in decibels. This is covered in Chap. 6. The Bluetooth receiver in Fig. 5-2 would normally have an output signal of about 300 mV rms.

Self-Test

Determine whether each statement is true or false.

1. An amplifier must be powered and have an input signal to develop a normal output signal.
2. An amplifier has a voltage gain of 50. If the input signal is 2 millivolts (mV), the output signal should be 50 mV.
3. The input signal to an amplifier is 1 mA. The output signal is 10 mA. The amplifier has a current gain of 10 W.
4. The input signal to an amplifier is 100 microvolts (μV), and its output signal is 50 mV, so its voltage gain is 500.
5. A step-up transformer has voltage gain, so it may be considered an amplifier.
6. All amplifiers have power gain.

5-2 Transistors

Transistors provide the power gain that is needed for most electronic applications. They also can provide voltage gain and current gain. There are several important types of transistors. One popular type is the *bipolar junction transistor* (BJT). Field-effect transistors are also widely used. Both types are covered here.

Bipolar junction transistors are similar to junction diodes, but one more junction is included. Figure 5-3 shows one way to make a transistor. A P-type semiconductor region is located between two N-type regions. The polarity of these regions is controlled by the valence of the materials used in the doping process. If you have forgotten this process and how it works, review the information in Chap. 2.

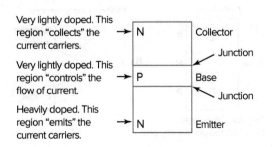

Fig. 5-3 NPN transistor structure.

Bipolar junction transistor

Emitter

Collector

Base

Collector-base junction

NPN transistor

PNP transistor

The transistor regions shown in Fig. 5-3 are named *emitter, collector,* and *base.* The *emitter* is very rich in current carriers. Its job is to send its carriers into the base region and then on to the collector. The *collector* collects the carriers. The emitter emits the carriers. The base acts as the control region. The *base* can allow none, some, or many of the carriers to flow from the emitter to the collector.

The transistor in Fig. 5-3 is *bipolar* because both holes (+) and electrons (−) will take part in the current flow through the device. The N-type regions contain free electrons, which are negative carriers. The P-type region contains free holes, which are positive carriers. Two (bi) polarities of carriers are present. Note that there are also two PN junctions in the transistor. It is a BJT.

The transistor shown in Fig. 5-3 would be classified as an *NPN transistor.* Another way to make a bipolar junction transistor is to make the emitter and collector of P-type material and the base of N-type material. This type would be classified as a *PNP transistor.* Figure 5-4 shows both possibilities and the schematic symbols for each. You should memorize the symbols. Remember that the emitter lead is always the one with the arrow. Also remember that if the arrow is *N*ot *P*ointing i*N*, the transistor is an NPN type.

The two transistor junctions must be biased properly. This is why you cannot replace an NPN transistor with a PNP transistor. The

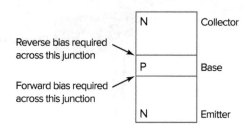

Fig. 5-5 Biasing the transistor junctions.

polarities would be wrong. Transistor bias is shown in Fig. 5-5. The *collector-base junction* must be *reverse-biased* for proper operation. In an NPN transistor, the collector will have to be positive with respect to the base. In a PNP transistor, the collector will have to be negative with respect to the base. PNP and NPN transistors are not interchangeable.

The *base-emitter junction* must be *forward-biased* to turn the transistor on, as shown in Fig. 5-5. This makes the resistance of the base-emitter junction very low as compared with the resistance of the collector-base junction. A forward-biased semiconductor junction has low resistance. A reverse-biased junction has high resistance. Figure 5-6 compares the two junction resistances.

The large difference in junction resistance makes the transistor capable of power gain. Assume that a current is flowing through the two resistances shown in Fig. 5-6. Power can be calculated using

$$P = I^2 \times R$$

The power gain from R_{BE} to R_{CB} could be established by calculating the power in each and dividing:

$$P_{\text{gain}} = \frac{I^2 \times R_{CB}}{I^2 \times R_{BE}}$$

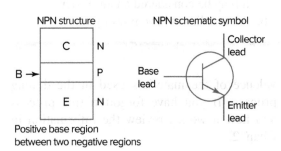

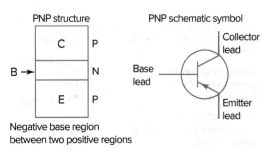

Fig. 5-4 Transistor structures and symbols.

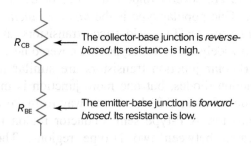

Fig. 5-6 Comparing junction resistances.

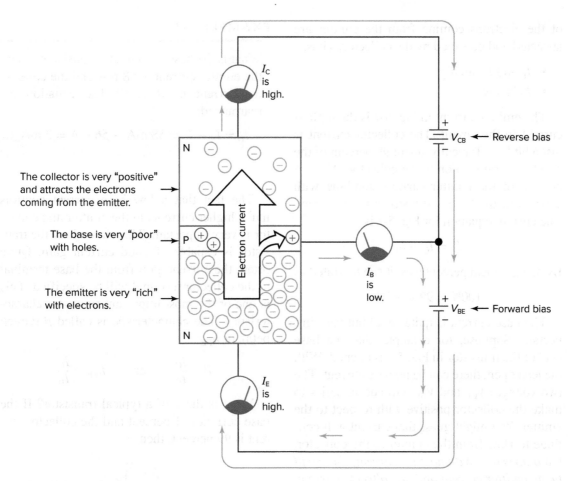

The collector is very "positive" and attracts the electrons coming from the emitter. →

The base is very "poor" with holes.

The emitter is very "rich" with electrons. →

I_C is high.

V_{CB} ← Reverse bias

Electron current

I_B is low.

V_{BE} ← Forward bias

I_E is high.

N
P
N

Fig. 5-7 NPN transistor currents.

If the current through R_{CB} happened to be equal to the current through R_{BE}, I^2 would cancel out and the power gain would be

$$P_{gain} = \frac{R_{CB}}{R_{BE}}$$

The currents in transistors are not equal, but they are very close. A typical value for R_{CB} might be 10 kΩ. It is high since the collector-base junction is reverse-biased. A typical value for R_{BE} might be 100 Ω. It is low because the base-emitter junction is forward-biased. The power gain for this typical transistor would be

$$P_{gain} = \frac{R_{CB}}{R_{BE}} = \frac{10 \times 10^3 \ \Omega}{100 \ \Omega}$$
$$= 100$$

Note: The units (Ω) cancel, and the gain is a pure number.

Perhaps the biggest puzzle is why the current through the reverse-biased junction is as high as the current through the forward-biased junction. Diode theory tells us to expect almost no current through a reverse-biased junction. This is true in a diode but not true in the collector-base junction of a transistor.

Figure 5-7 shows why the collector-base junction current is high. The collector-base voltage V_{CB} produces a reverse bias across the collector-base junction. The base-emitter voltage V_{BE} produces a forward bias across the base-emitter junction. If the transistor were simply two diode junctions, the results would be as follows:

- I_B and I_E would be high.
- I_C would be zero.

The base region of the transistor is very narrow (about 0.0025 cm or 0.001 in.). The base region is lightly doped. It has only a few free holes. It is not likely that an electron coming from the emitter will find a hole in the base with which to combine. With so few electron-hole combinations in the base region, the base current is *very small*. The collector is an N-type region but is charged positively by V_{CB}. Since the base is such a narrow region, the positive field of the collector is quite strong, and the great majority of the electrons coming from the emitter are attracted and collected by the collector. Thus,

- I_E and I_C are high.
- I_B is low.

The emitter current in Fig. 5-7 is the highest current in the circuit. The collector current is just a bit less. Typically, about 99 percent of the emitter carriers go on to the collector. About 1 percent of the emitter carriers combine with carriers in the base and become base current. The current equation for Fig. 5-7 is

$$I_E = I_C + I_B$$

By using typical percentages, it can be stated as

$$100\% = 99\% + 1\%$$

β (Greek beta)

The base current is quite small but *very* important. Suppose, for example, that the base lead of the transistor in Fig. 5-7 is opened. With the lead open, there can be no base current. The two voltages V_{CB} and V_{BE} will add in series to make the collector positive with respect to the emitter. You might guess that current will continue to flow from the emitter to the collector, but *it does not. With no base current, there will be no emitter current and no collector current.* The base-emitter junction *must* be forward-biased for the emitter to emit. Opening the base lead removes this forward bias. If the emitter is not emitting, there is nothing for the collector to collect. Even though the base current is very low, it must be present for the transistor to conduct from emitter to collector.

EXAMPLE 5-3

Determine the emitter current in a transistor when the base current is 1 mA and the collector current is 150 mA.

$$I_E = I_C + I_B = 150 \text{ mA} + 1 \text{ mA} = 151 \text{ mA}$$

ABOUT ELECTRONICS

Transistor Applications—Then and Now
- The earliest commercial transistors worked only at frequencies below 1 MHz.
- Transistors intended for harsh environments use metal, glass, and ceramic packages.

EXAMPLE 5-4

What is the base current in a transistor when the emitter current is 58 mA and the collector current is 56 mA? The equation is rearranged:

$$I_B = I_E - I_C = 58 \text{ mA} - 56 \text{ mA} = 2 \text{ mA}$$

The fact that a low base current controls much higher currents in the emitter and collector is very important. This shows how the transistor is capable of good current gain. Quite often, the current gain from the base terminal to the collector terminal will be specified. This is one of the most important transistor characteristics. The characteristic is called β (Greek beta), or h_{FE}:

$$\beta = \frac{I_C}{I_B} \qquad \text{or} \qquad h_{FE} = \frac{I_C}{I_B}$$

What is the β of a typical transistor? If the base current is 1 percent and the collector current is 99 percent, then

$$\beta = \frac{99\%}{1\%}$$
$$= 99$$

Note that the percent symbol cancels since it appears in both the numerator and the denominator. This is also the case if actual current readings are used. The units of current will cancel, leaving β as a pure number.

EXAMPLE 5-5

Find β for a transistor with a base current of 0.3 mA and a collector current of 60 mA.

$$\beta = \frac{I_C}{I_B} = \frac{60 \text{ mA}}{0.3 \text{ mA}} = 200$$

Don't forget to take *prefixes* such as milli and micro into account when using the β equation. For example, if a transistor has a collector current of 5 mA and a base current of 25 μA, its β is found by

$$\beta = \frac{I_C}{I_B} = \frac{5 \times 10^{-3} \text{A}}{25 \times 10^{-6} \text{A}}$$
$$= 200$$

The ampere units cancel. β is a pure number. Sometimes β is known and must be used to find either base current or collector current. If a transistor has a β of 150 and a collector current of 10 mA, how much base current is flowing? Rearranging the β equation and solving for I_B gives

$$I_B = \frac{I_C}{\beta}$$

$$= \frac{10 \times 10^{-3}A}{150}$$

$$= 66.7 \ \mu A$$

As another example, let's find the collector current in a transistor circuit with a β of 40 and a base current of 85 mA:

$$I_C = \beta \times I_B$$

$$= 40 \times 85 \ mA$$

$$= 3.4 \ A$$

Occasionally a current must be calculated before the current gain can be determined. Don't forget that the emitter current is the sum of the collector and base currents.

EXAMPLE 5-6

A transistor has an emitter current of 12.1 mA and a collector current of 12.0 mA. What is the β of this transistor? First, rearrange the current equation to find the base current:

$$I_B = I_E - I_C = 12.1 \ mA - 12.0 \ mA$$

$$= 0.1 \ mA$$

Then find β:

$$\beta = \frac{I_C}{I_B} = \frac{12 \ mA}{0.1 \ mA} = 120$$

The β of actual transistors varies greatly. Certain power transistors can have a β as low as 20. Small-signal transistors can have a β as high as 400. If you have to guess, a β of 150 can be used for small transistors and a β of 50 for power transistors.

The value of β varies among transistors with the same part number. A 2N2222 is a registered transistor. One manufacturer of this particular device lists a typical β range of 100 to 300. Thus, if three seemingly identical 2N2222 transistors are checked for β, values of 108, 167, and 256

could be obtained. It is *very* unlikely that they would check the same (especially if they are from different manufacturers or different production runs from the same manufacturer).

The value of β is important but unpredictable. Luckily, there are ways to use transistors that make the actual value of β less important than other, more predictable circuit characteristics. This will become clear in the next chapter. For now, focus on the idea that the current gain from the base terminal to the collector terminal tends to be high. Also, remember that the base current is small and controls the collector current.

Figure 5-8 shows what happens in a PNP transistor. Again, the base-emitter junction must be forward-biased for the transistor to be on. Note that V_{BE} is reversed in polarity when compared with Fig. 5-7. The collector-base junction of the PNP transistor must be reverse-biased. Note also that V_{CB} has been reversed in polarity. This is why PNP and NPN transistors are not interchangeable. If one were substituted for the other, both the collector-base and the base-emitter junction would be biased incorrectly.

Figure 5-8 shows the flow from emitter to collector as *hole current*. In an NPN transistor, it is *electron* current. The two transistor structures operate about the same in most ways. The emitter is very rich with carriers. The base is quite narrow and has only a few carriers. The collector is charged by the external bias source and attracts the carriers coming from the emitter. The major difference between PNP and NPN transistors is polarity.

The NPN transistor is more widely used than the PNP transistor. Electrons have better *mobility* than holes; that is, they can move more quickly through the crystal structure. This gives NPN transistors an advantage in high-frequency circuits where things have to happen quickly. Transistor manufacturers have more NPN types in their line. This makes it easier for circuit designers to choose the exact characteristics they need from the NPN group. Finally, it is often more convenient to use NPN devices in negative ground systems. Negative ground systems are more prevalent than positive ground systems.

You will find both types of transistors in use. Many electronic systems use both PNP and NPN transistors in the same circuit. It is very convenient to have both polarities available. This adds flexibility to circuit design.

Hole current

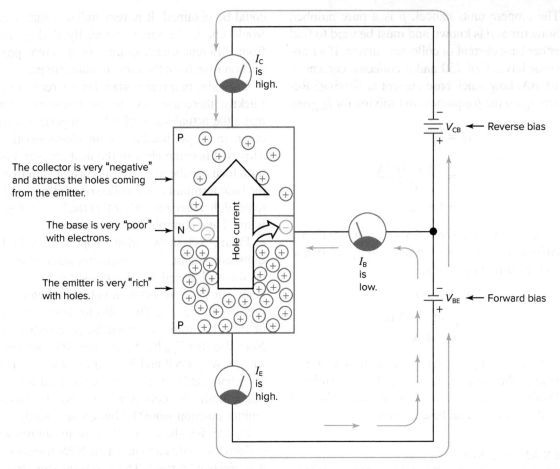

The collector is very "negative" and attracts the holes coming from the emitter.

The base is very "poor" with electrons.

The emitter is very "rich" with holes.

I_C is high.

V_{CB} ← Reverse bias

Hole current

I_B is low.

V_{BE} ← Forward bias

I_E is high.

Fig. 5-8 PNP transistor currents.

Self-Test

Determine whether each statement is true or false.

7. The emitter region of a junction transistor is heavily doped to have many current carriers.
8. A bipolar device may be connected in either direction and still give proper operation.
9. The collector-base junction must be forward-biased for proper transistor action.
10. A defective NPN transistor can be replaced with a PNP type.
11. Even though the collector-base junction is reverse-biased, considerable current can flow in this part of the circuit.
12. The base of a BJT is thin and lightly doped with impurities.
13. When I_B is equal to zero a BJT is off, and I_C will also be close to or equal to zero.
14. Base current controls collector current.
15. Base current is greater than emitter current.
16. Transistor β is measured in milliamperes.

17. 2N2222 transistors are manufactured to have a current gain of 222 from the base to the collector.
18. In a PNP transistor, the emitter emits holes and the collector collects them.
19. A PNP transistor is turned on by forward-biasing its base-emitter junction.

Solve the following problems.

20. A transistor has a base current of 500 μA and a β of 85. Find the collector current.
21. A transistor has a collector current of 1 mA and a β of 150. Find its base current.
22. A transistor has a base current of 200 μA and a collector current of 50 mA. Find its β.
23. A transistor has a collector current of 1 A and an emitter current of 1.01 A. Find its base current.
24. Find β for the transistor described in problem 23.

5-3 Characteristic Curves

As with diodes, transistor characteristic curves can provide much information. There are many types of transistor characteristic curves. One of the more popular types is the *collector family of curves*. An example of this type is shown in Fig. 5-9. The vertical axis shows collector current (I_C) and is calibrated in milliamperes. The horizontal axis shows collector-emitter bias (V_{CE}) and is calibrated in volts. Figure 5-9 is called a collector *family* since several volt-ampere characteristic curves are presented for the same transistor.

Figure 5-10 shows a circuit that can be used to measure the data points for a collector family of curves. Three meters are used to monitor base current I_B, collector current I_C, and collector-emitter voltage V_{CE}. To develop a graph of three values, one value can be held constant as the other two vary. This produces one curve. Then the constant value is set to a new level. Again, the other two values are changed and recorded. This produces the second curve. The process can be repeated as many times as required. For a collector family of curves, the constant value is the *base current*. The variable resistor in Fig. 5-10 is adjusted to produce the desired level of base current. Then the adjustable source is set to some value of V_{CE}. The collector current is recorded. Next, V_{CE} is changed to a new value.

Again, I_C is recorded. These data points are plotted on a graph to produce a volt-ampere characteristic curve of I_C versus V_{CE}. A very accurate curve can be produced by recording many data points. The next curve in the family is produced in exactly the same way but at a new level of base current.

The curves of Fig. 5-9 show some of the important characteristics of junction transistors. One detail is missing. It is the *collector breakdown voltage* which is designated V_{CEO}. It is listed in Table 5-1 under Maximum ratings. It is shown graphically in Fig. 5-34.

Most of Fig. 5-9 occupies what is called the *constant current region*. It also can be called the *active region* or the *linear region*. It serves when a transistor is used to amplify (make a signal larger). Notice that over most of the graph V_{CE} has little effect on the collector current. Examine the curve for $I_B = 20\ \mu$A. How much change in collector current can you see over the range from 2 to 18 V? It increases from 3 to 3.5 mA, for a change of only 0.5 mA. This is a 9 times increase in voltage. Ohm's law tells us to expect the current to increase 9 times. It would increase 9 times if the transistor was just a resistor. In a transistor, the base current has the major effect on collector current. The base current controls the collector current. When $I_B = 0$, the transistor is off or said to be *in cutoff*.

Collector family of curves

Base current

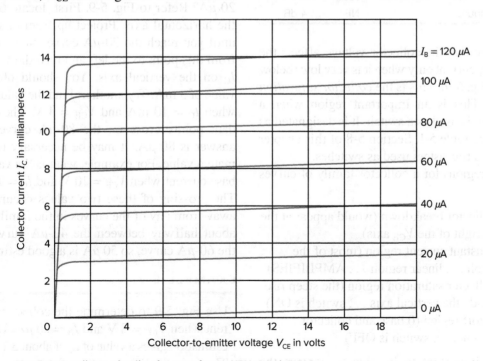

Fig. 5-9 A collector family of curves for an NPN transistor.

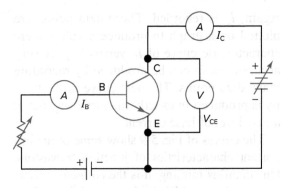

Fig. 5-I0 Circuit for collecting transistor data.

Table 5-I	Selected Specifications for the 2N2222A Bipolar Junction Transistor	

Parameter	Symbol	Value
Maximum ratings		
Collector-emitter voltage	V_{CEO}	40 V dc
Collector-base voltage	V_{CB}	75 V dc
Emitter-base voltage	V_{EB}	6 V dc
Collector current	I_C	800 mA dc
Total device dissipation (derate above 25°C)	P_D	1.8 W 12 mW/°C
Characteristics		
DC current gain	h_{FE}	100 to 300
AC current gain	h_{fe}	50 to 375
Gain-bandwidth product	f_T	300 MHz
Collector-emitter saturation	$V_{CE(sat)}$	0.3 V dc
Noise figure	NF	4 dB

Notice that the collector voltage affects the collector current only when it is very low (below 1 V in Fig. 5-9). This is the *collector saturation region*. This is an important region when a transistor is used as a switch. It is designated as $V_{CE(sat)}$ in Table 5-1. Section 5-8 of this chapter discusses transistors used as switches.

The regions for a collector family of curves are:

1. Collector breakdown (would appear at the far right of the V_{CE} axis)
2. Constant current region (most of the graph . . . linear region . . . AMPLIFIES)
3. Collector saturation region (the steep rise beside the vertical axis . . . switch is ON)
4. Cutoff region (0 base and collector currents . . . switch is OFF)

It is important to be able to convert the curves back into data points. For example, can

you find the value of I_C when $V_{CE} = 6$ V and $I_B = 20$ μA? Refer to Fig. 5-9. First, locate 6 V on the horizontal axis. Project up from this point until you reach the 20-μA curve. Now, project from this point to the left and read the value of I_C on the vertical axis. You should obtain a value of 3 mA. This is in the linear region. Try another: Find the value of I_B when $I_C = 10$ mA and $V_{CE} = 4$ V. These two data points cross on the 80-μA curve. The answer is 80 μA. This is also in the linear region. What if $I_C = 10$ mA and $V_{CE} = 0.3$ V? This is in the saturation region.

It may be necessary to estimate a value. For example, what is the value of base current when $V_{CE} = 10$ V and $I_C = 7$ mA in Fig. 5-9? The crossing of these two values occurs well away from any of the curves in the family. It is about halfway between the 40-μA curve and the 60-μA curve, so 50 μA is a good estimate. This is in the linear region. It might be necessary to guess. Here are two examples:

1. $I_B = 200$ μA and $I_C = 20$ mA: $V_{CE} = 1$ V: the transistor is in saturation
2. $V_{CE} = 25$ V and $I_C = 0$: $I_B = 0$: the transistor is in cutoff

It is important to be able to convert the curves back into data points. For example, can you read the value of I_C when $V_{CE} = 6$ V and $I_B = 20$ μA? Refer to Fig. 5-9. First, locate 6 V on the horizontal axis. Project up from this point until you reach the 20-μA curve. Now, project from this point to the left and read the value of I_C on the vertical axis. You should obtain a value of 3 mA. Try another: Find the value of I_B when $I_C = 10$ mA and $V_{CE} = 4$ V. These two data points cross on the 80-μA curve. The answer is 80 μA. It may be necessary to estimate a value. For example, what is the value of base current when $V_{CE} = 10$ V and $I_C = 7$ mA? The crossing of these two values occurs well away from any of the curves in the family. It is about halfway between the 40-μA curve and the 60-μA curve, so 50 μA is a good estimate.

EXAMPLE 5-7

Use Fig. 5-9 to determine the collector current when $V_{CE} = 8$ V and $I_B = 20$ μA. Using estimation gives a value of I_C of about 3.1 mA. This is in the linear (active) region.

EXAMPLE 5-8

Use the curves in Fig. 5-9 to find the emitter current when V_{CE} is 6 V and I_B is 100 μA. The collector curves do not show any emitter data, but emitter current can be found from base current and collector current. We already know the base current, so we inspect the curves to find the collector current. Figure 5-9 shows that $V_{CE} = 6$ V and $I_B = 100$ μA intersect at $I_C = 12$ mA. Thus,

$$I_E = I_C + I_B = 12 \text{ mA} + 100 \text{ } \mu\text{A}$$
$$= 12.1 \text{ mA}$$

This is in the active region.

The curves in Fig. 5-9 give enough information to calculate β. What is the value of β at $V_{CE} = 8$ V and $I_C = 8$ mA? The first step is to find the value of the base current. The two values intersect at a base current of 60 μA. Now, β can be calculated:

$$\beta = \frac{I_C}{I_B}$$
$$= \frac{8 \text{ mA}}{60 \text{ } \mu\text{A}}$$
$$= 133$$

Calculate β for the conditions of $V_{CE} = 12$ V and $I_C = 14$ mA. These values intersect at $I_B = 120$ μA:

$$\beta = \frac{14 \text{ mA}}{120 \text{ } \mu\text{A}}$$
$$= 117$$

The two prior calculations reveal another fact about transistors. Not only does β vary from transistor to transistor, but it also varies with I_C. Later, it will be shown that temperature also affects β.

Suppose for Fig. 5-11 that the base current is 1 mA. This is off the graph but can be interpreted. *First*, it is clearly in saturation and a low value of V_{CE} is expected. For example, the transistor in Table 5-1 lists $V_{CE(sat)}$ at 0.3 V. $I_{B(max)}$ is not specified. This value is not normally listed. The base current 1 mA will not damage the transistor. *Second*, it tells us that Beta is no longer applicable. Beta assumes linear operation and is not useful for saturation or cutoff. With a high base current what is important is if the collector current and device dissipations are at safe values. This will be clarified in Sec. 5-7 of this chapter.

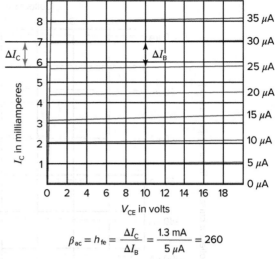

$$\beta_{ac} = h_{fe} = \frac{\Delta I_C}{\Delta I_B} = \frac{1.3 \text{ mA}}{5 \text{ } \mu\text{A}} = 260$$

Fig. 5-11 Calculating β_{ac} with characteristic curves.

There is another form of current gain from base to collector called β_{ac} or h_{fe}. Study the following equations to see how β_{ac} differs from what has already been discussed:

$$\beta_{dc} = h_{FE}$$
$$= \frac{I_C}{I_B}$$
$$\beta_{ac} = h_{fe}$$
$$= \frac{\Delta I_C}{\Delta I_B}\bigg|V_{CE}$$

The Greek delta symbol (Δ) means "change in" and the symbol | means that V_{CE} is to be held constant. Figure 5-11 shows the process. The collector-to-emitter voltage is constant at 10 V. The base current changes from 30 to 25 μA, for a ΔI_B value of 5 μA. Projecting to the left shows a corresponding change in collector current from 7.0 to 5.7 mA. This represents a ΔI_C value of 1.3 mA. Dividing gives a β_{ac} of 260.

There is no significant difference between β_{dc} and β_{ac} at low frequencies. This book emphasizes β_{dc}. The beta symbol with no subscript will designate *dc current gain*. Alternating current gain will be designated by β_{ac}.

EXAMPLE 5-9

Use Fig. 5-9 to obtain the data needed to calculate β_{ac} when $V_{CE} = 4$ V and I_B is changing from 20 to 40 μA.

$$\beta_{ac} = \frac{\Delta I_C}{\Delta I_B} = \frac{2.6 \text{ mA}}{20 \text{ } \mu\text{A}} = 130$$

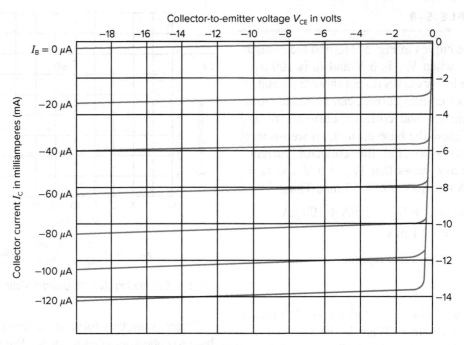

Fig. 5-12 A collector family of curves for a PNP transistor.

At high frequencies the ac current gain of BJTs starts to fall off. This effect limits the useful frequency range of transistors. The *gain-bandwidth product* is the frequency at which the ac current drain drops to 1. The symbol for gain-bandwidth product is f_T. This transistor specification is important in high-frequency applications. For example, the 2N5179 is a radio-frequency transistor and has an f_T of 1.4 GHz. The 2N3904 is a general-purpose transistor and has an f_T of 300 MHz. Thus, it would not be good practice to substitute a 2N3904 for a 2N5179 in a radio-frequency circuit.

It is standard practice to plot positive values to the right on the horizontal axis and up on the vertical axis. Negative values go to the left and down. A family of curves for a PNP transistor may be plotted on a graph as shown in Fig. 5-12. The collector voltage must be negative in a PNP transistor. Thus, the curves go to the left. The collector current is in the opposite direction, compared with an NPN transistor. Thus, the curves go down. However, curves for PNP transistors are sometimes drawn up and to the right. Either method is equally useful for presenting the collector characteristics.

Some shops and laboratories are equipped with a device called a *curve tracer* (an example is shown in Fig. 5-19). This device draws the characteristic curves on a cathode-ray tube or liquid crystal display (LCD). This is far more convenient than collecting many data points

and plotting the curves by hand. Curve tracers show NPN curves in the first quadrant (as in Fig. 5-9) and PNP curves in the third quadrant (as in Fig. 5-12). High-end (expensive) curve tracers are not commonplace.

The *transfer characteristic curves* shown in Fig. 5-13 are another example of how curves can be used to show the electrical characteristics of a transistor. Curves of this type show how one transistor terminal (the base) affects another (the collector). This is why they are called transfer curves. We know that base current controls collector current. Figure 5-13 shows how base-emitter voltage controls collector current. This is because the base-emitter bias sets the level of base current.

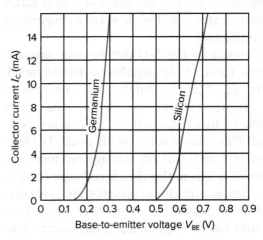

Fig. 5-13 Comparing silicon and germanium transistors.

Figure 5-13 also shows one of the important differences between silicon transistors and germanium transistors. Like diodes, germanium transistors turn on at a much lower voltage (approximately 0.2 V). The silicon device turns on near 0.6 V. These voltages are important to remember. They are reasonably constant and can be of great help in troubleshooting transistor circuits. They can also help a technician determine whether a transistor is made of silicon or germanium. Germanium transistors are rarely used now. They have been replaced by silicon devices because silicon works better at high temperatures.

Self-Test

Supply the missing word or number in each statement.

25. Refer to Fig. 5-9. Voltage $V_{CE} = 4$ V and current $I_C = 3$ mA. $I_B =$ _____.
26. Refer to Fig. 5-9. Current $I_B = 90$ μA and voltage $V_{CE} = 4$ V. $I_C =$ _____.
27. Refer to Fig. 5-9. Voltage $V_{CE} = 6$ V and current $I_C = 8$ mA. $\beta =$ _____.
28. Refer to Fig. 5-9. Current $I_B = 100$ μA and voltage $V_{CE} = 8$ V. $P_C =$ _____.

29. Refer to Fig. 5-9. V_{CE} is held constant at 4 V. I_B changes from 60 to 80 μA. $\beta_{ac} =$ _____.
30. Germanium transistors turn on when V_{BE} reaches _____ V.
31. Silicon transistors turn on when V_{BE} reaches _____ V.
32. Of the two semiconductor materials, _____ is the better conductor.

5-4 Transistor Data

Table 5-1 is a short list of transistor *parameters*. A parameter is any of a set of properties (i.e., electrical, mechanical, thermal) whose values determine the characteristics or behavior of something. One method of finding information about transistors is to use the Internet to conduct a *parametric search*. In the United States, companies such as *Mouser, Digi-Key, Allied,* and *Arrow* have websites that support this. For example, use a web browser and navigate to a supplier and select **BJT** as a parts category. There could be many thousands of choices at this point. Next, start to limit the search: **Semiconductors > Discrete > Transistors, Single, BJT** (at this point there might be 20,000 choices) > **Through Hole** (down to 9,588 choices), **NPN** > 5,093, $V_{CE(MAX)}$ 40V > 325, $I_{C(MAX)}$ 0.8A > 16, **TO-92** > 1. Starting at 20,000 choices we end up with only 1 transistor remaining! Obviously, your results will vary over a wide range depending on the parameters and suppliers that you use. With some practice you will find this to be a valuable method of finding parts.

Another approach is to use a search engine and enter a part number such as 2N2222A. You could have tens of thousands of hits. Select links that look promising. There will be many different data sheets available as PDF files. Table 5-1 is a *very brief* sample of what will be found on the data sheets. One item missing is the case style. When replacing a transistor it is often important to use the same type. A 2N2222A is available in the TO-92 (plastic) or the TO-18 (metal) package. The TO-18 allows more current and power dissipation. Table 5-1 lists 1.8 watts maximum power and that is for the metal case. The plastic package is rated for only 625 mW.

Occasionally there are oddities. A 2N2222A has the lead arrangement shown in Fig. 5-14 at the right (Style 1) and a P2N2222A has a different arrangement as shown in Fig. 5-14 at the left (Style 17).

Both are TO-92

C B E
Style 17

E B C
Style 1

Fig. 5-14 Case style variations.

Charles A Schuler

The Joint Electron Device Engineering Council (JEDEC) lists case styles for transistors ranging from TO-1 to TO-249 plus some SOT (small outline transistor) packages. Figure 5-15 shows a few samples.

Finding a replacement for a transistor might be easy. If the same part number is available this almost always works even if it is made by a different manufacturer. Once in a while, a transistor might be selected for beta or some other parameter. Another example is matched transistor pairs for audio amplifiers. When a substitute transistor must be used, there are some obvious rules:

1. Use the same polarity (NPN for NPN or PNP for PNP)
2. Use the same technology (BJT for BJT or FET for FET)
3. Use the same case style and lead arrangement.
4. Use equal or better ratings (V_{CEO}, I_C, P_D, f_T, NF)

Parameters like f_T and NF (noise figure) become important in radio frequency applications and low noise applications. f_T is the frequency where transistor gain goes to 0. One can't use a transistor with $f_T = 100$ MHz in an RF (radio frequency) amplifier designed to operate at 500 MHz. NF is a measure of how much internal noise is generated in a transistor. This should be as low as possible—especially when only very small signals are being amplified. It is a measure of the signal to noise ratio (SNR) at the input of the transistor compared to the SNR at the output:

$$NF = 10\,Log_{10}\frac{SNR_{in}}{SNR_{out}}$$

Chapter 6 will discuss how common logarithms and decibels (dBs) are used with signal ratios (input signal compared to output signal). For now, it is enough to state that an ideal transistor contributes no noise to the desired signal and has an NF = 0 dB. An extremely low noise transistor might have an NF = 1 dB. Table 5-1 shows NF = 4 dB. A 2N2222A is a switching transistor and is not noted for its low noise. We can surmise from this that it could be a poor choice for replacing an audio preamplifier that has to work with very low level signals.

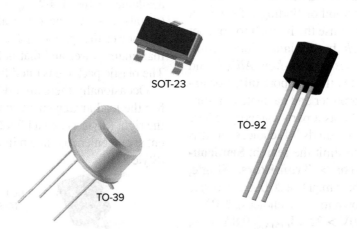

SOT-23

TO-92

TO-39

Fig. 5-15 Three transistor case styles.

Self-Test

Determine whether each statement is true or false.

33. Device manufacturers publish data sheets and data manuals for solid-state devices.

34. The order of the leads on a TO-92 package is always E - B - C.

35. A PNP transistor can be replaced with an NPN type if it is a general-purpose type.

36. Replacing a 2N2222 transistor made by Motorola with a 2N2222 transistor made by another company is not an exact replacement.

37. It is possible to choose a replacement transistor by considering polarity, semiconductor material, voltage and current levels, and circuit function.

5-5 Transistor Testing

One way to test transistors is to use a curve tracer. This technique is used by semiconductor manufacturers and by equipment makers to test incoming parts. Curve tracers are also used in design labs. Figure 5-19 at the end of this section shows an affordable curve tracer.

Another technique used at manufacturing and design centers is to place the transistor in a special fixture or test circuit. This is a *dynamic test* because it makes the device operate with real voltages and signals. This method of testing is often used for VHF and UHF transistors. Dynamic testing reveals power gain and *NF* under signal conditions. NF is a measure of a transistor's ability to amplify weak signals. Some transistors generate enough electrical noise to overpower a weak signal. These transistors are said to have a poor NF.

A few transistor types may show a gradual loss of power gain. Radio-frequency power amplifiers, for example, may use *overlay-type transistors*. These transistors can have over 100

separate emitters. They can suffer base-emitter changes that can gradually degrade power gain. Another problem is moisture, which can enter the transistor package and gradually degrade performance. Even though gradual failures are possible in transistors, they are *not* typical.

For the most part, *transistors fail suddenly and completely.* One or both junctions (the junction is the transition region between P and N) may short-circuit. An internal connection can break loose or burn out from an overload. This type of failure is easy to check. Most bad transistors can be identified with a few ohmmeter tests out-of-circuit or with voltmeter checks in-circuit.

A good transistor has two PN junctions. Both can be checked with an ohmmeter. As shown in Fig. 5-16, a PNP transistor is comparable to two diodes with a common cathode connection. The base lead acts as the common cathode. Figure 5-17 shows an NPN transistor as two diodes with a common anode connection. If two good diodes can be verified by ohmmeter tests, the transistor is probably good.

Dynamic test

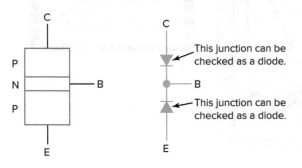

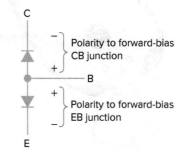

Fig. 5-16 PNP junction polarity.

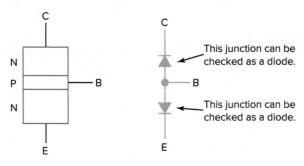

Fig. 5-17 NPN junction polarity.

The ohmmeter can also be used to identify the polarity (NPN or PNP) of a transistor and the three leads. This can be helpful when data are not available. Analog ohmmeters should be set to the $R \times 100$ range for testing most transistors. For DMMs, the diode function can be used.

The first step in testing transistors is to connect the ohmmeter leads across two of the transistor leads. If a lower resistance is indicated, the leads are across one of the diodes or else the transistor is shorted. To decide which is the case, reverse the ohmmeter leads. If the *transistor junction* is good, the ohmmeter will show a high resistance. If you happen to connect across the emitter and collector leads of a good transistor, the ohmmeter will show high resistance in both directions. A DMM might indicate "OL" for overrange. (Refer back to Fig. 3-14 to review junction testing with meters.) The reason is that two junctions are in the ohmmeter circuit. Study Figs. 5-16 and 5-17 and verify that with either polarity applied from emitter to collector, one of the diodes will be reverse-biased.

Once the emitter-collector connection is found, the base has been identified by the process of elimination. Now connect the negative lead of the ohmmeter to the base lead. Touch the positive lead to one and then the other of the two remaining leads. If a low resistance is shown, the transistor is a PNP type. Connect the positive lead to the base lead. Touch the negative lead to one and then the other of the two remaining leads. If a low resistance is shown, then the transistor is an NPN type.

Thus far, you have identified the base lead and the polarity of the transistor. Now it is possible to check the transistor for gain and to identify the collector and emitter leads. A 100,000-Ω resistor and an ohmmeter can check for gain.

The resistor will be used to provide the transistor with a small amount of base current. If the transistor has good current gain, the collector current will be much greater. The ohmmeter will indicate a resistance much lower than 100,000 Ω, and this proves that the transistor is capable of current gain. This check is made by connecting the ohmmeter across the emitter and collector leads at the same time that the resistor is connected across the collector and base leads. The technique is illustrated for both kinds of meters in Fig. 5-18. The DMM is on the diode function and the reading is higher than it is for a diode test. If you guess wrong and have the positive lead to the emitter and the

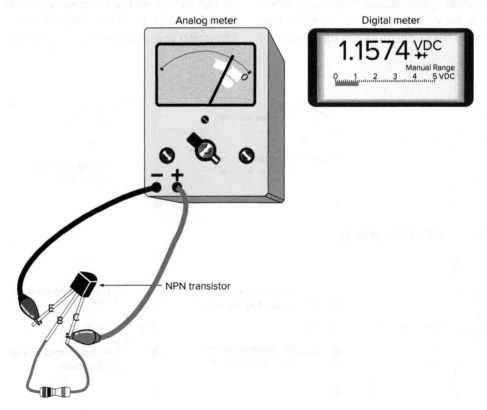

Analog meter   Digital meter

NPN transistor

Fig. 5-I8 Checking NPN gain.

negative lead to the collector, a low resistance reading will not be seen. The DMM will indicate OL (overload or overrange). Just remember that the resistor must be connected from the positive lead to the base when testing for gain in an NPN transistor. The emitter-collector combination showing the most gain (lowest resistance) is the correct connection. You will also be sure that the transistor has gain because of the low resistance reading or the high DMM reading with the diode test function selected.

Transistors have some *leakage current*. This is due to minority carrier action. One leakage current in a transistor is called I_{CBO}. (The symbol I stands for current, CB stands for the collector-base junction, and O tells us the emitter

is open.) This is the current that flows across the collector-base junction under conditions of reverse bias and with the emitter lead open. Another transistor leakage current is I_{CEO}. (The symbol I stands for current, CE stands for the collector-emitter terminals, and O tells us that the base terminal is open.) I_{CEO} is the largest leakage current. It is an amplified form of I_{CBO}:

$$I_{CEO} = \beta \times I_{CBO}$$

With the base terminal open, any current leaking across the reverse-biased collector-base junction will have the same effect on the base-emitter junction as an externally applied base current. With the base terminal open, there is no other place for the leakage current to go.

Leakage current

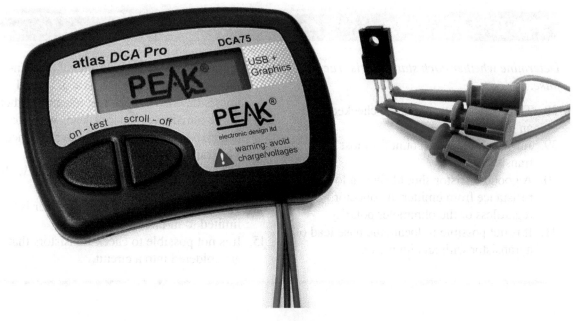

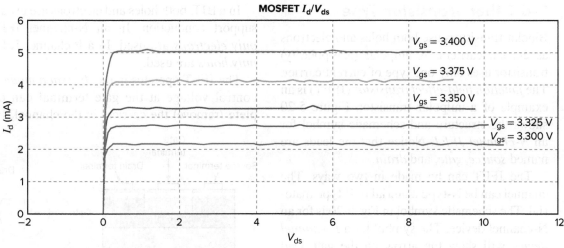

Fig. 5-19 Atlas DCA Pro curve tracer and analyzer.

Courtesy of Peak Electronic Design Ltd.

The transistor amplifies this leakage just as it would any base current:

$$I_C = \beta \times I_B$$

Silicon transistors have very low leakage currents. When ohmmeter tests are made, the ohmmeter should show an infinite resistance when the junctions are reverse-biased. Anything less may mean the transistor is defective. Germanium transistors have much greater leakage currents. This will probably show up as a high, but not infinite, reverse resistance. It will be most noticeable when checking from the emitter to the collector terminal. This is because I_{CEO} is an amplified version of I_{CBO}. It is not likely that technicians will encounter germanium transistors unless they are working on old circuitry.

In-circuit transistor testing can work but may not be straightforward. First, it must *never* be used on live circuits. Second, the ohmmeter might read low due to paths around the transistor. Third, more reliable methods based on voltage or signal path analysis are usually preferable. Later chapters cover the troubleshooting techniques most often used.

Figure 5-19 shows an out-of-circuit transistor analyzer that also displays characteristic curves on a USB connected computer. MOSFET transistors are covered in the next section. Transistor testers are common and might be an added feature of some instruments.

Self-Test

Determine whether each statement is true or false.

38. Transistor junctions can be checked with an ohmmeter.
39. Junction failures account for most bad transistors.
40. A good transistor should show a low resistance from emitter to collector, regardless of the ohmmeter polarity.
41. It is not possible to locate the base lead of a transistor with an ohmmeter.

42. Suppose that the positive lead of an ohmmeter is connected to the base of a good transistor. Also assume that touching either of the remaining transistor leads with the negative lead shows a moderate resistance. The transistor must be an NPN type.
43. It is possible to verify transistor gain with an ohmmeter.
44. Transistor testing with an ohmmeter is limited to in-circuit checks.
45. It is not possible to check transistors that are soldered into a circuit.

5-6 Other Transistor Types

Bipolar transistors use both holes and electrons as current carriers. A unipolar (one-polarity) transistor uses only one type of current carrier. The *junction field-effect transistor* (JFET) is an example of a unipolar transistor. Figure 5-20 shows the structure and schematic symbol for an *N-channel JFET*. Notice that the leads are named *source*, *gate*, and *drain*.

The JFET can be made in two ways. The channel can be N-type material or P-type material. The schematic symbol in Fig. 5-20 is for an N-channel device. The symbol for a *P-channel device* will show the arrow on the gate lead pointing out. Remember, pointing i*N* indicates an *N-channel device*.

In a BJT, both holes and electrons are used to support conduction. In an N-channel JFET, *only electrons* are used. In a P-channel JFET, *only holes* are used.

The JFET operates in the *depletion mode*. A control voltage at the gate terminal can deplete (remove) the carriers in the channel. For

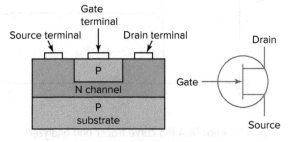

Fig. 5-20 An N-channel JFET.

Depletion mode

N-channel JFET

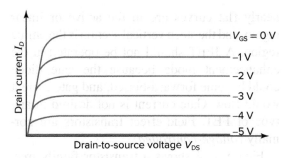

Fig. 5-21 Characteristic curves of a JFET.

example, the transistor in Fig. 5-20 will normally conduct from the source terminal to the drain terminal. The N channel contains enough free electrons to support the flow of current. If the gate is made negative, the free electrons can be *pushed out of the channel*. Like charges repel. This leaves the channel with fewer free carriers. The resistance of the channel is now much higher, and this tends to decrease the source and drain currents. In fact, if the gate is made negative enough, the device can be turned off and no current will flow.

Examine the curves of Fig. 5-21. Notice that as the voltage from gate to source ($-V_{GS}$) increases, the drain current I_D decreases. The curves in Fig. 5-21 are sometimes divided into three regions: (1) the ohmic region where the current ID increases rapidly from 0 to the bends, (2) the linear or active region where the curves are flat, and (3) the cutoff region where the device is off and I_D is 0. Compare a JFET with a BJT:

- A BJT is off (there is no collector current) until base current is provided.
- A JFET is on (drain current is flowing) until the gate voltage becomes high enough to remove the carriers from the channel.

These are *important* differences: (1) The bipolar device is current-controlled. (2) The unipolar device is voltage-controlled. (3) The bipolar transistor is normally off. (4) The JFET is normally on.

Will there be any *gate current* in the JFET? Check Fig. 5-20. The gate is made of P-type material. To control channel conduction, the gate is made negative. This reverse-biases the gate-channel diode. The gate current should be *zero* (there may be a very small leakage current).

There are also P-channel JFETs. They use P-type material for the channel and N-type material for the gate. The gate will be made positive to repel the holes in the channel. Again, this reverse-biases the gate-channel diode, and the gate current will be zero if the gate voltage is high. Since the polarities are opposite, N-channel JFETs and P-channel JFETs are not interchangeable.

Field-effect transistors (FETs) do not require any gate current for operation. This means the gate structure can be completely insulated from the channel. Thus, any slight leakage current resulting from minority carrier action is blocked. The gate can be made of metal. The insulation used is an oxide of silicon. This structure is shown in Fig. 5-22. It is called a *metal oxide semiconductor field-effect transistor* (MOSFET). The MOSFET can be made with a P channel or an N channel. Again, the arrow pointing i*N* (toward the center) tells us that the channel is N-type material.

MOSFET

Early MOSFETs were very delicate. The thin oxide insulator was easily damaged by excess voltage. The static charge on a technician's body could easily break down the gate insulator. These devices had to be handled very carefully. Their leads were kept shorted together until the device was soldered into the circuit. Special precautions were needed to safely make measurements in some MOSFET circuits. Today most MOSFET devices have built-in diodes to protect the gate insulator. If the gate voltage goes too high, the diodes turn on and safely discharge the

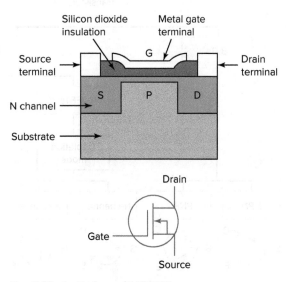

Fig. 5-22 An N-channel MOSFET.

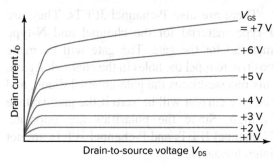

Fig. 5-23 Enhancement-mode characteristic curves.

potential. However, manufacturers still advise careful handling of MOSFET devices.

The gate voltage in a MOSFET circuit can be of either polarity since a diode junction is not used. This makes another mode of operation possible—the *enhancement mode*. An enhancement-mode device normally has no conductive channel from the source to the drain. It is a normally off device. The proper gate voltage will attract carriers to the gate region and form a conductive channel. The channel is *enhanced* (aided by gate voltage).

Figure 5-23 shows a family of curves for an N-channel enhancement-mode device. As the gate is made more positive, more electrons are attracted into the channel area. This enhancement improves channel conduction, and the drain current increases. When $V_{GS} = 0$, the drain current is 0. This is the cutoff region mentioned before. In Fig. 5-23, the cutoff region lies along the V_{DS} axis where I_D is zero. The flat or

Enhancement mode

nearly flat curves are in the active or linear region, and the steep vertical section is the ohmic region. A JFET should not be operated in the enhancement mode because the gate diode could become forward-biased, and gate current would flow. Gate current is not desired in any type of FET. Field-effect transistors are normally *voltage-controlled*.

Figure 5-24 shows a transistor family tree. Note that the enhancement-mode symbols use a broken line from the source to the drain. This is because the channel can be created or enhanced by applying the correct gate voltage. Field-effect transistors have some advantages over bipolar transistors that make the former attractive for certain applications. Their gate terminal does not require any current. This is a good feature when an amplifier with high input resistance is needed. This is easy to understand by inspecting Ohm's law:

$$R = \frac{V}{I}$$

Consider V to be a signal voltage supplied to an amplifier and I the current taken by the amplifier. In this equation, as I decreases, R increases. This means that an amplifier that draws very little current from a signal source has a high input resistance. Bipolar transistors are current-controlled. A bipolar amplifier must take a great deal more current from the signal source. As I increases, R decreases. Bipolar-junction transistor amplifiers have a low input resistance compared with FET amplifiers.

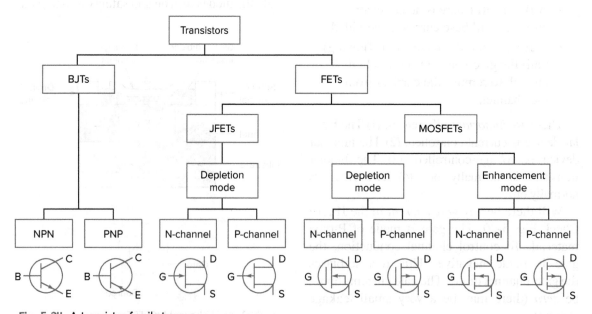

Fig. 5-24 A transistor family tree.

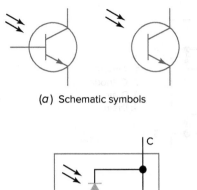

(a) Schematic symbols

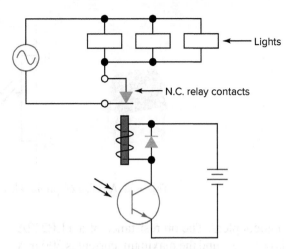

Fig. 5-26 Phototransistor-controlled lighting.

(b) Equivalent circuit

Fig. 5-25 Phototransistors.

So far we have looked at transistors that are current-controlled (BJTs) and voltage-controlled (FETs). What if a transistor could be controlled by something else? How about light? One can imagine uses for such a device. It turns out that bipolar junction transistors are *inherently light-sensitive*, and their packages are designed to eliminate this effect. *Phototransistors* are packaged differently to allow light to enter the crystal. Entry of the light energy creates hole-electron pairs in the base region and turns the transistor on. Thus, phototransistors can be controlled by light instead of by base current. In fact, some phototransistors are manufactured without a base lead, as shown by the schematic symbol in the middle of Fig. 5-25. Figure 5-25(b) shows the equivalent circuit for a phototransistor. You may assume that the collector is several volts positive with respect to the emitter. With no light entering the package, only a small current flows. It is typically on the order of 10 nanoamperes (nA) at room temperature. It is called the *dark current*. When light does enter, it penetrates the diode depletion region and generates carriers. The diode conducts and provides base current for the photo-transistor. The transistor has gain, so we can expect the collector current to be a great deal larger than the current flow in the diode in Fig. 5-25(b). A typical photo transistor might show 5 mA of collector current with a light input of 3 mW per square centimeter.

One possible application for a phototransistor is shown in Fig. 5-26. This circuit provides automatic lighting. With daylight conditions, the transistor conducts and holds the normally closed (NC) contacts of the relay open. This keeps the lights turned off. When night falls, the phototransistor dark current is too small to hold the relay in, and the contacts close and turn on the lights.

Phototransistors can also be used in *optoisolators* (also called optocouplers). Figure 5-27 shows the 4N35 optoisolator package, which houses a gallium arsenide, infrared-emitting diode and an NPN silicon phototransistor. The diode and transistor are optically coupled. Applying forward bias to the diode will cause it to produce infrared light and turn on the transistor. The 4N35 can safely withstand as much as 2,500 V across the input-to-output circuit, so its ability to isolate circuits is good. 4N35s are also available in SMT packages.

Photo MOSFET transistors are another possibility. Figure 5-28 shows a gallium arsenide, infrared-emitting diode optically coupled to a photo-MOSFET pair in a dual inline package (DIP). These photo relays have higher output current ratings than the BJT phototransistor-type

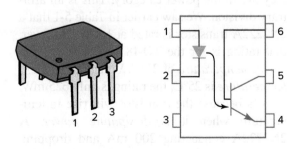

Fig. 5-27 4N35 optoisolator.

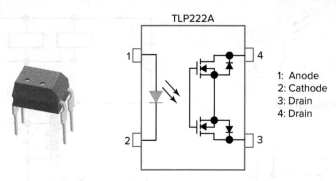

TLP222A

1: Anode
2: Cathode
3: Drain
4: Drain

Fig. 5-28 A TLP222A photo relay.

optocouplers. The on resistance of a TLP222A is only 2 Ω, and the maximum current is 500 mA and can be bidirectional: it can flow from pin 3 to pin 4 or from 4 to pin 3. It has a maximum isolation rating of 2,500 V, the same as the 4N35.

Determine whether each statement is true or false.

46. The schematic symbol for a phototransistor may or may not show a base lead.
47. Refer to Fig. 5-26. The current in the relay coil will increase as more light enters the transistor.
48. Refer to Fig. 5-27. Applying forward bias across pins 1 and 2 will allow current to flow from pin 4 to pin 5.
49. Bipolar junction transistors are bipolar devices.
50. The JFET is a unipolar device.
51. A depletion-mode transistor uses gate voltage to increase the number of carriers in the channel.

52. Bipolar transistors are current amplifiers, while unipolar transistors are voltage amplifiers.
53. It is possible to turn off an N-channel JFET with negative gate voltage.
54. In P-channel JFET circuits, the gate diode is normally forward-biased.
55. A MOSFET must be handled carefully to prevent breakdown of the gate insulator.
56. It is possible to operate a MOSFET in the enhancement mode.
57. The enhancement mode means that carriers are being pushed out of the channel by gate voltage.
58. The FET makes a better high-input-resistance amplifier than the bipolar type.

5-7 Power Transistors

Transistors can be divided into two broad categories: small-signal devices and power devices. When they must safely handle more than 1 W, they are in the power category. This is an arbitrary division. We saw earlier in Table 5-1 that a 2N2222A transistor is rated at 1.8 W. However, that rating is for the TO-18 version and for a *device temperature* of 25°C. When the *ambient temperature* is 25°C, the rating is only 625 mW. This is because the transistor will rise in temperature when it is *dissipating power*. A 2N2222A conducting 200 mA and dropping 9 V (that's 1.8 W) can burn your finger if you touch it. It will be operating well above 25°C

and will fail if it operates like that for a period of time (perhaps only seconds). When operated at 625 mW, it will still burn your finger (but not as badly) because it will reach a case temperature of around 90°C. Compare that with a 2N6288 power transistor, which will reach a case temperature of only about 55°C when dissipating 625 mW. Looking at the two cases shown in Fig. 5-29 makes it clear why the power transistor operates cooler. The 2N6288 is packaged in a TO-220 case, while a 2N2222A uses the TO-92.

Figure 5-29 shows that there is a significant difference in transistor case sizes. It also shows that the power transistor has a metal tab. This tab is often mechanically connected to a heat

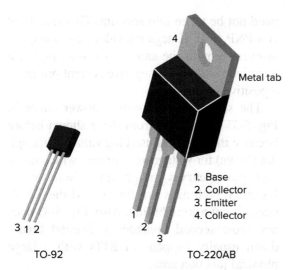

Fig. 5-29 A small-signal transistor and a power transistor.

TO-92

TO-220AB

Metal tab

1. Base
2. Collector
3. Emitter
4. Collector

sink. The heat sink is designed to conduct and transfer heat to the ambient environment, which will prevent the transistor from failing due to overtemperature. A 2N6288 is rated at 40 W maximum dissipation, but that's only if the heat sink maintains the case temperature at 25°C (77°F). That is not easily done; think of Florida in August. Power transistors are almost never operated at their maximum ratings. Figure 5-30 shows the power derating curve for the device.

EXAMPLE 5-10

What is the maximum safe power dissipation for a 2N6288 operating at a case temperature of 60°C? Consulting the graph shows about 28 W.

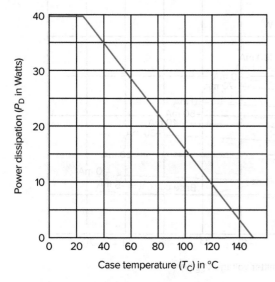

Fig. 5-30 Power derating curve for a 2N6288 transistor.

Heat is one of biggest factors in the failure of electronic devices, and that certainly includes transistors. Most have an upper temperature limit of 150°C, although some are rated at 200°C. Those are junction temperature ratings. If you burn a finger on a transistor, consider that the junction inside the case is a lot hotter! How hot will a transistor get? Power (dissipation) can be determined using

$$P_D = V_{DS} \times I_D \text{ (for FETs) or } P_C = V_{CE} \times I_C \text{ (for BJTs)}$$

Thus, as current and/or voltage increases, so does the power. This implies limits. Look at Fig. 5-31.

The most obvious limits are the maximum safe current and the maximum safe voltage. Exceeding can either damage or destroy a transistor. The other limit is set by the product of voltage and current. Thus, as V_{DS} increases, I_D will be less for the same power. The line between points 1 and 2 in Fig. 5-31 represents the maximum power limit for the device.

Bipolar junction transistors have an additional limitation, as shown in Fig. 5-32.

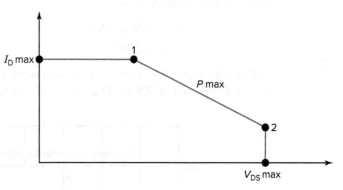

Fig. 5-31 Safe limits of FET operation.

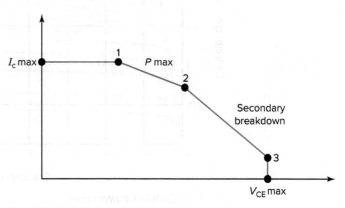

Fig. 5-32 Safe limits of BJT operation.

Notice the curve between points 2 and 3. It has a steeper slope than P_{max}, which implies an additional limit on the maximum safe power. With BJTs, higher currents and voltages can cause current flow to become confined to a small region of the crystal. A hot spot forms, and the crystal is damaged. This phenomenon does not exist with power FETs and is one of the reasons that FETs have become dominant in some applications. There is a related phenomenon called *secondary breakdown* or *second breakdown*, but it usually does not cause as many device failures. Although BJTs are often cheaper, power FETs are often more reliable.

Semiconductor manufacturers publish many kinds of graphs for their devices. Figure 5-33 shows the safe operating area for a PNP transistor. The maximum safe transistor dissipation for this particular transistor happens to be 7.5 W, and no operating point that falls to the right of the power curve would be safe. At $V_{CE} = 4$ V, the power curve crosses at a little less than 1.9 A on the I_C axis:

$$P_C = 4 \text{ V} \times 1.9 \text{ A}$$
$$= 7.6 \text{ W}$$

At $V_{CE} = 8$ V, the power curve crosses a bit above 0.9 A on the I_C axis:

$$P_C = 8 \text{ V} \times 0.9 \text{ A}$$
$$= 7.2 \text{ W}$$

All points along the power curve represent a $V_{CE} \times I_C$ product of 7.5 W. The negative values need not be taken into account. This transistor is a PNP type. If negative values are used, the answers remain the same since multiplying a negative voltage by a negative current produces a positive power value.

The shape of the constant power curve in Fig. 5-33 is different from those shown before because those graphs used log values (although not shown) for voltage and current, which made their power curves straight lines. Note that in Fig 5-33, both axes are linear, and the 7.5 W curve is not a straight line. Also, Fig. 5-33 does not show second breakdown. Second breakdown usually happens in BJTs with a large physical junction area.

EXAMPLE 5-11

Calculate the power dissipation for Fig. 5-33 when $V_{CE} = -10$ V and $I_B = -70$ mA. Begin by using the graph to find the collector current.

$$P_C = -10 \text{ V} \times -1.8 \text{ A} = +18 \text{ W}$$

Note: This exceeds the safe limit for this device.

If the collector characteristic curves are extended to include higher voltages, *collector breakdown* can be shown. Like diodes, transistors have limits to the amount of reverse bias that can be applied. BJTs have two junctions, and their breakdown ratings are more

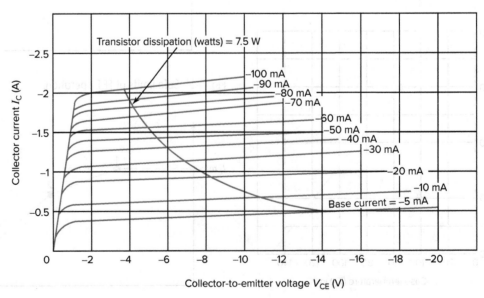

Fig. 5-33 Constant power curve.

complicated than those for diodes. Figure 5-34 shows a collector family of curves where the horizontal axis is extended to 140 V. When collector voltage becomes very high, it begins to control collector current, which is not desired. The base current is supposed to control the collector current. Transistors should not be operated near or over their maximum voltage ratings. As can be seen from Fig. 5-34, collector breakdown is not a fixed point as it is with diodes. It varies with the amount of base current. At 15 μA, the collector breakdown point is around 110 V. At 0 μA, it occurs near 130 V.

Figure 5-35 shows the safe operating area (SOA) curves for a power MOSFET. Both the continuous and peak drain currents are given as a function of drain-source voltage up to the breakdown limit of 55 V. The values are for an initial temperature of 25°C and a *single current pulse*.

EXAMPLE 5-12

Determine the maximum safe drain current for the device shown in Fig. 5-35 for a V_{DS} of 10 V for both dc and a 1-ms pulse, and compare the two. For dc, we find a maximum current of about 4 A, and for the pulse, it is about 55 A. The graph values are for a single pulse, hence the large difference compared to dc. Many circuits are operated in the digital mode (also called *switch mode*), so the pulse ratings are important.

Figure 5-36 shows the SOA curves for a 2N6284 Darlington power transistor. Pay close attention to the second breakdown region where the curves are solid and are shown at the right. Another factor for safe operation is the *bonding wire limit*. This occurs for pulse currents in excess of 40 A or a dc current of more than 20 A. A transistor is made of a small slab of semiconductor material called a *die*. The die is connected to the pins or terminals with small bonding wires. These wires can fuse (fail by open circuiting) at high currents.

EXAMPLE 5-13

Determine the maximum safe collector current for the device shown in Fig. 5-36 for a V_{CE} of 10 V for both dc and a 1-ms pulse, and compare the two. For dc, we find a maximum current of about 17 A, and for the pulse, it is about 40 A.

EXAMPLE 5-14

Is the device shown in Fig. 5-36 safe for a dc collector current of 5 A and a collector-to-emitter voltage of 40? No, that point is in the second breakdown region.

Figure 5-37 shows the internal circuit for a 2N6284 Darlington power transistor. The case is the TO-3 style, which is similar to

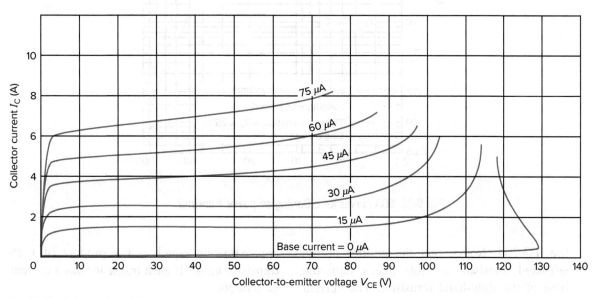

Fig. 5-34 Collector breakdown.

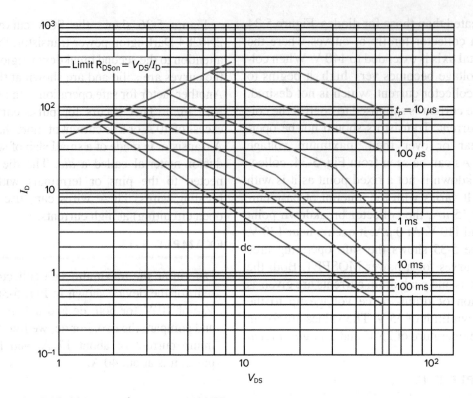

Fig. 5-35 SOA curves for a power MOSFET.

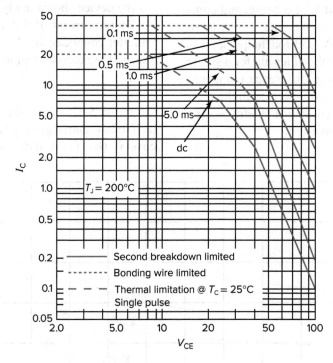

Fig. 5-36 SOA curves for a Darlington power transistor.

the TO-204. Notice that the emitter of the left-hand transistor controls (feeds into) the base of the right-hand transistor. The current gain from the B terminal to the C terminal is approximately equal to the product of both transistor gains. If each transistor has a current gain of 50:

$$h_{FE(both)} = h_{FE(1)} \times h_{FE(2)} = 50 \times 50 = 2,500$$

The high current gain of Darlingtons makes them easy to drive. The next section shows how four of them can be used to control a stepper motor.

An ohmmeter test of a device like the one shown in Fig. 5-37 can be misleading. If the ohmmeter did not turn on the series base-emitter junctions, then about 8 kΩ would be measured in both directions. There would be no indication of a good junction. Also, the collector and emitter leads would not check normally because of the added internal diode. Ohmmeter tests have limitations. Part numbers, schematics, and data sheets or manuals are needed.

In addition to extra components being placed inside transistor cases, there can be *parasitic components*. Figure 5-38 shows the internal structure of a power vertical metallic oxide semiconductor (VMOS) transistor. The name derives from the V-shaped channel. If the parasitic BJT turns on, it cannot be turned off

because the gate has no control over it. This phenomenon is known as *latchup*, which can lead to device destruction. The parasitic BJT might be turned on by a voltage drop across the P-type body region. To avoid latchup, the body and source are typically short-circuited within the device package.

Figure 5-39 shows the schematic symbol for the transistor shown in Fig. 5-38. The gate is insulated from the N channel. Note that the schematic symbol shows no electrical connection between the G terminal and the D or S terminal. Also note the diode across the S and D terminals. This is the integral body diode that can be seen by looking closely at Fig. 5-38. There is a parasitic diode between the source (which forms the P portion of the diode) and the drain (which forms the N portion).

The body diode is convenient in circuits that require a path for any possible reverse drain current (often called the *freewheeling current*).

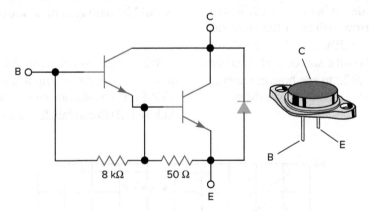

Fig. 5-37 2N6284 Darlington Power internal circuit and TO-3 case.

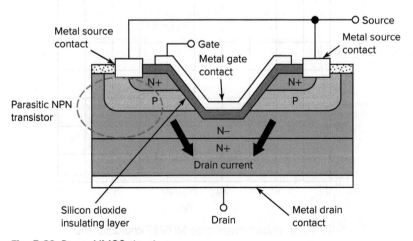

Fig. 5-38 Power VMOS structure.

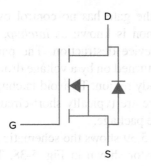

Fig. 5-39 Enhancement mode power N-channel MOSFET schematic symbol shows the integral body diode.

Inductive loads store energy, and when a MOSFET turns off, the body diode can provide a safe path for the discharge current.

EXAMPLE 5-15

Describe the differences between an N-channel enhancement mode power MOSFET schematic symbol and a P-channel type. The arrow and the diode are both reversed. With a P-channel, the arrow points to the right (as compared to Fig. 5-39), and the cathode of the diode connects to the source lead. You can refer back to Fig. 5-22 to see how the arrows differ, but the diodes are not shown there.

Figure 5-40 shows the V_{DS} versus I_D characteristic curves for an enhancement mode power MOSFET. The drain current is definitely enhanced by the gate voltage. As you can see, the current becomes high for gate voltages greater than 4 V. For much lower gate voltages, such as those near 0 V, the transistor will be off. As we learned before, enhancement mode transistors are normally off and must be turned on by applying gate voltage. Note that the specified pulse duration is short and specified at a very small duty cycle. Otherwise, the transistor would be destroyed by heat.

EXAMPLE 5-16

Calculate the power for Fig. 5-40 at $V_{DS} = 40$ V and $V_{GS} = 6$ V. This shows an operating point near 25 A:

$$P_D = 40 \text{ V} \times 25 \text{ A} = 1.125 \text{ kW}$$

That's a lot of power! With dc, a large pulse width, or a large duty cycle, the transistor would be damaged or destroyed.

With high-power transistors, the drive requirements become important. Both BJT and MOSFET transistors can handle high power. Although BJTs can fail due to second breakdown,

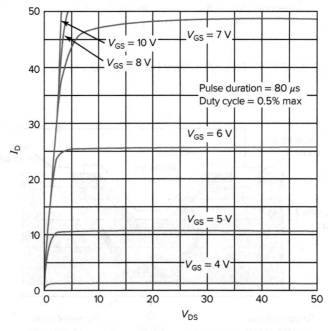

Fig. 5-40 Power enhancement-type MOSFET characteristic curves.

they are still good choices for some applications. MOSFETs are voltage-driven or voltage-controlled. This is an advantage in power devices. The driver only has to supply a changing voltage. In BJTs, the driver has to supply current and voltage. Darlingtons are easier to drive but also can fail from second breakdown, so MOSFETs might be preferred. However, MOSFETs still require some drive power due to their input capacitance. In switching circuits with fast rise times or in high-frequency designs, the capacitance means there will have to be input current. Recall that rapid voltage change causes significant current in capacitive circuits.

Yet another advantage of MOSFETs is that they don't have a problem with minority carrier storage that can limit how quickly a transistor can be turned off. In switching circuits, when the transistors are on, they are turned on hard (they are said to be *saturated*). To turn off an NPN transistor, for example, all the minority electrons in the base have to be cleared out before the device shuts off. There are lots of those in a saturated BJT. PNPs have the same issue.

Power BJTs are not as popular as they once were. They are still a viable technology, but MOSFETs are often better even though they may cost a bit more.

The insulated gate bipolar transistor (IGBT) is yet another choice. Figure 5-41 shows the structure and symbol for an IGBT. These devices can operate into the kilowatt region and are similar in structure to VMOS transistors. The major difference is that a P-type substrate is added at the bottom of the structure. This P-layer serves to lower the on-resistance of the device via a process known as *hole injection*. Holes from the added P layer move into the N region when the device is conducting. The added holes greatly improve the conductivity of the N channel (more current carriers mean better conductivity). Hole injection allows for very high current densities in IGBTs. Current density is rated in amperes per square millimeter (A/mm^2). Semiconductor manufacturers such as ON Semiconductor^{TM} sell unpackaged dies such as the NGTD21T65F2, which is about 20 mm^2 and rated at 200 A. With a high current density, a given device size can support more current flow. High current densities are important for power transistors.

Figure 5-42 shows the saturation curves and the case (package) for one IGBT. Saturation curves are used to predict how a device behaves when it is turned on very hard (saturated). The selected operating point in the green circle represents an R_{CE} value of only 8.33 mΩ:

$$R_{CE} = V_{CE}/I_C = 2 \text{ V}/240 \text{ A} = 8.33 \text{ m}\Omega$$

In switching circuits, the main idea is to keep the power *dissipated in a switch* as low possible. Here is a quick review of how power is calculated:

1. $P = V \times I$ (the definition equation)
2. $P = I^2R$ (as R_{CE} or R_{DS} approaches 0, so does the power)
3. $P = V^2/R$ (as V_{SAT} approaches 0, so does the power)

Equation 2 above comes into play with IGBTs and MOSFETs, and equation 3 comes

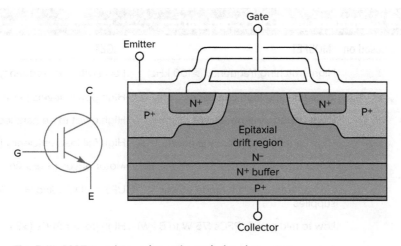

Fig. 5-41 IGBT transistor schematic symbol and structure.

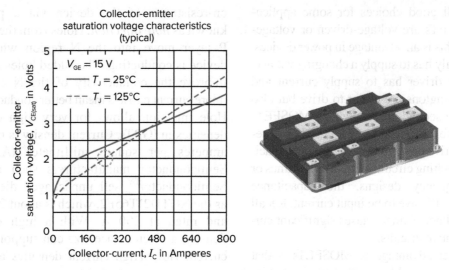

Collector-emitter saturation voltage characteristics (typical)

$V_{GE} = 15$ V

— $T_J = 25°C$

- - - $T_J = 125°C$

Collector-emitter saturation voltage, $V_{CE(sat)}$ in Volts (y-axis: 0 to 5)

Collector-current, I_C in Amperes (x-axis: 0, 160, 320, 480, 640, 800)

Fig. 5-42 IGBT collector current versus saturation voltage curves and case style.

into play with BJTs. V_{SAT}, the collector saturation voltage, should be as low as possible in switches. It is typically less than 0.4 V in power BJTs.

IGBTs can be compared to MOSFETs, as shown in Table 5-2.

EXAMPLE 5-17

Calculate the power dissipated in a BJT with $V_{SAT} = 0.35$ V and $I_C = 7$ A.

$$P_C = 0.35 \times 7 = 2.45 \text{ W}$$

EXAMPLE 5-18

Calculate the power dissipated in an IGBT with $R_{CE} = 8.33$ mΩ and $I_C = 7$ A.

$$P_C = 7^2 \times 8.33 \times 10^{-3} = 408 \text{ mW}$$

The thermal model of a transistor is shown in Fig. 5-43. Heat flow can be modeled as current flow. The model shows there are three thermal resistances: the thermal resistance of the junction to the case ($R\theta_{JC}$), the case to the heat sink ($R\theta_{CS}$), and the heat sink to the ambient ($R\theta_{SA}$). $R\theta_{JC}$ is due to the thermal resistance of the material used to mount the die to the case (solder). As the chip heats, the heat will move on to the case through the equivalent resistance ($R\theta_{JC}$). $R\theta_{CS}$ is the resistance for heat flow from the transistor case to the heat sink. Note that silicon grease and a mica washer are sometimes used to lower $R\theta_{CS}$. $R\theta_{SA}$ is the resistance for heat flow from the case to the ambient environment. Designers may have to choose a large metal heat sink or use fan cooling to reduce that resistance. Knowing the *total resistance* will allow you to calculate the *total temperature difference* just as total voltage can be calculated for a series electrical circuit when the flow is known.

Table 5-2 A Comparison of MOSFETs and IGBTs

Preferred device based on	MOSFET	IGBT
Conditions	High switching frequency (>100 kHz)	Low switching frequency (< 20 kHz)
	Wide line and load conditions	High power levels (>3 kW)
	dv/dt on the diode is limited	High dv/dt to be handled by the diode
	High light load efficiency is needed	High full load efficiency is needed
Applications	Motor drives (<250 W)	Motor drives (>250 W)
	Line operated switch mode power supplies	UPS and Welding H Bridge inverters
	Low to mid power PFCs (75 W to 3 kW)	High power PFCs (>3 kW)
	Solar inverters Battery charging	High power solar/wind inverters (>5 kW) Welding

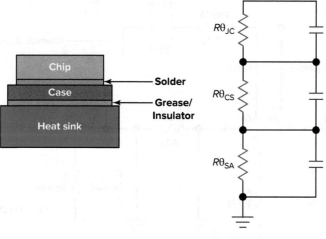

$$R\theta_{JA} = R\theta_{JC} + R\theta_{CS} + R\theta_{SA}$$

Fig. 5-43 Thermal equivalent circuit.

EXAMPLE 5-19

One particular 200-W power transistor in a TO-3 case has these thermal resistances (note that the unit θ has a dimension of °C/W). When a transistor is not on a heat sink, then there are only two thermal resistances:

$$\theta_{JC} = 0.875°C/W \text{ and } \theta_{CA} = 34°C/W$$

Assuming an ambient temperature of 25°C and a 200°C maximum junction temperature, determine the maximum safe power. In this example, θ_{JC} is not significant and can be ignored and the maximum dissipation can be found by considering only θ_{CA}. The temperature can go up by 175°C before the device will be in danger:

$$P_{MAX} = \text{Max temp rise}/\theta_{CA}$$
$$= 175°C/34°C/W = 5.15 \text{ W}$$

This should make it obvious why power transistors usually need heat sinks. The case to ambient thermal resistance is too high! Let's make another calculation based on a mica insulator plus a metal heat sink.

EXAMPLE 5-20

Find P_{MAX} for the data of the last example if a mica insulator rated at 0.4°C/W is used with a heat sink rated at 2.5°C/W.

θ_{JA} = the total thermal resistance from the junction to the ambient
$$= 0.875°C/W + 0.4°C/W + 2.5°C/W$$
$$= 3.78°C/W$$

$$P_{MAX} = \text{Max temp rise}/\theta_{CA}$$
$$= 175°C/3.78°C/W = 46.3 \text{ W}$$

This is much better than 5.15 W, so the usefulness of the heat sink has been demonstrated. The case to ambient thermal resistance has been eliminated.

The thermal model shown in Fig. 5-43 also shows capacitors; these model *thermal capacitance*. Thermal capacitance is equivalent to *thermal mass*. It takes time to charge a capacitor, and it takes time to raise or lower temperature based on thermal mass. People who cook know about thermal mass. It takes a lot longer to bring a quart of water to a boil than to do the same with a cup. Thermal mass is important in pulse circuits (they have smaller duty cycles). In a pulse circuit, the total thermal mass will help limit how hot it gets.

Figure 5-44 shows that thermal models can be used by circuit simulators. Transistor manufacturers make this information available so simulators can take advantage of it. In Fig. 5-44, the ambient temperature is modeled as a 25-V battery and the case to ambient thermal resistance as a 45-Ω resistor. Running the simulation will produce a voltage rise that is proportional to temperature.

The hardware and handling of power transistors are important. Mounting must use the correct fasteners, washers, and silicon grease (or insulators that do not require grease). Figure 5-45 shows the hardware that might be used. The washer between the screw head and

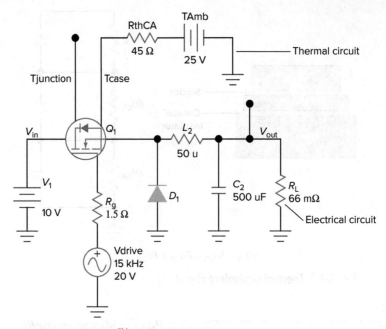

Fig. 5-44 A Multisim™ screen capture showing two circuits.

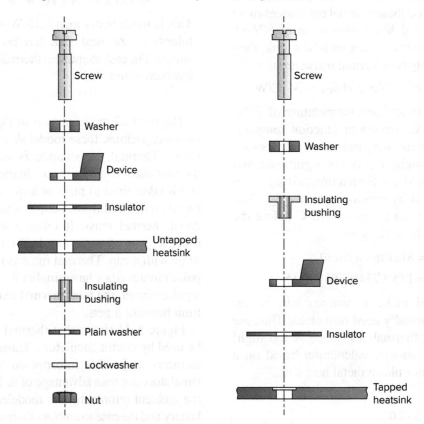

Fig. 5-45 Power transistor mounting hardware for two kinds of heat sinks.

the device package is necessary to avoid twisting (stressing) the package when the screw is tightened.

The handling of power devices is also important. Figure 5-46 shows that lead bending must be done in a way that does not stress the device.

When the transistor is heat-sink mounted, tighten the screw before soldering to avoid stressing the transistor leads.

Power transistors fail more often than small-signal transistors. The former normally run hot, which shortens the life. Technicians typically

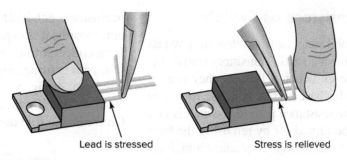

Lead is stressed   Stress is relieved

Fig. 5-46 Stressing transistor leads should be avoided.

replace more power transistors than they do small-signal transistors. Usually, an exact replacement is the best bet. If a substitution is required, the same type and polarity are mandated. The maximum ratings for voltage, current, and power should not be exceeded. Sometimes manufacturers recommend an upgraded device.

It *might* be possible to adapt a different type but *safety* and *reliability* could be compromised; therefore, this practice is not advised. When a power device is replaced, the mounting hardware and possibly the application of a special thermal compound (such as silicon grease) are important considerations.

Self-Test

Determine whether each statement is true or false.

59. All TO-220 transistors must be bolted to a heat sink.
60. Power transistors are derated for temperatures below 25°C.
61. Figure 5-30 shows second breakdown.
62. Dissipation increases with increases in voltage or current.
63. Power transistors are almost never operated at their maximum ratings.
64. $V_{DS(max)}$ can be safely exceeded as long as $P_{(max)}$ is not.
65. Secondary breakdown only occurs in VFETs.
66. Second breakdown only occurs at a current near the maximum.
67. The operating point at –12 V and –1 A in Fig. 5-33 represents negative power.
68. The data from question 67 represent an unsafe operating point.
69. Collector breakdown occurs at high I_C.
70. SOA means silicon on arsenide.
71. A transistor could be safe with a 20-A pulse yet be unsafe with a steady current of 5 A.

72. Duty cycle has no bearing on $P_{D(max)}$ or $I_{D(max)}$.
73. A Darlington transistor might have an h_{FE} of several thousand.
74. A large current can cause a transistor to go permanently open circuit.
75. A power Darlington cannot be tested with an ohmmeter.
76. A parasitic transistor will always cause latchup and circuit damage.
77. The body diode inside a power MOSFET can act as a freewheeling diode.
78. N-channel enhancement mode devices are turned off by applying about 10 V positive to the gate lead.
79. IGBTs need a large positive dc gate current to turn on.
80. A thermal circuit is usually modeled as a series electrical circuit.
81. Heat sinks should lower the resistance of a thermal circuit.
82. Heat sink temperature cannot be predicted by circuit simulation.
83. When remounting power transistors, it is OK to discard any washers that did not act as electrical insulators.

5-8 Transistors as Switches

The term "solid-state switch" refers to a switch that has no moving parts. Transistors lend themselves for use as switches because they can be turned on with a base current or a gate voltage to produce a low resistance path (the switch is on), or they can be turned off by removing the base current or the gate voltage to produce a high resistance (the switch is off). They are very widely applied because they are small, quiet, inexpensive, reliable, capable of high-speed operation, easy to control, and relatively efficient.

Figure 5-47 shows a typical application. It is a computer-controlled battery conditioner. It is used to determine the condition of rechargeable batteries. It automatically cycles a battery from charge to discharge to charge while it monitors both battery voltage and temperature. Modern rechargeable batteries are often expensive. Many require specific charging methods for maximum life.

Q_2 is on in Fig. 5-47 when the computer sends a signal to the charge/discharge control block to forward-bias its base-emitter junction. A typical circuit would apply about +5 V to the 1-kΩ base resistor. In switching applications, a transistor is either off or turned on very hard. In other words, it is expected to drop little voltage from collector to emitter when it is on. When a transistor is turned on hard, it is said to be saturated. So, in Fig. 5-47, when Q_2 is on, only the 10-Ω resistor limits current flow. We can use Ohm's law to find the discharge current:

$$I_{\text{discharge}} = \frac{12 \text{ V}}{10 \text{ }\Omega} = 1.2 \text{ A}$$

Let's estimate Q_2's base current, assuming that the NPN Darlington transistor has a current gain of 1,000:

$$I_B = \frac{I_C}{\beta} = \frac{1.2 \text{ A}}{1,000} = 1.2 \text{ mA}$$

Assuming the charge/discharge control circuit puts out a 5-V signal, we can calculate a value for Q_2's base resistor:

$$R_B = \frac{5 \text{ V}}{1.2 \text{ mA}} = 4.2 \text{ k}\Omega$$

Using this value of base resistor would result in *soft saturation*. Switching circuits use *hard saturation*. A 1-kΩ base resistor guarantees that the transistor will operate in saturation even though a transistor with less gain might

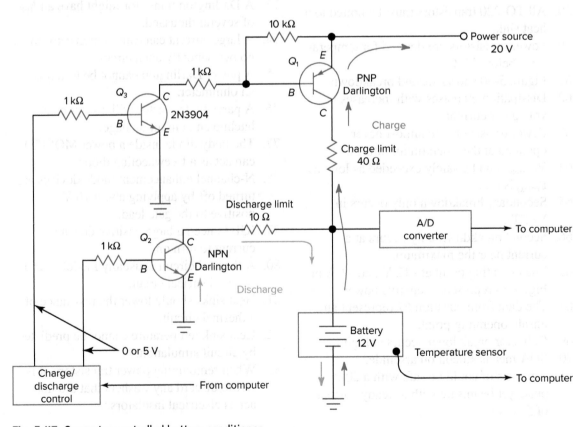

Fig. 5-47 Computer-controlled battery conditioner.

be used. A 4.7-kΩ base resistor would produce soft saturation, and in that case, a transistor with low gain would not achieve saturation. Soft saturation is not used in power-switching circuits because the load voltage and load current could be less than normal, and the switching transistor could get very hot. When a switching transistor is turned on but not saturated, it will drop several volts, and the power dissipation in the transistor will be abnormally high. Recall that

$$P_C = V_{CE} \times I_C$$

V_{CE} is very low in a saturated transistor, which keeps the collector dissipation low and the circuit efficiency high even with high values of collector current. *Troubleshooter's tip*: Failures of switching transistors can sometimes be attributed to not enough drive current or drive voltage.

When Q_2 is on in Fig. 5-47, Q_1 must be off since we can't charge and discharge at the same time. The computer will reverse the conditions when it is time to charge the battery. For charging, Q_1 must be on so that the 20-V power source is effectively connected to the battery through the 40-Ω resistor. Q_1 will be turned on hard, so once again the current can be calculated by viewing the transistor as a closed switch. This time, however, the battery

voltage must be subtracted from the source voltage:

$$I_{charge} = \frac{20\ V - 12\ V}{40\ \Omega} = 200\ mA$$

The computer, via the charge/discharge control circuit, turns on Q_3 when it is time to charge the battery. With Q_3 on, base current flows through the 1-kΩ resistor into the base of the PNP Darlington, which turns it on. As discussed before, both Q_3 and the PNP transistor are now in *hard saturation*. With Q_3 off, there is no path for base current, so the PNP transistor is off and no battery charging current flows. *Troubleshooter's tip*: In Fig. 5-47, the outputs from the charge/discharge control circuit should be either 5 V or 0 V, but never both 5 V at the same time.

Hard saturation

Figure 5-48 is a dusk-to-dawn controller without a mechanical relay or a phototransistor. This circuit uses a light-dependent resistor (LDR) as a sensor along with two transistors that act as switches. The LDR (R6) is made of a material that conducts better when exposed to light. When the sun comes up, Q_2 turns on, and its collector voltage drops to some low value, as does the gate voltage of Q_1. The load is now off (lights off at dawn). When darkness comes, the resistance of R6 goes up and Q_2 turns off. With Q_2 off, its collector voltage goes high, as does the gate of Q_1, turning it on; the load is now on (lights go on at dusk).

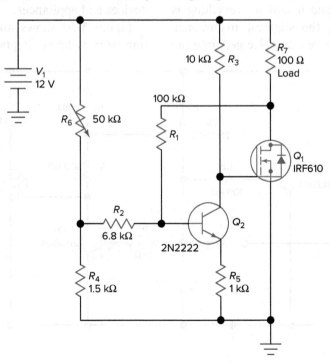

Fig. 5-48 Dusk-to-dawn control circuit.

The circuit in Fig. 5-48 is straightforward but has an added feature called *hysteresis*, which is sometimes needed in on-off control. Hysteresis can be defined as *having two trip points*. Resistor R_1 provides positive feedback from the drain of Q_1 (the output) to the base of Q_2 (the input), and that is the basis for two trip points. Driving the base voltage of Q_2 in a positive direction also drives the drain voltage of Q_1 in a positive direction. When an output is fed back to an input in a way that reinforces any change, the feedback is said to be *positive*. Circuits with positive feedback have hysteresis.

Why is hysteresis used? In a controller without it, *chatter* often results. Chatter is a sound that relays sometimes make. Relays are not used as often now, but the term *chatter* persists. What could happen in Fig. 5-48 since there is no relay to physically chatter? At dusk and dawn, the circuit will go into a mode where the load (the lights controlled by this circuit) will flicker or run dim. Chatter was bad in relays because it sometimes damaged the contact points and was annoying. Also, it can be difficult to keep the output of a dusk-to-dawn controller (the light) from entering the sensor, so *oscillation* can result. Chatter is oscillation.

How would a thermostat in a building with AC work with no hysteresis? There would be problems. As the set point is reached, the AC should shut down and it will if everything is working OK. But, the slightest disturbance (somebody walks by or line noise gets into the circuit or wind creates a warm draft) will call for cooling and the compressor will try to start up again. Not good! Compressors don't like to be immediately restarted.

Figure 5-49 is another on-off controller but this one is controlled by a push-button. Q_2 and Q_3 form a latch. When a latch is off it will stay off until some event triggers it on. The event here is initiated by pushing button S_1. This results in the discharge of C_1 into the base of Q_2. Q_2 will turn on and current will flow in R_1 and R_2. The current in R_1 causes a voltage across the base-emitter junction of Q_3 and it turns on. With Q_3 on, Q_2 stays on when S_1 is released. Q_2 and Q_3 are now **latched on**. The load is off because the gate voltage of Q_4 is at a low voltage. Q_1 is also on and C_1 discharges through R_7. With the capacitor discharged, the circuit is ready for the next press of the button (S_1) which will turn the load on.

Pushing the button a second time applies zero volts to the base circuit of Q_2, (because C_1 is currently discharged) turning it off. With Q_2 off, R_1 and R_2 pull up the gate voltage of Q_4 and it turns on which activates the load. Q_1 is now off so C1 can charge up again which makes it ready for the next cycle to turn Q_2 on again when the button is pushed again. Circuits like this are common now and have eliminated toggle, rocker and slide type switches in many devices and appliances.

Figure 5-50 shows another application for transistor switches. Stepper motors can be used

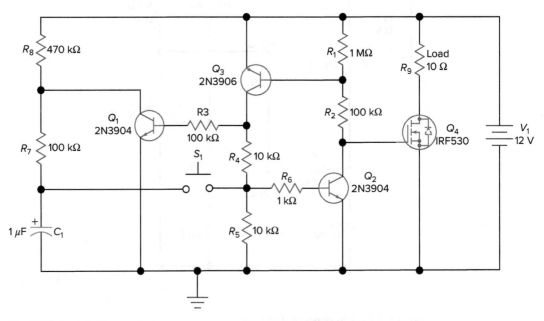

Fig. 5-49 Push button control.

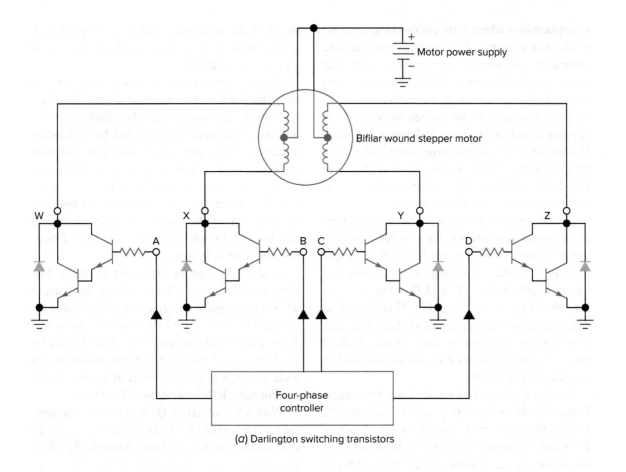

Motor power supply

Bifilar wound stepper motor

W   X   B C   Y   D   Z
A

Four-phase
controller

(a) Darlington switching transistors

Phase	Step 1	Step 2	Step 3	Step 4	Step 5
A					
B					
C					
D					

(b) Controller waveforms

(c) Stepper-motor appearance

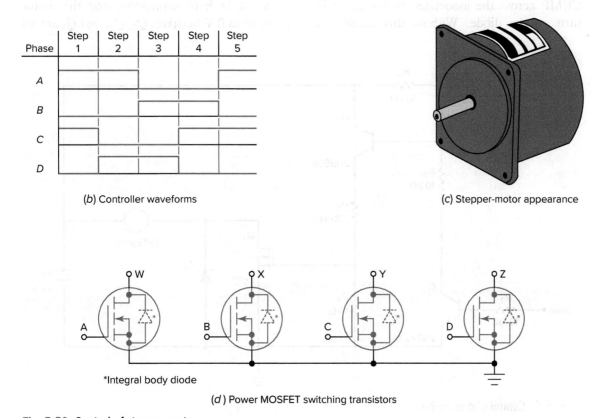

W   X   Y   Z
A   B   C   D

*Integral body diode

(d) Power MOSFET switching transistors

Fig. 5-50 Control of stepper motors.

in applications where tight control of speed and position is required. Such applications include computer disk drives, numerically controlled (automated) lathes and milling machines, and automated surface mount assembly lines. The shaft of a stepper motor moves in defined increments such as 1, 2, or 5 degrees per step. If a motor is a 1-degree type, then 180 pulses will move the shaft exactly one-half turn. As Fig. 5-50(b) shows, four groups of pulses are required with this particular motor.

A computer or a microprocessor sends precisely timed waveforms to the switching transistors to control the four motor leads: W, X, Y, and Z in Fig. 5-50(a). The control waveforms are shown as phases A, B, C, and D in Fig. 5-50(b); note that they are *rectangular*. This is typical when transistors are used as switches. They are either on or off (high or low). Figure 5-50(d) shows that power MOSFETs can also be used to control stepper motors.

Motors are inductive loads. They generate a large CEMF when they are switched off. Notice the protection diodes in Fig. 5-50(a). They can be separate diodes or built into the transistor, as was shown earlier (integral body diodes). When the transistor turns off, the CEMF across the associated motor coil will turn on the diode. Without this diode, the

drain of the switching transistor would break down from high voltage, and the transistor would be damaged.

Stepper motors are expensive and are not available for high-power applications. Variable-speed induction motors are often better choices, and they can also be controlled by solid-state switches. They are more like the ordinary motors found in home appliances. They are good choices when efficient control of motor speed is the main issue, especially in multiple horse-power applications. An electric vehicle can use solid-state switch control for efficient control of vehicle speed.

Figure 5-51 shows control of a dc motor. The circuit achieves on-off control by applying a +5-V drive signal to Q_3 to turn the motor on and a 0-V drive signal to turn the motor off. Also, pulse-width modulation (PWM) can be used to control the speed of the motor over a wide range. PWM is covered in Chaps. 8 and 14. For now, let's look at on-off control.

With +5 V of drive, Q_3 will be in hard saturation. This also allows Q_2 to turn on hard, as base current can now flow through R_3. With Q_2 saturated, R_4 and R_6 divide the 24-V supply and the gate of Q_1 goes to 16.3 V, which places it in hard saturation, and the motor runs. With 0 V of drive, Q_3, Q_2, and Q_1 are all

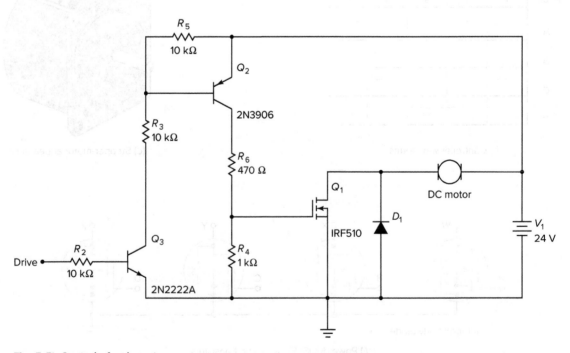

Fig. 5-51 Control of a dc motor.

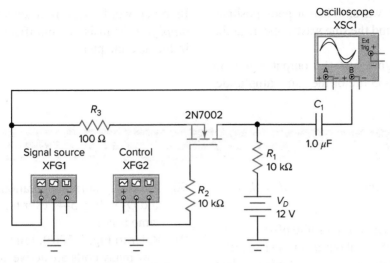

Fig. 5-52 MOSFET analog switch.

off and off the motor stops (the gate voltage is now zero). *Troubleshooter's tip*: Technicians will often measure Q_1's gate voltage with a meter or, in the case of a PWM controller, an oscilloscope.

Circuits such as the one in Fig. 5-51 are widely used, since power FETs are inexpensive, can have very low on-resistance, and operate efficiently at high frequencies (which makes them attractive in PWM applications).

So far, the transistor switches discussed have been used for turning loads on and off. There is another category called an *analog switch*, which is used to control the flow of analog signals. For example, analog switches can be used to select among different signal sources in a sound system (tuner, MP3 player, CD, DVD,

and so on). Generally, they are offered as integrated circuits, but it is possible to use enhancement-mode transistors to achieve this function. Figure 5-52 shows an example circuit, and Fig. 5-53 shows example waveforms.

In Fig. 5-52, the switching transistor is a 2N7002, which is an N-channel, enhancement-mode MOSFET. This device shows a low resistance from source to drain when a positive voltage is applied to the gate terminal. The signal source is connected to the source terminal, and the drain terminal provides the output signal. The control turns the transistor on by applying a voltage to the gate. Figure 5-53 shows the input signal in red and the output signal in blue. Note that the output signal is being switched on and off. The control signal in this

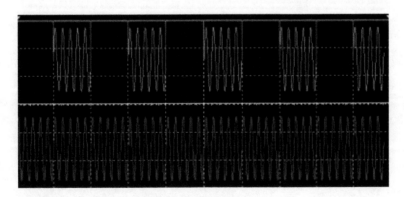

Fig. 5-53 Simulated analog switch waveforms.

Source: Multisim

case is a square wave. When it goes positive, the switch is on and the sine wave appears at the output.

The input signal and the output signal in Figs. 5-52 and 5-53 are of the same amplitude.

In other words, this is a switch and *not an amplifier*. Transistor amplifiers are discussed in the next chapter.

Self-Test

Answer the following questions with a short phrase or word.

84. The ideal switch has infinite off-resistance. What about its on-resistance?
85. An ideal switch dissipates no power when it is on. Why?
86. An ideal switch dissipates no power when it is off. Why?
87. What is happening to the battery in Fig. 5-47 when Q_3 is turned on by the computer?
88. What is happening to the battery in Fig. 5-47 when Q_2 is turned on by the computer?
89. What is happening to the battery in Fig. 5-47 when both lines coming out of the charge/discharge control block are at 0 V?

90. Assuming normal operation, are both Q_3 and Q_2 in Fig. 5-47 ever turned on at the same time?
91. Refer to Fig. 5-50(*b*). Ignoring transitions, how many coils are active at any given time?
92. Stepper motors have poor efficiency since they use power when they are standing still. Which section of Fig. 5-50 proves this?

Solve the following problems.

93. Determine a new value for the charge limit resistor in Fig. 5-47 to set the charging current to 1 A.
94. Determine a new value for the discharge limit resistor in Fig. 5-47 to set the discharge current to 0.5 A.

Chapter 5 Summary and Review

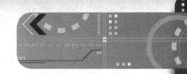

Summary

1. Gain is the basic function of any amplifier.
2. Gain can be calculated using voltage, current, or power. In all cases, the units cancel and gain is simply a number.
3. Power gain is the product of voltage gain and current gain.
4. The term *voltage amplifier* is often used to describe a small-signal amplifier.
5. The term *power amplifier* is often used to describe a large-signal amplifier.
6. Bipolar junction transistors are manufactured in two polarities: NPN and PNP. The NPN types are more widely applied.
7. In a BJT, the emitter emits the carriers, the base is the control region, and the collector collects the carriers.
8. The schematic symbol of an NPN transistor shows the emitter lead arrow *N*ot *P*ointing i*N*.
9. Normal operation of a BJT requires that the collector-base junction be reverse-biased and the base-emitter junction be forward-biased.
10. Most of the current carriers coming from the emitter cannot find carriers in the base region with which to combine. This tends to make the base current much less than the other currents.
11. The base is very narrow, and the collector bias attracts the carriers coming from the emitter. This tends to make the collector current almost as high as the emitter current.
12. Beta (β), or h_{FE}, is the current gain from the base terminal to the collector terminal. The value of β varies considerably, even among devices with the same part number.
13. Base current controls collector current and emitter current.
14. Emitters of PNP transistors produce holes. Emitters of NPN transistors produce electrons.
15. A collector characteristic curve is produced by plotting a graph of I_C versus V_{CE} with I_B at some fixed value.
16. Collector voltage has only a small effect on collector current over most of the operating range.
17. A power curve can be plotted on the graph of the collector family to show the safe area of operation.
18. Collector dissipation is the product of collector-emitter voltage and collector current.
19. Germanium transistors require a base-emitter bias of about 0.2 V to turn on. Silicon units need about 0.6 V.
20. Silicon transistors are much more widely used than germanium transistors.
21. Substitution guides provide the technician with needed information about solid-state devices.
22. The physical characteristics of a part can be just as important as the electrical characteristics.
23. Transistors can be tested with curve tracers, dynamic testers, ohmmeters, and with various in-circuit checks.
24. Most transistors fail suddenly and completely. One or both PN junctions may short or open.
25. An ohmmeter can check both junctions, identify polarity, identify leads, check gain, indicate leakage, and may even identify the transistor material.
26. Leakage current I_{CEO} is β times larger than I_{CBO}.
27. Phototransistors are biased on with light.
28. Phototransistors can be packaged with LEDs to form devices called optoisolators or optocouplers.
29. Bipolar transistors (NPN and PNP) use both holes and electrons for conduction.
30. Unipolar transistors (N-channel and P-channel types) use either electrons or holes for conduction.
31. A BJT is a normally off device. It is turned on with base current.
32. A JFET is a normally on device. It is turned off with gate voltage. This is called the depletion mode.
33. A MOSFET uses an insulated gate structure. Manufacturers make both depletion-type and enhancement-type MOSFETs.

34. An enhancement-mode MOSFET is a normally off device. It is turned on by gate voltage.
35. Field-effect transistors have a very high input resistance.
36. The abbreviations VFET and VMOS are used to refer to power field-effect transistors that have a vertical flow of current from source to drain.
37. Power FETs do not have some of the limitations of power bipolar transistors. The FETs are voltage-controlled, they are faster (no minority-carrier storage), they do not exhibit thermal runaway, and they are not prone to secondary breakdown.
38. Power FETs operate in the enhancement mode.
39. Transistors that are controlled by light are useful for applications such as dusk-to-dawn circuits.
40. Combining LEDs and transistors in the same package provides functions such as optoisolators and photo relays.

41. When a bipolar junction transistor is operated as a switch, there is going to be either no base current or a lot of base current.
42. A switching transistor is either turned on hard (saturated) or is off.
43. Ideally, switching is very efficient since an open switch shows no current for zero power dissipation, and a closed switch shows no voltage drop, which is another case of zero power dissipation.
44. In switching circuits, the control waveforms are often rectangular.
45. When inductive loads are being switched, some sort of protection device or circuit is needed because of the CEMF generated by the inductance.
46. The thermal resistance unit called θ has a dimension of °C/W.
47. Circuits with hysteresis have two trip points.

Related Formulas

Power gain: $P_{gain} = V_{gain} \times I_{gain}$ and

$$P_{gain} = \frac{R_{CB}}{R_{BE}}$$

Voltage gain: $A_V = \dfrac{V_{out}}{V_{in}}$

BJT current: $I_E = I_B + I_C$

BJT current gain: $\beta = \dfrac{I_C}{I_B}$ or $h_{FE} = \dfrac{I_C}{I_B}$

BJT ac gain: $\beta_{ac} = h_{fe} = \dfrac{\Delta I_C}{\Delta I_B}\Big|_{V_{CE}}$

Transistor dissipation: $P_C = V_{CE} \times I_C$ and $P_D = V_{DS} \times I_D$

Leakage current: $I_{CEO} = \beta \times I_{CBO}$

Darlington gain: $h_{FE(BOTH)} = h_{FE(1)} \times h_{FE(2)}$

Switch on dissipation: $P_{switch} = I^2 R_{switch(on)}$ or $V_{SAT} \times I_C$

$P_{MAX} = \text{max temp rise}/\theta$

Chapter Review Questions

Supply the missing word in each statement.

5-1. Small-signal amplifiers are usually called _____ amplifiers. (5-1)
5-2. Large-signal amplifiers are usually called _____ amplifiers. (5-1)
5-3. Bipolar junction transistors are made in two basic polarities: NPN and _____. (5-2)
5-4. Current flow in bipolar transistors involves two types of carriers: electrons and _____. (5-2)
5-5. The base-emitter junction is normally _____ biased. (5-2)

5-6. The collector-base junction is normally _____ biased. (5-2)
5-7. The smallest current in a BJT is normally the _____ current. (5-2)
5-8. In a normally operating BJT, the collector current is controlled mainly by the _____ current. (5-2)
5-9. Turning on an NPN BJT requires that the base be made _____ with respect to the emitter terminal. (5-2)

5-10. For proper operation, the base terminal of a PNP BJT should be _____ with respect to the emitter terminal. (5-2)

5-11. The emitter of a PNP transistor produces _____ current. (5-2)

5-12. The emitter of an NPN transistor produces _____ current. (5-2)

5-13. The symbol h_{FE} represents the _____ current gain of a transistor. (5-3)

5-14. The symbol h_{fe} represents the _____ current gain of a transistor. (5-3)

5-15. The equivalent symbol for h_{FE} is _____. (5-3)

5-16. The equivalent symbol for h_{fe} is _____. (5-3)

5-17. In testing bipolar transistors with an ohmmeter, a good diode indication should be noted at the collector-base and _____ junctions. (5-5)

5-18. In an ohmmeter test of a good BJT, the collector and emitter leads should check _____ regardless of meter polarity. (5-5)

5-19. A phototransistor's current is usually controlled by _____. (5-6)

5-20. Optocoupler is another name for _____. (5-6)

5-21. Refer to Fig. 5-21. As V_{GS} becomes more positive, drain current _____. (5-6)

5-22. An N-channel JFET uses _____ to support the flow of current. (5-6)

5-23. A P-channel JFET uses _____ to support the flow of current. (5-6)

5-24. Gate voltage in a JFET can remove carriers from the channel. This is known as the _____ mode. (5-6)

5-25. Gate voltage in a MOSFET can produce carriers in the channel. This is known as the _____ mode. (5-6)

5-26. A JFET is not normally operated in the enhancement mode because the gate diode may become _____ biased. (5-6)

5-27. The current flow in power field-effect transistors is _____ rather than lateral. (5-7)

5-28. Power bipolar transistors may be damaged by hot spots in the crystal caused by current crowding. This phenomenon is known as _____. (5-7)

5-29. Vertical metal oxide semiconductor transistors operate in the _____ mode. (5-7)

5-30. When the gate of an IGBT goes positive, RCE is expected to _____.

Chapter Review Problems

5-1. An amplifier provides a voltage gain of 20 and a current gain of 35. Find its power gain. (5-1)

5-2. An amplifier must give an output signal of 5 V peak-to-peak. If its voltage gain is 25, determine its input signal. (5-1)

5-3. If an amplifier develops an output signal of 8 V with an input signal of 150 mV, what is its voltage gain? (5-1)

5-4. A BJT has a base current of 25 μA and its $\beta = 200$. Determine its collector current. (5-2)

5-5. A BJT has a collector current of 4 mA and a base current of 20 μA. Find its β. (5-2)

5-6. A BJT has a $\beta = 250$ and a collector current of 3 mA. What is its base current? (5-2)

5-7. A bipolar transistor has a base current of 200 μA and an emitter current of 20 mA. What is the collector current? (5-2)

5-8. Find β for problem 5-7. (5-2)

5-9. Refer to Fig. 5-11. $V_{CE} = 10$ V and $I_B = 20$ μA. Find β. (5-3)

5-10. Refer to Fig. 5-12. $V_{CE} = -16$ V and $I_C = -7$ mA. Find I_B. (5-3)

5-11. Refer to Fig. 5-12. $I_B = -100$ μA and $V_{CE} = -10$ V. Find P_C. (5-3)

5-12. Refer to Fig. 5-13. The transistor is silicon and $V_{BE} = 0.65$ V. What is I_C? (5-3)

Critical Thinking Questions

5-1. If a transistor has a current gain of 100, how much current gain would be available by using three transistors? How would they be arranged?

5-2. You are looking at a collector family of characteristic curves on a curve tracer and you notice that the curves appear to be spreading (moving apart). What is happening?

5-3. Transistor heating is a big problem in some circuits. Today, it is becoming popular to operate transistors in a digital mode to alleviate the heat problem. Why?

5-4. Transistors are very popular, but an older technology based on *vacuum tubes* is still in use in very high-power applications such as large radio and television transmitters. Why? (*Hint:* Vacuum tubes can operate at thousands of volts.)

5-5. FETs are unipolar devices, and BJTs are bipolar devices. Will the future bring a new category of electronic devices that are tripolar?

5-6. When examining a piece of electronic equipment, why can't you assume that the transistors will all have three leads and the diodes will all have two leads?

Answers to Self-Tests

1. T	26. 11 mA	51. F	76. F
2. F	27. 133	52. T	77. T
3. F	28. 96 mW	53. T	78. F
4. T	29. 100	54. F	79. F
5. F	30. 0.2	55. T	80. T
6. T	31. 0.6	56. T	81. T
7. T	32. germanium	57. F	82. F
8. F	33. T	58. T	83. F
9. F	34. F	59. F	84. ideally, zero
10. F	35. F	60. F	85. ideally, it drops zero volts
11. T	36. F	61. F	
12. T	37. T	62. T	86. ideally, the current flow is zero
13. T	38. T	63. T	
14. T	39. T	64. F	87. it is charging
15. F	40. F	65. F	88. it is discharging
16. F	41. F	66. F	89. nothing (it is not charging or discharging)
17. F	42. T	67. F	
18. T	43. T	68. T	
19. T	44. F	69. F	90. no
20. 42.5 mA	45. F	70. F	91. two
21. 6.67 μA	46. T	71. T	92. (*b*)
22. 250	47. T	72. F	93. 8 Ω
23. 10 mA	48. T	73. T	94. 24 Ω
24. 100	49. T	74. T	
25. 20 μA	50. T	75. F	

CHAPTER 6

Introduction to Small-Signal Amplifiers

Learning Outcomes

This chapter will help you to:

6-1 *Calculate* decibel gain and loss. [6-1]

6-2 *Draw* a load line for a basic common-emitter amplifier. [6-2]

6-3 *Define* clipping in a linear amplifier. [6-2]

6-4 *Find* the operating point for a basic common-emitter amplifier. [6-2, 6-3]

6-5 *Determine* common-emitter amplifier voltage gain. [6-3]

6-6 *Identify* common-base and common-collector amplifiers. [6-4]

6-7 *Explain* the importance of impedance matching. [6-4]

6-8 *Discuss* SPICE and explain the importance of models. [6-5]

This chapter deals with gain. Gain is the ability of an electronic circuit to increase the level of a signal. As you will see, gain can be expressed as a ratio or as a logarithm of a ratio. Transistors provide gain. This chapter will show you how they can be used with other components to make amplifier circuits. You will learn how to evaluate a few amplifiers using some simple calculations. This chapter is limited to small-signal amplifiers. As mentioned before, these are often called voltage amplifiers.

6-1 Measuring Gain

Gain is the basic function of all amplifiers. It is a comparison of the signal fed into the amplifier with the signal coming out of the amplifier. Because of gain, we can expect the output signal to be greater than the input signal. Figure 6-1 shows how measurements are used to calculate the voltage gain of an amplifier. For example, if the input signal is 1 V and the output signal is 10 V, the gain is

Gain

$$\text{Gain} = \frac{\text{signal out}}{\text{signal in}} = \frac{10 \text{ V}}{1 \text{ V}} = 10$$

Note that the units of voltage cancel, and gain is a *pure number*. It is *not* correct to say that the gain of the amplifier is 10 V.

EXAMPLE 6-1

Calculate the gain of an amplifier if the output signal is 4 V and the input signal is 50 mV.

$$\text{Gain} = \frac{V_{\text{out}}}{V_{\text{in}}} = \frac{4 \text{ V}}{50 \times 10^{-3} \text{ V}} = 80$$

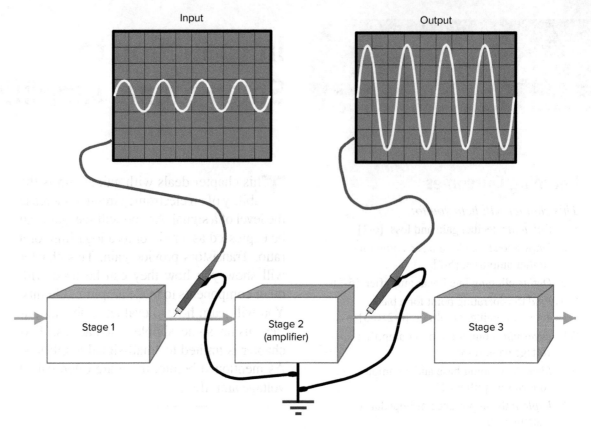

Input

Output

Stage 1

Stage 2
(amplifier)

Stage 3

Fig. 6-1 Measuring gain.

EXAMPLE 6-2

If an amplifier has a gain of 50, find its output signal amplitude when its input is 20 mV. Rearranging,

$$V_{out} = \text{gain} \times V_{in} = 50 \times 20 \text{ mV} = 1 \text{ V}$$

A circuit that has gain *amplifies*. The letter *A* is the general symbol for gain or amplification in electronics. A subscript can be added to specify the type of gain:

$$A_V = \frac{V_{out}}{V_{in}} = \text{voltage gain}$$

$$A_I = \frac{I_{out}}{I_{in}} = \text{current gain}$$

$$A_P = \frac{P_{out}}{P_{in}} = \text{power gain}$$

Voltage gain

Voltage gain A_V is used to describe the operation of small-signal amplifiers. Power gain A_P is used to describe the operation of large-signal amplifiers. If the amplifier in Fig. 6-1 were a power or large-signal amplifier, the gain would be based on watts rather than on volts. For

EXAMPLE 6-3

Find P_{in} for an amplifier with a power gain of 20 and an output power of 1 W. Rearranging,

$$P_{in} = \frac{P_{out}}{A_P} = \frac{1 \text{ W}}{20} = 50 \text{ mW}$$

example, if the input signal is 0.5 W and the output signal is 8 W, the power gain is

$$A_P = \frac{P_{out}}{P_{in}} = \frac{8 \text{ W}}{0.5 \text{ W}} = 16$$

Early work in electronics was in the communications area. The useful output of most circuits was audio for headphones or speakers. Thus, engineers and technicians needed a way to align circuit performance with human hearing. The human ear is *not linear* for audio power. It does not recognize intensity or loudness in the way a linear device does. For example, if you are listening to a speaker with 0.1-W input and the power suddenly increases to 1 W, you will notice that the sound has become louder. Then assume that the power suddenly increases again to 10 W. You will notice a second increase in loudness. The interesting thing is that you will probably rate the second increase in loudness as about *equal* to the first increase in loudness.

A linear detector would rate the second increase to be 10 times greater than the first. Let us see why:

- First increase from 0.1 to 1 W, which is a *0.9-W* linear change.
- Second increase from 1 to 10 W, which is a *9-W* linear change.

The second change is 10 times greater than the first:

$$\frac{9\text{ W}}{0.9\text{ W}} = 10$$

The loudness response of human hearing is *logarithmic*. Logarithms are therefore often used to describe the performance of audio systems. We are often more interested in the *logarithmic gain* of an amplifier than in its *linear gain*. Logarithmic gain is very convenient and widely applied. What started out as a convenience in audio work has now become the universal standard for amplifier performance. It is used in radio-frequency systems, video systems, and just about anywhere there is electronic gain.

Common logarithms are *powers (exponents) of 10*. For example,

$$10^{-3} = 0.001$$
$$10^{-2} = 0.01$$
$$10^{-1} = 0.1$$
$$10^{0} = 1$$
$$10^{1} = 10$$
$$10^{2} = 100$$
$$10^{3} = 1,000$$

The logarithm of 10 is 1. The logarithm of 100 is 2. The logarithm of 1,000 is 3. The logarithm of 0.01 is −2. Any positive number can be converted to a common logarithm. Logarithms can be found with a scientific calculator. Enter the number and then press the "log" key to obtain the common logarithm for the number.

EXAMPLE 6-4

Use a scientific calculator to find the common log of 2,138. Enter 2138 and then press the log key:

$$\text{Display} \approx 3.33$$

EXAMPLE 6-5

Use a scientific calculator to find the common log of 0.0316. Enter .0316 and then press the log key:

$$\text{Display} \approx -1.5$$

Or, enter 31.6, then press the EXP key, then enter 3, then press the ± key, and finally, press the log key.

Power gain is very often measured in *decibels (dB)*. The decibel is a logarithmic unit. Decibels can be found with this formula:

$$\text{dB power gain} = 10 \times \log_{10}\frac{P_{\text{out}}}{P_{\text{in}}}$$

Decibel (dB)

Gain in decibels is based on *common logarithms*. Common logarithms are based on 10. This is shown in the preceding equation as $\log_{10}$ (the base is 10). Hereafter the base 10 will be dropped, and log will be understood to mean $\log_{10}$.

Common logarithm

Logarithmic gain

Logarithms for numbers less than 1 are *negative*. This means that any part of an electronic system that produces *less* output than input will have a negative gain (−dB) when the above formula is used.

Let's apply the formula to the example given previously. The first loudness increase:

$$\text{dB } power\ gain = 10 \times \log\frac{1\text{ W}}{0.1\text{ W}}$$
$$= 10 \times \log 10$$

The logarithm of 10 (log 10) is 1, so

$$\text{dB power gain} = 10 \times 1 = 10$$

Thus, the first increase in level or loudness was equal to 10 dB. The second loudness increase:

$$\text{dB power gain} = 10 \times \log\frac{10\text{ W}}{1\text{ W}}$$
$$= 10 \times \log 10$$
$$= 10 \times 1 = 10$$

The second increase was also equal to 10 dB. Since the decibel is a logarithmic unit and because your hearing is logarithmic, the two 10-dB increases sound about the same. The average person can detect a change as small as 1 dB. Any change smaller than 1 dB would be very difficult for most people to hear.

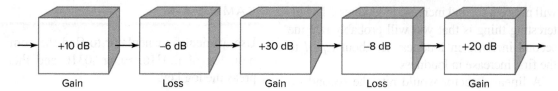

Fig. 6-2 Gain and loss in decibels.

Why has the decibel, which was developed for audio, come to be used in all areas of electronics where gain is important? The answer is that it is so convenient to work with. Figure 6-2 shows why. Five stages, or parts, of an electronic system are shown. Three of the stages show gain (+dB), and two show loss (−dB). To **dB voltage gain** evaluate the overall performance of the system shown in Fig. 6-2, all that is required is to add the numbers:

$$\text{Overall gain} = +10 - 6 + 30 - 8 + 20$$
$$= +46 \text{ dB}$$

When the gain or loss of individual parts of a system is given in decibels, it is very easy to evaluate the overall performance. This is why the decibel has come to be so widely used in electronics.

Figure 6-3 shows the same system with the individual stage performance stated as *ratios*. Now it is not so easy to evaluate the overall performance. The overall performance will be given by

$$\text{Overall gain} = \frac{10}{4} \times \frac{1,000}{6.31} \times 100$$
$$= 39,619.95$$

Notice that it is necessary to *multiply* for the gain stages and *divide* for the loss stages. When stage performance is given in decibels, gains are added and losses subtracted. The overall system performance is easier to determine using the dB system.

Figures 6-2 and 6-3 describe the same system. One has an overall gain of +46 dB, and the other has an overall gain of 39,619.65. The dB gain and the ratio gain should be the same:

$$\text{dB} = 10 \times \log 39,619.65$$
$$= 10 \times 4.60$$
$$= 46$$

The decibel is based on the ratio of the power output to the power input. It can also be used to describe the ratio of two voltages. The equation for finding *dB voltage gain* is slightly different from the one used for finding dB power gain:

$$\text{dB voltage gain} = 20 \times \log \frac{V_{\text{out}}}{V_{\text{in}}}$$

Notice that the logarithm is multiplied by 20 in this equation. This is because power varies as the *square* of the voltage:

$$\text{Power} = \frac{V^2}{R}$$

Power gain can therefore be written as

$$A_P = \frac{(V_{\text{out}})^2 / R_{\text{out}}}{(V_{\text{in}})^2 / R_{\text{in}}}$$

If R_{out} and R_{in} happen to be equal, they will cancel. Now power gain reduces to

$$A_P = \frac{(V_{\text{out}})^2}{(V_{\text{in}})^2} = \left(\frac{V_{\text{out}}}{V_{\text{in}}}\right)^2$$

Since the log of $V^2 = 2 \times \log$ of V, the logarithm can be multiplied by 2 to eliminate the need for squaring the voltage ratio:

$$\text{dB voltage gain} = 10 \times 2 \times \log \frac{V_{\text{out}}}{V_{\text{in}}}$$
$$= 20 \times \log \frac{V_{\text{out}}}{V_{\text{in}}}$$

The requirement that R_{in} and R_{out} be equal is often set aside for voltage amplifiers. It is

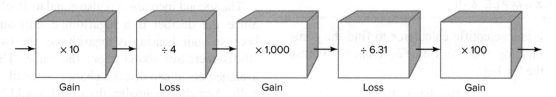

Fig. 6-3 Gain and loss in ratios.

important to remember that if the resistances are not equal, then the dB voltage gain will *not* be equal to the dB power gain. For example, suppose that an amplifier has a voltage gain of 50, an input resistance of 1 kΩ, and an output resistance of 150 Ω. Its dB voltage gain will be

$$A_V = 20 \times \log 50 = 34 \text{ dB}$$

The power gain of this same amplifier can be found by assigning some input voltage (the value does not matter). If we assign the input to be 1 V, then the output will be 50 V because of the stated voltage ratio. Now, power in and power out can be calculated:

$$P_{in} = \frac{V^2}{R} = \frac{1^2}{1,000} = 1 \text{ mW}$$

$$P_{out} = \frac{50^2}{150} = 16.7 \text{ W}$$

The power gain of the amplifier in decibels is

$$A_P = 10 \times \log \frac{16.7 \text{ W}}{1 \text{ mW}} = 42.2 \text{ dB}$$

Note that this does *not* equal the dB voltage gain.

As another example, suppose an amplifier has a voltage gain of 1, an input resistance of 50,000 Ω, and an output resistance of 100 Ω. The dB voltage gain of this amplifier is

$$A_V = 20 \times \log 1 = 0 \text{ dB}$$

Assuming a 1-V input signal,

$$P_{in} = \frac{1^2}{50,000} = 20 \text{ } \mu W$$

$$P_{out} = \frac{1^2}{100} = 10 \text{ mW}$$

The power gain of this amplifier in dB is

$$A_P = 10 \times \log \frac{10 \times 10^{-3}}{20 \times 10^{-6}} = 27 \text{ dB}$$

Table 6-1	Common Values for Estimating dB Gain and Loss	
Change	Power	Voltage
Multiplied by 2	+3 dB	+6 dB
Divided by 2	−3 dB	−6 dB
Multiplied by 10	+10 dB	+20 dB
Divided by 10	−10 dB	−20 dB

Once again, we see that the dB power gain is *not* equal to the dB voltage gain because R_{in} does not equal R_{out}. We see another interesting fact: *An amplifier may have no voltage gain yet offer a significant power gain.*

Technicians should have a feeling for gain and loss expressed in decibels. Often, a quick estimate is all that is required. Table 6-1 contains the common values used by technicians for making estimates.

EXAMPLE 6-6

A 100-W amplifier has a power gain of 10 dB. What input signal power is required to drive the amplifier to full output? Table 6-1 shows the multiplication (ratio) to be 10 for a power gain of 10 dB. The required input power is therefore one-tenth the desired output power:

$$P_{in} = \frac{100 \text{ W}}{10} = 10 \text{ W}$$

EXAMPLE 6-7

A transmitter feeds an antenna through a long run of coaxial cable. The transmitter develops 1 kW of output power and only 500 W reaches the antenna. What is the performance of the coaxial cable in dB? The 500 W is one-half of the input power. Therefore, the power has been divided by 2. Table 6-1 shows that this equals −3 dB. The performance of this cable can be verbalized in different ways:

1. The cable gain is −3 dB.
2. The cable loss is 3 dB.
3. The cable loss is −3 dB.

The first statement is technically correct. A negative dB gain means that there is actually a loss. The second statement is also technically correct. The word *loss* means that the value of 3 dB is to be preceded by a minus sign when used in system calculations. The third statement is *not* technically correct. Since the word *loss* means to precede the dB value with a minus sign, the result would be −(−3 dB) = +3 dB, which is a gain. A coaxial cable cannot produce a power gain. Double negatives should be avoided when describing dB losses.

EXAMPLE 6-8

The response of a low-pass filter is specified to be −6 dB at 5 kHz. A technician measures the filter output and finds 1 V at 1 kHz and notes that it drops to 0.5 V at 5 kHz. Is the filter working properly? Table 6-1 shows that a voltage division of 2 is equal to −6 dB. The filter is working properly.

EXAMPLE 6-9

An amplifier develops a 2-W output signal when its input signal is 100 mW. What is the power gain of this amplifier in dB? Find the ratio first:

$$\frac{2\ W}{0.1\ W} = 20$$

The value 20 is not in Table 6-1. However, it may be possible to *factor* a gain value into values that are in the table. A power gain of 20 can be broken down into a power gain of 10 (+10 dB) times a power gain of 2 (+3 dB). Add the dB gains:

$$\text{Gain} = 10\ \text{dB} + 3\ \text{dB} = 13\ \text{dB}$$

EXAMPLE 6-10

An amplifier has a voltage gain of 60 dB. If its input signal is 10 μV, what output signal can be expected? The table shows that a voltage gain of 20 dB produces a multiplication of 10; 60 dB = 3 × 20 dB. Thus a gain of 60 dB will multiply the signal by 10 three times:

$$V_{out} = V_{in} \times 10 \times 10 \times 10$$
$$= 10\ \mu V \times 1{,}000 = 10\ \text{mV}$$

Calculators with logarithms are inexpensive. Technicians are expected to be able to use calculators to find dB gain and loss. The following example is easy to work using a calculator.

EXAMPLE 6-11

If the input signal to a voltage amplifier is 350 mV and the output signal is 15 V, what is the performance of this amplifier in decibels?

$$\text{dB} = 20 \times \log \frac{V_{out}}{V_{in}}$$
$$= 20 \times \log \frac{15\ V}{0.35\ V}$$
$$= 20 \times \log 42.9$$
$$= 20 \times 1.63$$
$$= 32.6$$

The amplifier shows a voltage gain of 32.6 dB. The calculator manipulation is straightforward. First, divide the input signal into the output signal. Next, press the log key. Finally, multiply by 20.

A little algebraic manipulation will be required to solve some problems. The following example demonstrates this.

EXAMPLE 6-12

You are using an oscilloscope to measure a high-frequency waveform. The manufacturer of your oscilloscope specifies that its response is −3 dB at the frequency of measurement. If the screen shows a peak-to-peak value of 7 V, what is the actual value of the signal? Begin by plugging the known information into the dB equation for voltage:

$$-3 = 20 \times \log \frac{7\ V_{p-p}}{V_{in}}$$

Divide both sides of the equation by 20:

$$-0.15 = \log \frac{7\ V_{p-p}}{V_{in}}$$

Take the *inverse* log of both sides of the equation. This *removes* the log term from the right-hand side of the equation. For the left-hand side, you must find the inverse log of −0.15 using your calculator. On some calculators, you must use an INV key in conjunction with the log key. Press INV and then press log. On other calculators, you will find a key marked 10^X. Simply press this key. Perform the operation with -0.15 showing in the calculator display. The calculator should respond with 0.708. We are now at this point:

$$0.708 = \frac{7\,V_{p-p}}{V_{in}}$$

Rearrange and solve:

$$V_{in} = \frac{7\,V_{p-p}}{0.708} = 9.89\ \text{V peak-to-peak}$$

The dB system is sometimes misused. Absolute values are often given in decibel form. For example, you may have heard that the sound level of a musical group is 90 dB. This provides no information at all unless there is an agreed-upon reference level. One reference level used in sound is a pressure of 0.0002 dynes per square centimeter (dyn/cm^2) or 2×10^{-5} newtons per square meter (N/m^2). This reference pressure is equated to 0 dB, the *threshold of human hearing*. Now, if a second pressure is compared with the reference pressure, the dB level of the second pressure can be found. For example, a jet engine produces a sound pressure of 2,000 dyn/cm^2:

$$\text{Sound level} = 20 \times \log \frac{2,000}{0.0002} = 140\ \text{dB}$$

(The log is multiplied by 20 since sound power varies as the square of sound pressure.) In the average home there is a sound pressure of 0.063 dyn/cm^2, which can be compared with the reference level:

$$\text{Sound level} = 20 \times \log \frac{0.063}{0.0002} = 50\ \text{dB}$$

It is interesting to note that the decibel scale that places the threshold of human hearing at 0 dB places the threshold of *feeling* at 120 dB. You may have noticed that a very loud sound can be felt in the ear in addition to hearing it. An even louder sound will produce pain. The total dynamic range of hearing is 140 dB. Any sound louder than 140 dB (a jet engine) will not sound any louder to a person (although it would cause more pain).

Loudness is often measured using the *dBA scale*. This scale also places the threshold of hearing at 0 dB. The *A* refers to the *weighting*

used when making measurements. A filter tailors the frequency response to match how people hear. Tests have confirmed that *A weighting* closely matches what the instruments report to what people hear. The dBA scale is often used to determine if workers need hearing protection. For example, 90 dBA is the maximum safe work week (8 hours per day) exposure level (see Table 6-2). Some common levels are

Whisper	30 dBA
Conversation	60 dBA
Busy city street	80 dBA
Nearby auto horn	100 dBA
Nearby thunder	120 dBA

Table 6-2   OSHA Standard for Sound Exposure

Maximum Exposure per Day in Hours	Sound Level in dBA
8	90
6	92
4	95
3	97
2	100
1.5	102
1	105
0.5	110
0.25	115

Source: U.S. Department of Labor.

Trying to converse at a noisy party or an athletic event is difficult due to a poor signal-to-noise ratio (SNR). You can compensate by shouting which raises the desired signal and improves the SNR and the chances of being understood improves until everyone else also shouts.

SNR or S/N compares the level of a desired signal to the level of the noise. It is defined as the ratio of the signal power to the noise power and is often expressed in decibels. A ratio higher than 1:1 indicates more signal than noise. The larger this ratio the easier conversation becomes.

$$SNR_{dB} = 10 \log_{10} \frac{P_{signal}}{P_{noise}}$$

Studies show that for human speech at least 10 dB SNR is required for reasonable intelligibility. This can be lower for venues like parties

dBA scale

This EXTECH sound-level meter meets OSHA requirements.

Courtesy of FLIR Systems

because human intelligence comes into play (uses the context of what is being discussed) and the ability to filter based on sound direction can also help.

Amplifiers must deal with electrical noise, which is present in all devices and all circuits. Amplifiers are useful only if they don't contribute too much noise of their own. With weak (low amplitude) signals, low noise amplifiers are needed. Electronic amplifiers are built using resistors, transistors, capacitors, and so on. Each of those contributes some electrical noise. All devices generate thermal noise caused by the motion of charge carriers. It increases with temperature, which explains why cooling is used with some low-noise (weak-signal) amplifiers.

Noise figure is expressed in decibels and is a measure of how much a signal is degraded by a device.

$$\text{NF} = 10 \log_{10} \frac{\dfrac{\text{Signal}_{in}}{\text{Noise}_{in}}}{\dfrac{\text{Signal}_{out}}{\text{Noise}_{out}}}$$

As an example, let's say the input signal is 100 mV with 10 mV of noise and the output signal is 1 V with 200 mV of noise.

$$\text{NF} = 10 \log_{10} \frac{\dfrac{0.1}{0.01}}{\dfrac{1}{0.2}} = 3 \text{ dB}$$

An ideal amplifier would add no noise, but there is no such thing. But if there were:

$$\text{NF} = 10 \log_{10} \frac{\dfrac{0.1}{0.01}}{\dfrac{1}{0.1}} = 0 \text{ dB}$$

Some microwave amplifiers have noise figures less than 1 dB.

The *dBm scale* is widely applied in electronic communications. This scale places the 0-dB reference level at a power level of 1 mW. Signals and signal sources for radio frequencies and microwaves are often calibrated in dBm. With this reference level, a signal of 0.25 W is

$$\text{Power level} = 10 \times \log \frac{0.25}{0.001} = +24 \text{ dBm}$$

A 40-μW signal is

$$\text{Power level} = 10 \times \log \frac{40 \times 10^{-6}}{1 \times 10^{-3}}$$
$$= -14 \text{ dBm}$$

The dB scale can also be used to specify total harmonic distortion (THD). This is a measure of spectral purity and can be determined using a spectrum analyzer or a distortion meter. THD measurements are often mentioned for high-end audio equipment. They also can be used for the sine wave power supplied by the grid or by an inverter.

EXAMPLE 6-13

What is the dB distortion for an amplifier with 1% THD? Convert the percentage to a decimal and use the same methods we have been using.

$$\text{THD}_{(dB)} = 20 \times \log_{10}.01 = -40 \text{ dB}$$

Briefly, a sine wave (which has only one frequency) is fed into an amplifier, and the output is examined for harmonics (frequencies that should not be there). In the case of an amplifier, one possible cause for harmonic distortion is clipping, which is covered in the next section. In the case of 60 Hz power, the grid or an inverter supplies the signal to be analyzed. As discussed in the last chapter, capacitor input filters cause power line harmonics. THD can be defined by

$$\text{THD} = \sqrt{\frac{V_2^2 + V_3^2 + V_4^2 \cdots V_n^2}{V_1}} \times 100\%$$

where V_1 is the test signal voltage (or the 60-Hz line voltage), and the others are the harmonic voltages. Harmonics are integer multiples. For a 60 Hz line frequency, the harmonics are 120, 180, 240 Hz, and so on. Ideally, the numerator in the equation above is 0 and the THD is 0%. For humans, the threshold of THD detection is 1% for audio. Most people cannot hear that level of distortion or anything below it.

EXAMPLE 6-14

What is the percentage of THD for an inverter rated at −25 dB THD?

$$10 \times \text{inverse log} \left(\frac{-25}{20}\right) \times 100\%$$

Recall that the inverse log function is needed (10^X key on some calculators).

Self-Test

Determine whether each statement is true or false.

1. The ratio of output to input is called gain.
2. The symbol for voltage gain is A_V.
3. Human hearing is linear for loudness.
4. The dB gain or loss of a system is proportional to the common logarithm of the gain ratio.
5. The overall performance of a system is found by multiplying the individual dB gains.
6. The overall performance of a system is found by adding the ratio gains.
7. If the output signal is less than the input signal, the dB gain will be negative.
8. The voltage gain of an amplifier in decibels will be equal to the power gain in decibels only if $R_{in} = R_{out}$.
9. The dBm scale uses 1 μW as the reference level.

Solve the following problems.

10. A two-stage amplifier has a voltage ratio of 35 in the first stage and 80 in the second stage. What is the overall voltage ratio of the amplifier?

11. A two-stage amplifier has a voltage gain of 26 dB in the first stage and 38 dB in the second stage. What is the overall dB gain?
12. A two-way radio needs about 3 V of audio input to the speaker for good volume. If the receiver sensitivity is specified at 1 μV, what will the overall gain of the receiver have to be in decibels?
13. A 100-W audio amplifier is specified at −3 dB at 20 Hz. What power output can be expected at 20 Hz?
14. A transmitter produces 5 W of output power. A 12-dB power amplifier is added. What is the output power from the amplifier?
15. A transmitting station feeds 1,000 W of power into an antenna with an 8-dB gain. What is the effective radiated power of this station?
16. The manufacturer of an RF generator specifies its maximum output as +10 dBm. What is the maximum output power available from the generator in watts?

6-2 Common-Emitter Amplifier

Figure 6-4 shows a *common-emitter amplifier*. It is so named because the emitter of the transistor is *common* to both the input circuit and the output circuit. The input signal is applied across ground and the base circuit of the transistor. The output signal appears across ground and the collector of the transistor. Since the emitter is connected to ground, it is common to both signals, input and output.

The *configuration* of an amplifier is determined by which transistor terminal is used for signal input and which is used for signal output. The common-emitter configuration is one of three possibilities. The last section of this chapter discusses the other two configurations.

Common-emitter amplifier

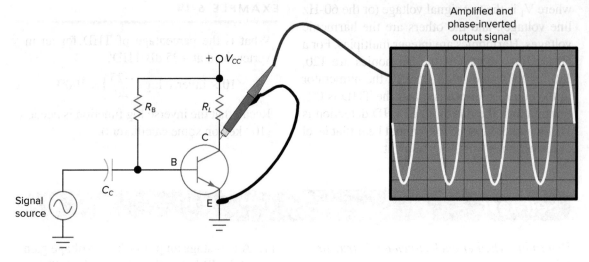

Fig. 6-4 A common-emitter amplifier.

There are two resistors in the circuit in Fig. 6-4. One is a *base bias resistor* R_B, and the other is a *collector load resistor* R_L. The base bias resistor is selected to limit the base current to some low value. The collector load resistor makes it possible to develop a voltage swing across the transistor (from collector to emitter). This voltage swing becomes the *output signal*.

C_C in Fig. 6-4 is called a *coupling capacitor*. Coupling capacitors are often used in amplifiers where only ac signals are important. A capacitor *blocks direct current*. Coupling capacitors may also be called *dc blocking capacitors*. *Capacitive reactance* is infinite at 0 Hz:

$$X_C = \frac{1}{2\pi fC}$$

As frequency f approaches that of direct current (0 Hz), capacitive reactance X_C approaches infinity.

A coupling capacitor may be required if the signal source provides a dc path. For example, the signal source could be a pickup coil in a microphone. This coil can have low resistance. Current flow takes the path of least resistance, and without a blocking capacitor, the direct current will flow in the coil instead of the transistor base circuit. Figure 6-5 shows this. The direct current flow has been diverted from the transistor base by the signal source.

The direct current through R_B is supposed to come from the base of the transistor, as shown in Fig. 6-6. There *must* be base current for the transistor to be on.

Coupling capacitor

Figure 6-6 has enough information to show how the amplifier operates. We will begin by finding the base current. Two parts can limit base current: R_B and the base-emitter junction.

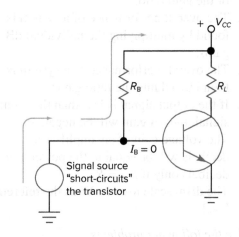

Fig. 6-5 The need for a coupling capacitor.

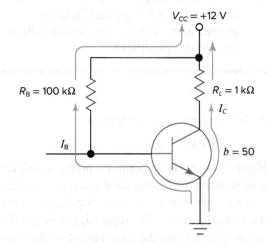

Fig. 6-6 Transistor circuit currents.

Resistor R_B has high resistance. The base-emitter junction is forward-biased so its resistance is low. Thus, R_B and the supply voltage are the major factors determining base current. By Ohm's law,

$$I_B = \frac{V_{CC}}{R_B} = \frac{12 \text{ V}}{100 \times 10^3 \text{ }\Omega}$$
$$= 120 \times 10^{-6} \text{ A}$$

It is possible to make a better approximation of base current by taking into account the drop across the base-emitter junction of the transistor. This drop is about 0.6 V for a silicon transistor. It is subtracted from the collector supply:

$$I_B = \frac{V_{CC} - 0.6 \text{ V}}{R_B}$$

Applying this to the circuit of Fig. 6-6,

$$I_B = \frac{12 \text{ V} - 0.6 \text{ V}}{100 \text{ k}\Omega} = 114 \times 10^{-6} \text{ A}$$

This shows that ignoring the transistor base-emitter drop does not produce a large error.

The base current in Fig. 6-6 is small. Since β is given, the collector current can now be found. We will use the first approximation of base current (120 μA) and β to find I_C:

$$I_C = \beta \times I_B = 50 \times 120 \times 10^{-6} \text{ A}$$
$$= 6 \times 10^{-3} \text{ A}$$

The collector current will be 6 mA. This current flows through load resistor R_L. The voltage drop across R_L will be

$$V_{R_L} = I_C \times R_L = 60 \times 10^{-3} \text{ A} \times 1 \times 10^3 \text{ }\Omega$$
$$= 6 \text{ V}$$

With a 6-V drop across R_L, the drop across the transistor will be

$$V_{CE} = V_{CC} - V_{R_L} = 12 \text{ V} - 6 \text{ V} = 6 \text{ V}$$

The calculations show the condition of the amplifier at its *static*, or resting, state. An input signal will cause the *static conditions* to change. Figure 6-7 shows why. As the signal source goes positive with respect to ground, the base current increases. The positive-going signal causes additional base current to flow onto the plate of the coupling capacitor. This is shown in Fig. 6-7(a). Figure 6-7(b) shows the input signal going negative. Current flows off the capacitor plate and up through R_B. This decreases the base current.

Static condition

As the base current increases and decreases, so does the collector current. This is because base current controls collector current. As the collector current increases and decreases, the voltage drop across the load resistor also increases and decreases. This means that the voltage drop across the transistor must also be changing. It does not remain constant at 6 V.

Figure 6-8 shows how the output signal is produced. A transistor can be thought of as a resistor from its collector terminal to its emitter terminal. The better the transistor conducts, the lower this resistor is in value. The poorer it

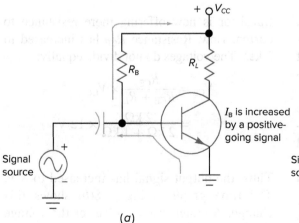

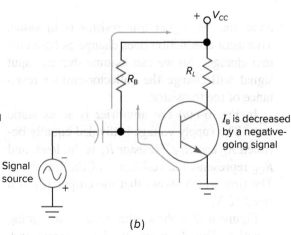

Fig. 6-7 The effect of the input signal on I_B.

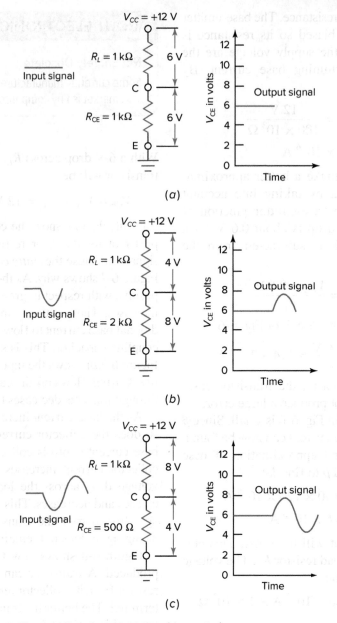

Fig. 6-8 How the output signal is created.

conducts, the higher this resistor is in value. Transistor conduction does change as base current changes. So we can assume that an input signal will change the collector-emitter resistance of the transistor.

In Fig. 6-8(a) the amplifier is at its static state. The supply voltage is divided equally between R_L and R_{CE}. Resistor R_L is the load, and R_{CE} represents the resistance of the transistor. The time graph shows that the output V_{CE} is a steady 6 V.

Figure 6-8(b) shows the input signal going negative. This decreases the base current and, in turn, decreases the collector current. The

transistor is now offering more resistance to current flow. Resistance R_{CE} has increased to 2 kΩ. The voltages do not divide equally:

$$V_{CE} = \frac{R_{CE}}{R_{CE} + R_L} \times V_{CC}$$

$$= \frac{2\ k\Omega}{2\ k\Omega + 1\ k\Omega} \times 12\ V$$

$$= 8\ V$$

Thus, the output signal has increased to 8 V. The time graph in Fig. 6-8(b) shows this change. Another way to solve for the voltage drop across the transistor would be to solve for the current:

$$I = \frac{V_{CC}}{R_L + R_{CE}}$$

$$= \frac{12 \text{ V}}{1 \text{ k}\Omega + 2 \text{ k}\Omega}$$

$$= 4 \times 10^{-3} \text{ A}$$

Now, this current can be used to calculate the voltage across the transistor:

$$V_{CE} = I \times R_{CE}$$

$$= 4 \times 10^{-3} \text{ A} \times 2 \times 10^3 \text{ } \Omega$$

$$= 8 \text{ V}$$

This agrees with the voltage found by using the ratio technique.

Figure 6-8(c) shows what happens in the amplifier circuit when the input signal goes positive. The base current increases. This makes the collector current increase. The transistor is conducting better, so its resistance has decreased. The output voltage V_{CE} is now

$$V_{CE} = \frac{0.5 \text{ k}\Omega}{0.5 \text{ k}\Omega + 1 \text{ k}\Omega} \times 12 \text{ V} = 4 \text{ V}$$

The time graph shows this change in output voltage.

Note that in Fig. 6-8 the output signal is *180 degrees out of phase* with the input signal. When the input goes negative [Fig. 6-8(b)], the output goes in a positive direction. When the input goes positive [Fig. 6-8(c)], the output goes in a negative direction (less positive). This is called *phase inversion*. It is an important characteristic of the common-emitter amplifier.

Phase inversion

The output signal should be a good replica of the input signal. If the input is a sine wave, the output should be a sine wave. When this is achieved, the amplifier is *linear*.

One thing that can make an amplifier nonlinear is too much input signal. When this occurs, the amplifier is *overdriven*. This will cause the output signal to show *distortion*, as shown in Fig. 6-9.

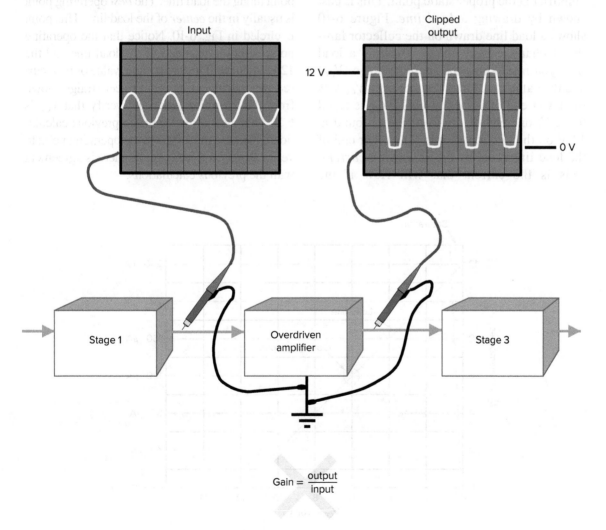

Fig. 6-9 An overdriven amplifier has a clipped output.

The waveform is clipped. V_{CE} cannot exceed 12 V. This means that the output signal will approach this limit and then suddenly stop increasing. Note that the positive-going part of the sine wave has been clipped off at 12 V. V_{CE} cannot go below 0 V. Note that the negative-going part of the signal clips at 0 V. 12 V and 0 V are the *limits* for this particular amplifier. It is not appropriate to calculate gain when the output is clipped. The equation in Fig. 6-9 is crossed out because it is not valid for nonlinear operation.

Clipping is a form of *distortion*. Such distortion in an audio amplifier will cause speech or music to sound bad. This is what happens when the volume control on a radio or stereo is turned up too high. One or more stages are overdriven and distortion results.

Clipping can be avoided by controlling the amplitude of the input and by operating the amplifier at the proper static point. This is best shown by drawing a *load line*. Figure 6-10 shows a load line drawn on the collector family of characteristic curves. To draw a load line, you must know the supply voltage (V_{CC}) and the value of the load resistor (R_L). V_{CC} sets the lower end of the load line. If V_{CC} is equal to 12 V, one end of the load line is found at 12 V on the horizontal axis. The other end of the load line is set by the *saturation current*. This is the current that will flow if the

collector-emitter resistance drops to zero. With this condition, only R_L will limit the flow. Ohm's law is used to find the saturation current:

$$I_{sat} = \frac{V_{CC}}{R_L} = \frac{12\ \text{V}}{1\ \text{k}\Omega}$$
$$= 12 \times 10^{-3}\ \text{A, or 12 mA}$$

This value of current is found on the vertical axis. It is the other end of the load line. As shown in Fig. 6-10, the load line for the amplifier runs between 12 mA and 12 V. These two values are the circuit *limits*. One limit is called *saturation* (12 mA in the example) and the other is called *cutoff* (12 V in the example). No matter what the input signal does, the collector current cannot exceed 12 mA, and the output voltage cannot exceed 12 V. If the input signal is too large, the output will be clipped at these points.

It is possible to operate the amplifier at any point along the load line. The *best* operating point is usually in the *center* of the load line. This point is circled in Fig. 6-10. Notice that the operating point is the *intersection* of the load line and the 120-μA curve. This is the same value of base current we calculated before. Project straight down from the operating point and verify that V_{CE} is 6 V. This also agrees with the previous calculation. Project to the left from the operating point to verify that I_C is 6 mA. Again, there is agreement with the previous calculations.

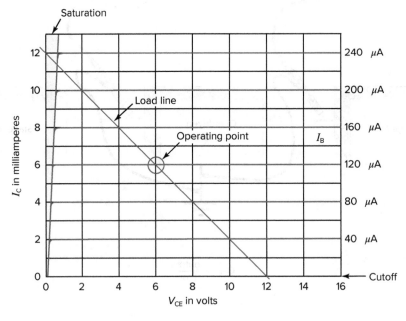

Fig. 6-10 Transistor amplifier load line.

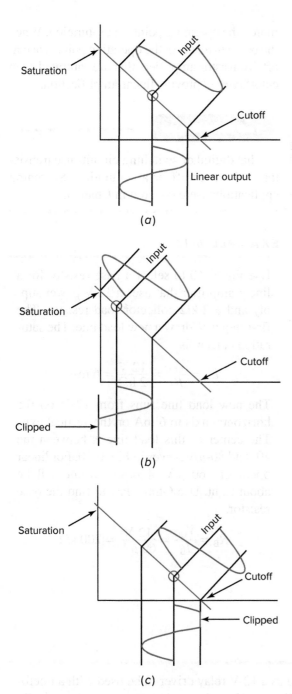

Fig. 6-11 Comparing amplifier operating points.

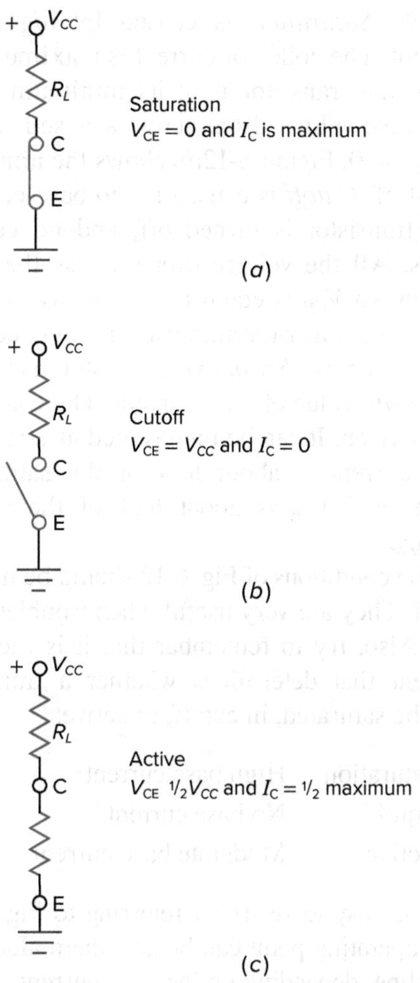

The load line does not really provide new information. It is helpful because it provides a visual or graphical view of circuit operation. For example, examine Fig. 6-11(a). The operating point is in the center of the load line. Notice that as the input goes positive and negative, the output signal follows with no clipping. Look at Fig. 6-11(b). The operating point is near the saturation end of the load line. The output signal is now clipped on the negative-going portion.

Saturation
$V_{CE} = 0$ and I_C is maximum

(a)

Cutoff
$V_{CE} = V_{CC}$ and $I_C = 0$

(b)

Active
V_{CE} $^1/_2 V_{CC}$ and $I_C = ^1/_2$ maximum

(c)

Fig. 6-12 Three possible amplifier conditions.

Figure 6-11(c) shows the operating point near the cutoff end of the load line. The signal is clipped on the positive-going portion. It should be obvious now why the best operating point is near the center of the load line. It allows the most output before clipping occurs.

Saturation and cutoff must be avoided in linear amplifiers. They cause signal distortion. Three possible conditions for an amplifier are shown in Fig. 6-12. Figure 6-12(a) indicates that a saturated amplifier is similar to a *closed* switch. *Saturation* is caused by *high* base current. The collector current is maximum because the transistor is at its minimum resistance. No voltage drops across a closed switch, so $V_{CE} = 0$. Figure 6-12(b) shows the amplifier at cutoff. *Cutoff* is caused by *no* base current. The transistor is turned off, and no current flows. All the voltage drops across the open switch, so V_{CE} is equal to the supply voltage. An active transistor amplifier lies between the two extremes. An *active* transistor has some *moderate* value of base current. The transistor is partly on. It can be represented as a resistor. The current is about half of the saturation value, and V_{CE} is about half of the supply voltage.

The conditions of Fig. 6-12 should be memorized. They are very useful when troubleshooting. Also, try to remember that it is the base current that determines whether a transistor will be saturated, in cutoff, or active:

Saturation	High base current
Cutoff	No base current
Active	Moderate base current

This is easy to verify by referring to Fig. 6-10. The operating point can be anywhere along the load line, depending on the base current. When the base current is moderate (120 μA), the operating point is active and in the center of the load line. When the base current is high (240 μA or more), the operating point is at saturation. When the base current is 0, the operating point is at cutoff. Transistor amplifiers that are saturated or in cutoff *cannot* provide linear amplification.

You May Recall

. . . that digital or switching circuits use transistors only at cutoff or at saturation. Switching applications were covered in Chap. 5.

EXAMPLE 6-15

Use Fig. 6-10 to select a base resistor for a linear amplifier that uses a 12-V power supply and a 2-kΩ collector load resistor. The first step is to draw a new load line. The saturation current is

$$I_{sat} = \frac{V_{CC}}{R_L} = \frac{12\ V}{2\ k\Omega} = 6\ mA$$

The new load line runs from 12 V on the horizontal axis to 6 mA on the vertical axis. The center of this load line is between the 40- and 80-μA curves in Fig. 6-10. For linear operation, 60 μA of base current will be about right. Use Ohm's law to find the base resistor:

$$R_B = \frac{V_{CC}}{I_B} = \frac{12\ V}{60\ \mu A} = 200\ k\Omega$$

EXAMPLE 6-16

Choose a value for R_B if the transistor will serve as a 12-V relay driver to be used with a microcomputer with an output high of 3.3 V. We will need the relay coil current. This is typically 100 mA for small 12-V relays. Next, we will need the minimum value of h_{FE}. This is typically 50, so the base current will be

$$I_B = I_C/\beta = 100\ mA/50 = 2\ mA$$

R_B will be connected from the output of the computer to the base of the driver transistor and is found with

$$R_B = (3.3\ V - 0.7\ V)/2\ mA = 1.3\ k\Omega$$

That value will not guarantee *hard saturation* so the resistor should be smaller, typically around 270 Ω. Note: If the microcomputer high output current is limited, a Darlington transistor relay driver can be used.

Self-Test

Determine whether each statement is true or false.

17. In a common-emitter amplifier, the input signal is applied to the collector.
18. In a common-emitter amplifier, the output signal is taken from the emitter terminal.
19. A coupling capacitor allows ac signals to be amplified but blocks direct current.
20. Common-emitter amplifiers show a 180-degree phase inversion.
21. Overdriving an amplifier causes the output to be clipped.
22. The best operating point for a linear amplifier is at saturation.

Solve the following problems.

23. Refer to Fig. 6-6. Change the value of R_B to 75 kΩ. Do not take V_{BE} into account. Find I_B.
24. With R_B changed to 75 kΩ in Fig. 6-6, what is the new value of collector current?

25. With R_B changed to 75 kΩ in Fig. 6-6, what is V_{R_L}?
26. With R_B changed to 75 kΩ in Fig. 6-6, what is V_{CE}?
27. Refer to Fig. 6-10. Find the new operating point on the load line using your answer from problem 23, and project down to the voltage axis to find V_{CE}.
28. Refer to Fig. 6-10. Project to the left from the new operating point and find I_C.
29. Refer to Fig. 6-6. Change the value of R_B to 50 kΩ. Do not correct for V_{BE} and determine the base current, the collector current, the voltage drop across the load resistor, and the voltage drop across the transistor. Is the transistor in saturation, in cutoff, or in the linear range?
30. Refer to Figs. 6-6 and 6-10. If V_{CC} is changed to 10 V, determine both end points for the new load line.

6-3 Stabilizing the Amplifier

Figure 6-13 shows a common-emitter amplifier that is the same as the one shown in Fig. 6-6 with one important exception: The transistor has a β of 100. If we analyze the circuit for base current, we get

$$I_B = \frac{V_{CC}}{R_B} = \frac{12 \text{ V}}{100 \text{ k}\Omega} = 120 \ \mu\text{A}$$

This is the same value of base current that was calculated before. The collector current, however, is greater:

$$I_C = \beta \times I_B = 100 \times 120 \ \mu\text{A} = 12 \text{ mA}$$

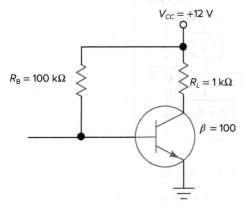

Fig. 6-13 Amplifier with a higher β transistor.

This is twice the collector current of Fig. 6-6. Now, we can solve for the voltage drop across R_L:

$$V_{R_L} = I_C \times R_L = 12 \text{ mA} \times 1 \text{ k}\Omega = 12 \text{ V}$$

And the voltage drop across the transistor is

$$V_{CE} = V_{CC} - V_{R_L} = 12 \text{ V} - 12 \text{ V} = 0 \text{ V}$$

There is no voltage across the transistor. The transistor amplifier is in *saturation*. A saturated transistor is *not* capable of linear amplification. The circuit of Fig. 6-13 would produce severe clipping and distortion.

The only change from Fig. 6-6 to Fig. 6-13 is the β of the transistor. The value of β can vary widely among transistors with the same part number. The 2N3904 is a general-purpose NPN transistor. If you consult the data sheet for this device, you will find that h_{FE} (β) is listed as 100 minimum and 300 maximum. This means that amplifiers like the one shown in Fig. 6-13 cannot be used with this transistor or with any other general-purpose transistor because of β variation. Amplifiers that are β sensitive or β dependent are not practical. It's possible to adjust the value of R_B for the actual β value of each device, but that's not practical. However, it will be instructive to follow the process. We know that a collector current of 6 mA is the center of the load line when $V_{CC} = 12$ V and $R_L = 1$ kΩ. Let's calculate a value for R_B, assuming a β of 200 and assuming that we want to operate the amplifier at the center of the load line:

$$I_B = \frac{I_C}{\beta} = \frac{6 \text{ mA}}{200} = 30 \text{ }\mu A$$

$$R_B = \frac{V_{CC}}{I_B} = \frac{12 \text{ V}}{30 \text{ }\mu A} = 400 \text{ k}\Omega$$

β varies with temperature, so this adds another problem in that the circuit of Fig. 6-13 would not maintain the same operating point over a

EXAMPLE 6-17

Find a better estimate of the conditions for the amplifier shown in Fig. 6-13 by taking the base-emitter junction voltage into account:

$$I_B = \frac{V_{CC} - V_{BE}}{R_B} = \frac{12 \text{ V} - 0.7 \text{ V}}{100 \text{ k}\Omega}$$

$$= 113 \text{ }\mu A$$

$$I_C = \beta \times I_B = 100 \times 113 \text{ }\mu A = 11.3 \text{ mA}$$

$$V_{RL} = I_C \times R_L = 11.3 \text{ mA} \times 1 \text{ k}\Omega = 11.3 \text{ V}$$

$$V_{CE} = V_{CC} - V_{RL} = 12 \text{ V} - 11.3 \text{ V} = 0.7 \text{ V}$$

Note: The better estimate does not make a large difference. V_{CE} is very close to saturation, and the amplifier is still in trouble.

EXAMPLE 6-18

Solve Fig. 6-13 assuming a current gain of 200. The base current remains the same. Using the value of 120 μA found earlier,

$$I_C = \beta \times I_B = 200 \times 120 \text{ }\mu A = 24 \text{ mA}$$

$$V_{RL} = I_C \times R_L = 24 \text{ mA} \times 1 \text{ k}\Omega = 24 \text{ V}$$

Note: V_{RL} can't be larger than V_{CC}. When this happens, the amplifier is in hard saturation, and the actual collector current is limited by Ohm's law:

$$I_C = \frac{V_{CC}}{R_L} = \frac{12 \text{ V}}{1 \text{ k}\Omega} = 12 \text{ mA}$$

The collector current in Fig. 6-13 cannot exceed 12 mA, regardless of β.

wide temperature range. We need a circuit design that is not sensitive to β. Such designs are often called β-independent, and they work well over a range of beta values and temperature.

It is possible to significantly improve a common-emitter amplifier by adding two resistors: one in the emitter circuit and another from base to ground. These resistors can make an amplifier less sensitive to β and to temperature changes. The additional resistors create some additional voltage drops. Figure 6-14 shows the drops most commonly used for analyzing and troubleshooting transistor amplifiers.

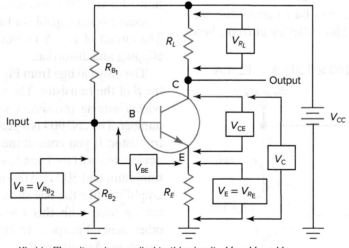

Kirchhoff's voltage law applied to this circuit: $V_B = V_{BE} + V_E$
$$V_{CC} = V_{RL} + V_{CE} + V_E$$
$$V_C = V_{CE} + V_E = V_{CC} - V_{RL}$$

Fig. 6-14 Common labels for voltage drops in a transistor amplifier.

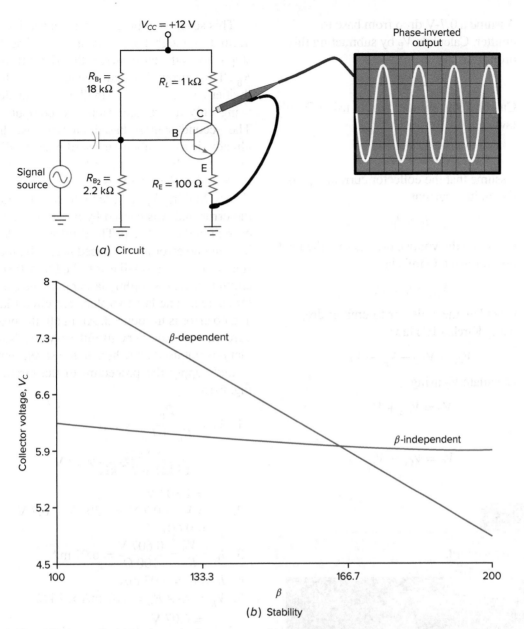

(a) Circuit

(b) Stability

Fig. 6-15 A practical (β-independent) common-emitter amplifier.

Beginning at the left in Fig. 6-14, V_B refers to the drop from the base terminal of the transistor to ground. This is an example of *single-subscript notation*. When single subscripts (the B in V_B) are used, ground is the reference point. V_B is measured from the base to ground. V_{BE} is an example of double-subscription notation. When double subscripts (B and E in V_{BE}) are used, the voltage drop is measured from one subscript point to the other subscript point. Another possibility is the drop across a part with two leads, such as a resistor. An example is V_{R_L} in Fig. 6-14, which specifies the drop across the load resistor. Study this system of notation for

voltage drops carefully. It will help you analyze and troubleshoot circuits.

Figure 6-15 shows an amplifier with all of the resistor values and the supply voltage specified. This information will allow us to analyze the circuit using the following steps:

1. Calculate the voltage drop across R_{B_2}. This is also called the base voltage, or V_B. The two base resistors form a voltage divider across the supply V_{CC}. The voltage divider equation is

$$V_B = \frac{R_{B_2}}{R_{B_1} + R_{B_2}} \times V_{CC}$$

2. Assume a 0.7-V drop from base to emitter. Calculate V_E by subtracting this drop from V_B:

$$V_E = V_B - 0.7$$

3. Calculate the emitter current using Ohm's law:

$$I_E = \frac{V_E}{R_E}$$

4. Assume that the collector current equals the emitter current:

$$I_C = I_E$$

5. Calculate the voltage drop across the load resistor using Ohm's law:

$$V_{R_L} = I_C \times R_L$$

6. Calculate the collector-to-emitter drop using Kirchhoff's law:

$$V_{CE} = V_{CC} - V_{R_L} - V_E$$

7. Calculate V_C using

$$V_C = V_{CE} + V_E$$

or

$$V_C = V_{CC} - V_{R_L}$$

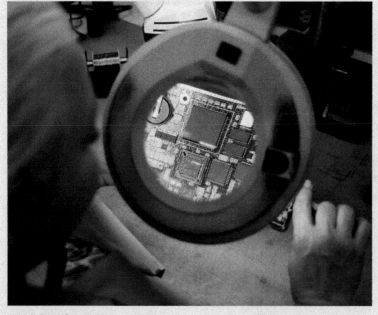
This seven-step process is not exact, but it is accurate enough for practical work. The first step ignores the base current that flows through R_{B_1}. This current is usually about one-tenth the divider current. Thus, a small error is made by using only the resistor values to compute V_B. The second step is based on what we have already learned about forward-biased silicon junctions. However, to get better accuracy, 0.7 V, rather than 0.6 V, is used for silicon transistors. Making V_{BE} a little high tends to reduce the error that was caused by ignoring the base current in the first step. The third step is Ohm's law, and no error is introduced here. The fourth step introduces a small error. We know that the emitter current is slightly larger than the collector current. The last two steps are circuit laws, and no error is introduced. All in all, the procedure can provide very useful answers. Notice that β is not used anywhere in the seven steps.

Let's apply the procedure to the circuit in Fig. 6-15:

1. $V_B = \dfrac{R_{B_2}}{R_{B_1} + R_{B_2}} \times V_{CC}$

   $= \dfrac{2.2\ \text{k}\Omega}{18\ \text{k}\Omega + 2.2\ \text{k}\Omega} \times 12\ \text{V}$

   $= 1.307\ \text{V}$

2. $V_E = V_B - 0.7\ \text{V} = 1.307\ \text{V} - 0.7\ \text{V}$
   $= 0.607\ \text{V}$

3. $I_E = \dfrac{V_E}{R_E} = \dfrac{0.607\ \text{V}}{100\ \Omega} = 6.07\ \text{mA}$

4. $I_C = I_E = 6.07\ \text{mA}$

5. $V_{R_L} = I_C \times R_L = 6.07\ \text{mA} \times 1\ \text{k}\Omega$
   $= 6.07\ \text{V}$

6. $V_{CE} = V_{CC} - V_{R_L} - V_E$
   $= 12\ \text{V} - 6.07\ \text{V} - 0.607\ \text{V}$
   $= 5.32\ \text{V}$

7. $V_C = V_{CE} + V_E = 5.32\ \text{V} + 0.607\ \text{V}$
   $= 5.93\ \text{V}$

Since the collector-emitter voltage is nearly half the supply voltage, we can assume the circuit will make a good linear amplifier. The circuit will work well with any reasonable value of β and will be stable over a wide temperature range.

The graph shown in Fig. 6-15(b) shows the performance of the amplifier with transistor gain ranging from 100 to 200. Note that there is only a minor change in the performance of the β-independent amplifier. For contrast, the performance of the β-dependent circuit of Fig. 6-13 is also shown. The graph was produced using a circuit simulator.

So far the discussions and examples have been concerned with the *dc analysis* of transistor amplifiers. The dc conditions include all of the static currents and voltage drops. An *ac analysis* of an amplifier will allow us to determine its *voltage gain*.

EXAMPLE 6-19

Modify the circuit shown in Fig. 6-15 so that V_{CE} is half the supply voltage. There are many ways to accomplish this. Perhaps the easiest approach is to change the value of R_L, since doing so will not affect the collector current or the emitter voltage. Begin by rearranging the Kirchhoff voltage equation to find the value of V_{R_L} with V_{CE} equal to half the supply:

$$V_{R_L} = V_{CC} - V_{CE} - V_E$$
$$V_{R_L} = 12\text{ V} - 6\text{ V} - 0.607\text{ V} = 5.39\text{ V}$$

Find the new value for R_L with Ohm's law:

$$R_L = \frac{V_{R_L}}{I_C} = \frac{5.39\text{ V}}{6.07\text{ mA}} = 888\ \Omega$$

Voltage gain is an ac amplifier characteristic and one of the most important for small-signal amplifiers. It is the easiest form of gain to measure. For example, an oscilloscope can be used to look at the input signal and then the output signal. Dividing the input into the output will give the gain. Current gain and power gain are not as easy to measure.

Bipolar junction transistors are current amplifiers. A changing base current will produce an output signal. However, as the input signal *voltage* changes, the input signal *current* will also change. In other words, it is the input signal voltage that controls the input signal current. This makes it possible to discuss, calculate, and measure signal voltage gain even though BJTs are current-controlled.

The first step in calculating voltage gain is to estimate the *ac resistance* of the transistor emitter. You can't see this resistance on a schematic diagram because it is inside the device. The symbol used to represent it is r_E. In electronics, a lowercase "r" is often used to

represent an ac resistance. r_E, the ac resistance of the emitter, is a function of the dc emitter current and is estimated with

$$r_E = \frac{25\text{ mV}}{I_E}$$

Since the numerator is in millivolts, the formula can be applied directly when the emitter current is in milliamperes. However, if the emitter current is in amperes, change the numerator to volts (25 mV = 0.025) or change the emitter current to milliamperes.

We have already solved the circuit in Fig. 6-15 for the dc emitter current, so we can estimate r_E:

$$r_E = \frac{25\text{ mV}}{6.07\text{ mA}} = 4.12\ \Omega$$

In actual circuits and with higher temperatures, r_E tends to be higher. It can be estimated with

$$r_E = \frac{50\text{ mV}}{I_E}$$

So, for the circuit in Fig. 6-15, r_E could be as high as

$$r_E = \frac{50\text{ mV}}{6.07\text{ mA}} = 8.24\ \Omega$$

Knowing r_E allows the voltage gain to be found from

$$A_V = \frac{R_L}{R_E + r_E}$$

Let's use this formula to find the voltage gain for the circuit of Fig. 6-15:

$$A_V = \frac{1,000\ \Omega}{100\ \Omega + 4.12\ \Omega} = 9.6$$

The amplifier will have a voltage gain of 9.6. If the input signal is 1 V peak-to-peak, then the output signal should be 9.6 V peak-to-peak. If the input signal is 2 V peak-to-peak, then the output will be clipped. It is *not possible* to exceed the supply voltage in peak-to-peak output in this type of amplifier. The calculated gain will hold true *only* if the amplifier is operating in a linear fashion.

Sometimes, a higher gain is needed. It is possible to improve the gain by adding an *emitter bypass capacitor*. Figure 6-16 shows one across the 100 Ω emitter resistor. The capacitor is chosen to have a low reactance. It acts as a short circuit at signal frequencies and effectively grounds the emitter. It is an open circuit at 0 Hz,

DC analysis

AC analysis

Voltage gain

Emitter bypass capacitor

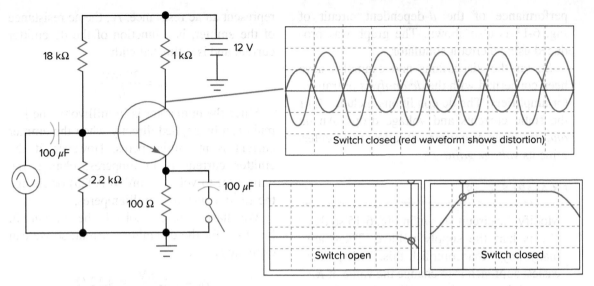

Fig. 6-16 Multisim™ circuit, waveforms, and Bode plots.

so it has no effect on the dc circuit. Said another way, the *ac signal is bypassed* around R_E. Since R_E is bypassed, when the switch is closed, the voltage gain is set by R_L and r_E:

$$A_V = \frac{R_L}{r_E} = \frac{1{,}000\ \Omega}{4.12\ \Omega} = 243 = 48\ \text{dB}$$

That is a big difference. Without the capacitor, the gain is only 9.6 or about 20 dB. We know the actual value of r_E is going to be a bit higher, so the actual gain will be somewhat less. The circuit simulator predicts about 45 dB. With the switch open, the circuit simulator predicts 19.52 dB gain.

Emitter bypassing affects the input impedance of the amplifier, its frequency range, and its distortion. These effects can make bypassing a poor choice. Therefore, circuit designers may choose to develop a gain of 100 by using two amplifier stages, each having a gain of 10.

Looking again at Fig. 6-16, note the distortion in the red waveform. The positive sine peaks are compressed. This is the collector signal with the switch closed. The sine wave has harmonic distortion. It's not badly clipped but the simulator reports a THD of 12.3% which is too high for quality audio. The THD is only 0.032% when the switch is open. The Bode plot shows another issue: the bandwidth has suffered. It decreases from 46 to 17 MHz. Note the striking difference in low-frequency performance. Emitter bypassing might give "cheap" gain but there are three costs: (1) more distortion, (2) less bandwidth, (3) lower input impedance. Emitter bypassing is more useful in some RF circuits. Chapter 7 covers these ideas in more detail. It also shows how to cascade two or more amplifiers to get more gain without the problems emitter bypassing causes.

Self-Test

Solve the following problems.

31. Refer to Fig. 6-15. If $V_B = 1.5$ V and $V_{BE} = 0.7$ V, what is the voltage drop across R_E?
32. Refer to Fig. 6-15. If $V_{CC} = 10$ V, $V_{R_L} = 4.4$ V, and $V_{R_E} = 1.2$ V, what is V_{CE}?
33. Using the data from problem 32, find V_C.

Problems 34 through 40 refer to Fig. 6-15 with these changes: $R_{B_2} = 1.5\ k\Omega$ and $R_L = 2{,}700\ \Omega$

34. Calculate V_B.
35. Calculate I_E.
36. Calculate V_{R_L}.
37. Calculate V_{CE}.
38. Is the amplifier operating in the center of the load line?
39. Calculate A_V.
40. Calculate the range for A_V if an emitter bypass capacitor is added to the circuit.

6-4 Other Configurations

The common-emitter configuration is a very popular circuit. It serves as the basis for most linear amplifiers. However, for some circuit conditions, one of two other configurations could be a better choice.

Amplifiers have many characteristics. Among these is *input impedance*. The input impedance of an amplifier is the loading effect it will present to a signal source. Figure 6-17 shows that when a signal source is connected to an amplifier, that source sees a load, not an amplifier. The load seen by the source is the input impedance of the amplifier.

Signal sources vary widely. An antenna is the signal source for a radio receiver. An antenna might have an impedance of 50 Ω. A microphone is the signal source for a public address system. A microphone might have an impedance of 100,000 Ω. Every signal source has a *characteristic impedance*.

The situation can be stated simply: For the best power transfer, the source impedance should *equal* the amplifier input impedance. This is called *impedance matching*. Figure 6-18 shows why impedance matching gives the best power transfer. In Fig. 6-18(a), a 60-V signal source has an impedance of 15 Ω. This impedance (Z_G) is drawn as an external resistor in series with the generator (signal source). Since

the generator impedance does act in series, the circuit is a good model, as shown. The load impedance in Fig. 6-18(a) is also 15 Ω. Thus, we have an *impedance match*. Let us see how much power is transferred to the load. We begin by finding the current flow:

$$I = \frac{V}{Z} = \frac{60\ V}{15\ \Omega + 15\ \Omega} = 2\ A$$

The power dissipated in the load is

$$P = I^2 \times Z_L = (2\ A)^2 \times 15\ \Omega = 60\ W$$

A 15-Ω load will dissipate 60 W. Figure 6-18(b) shows the same source with a load of 5 Ω. Solving this circuit gives

$$I = \frac{60\ V}{15\ \Omega + 5\ \Omega} = 3\ A$$

$$P = (3\ A)^2 \times 5\ \Omega = 45\ W$$

Note that the dissipation is less than maximum when the load impedance is less than the source impedance. Figure 6-18(c) shows the load impedance at 45 Ω. Solving this circuit gives

$$I = \frac{60\ V}{15\ \Omega + 45\ \Omega} = 1\ A$$

$$P = (1\ A)^2 \times 45\ \Omega = 45\ W$$

The dissipation is *less* than maximum when the load impedance is greater than the source impedance. Maximum load power will occur *only* when the impedances are matched.

The common-emitter configuration typically has an input impedance of around 1,000 Ω. The actual value depends on both the transistor used and the other parts in the amplifier. This may or may not be a desirable input impedance. It depends on the signal source.

Figure 6-19 shows a *common-collector amplifier*. It is so named because the collector terminal is common to the input and the output circuits. At first glance, this circuit may seem to be the

Input impedance

Characteristic impedance

Impedance matching

Common-collector amplifier

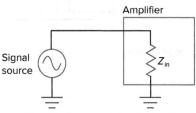

Amplifier

The signal source is loaded by Z_{in}, the input impedance of the amplifier

Fig. 6-17 Amplifier loading effect.

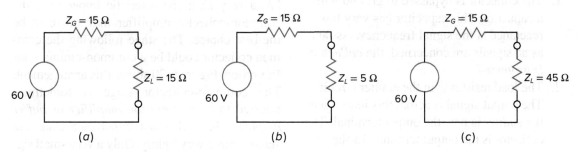

(a) (b) (c)

Fig. 6-18 The need for impedance matching.

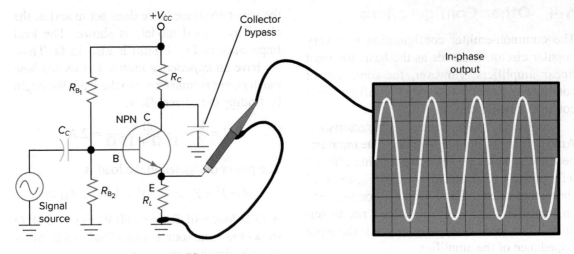

Fig. 6-19 A common-collector amplifier.

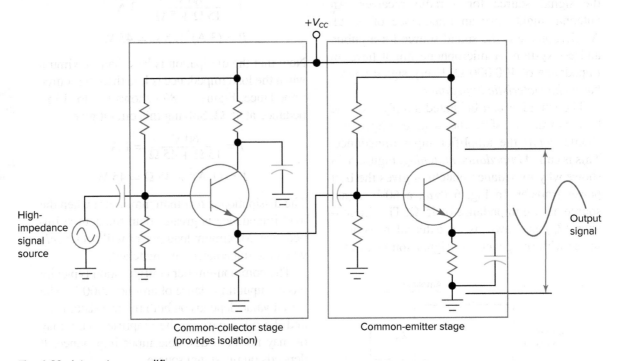

Fig. 6-20 A two-stage amplifier.

same as a common-emitter amplifier. There are two very important differences, however:

1. The collector is bypassed to ground with a capacitor. This capacitor has very low reactance at the signal frequency. As far as ac signals are concerned, the *collector is grounded*.

2. The load resistor is in the emitter circuit. The output signal is across this load. Thus, the *emitter* is now the output terminal. The collector is the output terminal for the common-emitter configuration.

Isolation amplifier

The common-collector amplifier can have a very high input impedance, as much as several hundred thousand ohms. If the signal source has a very high characteristic impedance, the common-collector amplifier may prove to be the best choice. The stage following the common collector could be a common-emitter configuration. Figure 6-20 shows this arrangement. The common-collector stage is sometimes referred to as an *isolation amplifier* or *buffer amplifier*. Its high input impedance loads the signal source very lightly. Only a very small signal current will flow. Thus, the signal source

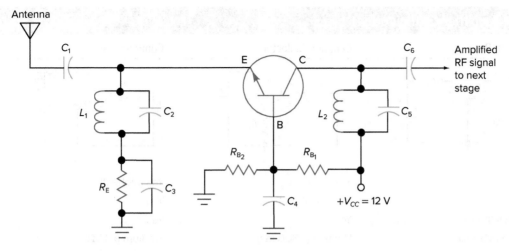

Fig. 6-21 A common-base RF amplifier.

has been isolated (buffered) from the loading effects of the rest of the circuit.

In addition to a high input impedance, the common-collector amplifier has some other important characteristics. It is not capable of giving any voltage gain. The output signal will always be less than the input signal as far as voltage is concerned. The current gain is very high. There is also a moderate power gain. There is no phase inversion in the common-collector amplifier. As the signal source drives the base terminal in a positive direction, the output (emitter terminal) also goes in a positive direction. The fact that the output *follows* the input has led to a second name for this amplifier. It is frequently called an *emitter follower*.

Emitter followers are also noted for their low *output impedance*. This is an advantage when a signal must be supplied to a low-impedance load. For example, a speaker typically has an impedance of 4 to 8 Ω. An emitter follower can drive a speaker reasonably well, whereas the common-emitter amplifier cannot.

The last circuit to be discussed is the *common-base amplifier*. This configuration has its base terminal common to both the input and output signals. It has a very low *input impedance*, on the order of 50 Ω. Therefore, it is useful only with low-impedance signal sources. It is a good performer at radio frequencies.

Figure 6-21 is a schematic diagram of a common-base RF amplifier. It is designed to amplify weak radio signals from the antenna circuit. The antenna impedance is low, on the order of 50 Ω. This makes a good impedance match from the antenna to the amplifier. The base is

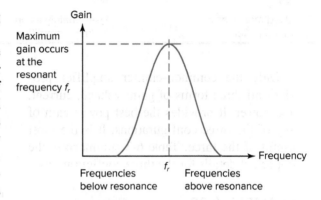

Fig. 6-22 Frequency response of the tuned RF amplifier.

grounded at the signal frequency by C_4. The signal is fed into the emitter terminal of the transistor, and the amplified output signal is taken from the collector. Circuits L_1C_2 and L_2C_5 are for *tuning*. They are *resonant* at the desired frequency of operation. This allows the amplifier to reject other frequencies that could cause interference. Coil L_2 and capacitor C_5 form the collector load for the amplifier. This load will be a high impedance at the resonant frequency. This makes the voltage gain high for this frequency. Other frequencies will have less gain through the amplifier. Figure 6-22 shows the gain performance for a tuned RF amplifier of this type.

The common-base amplifier is not capable of providing any current gain. The input current will always be more than the output current. This is because the emitter current is always highest in BJTs. The amplifier is capable of a large voltage gain. It is also capable of power gain. As with the emitter-follower configuration, it does not invert the signal (no phase inversion).

Emitter follower

Common-base amplifier

Table 6-3 Summary of Amplifier Configurations

	Common Base	Common Collector	Common Emitter
Basic circuit (showing signal source and load R_L)			
Power gain	Yes	Yes	Yes (highest)
Voltage gain	Yes	No (less than 1)	Yes
Current gain	No (less than 1)	Yes	Yes
Input impedance	Lowest ($\approx$50 Ω)	Highest ($\approx$300 kΩ)	Medium ($\approx$1 kΩ)
Output impedance	Highest ($\approx$1 MΩ)	Lowest ($\approx$300 Ω)	Medium ($\approx$50 kΩ)
Phase inversion	No	No	Yes
Application	Used mainly as an RF amplifier	Used mainly as an isolation amplifier	Universal—works best in most applications

Only the common-emitter amplifier provides all three forms of gain: voltage, current, and power. It provides the best power gain of any of the three configurations. It is the most useful of the three. Table 6-3 summarizes the important details for the three configurations.

EXAMPLE 6-20

Find the emitter current and V_{CE} for Fig. 6-21 if R_{B_1} = 10 kΩ, R_{B_2} = 2.2 kΩ, and R_E = 470 kΩ. Although this circuit looks very different, it can be analyzed with the same approach that was used for Fig. 6-15. As before, find the base voltage:

$$V_B = \frac{R_{B_2}}{R_{B_1} + R_{B_2}} \times V_{CC}$$

$$= \frac{2.2 \text{ k}\Omega}{10 \text{ k}\Omega + 2.2 \text{ k}\Omega} \times 12 \text{ V} = 2.16 \text{ V}$$

Subtract for the base-emitter drop:

$$V_E = V_B - 0.7 \text{ V} = 2.16 \text{ V} - 0.7 \text{ V}$$
$$= 1.46 \text{ V}$$

Note: The dc resistance of coil L_1 is very low and will have no effect on the dc emitter current. Calculate the emitter current:

$$I_E = \frac{V_E}{R_E} = \frac{1.46 \text{ V}}{470 \text{ }\Omega} = 3.11 \text{ mA}$$

Note: The dc resistance of coil L_2 is very low and will have no effect on the dc voltage drop across the transistor. Find the drop across the transistor:

$$V_{CE} = V_{CC} - V_E = 12 \text{ V} - 1.46 \text{ V} = 10.5 \text{ V}$$

So far, only NPN transistor amplifiers have been shown. Everything that has been discussed for NPN circuits applies to PNP circuits, with the exception of *polarity*. Figure 6-23 shows a PNP amplifier. Note that the supply V_{CC} is *negative*. If you compare this with the NPN amplifiers shown in this chapter, you will find that they are energized with a positive supply.

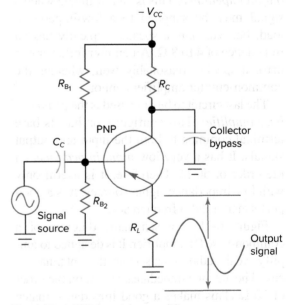

Fig. 6-23 A PNP transistor amplifier.

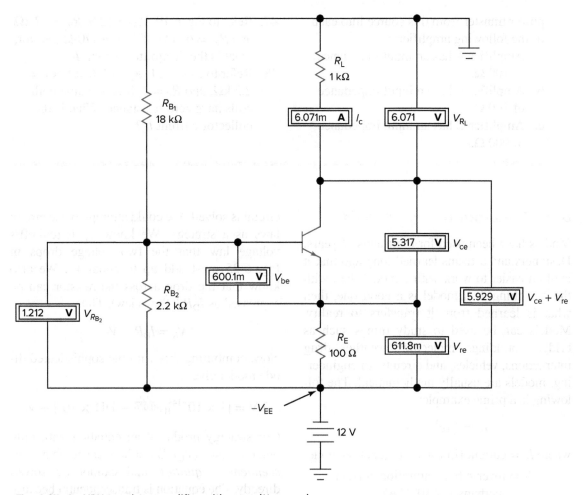

Fig. 6-24 An NPN transistor amplifier with an emitter supply.

NPN circuits *can* be energized with a negative supply. When this is done, the supply terminal is named V_{EE} as opposed to V_{CC}. This is shown in Fig. 6-24. Study this circuit and compare it with Fig. 6-15 to verify that both transistors are properly biased. You should determine in both circuits that the collector is reverse-biased and that the base-emitter junction is forward-biased.

Remember, these bias conditions must be met for a transistor to serve as a linear amplifier.

Figure 6-24 is a printout from a circuit simulator. Note how closely the meter readings agree with the values that were found earlier for Fig. 6-15. Some of the meter labels are different to reflect the change in ground reference for this circuit.

Self-Test

Determine whether each statement is true or false.

41. The configuration of an amplifier can be determined by inspecting which transistor terminals are used for input and output.
42. For maximum power transfer, the source resistance must be equal to the load resistance.
43. The term *emitter follower* is applied to common-emitter amplifiers.

44. The common-collector configuration is the best choice for matching a high-impedance source to a low-impedance load.
45. The amplifier shown in Fig. 6-23 is in the common-emitter configuration.

Solve the following problems.

46. A signal source has an impedance of 300 Ω. It develops an output of 1 V. Calculate the

power transfer from this source into each of the following amplifiers:

a. Amplifier A has an input impedance of 100 Ω.
b. Amplifier B has an input impedance of 300 Ω.
c. Amplifier C has an input impedance of 900 Ω.

47. Refer to Fig. 6-19. $V_{CC} = 12$ V, $R_{B_1} = 47$ kΩ, and $R_{B_2} = 68$ kΩ. If R_L is a 470-Ω resistor, what is the dc emitter current I_E?
48. Refer to Fig. 6-21. $R_{B_1} = 5.6$ kΩ, $R_{B_2} = 2.2$ kΩ, and $R_E = 270$ Ω. Assume both coils have zero resistance. What is the collector current I_C?

6-5 Simulation and Models

Models have been used for thousands of years. Designers and artisans learned long ago that it is often easier to work with a model than with the real thing. If a model is a good one, then what is learned from it transfers to reality. Models can be used to study things such as bridges, buildings, molecules, weather, drug interactions, vehicles, and circuits. In engineering, models are usually mathematical. The following is a prime example:

$$I_D = I_S[e^{V_D/V_T} - 1]$$

Transcendental equation

where I_D = current flow in a PN-junction diode

I_S = reverse bias saturation current (perhaps 1×10^{-14} A)

$e \cong 2.718$

Iteration

V_D = voltage across the PN junction

V_T = thermal voltage (0.026 V at 27°C)

Technicians seldom use models such as this. Here is why: Look at the circuit in Fig. 6-25. It can be solved easily with a simplified model that fixes the drop across a forward-biased diode at 0.7 V. The sophisticated model is not easy to use because we don't know V_D until the

circuit is solved. We could attempt to use circuit laws as a strategy. We know by Kirchhoff's voltage law that the two voltage drops in Fig. 6-25 must add up to equal V_S. We also know that the drop across the resistor can be expressed as $I_D R$ (Ohm's law). Thus,

$$V_S = I_D R + V_D$$

Now, combining this with the sophisticated diode model gives

$$3 \text{ V} = [1 \times 10^{-14}][e^{\frac{V_D}{0.026}} - 1][1 \times 10^3] + V_D$$

Our strategy produced an equation with only one unknown (V_D). But it happens to be a *transcendental equation* and cannot be solved directly. The equation is transcendental because of the exponent ($V_D/0.026$). Computers (and people) can use a process of repeated guessing called *iteration* to solve such equations. One keeps plugging values for V_D into the right-hand side of the equation and solving until the result is close to 3 V. Iteration doesn't always work. If you have ever used software that reported "failed to converge," the equation did not balance after a number of attempts. Yes, computers can be programmed to "know" when to give up. When software iteration fails, it may be possible to change the simulation tolerance or the iteration limits to obtain a solution.

By the way, if you want to try iteration for the diode problem, use a calculator with an e^x key. Solve the fractional exponent first by using a guess value of 0.7 V for V_D, divide by 0.026, then press the e^x key, subtract 1, multiply by the saturation current, multiply by the resistor value, and then add 0.7 V. *Hint:* 0.68 V is a better guess value for V_D in this case.

Computers are tireless, and people are not. Computers can do millions of calculations in a second. Computers rarely make mistakes.

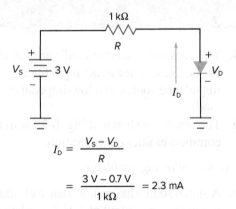

Fig. 6-25 A diode circuit solved using a simple model.

$$I_D = \frac{V_S - V_D}{R}$$

$$= \frac{3 \text{ V} - 0.7 \text{ V}}{1 \text{ k}\Omega} = 2.3 \text{ mA}$$

Computers are very good at simulation using complex models. IBM developed one of the first circuit simulators about 1960 and called it ECAP (Electronic Circuit Analysis Program). After ECAP came SPICE (Simulation Program with Integrated Circuit Emphasis). SPICE was developed at the University of California at Berkeley in the early 1970s. This program is in the public domain and has become a standard for electronic simulation software. Like ECAP, SPICE originally ran on mainframe computers that were available only in governmental agencies, large companies, and some universities. Today, relatively inexpensive personal computers have more computing power than the old mainframes did.

Figure 6-26 shows that the SPICE model is important. The 1N4001 is a general-purpose silicon diode (often used as a rectifier), and the 1N4148 is a high-speed silicon switching diode. The meter readings (diode forward voltage drops) are significantly influenced by the models for the diodes.

EXAMPLE 6-21

Determine the diode currents for Fig. 6-26.

Begin by finding the drops across the resistors:

$$V_{1\,k\Omega} = 3\text{ V} - 0.513\text{ V} = 2.487\text{ V and}$$
$$V_{1\,k\Omega} = 3\text{ V} - 0.622\text{ V} = 2.378\text{ V}$$

Use Ohm's law:

$$I = \frac{2.487\text{ V}}{1{,}000\text{ }\Omega} = 2.487\text{ mA}\quad\text{and}$$
$$I = \frac{2.378\text{ V}}{1{,}000\text{ }\Omega} = 2.378\text{ mA}$$

SPICE development continues to evolve, with more models, more features, and more versions adapted to personal computers. It's very likely that most electronic workers will use some type of circuit simulator. Even some electronic hobbyists use simulators. It is worth knowing whether a given simulator is based on SPICE. If it is, chances are that it will be compatible with other simulators and with device models that have been developed for SPICE. Device models are important. As an example, the reverse-bias saturation current for a real diode varies according to its physical size and how it is doped. The range of practical values for I_S is from 10^{-15} to 10^{-13} A. I_S doubles for approximately every 5°C rise in temperature (this is one way that circuit simulators deal with temperature change). Thus, accurate SPICE models can predict how various real parts perform in circuits and what effects temperature will have.

Simulation software allows things to be investigated that would take too long or cost too much by using other means. For example, a *parameter sweep* can allow a circuit to be solved for a range

Parameter sweep

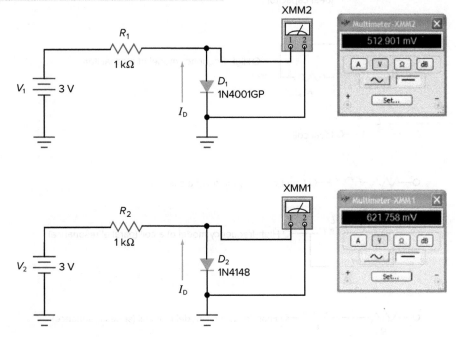

Fig. 6-26 The SPICE model is important.

of values. This is how the graph in Fig. 6-15(*b*) was prepared. Transistor current gain β was swept over a range of 100 to 200. Temperature sweeps are another example. It's easy to solve a diode circuit for 10 different temperatures. Imagine doing this using a real circuit. Circuit simulation offers the following advantages:

- Savings in time and money
- A dynamic learning environment
- An easy way to try ideas (called "what if?")
- A safe way of working with high-energy circuits
- A way to conduct investigations that would not be practical using other methods
- The ability to do an independent check on circuit designs (detect and correct mistakes)
- A way to improve designs (reliability, cost, etc.)

Modeling is not perfect. If a model is not appropriate, it does not represent reality. Also, simulators must use guessing and iteration to solve certain kinds of problems. Circuit simulators have some limitations, which may include

- Inaccuracy in high-frequency simulator behavior
- Failure to check for failure modes (voltage breakdown, temperature breakdown, etc.)
- Inability to simulate some circuits (a solution is not reached)
- Inability to simulate some circuits accurately (solution is not realistic)
- Inability to provide a particular type of analysis
- Unavailability of needed device models

Let's examine some things that occur at high frequencies. Figure 6-27 shows that, at high frequencies, basic components are more complicated. Why does this happen? There are so-called stray inductances and stray

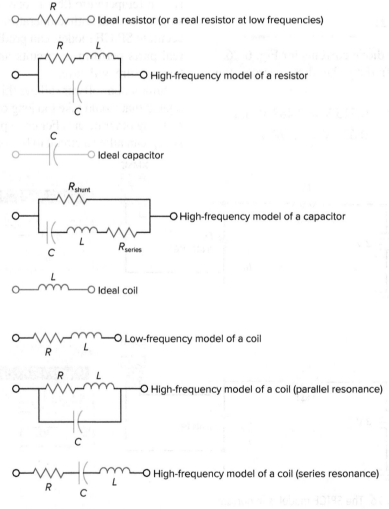

Fig. 6-27 Device models.

capacitances in all real devices. If a device has wire leads, they produce stray inductance. If a resistor has a film deposited on an insulating body, there is a stray capacitance between the film and the body. These stray effects are very small. They are in the nanohenry (nH) and picofarad (pF) ranges. But at very high frequencies, these stray effects should not be ignored.

What about capacitors? They often have wire leads. The leads have a small inductance, but it becomes significant at high frequencies. High-frequency designs often use leadless parts (like chip capacitors and resistors) to reduce stray inductance.

Consider an inductor or a coil. It has turns of wire that are close to each other. There is a small capacitance from turn to turn. The overall effect is called the *distributed capacitance* of the coil. Distributed capacitance is small and can be ignored at low frequencies. At high frequencies, it becomes significant and can make a coil act like a resonant circuit. The wire used to wind a coil has both a low-frequency resistance and a different high-frequency resistance caused by *skin effect*. You may recall that skin effect means that most of the high-frequency current flow is confined to the skin of a conductor.

Distributed capacitance

Skin effect

Self-Test

Answer the following with a short phrase or a word.

49. List some things that can be simulated with computers.
50. Give an example of a mathematical model.
51. What does the acronym SPICE represent?
52. Give an example of a device model.
53. What is another name for repeated guessing?
54. What does *failure to converge* mean?
55. What is a parameter sweep good for?
56. Why are ideal models OK at low frequencies?

Chapter 6 Summary and Review

Summary

1. Amplifier gain is determined by dividing the output by the input.
2. Gain is specified as a voltage ratio, a current ratio, a power ratio, or as the logarithm of a ratio.
3. When each part of a system is specified in decibel gain or loss, the overall performance can be obtained by simply adding all gains and subtracting all losses.
4. When each part of a system is specified in ratios, the overall performance is obtained by multiplying all gains and dividing by all losses.
5. The decibel is based on power gain or loss. It can be adapted to voltage gain or loss by assuming the input and output resistances to be equal. When they are not equal, the dB voltage gain does not equal the dB power gain.
6. In a common-emitter amplifier, the emitter of the transistor is common to both the input signal and the output signal.
7. The collector load resistor in a common-emitter amplifier allows the output voltage swing to be developed.
8. The base bias resistor limits base current to the desired steady or static level.
9. The input signal causes the base current, the collector current, and the output voltage to change.
10. The common-emitter amplifier produces a 180-degree phase inversion.
11. One way to show amplifier limits is to draw a load line. Linear amplifiers are operated in the center of the load line.
12. A saturated transistor can be compared with a closed switch. The voltage drop across it will be zero (or very low).
13. A cutoff transistor can be compared with an open switch. The voltage drop across it will be equal to the supply voltage.
14. A transistor set up for linear operation should be between saturation and cutoff. The voltage drop across it should be about half the supply voltage.
15. For a transistor amplifier to be practical, it cannot be too sensitive to β.
16. A practical and stable amplifier circuit uses a voltage divider to set the base voltage and a resistor in the emitter lead.
17. The voltage gain of a common-emitter amplifier is set by the load resistance and the emitter resistance.
18. The common-emitter amplifier is the most popular of the three possible configurations.
19. The best transfer of signal power into an amplifier occurs when the source impedance matches the amplifier input impedance.
20. The common-collector, or emitter-follower, amplifier has a very high input impedance and a low output impedance.
21. The common-collector amplifier has a voltage gain of less than 1.
22. Because of its high input impedance, the common-collector amplifier makes a good isolation amplifier.
23. The common-base amplifier has a very low input impedance.
24. The common-base amplifier is used mainly as an RF amplifier.
25. The common-emitter amplifier is the only configuration that gives both voltage and current gain. It has the best power gain.
26. Any of the three amplifier configurations can use either NPN or PNP transistors. The major difference is in polarity.
27. When the collector circuit of an amplifier is powered, the supply point is called V_{CC}.
28. When the emitter circuit of an amplifier is powered, the supply point is called V_{EE}.
29. A lot can be learned by working with realistic models of real things, like circuits.
30. In engineering and technology, mathematical models are popular.
31. Although some equations cannot be solved directly, many of them yield to iteration, which is a process of repeated attempts using guess values.

32. Computers are excellent tools for working with mathematical models and iterative solutions.

33. SPICE, developed by the University of California at Berkeley, is the most popular electronic simulation software.

34. SPICE models are available for many electronic devices.

35. Although circuit simulators are very useful and powerful, they have limitations.

Related Formulas

Gain: $\text{Gain} = \dfrac{\text{signal out}}{\text{signal in}}$; $A_V = \dfrac{V_{\text{out}}}{V_{\text{in}}}$;

$$A_I = \dfrac{I_{\text{out}}}{I_{\text{in}}}; A_P = \dfrac{P_{\text{out}}}{P_{\text{in}}}$$

dB gain: $\text{Gain} = 20 \times \log \dfrac{V_{\text{out}}}{V_{\text{in}}}$;

$$\text{Gain} = 10 \times \log \dfrac{P_{\text{out}}}{P_{\text{in}}}$$

Base current (β-dependent circuit):

$$I_B = \dfrac{V_{CC} - 0.6 \text{ V}}{R_B}$$

Collector current (β-dependent circuit):

$$I_C = \beta \times I_B$$

Collector-emitter voltage (β-dependent circuit):

$$V_{CE} = V_{CC} - I_C R_L$$

Saturation current (β-dependent circuit):

$$I_{\text{sat}} = \dfrac{V_{CC}}{R_L}$$

Base voltage (β-independent circuit):

$$V_B = \dfrac{R_{B_2}}{R_{B_1} + R_{B_2}} \times V_{CC}$$

Emitter voltage (β-independent circuit):

$$V_E = V_B - 0.7 \text{ V}$$

Emitter current (β-independent circuit):

$$I_E = \dfrac{V_E}{R_E}$$

Collector current (either circuit):

$$I_C \approx I_E$$

V_{R_L} (either circuit):

$$V_{R_L} = I_C \times R_L$$

V_{CE} (β-independent circuit):

$$V_{CE} = V_{CC} - V_{R_L} - V_E$$

V_C (β-independent circuit):

$$V_C = V_{CE} + V_E = V_{CC} - V_{R_L}$$

AC emitter resistance: $r_E = \dfrac{0.025}{I_E}$ or

$$r_E = \dfrac{.05}{I_E}$$

Voltage gain: $A_V = \dfrac{R_L}{R_E + r_E}$ or

$$A_V = \dfrac{R_L}{r_E}$$

Chapter Review Questions

Supply the missing word in each statement.

6-1. A_V is the symbol for _____. (6-1)

6-2. A_P is the symbol for _____. (6-1)

6-3. Common logarithms are powers of _____. (6-1)

6-4. If the signal out is less than the signal in, the dB gain will be a _____ number. (6-1)

6-5. The sensitivity of human hearing to loudness is not linear but _____. (6-1)

6-6. In a common-emitter amplifier, the signal is fed to the base circuit of the transistor, and the output is taken from the _____. (6-2)

6-7. Refer to Fig. 6-4. The component that prevents the signal source from bypassing the base-emitter direct current flow is _____. (6-2)

6-8. Refer to Fig. 6-4. The component that allows the amplifier to develop an output voltage signal is _____. (6-2)

6-9. Refer to Fig. 6-4. If R_B opens (infinite resistance), then the transistor will operate in _____. (6-2)

6-10. Refer to Fig. 6-6. As an input signal drives the base in a positive direction, the collector will change in a _____ direction. (6-2)

6-11. Refer to Fig. 6-6. As an input signal drives the base in a positive direction, the collector current should _____. (6-2)

6-12. Clipping can be avoided by controlling the input signal and by operating the amplifier at the _____ of the load line. (6-2)

6-13. Refer to Fig. 6-10. The base current is zero. The amplifier will be in _____. (6-2)

6-14. Refer to Fig. 6-10. The base current is 300 μA. The amplifier will be in _____. (6-2)

6-15. A technician is troubleshooting an amplifier and measures V_{CE} to be near 0 V. Voltage V_{CC} is normal. The transistor is operating in _____. (6-2)

6-16. Refer to Fig. 6-13. This amplifier circuit is not practical because it is too sensitive to temperature and to _____. (6-3)

6-17. A signal source has an impedance of 50 Ω. For best power transfer, an amplifier designed for this signal source should have an input impedance of _____. (6-4)

6-18. Refer to Fig. 6-19. As the base is driven in a positive direction, the emitter will go in a _____ direction. (6-4)

6-19. Refer to Fig. 6-19. This configuration is noted for a high input impedance and a _____ output impedance. (6-4)

6-20. An amplifier is needed with a low input impedance for a radio-frequency application. The best choice is probably the common _____ configuration. (6-4)

6-21. The only amplifier that produces a 180-degree phase inversion is the _____ configuration. (6-4)

6-22. An amplifier is needed with a moderate input impedance and the best possible power gain. The best choice is probably the common _____ configuration. (6-4)

6-23. An amplifier is needed to isolate a signal source from any loading effects. The best choice is probably the common _____ configuration. (6-4)

6-24. Refer to Fig. 6-24. If this circuit were designed for a PNP transistor, then V_{EE} would have to be _____ with respect to ground. (6-4)

Chapter Review Problems

6-1. The signal fed into an amplifier is 100 mV, and the output signal is 8.5 V. What is A_V? (6-1)

6-2. What is the dB voltage gain in problem 6-1?

6-3. If $R_{in} = R_{out}$ in problem 6-1, what is the dB power gain? (6-1)

6-4. An amplifier with a power gain of 6 dB develops an output signal of 20 W. What is the signal input power? (6-1)

6-5. A 1,000-W transmitter is fed into a coaxial cable with a 2-dB loss. How much power reaches the antenna? (6-1)

6-6. A two-stage amplifier has a gain of 40 in the first stage and a gain of 18 in the second stage. What is the overall ratio gain? (6-1)

6-7. A two-stage amplifier has a gain of 18 dB in the first stage and 22 dB in the second stage. What is the overall dB gain? (6-1)

6-8. An oscilloscope has a frequency response that is −3 dB at 50 MHz. A 10-V peak-to-peak, 50-MHz signal is fed into the oscilloscope. What voltage will the oscilloscope display? (6-1)

6-9. The signal coming from a microwave antenna is rated at −90 dBm. What is the level of this signal in watts? (6-1)

6-10. Refer to Fig. 6-4. Assume $R_B = 100$ kΩ and $V_{CC} = 10$ V. Do not correct for V_{BE}, and find I_B. (6-2)

6-11. Refer to Fig. 6-6. Do not correct for V_{BE}. Assume that $\beta = 80$. Determine V_{CE}. (6-2)

6-12. Refer to Fig. 6-10. Assume the base current is 180 μA. Find I_C. (6-2)

6-13. Refer to Fig. 6-10. If the base current is 200 μA, what is V_{CE}? (6-2)

Chapter Review Problems...continued

6-14. Refer to Fig. 6-15(a). Assume that R_L is 1,500 Ω. Calculate V_{CE}. (6-3)

6-15. Refer to Fig. 6-15(a). Find the maximum value for I_E. (6-3)

6-16. Refer to Fig. 6-16. Assume the emitter current to be 5 mA. The voltage gain could be as high as _____. (6-3)

6-17. Find A_V for the data in problem 6-16 if the emitter bypass capacitor is open (a common defect in electrolytics). (6-3)

6-18. Refer to Fig. 6-23. $V_{CC} = -20$ V, $R_{B_1} = R_{B_2} = 10$ kΩ, $R_L = 1$ kΩ, and $R_C = 100$ Ω. Find V_B, V_E, I_E, R_{R_C}, and V_{CE}. (6-4)

Critical Thinking Questions

6-1. Is there any advantage to human hearing being logarithmic?

6-2. Suppose an amplifier is defective and no matter what the input signal is, the output is always zero. Can the performance of this amplifier be expressed using decibels?

6-3. You are approached by an inventor who wants you to invest money in a new development called an *energy amplifier*. Why should you be extremely cautious?

6-4. We know that amplifiers can make sounds louder. Can they also improve the quality of sound?

6-5. A transistor has an operating point at the center of the load line. Assuming no clipping, will this transistor run at a different temperature when it is amplifying signals as compared with when it is not fed any input signal?

6-6. In some cases, gain is needed but a phase inversion is not acceptable. Can the common-emitter configuration be used in these cases?

Answers to Self-Tests

1. T
2. T
3. F
4. T
5. F
6. F
7. T
8. T
9. F
10. 2,800
11. 64 dB
12. 130 dB
13. 50 W
14. 79.2 W
15. 6.31 kW
16. 10×10^{-3} W
17. F
18. F
19. T
20. T
21. T

22. F
23. 160 μA
24. 8 mA
25. 8 V
26. 4 V
27. $V_{CE} = 4$ V
28. $I_C = 8$ mA
29. $I_B = 240$ μA
    $I_C = 12$ mA
    $V_{R_L} = 12$ V
    $V_{CE} = 0$ V
    saturation
30. $V_{CE(cutoff)} = 10$ V
    $I_{sat} = 10$ mA
31. 0.8 V
32. 4.4 V
33. 5.6 V
34. 0.923 V
35. 2.23 mA
36. 6.02 V
37. 5.75 V

38. very close to it
39. 24
40. 120 to 241
41. T
42. T
43. F
44. T
45. F
46. a. 0.625 mW
    b. 0.833 mW
    c. 0.625 mW
47. 13.6 mA
48. 9.94 mA
49. weather, circuits, physical systems, etc.
50. Ohm's law, the diode equation, etc.
51. simulation program with integrated circuit emphasis

52. data for an actual device, including its temperature performance, its saturation current, etc.
53. iteration
54. iteration failed to produce an answer close enough to a required value
55. investigating the effect of a device parameter or temperature
56. the stray inductance and capacitance are too small to be significant

More about Small-Signal Amplifiers

Learning Outcomes

This chapter will help you to:

7-1 *Identify* the standard methods of signal coupling and list their characteristics. [7-1]

7-2 *Calculate* the input impedance of common-emitter amplifiers. [7-2]

7-3 *Find* voltage gain in cascade amplifiers. [7-2]

7-4 *Draw* a signal load line for a common-emitter amplifier. [7-2]

7-5 *Solve* FET amplifier circuits. [7-3]

7-6 *Identify* negative feedback and list its effects. [7-4]

7-7 *Determine* the frequency response of a common-emitter amplifier. [7-5]

7-8 *Identify* positive feedback and list its effects. [7-6]

7-9 *Define* hysteresis. [7-6]

A single stage of amplification is often not enough. This chapter covers multistage amplifiers and the methods used to transfer signals from one stage to the next. It also covers field-effect transistor amplifiers. The FET has certain advantages that make it attractive for some amplifier applications. This chapter also discusses negative feedback and frequency response.

7-1 Amplifier Coupling

Coupling refers to the method used to transfer the signal from one stage to the next. There are three basic types of amplifier coupling: capacitive, direct, and transformer.

Capacitive coupling is useful when the signals are *alternating current*. Coupling capacitors are selected to have a low reactance at the lowest signal frequency. This gives good performance over the frequency range of the amplifier. Any dc component will be blocked by a coupling capacitor.

Figure 7-1 shows why it is important to block the dc component in a multistage amplifier. Transistor Q_1 is the first gain stage. Its static collector voltage is 7 V. This is measured from ground to the collector terminal. Transistor Q_2 in Fig. 7-1 has a static base potential of 3 V. This is measured from ground to the base terminal. Because the grounds are common, it is easy to calculate the voltage from the collector of transistor Q_1 to the base of Q_2:

$$V = 7\text{ V} - 3\text{ V} = 4\text{ V}$$

There is 4 V dc across capacitor C_2.

What would happen in Fig. 7-1 if C_2 shorted? The collector of Q_1 and the base of Q_2 would show the *same* voltage with respect to ground. This would greatly change the operating point of Q_2. The base voltage of Q_2 would be higher than 3 V. The increase in base voltage would

Capacitive coupling

186 Small-Signal Amplifiers Chapter 7

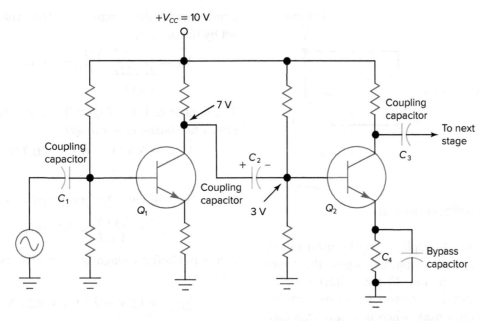

Fig. 7-1 A capacitively coupled amplifier.

drive Q_2 into saturation. It would no longer be capable of linear operation.

Coupling capacitors used in transistor circuits are often of the electrolytic type. This is especially true in low-frequency amplifiers. High values of capacitance are needed to pass the signals with little loss. *Polarity* is an important factor when working with electrolytic capacitors. Again, refer to Fig. 7-1. The collector of Q_1 is 4 V more positive than the base of Q_2. C_2 *must* be installed with the polarity shown.

Capacitive coupling is widely applied in electronic amplifiers that process ac signals. Some applications, however, require operation down to dc (0 Hz). Electronic instruments, such as oscilloscopes and meters, often have to respond to direct current. The amplifiers in these instruments *cannot* use capacitive coupling.

Direct coupling does work at 0 Hz (direct current). A direct-coupled amplifier uses wire or some other dc path between stages. Figure 7-2 shows a direct-coupled amplifier. Notice that the emitter of Q_1 is directly connected to the base of Q_2. An amplifier of this type will have to be designed so that the static terminal voltages are compatible with each other. In Fig. 7-2, the emitter voltage of Q_1 will be the same as the base voltage of Q_2.

Temperature sensitivity can be a problem in direct-coupled amplifiers. As temperature goes up, β and leakage current increase. This tends to shift the static operating point of an amplifier.

When this happens in an early stage of a dc amplifier, all of the following stages will *amplify* the temperature drift. In Fig. 7-2, assume the temperature has gone up. This will make Q_1 conduct more current. More current will flow through Q_1's emitter resistor, increasing its voltage drop. The base of Q_2 now sees more voltage, so it is turned on harder. If a third and fourth stage follow, even a slight shift in the operating point of Q_1 may cause the fourth stage to be driven out of the linear range of operation.

Direct coupling a few stages is not difficult. It may be the least expensive way to get the gain

Polarity

Direct coupling

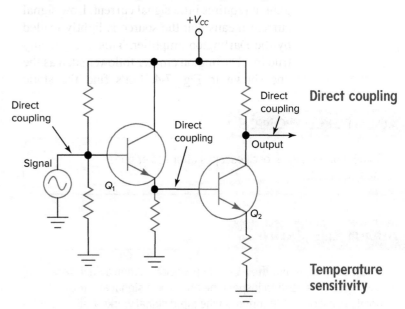

Fig. 7-2 A direct-coupled amplifier.

Temperature sensitivity

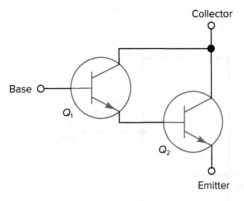

Collector

Base

Q_1

Q_2

Emitter

Fig. 7-3 A Darlington amplifier.

needed. Direct coupling may be used in sections of an audio amplifier where the lowest frequency is around 20 Hz. Direct coupling provides good low-frequency response, and it is used in audio work when it is less expensive than another coupling method.

Darlington circuit   A *Darlington circuit* is another example of direct coupling. You should recall that Darlington pairs may be packaged as a single device or formed by connecting two devices, as shown in Fig. 7-3. The current gain for Fig. 7-3 is approximately equal to the product of the individual betas:

$$\text{Current gain} = A_I \approx \beta_1 \times \beta_2$$

If each transistor has a β of 100, then

$$A_I \approx 100 \times 100 = 10,000$$

The Darlington circuit is a good choice when a high current gain or a high input impedance is required. Since the circuit has such high current gain, it requires little signal current. Low signal current means that the source is lightly loaded by the Darlington amplifier. This is especially true in a Darlington emitter follower such as the one shown in Fig. 7-4. Let's find the static

conditions for this circuit. Q_1's base voltage is set by the divider:

$$V_{B_1} = \frac{220 \text{ k}\Omega}{220 \text{ k}\Omega + 470 \text{ k}\Omega} \times 12 \text{ V}$$
$$= 3.83 \text{ V}$$

The emitter resistor of Q_2 will see this voltage *minus two base-emitter drops:*

$$V_{E(Q_2)} = 3.83 \text{ V} - 0.7 \text{ V} - 0.7 \text{ V}$$
$$= 2.43 \text{ V}$$

Ohm's law will give Q_2's emitter current:

$$I_{E(Q_2)} = \frac{2.43 \text{ V}}{1 \text{ k}\Omega} = 2.43 \text{ mA}$$

And Kirchhoff's voltage law will give the drop across Q_2:

$$V_{CE(Q_2)} = 12 \text{ V} - 2.43 \text{ V} = 9.57 \text{ V}$$

EXAMPLE 7-1

Modify the circuit shown in Fig. 7-4 so that $V_{CE(Q_2)}$ is half the supply voltage. This means that the emitter of Q_2 should be at 6 V. We can change the voltage divider to produce a voltage that is two base-emitter drops above 6 V:

$$V_{\text{divider}} = 6 \text{ V} + 0.7 \text{ V} + 0.7 \text{ V} = 7.4 \text{ V}$$

A new value for the 470-kΩ resistor of the divider can be found by solving the following equation for R:

$$7.4 \text{ V} = \frac{220 \text{ k}\Omega}{220 \text{ k}\Omega + R} \times 12 \text{ V}$$

Multiply both sides by the denominator:

$$7.4 \text{ V} \times (220 \text{ k}\Omega + R) = 220 \text{ k}\Omega \times 12 \text{ V}$$

Divide both sides by 12 V:

$$0.617(220 \text{ k}\Omega + R) = 220 \text{ k}\Omega$$

Multiply:

$$136 \text{ k}\Omega + 0.617R = 220 \text{ k}\Omega$$

Subtract 136 kΩ from both sides:

$$0.617R = 84 \text{ k}\Omega$$

Divide both sides by 0.617:

$$R = 136 \text{ k}\Omega$$

Replace the 470-kΩ resistor in Fig. 7-4 with a 130-kΩ resistor, which is the closest standard value.

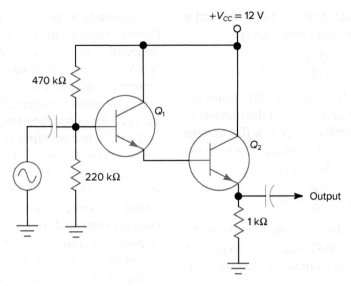

Fig. 7-4 A Darlington emitter follower.

Figure 7-5 shows a *transformer-coupled* amplifier. The transformer serves as the collector load for the transistor and as the coupling device to the amplifier load. The advantage of transformer coupling is easy to understand if we examine its *impedance-matching* property. The *turns ratio* of a transformer is given by

$$\text{Turns ratio} = \frac{N_P}{N_S}$$

where N_P = number of primary turns and N_S = number of secondary turns

The *impedance ratio* is given by

$$\text{Impedance ratio} = (\text{turns ratio})^2$$

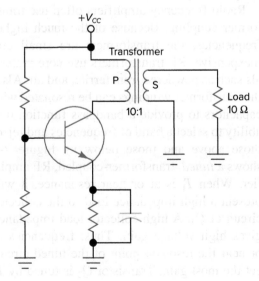

Fig. 7-5 A transformer-coupled amplifier.

If the transformer in Fig. 7-5 has 100 primary turns and 10 secondary turns, its turns ratio is

$$\text{Turns ratio} = \frac{100}{10} = 10$$

and its impedance ratio is

$$\text{Impedance ratio} = 10^2 = 100$$

This means that the load seen by the collector of the transistor will be 100 times the impedance of the actual load. If the load is 10 Ω, the collector will see $100 \times 10\ \Omega = 1\ \text{k}\Omega$.

The output impedance of common-emitter amplifiers is much higher than 10 Ω. When the amplifier must deliver signal energy to such a low impedance, a matching transformer will greatly improve the power transfer to the load. The collector load of 1,000 Ω in Fig. 7-5 is high enough to provide good voltage gain. If we assume that the emitter current is 5 mA, we have enough information to calculate the gain. First, we must estimate the ac resistance of the emitter as discussed in the preceding chapter:

$$r_E = \frac{25\ \text{mV}}{5\ \text{mA}} = 5\ \Omega$$

The emitter resistor is bypassed, so the voltage gain will be given by

$$A_V = \frac{R_L}{r_E}$$

There is no load resistor in Fig. 7-5, but there is a *transformer-coupled* 10-Ω load. This load is *transformed* to 1 kΩ by the transformer, and it will set the voltage gain along with r_E:

$$A_V = \frac{1,000 \ \Omega}{5 \ \Omega} = 200$$

Does the external 10-Ω load see 200 times the signal voltage sent to the base of the transistor? *No*, because the transformer gives a 10:1 voltage *step-down*. Therefore, the voltage gain from the base circuit to the 10-Ω load is

$$A_V = \frac{200}{10} = 20$$

This is still better than we can do without the transformer. If the 10-Ω load were connected directly in the collector circuit as a load resistor, the gain would be

$$A_V = \frac{10 \ \Omega}{5 \ \Omega} = 2$$

Obviously, the transformer does quite a bit to improve the voltage gain of the circuit.

EXAMPLE 7-2

Calculate the voltage gain for Fig. 7-5 using the conservative estimate for the ac emitter resistance. Assume that the dc emitter current is 5 mA. The conservative estimate for ac emitter resistance uses 50 mV, so the voltage gain will be 10, which is half that calculated when using 25 mV. The emitter resistance using 50 mV:

$$r_E = \frac{50 \ \text{mV}}{5 \ \text{mA}} = 10 \ \Omega$$

The transformed collector load is 1,000 Ω, and the gain to the collector circuit is

$$A_V = \frac{1,000 \ \Omega}{10 \ \Omega} = 100$$

The signal voltage across the 10-Ω load is stepped down and so is the gain:

$$A_V = \frac{100}{10} = 10$$

Vacuum-tube power amplifiers used transformer coupling to match the relatively high impedance of the plate circuits to the loudspeakers. Now, solid-state amplifiers use the common-collector configuration with its low output impedance to drive the speakers without the need for impedance-matching transformers.

Transformer coupling is used in distributed sound systems. These installations are called *constant-voltage systems* (and sometimes 70.7-V systems in the United States). The amplifier output is stepped up in voltage so that less audio current is required for a given power level. The step-up transformer can be in the amplifier enclosure or exist as a separate unit. The higher voltage and smaller current allows the use of smaller conductors, and that provides considerable savings in a physically large structure or building. The distributed audio signal reaches 70.7 V when the amplifier is producing its maximum output. Each speaker in the system contains, or is paired with, a step-down transformer to match the relatively high impedance of the distribution system to the 4- or 8-Ω impedances, which are common. The step-down transformers often have taps so that the volume at each speaker can be controlled. Other voltages can be used, and 70.7 is common in the United States. The 70.7-V standard originated with a Underwriter's Laboratories (UL) requirement that the conductors for any distributed voltage over 100-V peak be located in conduit. Conduit is expensive, so distributed sound systems were standardized at 70.7 V_{rms} (100 V_{p-p}) when operating at full power.

Radio-frequency amplifiers often use transformer coupling. Because of the much higher frequencies, the transformers are small and inexpensive. RF transformers use core materials such as powdered iron, ferrite, and air. Also, the transformer windings can be resonated with capacitors to provide a bandpass function (the ability to select a band of frequencies and reject those above and those below it). Figure 7-6 shows a *tuned*, transformer-coupled, RF amplifier. When T_1 is at or near resonance, it will present a high impedance load to the collector circuit of Q_1. A high collector load impedance gives high voltage gain. Thus, frequencies at or near the resonant point of the tuned circuit get the most gain. Transistor Q_2 is tuned by T_2 in Fig. 7-6. Using additional tuned stages

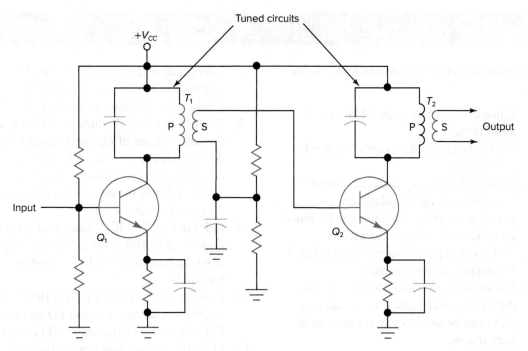

Tuned circuits

$+V_{CC}$

T_1

P   S

Input

Q_1

T_2

P   S   Output

Q_2

Fig. 7-6 A tuned RF amplifier.

improves the *selectivity* of amplifiers of this type. Selectivity is the ability to reject unwanted frequencies.

The transformers in Fig. 7-6 also provide an impedance match. Transformer T_1 normally has more primary turns than secondary turns. This matches the high collector impedance of Q_1 to the lower input impedance of Q_2.

The secondary of T_1 delivers the ac signal to the base of Q_2. It also provides the dc base voltage for Q_2. Figure 7-6 shows that the base divider network for Q_2 is connected to the bottom of T_1's secondary. The secondary winding has a low dc resistance, so the voltage at the junction of the divider resistors will also appear at the base of Q_2. Note that a *bypass* capacitor provides a *signal ground* for the bottom of the secondary winding of T_1. Without this capacitor,

signal current would flow in the divider network, and much of the signal energy would be dissipated (lost).

Knowing the function of circuit components is important for component-level troubleshooting. For example, in Fig. 7-6 if the bypass capacitor just discussed *opens*, the symptom is loss of gain because much of the signal will be dissipated in the divider network. On the other hand, if the capacitor *shorts*, the amplifier may pass no signal because there won't be any base bias for Q_2 and it will be in *cutoff*. Also, if the capacitor opens, the fault *cannot* be found by checking dc voltages, but if the capacitor shorts, the fault *can* be found by checking dc voltages.

Three methods of coupling have been presented. Table 7-1 summarizes some of the important points for each method discussed.

Selectivity

Table 7-1	Summary of Coupling Methods		
	Capacitor Coupling	Direct Coupling	Transformer Coupling
Response to direct current	No	Yes	No
Provides impedance match	No	No	Yes
Advantages	Easy to use. Terminals at different dc levels can be coupled.	Simplicity when a few stages are used.	High efficiency. Can be tuned to make a selective amplifier.
Disadvantages	May require high values of capacity for low-frequency work.	Difficult to design for many stages. Temperature sensitivity.	Cost, size, and weight can be a problem.

Determine whether each statement is true or false.

1. Capacitive coupling cannot be used in dc amplifiers.
2. Transformer coupling cannot be used in dc amplifiers.
3. A shorted coupling capacitor cannot be found by making dc voltage checks.
4. An open coupling capacitor can be found by making dc voltage checks.
5. A shorted bypass capacitor can be found by making dc voltage checks.
6. If a signal source and a load have two different impedances, transformer coupling can be used to achieve an impedance match.

Solve the following problems.

7. Refer to Fig. 7-1. A coupling capacitor should present no more than one-tenth the impedance of the load it is working into. If the second stage has an input impedance of 2 kΩ, and the circuit must amplify frequencies as low as 20 Hz, what is the minimum value for C_2?
8. Refer to Fig. 7-1. Assume C_2 shorts and the base voltage of Q_2 increases to 6 V. Also assume that Q_2 has a load resistor of 1,200 Ω and an emitter resistor of 1,000 Ω. Solve the circuit and prove that Q_2 goes into saturation.
9. Refer to Fig. 7-3. If Q_1 has a β of 50 and Q_2 has a β of 100, what is the current gain from the base terminal to the emitter terminal?
10. Refer to Fig. 7-4. Assume that the 220-kΩ resistor is changed to a 330-kΩ resistor. Find the current flow in the 1-kΩ resistor.
11. Find the voltage drop from collector to emitter for Q_2 for the data given in problem 10.
12. Refer to Fig. 7-5. Assume the turns ratio is 14:1 (primary to secondary). What load does the collector of the transistor see?

7-2 Voltage Gain in Coupled Stages

Figure 7-7(a) shows a common-emitter amplifier driven by a 100-mV signal source. This particular signal source has an internal impedance of 10 kΩ. Signal sources with high internal impedances deliver only a fraction of their output capability when connected to amplifiers with moderate input impedances. As Fig. 7-7(b) shows, the internal impedance of the source and the input impedance of the amplifier form a *voltage divider*. To find the actual signal voltage delivered to the transistor amplifier, the voltage divider equation is used:

$$V = \frac{6.48\ k\Omega}{10\ k\Omega + 6.48\ k\Omega} \times 100\ mV$$

$$= 39.3\ mV$$

This calculation demonstrates why it is sometimes important to know the input impedance of an amplifier.

EXAMPLE 7-3

How much signal would be delivered to the amplifier in Fig. 7-7 if the signal source had an internal impedance of only 50 Ω?

$$V = \frac{6.48\ k\Omega}{50\ \Omega + 6.48\ k\Omega} \times 100\ mV$$

$$= 99.2\ mV$$

This demonstrates that amplifier loading has little effect when the signal source impedance is relatively low.

Finding the *input impedance* of a common-emitter amplifier is detailed in Fig. 7-8. As Fig. 7-8(a) shows, the total ac signal current divides into *three paths*. The power supply point is marked + and is at *ground* potential as far as ac signals are concerned. Power supplies normally have a very low impedance for ac signals. The top of R_{B_1} is effectively at signal

Input impedance

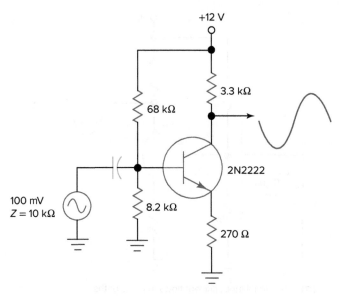

+12 V

3.3 kΩ

68 kΩ

2N2222

100 mV
Z = 10 kΩ

8.2 kΩ

270 Ω

(a) A common-emitter amplifier

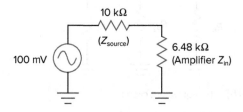

10 kΩ

(Z$_{source}$)

100 mV

6.48 kΩ
(Amplifier Z$_{in}$)

(b) The equivalent input circuit

Fig. 7-7 An amplifier input loads a signal source.

ground. So both base bias resistors plus the transistor itself load the ac signal source. Figure 7-8(b) shows the equivalent circuit. If all three of these loads are known, the input impedance of the amplifier can be found by using the reciprocal equation normally used for parallel resistors.

Figure 7-7 will be used as an example for determining the input impedance of a common-emitter amplifier. The dc conditions are solved first by the approach from the previous chapter:

$$V_B = \frac{8.2\ k\Omega}{68\ k\Omega + 8.2\ k\Omega} \times 12\ V = 1.29\ V$$

$$V_E = 1.29\ V - 0.7\ V = 0.591\ V$$

$$I_E = \frac{0.591\ V}{270\ \Omega} = 2.19\ mA$$

$$r_E = \frac{25\ mV}{2.19\ mA} = 11.4\ mA$$

The input resistance of the transistor itself can now be determined. To avoid confusion, we will use the symbol r_{in} for this resistance and later the symbol Z_{in} to represent the input impedance of the overall amplifier. Input resistance r_{in} is found by multiplying β times the sum of the unbypassed emitter resistances. R_E is *not* bypassed in Fig. 7-7, so it must be used:

$$r_{in} = \beta(R_E + r_E)$$
$$= 200(270\ \Omega + 11.4\ \Omega)$$
$$= 56.3\ k\Omega$$

Note: A good estimate of β for a 2N2222 transistor is 200. Also note that r_E could have been ignored without significantly affecting the result:

$$r_{in} = 200 \times 270\ \Omega = 54\ k\Omega$$

The result of 56.3 kΩ is more accurate. The more accurate approach requires that r_E be known. When the emitter resistor is bypassed with a capacitor, r_E *must* be used because the emitter resistor (R_E) is eliminated from the calculation just as it is when solving for voltage gain. If a bypass capacitor were connected across the 270-Ω emitter resistor in Fig. 7-7(a), then the result would be

$$r_{in} = 200 \times 11.4\ \Omega = 2.28\ k\Omega$$

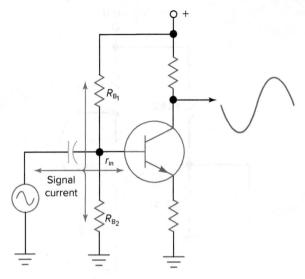

(a) The signal input current flows in three paths

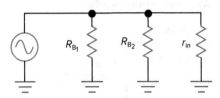

(b) The equivalent circuit

Fig. 7-8 Amplifier input impedance.

We are now prepared to find the input impedance of the amplifier shown in Fig. 7-7 using the standard reciprocal equation and the more accurate value of r_{in} for the unbypassed condition:

$$Z_{in} = \frac{1}{1/R_{B_1} + 1/R_{B_2} + 1/r_{in}}$$

$$= \frac{1}{1/68 \text{ k}\Omega + 1/8.2 \text{ k}\Omega + 1/56.3 \text{ k}\Omega}$$

$$= 6.48 \text{ k}\Omega$$

If the 270-Ω emitter resistor in Fig. 7-7 *is* bypassed, the input impedance *drops* to 1.74 kΩ. You are encouraged to verify this by combining 2.28 kΩ with the two base resistors. Because it may be desirable to have the input impedance of amplifiers as high as possible, you can see that emitter bypassing must be avoided in some applications.

Parallel equivalent

Cascade

Let's apply what we have learned to the cascade amplifier shown in Fig. 7-9. *Cascade* means that the output of one stage is connected to the input of the next. Knowing how to calculate the input impedances of the amplifiers will allow us to find the overall gain of this circuit from the source to the 680-Ω load resistor and

the amplitude of the output signal. We begin by solving the *second* stage for its dc conditions:

$$V_B = \frac{3.9 \text{ k}\Omega}{27 \text{ k}\Omega + 3.9 \text{ k}\Omega} \times 12 \text{ V} = 1.51 \text{ V}$$

$$V_E = 1.51 \text{ V} - 0.7 \text{ V} = 0.815 \text{ V}$$

$$I_E = \frac{0.815 \text{ V}}{100 \text{ }\Omega} = 8.15 \text{ mA}$$

$$r_E = \frac{25 \text{ mV}}{8.15 \text{ mA}} = 3.07 \text{ }\Omega$$

Knowing r_E allows us to find the voltage gain for the second stage. The voltage gain in common-emitter amplifiers is found by dividing the collector load by the resistance in the emitter circuit. However, when the output signal is loaded by another resistor such as the 680-Ω resistor in Fig. 7-9, then the total collector load is the *parallel equivalent* of the collector resistor and the other load resistor. Once again, we see that the power supply point is a ground as far as the ac signal is concerned. So the second transistor in Fig. 7-9 supplies signal current to both the 1-kΩ collector resistor and the 680-Ω resistor. Using the product-over-sum technique to find the parallel equivalent of these two resistors gives:

$$R_P = \frac{1 \text{ k}\Omega \times 680 \text{ k}\Omega}{1 \text{ k}\Omega + 680 \text{ k}\Omega} = 405 \text{ }\Omega$$

The voltage gain is found next:

$$A_V = \frac{R_P}{R_E + r_E}$$

$$= \frac{405 \text{ }\Omega}{100 \text{ }\Omega + 3.07 \text{ }\Omega} = 3.93$$

EXAMPLE 7-4

Calculate the gain for the second stage of Fig. 7-9 using the conservative estimate for ac emitter resistance (use 50 mV). The calculated gain won't change much because the emitter resistor is not bypassed. Find the ac emitter resistance:

$$r_E = \frac{50 \text{ mV}}{8.15 \text{ mA}} = 6.13 \text{ }\Omega$$

Determine the voltage gain:

$$A_V = \frac{R_P}{R_E + r_E} = \frac{405 \text{ }\Omega}{100 \text{ }\Omega + 6.13 \text{ }\Omega} = 3.82$$

This is not much different than 3.93.

Once again, we can determine that ignoring r_E will not significantly change the result. The voltage gain calculates to 4.05 when this approach is taken ($^{405}/_{100}$). If the 100-Ω emitter resistor is bypassed, then r_E *must* be used since R_E is eliminated:

$$A_{V(\text{bypassed})} = \frac{405}{3.07} = 132$$

Adding a load to a transistor amplifier changes the way its gain is calculated. Amplifier gain is *always decreased* by loading. The question now is, what does the second stage in Fig. 7-9 do to the first stage? The answer is that *it loads it*. Therefore to find the first-stage voltage gain, we must first find the input impedance of the second stage. We begin by finding the input resistance of the second transistor itself:

$$r_{in} = 200(100 \text{ }\Omega + 3.07 \text{ }\Omega) = 20.6 \text{ k}\Omega$$

The input impedance is calculated next:

$$Z_{in} = \frac{1}{1/27 \text{ k}\Omega + 1/3.9 \text{ k}\Omega + 1/20.6 \text{ k}\Omega}$$

$$= 2.92 \text{ k}\Omega$$

This 2.92-kΩ input impedance of the second stage acts in *parallel* with the 3.3-kΩ collector resistor of the first stage in Fig. 7-9:

$$R_P = \frac{3.3 \text{ k}\Omega \times 2.92 \text{ k}\Omega}{3.3 \text{ k}\Omega + 2.92 \text{ k}\Omega} = 1.55 \text{ k}\Omega$$

Therefore, the gain of the first stage is

$$A_V = \frac{1,550}{270 + 11.4} = 5.51$$

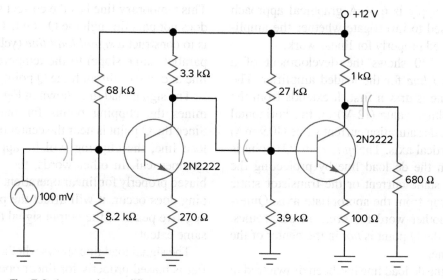

Fig. 7-9 A cascade amplifier.

The *overall gain* of the two-stage amplifier shown in Fig. 7-9 is found by multiplying the individual gains:

$$A_{V(\text{total})} = A_{V_1} + A_{V_2}$$
$$= 5.51 \times 3.93 = 21.7$$

Since no output impedance is specified for the signal source, we will assume that it is an ideal voltage source (has zero internal impedance). So, all of the signal is delivered to the first stage, and the output signal is

$$V_{\text{out}} = 100 \text{ mV} \times 21.7 = 2.17 \text{ V}$$

You may have become used to the idea that the quiescent (static) collector-to-emitter voltage should be about half the supply voltage in a linear amplifier. This is *not* the case when *RC*-coupled amplifiers are loaded, as the second stage in Fig. 7-9 is loaded by the 680-Ω resistor.

We have already solved for most of the dc conditions for the second stage of Fig. 7-9. We found that the emitter voltage (V_E) is 0.815 V and that the emitter current is 8.15 mA. Assuming that the collector current is equal to the emitter current, Ohm's law will give us the drop across the collector resistor:

$$V_{RL} = 8.15 \text{ mA} \times 1 \text{ k}\Omega = 8.15 \text{ V}$$

Kirchhoff's voltage law will give us the transistor static drop:

$$V_{CE} = V_{CC} - V_{RL} - V_E$$
$$= 12 \text{ V} - 8.15 \text{ V} - 0.815 \text{ V}$$
$$= 3.04 \text{ V}$$

Half the supply is 6 V. A graphical approach will be used to investigate whether the amplifier is biased properly for linear work.

Signal load line

Figure 7-10 shows the development of a *signal load line* for the loaded amplifier. The dc load line is drawn first. It extends from the supply voltage value (12 V) on the horizontal axis to the dc saturation current value (10.9 mA) on the vertical axis. The *quiescent* (Q) point is located on the dc load line by projecting the transistor static current or the transistor static voltage drop from the appropriate axis. Quiescent is another word for *static* in electronics. Note that the Q point is *not* in the center of the dc load line.

When the dc load line has been drawn (red in Fig. 7-10) and the Q point has been located on

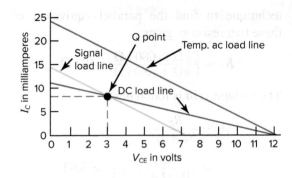

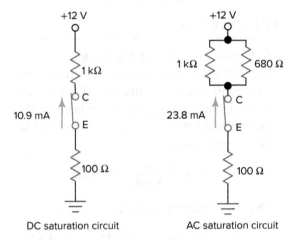

Fig. 7-10 Development of the signal load line.

it, it is time to draw a *temporary ac load line* (blue). Figure 7-10 shows that the ac saturation circuit is different from the dc saturation circuit. Once again, we find that the collector resistor and the 680-Ω resistor act in parallel. This makes the ac saturation current higher than the dc saturation current. A temporary ac load line is drawn from the supply voltage value (12 V) to the ac saturation current value (23.8 mA). This temporary line has the correct slope, but it does not pass through the Q point. The last step is to construct a *signal load line* (yellow) that is parallel (same slope) to the temporary ac load line and passes through the Q point.

The signal load line shown in Fig. 7-10 determines the clipping points for the amplifier. Since the Q point is near the center of the signal load line, the clipping will be approximately symmetrical. In other words, the amplifier is biased properly for linear operation. When clipping does occur, it will affect the positive and negative peaks of the output signal to about the same extent.

The signal load line shows whether an amplifier is biased properly for linear operation. It is often a requirement to have the Q point in the

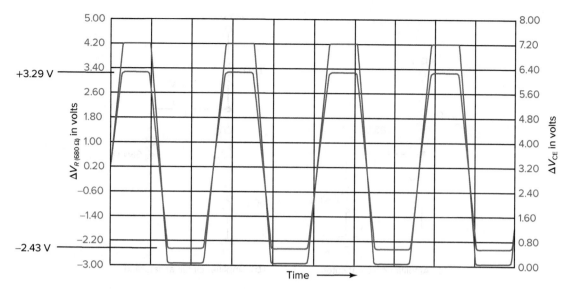

Fig. 7-11 Driving the amplifier output into clipping.

center of the signal load line. The signal load line shows the maximum peak-to-peak swing of V_{CE}. The output swing across the 680-Ω load will be less for Fig. 7-9 because of signal voltage drop across the 100-Ω emitter resistor. Figure 7-11 shows the performance of the amplifier when its output is driven into clipping. The larger of the two waveforms (red) shows the swing of V_{CE} and is a bit more than 7 V_{p-p}. This agrees with the signal load line in Fig. 7-10.

An analysis of the clipping points of the amplifier will provide more insight as to how the circuit operates. There are two clipping points: cutoff and saturation. Figure 7-12(a) shows what the output circuit looks like at cutoff. The transistor is off, so all that must be considered is the supply voltage, R_L, the 680-Ω load, and the output coupling capacitor, which has a charge of 3.86 V. This charge is equal to the quiescent collector voltage. We have already solved the output circuit for the quiescent values of V_E and V_{CE}. The quiescent collector voltage is found by adding them (Kirchhoff's voltage law):

$$V_C = V_E + V_{CE} = 0.815 \text{ V} + 3.04 \text{ V}$$
$$= 3.86 \text{ V}$$

If the time constant of the output circuit is relatively long compared with the signal period, the capacitor will maintain a constant voltage. This is why the cutoff circuit shown in Fig. 7-12(a) shows a 3.86-V battery. In many cases, a charged capacitor can be viewed as a battery. Note that the battery voltage opposes the +12-V supply.

The voltage drop across the 680-Ω resistor can be found with the voltage divider equation:

$$V = (+12 \text{ V} - 3.86 \text{ V}) \times \frac{680 \text{ }\Omega}{1 \text{ k}\Omega + 680 \text{ }\Omega}$$
$$= +3.29 \text{ V}$$

This verifies the positive clipping point shown in Fig. 7-11.

Figure 7-12(b) shows the output circuit at saturation. It is more complicated because it is a multiple-source circuit.

> **You May Recall**
>
> . . . that one way to solve multiple-source circuits is to use the *superposition theorem*.

Superposition theorem

The steps are:

1. Replace every voltage source but one with a short circuit.
2. Calculate the magnitude and direction of the current through each resistor in the temporary circuit produced by step 1.
3. Repeat steps 1 and 2 until each voltage source has been used as an active source.
4. Algebraically add all the currents from step 2.

Figure 7-12(c) shows the steps. The result is a current *down* through the 680-Ω resistor of 3.58 mA. The top of the resistor is now negative with respect to ground:

$$V = -3.58 \text{ mA} \times 680 \text{ }\Omega = -2.43 \text{ V}$$

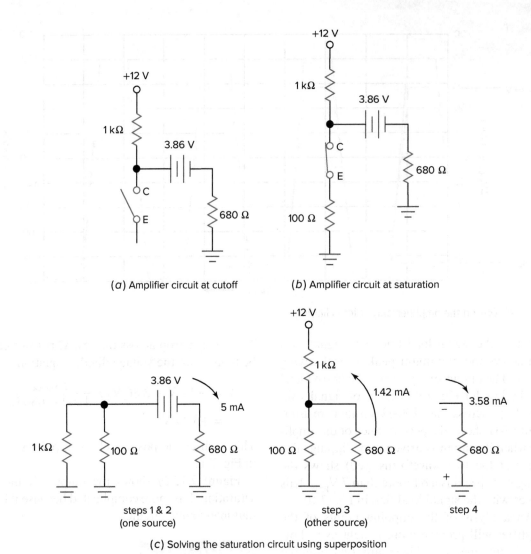

(a) Amplifier circuit at cutoff

(b) Amplifier circuit at saturation

(c) Solving the saturation circuit using superposition

Fig. 7-12 Verifying the load waveform of Fig. 7-11.

This verifies the negative limit of the output swing across the 680-Ω load resistor as shown in Fig. 7-11. The maximum signal swing across the 680-Ω resistor is the difference between both limits:

$$V_{out,\,max} = 3.29 - (-2.43) = 5.72 \text{ V}_{p-p}$$

This is quite a bit less than the V_{CE} swing due to the loss across the 100-Ω emitter resistor.

Amplifier clipping can cause problems. As Fig. 7-11 shows, the clipped output starts to look like a square wave. Square waves have high-frequency components called *harmonics*. Harmonics can damage components. For example, it is possible to burn out the high-frequency speakers in some stereo systems by playing them too loudly. This results in clipping and the generation of high-frequency harmonics. Another problem

caused by clipping and harmonics is interference. A transmitter that is driven into clipping will generate harmonic frequencies that will interfere with other communications channels.

Figure 7-13(a) shows the spectral output of a lab signal generator. Most of the energy appears at only one frequency, 1 kHz. There is also some energy at 3, 5, and 9 kHz, which are odd harmonics. There is no such thing as a perfect sine wave in the real world. Laboratories often use signal, or function, generators to provide sine signals. Even the best of these generators cannot provide a pure sine wave. The instrument used for Fig. 7-13(a) has 0.008 percent total harmonic distortion (THD). Some very expensive signal generators can achieve 0.0003 percent THD. Older lab equipment was often specified at 1 percent THD.

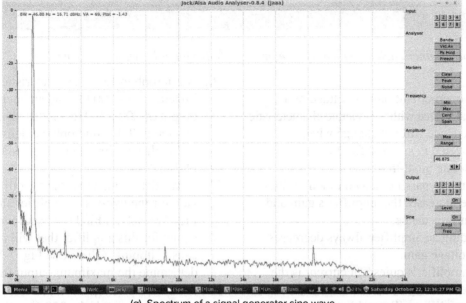

(a) Spectrum of a signal generator sine wave

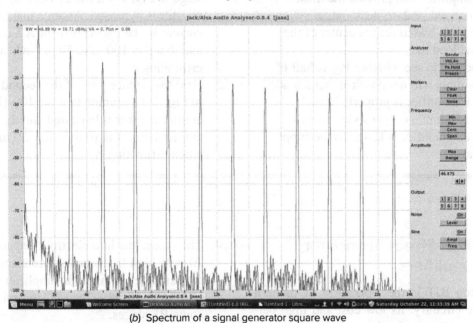

(b) Spectrum of a signal generator square wave

Fig. 7-13 The spectra of two signals: sine and square.

(a–b) Source: www.linuxaudio.org

Figure 7-13(b) shows the frequency spectrum of a square wave. It contains significant energy at the odd harmonics (3, 5, 7 kHz, and so on) of the 1-kHz fundamental signal. A perfect square wave has harmonics that extend to infinity and a THD of 48.3 percent. This means that almost half of its energy lies in the harmonics. Practical square waves have less than 48.3 percent THD but still have a high percentage.

The personal computers (PCs) of today usually have good-quality sound cards, and that is how the spectra of Fig. 7-13 were obtained. The software

used is shareware. It is amazing how measurements can sometimes be made using low-cost or free software. It was not that long ago that measuring THD with reasonable accuracy at the 1 percent level or below required heavy and expensive test equipment. This expensive equipment may still be required. One example is working with extremely low distortion. A THD of 0.0003 percent translates to −110 dB, which cannot normally be measured with PC sound cards. High-end audio analyzers costing tens of thousands of dollars are available, but very few labs have them.

Determine whether each statement is true or false.

13. The open-circuit output voltage of a signal source with a moderate characteristic impedance will not change when connected to an amplifier that has a moderate input impedance.
14. Emitter bypassing in a common-emitter stage increases the amplifier's gain and input impedance.
15. Loading an amplifier always decreases its voltage gain.
16. An amplifier's quiescent current is the same as its static current.
17. An amplifier will provide the most undistorted peak-to-peak output swing when it is biased for the center of the signal load line.
18. Checking to see whether V_{CE} is half of the supply will not confirm that a loaded linear amplifier is properly biased.
19. The maximum output swing from a loaded amplifier is less than the supply voltage.

Solve the following problems.

20. A microphone has a characteristic impedance of 100 kΩ and an open-circuit output of 200 mV. How much signal voltage will this microphone deliver to an amplifier with an input impedance of 2 kΩ?
21. It was determined that the overall voltage gain for the two-stage amplifier shown in Fig. 7-9 is 21.7. Find the maximum input signal for this amplifier that will not cause clipping. (*Hint:* Use Fig. 7-10 to determine the maximum output first.)
22. Find the input impedance for the first stage in Fig. 7-9 if the 270-Ω emitter resistor is bypassed. Assume that the current in the 270-Ω resistor is 5 mA and that $\beta = 100$. Use 50 mV when estimating r_E.
23. Find the voltage gain of the second stage in Fig. 7-9 assuming that the 100-Ω emitter resistor is bypassed and the emitter current is 10 mA. Use 50 mV when estimating r_E.

7-3 Field-Effect Transistor Amplifiers

The silicon BJT is the workhorse of modern electronic circuitry. Its low cost and high performance make it the best choice for most applications. Field-effect transistors do, however, offer certain advantages that make them a better choice in some circuits. Some of these advantages are as follows:

Common-source amplifier

1. They are voltage-controlled amplifiers. Because of this, their input impedance is very high.
2. They have a low noise output. This makes them useful as preamplifiers when noise must be very low because of high gain in the following stages.
3. They have better linearity. This makes them attractive when distortion must be minimized.

4. They have low interelectrode capacitance. At very high frequencies, interelectrode capacitance can make an amplifier work poorly. This makes the FET desirable in some RF stages.
5. They can be manufactured with two gates. The second gate is useful for gain control or the application of a second signal.

Figure 7-14(*a*) shows an FET *common-source amplifier*. The source terminal is common to both the input and the output signals. This circuit is similar to the BJT common-emitter configuration. The supply voltage V_{DD} is positive with respect to ground. The current will flow from ground, through the N channel, through the load resistor, and into the positive end of the power supply. Note that a bias supply V_{GS} is applied across the gate-source junction. The polarity of this bias supply is arranged to

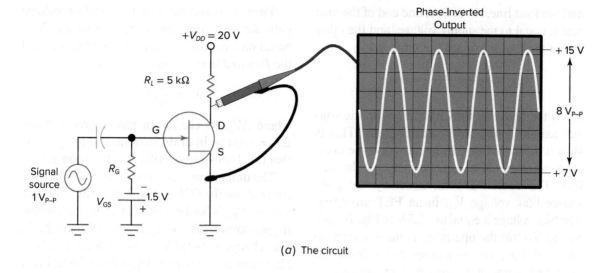

Phase-Inverted Output

+ 15 V

8 V_P-P

+ 7 V

(a) The circuit

+V_{DD} = 20 V

R_L = 5 kΩ

G D
S

Signal source 1 V_P-P

R_G

V_{GS} 1.5 V

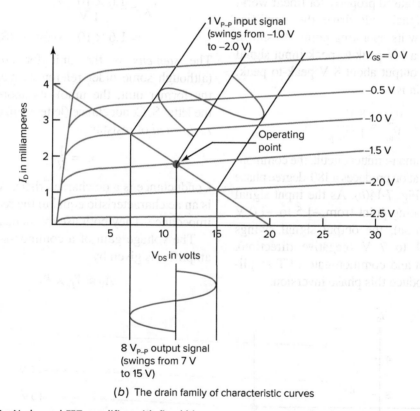

1 V_P-P input signal (swings from –1.0 V to –2.0 V)

V_{GS} = 0 V
–0.5 V
–1.0 V
–1.5 V
–2.0 V
–2.5 V

Operating point

I_D in milliamperes

V_{DS} in volts

8 V_P-P output signal (swings from 7 V to 15 V)

(b) The drain family of characteristic curves

Fig. 7-14 An N-channel FET amplifier with fixed bias.

reverse-bias the junction. Therefore, we may expect the gate current to be zero.

The gate resistor R_G in Fig. 7-14(a) will normally be a very high value [around 1 megohm (MΩ)]. It will not drop any dc voltage because there is no gate current. Using a large value of R_G keeps the input impedance high. If R_G = 1 MΩ, the signal source sees an impedance of 1 MΩ. This

high input impedance is ideal for amplifying high-impedance signal sources. At very high frequencies, other effects can lower this impedance. At low frequencies, the amplifier input impedance is simply equal to the value of R_G.

The load resistor in Fig. 7-14(a) allows the circuit to produce a voltage gain. Figure 7-14(b) shows the characteristic curves for the transistor and the load line. As before, one end of the load line is equal to the supply voltage, and the other end of the load line is set by Ohm's law:

$$I_{sat} = \frac{V_{DD}}{R_L} = \frac{20\ V}{5\ k\Omega} = 4\ mA$$

Thus, the load line runs from 20 V on the voltage axis to 4 mA on the current axis. This is shown in Fig. 7-14(b). In linear work, the operating point should be near the center of the load line. The operating point is set by the gate-source bias voltage V_{GS} in an FET amplifier. The bias voltage is equal to −1.5 V in Fig. 7-14(a). In Fig. 7-14(b) the operating point is shown on the load line. By projecting down from the operating point, it is seen that the resting or static voltage across the transistor will be about 11 V. This is roughly half the supply voltage, so the transistor is biased properly for linear work.

An input signal will drive the amplifier above and below its operating point. As shown in Fig. 7-14(b), a 1-V peak-to-peak input signal will swing the output about 8 V peak-to-peak. The voltage gain is

$$A_V = \frac{V_{out}}{V_{in}} = \frac{8\ V_{p-p}}{1\ V_{p-p}} = 8$$

As with the common-emitter circuit, the common-source configuration produces a 180-degree phase shift. Look at Fig. 7-14(b). As the input signal shifts the operating point from −1.5 to −1.0 V (positive direction), the output signal swings from about 11 to 7 V (negative direction). Common-drain and common-gate FET amplifiers do not produce this phase inversion.

There is a second way to calculate voltage gain for the common-source amplifier. It is based on a characteristic of the transistor called the *forward transfer admittance Y_{fs}*:

$$Y_{fs} = \frac{\Delta I_D}{\Delta V_{GS}}\ |V_{DS}$$

where ΔV_{GS} = change in gate-source voltage, ΔI_D = change in drain current, and $|V_{DS}$ means that the drain-source voltage is held constant.

The drain family of curves can be used to calculate Y_{fs} for the FET. In Fig. 7-15 the drain-source voltage V_{DS} is held constant at 11 V. The change in gate-source voltage V_{GS} is from −1.0 to −2.0 V. The change is (−1.0 V) − (−2.0 V) = 1 V. The change in drain current ΔI_D is from 2.6 to 1 mA for a change of 1.6 mA. We can now calculate the forward transfer admittance of the transistor:

$$Y_{fs} = \frac{1.6 \times 10^{-3}\ A}{1\ V}$$

$$= 1.6 \times 10^{-3}\ \text{siemens (S)}$$

The *siemens* is the unit for *conductance* (although some older references may still use the former unit, the mho). Its abbreviation is the letter S. Conductance (letter symbol G) is the *reciprocal* of resistance:

$$G = \frac{1}{R}$$

Conductance is a dc characteristic. Admittance is an ac characteristic equal to the reciprocal of impedance. They both use the siemens unit.

The voltage gain of a common-source FET amplifier is given by

$$A_V \approx Y_{fs} \times R_L$$

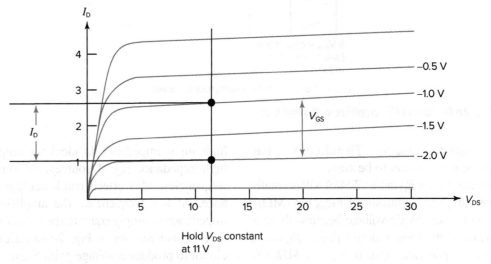

Fig. 7-15 Calculating forward transfer admittance.

For the circuit in Fig. 7-14, the voltage gain will be

$$A_V = 1.6 \times 10^{-3} \text{ S} \times 5 \times 10^3 \text{ } \Omega = 8$$

Note: Since the siemens and ohm units have a *reciprocal* relationship, the units *cancel* and the gain is just a number as always.

A voltage gain of 8 agrees with the graphical solution in Fig. 7-14(*b*). One advantage of using the gain equation is that it makes it easy to calculate the voltage gain for different values of load resistance. It will not be necessary to draw additional load lines. If the load resistance is changed to 8.2 kΩ, the voltage gain will be

$$A_V = 1.6 \times 10^{-3} \text{ S} \times 8.2 \times 10^3 \text{ } \Omega$$
$$= 13.12$$

With a load resistance of 5 kΩ, the circuit gives a voltage gain of 8. With a load resistance of 8.2 kΩ, the circuit gives a voltage gain slightly over 13. This shows that gain is *directly* related to load resistance. This was also the case in the *common-emitter* BJT amplifier circuits. Remember this concept because it is valuable for understanding and troubleshooting amplifiers.

Figure 7-16 shows an improvement for the common-source amplifier. The bias supply V_{GS} has been eliminated. Instead, we find resistor R_S in the source circuit. As the source current flows through this resistor, voltage will drop across it. This voltage drop will serve to bias the gate-source junction of the transistor. If the desired bias voltage and current are known, it is an easy matter to calculate the value of the source resistor. Since the drain current and the source current are equal,

$$R_S = \frac{V_{GS}}{I_D}$$

If we assume the same operating conditions as in the circuit of Fig. 7-14(*a*), the gate-bias

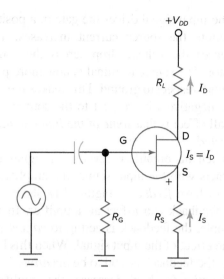

Fig. 7-16 An N-channel FET amplifier using source bias.

voltage should be −1.5 V (the sign is not used in the calculation). The source resistor should be

$$R_S = \frac{1.5 \text{ V}}{1.9 \text{ mA}} = 790 \text{ } \Omega$$

Note that the value of current used in the calculation is about half the saturation current. Check Fig. 7-14(*b*) and verify that this value of drain current is near the center of the load line.

The circuit of Fig. 7-16 is called a *source bias* circuit. The bias voltage is produced by source current flowing through the source resistor. The drop across the resistor makes the source terminal *positive* with respect to ground. The gate terminal is at ground potential. There is no gate current and therefore no drop across the gate resistor R_G. Thus, the source is positive with respect to the gate. To say it another way, the gate is negative with respect to the source. This accomplishes the same purpose as the separate supply V_{GS} in Fig. 7-14(*a*).

Source bias is much simpler than using a separate bias supply. The voltage gain does suffer, however. To see why, examine Fig. 7-17.

Source bias

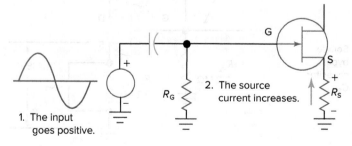

1. The input goes positive.

2. The source current increases.

3. The voltage drop across R_S increases.

4. Increased voltage drop across R_S makes the gate more negative with respect to the source.

5. Some of the input signal is canceled.

Fig. 7-17 Source feedback.

As the input signal drives the gate in a positive direction, the source current increases. This increases the voltage drop across the source resistor. The source terminal is now more positive with respect to ground. This makes the gate more negative with respect to the source. The overall effect is that *some of the input signal is canceled*.

When an amplifier develops a signal that interacts with the input signal, the amplifier is said to have *feedback*. Figure 7-17 shows one way feedback can affect an amplifier. In this example, the feedback is acting to *cancel* part of the effect of the input signal. When this happens, the feedback is said to be *negative*.

Negative feedback decreases the amplifier's gain. It is also capable of increasing the frequency range of an amplifier. Negative feedback may be used to decrease an amplifier's distortion. So, negative feedback is not good or bad—it is a mixture. If maximum voltage gain is required, the negative feedback will have to be eliminated. In Fig. 7-18 the source resistor is shunted with a *source bypass capacitor*. This capacitor will eliminate the negative feedback and increase the gain. It is selected to have low reactance at the signal frequency. It will prevent the source terminal voltage from swinging with the increases and decreases in source current. It has pretty much the same effect as the emitter bypass capacitor in the common-emitter amplifier studied before.

When we studied BJTs, we found them to have an unpredictable β. This made it necessary to investigate a circuit that was not as sensitive to β. Field-effect transistors also have characteristics

that vary widely from unit to unit. It is necessary to have circuits that are not as sensitive to certain device characteristics.

The circuit in Fig. 7-14(*a*) is called *fixed bias*. This circuit will work well only if the transistor has predictable characteristics. The fixed-bias circuit usually is not a good choice. The circuit in Fig. 7-16 uses source bias. It is much better and allows the transistor characteristics to vary. If, for example, the transistor tended toward more current, the source bias would automatically increase. More bias would reduce the current. Thus, we can see that the source resistor stabilizes the circuit.

The greater the source resistance, the more stability we can expect in the operating point. But too much source resistance could create too much bias, and the circuit will operate too close to cutoff. If there were some way to offset this effect, a better circuit would result. Figure 7-19 shows a way. This circuit uses *combination bias*. The bias is a combination of a fixed positive voltage applied to the gate terminal and source bias. The positive voltage is set by a voltage divider. The divider network is made up of R_{G_1} and R_{G_2}. These resistors will usually be high in value to maintain a high input impedance for the amplifier.

The combination-bias circuit can use a larger value for R_S, the source resistor. The bias voltage V_{GS} will not be excessive because a positive, fixed voltage is applied to the gate. This fixed, positive voltage will reduce the effect of the voltage drop across the source resistor.

A few calculations will show how the combination-bias circuit works. Assume in Fig. 7-19 that the desired source current is to

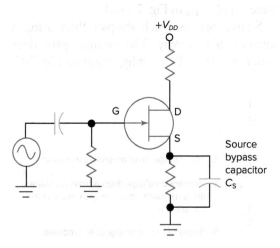

Fig. 7-18 Adding the source bypass capacitor.

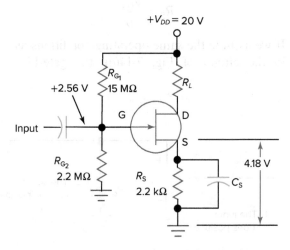

Fig. 7-19 An FET amplifier using combination bias.

be 1.9 mA. The voltage drop across R_S is

$$V_{R_S} = 1.9 \text{ mA} \times 2.2 \text{ k}\Omega$$
$$= 4.18 \text{ V}$$

Next, the voltage divider drop across R_{G_2} sets V_G at

$$V_G = \frac{2.2 \text{ M}\Omega}{2.2 \text{ M}\Omega + 15 \text{ M}\Omega} \times 20 \text{ V}$$
$$= 2.56 \text{ V}$$

Both of the above voltages are positive with respect to ground. The source voltage is *more* positive. The gate is therefore *negative* with respect to the source by an amount of

$$V_{GS} = 2.56 \text{ V} - 4.18 \text{ V} = -1.62 \text{ V}$$

In measuring bias voltage in circuits such as the one in Fig. 7-19, remember that V_G and V_{GS} are different. The gate voltage V_G is measured with respect to ground. The gate-source voltage V_{GS} is measured with respect to the source. Also, don't forget to take the loading effect of your meter into account. When measuring V_G in a high-impedance circuit like Fig. 7-19, a meter with a high input impedance is required for reasonable results.

Figure 7-20 shows a P-channel JFET amplifier. The supply voltage is *negative* with respect to ground. Note the direction of the source current I_S. The voltage drop produced by this current will reverse-bias the gate-source diode. This is proper, and no gate current will flow through R_G.

EXAMPLE 7-5

Determine the gate-source bias for Fig. 7-20 if the drain current is 2 mA and the source resistor is 860 Ω. Is there any difference when this bias is compared with Fig. 7-16? The bias is found using Ohm's law, and it is understood that the source current is equal to the drain current:

$$V_{GS} = I_S \times R_S = 2 \text{ mA} \times 860 \ \Omega = 1.72 \text{ V}$$

V_{GS} is positive in Fig. 7-20 and negative in Fig. 7-16.

It is possible to have linear operation with *zero bias*, as shown in Fig. 7-21. You should recognize the transistor as a MOSFET. The gate is

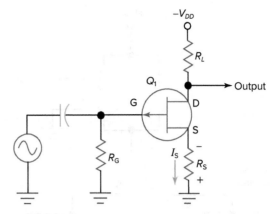

Fig. 7-20 A P-channel JFET amplifier.

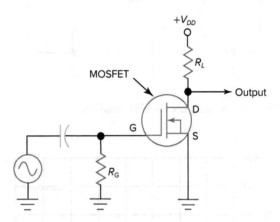

Fig. 7-21 A zero-bias MOSFET amplifier.

insulated from the source in this type of transistor. This prevents gate current regardless of the gate-source polarity. As the signal goes positive, the drain current will increase. As the signal goes negative, the drain current will decrease. This type of transistor can operate in both the *enhancement* and the *depletion* modes. This is *not* true with junction FETs. The zero-bias circuit in Fig. 7-21 is restricted to depletion-mode MOSFETs for linear work.

Figure 7-22 shows the schematic for a *dual-gate MOSFET* amplifier. The circuit uses tuned transformer coupling. Good gain is possible at frequencies near the resonant point of the transformers. The signal is fed to gate 1 (G_1) of the MOSFET. The output signal appears in the drain circuit. The gain of this amplifier can be controlled over a large range by gate 2 (G_2).

The graph in Fig. 7-23 shows a typical gain range for this type of amplifier. Note that maximum power gain, 20 dB, occurs when gate 2 is positive with respect to the source by about 3 V. At zero bias, the gain is only about 5 dB. As gate 2 goes negative with respect to the

Dual-gate MOSFET

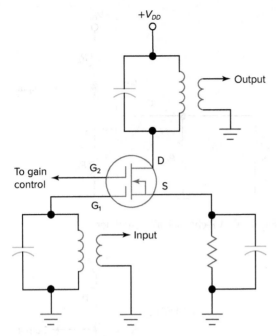

Fig. 7-22 A dual-gate MOSFET amplifier.

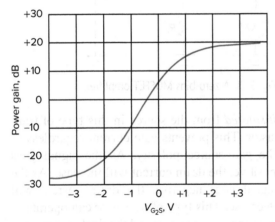

Fig. 7-23 The effect of V_{G_2} on gain.

source, the gain continues to drop. The minimum gain is about −28 dB. Of course, −28 dB represents a large loss.

The total range of gain for the amplifier is from +20 to −28 dB, or 48 dB. This means a power ratio of about 63,000:1. Thus, with the proper control voltage applied to G_2, the circuit in Fig. 7-22 can keep a constant output over a tremendous range of input levels. Amplifiers of this type are used in applications such as communications where a wide range of signal levels is expected.

Field-effect transistors have some very good characteristics. However, bipolar transistors usually cost less and give much better voltage gains. This makes the bipolar transistor the best choice for most applications.

Field-effect transistors are used when some special feature is needed. For example, they are a good choice if a very high input impedance is required. When used, FETs are generally found in the first stage or two of a linear system.

Examine Fig. 7-24. Transistor Q_1 is a JFET, and Q_2 is a BJT. Thanks to the JFET, the input impedance is high. Thanks to the bipolar device, the power gain is good and the cost is reasonable. This is typical of the way circuits are designed to have the best performance for the least cost.

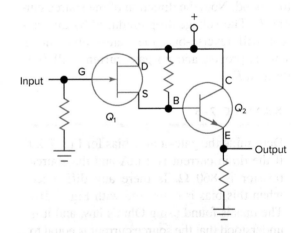

Fig. 7-24 Combining FET and bipolar devices.

Self-Test

Determine whether each statement is true or false.

24. Voltage-controlled amplifiers usually have a lower input impedance than current-controlled amplifiers.

25. Gate current is avoided in JFET amplifiers by keeping the gate-source junction reverse-biased.

26. A separate gate supply, such as the one shown in Fig. 7-14(a), is the best way to keep the gate junction reverse-biased.
27. The voltage gain of an FET amplifier is given in siemens.
28. Source bias tends to stabilize FET amplifiers.
29. Dual-gate MOSFETs are not used as linear amplifiers.

Solve the following problems.

30. Refer to Fig. 7-14(a). Assume $I_D = 3$ mA. Find V_{DS}.
31. Refer to Fig. 7-14(b). Assume $V_{GS} = 0$ V. Where would the transistor be operating?

32. Refer to Fig. 7-14(b). Assume $V_{GS} = -3.0$ V. Where would the transistor be operating?
33. Refer to Fig. 7-16. If $I_D = 1.5$ mA and $R_S = 1,000$ Ω, what is the value of V_{GS}?
34. Refer to Fig. 7-19. Assume $V_{DD} = 15$ V and $I_D = 2$ mA. What is the value of V_{GS}? What is the gate polarity with respect to the source?
35. Refer to Fig. 7-24. In what circuit configuration is Q_1 connected? Q_2?
36. Refer to Fig. 7-24. If the input is going in a positive direction, in what direction will the output go?

7-4 Negative Feedback

Figure 7-25 shows two block diagrams. In Fig. 7-25(a), block *A* represents the *open-loop gain* of the amplifier. This is the gain with *no negative feedback*. As an example, perhaps *A* = 50. With negative feedback, it is the *closed-loop gain* that determines V_{out}, and this gain is always less than *A*. So if *A* = 50, the closed-loop gain must be less than 50. Block *B* represents the feedback circuit that returns some of the output signal back to the input. The *feedback ratio* tells us how much of the output

Closed-loop gain

Open-loop gain

Feedback ratio

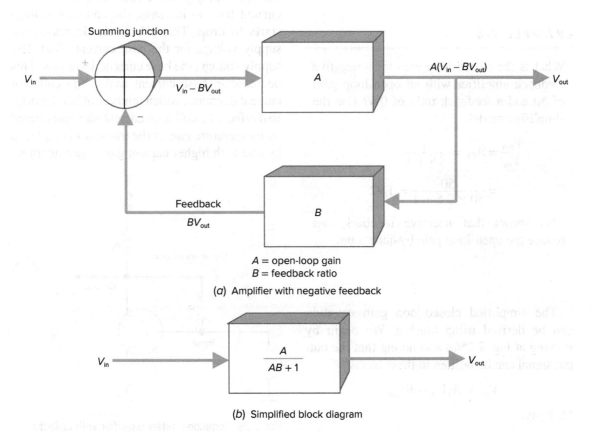

(a) Amplifier with negative feedback

(b) Simplified block diagram

Fig. 7-25 Negative feedback block diagrams.

is returned. If block B is a voltage divider formed with two equal resistors, then $B = 0.5$. The summing junction in Fig. 7-25(a) is where the feedback and V_{in} are put together (combined).

Figure 7-25(b) shows that it is possible to simplify the circuit. The simplification makes it easy to find the closed-loop gain.

Negative feedback must lower the gain of an amplifier, so it doesn't seem to provide any advantage. It is an excellent trade-off in many cases. Here are the reasons why negative feedback is used:

Collector feedback

DC feedback

1. *Stabilize an amplifier:* Make the gain and/or the operating point independent of device characteristics and temperature.
2. *Increase the bandwidth of an amplifier:* Make it provide useful gain over a broader range of frequencies.
3. *Improve the linearity of an amplifier:* Decrease the amount of signal distortion.
4. *Improve the noise performance of an amplifier:* Make the amplifier quieter.
5. *Change amplifier impedances:* Raise or lower the input or output impedance.

EXAMPLE 7-6

What is the closed-loop gain for a negative feedback amplifier with an open-loop gain of 50 and a feedback ratio of 0.5? Use the simplified model:

$$\frac{V_{out}}{V_{in}} = A_{CL} = \frac{A}{AB + 1}$$

$$= \frac{50}{50 \times 0.5 + 1} = 1.92$$

This shows that negative feedback can reduce the open-loop gain by quite a bit.

The simplified closed-loop gain equation can be derived using algebra. We begin by looking at Fig. 7-25(a) and noting that the output signal can be written in these terms:

$$V_{out} = A(V_{in} - BV_{out})$$

Multiply:

$$V_{out} = AV_{in} - ABV_{out}$$

Divide each term by V_{out}:

$$1 = \frac{AV_{in}}{V_{out}} - AB$$

Add AB to both sides:

$$AB + 1 = \frac{AV_{in}}{V_{out}}$$

Divide both sides by A:

$$\frac{AB + 1}{A} = \frac{V_{in}}{V_{out}}$$

Invert both sides:

$$\frac{A}{AB + 1} = \frac{V_{out}}{V_{in}} = A_{CL}$$

Figure 7-26 shows a common-emitter amplifier with *collector feedback*. Resistor R_F provides feedback from the collector terminal (the output) back to the base terminal (the input). Feedback resistor R_F provides both dc and ac feedback. The *dc feedback* stabilizes the amplifier's operating point. If temperature increases, the gain of the transistor increases and more collector current will flow. However, in Fig. 7-26, R_F is connected to the collector and not to the V_{CC} supply point. Thus, when the collector current tries to increase, the collector voltage starts to drop. This effectively decreases the supply voltage for the base current. With less supply voltage, the base current decreases. This decrease in base current makes the collector current decrease, which tends to offset the original effect of a collector current increase caused by temperature rise. If the transistor is replaced by one with higher current gain, once again R_F

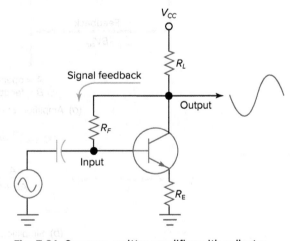

Fig. 7-26 Common-emitter amplifier with collector feedback.

would help to stabilize the circuit. Direct current feedback in an amplifier is helpful for keeping the operating point near its desired value. The emitter resistor in Fig. 7-26 provides additional feedback and improves stability of the operating point even more.

In Fig. 7-26, R_F also provides *ac feedback*. When the amplifier is driven with a signal, a larger out-of-phase signal appears at the collector of the transistor. This signal feeds back and reduces the ac current gain of the amplifier and decreases its input impedance. Suppose, for example, that the voltage gain of the amplifier is 50 and that R_F is a 100-kΩ resistor. Any voltage change at the signal source would cause the collector end of R_F to go 50 times in the opposite direction. This means that the *ac signal current* flow in R_F would be *50 times greater* than it would be if the resistor were connected to V_{CC}. The signal current in R_F is proportional to the signal voltage across it. The gain of the amplifier is 50; therefore, the signal current must be 50 times the value that would be predicted by the signal voltage and the value of R_F. It also means that R_F loads the signal source as if it were ¹⁄₅₀ of its actual value, or 2 kΩ in this example. Therefore, we see that the ac feedback in Fig. 7-26 causes extra signal current to flow in the input circuit. The current gain of the amplifier has been decreased, and its input impedance has been decreased. The voltage gain has not been changed by R_F.

The ac feedback may not be desirable. Figure 7-27 shows how it can be eliminated. The feedback resistor has been replaced with two resistors, R_{F_1} and R_{F_2}. The junction of these

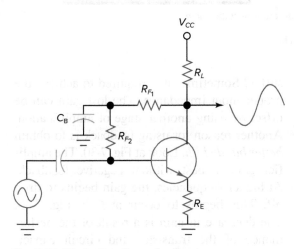

Fig. 7-27 Eliminating the ac feedback.

two resistors is bypassed to ground with capacitor C_B. This capacitor is chosen to have a very low reactance at the signal frequencies. It acts as a short circuit and prevents any ac feedback from reaching the base of the transistor. Now the input impedance and the current gain of the amplifier are both higher than in Fig. 7-26. The bypass capacitor in Fig. 7-27 has *no effect* on the dc feedback. Capacitive reactance is infinite at 0 Hz; therefore, the amplifier has the same operating point stability as discussed for Fig. 7-26.

In many cases, an emitter resistor provides enough feedback. In Fig. 7-28, the emitter resistor R_E provides both dc and ac feedback. The dc feedback acts to stabilize the operating point of the amplifier. Emitter resistor R_E is not bypassed, and a signal appears across it when the amplifier is driven. This signal is in phase with the input signal. If the signal source drives the base in a positive direction, the signal across R_E will also cause the emitter to go in a positive direction. This action reduces the base-emitter signal voltage and decreases the voltage gain of the amplifier. It also increases the input impedance of the amplifier as discussed earlier in this chapter. We know that R_E can be bypassed for improved voltage gain at the cost of decreased input impedance.

When very high input impedances are required, the circuit in Fig. 7-29 may be used. It is often called a *bootstrap circuit*. Capacitor C_B and resistor R_B provide a feedback path that works to increase the input impedance of the amplifier. An in-phase signal is developed across the unbypassed part of the emitter resistor. This signal is coupled by C_B to the right end of R_B. Since it is in phase with the input signal, it decreases the signal current flowing in R_B. For example, if the feedback signal at the right end of R_B were equal to the signal supplied by the source, there would be no signal voltage difference across R_B, and no signal current could flow in it. This would make R_B appear as an *infinite impedance* as far as signal currents are concerned. In actual circuitry, the feedback signal is lower in amplitude than the input signal; therefore, there is some signal current flow in R_B. However, for the input signal, R_B appears to be many times greater in impedance than its value in ohms. Since R_B is in series with the dc bias divider, it effectively isolates the signal

AC feedback

Bootstrap circuit

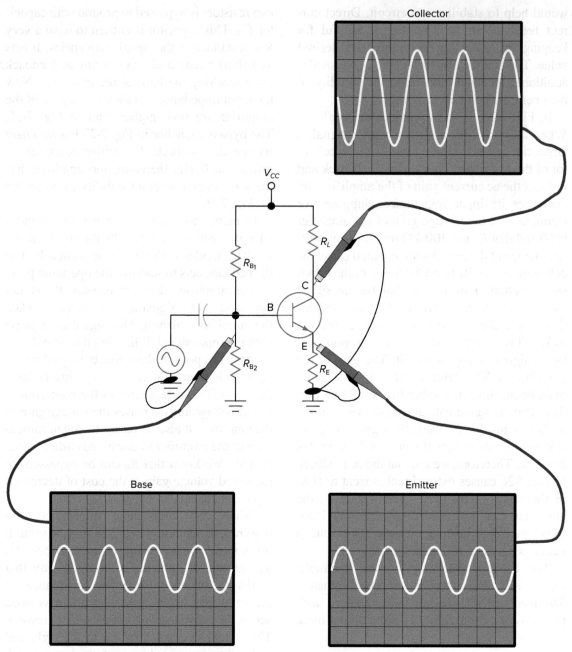

The base and emitter signals are almost the same.

Fig. 7-28 Common-emitter amplifier with emitter feedback.

source from the two bias resistors. A bootstrap amplifier, such as the one shown in Fig. 7-29, will have an input impedance of several hundred thousand ohms, whereas the circuit shown in Fig. 7-28 would be on the order of several thousand ohms.

So far we have seen that negative feedback in an amplifier can decrease current gain or voltage gain. We have seen that it can lower or raise the input impedance of the amplifier. Why use feedback if gain (current or voltage) must

suffer? Sometimes it is required to achieve the proper input impedance. The lost gain can be offset by using another stage of amplification. Another reason for using feedback is to obtain *better bandwidth*. Look at Fig. 7-30. The amplifier gain is best without negative feedback. At higher frequencies, the gain begins to drop off. This begins to occur at f_1 in Fig. 7-30. The decrease in gain is a result of the performance of the transistor and circuit capacitance. All transistors show less gain as the

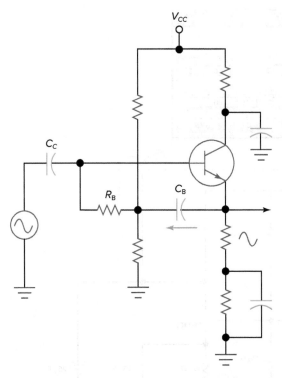

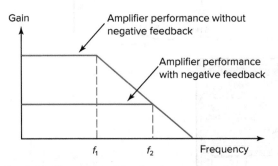

Fig. 7-29 Bootstrapping for high input impedance.

Fig. 7-30 The effect of feedback on gain and bandwidth.

frequency increases. The reactance of a capacitor decreases as the frequency increases. This decreasing reactance loads the amplifier, and the gain drops.

Now, look at the performance of the amplifier in Fig. 7-30 with negative feedback. The gain is much less at the lower frequencies, but it does not start dropping off until f_2 is reached. Frequency f_2 is much higher than f_1. With negative feedback, the amplifier provides less gain, but the gain is constant over a wider frequency range. The loss of gain in one stage can easily be overcome by adding another stage of amplification.

Negative feedback also *reduces noise and distortion*. Suppose the signal picks up some

noise or distortion in the amplifier. This appears on the output signal. Some of the output signal is fed back to the input. The noise or distortion will be placed on the input signal in an opposite way. Remember, the feedback is out of phase with, or opposite to, the input. Much of the noise and distortion is canceled in this way. By intentionally distorting the input signal, opposite to the way the amplifier distorts it, the circuit becomes more linear.

EXAMPLE 7-7

Verify the negative feedback gain in Table 7-2. A look at Fig. 7-31 shows that the negative feedback is applied to Q_1's source resistor, which forms a voltage divider with the 5.1-kΩ resistor. The feedback ratio is found with the familiar voltage divider equation:

$$B = \frac{180 \ \Omega}{180 \ \Omega + 5.1 \text{ k}\Omega} = 0.0341$$

Now, we can apply the closed-loop gain equation that was developed earlier. The open-loop gain value is found in Table 7-2 ($A_{OL} = 240$):

$$A_{CL} = \frac{A}{AB + 1}$$

$$= \frac{240}{(240)(0.0341) + 1} = 26.1$$

This agrees with the value in the table. The decrease in gain is the price paid for low distortion and wide bandwidth.

Figure 7-31 shows a cascade amplifier with switchable negative feedback. The feedback is for alternating current only because the 25-μF coupling capacitor blocks direct current. The top oscilloscope display shows the performance with negative feedback. Notice that both waveforms are triangular, with no apparent distortion. The bottom display shows performance without negative feedback. The output triangle wave is distorted. Table 7-2 compares performance with and without feedback. Notice that the *bandwidth* is also much better with negative feedback: bandwidth $= f_H - f_L$.

Triangle waves are often used when checking for clipping or other forms of distortion. Sine waves are not as useful for distortion checks. It is much easier to notice deviations from straight lines.

Noise and distortion

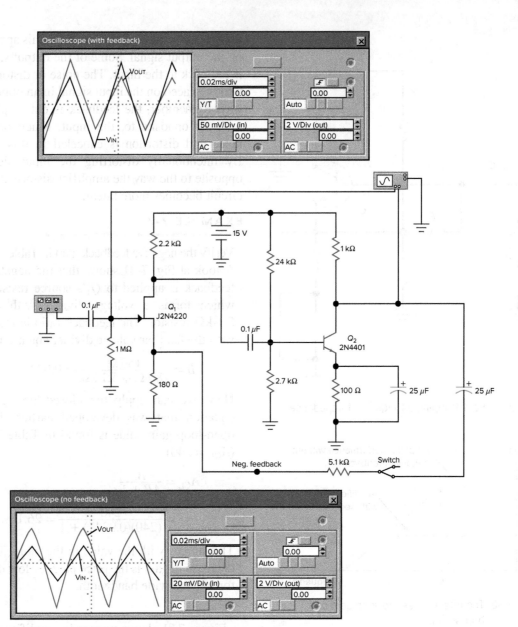

Fig. 7-31 A two-stage amplifier with switchable negative feedback.

Table 7-2	Performance with and without Feedback	
	With Feedback	Without Feedback
A_V	26 (28 dB)	240 (48 dB)
f_L	150 Hz	1.2 kHz
f_H	1.9 MHz	190 kHz
Distortion	Low	High

Figure 7-32 shows an NPN-PNP amplifier with negative feedback. The PNP transistor shares its collector circuit with the emitter circuit of the NPN transistor. There is both dc and ac feedback in this amplifier.

EXAMPLE 7-8

Find the dc conditions for the amplifier shown in Fig. 7-32. Begin with the base voltage at the input:

$$V_{B(NPN)} = \frac{5.1\ k\Omega}{5.1\ k\Omega + 27\ k\Omega} \times 12\ V$$

$$= 1.91$$

$$I_{(560\ \Omega)} = \frac{1.91\ V - 0.7\ V}{560\ \Omega} = 2.16\ mA$$

This is *not* the current flow in the NPN transistor because both transistors share this resistor. Notice in Fig. 7-32 that the base-emitter junction of the PNP transistor is in *parallel* with a 680-Ω resistor. As always, it is reasonable to assume a 0.7-V drop, so

$$I_{680\,\Omega} = \frac{0.7\text{ V}}{680\ \Omega} = 1.03\text{ mA}$$

We usually assume that $I_E = I_C$, so it's possible to subtract this current from 2.16 mA to determine the PNP transistor current:

$$I_{C(PNP)} = I_{E(PNP)} = 2.16\text{ mA} - 1.03\text{ mA}$$

$$= 1.13\text{ mA}$$

Finish by finding the dc output voltage of the amplifier. First, the drop across the 4.7-kΩ resistor is

$$V_{4.7\,k\Omega} = 1.13\text{ mA} \times 4.7\text{ k}\Omega = 5.31\text{ V}$$

$$V_{out} = V_{560\,\Omega} + V_{4.7\,k\Omega}$$

$$= 1.21\text{ V} + 5.31\text{ V} = 6.52\text{ V}$$

The fact that this voltage is close to half the supply voltage tells us that the circuit is biased for linear operation. The method used here is based on several assumptions, and there is some error. As Fig. 7-32 shows, a circuit simulator reports different dc conditions—but not drastically different.

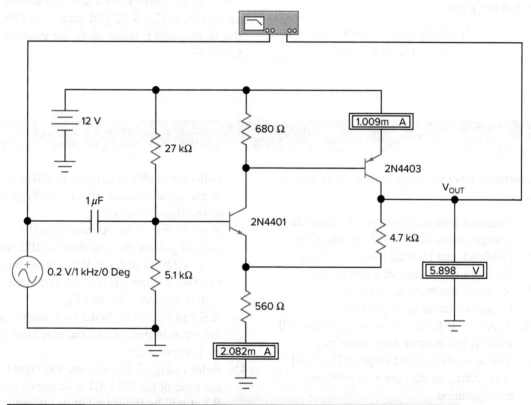

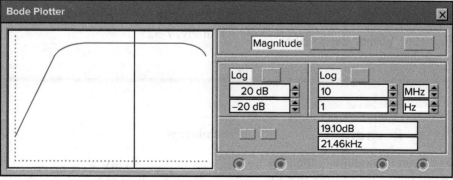

Fig. 7-32 NPN-PNP amplifier with series feedback.

EXAMPLE 7-9

Find the dB closed-loop gain for Fig. 7-32 and compare it with the simulation. Begin by finding the gain of each stage. The emitter resistor for the NPN transistor is large and is not bypassed, so the gain of this stage is

$$A_{V(\text{NPN})} = \frac{R_L}{R_E} = \frac{680 \ \Omega}{560 \ \Omega} = 1.21$$

The PNP transistor has no emitter resistor, so to find the gain, we need r_E first:

$$r_E = \frac{25}{I_E} = \frac{25}{1.13} = 22.1 \ \Omega$$

$$A_{V(\text{PNP})} = \frac{R_L}{R_E} = \frac{4.7 \ k\Omega}{22.1 \ \Omega} = 213$$

The open-loop gain for Fig. 7-32 is the product of the gains:

$$A_{\text{OL}} = A_{V(\text{NPN})} \times A_{V(\text{PNP})}$$

$$= 1.21 \times 213 = 258$$

The closed-loop gain is less because of the feedback. The feedback ratio is set by the voltage division produced by the 560-Ω and 4.7-kΩ resistors:

$$B = \frac{560 \ \Omega}{560 \ \Omega + 4.7 \ \Omega} = 0.106$$

The closed-loop gain is

$$A_{\text{CL}} = \frac{A}{AB + 1}$$

$$= \frac{258}{(258)(0.106) + 1} = 9.11$$

Convert to dB:

$$= 20 \times \log 9.11 = 19.2 \ \text{dB}$$

This agrees closely with the simulation of the circuit in Fig. 7-32 (the gain is shown on the Bode plotter. Bode plots are covered in Chap. 9).

Self-Test

Determine whether each statement is true or false.

37. Negative feedback always decreases the voltage or current gain of an amplifier.
38. Feedback can be used to raise or lower the input impedance of an amplifier.
39. Negative feedback decreases the frequency range of amplifiers.
40. Negative dc feedback in an amplifier will make it less temperature-sensitive.
41. The second amplifier stage in Fig. 7-31 is operating in the common-collector configuration.

Solve the following problems.

42. Refer to Fig. 7-26. Assume the voltage gain of the amplifier is 50 and the

collector feedback resistor is 100 kΩ. What signal loading effect will R_F present to the signal source?
43. Refer to Fig. 7-28. Assume that the current gain of the transistor is 100 and R_E is a 220-Ω resistor. What is the base-to-ground impedance r_{in} for the input signal, ignoring R_{B_1} and R_{B_2}?
44. Refer to Fig. 7-28. What will happen to the input impedance of the amplifier if R_E is bypassed?
45. Refer to Fig. 7-32. Assume the signal at the base of the 2N4401 is negative-going. What will be the signal at its collector?
46. What is the configuration of the 2N4403 in Fig. 7-32?

7-5 Frequency Response

Figure 7-33 shows a common-emitter amplifier. A dc analysis of this circuit determines that the emitter current is 8.38 mA. The ac resistance of the emitter is

$$r_E = \frac{50 \ \text{mV}}{8.38 \ \text{mA}} = 5.96 \ \Omega$$

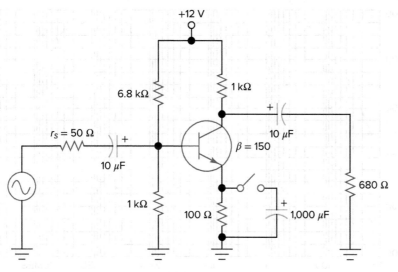

Fig. 7-33 A common-emitter amplifier.

With the emitter-bypass switch open, the voltage gain of the amplifier is

$$A_V = \frac{1\ \text{k}\Omega \,\|\, 680\ \Omega}{100\ \Omega + 5.96\ \Omega} = 3.82$$

To get a more precise gain estimate, the internal resistance of the signal source must be taken into account. The amplifier's input impedance causes a loading effect and some loss of signal voltage across r_S in Fig. 7-33. For this loss to be found, the input impedance of the amplifier must be known. The first step is to find the input resistance of the transistor:

$$r_{\text{in}} = \beta(R_E + r_E)$$
$$= 150(100\ \Omega + 5.96\ \Omega) = 15.9\ \text{k}\Omega$$

The input impedance of the amplifier is determined next:

$$Z_{\text{in}} = \cfrac{1}{\cfrac{1}{R_{B_1}} + \cfrac{1}{R_{B_2}} + \cfrac{1}{r_{\text{in}}}}$$
$$= \cfrac{1}{\cfrac{1}{6.8\ \text{k}\Omega} + \cfrac{1}{1\ \text{k}\Omega} + \cfrac{1}{15.9\ \text{k}\Omega}} = 826\ \Omega$$

Some fraction of the signal source voltage in Fig. 7-33 will reach the base of the transistor. Using the voltage divider equation gives

$$V_{\text{fraction}} = \frac{826}{826 + 50} = 0.943$$

So, the unbypassed voltage gain of the common-emitter amplifier in Fig. 7-33 is

$$A_V = 0.943 \times 3.82 = 3.60$$

And the dB gain is

$$A_V = 20 \times \log_{10} 3.60 = 11.1\ \text{dB}$$

Figure 7-34 shows the gain-versus-frequency response for the amplifier. The lower curve is for the unbypassed condition. In the *midband* of the amplifier, the gain is 11.1 dB. Note that the gain curve starts to drop at frequencies above and below the midband.

The curves in Fig. 7-34 are plotted on a *semi-log graph*. The frequency axis is logarithmic, and the vertical axis is linear. If both axes were logarithmic, the graph would be called *log-log*. One reason for using a logarithmic axis is to obtain good resolution over a large range.

Frequency response graphs show the range over which an amplifier is useful. Generally, the *bandwidth* of an amplifier is the range of frequencies where the gain of the amplifier is within 3 dB of its midband gain. An examination of the lower curve of Fig. 7-34 shows that the gain is down 3 dB around frequencies of 20 Hz and 40 MHz. So the amplifier is useful from 20 Hz to 40 MHz, and the bandwidth is just a little less than 40 MHz.

The point at which an amplifier's gain drops 3 dB from its best gain is sometimes called a *break frequency*. The lower break frequency is caused by the capacitors in Fig. 7-33. In the midband of the amplifier, the capacitors can be viewed as ac short circuits. At a break frequency, a capacitor has a reactance *equal* to the

Break frequency

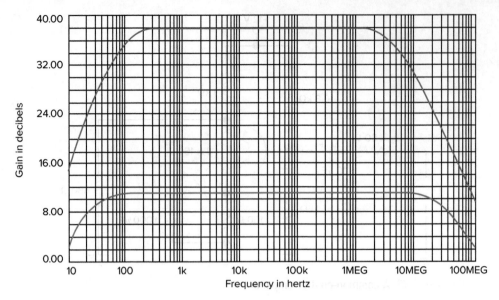

Fig. 7-34 Frequency response of common-emitter amplifier.

equivalent resistance that it is coupling or bypass-ing. In other words, at the break frequency,

$$X_C = R_{eq}$$

The capacitive reactance at the break frequency (f_b) is

$$X_C = \frac{1}{2\pi f_b C}$$

Substituting:

$$R_{eq} = \frac{1}{2\pi f_b C}$$

Solving for f_b:

$$f_b = \frac{1}{2\pi R_{eq} C}$$

The preceding equation shows that it is not dif-ficult to calculate a break frequency if the equiv-alent resistance and capacitance are known.

The 10-μF output coupling capacitor in Fig. 7-33 will cause a 3-dB drop in gain at some frequency. This capacitor couples into 680 Ω. It is fed by the collector circuit of the amplifier, which has an output impedance of 1 kΩ (it is equal to R_L). The equivalent resistance loading the output capacitor is

$$R_{eq} = 1\ \text{k}\Omega + 680\ \Omega = 1.68\ \text{k}\Omega$$

The break frequency is

$$f_b = \frac{1}{2\pi R_{eq} C}$$

$$= \frac{1}{6.28 \times 1.68\ \text{k}\Omega \times 10 \times 10^{-6}\ \text{F}}$$

$$= 9.47\ \text{Hz}$$

The other 10-μF capacitor in Fig. 7-33 cou-ples into the amplifier. The input impedance has already been calculated to be 826 Ω. The internal resistance of the signal source is 50 Ω. The equivalent resistance for the input circuit is

$$R_{eq} = 50\ \Omega + 826\ \Omega = 876\ \Omega$$

The break frequency is

$$f_b = \frac{1}{6.28 \times 876\ \Omega \times 10 \times 10^{-6}\ \text{F}}$$

$$= 18.2\ \text{Hz}$$

This is higher than the break frequency for the output circuit. The highest number is the one used to determine the low-frequency response. In this case, the break frequencies are close (9.47 and 18.2 Hz). When this happens, the actual break point of the amplifier will be higher than the highest individual break frequency. The circuit in Fig. 7-33 has a break frequency around 24 Hz.

The upper break frequency of a common-emitter amplifier such as the one shown in Fig. 7-33 is partly determined by capacitances that do not show on the schematic. Transistors have internal junction capacitances that act to bypass high frequencies. This topic is covered in some detail in Chap. 9.

The performance of the common-emitter amplifier in Fig. 7-33 changes quite a bit when the switch is closed. The switch connects an emitter bypass capacitor. This capacitor raises the gain, decreases the input impedance of the amplifier, and decreases the bandwidth of the amplifier. The gain with R_E bypassed is

$$A_V = \frac{1 \text{ k}\Omega \,\|\, 680 \,\Omega}{5.96 \,\Omega} = 67.9$$

The input impedance with R_E bypassed:

$$r_{in} = \beta \times r_E = 150 \times 5.96 \,\Omega = 894 \,\Omega$$

$$Z_{in} = \frac{1}{1/R_{B_1} + 1/R_{B_2} + 1/r_{in}}$$

$$= \frac{1}{1/6.8 \text{ k}\Omega + 1/1 \text{ k}\Omega + 1/894 \,\Omega}$$

$$= 441 \,\Omega$$

As before, using the voltage divider equation,

$$V_{fraction} = \frac{441}{441 + 50} = 0.898$$

So, the bypassed voltage gain of the common-emitter amplifier in Fig. 7-33 is

$$A_V = 0.898 \times 67.9 = 61.0$$

And the dB gain is

$$A_V = 20 \times \log_{10} 61.0 = 35.7 \text{ dB}$$

The higher curve in Fig. 7-34 shows the mid band gain to be 38 dB. We have used the conservative value of 50 mV to find the ac emitter resistance. The actual gain of the amplifier tends to be somewhat higher.

The breakpoint for the output coupling capacitor remains the same (9.47 Hz). The break point changes for the input coupling capacitor because the input impedance of the amplifier is lower:

$$R_{eq} = 50 \,\Omega + 441 \,\Omega = 491 \,\Omega$$

The break frequency is now

$$f_b = \frac{1}{6.28 \times 491 \,\Omega \times 10 \times 10^{-6} \text{ F}}$$

$$= 32.4 \text{ Hz}$$

Another lower break frequency is caused by the emitter bypass capacitor. We must find the equivalent resistance bypassed by this capacitor. This resistance is partly determined by the base circuit, as viewed from the emitter terminal. This is sort of a backward view through the

amplifier, and we find that resistors appear smaller by a factor equal to β:

$$r_{EB} = \frac{r_S \,\|\, R_{B_1} \,\|\, R_{B_2}}{\beta}$$

$$= \frac{50 \,\Omega \,\|\, 6.8 \text{ k}\Omega \,\|\, 1 \text{ k}\Omega}{150}$$

$$= 0.315 \,\Omega$$

This resistance is in series with r_E, and that combination is in parallel with R_E:

$$R_{eq} = (r_{EB} + r_E) \,\|\, R_E$$

$$= (0.315 \,\Omega + 5.96 \,\Omega \,\|\, 100 \,\Omega)$$

$$= 5.90 \,\Omega$$

$$f_b = \frac{1}{6.28 \times 5.9 \,\Omega \times 1,000 \times 10^{-6} \text{ F}}$$

$$= 27.0 \text{ Hz}$$

As Fig. 7-34 shows, the lower break frequency is about 80 Hz. Because the three break frequencies are so close to each other (9.47, 32.4, and 27.0 Hz), the actual breakpoint is greater than the highest of the three.

Please notice that the bandwidth suffers in Fig. 7-34 when the gain is increased by the addition of the emitter bypass capacitor. As discussed in the preceding section, the negative feedback provided by using emitter feedback makes the amplifier useful over a wider range of frequencies.

The upper break frequency in Fig. 7-34 drops from 40 to 4 MHz when the capacitor is added. This is because the increased gain multiplies the effect of the internal capacitance of the transistor. When designers need very-wide-band amplifiers, they often keep the gain in any one stage at a moderate value and add an additional stage or two to obtain the overall gain required.

Figure 7-35 shows a *cascode amplifier*. This is a special cascade arrangement in which a common-emitter stage is direct-coupled to a common-base stage. Cascode amplifiers extend the frequency limitations of common-emitter amplifiers by decreasing the effect of one of the internal capacitances in transistors. This allows extended bandwidth. Cascode amplifiers are used in radio-frequency applications.

Common-base amplifiers have extended bandwidth, but they also have a very low input impedance. The cascode arrangement is suited to situations in which the extended frequency

Cascode amplifier

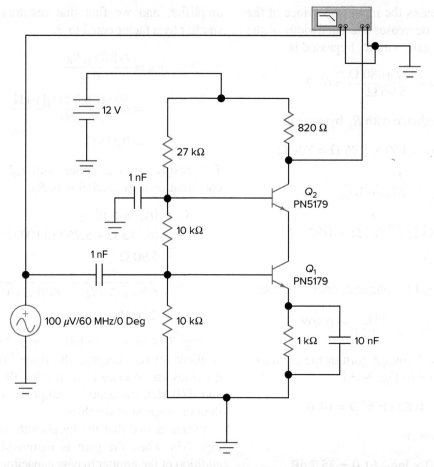

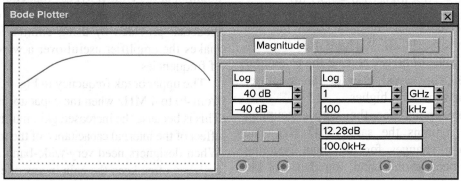

Fig. 7-35 A cascode amplifier.

performance of a common-base amplifier is desired along with the higher input impedance of a common-emitter amplifier. In Fig. 7-35, Q_1 is the common-emitter input amplifier, and Q_2 is the common-base output amplifier. The cascode arrangement provides high-gain (almost 33 dB), wide bandwidth (almost 300 MHz), and an input impedance in the kilohm range.

Self-Test

Solve the following problems.

47. Suppose the amplifier in Fig. 7-33 has an input impedance of 600 Ω and a midband voltage gain of 10. Find the midband output voltage from the amplifier if the signal source has an impedance of 600 Ω and develops 100 mV$_{\text{p-p}}$.

48. Find the lower break frequency for the input circuit in question 47 if the input coupling capacitor is 0.1 μF.
49. What is the midband dB gain of the amplifier described in question 47?
50. What is the gain of the amplifier in question 47 at its break frequency?
51. Select an input coupling capacitor for the amplifier described in question 47 that will change its break frequency to 10 Hz.
52. An amplifier has three capacitors. Calculations reveal break frequencies of 10, 15, and 150 Hz. What is the lower break frequency for the entire amplifier?
53. An amplifier has three capacitors. Calculations reveal break frequencies of 135, 140, and 150 Hz. What is the lower break frequency for the entire amplifier?
54. Determine the dc conditions for Fig. 7-35.

7-6 Positive Feedback

Positive feedback is the opposite of negative feedback. Some of the output is fed back to the input so as to add to or enhance the input signal. It can be used to

- Decrease bandwidth
- Increase gain
- Create a signal via oscillation (covered in Chap. 11)
- Change amplifier impedances
- Reduce the effect of noise

Here we will cover only the last item listed above.

Transistors used as switches were covered in Chap. 5. In the circuits discussed there, the input to the switch varied over a relatively large range between the off and on threshold conditions (perhaps 0 V for off and 5 V for on). Suppose a switch is needed that changes from off to on over a very narrow range (perhaps from 0.7 V for off to 0.6 V for on). Figure 7-36 shows a two-stage transistor switching circuit with a very small difference between the input thresholds for off and on.

The gain of the two stages in Fig. 7-36 makes the threshold very sharp (well defined). When the control voltage reaches about 0.7 V, U_1 turns on, and its collector voltage drops to around 0.2 V as it goes into hard saturation. Thus, the base-emitter circuit of U_2 is no longer forward-biased, and it turns off. Figure 7-37 shows the control voltage (white) ramping in a positive direction until the threshold is reached, and then the collector voltage of U_2 (yellow) rises rapidly from 0 to 12 V. Later, when the control voltage drops below the

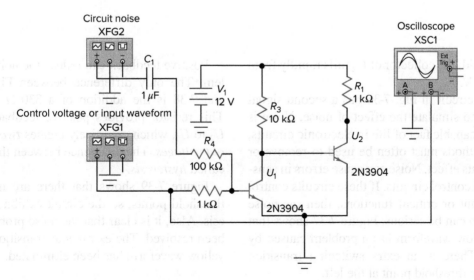

Fig. 7-36 Switching (or wave-shaping) circuit.

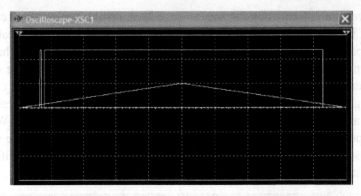

Fig. 7-37 Input and output waveforms for the transistor switching circuit.

Source: Multisim

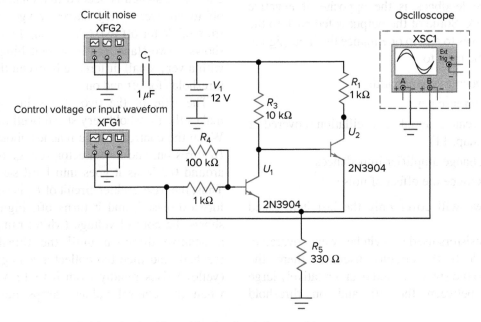

Fig. 7-38 Switching circuit with positive feedback (hysteresis).

threshold, the collector of U_2 falls rapidly from 12 to 0 V.

The circuit in Fig. 7-36 has a second signal source to simulate the effect of noise. Noise is an inescapable fact of life in electronic circuits, and methods must often be used to remove or reduce its effect. Noise can cause errors in sensing and control circuits. If these circuits control important or critical functions, then the noise problem can be serious. Figure 7-37 shows that the yellow waveform has a problem caused by noise. There is an extra switching transition near the threshold point at the left.

Positive feedback can reduce the noise problem. The only difference between Figs. 7-36 and 7-38 is the addition of a 330-Ω resistor. This resistor provides positive feedback from U_2 to U_1, which effectively creates *two threshold voltages*. The difference between the two is called *hysteresis*.

Figure 7-39 shows that there are now two threshold points, so the circuit exhibits hysteresis. Also, it is clear that the noise problem has been resolved. The extraneous transition in the yellow waveform has been eliminated.

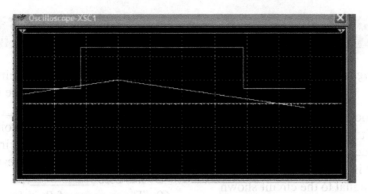

Fig. 7-39 Input and output waveforms for the switching circuit with hysteresis.

Source: Multisim

EXAMPLE 7-10

Determine the threshold voltages for Fig. 7-36, and corroborate them with the white waveform in Fig. 7-37, assuming that the oscilloscope displays 5 V per division.

Because there is no feedback (the emitters are grounded), the threshold voltages are simply $V_{BE} = 0.7$, since the transistors are silicon. This agrees with the white waveform in Fig. 7-37.

EXAMPLE 7-11

Determine the threshold voltages for Fig. 7-38, and corroborate them with the white waveform in Fig. 7-39, assuming that the oscilloscope is set at 5 V per division.

Starting with the case where U_2 is on, the voltage divider formed by R_1 and R_5 and the 12-V supply determine V_E for both transistors (we can ignore V_{CE} for U_2 as it is in hard saturation),

$$V_E = 12 \text{ V} \times \frac{330 \text{ }\Omega}{1,330 \text{ }\Omega} = 3 \text{ V}$$

$V_{BE(U_1)} = 3 \text{ V} + 0.7 \text{ V} = 3.7 \text{ V}$ (which agrees with the white waveform in Fig. 7-39, the left trip point)

Continuing with the case when U_1 is on, the voltage divider formed by R_3 and R_5 and the

12-V supply determines V_E for both transistors (as before, we can ignore V_{CE} for U_1 as it is in hard saturation):

$$V_E = 12 \text{ V} \times \frac{330 \text{ }\Omega}{10,330 \text{ }\Omega} = 0.4 \text{ V}$$

$V_{BE(U_1)} = 0.4 \text{ V} + 0.7 \text{ V} = 1.1 \text{ V}$ (which agrees with the white waveform in Fig. 7-39, the right trip point)

EXAMPLE 7-12

Determine the hysteresis voltages for Figs. 7-36 and 7-38.

The hysteresis voltage is zero for Fig. 7-36 because both thresholds are 0.7 V and for Fig. 7-38:

$$V_{hysteresis} = 3.7 \text{ V} - 1.1 \text{ V} = 2.6 \text{ V}$$

Circuits like the ones discussed here are sometimes called wave shapers. Regardless of the input waveform, assuming it is large enough to cross the thresholds, the output waveform will be rectangular. With hysteresis, the output waveform will be free of extra transitions, assuming that the noise voltage is less than the hysteresis voltage. Operational amplifiers and/or comparators can be operated with positive feedback to provide hysteresis. This will be discussed in Chap. 9.

Supply the missing word or words in each statement.

55. When feedback is used to decrease bandwidth, it is _____.

56. When feedback is used to create hysteresis, it is _____.

57. If the input signal to the circuit shown in Fig. 7-38 is changed from triangular to sinusoidal, the output signal will be _____.

58. A transistor switching circuit with positive feedback will not be subject to noise provided that the _____ voltage is greater than the noise voltage.

59. A transistor switching circuit with identical on and off threshold voltages has zero _____.

60. The purpose of R_5 in Fig. 7-38 is to provide _____ _____ from U_2 to U_1.

Chapter 7 Summary and Review

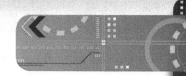

Summary

1. There are three basic types of amplifier coupling: capacitive, direct, and transformer.
2. Capacitor coupling is useful in ac amplifiers because a capacitor will block direct current and allow the ac signal to be coupled.
3. Electrolytic coupling capacitors must be installed with the correct polarity.
4. Direct coupling provides dc gain. It can be used only when the dc terminal voltages are compatible.
5. A Darlington amplifier uses direct coupling. Darlington transistors have high current gain.
6. Transformer coupling gives the advantage of impedance matching.
7. The impedance ratio of a transformer is the square of its turns ratio.
8. Radio-frequency transformers can be tuned to give selectivity. Those frequencies near the resonant frequency will receive the most gain.
9. Loading a signal source will reduce its voltage output. This effect is often most noticeable with high-impedance signal sources.
10. In multistage amplifiers, each stage is loaded by the input impedance of the next stage.
11. When an amplifier is loaded, its voltage gain decreases.
12. Emitter bypassing in a common-emitter amplifier increases voltage gain but lowers the amplifier's input impedance.
13. Loaded amplifiers should be biased at the center of the signal load line for optimal linear output swing.
14. The maximum output swing from a loaded amplifier is less than the supply voltage. Its value can be found by drawing a signal load line or analyzing the output circuit.
15. It is necessary to reverse-bias the gate-source junction for linear operation with junction field-effect transistors.
16. Because there is no gate current, the input impedance of FET amplifiers is very high.
17. The voltage gain in an FET amplifier is approximately equal to the product of the load resistance and the forward transfer admittance of the transistor.
18. More load resistance means more voltage gain.
19. Fixed bias, in FET amplifiers, is not desirable because the characteristics of the transistor vary quite a bit from unit to unit.
20. Source bias tends to stabilize an FET amplifier and make it more immune to the characteristics of the transistor.
21. Combination bias uses fixed and source bias to make the circuit even more stable and predictable.
22. A common-source FET amplifier using source bias must use a source bypass capacitor to realize maximum voltage gain.
23. The dual-gate MOSFET amplifier is capable of a tremendous range of gain by applying a control voltage to the second gate.
24. When the feedback tends to cancel the effect of the input to an amplifier, that feedback is negative. Another way to identify negative feedback is that it is out of phase with the input.
25. Direct current negative feedback can be used to stabilize the operating point of an amplifier.
26. Negative signal feedback may be used to decrease the current gain or the voltage gain of an amplifier.
27. Negative signal feedback may be used to decrease or increase the input impedance of an amplifier.
28. Negative signal feedback increases the bandwidth of an amplifier. It also reduces noise and distortion.

Related Formulas

Darlington current gain: $A_1 = \beta_1 \times \beta_2$

Transformers: Turns ratio $= \dfrac{N_P}{N_S}$

Impedance ratio $= $ (turns ratio)^2

AC emitter resistance: $r_E = \dfrac{25\ \text{mV}}{I_E}$

AC voltage gain: $A_V = \dfrac{R_L}{r_E}$ (emitter at ac ground)

(for common emitter amp)

AC input resistance: $r_{in} = \beta(R_E + r_E)$
(R_E not bypassed)

Amplifier input impedance:

$$Z_{in} = \dfrac{1}{1/R_{B_1} + 1/R_{B_2} + 1/r_{in}}$$

Parallel load: $R_P = \dfrac{R_1 \times R_2}{R_1 + R_2}$

Voltage gain: $A_V = \dfrac{R_P}{R_E + r_E}$ or $\dfrac{R_P}{r_E}$

(parallel load)

Cascade gain: $A_{V(total)} = A_{V_1} \times A_{V_2}$

Forward transfer admittance:

$$Y_{fs} = \dfrac{\Delta I_D}{\Delta V_{GS}}\bigg| V_{DS}$$

Voltage gain: $A_V \approx Y_{fs} \times R_L$

Voltage gain: $A_V = \dfrac{A}{AB + 1}$

(with negative feedback)

Break frequency: $f_b = \dfrac{1}{2\pi R_{eq}C}$

Chapter Review Questions

Determine whether each statement is true or false.

7-1. When electrolytic capacitors are used as coupling capacitors, they may be installed without checking polarity. (7-1)

7-2. Capacitors couple alternating current and block direct current. (7-1)

7-3. If an early stage in a direct-coupled amplifier drifts with temperature, the drift is amplified by following stages. (7-1)

7-4. A Darlington transistor has three leads but contains two BJTs. (7-1)

7-5. A transformer can match a high-impedance collector circuit to a low-impedance load. (7-1)

7-6. Refer to Fig. 7-6. This amplifier will provide a little gain at 0 Hz. (7-1)

7-7. Amplifier input impedance can be ignored when the signal source is ideal (0 internal impedance). (7-1)

7-8. Refer to Fig. 7-9. It can be determined that the input impedance of the first amplifier cannot be greater than 8.2 kΩ by inspection. (7-2)

7-9. Refer to Fig. 7-10. The Q point is also called the operating point. (7-2)

7-10. Refer to Fig. 7-10. The amplifier can develop a maximum output swing of 12 V peak-to-peak. (7-2)

7-11. Refer to Fig. 7-14(*b*). If $V_{GS} = -2.5$ V, the positive-going portion of the output signal will be severely clipped. (7-3)

7-12. Refer to Fig. 7-16. Increasing the value of R_L should increase the voltage gain of the amplifier. (7-3)

7-13. Refer to Fig. 7-18. The effect of capacitor C_S is to increase the voltage gain. (7-3)

7-14. Refer to Fig. 7-20. Transistor Q_1 is in the source follower configuration. (7-3)

7-15. Refer to Fig. 7-21. The bias V_{GS} is -1.5 V. (7-3)

7-16. Refer to Figs. 7-22 and 7-23. To decrease the gain of the amplifier, G_2 must be made more negative with respect to the source. (7-3)

7-17. Refer to Fig. 7-24. The input terminal and the output terminal should be 180 degrees out of phase. (7-4)

7-18. Negative feedback tends to cancel the input signal. (7-4)

7-19. Negative feedback increases the voltage gain of the amplifier but at the expense of reduced bandwidth. (7-4)

7-20. Negative feedback improves amplifier linearity. (7-4)

7-21. Negative dc feedback can be used to stabilize the amplifier operating point. (7-4)

7-22. Refer to Fig. 7-33. There is more amplifier distortion when the switch is closed. (7-5)

7-23. Refer to Fig. 7-33. There is wider frequency response when the switch is open. (7-5)

Chapter Review Problems

7-1. Refer to Fig. 7-4. Assume that the 220-kΩ resistor is changed to 470 kΩ. Find the base voltage for Q_1. (7-1)

7-2. Using the data from problem 7-1, find the emitter voltage of Q_1. (7-1)

7-3. Using the data from problem 7-1, find the emitter voltage of Q_2. (7-1)

7-4. Using the data from problem 7-1, find the emitter current for Q_2. (7-1)

7-5. Using the data from problem 7-1, find V_{CE} for Q_2.

7-6. Using the data from problem 7-1, find Z_{in} for Q_1. (*Hint:* Because of the high current gain and the 1-kΩ emitter resistor, r_{in} for Q_1 is so high in circuits of this type that it can be ignored.) (7-1)

7-7. Refer to Fig. 7-3. Each transistor has a β of 80. What is the overall current gain of the pair? (7-1)

7-8. Refer to Fig. 7-5. The secondary has 5 turns, and the primary has 200 turns. What is the turns ratio of the transformer? (7-1)

7-9. Use the data from problem 7-8 and find the peak-to-peak signal across the load if the collector signal is 40 V peak-to-peak. (7-1)

7-10. Use the data from problem 7-8 and find the collector load if the load resistor at the transformer secondary is 4 Ω. (7-1)

7-11. Refer to Fig. 7-6. The inductance of the transformer primaries is 100 μH. The capacitors across the primaries are both 680 pF. At what frequency will the gain of the amplifier be greatest? (7-1)

7-12. Refer to Fig. 7-14. If $R_L = 1,000$ Ω, where will the load line terminate on the vertical axis? (7-3)

7-13. Refer to Fig. 7-14. If $V_{DD} = 12$ V, where will the load line terminate on the horizontal axis? (7-3)

7-14. An FET drain swings 2 mA, with a gate swing of 1 V. What is the forward transfer admittance for this FET? (7-3)

7-15. An FET has a forward transfer admittance of 4×10^{-3} S. It is to be used in the common-source configuration with a load resistor of 4,700 Ω. What voltage gain can be expected? (7-3)

7-16. Refer to Fig. 7-16. Assume a source current of 10 mA and a source resistor of 100 Ω. What is the value of V_{GS}? (7-3)

7-17. Refer to Fig. 7-18. It is desired that $V_{GS} = -2.0$ V at $I_D = 8$ mA. What should be the value of the source resistor? (7-3)

7-18. Refer to Fig. 7-28. V_{CC} is 15 V, R_L is 1.2 kΩ, R_{B_1} is 22 kΩ, R_{B_2} is 4.7 kΩ, R_E is 470 Ω, and β is 150. Find Z_{in}. (7-4)

7-19. Using the data from problem 7-18, find A_V. (7-4)

7-20. Using the data from problem 7-18, find the amplifier output signal, assuming that the signal source develops an open-circuit output of 1 V peak-to-peak and has a characteristic impedance of 10 kΩ. (7-4)

Critical Thinking Questions

7-1. List some advantages and disadvantages for a direct-coupled audio amplifier.

7-2. A transformer can match impedances for best power transfer. Are there any mechanical analogies for this fact?

7-3. Can you think of any other methods of signal coupling that are different than the ones discussed in this chapter?

7-4. The gain of an amplifier tends to drop when it is loaded. Are there any analogies in the mechanical world?

7-5. On the basis of what you have learned about negative feedback, what do you think positive feedback would do to an amplifier?

7-6. Why is it not possible for any amplifier to have infinite bandwidth?

 ## Answers to Self-Tests

1. T
2. T
3. F
4. F
5. T
6. T
7. 39.8 μF
8. $I_E = 5.3$ mA, $V_{RL} =$ 6.36 V, $V_{RE} + V_{RL} >$ V_{CC} Amplifier is in saturation
9. 5,000
10. 3.55 mA
11. 8.45 V
12. 1.96 kΩ
13. F
14. F
15. T

16. T
17. T
18. T
19. T
20. 3.92 mV
21. 0.264 V peak-to-peak
22. 880 Ω
23. 81
24. F
25. T
26. F
27. F
28. T
29. F
30. 5 V
31. saturation
32. cutoff
33. −1.5 V

34. −2.48 V; negative
35. common drain (source follower) common collector (emitter follower)
36. positive
37. T
38. T
39. F
40. T
41. F
42. 2 kΩ
43. 22 kΩ
44. It will decrease.
45. positive-going
46. common emitter
47. 500 mV$_{p-p}$
48. 1.33 kHz
49. 20 dB

50. 17 dB (0.707×10)
51. 13.3 μF
52. 150 Hz
53. greater than 150 Hz
54. For Q_1: $V_B = 2.55$, $V_E = 1.85$, $I_E = 1.85$ mA, $V_C = 4.41$, $V_{CE} = 2.56$; for Q_2: $V_C = 10.5$, $V_E = 4.41$, $V_{CE} = 6.09$
55. positive
56. positive
57. rectangular
58. hysteresis
59. hysteresis
60. positive feedback

CHAPTER 8

Large-Signal Amplifiers

Learning Outcomes

This chapter will help you to:

8-1 *Calculate* amplifier efficiency. [8-1, 8-2, 8-3]

8-2 *Identify* the class of amplifier operation. [8-1]

8-3 *Recognize* crossover distortion in push-pull amplifiers. [8-3]

8-4 *Explain* the operation of complementary symmetry amplifiers. [8-4]

8-5 *Describe* tank circuit action in class C amplifiers. [8-5]

8-6 *Explain* how class D amplifiers work. [8-6]

This chapter introduces the idea of efficiency in an amplifier. An efficient amplifier delivers a large part of the power it receives from the supply as a useful output signal. Efficiency is most important when large amounts of signal power are required. It will be shown that amplifier efficiency is related to how the amplifier is biased. It is possible to make large improvements in efficiency by moving the operating point away from the center of the load line.

8-1 Amplifier Class

All amplifiers are power amplifiers. However, those operating in the early stages of the signal processing system deal with small signals. These early stages are designed to give good voltage gain. Since voltage gain is the most important function of these amplifiers, they are called voltage amplifiers. Figure 8-1 is a block diagram of a simple audio amplifier. The Bluetooth receiver produces a small signal in the millivolt range. The first stage amplifies this audio signal, and it becomes larger. The last stage produces a much larger signal. It is called a power amplifier.

A power amplifier is designed for good power gain. It must handle large voltage and current swings. These high voltages and currents make the power high. It is very important to have good *efficiency* in a power amplifier. An efficient power amplifier delivers the most signal power for the dc power it takes from the supply. Look at Fig. 8-2. Note that the job of the power amplifier is to change dc power into signal power. Its efficiency is given by

Efficiency

$$\text{Efficiency} = \frac{\text{signal power output}}{\text{dc power input}} \times 100\%$$

The power amplifier in Fig. 8-2 produces 8 W of signal power output. Its power supply

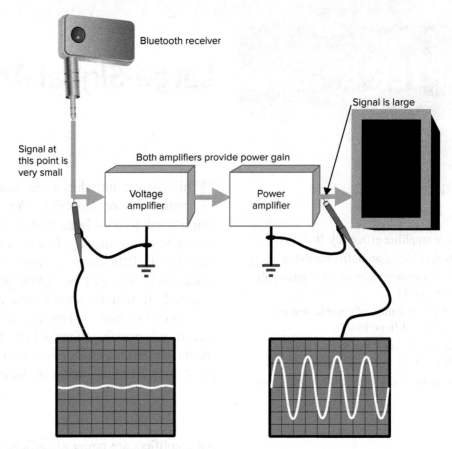

Fig. 8-1 A typical amplifier.

develops 16 V, and the amplifier draws 1 A. The dc power input to the amplifier is

$$P = V \times I = 16\ \text{V} \times 1\ \text{A} = 16\ \text{W}$$

The efficiency of the amplifier is

$$\text{Efficiency} = \frac{8\ \text{W}}{16\ \text{W}} \times 100\% = 50\%$$

Efficiency is very important in high-power systems. For example, assume that 100 W of audio power is required in a music amplifier. Also assume that the power amplifier is only 10 percent efficient. What kind of a power supply would be required? The power supply

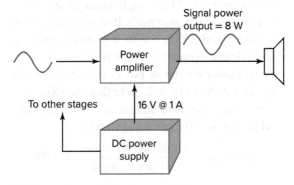

Fig. 8-2 Comparing the output signal with the dc power input.

would have to deliver 1,000 W to the amplifier! A 1-kilowatt (kW) power supply is a large, heavy, and expensive item. Heat would be another problem in this music amplifier. Of the 1-kW input, 900 W would become heat. This system would probably need a cooling fan.

EXAMPLE 8-1

What is the efficiency of an amplifier that draws 2 A from a 40-V supply when delivering 52 W of signal power? How many watts are dissipated in this amplifier? Find the dc input power first:

$$P_{\text{in}} = V \times I = 40\ \text{V} \times 2\ \text{A} = 80\ \text{W}$$

Find the efficiency:

$$\text{Efficiency} = \frac{P_{\text{out}}}{P_{\text{in}}} \times 100\% = \frac{52\ \text{W}}{80\ \text{W}} \times 100\%$$
$$= 65\%$$

The dissipation (heat) is the difference between input and output:

$$\text{Dissipation} = P_{\text{in}} - P_{\text{out}} = 80\ \text{W} - 52\ \text{W}$$
$$= 28\ \text{W}$$

The amplifier circuits covered in previous chapters have been *class A*. Class A amplifiers operate at the center of the load line, as shown in Fig. 8-3. The operating point is class A. This gives the maximum possible output swing without clipping. The output signal is a good replica of the input signal. This means that distortion is low. This is the greatest advantage of class A operation.

Class A

EXAMPLE 8-2

A stereo automotive audio amplifier is rated at 70 W of output per channel at an efficiency of 60 percent. How much current will this amplifier require when it is delivering rated output? Automotive electrical systems operate on 12 V. The first step is to find the input power, and then the current can be determined. Rearranging the efficiency formula gives

$$P_{in} = \frac{P_{out}}{\text{efficiency}} = \frac{140 \text{ W}}{0.6} = 233 \text{ W}$$

Rearranging the power formula gives

$$I = \frac{P}{V} = \frac{233 \text{ W}}{12 \text{ V}} = 19.4 \text{ A}$$

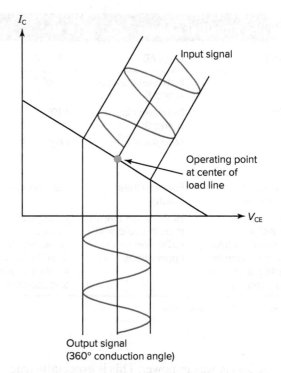

Fig. 8-3 Class A operating point.

Figure 8-4 shows another class of operation. The operating point is at *cutoff* on the load line. This is done by running the base-emitter junction of the transistor with *zero bias*. Zero bias means that only half the input signal will be amplified. Only that half of the signal that can turn on the base-emitter diode will produce any

output signal. The transistor conducts for *half* of the input cycle. A *class B* amplifier is said to have a *conduction angle* of 180 degrees. Class A amplifiers conduct for the entire input cycle. They have a conduction angle of 360 degrees.

Class B

Conduction angle

What is to be gained by operating in class B? Obviously, we have a distortion problem that was not present in class A. In spite of the distortion, class B is useful because it gives better efficiency. Biasing an amplifier at cutoff saves power.

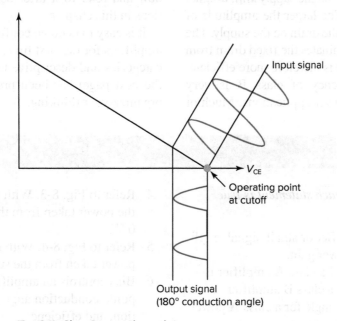

Fig. 8-4 Class B operating point.

Table 8-1 Summary of Amplifier Classes

	Class A	Class AB	Class B	Class C	Class D
Efficiency	50%*	Between classes A and B	78.5%*	100%*	100%
Conduction angle	360°	Between classes A and B	180°	Small (approx. 90°)	Not applicable (these use PWM)
Distortion	Low	Moderate	High	Extreme	Moderate after switching frequency is filtered out
Bias (emitter-base)	Forward (center of load line)	Forward (near cutoff)	Zero (at cutoff)	Reverse (beyond cutoff)	Not applicable
Applications	Practically all small-signal amplifiers. A few moderate power amplifiers in audio applications.	High-power stages in both audio and radio-frequency applications.	High-power stages— generally not used in audio applications due to distortion.	Generally limited to radio-frequency applications. Tuned circuits remove much of the extreme distortion.	Used in power amplifiers where efficiency is a prime concern.

*Theoretical maximums. Cannot be achieved in practice.

Class A wastes power. This is especially true at very low signal levels. The class A operating point is in the center of the load line. This means that about half the supply voltage is dropped across the transistor. The transistor is conducting half the saturation current. This voltage drop and current produce a power loss in the transistor. This power loss is *constant* in class A. There is a drain on the power supply even if no signal is being amplified.

The class B amplifier operates at *cutoff*. The transistor current is zero. Zero current means 0 W. There is no drain on the supply *until* a signal is being amplified. The larger the amplitude of the signal, the larger the drain on the supply. The class B amplifier eliminates the fixed drain from the power supply and is therefore more efficient.

The better efficiency of class B is very important in high-power applications. Much of the distortion can be eliminated by using two transistors: each amplifies one-half of the signal. Such circuits are a bit more complicated, but the improved efficiency is worth the effort.

There are also *class AB* and *class C* amplifiers. Again, it is a question of bias. *Bias controls the operating point, the conduction angle, and the class of operation.* With class D amplifiers, it's not a question of bias but a totally different mode of operation called pulse-width modulation. Table 8.1 summarizes the important features of the amplifier classes. Study this table now and refer to it after completing later sections in this chapter.

It is easy to become confused when studying amplifiers for the first time. There are so many categories and descriptive terms. Table 8.2 (on the next page) has been prepared to help you organize your thinking.

Self-Test

Determine whether each statement is true or false.

1. A voltage amplifier or small-signal amplifier gives no power gain.
2. The efficiency of a class A amplifier is less than that of a class B amplifier.
3. The conduction angle for a class A power amplifier is 180 degrees.
4. Refer to Fig. 8-3. With no input signal, the power taken from the supply will be 0 W.
5. Refer to Fig. 8-4. With no input signal, the power taken from the supply will be 0 W.
6. Bias controls an amplifier's operating point, conduction angle, class of operation, and efficiency.

Table 8-2 Amplifier Characteristics

	Explanations and Examples
Voltage amplifiers	Voltage amplifiers are small-signal amplifiers. They can be found in the early stages in the signal system. They are often designed for good voltage gain. An audio preamplifier would be a good example of a voltage amplifier.
Power amplifiers	Power amplifiers are large-signal amplifiers. They can be found late in the signal system. They are designed to give power gain and reasonable efficiency. The output stage of an audio amplifier would be a good example of a power amplifier.
Configuration	The configuration of an amplifier tells how the signal is fed to and taken from the amplifying device. For bipolar transistors, the configurations are common-emitter, common-collector, and common-base. For field-effect transistors, the configurations are common-source, common-drain, and common-gate.
Coupling	How the signal is transferred from stage to stage. Coupling can be capacitive, direct, or transformer.
Applications	Amplifiers may be categorized according to their use. Examples are audio amplifiers, video amplifiers, RF amplifiers, dc amplifiers, bandpass amplifiers, and wide-band amplifiers.
Classes	This category refers to how the amplifying device is biased in the case of classes A, B, AB, and C. Class D amplifiers use a different mode of operation called pulse-width modulation. Voltage amplifiers are usually biased for class A operation. For improved efficiency, power amplifiers may use class B, AB, or C operation. Class D amplifiers are used when efficiency is a prime consideration.

Solve the following problems.

7. Refer to Fig. 8-2. Suppose the power supply is rated at 20 V. What is the efficiency of the power amplifier?

8. A certain amplifier is producing an output power of 100 W. Its efficiency is 60 percent. How much power will the amplifier take from the supply?

8-2 Class A Power Amplifiers

The *class A* power amplifier operates near the center of the load line. It is not highly efficient, but it does offer low distortion. It is also the most simple design.

Figure 8-5 shows a class A power amplifier. We will use a load line to see how much signal power can be produced. The load line will be set by the supply voltage V_{CC} and the saturation current:

$$I_{sat} = \frac{V_{CC}}{R_{load}} = \frac{16\ V}{80\ \Omega} = 0.2\ A\ or\ 200\ mA$$

The load line will run from 16 V on the horizontal axis to 200 mA on the vertical axis.

Next, we must find the operating point for the amplifier. Solving for the base current, we get

$$I_B = \frac{V_{CC}}{R_B} = \frac{16\ V}{16\ k\Omega} = 1\ mA$$

The transistor has a β of 100. The collector current will be

$$I_C = \beta \times I_B$$
$$= 100 \times 1\ mA = 100\ mA$$

The load line can be seen in Fig. 8-6 with the 100-mA operating point.

The amplifier can be driven to the load-line limits before clipping occurs. The maximum

Class A

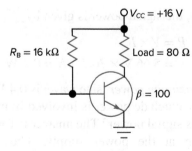

Fig. 8-5 Class A power amplifier.

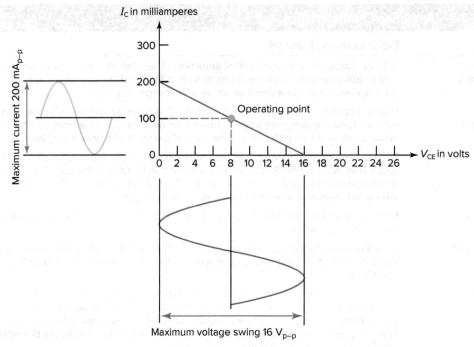

Fig. 8-6 Class A load line.

voltage swing will be 16 V peak-to-peak. The maximum current swing will be 200 mA peak-to-peak. Both of these maximums are shown in Fig. 8-6.

We now have enough information to calculate the *signal power*. The peak-to-peak values must be converted to rms values. This is done by

$$V_{rms} = \frac{V_{p-p}}{2} \times 0.707$$

$$= \frac{16\ V}{2} \times 0.707 = 5.66\ V$$

Next, the rms current is

$$I_{rms} = \frac{I_{p-p}}{2} \times 0.707$$

$$= \frac{200\ mA}{2} \times 0.707 = 70.7\ mA$$

Finally, the signal power is given by

$$P = V \times I$$

$$= 5.66\ V \times 70.7\ mA = 0.4\ W$$

The *maximum power* (sine wave) is 0.4 W.

How much dc power is involved in producing this signal power? The answer is found by looking at the power supply. The supply develops 16 V. The current taken from the

supply must also be known. The base current is small enough to ignore. The *average* collector current is 100 mA. Therefore, the *average* power is

$$P = V \times I$$

$$= 16\ V \times 100\ mA = 1.6\ W$$

The amplifier takes 1.6 W from the power supply to produce a signal power of 0.4 W. The efficiency of the amplifier is

$$Efficiency = \frac{P_{ac}}{P_{dc}} \times 100\%$$

$$= \frac{0.4\ W}{1.6\ W} \times 100\% = 25\%$$

The class A amplifier shows a maximum efficiency of 25 percent. This occurs *only* when the amplifier is driven to its maximum output. The efficiency is *less* when the amplifier is not driven hard. The 1.6 W in the above equation is *fixed*. As the drive decreases, the efficiency drops. With *no drive*, the efficiency drops to zero. An amplifier of this type would be a poor choice for high-power applications. The power supply would have to produce *four times* the required signal power. Three-fourths of this power would be wasted as heat. The transistor would probably require a large heat sink.

EXAMPLE 8-3

The amplifier shown in Fig. 8-5 is producing a peak-to-peak sine wave output of 8 V. Determine its efficiency. First, find the rms signal voltage:

$$V_{rms} = \frac{V_{p-p}}{2} \times 0.707 = \frac{8}{2} \times 0.707 = 2.83 \text{ V}$$

We could inspect Fig. 8-6 to determine the peak-to-peak signal current and convert it to rms to find the signal power. However, it is easier to use the power formula to calculate the output power directly:

$$P = \frac{V^2}{R} = \frac{2.83^2}{80} = 0.1 \text{ W}$$

The dc input power doesn't change in a class A amplifier, so the efficiency is

$$\text{Efficiency} = \frac{P_{ac}}{P_{dc}} \times 100\%$$

$$= \frac{0.1 \text{ W}}{1.6 \text{ W}} \times 100\% = 6.25\%$$

EXAMPLE 8-4

What is the dissipation in the amplifier in Example 8-3? What is its dissipation with no input signal? Dissipation is the difference between input and output:

$$\text{Dissipation} = P_{in} - P_{out} = 1.6 \text{ W} - 0.1 \text{ W}$$

$$= 1.5 \text{ W}$$

With no input signal, there is no useful output:

$$\text{Dissipation} = P_{in} - P_{out} = 1.6 \text{ W} - 0 \text{ W}$$

$$= 1.6 \text{ W}$$

One reason why the class A amplifier is so wasteful is that dc power is dissipated in the load. A big improvement is possible by removing the load from the dc circuit. Figure 8-7 shows how to do this. The transformer will couple the signal power to the load. Now, there is no direct current flow in the load. *Transformer coupling* allows the amplifier to produce *twice* as much signal power.

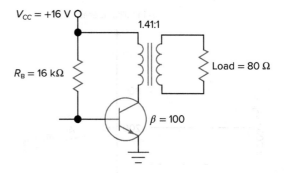

Fig. 8-7 Class A power amplifier with transformer coupling.

Figure 8-7 shows the same supply voltage, the same bias resistor, the same transistor, and the same load as in Fig. 8-5. The only difference is the coupling transformer. The dc conditions are now quite different. The transformer primary will have very low resistance. This means that all the supply voltage will drop across the transistor at the operating point.

The dc load line for the transformer-coupled amplifier is shown in Fig. 8-8. It is *vertical*. The operating point is still at 100 mA. This is because the base current and β have not changed. The change is the absence of the 80-Ω dc resistance in series with the collector. All the supply voltage now drops across the transistor.

Actually, the load line will not be perfectly vertical. The transformer and even the power supply always have a little resistance. However, the dc load line is very steep. We cannot show any output swing from this load line.

There is a second load line in a transformer-coupled amplifier. It is the result of the ac load in the collector circuit and is called the ac load line. The ac load is not 80 Ω in the collector circuit of Fig. 8-7. The transformer is a step-down type. Remember, transformer impedance ratio is equal to the *square of the turns ratio*. Therefore, the ac load in the collector circuit will be

$$\text{Load}_{ac} = (1.41)^2 \times 80 \ \Omega = 160 \ \Omega$$

Notice that the ac load line in Fig. 8-8 runs from 32 V to 200 mA. This satisfies an impedance of

$$Z = \frac{V}{I}$$

$$= \frac{32 \text{ V}}{200 \text{ mA}} = 160 \ \Omega$$

Also notice that the ac load line passes through the operating point. The dc load line and the ac

Transformer coupling

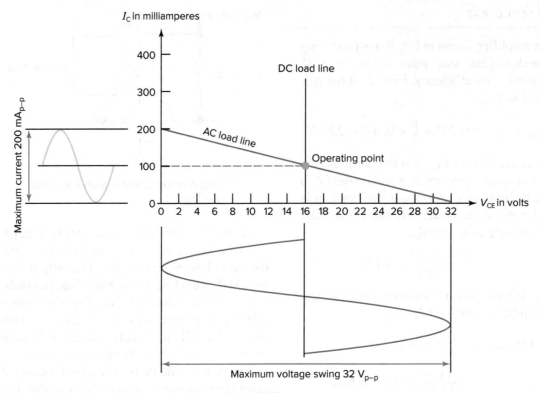

I_C in milliamperes

400

300

DC load line

200

AC load line

Operating point

100

Maximum current 200 mA$_{p-p}$

0

0 2 4 6 8 10 12 14 16 18 20 22 24 26 28 30 32

V_{CE} in volts

Maximum voltage swing 32 V$_{p-p}$

Fig. 8-8 Load lines for the transformer-coupled amplifier.

load line must always pass through the same operating point.

How does the ac load line extend to 32 V? This is *twice* the supply voltage! There are two ways to explain this. First, it *must* extend to 32 V if it is to pass through the operating point and satisfy a slope of 160 Ω. Second, a transformer is a type of inductor. When the field collapses, a voltage is generated. This voltage adds in series with the supply voltage. Thus, V_{CE} can swing to *twice* the supply voltage in a transformer-coupled amplifier.

Compare Fig. 8-8 with Fig. 8-6. The output swing doubles with transformer coupling. It is safe to assume that the output power also doubles. The *dc power input to the amplifier has not changed*. The supply voltage is still 16 V, and the average current is still 100 mA. Transformer coupling the class A amplifier provides twice as much signal power for the same dc power input. The maximum efficiency of the transformer-coupled class A amplifier is

$$\text{Efficiency} = \frac{P_{ac}}{P_{dc}} \times 100\%$$

$$= \frac{0.8 \text{ W}}{1.6 \text{ W}} \times 100\% = 50\%$$

Remember, however, that this efficiency is reached only at maximum signal level. The efficiency is less for smaller signals and drops to zero when the amplifier is not driven with a signal.

An efficiency of 50 percent may be acceptable in some applications. Class A power amplifiers are sometimes used in medium-power applications (up to 5 W or so). However, the transformer can be an expensive component. For example, in a high-quality audio amplifier, the output transformer can cost more than all the other amplifier parts combined! So for high-power and high-quality amplifiers, something other than class A is usually a better choice.

Our efficiency calculations have ignored some losses. First, we have ignored the *saturation voltage* of the transistor. In practice, V_{CE} cannot drop to 0 V. A power transistor might show a saturation of 0.7 V. This would have to be subtracted from the output swing. Second, we ignored transformer loss in the transformer-coupled amplifier. Transformers are not 100 percent efficient. Small, inexpensive transformers may be only 75 percent efficient at low audio frequencies. The calculated efficiencies of 25 and 50 percent are *theoretical maximums*. They are not realized in actual circuits.

Another problem with the class A circuit is the fixed drain on the power supply. Even when no signal is being amplified, the drain on the supply in our example was fixed at 1.6 W. Most power amplifiers must handle signals that change in level. An audio amplifier, for example, will handle a broad range of volume levels. When the volume is low, the efficiency of class A is quite poor.

EXAMPLE 8-5

What is the best efficiency for the amplifier shown in Fig. 8-7 if the efficiency of the transformer is 80 percent? The overall efficiency of a system is the product of the individual efficiencies:

$$\text{Efficiency} = 0.5 \times 0.8 \times 100\% = 40\%$$

Self-Test

Solve the following problems.

9. Refer to Fig. 8-5. The current gain of the transistor is 120. Calculate the power dissipated in the transistor with no input signal.
10. Refer to Fig. 8-6. The operating point is at $V_{CE} = 12$ V. Calculate the power dissipated in the transistor with no input signal.
11. Refer to Fig. 8-7. The transformer has a turns ratio from primary to secondary of 3:1. What load does the collector of the transistor see?
12. Refer to Fig. 8-7. The transformer has a turns ratio of 4:1. An oscilloscope shows a collector sinusoidal signal of 30 V peak-to-peak. What will the amplitude of the

signal be across the 80-Ω load? What will be the rms signal power delivered to the load?

Determine whether each statement is true or false.

13. Transformer coupling the output does not improve the efficiency of a class A amplifier.
14. Refer to Fig. 8-8. The dc load line is very steep because the dc resistance of the output transformer primary winding is so low.
15. In practice, it is possible to achieve 50 percent efficiency in class A by using transformer coupling.

8-3 Class B Power Amplifiers

The *class B* amplifier is biased at *cutoff*. No current will flow until an input signal provides the bias necessary to turn on the transistor. This eliminates the large fixed drain on the power supply. The efficiency is much better. Only *half the input* signal is amplified, however. This produces extreme distortion. A single class B transistor would not be useful in audio work. The sound would be horrible.

Two transistors can be operated in class B. One can be arranged to amplify the positive-going portion of the input and the other to amplify the negative-going portion. Combining

the two halves, or portions, will reduce much of the distortion. Two transistors operating in this way are said to be in *push-pull*.

Figure 8-9 shows a class B push-pull power amplifier. Two transformers are used. Transformer T_1 is called the *driver transformer*. It provides Q_1 and Q_2 with signal drive. Transformer T_2 is called the *output transformer*. It combines the two signals and supplies the output to the load. Notice that both transformers have one winding that is center-tapped.

With no signal input, there will not be any current flow in Fig. 8-9. Both Q_1 and Q_2 are *cut off*. There is no dc supply to turn on the base-emitter junctions. When the input

Class B

Push-pull

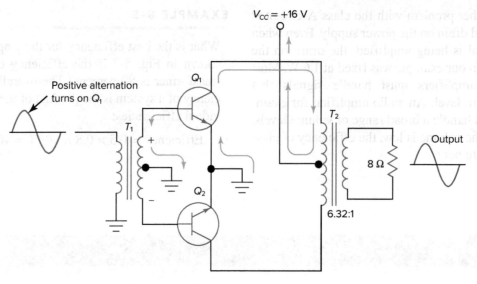

Fig. 8-9 Class B push-pull power amplifier with Q_1 turned on.

signal produces the secondary polarity in T_1 as shown, Q_1 is turned on. Current will flow through half of the primary of T_2. Since the primary current is changing in the output transformer, a signal will appear across the secondary. The positive-going portion of the input signal has been amplified and appears across the load.

When the signal reverses polarity, Q_1 is cut off and Q_2 turns on. This is shown in Fig. 8-10. Current will flow through the other

half of the primary of T_2. This time the current is flowing up through the primary. When Q_1 was on, the current was flowing down through the primary winding. This current *reversal* produces the negative alternation across the load. By operating two transistors in push-pull, much of the distortion has been eliminated. The circuit amplifies almost the entire input signal.

We can use graphs to show the output swing and efficiency for the class B push-pull

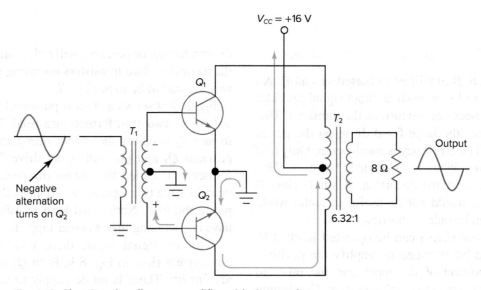

Fig. 8-10 Class B push-pull power amplifier with Q_2 turned on.

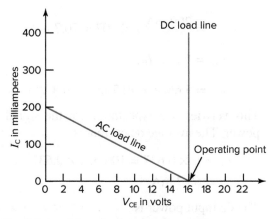

Fig. 8-II Load lines for the class B amplifier.

amplifier. Figure 8-11 shows the dc and ac load lines for the push-pull circuit. The dc load line is vertical. There is very little resistance in the collector circuit. The ac load line slope is set by the *transformed load* in the collector circuit.

Transformer T_2 shows a turns ratio of 6.32:1. This turns ratio determines the ac load that will be seen in the collector circuit. Only half the transformer primary is conducting at any time. Therefore, only half the turns ratio will be used to calculate the impedance ratio:

$$\frac{6.32}{2} = 3.16$$

Now, the collector load will be equal to the square of half the turns ratio times the load resistance:

$$\text{Load}_{ac} = (3.16)^2 \times 8\ \Omega = 80\ \Omega$$

Each transistor sees an ac load of 80 Ω. The load line of Fig. 8-11 runs from 16 V to 200 mA. This satisfies a slope of 80 Ω:

$$R = \frac{V}{I} = \frac{16\ \text{V}}{0.2\ \text{A}} = 80\ \Omega$$

Figure 8-11 is correct but shows only one transistor.

There is another way to graph a push-pull circuit, as shown in Fig. 8-12. This graph allows the entire output swing to be shown. The output voltage swings 32 V peak-to-peak. This must be converted to an rms value:

$$V_{\text{rms}} = \frac{V_{\text{p-p}}}{2} \times 0.707$$

$$= \frac{32\ \text{V}}{2} \times 0.707 = 11.31\ \text{V}$$

Next, the rms current is found:

$$I_{\text{rms}} = \frac{I_{\text{p-p}}}{2} \times 0.707$$

$$= \frac{400\ \text{mA}}{2} \times 0.707 = 141.4\ \text{mA}$$

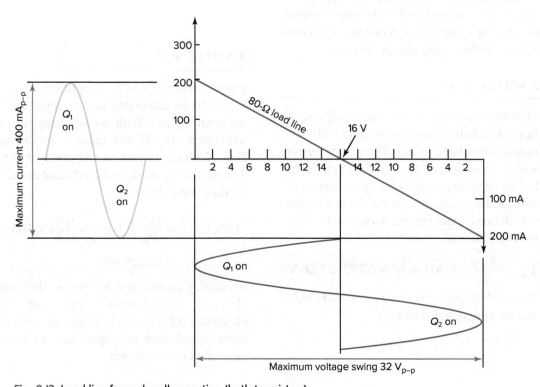

Fig. 8-I2 Load line for push-pull operation (both transistors).

Finally, the signal power is

$$P = V \times I$$

$$= 11.31 \text{ V} \times 141.4 \text{ mA} = 1.6 \text{ W}$$

To find the efficiency of the class B push-pull circuit, we will need the dc input power. The supply voltage is 16 V. The supply current varies from 0 to 200 mA. As in class A, the average collector current is what we need:

$$I_{av} = I_p \times 0.637 = 200 \text{ mA} \times 0.637$$

$$= 127.4 \text{ mA}$$

The average input power is

$$P = V \times I = 16 \text{ V} \times 127.4 \text{ mA} = 2.04 \text{ W}$$

The class B push-pull amplifier takes 2.04 W from the supply to give a signal output of 1.6 W. The efficiency is

$$\text{Efficiency} = \frac{P_{ac}}{P_{dc}} \times 100\%$$

$$= \frac{1.6 \text{ W}}{2.04 \text{ W}} \times 100\% = 78.5\%$$

The best efficiency for class A is 50 percent. The best efficiency for class B is 78.5 percent. This improved efficiency makes the class B push-pull circuit attractive for high-power applications. For smaller signals, the class B amplifier takes less from the power supply. The 2.04-W factor is not fixed in the above calculation. As the input signal decreases, the power demand on the supply also decreases.

EXAMPLE 8-6

Find the efficiency for the push-pull amplifier in Fig. 8-10 when it is driven to half of its maximum voltage swing. The power output will decrease to one-fourth of maximum, or 0.4 W, because power varies as the square of voltage. However, let's be certain and do the calculations. Half voltage swing is 16 V_{p-p} for Fig. 8-10 and

$$V_{rms} = \frac{V_{p-p}}{2} \times 0.707 = \frac{16}{2} \times 0.707 = 5.66 \text{ V}$$

The peak-to-peak current is now 200 mA, and the rms signal current is

$$I_{rms} = \frac{I_{p-p}}{2} \times 0.707$$

$$= \frac{200 \text{ mA}}{2} \times 0.707 = 70.7 \text{ mA}$$

$$P_{ac} = V_{rms} \times I_{rms}$$

$$= 5.66 \text{ V} \times 70.7 \text{ mA} = 0.4 \text{ W}$$

This verifies our expectation for the signal power. The average dc current is

$$I_{av} = I_p \times 0.637 = 100 \text{ mA} \times 0.637$$

$$= 63.7 \text{ mA}$$

The dc input power is

$$P_{dc} = 16 \times 63.7 \text{ mA} = 1.02 \text{ W}$$

Please notice that this is half of what it was when the amplifier was fully driven. The efficiency is also half of the fully driven value:

$$\text{Efficiency} = \frac{P_{ac}}{P_{dc}} \times 100\%$$

$$= \frac{0.4 \text{ W}}{1.02 \text{ W}} \times 100\% = 39.2\%$$

Efficiency decreases when the class B amplifier is not fully driven. However, this amplifier is more efficient than a class A amplifier driven to half of its maximum output swing.

EXAMPLE 8-7

Can the efficiency of the amplifier shown in Fig. 8-10 be calculated for the condition of no input signal? With no input signal, the transistors are off and there is no current flow. With no current, the dc power is zero. The equation cannot be solved since division by zero is not defined:

$$\text{Efficiency} = \frac{P_{out}}{P_{in}} \times 100\% = \frac{0 \text{ W}}{0 \text{ W}} \times 100\%$$

$$= \text{undefined}$$

Efficiency cannot be calculated in this case. However, we can reach a conclusion: The efficiency of a class B amplifier with no input signal does not equal zero as in the case of a class A device.

Class A power transistors require a *high wattage rating*. The reason is that the transistors are always biased on to half of the saturation current. For example, to build a 100-W class A amplifier, the transistor will need at least a 200-W rating. This is based on

$$\text{Efficiency} = \frac{P_{ac}}{P_{dc}} \times 100\%$$

$$= \frac{100 \text{ W}}{200 \text{ W}} \times 100\% = 50\%$$

Look at the above equation: 200 W goes into the transistor; 100 W comes out as signal power. The 100 W difference *heats* the transistor. What if the signal input is zero? The signal output is zero; yet *200 W still goes into the transistor* and changes to heat.

The wattage rating needed for class B at a given power level is only *one-fifth* that needed for class A. To build a 100-W amplifier requires a 200-W transistor in class A. In class B,

$$\frac{200}{5} = 40 \text{ W}$$

Two 20-W transistors operating in push-pull would provide 100 W output. Two 20-W transistors cost quite a bit less than one 200-W transistor. This is a marked advantage of class B push-pull over class A in high-power amplifiers.

The size of the heat sink is another factor. A transistor rating is based on some safe temperature. In high-power work, the transistor is mounted on a device that carries off the heat. A class B design will need only one-fifth the heat sink capacity for a given amount of power.

There is a very strong case for using class B in high-power work. However, there is still too much distortion for some applications. The push-pull circuit eliminates quite a bit of distortion, but some remains. The problem is called *crossover distortion*.

The base-emitter junction of a transistor behaves like a diode. It takes about 0.6 V to turn on the base-emitter junction in a silicon transistor. This means that the first 0.6 V of input signal in a class B push-pull amplifier *will not be amplified*. The amplifier has a *dead band* of about 1.2 V. The base-emitter

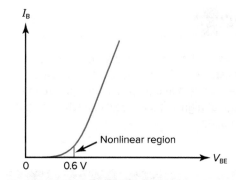

Fig. 8-13 Characteristic curve for a base-emitter junction.

junction is also very nonlinear near the turn-on point. Figure 8-13 shows the characteristic curve for a typical base-emitter junction. Note the curvature near the 0.6-V forward-bias region. As one transistor is turning off and the other is coming on in a push-pull design, this curvature distorts the output signal. The dead band and the nonlinearity make the class B push-pull circuit unacceptable for some applications.

The effect of crossover distortion on the output signal is shown in Fig. 8-14(*a*). It happens as the signal is *crossing over* from one

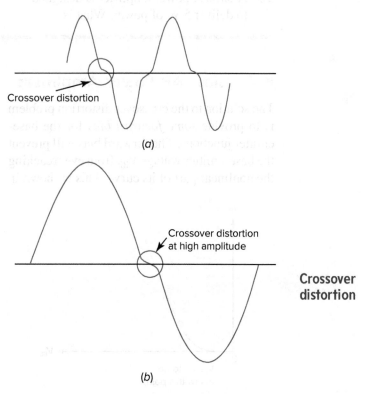

Fig. 8-14 Crossover distortion in the output signal.

Crossover distortion

Dead band

transistor to the other. Crossover distortion is very noticeable when the input signal is small. In fact, if the input signal is 1 V_{p-p} or less, there won't be any output with silicon transistors. As shown in Fig. 8-14(*b*), the distortion is less noticeable for large signals. This can be a valuable clue when troubleshooting.

Self-Test

Determine whether each statement is true or false.

16. Refer to Fig. 8-9. Transistors Q_1 and Q_2 operate in parallel.
17. Refer to Fig. 8-10. Transistors Q_1 and Q_2 will never be on at the same time.
18. Crossover distortion is caused by the nonlinearity of the base-emitter junctions in the transistors.

Solve the following problems.

19. Refer to Fig. 8-10. Transformer T_2 has a turns ratio of 20:1. What is the load seen by the collector of Q_1? Q_2?
20. A class A power amplifier is designed to deliver 5 W of power. What is

dissipated in the transistor at zero signal level?
21. A class B push-pull amplifier is designed to deliver 10 W of power. What is the most power that must be dissipated by each transistor?
22. Refer to Fig. 8-12. Assume the amplifier is being driven to only half its maximum swing. Calculate the rms power output.
23. Refer to Fig. 8-12. Assume the amplifier is driven to half its maximum swing. Calculate the average power input.
24. Refer to Fig. 8-12. Assume the amplifier is driven to half its maximum swing. Calculate the efficiency of the amplifier.

8-4 Class AB Power Amplifiers

Class AB

The solution to the crossover distortion problem is to provide *some forward bias* for the base-emitter junctions. The forward bias will prevent the base-emitter voltage V_{BE} from ever reaching the nonlinear part of its curve. This is shown in

Fig. 8-15. The forward bias is small and results in a *class AB* amplifier. It has characteristics between class A and class B.

The operating point for class AB is shown in Fig. 8-16. Note that class AB operates *near cutoff.*

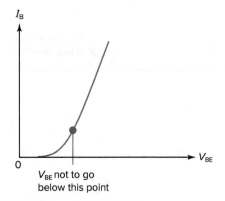

Fig. 8-15 Minimum value of V_{BE} to prevent crossover distortion.

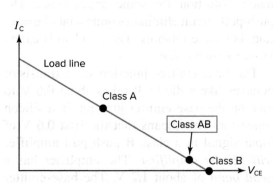

Fig. 8-16 Class AB operating point.

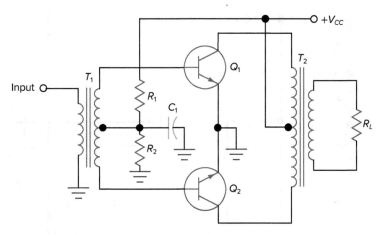

Fig. 8-17 Class AB push-pull power amplifier.

Figure 8-17 is a class AB push-pull power amplifier. Resistors R_1 and R_2 form a voltage divider to forward-bias the base-emitter junctions. The bias current flows through both halves of the secondary of T_1. Capacitor C_1 grounds the center tap for ac signals. Without it, signal current will flow in R_1 and R_2, and a large part of the signal energy will be wasted (dissipated as heat).

A class AB amplifier does not have as much efficiency as a class B amplifier, but its efficiency is better than that of a class A design. It is a compromise class that provides minimum distortion and reasonable efficiency. It is the most popular class for high-power audio work. Amplifiers such as the one shown in Fig. 8-17 are popular in modest power audio applications.

Now that the distortion problem is solved for push-pull amplifiers, it is time to look at the transformers. For high-power and high-quality work, the transformers are too expensive. They can be eliminated.

EXAMPLE 8-8

Find a value for R_1 in Fig. 8-17 assuming class AB bias, a supply of 12 V, and a value for R_2 of 1 kΩ. Use the voltage divider equation and assume that a 0.6-V drop is desired for class AB:

$$0.6 \text{ V} = \frac{1 \text{ k}\Omega}{1 \text{ k}\Omega + R_1} \times 12 \text{ V}$$

Eliminate the fraction by multiplying both sides by the denominator:

$$0.6 \text{ V}(1 \text{ k}\Omega + R_1) = 1 \text{ k}\Omega \times 12 \text{ V}$$

Divide both sides by 12 V:

$$0.05(1 \text{ k}\Omega + R_1) = 1 \text{ k}\Omega$$

Multiply:

$$50 \text{ }\Omega + 0.05 \text{ } R_1 = 1 \text{ k}\Omega$$

Subtract 50 Ω from both sides:

$$0.05 \text{ } R_1 = 950 \text{ }\Omega$$
$$R_1 = 19 \text{ k}\Omega$$

Driver transformers can be eliminated by using a combination of transistor polarities. A positive-going signal applied to the base of an NPN transistor tends to turn it on. A positive-going signal applied to the base of a PNP transistor tends to turn it off. This means that push-pull operation can be obtained without a center-tapped driver transformer.

Output transformers can be eliminated by using a different amplifier configuration. The emitter-follower (common-collector) amplifier is noted for its low output impedance. This allows good matching to low-impedance loads such as loudspeakers.

Figure 8-18 shows an amplifier design that eliminates the transformers. Transistor Q_1 is an NPN transistor, and Q_2 is a PNP transistor. Push-pull operation is realized without a center-tapped driver transformer. Notice that the load is capacitively coupled to the emitter leads of the transistors. The transistors are operating as *emitter followers*.

The circuit of Fig. 8-18 is known as a *complementary symmetry amplifier*. The transistors are *electrical complements*. One is NPN, and the other is PNP. The curves in Fig. 8-19 show the

Complementary symmetry amplifier

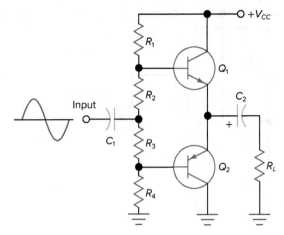

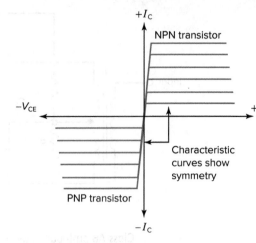

Fig. 8-18 A complementary symmetry amplifier.

Fig. 8-19 NPN-PNP symmetry.

symmetrical characteristics of NPN and PNP transistors. Good matching of characteristics is important in the complementary symmetry amplifier. For this reason, transistor manufacturers offer NPN-PNP pairs with good symmetry.

EXAMPLE 8-9

Select values for R_2 and R_3 for Fig. 8-18 assuming a 12-V supply and that R_1 and R_4 are each 10 kΩ. For class AB, we can assume a 0.6-V drop for *each* transistor or a 1.2-V drop for *both*. Set up the divider formula to find the total value of R_2 and R_3 (R_T) first:

$$1.2 \text{ V} = \frac{R_T}{20 \text{ k}\Omega + R_T} \times 12 \text{ V}$$

Solve as before (the algebra is similar to Example 8-8) to find that $R_T = 2.22$ kΩ. Each resistor should be half of that amount, or 1.11 kΩ.

Quasi-complementary symmetry amplifier

Figure 8-20 shows the output signal in a complementary symmetry amplifier when the input signal goes positive. Transistor Q_1, the NPN transistor, is turned on. Transistor Q_2, the PNP transistor, is turned off. Current flows through the load, through C_2, and through Q_1 into the power supply. This current charges C_2 as shown. Notice that there is *no phase inversion* in the amplifier. This is to be expected in an emitter follower.

When the input signal goes negative, the signal flow is as shown in Fig. 8-21. Now Q_1 is off and Q_2 is on. This causes C_2 to discharge as shown. Again, the output is in phase with the input. Capacitor C_2 is usually a large capacitor (1,000 μF or so). This is necessary for good low-frequency response with low values of R_L.

Another possibility is shown in Fig. 8-22. This is known as a *quasi-complementary symmetry amplifier*. The output transistors Q_3 and Q_4 are not complementary. They are both NPN

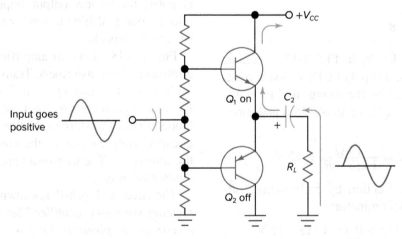

Fig. 8-20 A positive-going signal in a complementary symmetry amplifier.

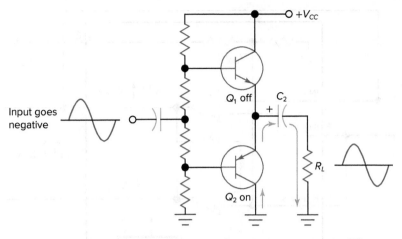

Fig. 8-21 A negative-going signal in a complementary symmetry amplifier.

types. The *driver transistors* Q_1 and Q_2 are complementary. A positive-going input signal will turn on Q_1, the NPN driver, and it will turn off Q_2, the PNP driver. This results in a push-pull action since the drivers supply the base current for the output transistors. Again, no center-tapped transformer is necessary in the output.

Notice the diodes used in the bias network of Fig. 8-22. These diodes provide *temperature compensation*. Transistors tend to conduct more current as temperature goes up. This is undesirable. The drop across a conducting diode *decreases* with an increase in temperature. If the diode drop is part of the bias voltage for the amplifier, compensation results. The decreasing voltage across the diode will lower the amplifier current. Thus, the operating point is more stable with this arrangement.

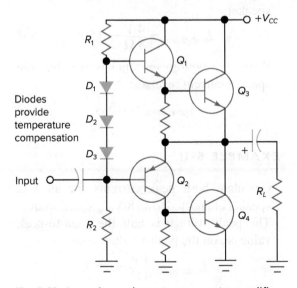

Diodes provide temperature compensation

Fig. 8-22 A quasi-complementary symmetry amplifier.

Integrated circuits are available for power applications. Figure 8-23 shows the TPA4861 1-W *power amplifier* made by Texas Instruments. It's an audio amplifier and is intended for battery-powered devices that operate at 3.3 or 5 V, such as notebook computers. This IC uses a *bridge tied load* (BTL) circuit for improved output power. BTL amplifiers are also called full-bridge amplifiers. The relatively low power supply voltage limits the amount of power that can be delivered, but the BTL circuit helps a lot. In a ground tied load circuit (like the one in Fig. 8-22), assuming an 8-Ω load and a 5-V supply, the most output power is only 250 mW. The best peak-to-peak output swing is usually about 1 V less than the supply. So with a 4-V peak-to-peak swing,

Temperature compensation

$$V_{\text{out(rms)}} = \frac{4\,V_{\text{p-p}}}{2} \times 0.707 = 1.414$$

$$P_{\text{out}} = \frac{V^2}{R_L} = \frac{1.414^2}{8} = 250 \text{ mW}$$

In Fig. 8-23(*a*) there are two outputs, and the swing across the load is two times the peak-to-peak swing of each output. Using the same load and supply voltage as above gives

$$V_{\text{out(rms)}} = \frac{8\,V_{\text{p-p}}}{2} \times 0.707 = 2.83$$

$$P_{\text{out}} = \frac{V^2}{R_L} = \frac{2.83^2}{8} = 1 \text{ W}$$

This is four times the output power that is available using a ground tied load. Remember, power varies as the square of voltage, so doubling voltage will quadruple power for any fixed load value.

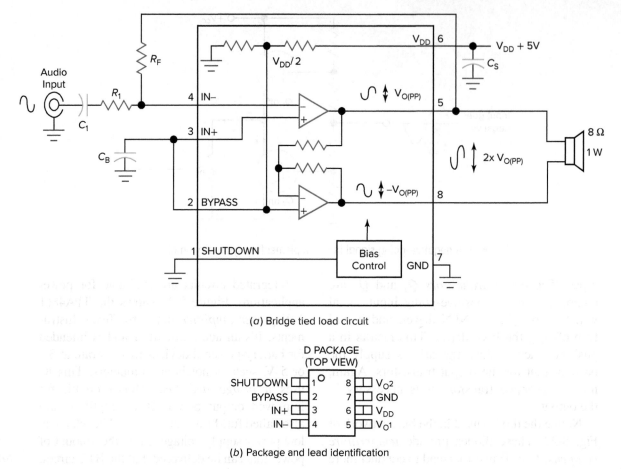

(a) Bridge tied load circuit

D PACKAGE
(TOP VIEW)

SHUTDOWN	1	8	V_O2
BYPASS	2	7	GND
IN+	3	6	V_{DD}
IN−	4	5	V_O1

(b) Package and lead identification

Fig. 8-23 Texas Instruments TPA4861 amplifier.

Why is the load swing in Fig. 8-23(a) twice the peak output of one amplifier? Notice that the amplifiers operate out of phase. The triangle symbols represent amplifiers. Since it's an IC, we need not be concerned about what's inside. When a signal is applied to a negative input, the amplifier phase inverts. When a signal is applied to a positive input, the output is in-phase. The top amplifier output [pin 5 in Fig. 8-23(a)] drives one end of the load and also drives the negative input of the bottom amplifier. So the output at pin 8 is 180 degrees out of phase from the output at pin 5. When one end of the load is driven in a positive direction, the other end is being driven in a negative direction.

At one peak, the speaker voltage in Fig. 8-23(a) is +4 V at the top terminal and 0 V at the bottom. At the other peak it is +4 V at the bottom speaker terminal and 0 V at the top. The same peak-to-peak ac flows in the speaker as it would if it was connected to an 8-V peak-to-peak signal source.

EXAMPLE 8-10

Calculate the peak speaker currents for Fig. 8-23(a). When the top speaker terminal is positive, the current flows up through the speaker:

$$I_{peak(down)} = \frac{4\ V}{8\ \Omega} = 0.5\ A$$

When the bottom terminal is positive, the speaker current reverses:

$$I_{peak(down)} = 0.5\ A$$

EXAMPLE 8-11

Calculate both peak currents for an 8-Ω speaker connected to an 8-V_{p-p} signal source. The peak voltage is half the peak-to-peak value, so on the positive alternation,

$$I_{peak(up)} = \frac{4\ V}{8\ \Omega} = 0.5\ A$$

On the negative alternation, the current flow in the speaker reverses:

$$I_{peak(down)} = 0.5 \text{ A}$$

Bridged amplifiers offer another important advantage. They eliminate the need for output coupling capacitors (like the one shown in Fig. 8-22). This is because there is no dc voltage that has to be blocked. In Fig. 8-22, the amplifier output has a dc component that is equal to half the power supply. You will notice in Fig. 8-23(a) that the speaker is directly connected to the outputs. Since output coupling capacitors have to be quite large for good low frequency response, this is an important advantage. Bridged amplifiers are also used in vehicles and can be made with discrete component designs (they are not limited to ICs).

Class AB continues to serve well, but there are two variations. One is to use additional power rails when the needed power level exceeds the clipping level. These rails activate when the output signal peaks would otherwise exceed the maximum output voltage available from a class AB amplifier's single supply rail. Class G amplifiers employ several power rails at discrete voltage steps and switch between them as needed. Instead of providing multiple discrete rails, class H amplifiers track the input signal level and vary the voltage on the supply rails as needed.

Self-Test

Answer the following questions.

25. Is the efficiency of class AB better than that of class A but not as good as that of class B?
26. Refer to Fig. 8-17. Assume that C_1 shorts. In what class will the amplifier operate?
27. Refer to Fig. 8-17. Assume that Q_1 and Q_2 are running very hot. Could C_1 be shorted? Why or why not?
28. Refer to Fig. 8-17. Assume that Q_1 and Q_2 are running very hot. Could R_2 be open? Why or why not?
29. Refer to Fig. 8-18. An input signal drives C_1 in a positive direction. In what direction will the top of R_L be driven?
30. Refer to Fig. 8-18. An input signal drives C_1 in a positive direction. Which transistor is turning off?
31. Refer to Fig. 8-18. Voltage $V_{CC} = 20$ V. With no input signal, what should the voltage be at the emitter of Q_1? At the base of Q_1? At the base of Q_2? (*Hint:* The transistors are silicon.)

8-5 Class C Power Amplifiers

Class C amplifiers are biased *beyond cutoff.* Figure 8-24 is a class C amplifier with a negative supply voltage V_{BB} applied to the base circuit. This negative voltage reverse-biases the base-emitter junction of the transistor. The transistor will not conduct until the input signal overcomes this reverse bias. This happens for only a small part of the input cycle. The transistor conducts for only a small part (90 degrees or less) of the input waveform.

As shown in Fig. 8-24, the collector-current waveform is not a whole sine wave. It is not even half a sine wave. This extreme distortion means the class C amplifier *cannot* be used for audio work. Class C amplifiers are used at *radio frequencies.*

Figure 8-24 shows the *tank circuit* in the collector circuit of the class C amplifier. This tank circuit restores the sine wave input signal. Note that a sine wave is shown across R_L. Tank circuits can restore sine wave signals but not rectangular waves or complex audio waves.

Tank circuit action is explained in Fig. 8-25. The collector-current pulse charges the capacitor [Fig. 8-25(a)]. After the pulse, the capacitor discharges through the inductor [Fig. 8-25(b)]. Energy is stored in the field around the inductor. When the capacitor discharges to zero, the field collapses and keeps the current flowing

Class C

Tank circuit

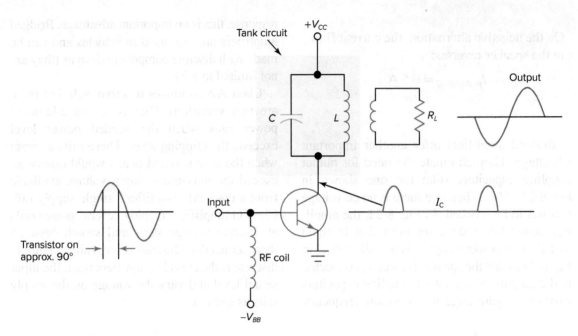

Fig. 8-24 A class C amplifier.

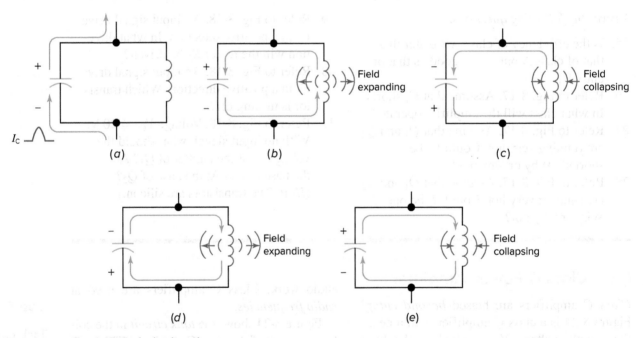

Fig. 8-25 Tank circuit action.

[Fig. 8-25(c)]. This charges the capacitor again, but note that the polarity is opposite. After the field has collapsed, the capacitor again begins discharging through the inductor [Fig. 8-25(d)]. Note that current is now flowing in the opposite direction, and the inductor field is expanding. Finally, the inductor field begins to collapse, and the capacitor is charged again to its original polarity [Fig. 8-25(e)].

Tank circuit action results from a capacitor discharging into an inductor that later discharges into the capacitor, and so on. Both the capacitor and the inductor are energy storage devices. As the energy transfers from one to the other, a sine wave is produced. Circuit loss (resistance) will cause the sine wave to decrease gradually. This is shown in Fig. 8-26(a); the wave is called a *damped sine wave*. By pulsing

Damped sine wave

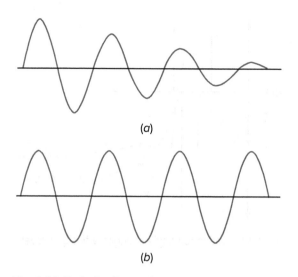

(a)

(b)

Fig. 8-26 Tank circuit waveforms.

the tank circuit every cycle, the sine wave amplitude can be kept constant. This is shown in Fig. 8-26(b). In a class C amplifier, the tank circuit is recharged by a collector-current pulse every cycle. This makes the sine wave output constant in amplitude.

The values of inductance and capacitance are important in a class C amplifier tank circuit. They must *resonate* at the frequency of the input signal. The equation for resonance is

$$f_r = \frac{1}{2\pi\sqrt{LC}}$$

What is the resonant frequency of a tank circuit that has 100 pF of capacitance and 1 μH of inductance? Substituting the values into the equation, we get

$$f_r = \frac{1}{6.28\sqrt{1 \times 10^{-6} \times 100 \times 10^{-12}}}$$

$$= 15.9 \times 10^6 \text{ Hz}$$

The resonant frequency is 15.9 MHz.

In some cases, the tank circuit is tuned to resonate at two or three times the frequency of the input signal. This produces an output signal two or three times the frequency of the input signal. Such circuits are called *frequency multipliers*. They are commonly used where high-frequency signals are needed. For example, suppose that a 150-MHz two-way transmitter is being designed. It is often easier to initially develop a lower frequency. The lower frequency can be multiplied up to the working frequency. Figure 8-27 shows the block diagram for such a transmitter.

Frequency multipliers

The class C amplifier is the *most efficient* of all the analog amplifier classes. Its high efficiency is shown by the waveforms in Fig. 8-28. The top waveform is the input signal. Only the positive peak of the input forward-biases the base-emitter junction. This occurs at 0.6 V in a silicon transistor. Most of the input signal falls below this value because of the negative bias (V_{BE}). The middle waveform in Fig. 8-28 is the collector current I_C. The collector current is in the form of narrow pulses. The bottom waveform is the output signal. It is sinusoidal because of tank-circuit action. Note that the collector-current pulses occur when the output waveform is near zero. This means that little power will be dissipated in the transistor:

Resonate

$$P_C = V_{CE} \times I_C = 0 \times I_C = 0 \text{ W}$$

If no power is dissipated in the transistor, it must all become signal power. This leads to the conclusion that the class C amplifier is 100 percent efficient. Actually, power is dissipated in the transistor. Voltage V_{CE} is low but not zero when the transistor is conducting. The tank circuit will also cause some loss. Practical class C

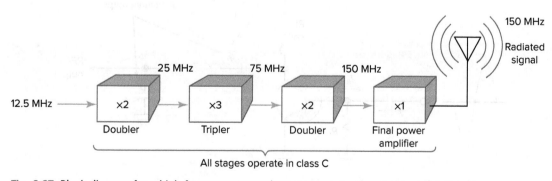

Fig. 8-27 Block diagram for a high-frequency transmitter.

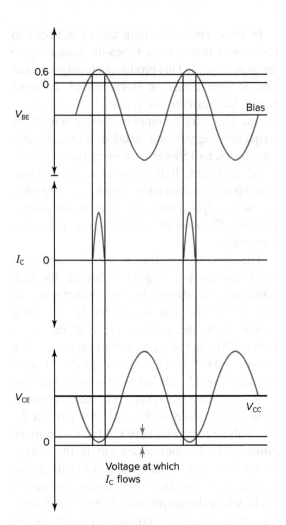

Fig. 8-28 Class C amplifier waveforms.

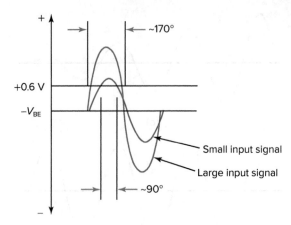

Fig. 8-30 Conduction angle changes with signal level.

Signal bias

power amplifiers can achieve efficiencies as high as 85 percent. Class C amplifiers are very popular at radio frequencies where tank circuits can restore the sine wave signal.

Practical class C power amplifiers seldom use a fixed bias supply for the base circuit.

A better way to do it is to use *signal bias*. This is shown in Fig. 8-29. As the input signal goes positive, it forward-biases the base-emitter junction. Base current I_B flows as shown. The current charges C_1. Resistor R_1 discharges C_1 between positive peaks of the input signal. Resistor R_1 cannot completely discharge C_1, and the remaining voltage across C_1 acts as a bias supply. Capacitor C_1's polarity reverse-biases the base-emitter junction.

One of the advantages of signal bias is that it is self-adjusting according to the level of the input signal. Class C amplifiers are designed for a small conduction angle to heighten the efficiency. If an amplifier uses fixed bias, the conduction angle will increase if the amplitude of the input signal increases. Figure 8-30 shows why. Two conduction angles can be seen for the fixed bias of $-V_{BE}$ shown on the graph. The angle is approximately 90 degrees for the small signal and approximately 170 degrees for the large signal. A 170-degree conduction

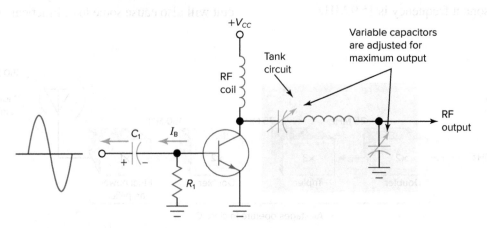

Fig. 8-29 Class C amplifier using signal bias.

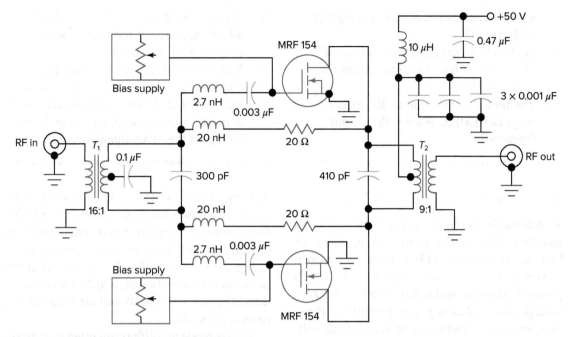

Fig. 8-31 Motorola I-kW RF power amplifier.

angle is too large, and the amplifier efficiency would suffer. The amplifier might also overheat since the average current flow would be much greater. Signal bias overcomes this problem because the conduction angle tends to remain constant. For example, if the input signal in Fig. 8-29 becomes larger, the average charge on C_1 will increase. This will increase the reverse bias $-V_{BE}$. More reverse bias means a smaller conduction angle. The signal bias circuit automatically adjusts to changes in the amplitude of the input signal and tends to keep the conduction angle constant.

Figure 8-29 also shows a different type of tank circuit. It is known as an *L network*. It matches the impedance of the transistor to the load. Radio-frequency power transistors often have an output impedance of about 2 Ω. The standard load impedance in RF work is 50 Ω. Thus, the *L* network is necessary to match the transistor to 50 Ω.

Figure 8-31 shows a high-power RF amplifier using VFETs or power MOSFETs. The amplifier is a push-pull design and develops over 1 kW of output power over the range of 10 to 90 MHz. Its power gain ranges from 11 to 14 dB over this frequency range.

The circuit in Fig. 8-31 uses negative feedback to achieve such a wide frequency range. The 20-nH inductors and 20-Ω resistors feed a part of the drain signal back into the gate circuit of each transistor. The transformers T_1 and T_2 are special wideband, ferrite core types. These also contribute to the wide frequency range of the amplifier.

It is possible to improve class C amplifiers by using the transistors as switches. Class E and class F amplifiers, which are often used at microwave frequencies, use matching networks designed so the transistor voltage peaks occur at minimum current. This decreases transistor dissipation and gives higher efficiency.

L network

Large-Signal Amplifiers **Chapter 8** 247

expected across R_L, assuming a high-Q tank circuit?

34. Is class C more efficient than class B?
35. Does class C have the smallest conduction angle?
36. The input frequency to an RF tripler stage is 10 MHz. What is the output frequency?
37. A tank circuit uses a 6.8-μH coil and a 47-pF capacitor. What is the resonant frequency of the tank?
38. Refer to Fig. 8-29. Assume that the amplifier is being driven by a signal and the voltage at the base of the transistor is negative. What should happen to the base voltage if the drive signal increases?

8-6 Switch-Mode Amplifiers

Switch-mode amplifiers operate transistors as switches. This makes them very efficient (as high as 95 percent). This is because an open switch has zero current and therefore zero power dissipation, and a closed switch has zero voltage drop and zero power dissipation. However, in practice there will be some small voltage drop across a transistor even when it is turned on very hard; thus, there is some dissipation but far less than that found in the amplifiers presented so far in this chapter.

Figure 8-32 is a log-log graph of efficiency versus power output. The three amplifiers are identical in terms of their maximum power output (10 W). The theoretical efficiency of class A reaches 25 percent when driven to maximum output, and it is 78.5 percent for class B. These were calculated earlier in this chapter. Look at the switch-mode curve (*class D*), which reaches an efficiency of 90 percent at full output. In typical use, the class D amplifier is much better

Class D

than the others. A 10-W amplifier rarely operates at that power level. A rule of thumb is that, on average, a properly sized audio amplifier operates at one-tenth of its maximum power rating. At an output of 1 W, the class D amplifier is 80 percent efficient, which is much better than 30 percent (class B) and far better than 3 percent (class A).

Switch-mode amplifiers are often used where

- High powers are required.
- Devices are battery operated.
- Devices must be as small as possible.
- Devices must be as lightweight as possible.
- Heat sinks are not desirable (often not needed up to 10 W).

Actual applications include home theater systems, automobile sound systems, large auditorium systems, and subwoofers. They are generally considered inferior to the best audio amplifiers due to noise and distortion. However,

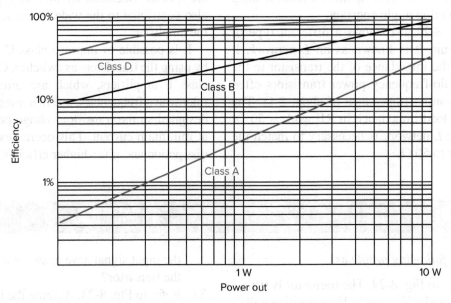

Fig. 8-32 Efficiency curves for three amplifier classes.

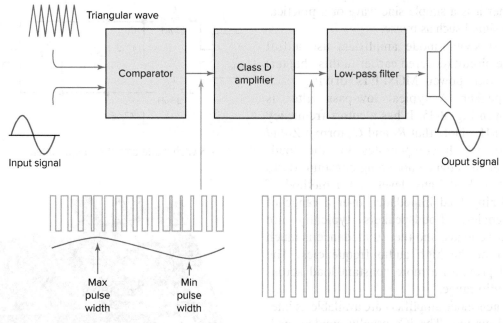

Fig. 8-33 Block diagram of a switch-mode amplifier.

they have been steadily improving and now rival high-end amplifiers. In any case, the loudspeaker is usually the major source of distortion. Noise can be a problem. Electromagnetic interference (EMI) might exclude switch-mode amplifiers from some applications.

Figure 8-33 shows a block diagram for a switch-mode amplifier. The audio signal is represented by a sine wave. It is *compared* to a triangle wave and converted to a pulse-width modulation signal. Comparators are covered in Chap. 9. The PWM signal is full on, full off, or in rapid transition between the two. Thus, the class D amplifier transistors are driven full on or full off for very low power loss in the transistors. The signal usually must be passed through a low-pass filter to smooth the waveform before sending it to the loudspeaker. The triangle wave

operates at a very high frequency to make the job of the low-pass filter easier. Switch-mode amplifiers usually operate at switching frequencies far above the highest audio frequency (20 kHz). The triangular signal in Fig. 8-33 will typically be hundreds of kilohertz.

Figure 8-34 superimposes the waveforms of the amplifier in Fig. 8-33. The triangle wave is compared to the sine wave. At the moment the two instantaneous amplitudes are close to equal, the comparator output is ready to switch from low to high or from high to low. Note that the pulse width of the comparator output signal is proportional to the amplitude of the sine wave. Higher amplitudes yield larger pulse widths, and lower amplitudes yield smaller pulse widths. Low-pass filtering the PWM signal will restore the shape of the input signal,

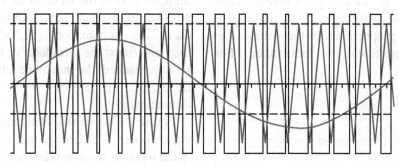

Fig. 8-34 Switch-mode amplifier waveforms.

whether it is a simple sine wave or a practical audio signal such as music.

Most switch-mode amplifiers use a full bridge circuit (covered earlier in this chapter) with four power MOSFETs driving each loudspeaker. A typical low-pass filter is shown in Fig. 8-35. It has a cutoff frequency of 27 kHz. Note that R_1 and C_4 form a *Zobel circuit*, which compensates for the loudspeaker impedance not being constant. (Otto Zobel of Bell Labs developed a method of equalizing load impedance over a range of frequencies.) Loudspeakers typically are inductive loads, and their impedance is much higher at the high audio frequencies. The Zobel provides a more constant load across the audio range.

Switch-mode amplifiers are available as integrated circuits. The ICs usually employ feedback to improve distortion and to increase the power supply rejection ratio (PSRR). Without feedback and compensation, the PSRR is 0 dB, and with feedback, it can be as high as 80 dB. Since they use linear feedback techniques, ICs are not true digital amplifiers. Figure 8-36 shows a 500-W stereo audio amplifier by Class D Audio. It measures only $5\frac{1}{2} \times 4\frac{1}{4}$ in. and requires a $\pm$ 65-V dc power supply. The THD is rated at 0.02 percent and the efficiency at over 90 percent.

Some class D amplifiers with more modest power ratings are available with no output inductors or with simple ferrite bead inductors. Since the switching frequencies are so high, the loudspeakers can sometimes serve as low-pass filters. However, EMI can occur, especially with long speaker leads. An external filter may be needed in some cases.

Switch-mode power amplifiers are also used with inductive loads, such as motors. In Fig. 8-37(*b*), the switch has been closed, and the inductor is charging. Current flow increases linearly with time. The actual rate of current rise is proportional to the voltage and inversely proportional to inductance. For

Core saturation

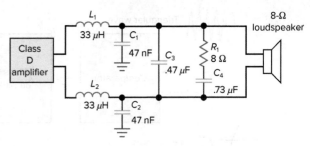

Fig. 8-35 Switch-mode amplifier output filter.

Fig. 8-36 Model SDS-500 from Class D Audio.
Courtesy of Class D Audio LLC.

example, in the case of a 100-V supply and a 100-mH inductor,

$$\frac{\Delta I}{\Delta T} = \frac{V}{L} = \frac{100}{0.1} = 1{,}000 \text{ A/s}$$

One thousand amperes is quite a current! However, if the switch is closed for only 1 ms, the current flow will reach 1 A. In other words, in a PWM amplifier the switch (or transistor) is on for a relatively short period of time. Another reason why the devices are only on briefly is core saturation. If the current flow becomes too high, the core of the inductor (or motor) will not be able to conduct any additional magnetic flux. This phenomenon is called *core saturation* and must be avoided. If the saturation point is reached, the inductance decreases drastically, and the rate of current rise increases dramatically.

Figure 8-37(*c*) shows what happens when the switch opens. The energy that has been stored in the inductor must be released. The field starts to collapse and maintains the inductor current

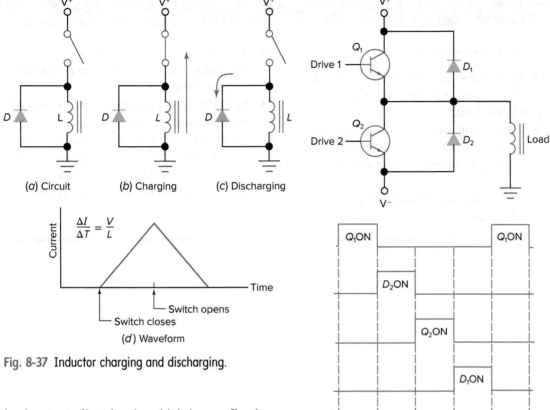

(a) Circuit   (b) Charging   (c) Discharging

$$\frac{\Delta I}{\Delta T} = \frac{V}{L}$$

(d) Waveform

Fig. 8-37 Inductor charging and discharging.

Fig. 8-38 Pulse-width modulation circuit and waveforms.

Freewheeling diodes

in the same direction in which it was flowing when it was charging. That is the purpose of the diode. It becomes forward-biased when the switch opens and allows the current to decay back to zero, as shown in Fig. 8-37(d). Diodes used in this way are called *freewheeling diodes*. Without the diode, an extremely high voltage would be generated at the moment when the switch opens. This would cause arcing or device damage in the case of a solid-state switch. Power field-effect transistors may not need an external freewheeling diode since they have an integral body diode that serves the same purpose.

Figure 8-38 shows a more practical arrangement. It is capable of providing load current in both directions. When Q_1 is on, the inductor current is to the positive supply. When Q_2 is on, the inductor current flows from the negative supply. Q_1 and Q_2 must never be on at the same time since this would provide a short-circuit path from $V-$ to $V+$.

Figure 8-38 also shows some possible waveforms. Starting at the top, Q_1 is on for a period of time. The inductor current increases from zero during this interval. When Q_1 switches off, D_2 is forward-biased, and current continues to flow until the inductor is discharged. Then Q_2

is switched on, and the load current again begins to increase, but now in an opposite direction. The negative supply is now providing the current. When Q_2 switches off, D_1 comes on and discharges the inductor. The load waveform shows that the average current is zero. Stated another way, there is no dc component in the load current.

If the positive and negative supplies in Fig. 8-36 have output filter capacitors, they are charged when the diodes are conducting. Or, if the supply uses rechargeable batteries, the batteries are charged by the diodes. This action is called *regeneration*. The energy stored in the inductors is returned and not wasted as heat. Electric vehicles might also use *regenerative braking*. This takes place when drive motors are turned into generators to slow down the vehicle.

Regeneration

Answer the following questions.

39. A digital amplifier applies 50-V pulses to a 150-mH load. Find the rate of current rise in the load.

40. When a digital amplifier is used with an inductive load, the output must not be on long enough to cause magnetic core _____.

41. The freewheeling components in Fig. 8-37 are _____.

42. The efficiency of digital amplifiers is significantly _____ than the efficiency of linear amplifiers.

43. In Fig. 8-38, transistors Q_1 and Q_2 should not be turned on at the _____.

44. When PWM is used to produce a sinusoidal load current, the digital switching frequency should be significantly _____ than the sinusoidal frequency.

Chapter 8 Summary and Review

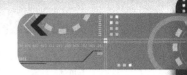

Summary

1. All amplifiers are technically power amplifiers. Only those that handle large signals are called power amplifiers.
2. The power amplifier is usually the last stage in the signal processing system.
3. Power amplifiers should be efficient. Efficiency is a comparison of the signal power output to the dc power input.
4. Poor efficiency in a power amplifier means the power supply will have to be larger and more expensive. It also means that the amplifier will convert too much electrical energy to heat.
5. Class A amplifiers operate at the center of the load line. They have low distortion and a conduction angle of 360 degrees.
6. Class B operates at cutoff. The conduction angle is 180 degrees.
7. Class B amplifiers do not present a fixed drain on the power supply. The drain is zero with no input signal.
8. Class B is more efficient than class A.
9. Bias controls the operating point and the class of operation in analog amplifiers.
10. The maximum theoretical efficiency for class A operation is 25 percent. With transformer coupling, it is 50 percent.
11. In a transformer-coupled amplifier, the impedance ratio is equal to the square of the turns ratio.
12. The fixed drain on the power supply is a major drawback with class A circuits. Efficiency is very poor when signals are small.
13. A single class B transistor will amplify half the input signal.
14. Two class B transistors can be operated in push-pull.
15. The maximum theoretical efficiency for class B is 78.5 percent.
16. A class B amplifier draws less current from the supply for smaller signals.
17. For a given output power, class B transistors will require only one-fifth of the power rating needed for class A.
18. The biggest drawback to class B push-pull is crossover distortion.
19. Crossover distortion can be eliminated by providing some forward bias for the base-emitter junctions of the transistors.
20. Class AB amplifiers are forward-biased slightly above cutoff.
21. Class AB operation is popular for high-power audio work.
22. Push-pull operation can be obtained without center-tapped transformers by using a PNP-NPN pair.
23. An amplifier that uses a PNP-NPN pair for push-pull operation is called a complementary symmetry amplifier.
24. Complementary pairs have symmetrical characteristic curves.
25. Diodes may be used to stabilize the operating point in power amplifiers.
26. Bridged amplifiers quadruple the maximum output power for any given supply voltage and load resistance, and they eliminate the output coupling capacitor.
27. Class C amplifiers are biased beyond cutoff.
28. The conduction angle for class C is around 90 degrees.
29. Class C amplifiers have too much distortion for audio work. They are useful at radio frequencies.
30. A tank circuit can be used to restore a sine wave signal in a class C amplifier.
31. The tank circuit should resonate at the signal frequency. In a frequency multiplier, the tank resonates at some multiple of the signal input frequency.
32. The class C amplifier has a maximum theoretical efficiency of 100 percent. In practice, it can reach about 85 percent.
33. Switch-mode power amplifiers may also be called class D or digital power amplifiers.
34. Switch-mode amplifiers often use pulse-width modulation and are noted for their high efficiency.

Related Formulas

Efficiency $= \dfrac{P_{out}}{P_{in}} \times 100\% = \dfrac{P_{ac}}{P_{dc}} \times 100\%$

Average values: $I_{av} = I_p \times 0.637$
(sine wave)

Dissipation $= P_{in} - P_{out}$

$$P_{av} = P_{dc} = V_{dc} \times I_{av}$$

RMS values: $V_{rms} = \dfrac{V_{p-p}}{2} \times 0.707$
(sine waves)

Resonance: $f_r = \dfrac{1}{2\pi\sqrt{LC}}$

$$I_{rms} = \dfrac{I_{p-p}}{2} \times 0.707$$

Collector dissipation: $P_C = V_{CE} \times I_C$

Signal (ac) power: $P_{ac} = V_{rms} \times I_{rms}$

Inductor rate of current change: $\dfrac{\Delta I}{\Delta T} = \dfrac{V}{L}$

Chapter Review Questions

Answer the following questions.

8-1. Refer to Fig. 8-1. In which of the three stages is efficiency the most important? (8-1)

8-2. An amplifier delivers 60 W of signal power. Its power supply develops 28 V, and the current drain is 4 A. What is the efficiency of the amplifier? (8-1)

8-3. An amplifier has an efficiency of 45 percent. It is rated at 5 W of output. How much current will it draw from a 12-V battery when delivering its rated output? (8-1)

8-4. Which class of amplifier produces the least distortion? (8-1)

8-5. What is the conduction angle of a class B amplifier? (8-1)

8-6. The operating point for an amplifier is at the center of the load line. What class is the amplifier? (8-2)

8-7. Refer to Fig. 8-7. What is the maximum theoretical efficiency for this circuit? At what signal level is this efficiency achieved? (8-2)

8-8. What will happen to the efficiency of the amplifier in Fig. 8-7 as the signal level decreases? (8-2)

8-9. Refer to Fig. 8-7. What turns ratio will be required to transform the 80-V load to a collector load of 1.28 kΩ? (8-2)

8-10. A class A transformer-coupled amplifier operates from a 9-V supply. What is the maximum peak-to-peak voltage swing at the collector? (8-2)

8-11. Refer to Fig. 8-9. What would have to be done to V_{CC} so the circuit could use PNP transistors? (8-3)

8-12. Refer to Fig. 8-9. With zero signal level, how much current will be taken from the 16-V supply? (8-3)

8-13. Refer to Fig. 8-9. What is the phase of the signal at the base of Q_1 as compared with the base of Q_2? What component causes this? (8-3)

8-14. Refer to Fig. 8-9. Assume the peak-to-peak sine wave swing across the collectors is 24 V. The transformer is 100 percent efficient. Calculate
   a. V_{p-p} across the load (do not forget to use half the turns ratio)
   b. V_{rms} across the load
   c. P_L (load power) (8-3)

8-15. Calculate the minimum wattage rating for each transistor in a push-pull class B amplifier designed for 100-W output. (8-3)

8-16. Calculate the minimum wattage rating for a class A power transistor that is transformer-coupled and rated at 25-W output. (8-3)

8-17. At what signal level is crossover distortion most noticeable? (8-3)

8-18. Refer to Fig. 8-17. Which two components set the forward bias on the base-emitter junctions of Q_1 and Q_2? (8-4)

8-19. Refer to Fig. 8-17. There is no input signal. Will the amplifier take any current from the power supply? (8-4)

8-20. Refer to Fig. 8-17. What will happen to the current taken from the power supply as the signal level increases? (8-4)

8-21. Refer to Fig. 8-18. What is the configuration of Q_1? (8-4)

8-22. Refer to Fig. 8-18. What is the configuration of Q_2? (8-4)

8-23. Refer to Fig. 8-21. When the input signal goes negative, what supplies the energy to the load? (8-4)

8-24. The major reason for using class AB in a push-pull amplifier is to eliminate distortion. What name is given to this particular type of distortion? (8-4)

8-25. Refer to Fig. 8-22. Assume that a signal drives the input positive. What happens to the current flow in Q_1 and Q_3? (8-4)

8-26. Refer to Fig. 8-22. Assume that a signal drives the input positive. What happens at the top of R_L? (8-4)

8-27. Refer to Fig. 8-22. Assume that Q_2 is turned on harder (conducts more). What should happen to Q_4? (8-4)

8-28. Refer to Fig. 8-22. Which two transistors operate in complementary symmetry? (8-4)

8-29. Which nondigital amplifier class has the best efficiency? (8-5), (8-6)

8-30. Refer to Fig. 8-24. What allows the output signal across the load resistor to be a sine wave? (8-5)

8-31. Refer to Fig. 8-24. What makes the conduction angle of the amplifier so small? (8-5)

8-32. Refer to Fig. 8-29. What does the charge on C_1 accomplish? (8-5)

8-33. Refer to Fig. 8-29. What two things does the tank circuit accomplish? (8-5)

8-34. Refer to Fig. 8-29. What will happen to the reverse bias at the base of the transistor if the drive level increases? (8-5)

Chapter Review Problems

8-1. Determine the efficiency of an amplifier that produces 18 W of output when drawing 300 mA from a 200-V supply. (8-1)

8-2. Determine the dissipation for the amplifier in problem 8-1. (8-1)

8-3. An amplifier provides 130 W, is 40 percent efficient, and operates from a 24-V supply. How much current will it draw from the supply when supplying maximum output? (8-1)

8-4. Refer to Fig. 8-5. Change the supply voltage to 24 V. Determine the maximum, nondistorted signal power, the quiescent supply current, and the efficiency when fully driven. (8-2)

8-5. Find the efficiency for the circuit in problem 8-4 when the amplifier is driven to one-fourth of its maximum value. (8-2)

8-6. Refer to Fig. 8-7. Change the turns ratio to 2.5:1. What is the ac load presented to the collector of the transistor? (8-2)

8-7. Determine the rms load voltage for problem 8-6 when the collector signal is 15 V peak-to-peak. (8-2)

8-8. Refer to Fig. 8-10. Change the turns ratio to 10:1. What signal load is seen by the collectors of the transistors? (8-3)

8-9. What is the minimum transistor power rating for a class A, 20-W, transformer-coupled amplifier? (8-3)

8-10. Suppose the circuit in problem 8-9 is replaced with a class B push-pull design. What is the required power rating for each transistor? (8-3)

8-11. Refer to Fig. 8-17. Assume that V_{BE} is 0.6 V, the supply is 20 V, and R_1 is 4.7 kΩ. Calculate the required value for R_2. (8-4)

8-12. A ground-tied load amplifier operates from a 12-V supply. Determine the maximum power that it can deliver to a 4-Ω load (do not correct for any transistor loss). (8-4)

8-13. A bridge-tied load amplifier operates from a 12-V supply. Determine the maximum power that it can deliver to a 4-Ω load (do not correct for any transistor loss). (8-4)

8-14. A class C amplifier tank circuit has a 0.2-μH inductor and a 22-pF capacitor. What is its resonant frequency? (8-5)

8-15. Determine the rate of current rise if a 35-mH inductor has 240 V across it. (8-6)

8-16. Assume, in problem 8-15, that the voltage has just been applied across the inductor. What is the current value 200 ms later? What will a graph from 0 s to 200 ms look like for this circuit? (8-6)

Critical Thinking Questions

8-1. Why can't the theoretical efficiency of an amplifier exceed 100 percent?

8-2. Can you identify any problem that could occur when the power transistors in a push-pull amplifier are poorly matched?

8-3. Can you think of any way to alleviate the problem identified in question 8-2?

8-4. Why is the failure rate for power amplifiers greater than that or small-signal amplifiers?

8-5. Amplifiers capable of power output levels in excess of a kilowatt are often based on vacuum tube technology. Why?

8-6. The diodes used to thermally compensate transistor power amplifiers are sometimes physically mounted to the same heat sink that is used to cool the transistors. Why?

8-7. Suppose you are working on an RF power amplifier similar to the one shown in Fig. 8-29. The amplifier repeatedly blows the V_{CC} fuse. What could be wrong?

Answers to Self-Tests

1. F
2. T
3. F
4. F
5. T
6. T
7. 40 percent
8. 167 W
9. 0.768 W
10. 0.6 W
11. 720 Ω
12. 7.5 V peak-to-peak; 88 mW
13. F
14. T

15. F
16. F
17. T
18. T
19. 800 Ω; 800 Ω
20. 10 W
21. 2 W
22. 0.4 W
23. 1.02 W
24. 39.2 percent (Note: This is half the efficiency achieved for driving the amplifier to its maximum swing.)

25. yes
26. class B
27. No, because this would remove forward bias and tend to make them run cooler.
28. Yes, because this would increase forward bias.
29. positive
30. Q_2
31. 10, 10.7, and 9.3 V
32. 6.6 to 6.7 V
33. sine wave

34. yes
35. yes
36. 30 MHz
37. 8.9 MHz
38. It should increase (go more negative).
39. 333 A/s
40. saturation
41. diodes
42. better
43. same time
44. greater

CHAPTER 9

Operational Amplifiers

Learning Outcomes

This chapter will help you to:

9-1 *Predict* the phase relationships in differential amplifiers. [9-1]

9-2 *Determine* the CMRR for differential amplifiers. [9-2]

9-3 *Calculate* the power bandwidth for operational amplifiers. [9-3]

9-4 *Find* voltage gain for operational amplifiers [9-4]

9-5 *Determine* the small-signal bandwidth for operational amplifiers. [9-5]

9-6 *Identify* various applications for operational amplifiers. [9-6]

9-7 *Discuss* the operation and application of comparators. [9-7]

Thanks to integrated-circuit technology, differential and operational amplifiers are inexpensive, offer excellent performance, and are easy to apply. This chapter deals with the theory and characteristics of these amplifiers. It also covers some of their many applications.

9-1 The Differential Amplifier

An amplifier can be designed to respond to the *difference* between two input signals. Such an amplifier has *two inputs* and is called a *difference,* or *differential,* amplifier. Figure 9-1 shows the basic arrangement. The $-V_{EE}$ supply provides forward bias for the base-emitter junctions, and $+V_{CC}$ reverse-biases the collectors. Such supplies are called *dual supplies* or *bipolar supplies.* Two batteries can be used to form a bipolar supply, as in Fig. 9-2. Figure 9-3 shows a bipolar rectifier circuit.

A differential amplifier can be driven at *one* of its inputs. This is shown in Fig. 9-4. An output signal will appear at *both* collectors. Assume that the input drives the base of Q_1 in a positive direction. The conduction in Q_1 will increase since it is an NPN device. More voltage will drop across Q_1's load resistor because of the increase in current. This will cause the collector of Q_1 to go less positive. Thus, an inverted output is available at the collector of Q_1.

In Fig. 9-4, Q_1 acts as a *common-emitter* amplifier, and that's why an inverted signal appears at its collector. However, it also serves as an e*mitter follower* and drives the emitter of Q_2. Q_2 acts as a *common-base* amplifier because its emitter is the input and its collector is the output. The emitter-follower and common-base configurations do not produce a phase inversion, which is why the signal at the collector of Q_2 is in phase with the signal source.

Difference or differential amplifier

Bipolar supply

257

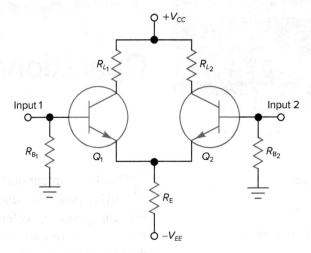

Fig. 9-I A differential amplifier.

The differential amplifier in Fig. 9-4 could also be driven at the right-hand side. In other words, the signal source could be disconnected from the base of Q_1 and connected to the base of Q_2. If this is done, then an in-phase signal would be available at the collector of Q_1, and an out-of-phase signal would be found at the collector of Q_2.

As can be seen in Fig. 9-4, both *inverted* (out-of-phase) and *noninverted* (in-phase) outputs are available. These appear from ground to either collector terminal and may be called *single-ended outputs*. A *differential output* is also available. This output is taken from the collector of Q_1 to the collector of Q_2. The differential output has twice the swing of either single-ended output. If, for example, Q_1's collector has gone 2 V negative and Q_2's collector has gone 2 V positive, the difference is $(+2) - (-2) = 4$ V.

The amplifier can also be *driven differentially*, as shown in Fig. 9-5. The advantage to this connection is that hum and noise can be significantly reduced. Power-line hum is a common problem in electronics, especially where high-gain amplifiers are involved. The 60-Hz

Single-ended output

Differential output

Common-mode signal

power circuits radiate signals that are picked up by sensitive electronic circuits. If the hum is common to both inputs (same phase), it will be rejected.

Figure 9-6 shows how hum can affect a desired signal. The result is a noisy signal of poor quality. The hum and noise can be stronger than the desired signal.

Refer to Fig. 9-7. A noisy differential signal is shown. Note that the phase of the hum signal is common. The hum goes positive to both inputs at the same time. Later, both inputs see a negative-going hum signal. This is called a *common-mode signal*. Common-mode signals are attenuated (made smaller) in differential amplifiers.

Here is what happens when the differential signal shown in Fig. 9-7 is applied to the circuit in Fig. 9-5. The blue signals are out of phase. They will be amplified because they represent a difference input to the amplifier. The hum signals (black) are in phase, and they represent no difference to the input of the amplifier and will not be amplified. As shown at the bottom of Fig. 9-7, the common-mode hum is rejected.

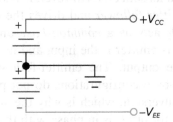

Fig. 9-2 A battery dual supply.

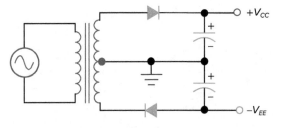

Fig. 9-3 A rectifier dual supply.

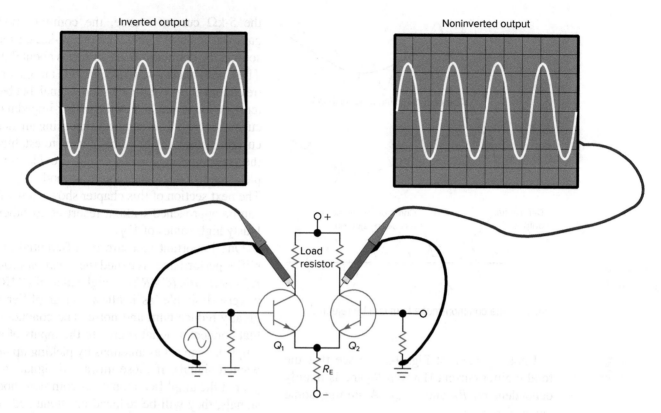

Fig. 9-4 Driving a differential amplifier at one input.

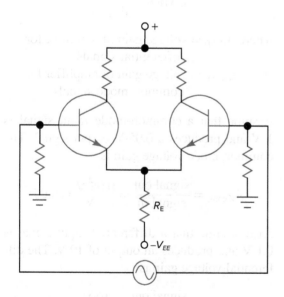

Fig. 9-5 Driving the amplifier differentially.

Understanding common-mode rejection is greatly assisted by assuming a *constant total emitter current*. If the *total* emitter current is constant, then both transistors *cannot* have increasing current at the same time because that would demand an increase in the total current. Nor can both transistors show decreasing current at the same time, as that would demand a decrease in the total current. Thus, common-mode signals won't affect the amplifier or produce any output signal because they drive both amplifier inputs in the same direction at the same time. On the other hand, a differential signal can affect the amplifier and create an output, since one transistor current can increase as the other decreases, even though the total current is constant.

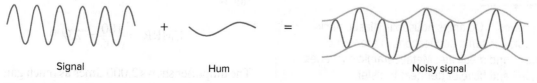

Fig. 9-6 Hum voltage can add to a signal.

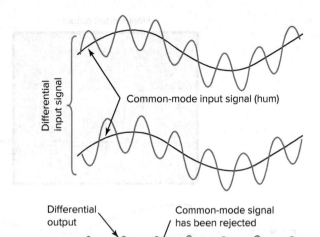

Fig. 9-7 The common-mode hum can be rejected.

Common-mode rejection ratio (CMRR)

Looking again at Fig. 9-5, we see that the total emitter current flows in R_E and is mainly dependent on R_E and $-V_{EE}$. Assuming some values for each:

$$I_{E(total)} = \frac{V_{EE}}{R_E} = \frac{10 \text{ V}}{5 \text{ k}\Omega} = 2 \text{ mA}$$

The same total emitter current can also be established by using much higher values of V_{EE} and R_E:

$$I_{E(total)} = \frac{V_{EE}}{R_E} = \frac{100 \text{ V}}{50 \text{ k}\Omega} = 2 \text{ mA}$$

Such a high value for V_{EE} is not practical but serves here to illustrate a point.

Differential amplifiers like the one shown in Fig. 9-5 provide better common-mode rejection with higher values of R_E. Why? You may recall that an ideal current source provides a constant current and has infinite resistance. Using a 50-kΩ resistor makes the total emitter current more constant and the common-mode rejection better. The 5-kΩ and 50-kΩ biasing schemes were both tested using a circuit simulator. With

the 5-kΩ current source, the common-mode gain was about 0.5, and with the 50-kΩ current source, the common-mode gain was about 0.05. Thus, both biasing schemes provide attenuation (rejection) of the commode-mode signal, but better rejection is realized by the higher-impedance current source. This suggests that using an ideal current source (infinite impedance) to establish the total emitter current would provide complete rejection of the common-mode signal. The next section of this chapter shows how this can be approached without resorting to ridiculously high values of V_{EE}.

One important measure of differential amplifier performance is called the *common-mode rejection ratio (CMRR)*. A high value of CMRR is very desirable, as it allows an amplifier to greatly reduce hum and noise. The conductors that connect a signal source to the inputs of an amplifier can act as antennas by picking up unwanted signals. If those unwanted signals appear at the amplifier's inputs as common-mode signals, they will be reduced or attenuated according to

$$\text{CMRR} = \frac{A_{V(dif)}}{A_{V(com)}}$$

where $A_{V(dif)}$ = voltage gain of amplifier for differential signals
$A_{V(com)}$ = voltage gain of amplifier for common-mode signals

Assume that a common-mode input signal is 1 V and produces a 0.05-V output signal. The common-mode voltage gain is

$$A_{V(com)} = \frac{\text{signal out}}{\text{signal in}} = \frac{0.05 \text{ V}}{1 \text{ V}} = 0.05$$

Also assume that a differential input signal is 0.1 V and produces an output of 10 V. The differential voltage gain is

$$A_{V(dif)} = \frac{\text{signal out}}{\text{signal in}} = \frac{10 \text{ V}}{0.1 \text{ V}} = 100$$

The common-mode rejection ratio of the amplifier is

$$\text{CMRR} = \frac{100}{0.05} = 2,000$$

The amplifier shows 2,000 times as much gain for differential signals as it does for common-mode

ABOUT ELECTRONICS

Plastic Op-Amp Packages Op amps housed in plastic packages are not acceptable for some applications in areas such as aerospace, the military, and medicine. Metal/ceramic packages are hermetically sealed and have better heat transfer.

signals. The CMRR is usually specified in decibels:

$$CMRR_{(dB)} = 20 \times \log 2{,}000 = 66 \text{ dB}$$

Some differential amplifiers have common-mode rejection ratios over 100 dB. They are very effective in rejecting common-mode signals.

EXAMPLE 9-1

An amplifier has a differential gain of 40 dB and a common-mode gain of −26 dB. What is the CMRR for this amplifier? When the differential and common-mode gains are expressed in decibels, the CMRR is found by subtracting:

$$CMRR = 40 \text{ dB} - (-26 \text{ dB}) = 66 \text{ dB}$$

Self-Test

Solve the following problems.

1. Refer to Fig. 9-1. What name is given to the energy source marked $+V_{CC}$ and $-V_{EE}$?
2. Refer to Fig. 9-1. Assume that input 1 and input 2 are driven 1 V positive. Ideally, what will happen at the collector terminals?
3. Refer to Fig. 9-1. Assume that a signal appears at input 1 and drives it positive. In what direction will the collector of Q_1 be driven? The collector of Q_2?
4. When a signal drives input 2 in Fig. 9-1, why is there a collector voltage change at Q_1?
5. Assume that in Fig. 9-1 a signal drives input 1 and produces an output at the collector of Q_1. This output measures 2 V peak-to-peak with respect to ground.

What signal amplitude should appear across the two collectors?
6. Refer to Fig. 9-2. Both batteries are 12 V. What is V_{CC} with respect to ground? What is V_{EE} with respect to ground? What is V_{CC} with respect to V_{EE}?
7. Refer to Fig. 9-4. What reference point is used to establish the inverted output and the noninverted output?
8. Refer to Fig. 9-5. Assume the signal source supplies 120 mV. The differential output signal is 12 V. Calculate the differential voltage gain of the amplifier.
9. Refer to Fig. 9-5. The differential voltage gain is 80. A common-mode hum voltage of 80 mV is applied to both inputs. The differential hum output is 8 mV. Calculate the CMRR for the amplifier.

9-2 Differential Amplifier Analysis

The properties of differential amplifiers can be demonstrated by working through the dc and ac conditions of a typical circuit. Figure 9-8 shows a circuit with all the values necessary to determine the dc and ac conditions.

Analyzing circuits like the one in Fig. 9-8 is made easier by making a few assumptions. One assumption is that the base leads of the transistors are at ground potential. This is reasonable since the base currents are very small, which makes the drops across R_{B1} and R_{B2} close to 0 V. The next assumption is that the transistors are turned on. If the bases are at 0 V, then the

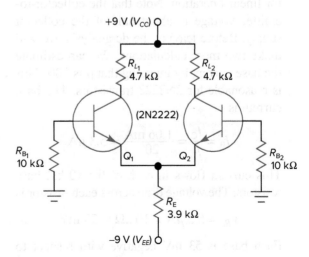

Fig. 9-8 A differential amplifier circuit.

emitters must be at −0.7 V. This condition is required to forward-bias the base-emitter junctions and turn on the transistors. Saying that the emitter is −0.7 V with respect to the base is the same as saying that the base is +0.7 V with respect to the emitter. This satisfies the bias requirements for NPN transistors.

Now that we have made our basic assumptions, we can begin the dc analysis. Knowing the voltage at both ends of R_E allows us to find its drop:

$$V_{R_E} = -9 \text{ V} - (-0.7 \text{ V}) = -8.3 \text{ V}$$

Knowing that there is a drop of 8.3 V (we now discard its sign) allows us to solve for the current flow in the emitter resistor:

$$I_{R_E} = \frac{V_{R_E}}{R_E} = \frac{8.3 \text{ V}}{3.9 \text{ k}\Omega} = 2.13 \text{ mA}$$

Assuming balance, each transistor will support half of this current. The emitter current for each transistor is

$$I_E = \frac{2.13 \text{ mA}}{2} = 1.06 \text{ mA}$$

As usual, we assume the collector currents to be equal to the emitter currents. The drop across each load resistor is

$$V_{R_L} = 1.06 \text{ mA} \times 4.7 \text{ k}\Omega = 4.98 \text{ V}$$

V_{CE} is found by Kirchhoff's voltage law:

$$V_{CE} = V_{CC} - V_{R_L} - V_E$$
$$= 9 - 4.98 - (-0.7) = 4.72 \text{ V}$$

This dc analysis shows that the dc conditions of the differential amplifier in Fig. 9-8 are good for linear operation. Note that the collector-to-emitter voltage is about half of the collector supply. Before leaving the dc analysis, we will make two more calculations. We can estimate the base current by guessing that β is 200. This is reasonable for 2N2222 transistors. The base current is

$$I_B = \frac{I_C}{\beta} = \frac{1.06 \text{ mA}}{200} = 5.3 \text{ }\mu\text{A}$$

This current flows in each of the 10-kΩ base resistors. The voltage drop across each resistor is

$$V_{R_B} = 5.3 \text{ }\mu\text{A} \times 10 \text{ k}\Omega = 53 \text{ mV}$$

Each base is 53 mV negative with respect to ground. Remember that base current flows out

of an NPN transistor. This direction of flow makes the bases in Fig. 9-8 slightly negative with respect to ground. The 53-mV value is very small, so the initial assumption was valid.

We are now prepared to do an *ac analysis* of the circuit. The first step is to find the ac resistance of the emitters:

$$r_E = \frac{50}{I_E} = \frac{50}{1.06} = 47 \text{ }\Omega$$

You may recall that ac emitter resistance can be estimated by using a 25- or a 50-mV drop. The higher estimate is more accurate for circuits like the one in Fig. 9-8.

Knowing r_E allows us to find the voltage gain for the differential amplifier. There are actually two gains to find: (1) the differential voltage gain and (2) the common-mode voltage gain. Figure 9-9 shows an ac equivalent circuit that is appropriate when the amplifier is driven at one input. Notice that r_E is shown in both emitter circuits. The differential voltage gain (A_D) is equal to the collector load resistance divided by 2 times r_E.

In Fig. 9-9, very little signal current flows in R_E, so it does not appear in the voltage gain equation. Q_1 is driven at its base by the signal source. Its emitter signal current must flow through its 47 Ω of ac emitter resistance. This emitter signal also drives the emitter of Q_2 and must overcome its 47 Ω of ac emitter resistance. Q_2 acts as a common-base amplifier in this circuit and is driven at its emitter by the emitter of Q_1. This is why the denominator of the gain equation contains $2 \times r_E$ (the two 47-Ω resistances are acting in series for signal current). R_E is much larger than the ac emitter resistances, and its effect is small enough to be ignored. In circuits of this type, the signal current in R_E is only about 1 percent of the signal current in the transistors.

The base resistors in Fig. 9-9 can also affect the differential voltage gain. When these resistors are small, they can be ignored. If the resistors are large, they will reduce the gain. The reason this happens is that the signal current flows in the base-emitter circuit, so the base resistor offers additional opposition. However, the effect of the base resistor is reduced by the current gain of the transistor. When viewed from the emitter, the base resistor appears smaller to the ac signal current. So, if the base resistors in

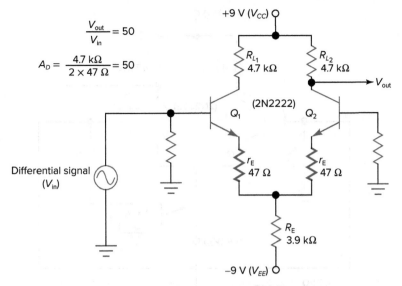

$$\frac{V_{out}}{V_{in}} = 50$$

$$A_D = \frac{4.7\ \text{k}\Omega}{2 \times 47\ \Omega} = 50$$

Fig. 9-9 AC equivalent circuit for differential signal gain.

Fig. 9-9 are fairly large, say 10 kΩ, the ac base resistance is found by

$$r_B = \frac{R_B}{\beta} = \frac{10\ \text{k}\Omega}{200} = 50\ \Omega$$

The ac base resistance decreases the differential gain:

$$
\begin{aligned}
A_{V(\text{dif})} &= \frac{R_L}{(2 \times r_E) + r_B} \\
&= \frac{4.7\ \text{k}\Omega}{(2 \times 47\ \Omega) + 50\ \Omega} \\
&= 32.6
\end{aligned}
$$

The base resistance is used only once in the gain equation (not multiplied by 2) because the signal source is applied directly to one base. In Fig. 9-9, only the base resistor on the right affects the signal current.

Considering the ac base resistance can provide a more accurate estimate of *differential gain*. However, it may not be necessary. Since we used the conservative 50-mV value for estimating r_E, the gain will probably be closer to 50 for the circuit in Fig. 9-9. Designers often use a very conservative approach to ensure that the actual circuit gain will be at least as high as their calculated value. Too much gain is an easier problem to solve than not enough gain.

A differential gain of 50 is very respectable. As we will see, the common-mode gain is much less. Figure 9-10 shows the ac equivalent circuit for *common-mode gain*. Here, the 47-Ω emitter resistances of the transistors are eliminated. They are so small compared with 7.8 kΩ that they can be ignored. R_E is physically a 3.9-kΩ resistor. However, it appears to be twice that value because it supports both transistor currents. As discussed in the first section of this chapter, the ideal situation is a constant total emitter current. However, this is not the case with Fig. 9-10. A common-mode signal will change the total emitter current because the impedance is not infinite. In the case where a common-mode signal is driving both bases in a positive direction, both transistors are turned on harder. R_E must support twice the increase in current that it would if it were serving just a single transistor. The output signal is taken from Q_2 in Fig. 9-10. As far as Q_2 is concerned, its emitter is loaded by 7.8 kΩ. This large resistance makes the common-mode gain less than 1:

$$
\begin{aligned}
A_{CM} &= \frac{4.7\ \text{k}\Omega}{7.8\ \text{k}\Omega} \\
&= 0.603
\end{aligned}
$$

Our ac analysis of the differential amplifier has shown a differential gain of 50 and a common-mode gain of 0.603. The ratio is

$$\frac{50}{0.603} = 82.9$$

Common-mode gain

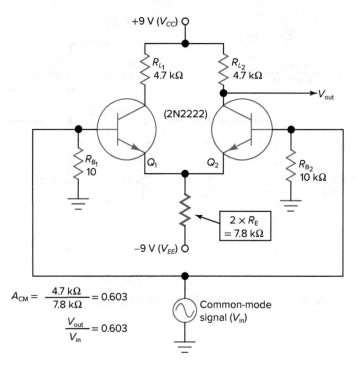

$$A_{CM} = \frac{4.7\ k\Omega}{7.8\ k\Omega} = 0.603$$

$$\frac{V_{out}}{V_{in}} = 0.603$$

Fig. 9-10 AC equivalent circuit for common-mode signal gain.

We can expect this differential amplifier to produce almost 83 times more gain for a differential signal than for a common-mode signal. This will go a long way toward eliminating noise and hum in many applications. The decibel CMRR is

$$CMRR = 20 \times \log 82.9 = 38.4\ dB$$

EXAMPLE 9-2

What is the CMRR for the circuit in Fig. 9-10 if V_{EE} is 95 V and R_E is 45 kΩ? First, we should determine if this will change the differential gain by calculating the total emitter current. With such a high value of V_{EE}, it will be safe to ignore the 0.7-V drop across the base-emitter junctions:

$$I_{RE} = \frac{V_{EE}}{R_E} = \frac{95\ V}{45\ k\Omega} = 2.11\ mA$$

This is almost the same total emitter current as before, so r_E remains the same and so does the differential gain. Next, find the common-mode gain:

$$A_{V(COM)} = \frac{R_L}{2 \times R_E} = \frac{4.7\ k\Omega}{2 \times 45\ k\Omega} = 0.0522$$

Current source

Finally, the CMRR is

$$CMRR = 20 \times \log \frac{50}{0.0522} = 59.6\ dB$$

Thus we find an improvement of 59.6 dB − 38.4 dB = 21.2 dB. Increasing R_E increases the CMRR by 21.2 dB and improves the amplifier's ability to reject unwanted common-mode signals.

Figure 9-11 shows a practical way to obtain a high CMRR. R_E has been replaced with a *current source* consisting of two resistors, a zener diode, and transistor Q_3. The zener diode is biased by the −9-V supply. The 390-Ω resistor limits the zener current. The zener cathode is 5.1 V positive with respect to its anode. This drop forward-biases the base-emitter circuit of Q_3. If we subtract for V_{BE}, we can determine the current flow in the 2.2-kΩ resistor:

$$I = \frac{5.1\ V - 0.7\ V}{2,200\ \Omega} = 2\ mA$$

The emitter current of Q_3 is 2 mA. We can make the usual assumption that the collector current is equal to the emitter current. Thus, the

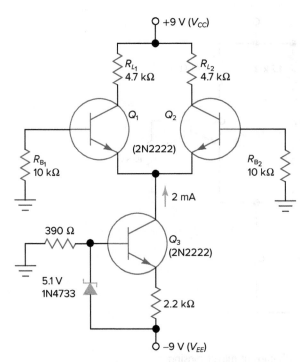

Fig. 9-II A differential amplifier with current source biasing.

collector of Q_3 in Fig. 9-11 supplies 2 mA to the emitters of the differential amplifier.

A current source such as the one shown in Fig. 9-11 has a very high ac resistance. This resistance is a function of Q_3's ac collector and ac emitter resistances and the 2.2-kΩ emitter resistor.

The ac collector resistance of small transistors typically ranges from 50 to 200 kΩ. As Fig. 9-12 shows, the collector curve is relatively flat. The collector current changes only a small amount over the 20-V range of the graph. The ac collector resistance can be found from the graph by using Ohm's law. The graph shows that the collector current change is 0.2 mA for a collector-to-emitter change of 20 V:

$$r_C = \frac{\Delta V_{CE}}{\Delta I_C} = \frac{20\ V}{0.2\ mA} = 100\ k\Omega$$

The ac emitter resistance is estimated with the familiar:

$$r_E = \frac{50\ mV}{I_E} = \frac{50\ mV}{2\ mA} = 25\ \Omega$$

The following formula can be used to estimate the ac resistance of a constant current source, such as the one shown in Fig. 9-11:

$$r_{EE} = r_C \times \left(1 + \frac{R_E}{r_E}\right)$$

$$= 100\ k\Omega \times \left(1 + \frac{2.2\ k\Omega}{25}\right)$$

$$= 8.9\ M\Omega$$

This rather high value of ac resistance makes the common-mode gain of the amplifier in Fig. 9-11 very small:

$$A_{V(com)} = \frac{R_L}{2 \times r_{EE}}$$

$$= \frac{4.7\ k\Omega}{2 \times 8.9\ M\Omega}$$

$$= 0.264 \times 10^{-3}$$

The current source biases the amplifier in Fig. 9-11 at about the same level of current as the circuit in Fig. 9-8. Therefore, the differential gain will be about the same (50). The CMRR for Fig. 9-11 is quite large:

$$CMRR = 20 \times \log \frac{50}{0.264 \times 10^{-3}} = 106\ dB$$

In practice, it is difficult to achieve such a high CMRR. However, the circuit in Fig. 9-11 is substantially better than the circuit in Fig. 9-8. When the CMRR must be optimized, matched components and laser-trimmed components can be used. Integrated circuit amplifiers with differential inputs usually have good CMRRs because the transistors and resistors tend to be well matched and track thermally (change temperature by the same amount).

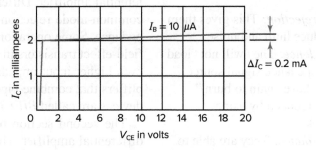

Fig. 9-I2 Typical collector curve for a 2N2222 transistor.

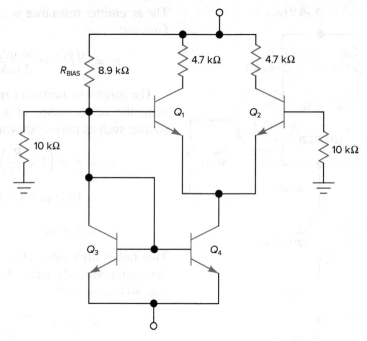

Fig. 9-13 A differential amplifier with current mirror biasing.

Figure 9-13 shows another way to bias a differential amplifier. Q_3 and Q_4 act as a *current mirror.* This term is used because the current in Q_3 is mirrored in Q_4. Current mirrors are common in integrated circuits, where it is much less expensive to form additional transistors than a zener diode. The two supplies and the bias resistor set the current in Q_3 and Q_4. With 9-V supplies, the current in the differential amplifier is about the same as the one in Fig. 9-11.

Self-Test

Solve the following problems.

10. Use Fig. 9-8 as a guide with the following changes and assumptions: the transistors are identical and $\beta = 200$, $R_{L_1} = R_{L_2} = 10$ kΩ, $R_E = 8.9$ kΩ, $R_{B_1} = R_{B_2} = 100$ kV. Use 50 mV to estimate r_E. Find: I_{R_E}, $I_{E(Q_1)}$, $I_{E(Q_2)}$, $V_{CE(Q_1)}$, $V_{CE(Q_2)}$, $V_{B(Q_1)}$, $V_{B(Q_2)}$, $A_{V(\text{dif})}$, $A_{V(\text{com})}$, and CMRR.

9-3 Operational Amplifiers

Operational amplifiers (op amps) use differential input stages. They have characteristics that make them very useful in electronic circuits. Some of their characteristics are as follows:

1. *Common-mode rejection:* This gives them the ability to reduce hum and noise.
2. *High input impedance:* They will not "load down" a high-impedance signal source.
3. *High gain:* They have "gain to burn," which is usually reduced by using negative feedback.
4. *Low output impedance:* They are able to deliver a signal to a low-impedance load.

No single-stage amplifier circuit can rate high in all the above characteristics. An operational amplifier is actually a combination of several amplifier stages. Refer to Fig. 9-14. The first section of this multistage circuit is a differential amplifier. Differential amplifiers have common-mode rejection and a high input impedance. Some operational amplifiers may use field-effect transistors in this first section for an even higher input impedance. Operational amplifiers that combine bipolar devices with FET devices are called *BIFET* op amps.

The second section of Fig. 9-14 is another differential amplifier. This allows the differential output of the first section to be used. This

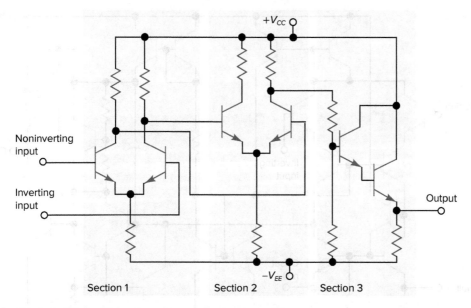

Fig. 9-I4 The major sections of an operational amplifier.

provides the best common-mode rejection and differential voltage gain.

The third section of Fig. 9-14 is a common-collector, or emitter-follower, stage. This configuration is known for its low output impedance. Notice that the output is a single terminal. No differential output is possible. This is usually referred to as a *single-ended output*. Most electronic applications require only a single-ended output.

A single-ended terminal can show only one phase with respect to ground. This is why Fig. 9-14 shows one input as *noninverting* and the other as *inverting*. The noninverting input will be in phase with the output terminal. The inverting input will be 180 degrees out of phase with the output terminal.

Figure 9-15 shows the amplifier in a simplified way. Notice the triangle. Electronic diagrams often use triangles to represent amplifiers.

Single-ended output

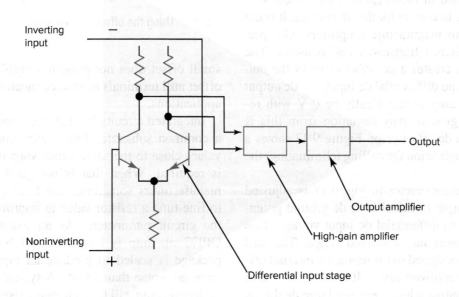

Fig. 9-I5 A simplified way of showing an operational amplifier.

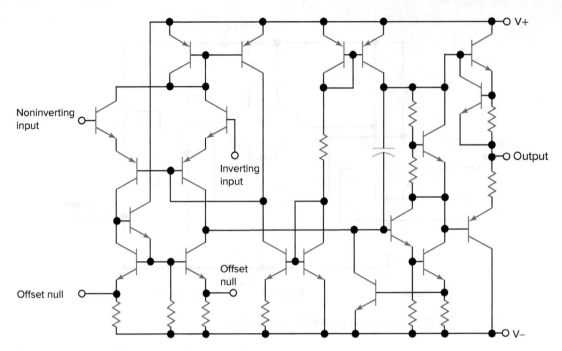

Fig. 9-16 A schematic of an operational amplifier.

Also notice that the inverting input is marked with a minus (−) sign and that the noninverting input is marked with a plus (+) sign. This is standard practice.

Figure 9-16 shows the schematic diagram of a common integrated-circuit op amp. This device has a noninverting input, an inverting input, and a single-ended output. It also has two terminals marked *offset null*. These terminals can be used in those applications where it is necessary to correct for dc offset error. It is not possible to manufacture amplifiers with perfectly matched transistors and resistors. The mismatch creates a dc offset error in the output. With no differential dc input, the dc output of an op amp should ideally be 0 V with respect to ground. Any deviation from this is known as dc offset error. Figure 9-17 shows a typical application for nulling (eliminating) the offset.

The potentiometer in Fig. 9-17 is adjusted so the output terminal is at dc ground potential with no differential dc input voltage. This potentiometer has a limited range. The null circuit is designed to overcome an internal offset in the millivolt range. It is not designed to null the output when there is a large dc differential input applied to the op amp by external circuit conditions. In many applications, a

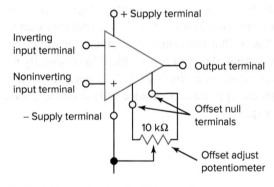

Fig. 9-17 Using the offset null terminals.

small offset does not present a problem. The offset null terminals are not connected in such applications.

Integrated circuits form all components into a common substrate. This makes component values close to the same value when matching is required. When that is not good enough, manufacturers sometimes use laser trimming to fine-tune a resistor value to improve one of the circuit parameters. As an example, the OP177 op amp is laser trimmed before the package is sealed to produce an input offset error no worse than 25 μV. A typical garden-variety op amp will have an input offset error of several millivolts. With such a low offset error, we know the amplifier output will have no dc

Offset null

offset when there is no dc input signal. In other words, the amplifier is almost perfectly balanced. The OP177 is classified as an ultra-precision amplifier and is more expensive than most op amps.

A very wide range of integrated-circuit operational amplifiers are available. The technologies used in manufacturing them include bipolar junction transistors, field-effect transistors (FET), and complementary metal oxide semiconductors. Some op amps combine device types. Examples include BIFET (bipolar field-effect transistors) and BICMOS (bipolar complementary metal oxide semiconductors). Some op amps are specialized for particular applications, such as low current drain for battery-powered devices or wide bandwidth for high-speed applications. The following list is for a general-purpose op amp such as the LM741C:

- Voltage gain: 200,000 (106 dB)
- Output impedance: 75 Ω
- Input impedance: 2 MΩ
- CMRR: 90 dB
- Offset adjustment range: ± 15 mV
- Output voltage swing: ± 13 V
- Small-signal bandwidth: 1 MHz
- Slew rate: 0.5 V/μs

The last characteristic in the list is *slew rate*. This is the maximum rate of change for the output voltage of an op amp. Figure 9-18 shows what happens when the input voltage changes suddenly. The output cannot produce an instantaneous voltage change. It *slews* (changes) a certain number of volts in a given period of time. The unit of time used to rate op amps is the microsecond (1 μs = 1 $\times$ 10^{-6} s). Some op amps have slew rates as low as 0.04 V/μs, and others are rated at 70 V/μs.

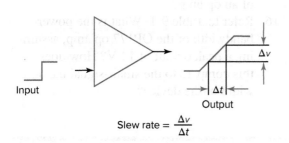

Slew rate = $\dfrac{\Delta v}{\Delta t}$

Fig. 9-18 Op-amp response to a sudden input change.

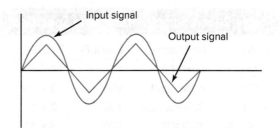

Fig. 9-19 Slew-rate distortion.

Slew rate is an important consideration for high-frequency operation. High frequency means rapid change. An op amp may not be able to slew fast enough to reproduce its input signal. Figure 9-19 shows an example of slew-rate *distortion*. Note that the input signal is sinusoidal and the output signal is triangular. The output signal from a linear amplifier is supposed to have the same shape as the input signal. Any deviation is called distortion.

Distortion

In addition to causing distortion, slew rate may prevent an op amp from producing its full output voltage swing. Large output signals are more likely to be limited than small signals. So, the factors are signal frequency, output swing, and the slew-rate specification of the op amp. The following equation predicts the maximum frequency of operation for sinusoidal input signals:

$$f_{max} = \frac{SR}{6.28 \times V_P}$$

Slew rate

where SR is the slew rate (V/μs), and V_P is the peak output.

A general-purpose op amp like the LM741C can produce a maximum peak output swing of 13 V when powered by a 15-V supply. Let's see what the maximum sine wave frequency is if its slew rate is 0.5 V/μs:

$$f_{max} = \frac{0.5 \text{ V}/\mu\text{s}}{6.28 \times 13 \text{ V}}$$

$$= \frac{1}{6.28 \times 13 \text{ V}} \times \frac{0.5 \text{ V}}{1 \times 10^{-6} \text{ s}}$$

$$= 6.12 \text{ kHz}$$

The voltage units cancel, and the expression reduces to the reciprocal of time, which is equal to frequency. Thus, 6.12 kHz may be called the *power bandwidth* of the op amp. Two things will happen if a sinusoidal input signal is significantly greater than 6.12 kHz and

Power bandwidth

Table 9-1   A Sample of Op-Amp Specifications

Device	Description	Gain (dB)	Z_{in} (Ω)	CMRR (dB)	Bandwidth (MHz)	SR (V/μs)	Supply (V)
LF412	Dual JFET	106	10^{12}	100	4	10	±18
LM318	Wide Band	106	3×10^6	100	15	50	±20
LM741	Gen Purpose	106	2×10^6	95	1.5	0.7	±15
MCP619	Bi-CMOS	100	3×10^6	100	0.19	0.08	5.5
OP177	Precision	140	45×10^6	130	0.6	0.3	±15
OPA134	Audio	120	10^{13}	100	8	20	±18
OPA151	Audio Power	97	10^{12}	113	0.055 (power)	10	±40
OPA127	CMOS	120	10^{11}	94	20	30	12
TLC2274	Rail to Rail	90	10^{12}	70	2.2	3.6	8

is large enough to drive the output to 13 V peak: (1) the output signal will show distortion (as in Fig. 9-19) and (2) the peak output swing will be less than 13 V.

A typical op amp like the LM741C has a *small-signal* bandwidth of 1.5 MHz. Large, high-frequency signals will be slew-rate-limited. The power bandwidth of an operational amplifier is less than its small-signal bandwidth. Table 9-1 lists several types of op amps along with a few of their specifications.

There are many hundreds of popular operational amplifiers. Table 9-1 gives only a glimpse of them. Online selection guides can be very helpful for choosing one.

EXAMPLE 9-3

Calculate the power bandwidth of a high-speed op amp with a slew rate of 70 V/μs when the output signal is 20 V_{p-p}. Apply the formula:

$$f_{max} = \frac{70 \text{ V}/\mu s}{6.28 \times 10 \text{ V}}$$

$$= \frac{1}{6.28 \times 10 \text{ V}} \times \frac{70 \text{ V}}{1 \times 10^{-6} \text{ s}}$$

$$= 1.11 \text{ MHz}$$

Self-Test

Solve the following problems.

11. Refer to Fig. 9-13. Which section of the amplifier (1, 2, or 3) operates as an emitter follower to produce a low output impedance?
12. Refer to Fig. 9-13. A signal is applied to the inverting input terminal. What is the phase of the signal that appears at the output terminal as compared with the input signal?
13. Refer to Fig. 9-13. Is the output of the amplifier differential or single-ended?

14. What is the name of the terminals used to null the effect of internal dc imbalance in an op amp?
15. What are two possible output effects if an input signal exceeds the power bandwidth of an op amp?
16. Refer to Table 9-1. What is the power bandwidth of the OP177 op amp, assuming a peak output of 14 V? How does this compare to the small-signal bandwidth of the device?

9-4 Setting Op-Amp Gain

A general-purpose op amp has an *open-loop* voltage gain of 200,000. Open-loop means *without feedback*. Op amps are usually operated *closed-loop*. The output, or a part of it, is fed back to the inverting (−) input. This is *negative* feedback, which reduces the gain and increases the bandwidth of the amplifier.

Figure 9-20 shows a closed-loop operational amplifier circuit. The output is fed back to the inverting input. The input signal drives the noninverting (+) input. The circuit is easy to analyze if we make an assumption: There is no difference in voltage across the op-amp inputs. What is the basis for this assumption? Considering a typical gain of 200,000, the assumption is reasonable. For example, if the output is at its maximum positive value, say 10 V, the differential input is only

$$V_{in(dif)} = \frac{V_{out}}{A_V} = \frac{10 \text{ V}}{2 \times 10^5} = 50 \ \mu\text{V}$$

Fifty microvolts is close to zero, so the assumption is valid. This is a *key point* for understanding op-amp circuits. The differential gain is *so large* that the differential input voltage can be assumed to be zero when making many practical calculations.

Now let's apply the assumption to the circuit shown in Fig. 9-20. The feedback will eliminate any voltage difference across the input terminals. If the input signal swings 1 V positive, the output terminal will do exactly the same. Since the output is fed back to the inverting input, both inputs will be at +1 V, and the differential input will be 0. If the input signal

swings 5 V negative, the output terminal will do exactly the same. Again, the differential input is 0 because of the feedback. It should be clear that in Fig. 9-20 the output *follows* the input signal. In fact, this circuit is called a *voltage follower*. $V_{out} = V_{in}$, so the circuit has a voltage gain of 1.

At first glance, an amplifier with a gain of 1 may seem to be no better than a piece of wire! However, such an amplifier can be useful if it has a high input impedance and a low output impedance. The input impedance of the voltage follower in Fig. 9-20 is approximately equal to the input resistance of the op amp times the open-loop gain: $Z_{in(CL)} \approx$ 2 MΩ × 200,000 ≈ 400 GΩ (for a 741 op amp). The output impedance of a voltage follower is approximately equal to the basic output impedance of the op amp divided by its open-loop gain. Because the open-loop gain is so high, the output impedance is 0 Ω for practical purposes:

$$Z_{out(CL)} \approx \frac{75 \ \Omega}{200 \times 10^3} = 0.375 \text{ m}\Omega$$

An amplifier that has an input impedance of 400 GΩ and an output impedance near 0 Ω makes an excellent *buffer*. Buffer amplifiers are used to isolate signal sources from any loading effects. They are also useful when working with signal sources that have rather high internal impedances.

Figure 9-21 shows an op-amp circuit that has a voltage gain greater than 1. The actual value of the gain is easy to determine. R_1 and the feedback resistor R_F form a voltage divider for the output voltage. The divided output voltage must be equal to the input voltage to satisfy the assumption that the differential input voltage is 0:

$$V_{in} = V_{out} \times \frac{R_1}{R_1 + R_F}$$

Dividing both sides by V_{out} and inverting gives

$$A_V = \frac{V_{out}}{V_{in}}$$

$$= \frac{R_1 + R_F}{R_1} = 1 + \frac{R_F}{R_1}$$

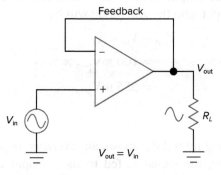

Feedback

V_{out}

R_L

V_{in}

$V_{out} = V_{in}$

Fig. 9-20 An op amp with negative feedback.

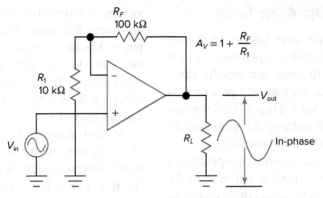

$$A_V = 1 + \frac{R_F}{R_1}$$

Fig. 9-21 A noninverting circuit with gain.

Let's apply this gain equation to Fig. 9-21:

$$A_V = 1 + \frac{R_F}{R_1}$$

$$= 1 + \frac{100 \text{ k}\Omega}{10 \text{ k}\Omega} = 11$$

Noninverting amplifier

The circuit of Fig. 9-21 is a *noninverting amplifier*. The input signal is applied to the + input of the op amp. An ac output signal will be in phase with the input signal. A dc input signal will create a dc output signal of the same polarity. For example, if the input is −1 V, the output will be −11 V (−1 V × 11 = −11 V).

A *transimpedance amplifier* converts current to voltage. If R_1 in Fig. 9-21 is replaced with a photodiode, it will produce an output voltage proportional to the diode current:

$$V_O = R_F \times I_D$$

Transimpedance amplifiers are also called current-to-voltage converters.

Figure 9-22 shows another model for amplifiers with negative feedback. This model was presented in Chap. 7.

EXAMPLE 9-4

Calculate the closed-loop gain for Fig. 9-21 by using the other model, assuming open-loop gains of 200,000 and 50,000. The feedback ratio B in Fig. 9-21 is determined by R_F and R_1, which form a voltage divider:

Inverting amplifier

$$B = \frac{R_1}{R_F + R_1} = \frac{1 \text{ k}\Omega}{10 \text{ k}\Omega + 1 \text{ k}\Omega} = 0.091$$

Applying the other model for open-loop gain $A = 200,000$ gives

$$A_{CL} = \frac{A}{AB + 1} = \frac{200,000}{(200,000)(0.091) + 1} \approx 11$$

Applying the other model for A = 50,000,

$$A_{CL} = \frac{A}{AB + 1} = \frac{50,000}{(50,000)(0.091) + 1} \approx 11$$

Please notice two important facts: (1) both models produce the same result and (2) the circuit is not sensitive to open-loop gain because of the negative feedback.

EXAMPLE 9-5

Determine the output signal (amplitude and phase) for Fig. 9-21 if R_1 is changed to 22 kΩ and V_{in} is 100 mV$_{p-p}$. First, determine the gain for the amplifier:

$$A_V = 1 + \frac{R_F}{R_1} = 1 + \frac{100 \text{ k}\Omega}{22 \text{ k}\Omega} = 5.55$$

The output signal will be in phase with the input, and the amplitude will be

$$V_{out} = V_{in} \times A_V$$

$$= 100 \text{ mV}_{p-p} \times 5.55$$

$$= 555 \text{ mV}_{p-p}$$

Figure 9-23(a) shows an *inverting amplifier*. Here, the signal is fed to the − input of the op amp. The output signal will be 180 degrees out of phase with the input signal.

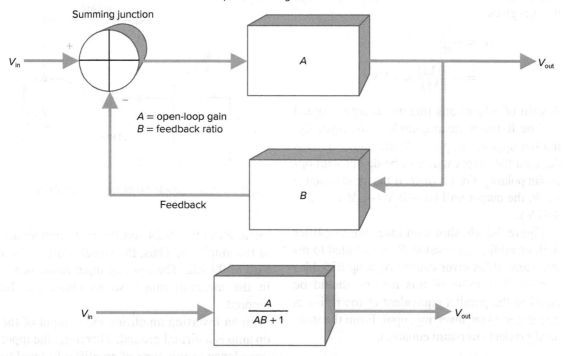

Summing junction

Amplifier with negative feedback

V_{in}

A

V_{out}

A = open-loop gain
B = feedback ratio

Feedback

B

V_{in}

$\dfrac{A}{AB+1}$

V_{out}

Fig. 9-22 Another model of negative feedback.

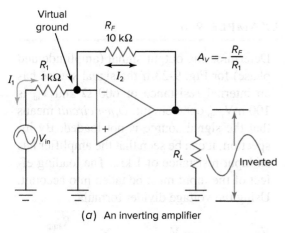

Virtual ground

R_F
10 kΩ

R_1
1 kΩ

I_1

I_2

$A_V = -\dfrac{R_F}{R_1}$

V_{in}

R_L

Inverted

(a) An inverting amplifier

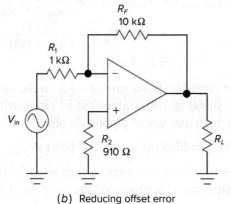

R_F
10 kΩ

R_1
1 kΩ

V_{in}

R_2
910 Ω

R_L

(b) Reducing offset error

Fig. 9-23 Inverting amplifier circuits.

The gain equation is a little different for the inverting circuit. As Fig. 9-23(a) shows, the noninverting input is at ground potential. Therefore the inverting input is also at ground potential because we again can assume that there is no difference across the inputs. The inverting input is known as a *virtual ground*. With the right end of R_1 effectively grounded (it is connected to the virtual ground), any input signal will cause a current to flow in R_1. According to Ohm's law,

Virtual ground

$$I_1 = \frac{V_{in}}{R_1}$$

Any output signal will cause a current to flow in R_F:

$$I_2 = \frac{-V_{out}}{R_F}$$

V_{out} is negative here because the amplifier *inverts*. The current into or out of the − terminal of the op amp is so small that it is effectively 0. Thus, $I_2 = I_1$, and by substitution

$$\frac{-V_{out}}{R_F} = \frac{V_{in}}{R_1}$$

Rearranging gives

$$A_V = -\frac{V_{out}}{V_{in}} = \frac{R_F}{R_1}$$

Applying the inverting gain equation to Fig. 9-23(a) gives

$$A_V = -\frac{R_F}{R_1}$$

$$= -\frac{10 \text{ k}\Omega}{1 \text{ k}\Omega} = -10$$

A gain of −10 means that an ac output signal will be 10 times the amplitude of the input signal but opposite in phase. If the input signal is dc, then the output will also be dc but with opposite polarity. For example, if the input signal is −1 V, the output will be +10 V (−1 V × −10 = +10 V).

Figure 9-23(b) shows an inverting amplifier with an additional resistor. R_2 is included to reduce any offset error caused by amplifier bias current. The value of this resistor should be equal to the parallel equivalent of the resistors connected to the inverting input. From the standard product-over-sum equation,

$$R_2 = \frac{R_1 \times R_F}{R_1 + R_F}$$

$$= \frac{1 \text{ k}\Omega \times 10 \text{ k}\Omega}{1 \text{ k}\Omega + 10 \text{ k}\Omega} = 909 \ \Omega$$

The closest standard value is 910 Ω. The amplifier bias currents will find the same effective resistance at both inputs. This will equalize the resulting dc voltage drops and eliminate any dc difference between the inputs caused by bias currents. One manufacturer of the 741 op amp lists the typical input bias current at 80 nA and the maximum value at 500 nA at room temperature.

The addition of R_2 in Fig. 9-23(b) does *not* substantially affect the signal voltage gain or the virtual ground. The current flowing in R_2 is so small that the drop across it is effectively 0. For example, if we use 80 nA and 910 Ω,

$$V = 80 \times 10^{-9} \text{ A} \times 910 \ \Omega$$

$$= 72.8 \ \mu\text{V}$$

Therefore, the noninverting input is still effectively at ground potential, and the inverting input is still a virtual ground.

AC-coupled

Figure 9-24 shows an *ac-coupled* noninverting amplifier. This situation mandates the use of R_2 to provide a dc path for the input bias current. To minimize any offset effect, R_2 is again chosen to be equal to the parallel equivalent of the resistors connected to the other op-amp

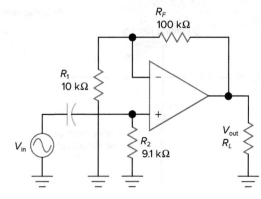

Fig. 9-24 An ac-coupled noninverting amplifier.

input. R_2 in Fig. 9-24 sets the input impedance of the amplifier. Thus, the signal source sees a load of 9.1 kΩ. The op-amp input resistance is in the megohm range, so its effect can be ignored.

In an inverting amplifier, the − input of the op amp is a virtual ground. Therefore, the input impedance of this type of amplifier is equal to the resistor connected between the signal source and the inverting input. The signal source in Fig. 9-23 sees a load of 1 kΩ.

EXAMPLE 9-6

Determine the output signal (amplitude and phase) for Fig. 9-23 if the signal source has an internal resistance of 600 Ω and V_{in} is 100 mV$_{p-p}$ open circuit. *Open circuit* means that the signal source is not loaded. By inspection, it can be seen that the amplifier has an input resistance of 1 kΩ. The loading effect of the input must be taken into account. Using the voltage divider formula,

$$V_{in(\text{closed circuit})} = V_{in(\text{open circuit})} \times \frac{R_{amp}}{R_{amp} + R_{source}}$$

$$= 100 \text{ mV}_{p-p} \times \frac{1 \text{ k}\Omega}{1 \text{ k}\Omega + 600 \ \Omega}$$

$$= 62.5 \text{ mV}_{p-p}$$

The amplifier has a gain of −10, so the output signal is 180 degrees out of phase with the input and has an amplitude of

$$V_{out} = 62.5 \text{ mV}_{p-p} \times 10 = 625 \text{ mV}_{p-p}$$

The negative gain has been accounted for by expressing the phase relationship as 180 degrees.

All op-amp circuits have limits. Two of these limits are set by the *rail voltages*. A rail is simply another name for the power supply in an op-amp circuit. If a circuit is powered by ±12 V, the positive rail is +12 V, and the negative rail is −12 V. The rail voltages cannot be exceeded by the output. In fact, the output voltage is usually limited to at least 1 V less than the rail. The most output that can be expected from an op amp powered by ±12 V is about ±11 V.

Suppose you are asked to calculate the output voltage for an inverting amplifier with a gain of −50 and an input signal of 500 mV dc. The supply is specified at ±15 V:

$$V_{out} = V_{in} \times A_V$$
$$= 500 \text{ mV} \times -50$$
$$= -25 \text{ V}$$

This output is *not* possible. The amplifier will *saturate* within about 1 V of the negative rail. The output will be about −14 V dc.

As another example, find the peak-to-peak output voltage for an op amp with a gain of 100. Assume a ±9 V supply and an ac input signal of 250 mV peak-to-peak:

$$V_{out} = V_{in} \times A_V$$
$$= 100 \times 250 \text{ mV peak-to-peak}$$
$$= 25 \text{ V peak-to-peak}$$

This output *cannot* be achieved. The maximum output swing will be from about −8 V to about +8 V, which is 16 V peak-to-peak. The output signal will be *clipped* in cases like this.

The graph in Fig. 9-25 shows gain versus frequency for a typical integrated-circuit op amp. Graphs of this type are known as *Bode plots*. They show how gain decreases as frequency increases. Notice in Fig. 9-25 that the open-loop performance curve shows a *break frequency* at about 7 Hz. This frequency is designated as f_b. The gain will *decrease at a uniform rate* as frequency is *increased* beyond the break frequency. Most op amps show a gain decrease of *20 dB per decade* above f_b.

Check the open-loop gain in Fig. 9-25 at 10 Hz and note that it is 100 dB. A *decade* increase in frequency means an increase of 10 times. Now check the gain at 100 Hz and

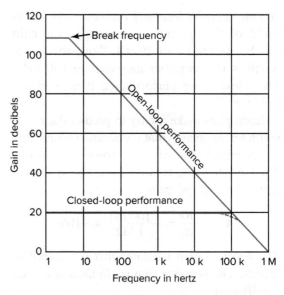

Fig. 9-25 Typical op-amp Bode plot.

verify that it drops to 80 dB. The loss in gain is 100 − 80 dB = 20 dB. Beyond the f_b, gain drops at 20 dB per decade.

Bode plots are approximate. Figure 9-26 shows that the actual performance of an amplifier is 3 dB less at f_b. This is the point of worst error, and Bode plots are accurate for frequencies significantly higher or lower than f_b. To find the true gain at f_b, subtract 3 dB.

The open-loop gain shown in Fig. 9-25 indicates a break frequency lower than 10 Hz. This is a Bode plot, so we know that the gain is already 3 dB less at this point. The gain of the general-purpose op amp begins to decrease around 5 Hz. Obviously, it is not a wideband amplifier when operated open-loop. Op amps are usually operated closed-loop,

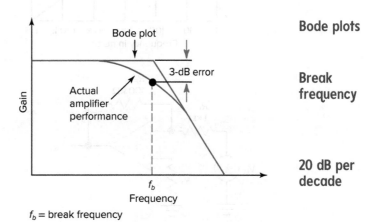

Fig. 9-26 Bode plot error is greatest at the break frequency.

Rail voltages

Saturate

Clipped

Bode plots

Break frequency

20 dB per decade

and the *negative feedback increases the bandwidth* of the op amp. For example, the gain can be decreased to 20 dB. Now, the bandwidth of the amplifier increases to 100 kHz. This closed-loop performance is also shown in Fig. 9-25.

Bode plots make it easy to predict the bandwidth for an op amp that is operating with negative feedback. Figure 9-27 shows an example. The first step is to find the closed-loop voltage gain. The appropriate equation is

$$A_V = -\frac{R_F}{R_1} = -\frac{100 \text{ k}\Omega}{1 \text{ k}\Omega} = -100$$

The negative gain indicates that the amplifier inverts. The negative sign is *eliminated* to find the dB gain:

$$A_V = 20 \times \log 100 = 40 \text{ dB}$$

The dB gain is located on the vertical axis of the Bode plot. Projecting to the right produces

an intersection with the open-loop plot at 10 kHz. This is f_b (the break frequency), and the bandwidth of the amplifier is 10 kHz. Above f_b, the gain drops at 20 dB per decade. So the gain will be 40 dB − 20 dB = 20 dB at 100 kHz. The gain at f_b is down 3 dB, and 40 dB − 3 dB = 37 dB at 10 kHz.

Earlier in this chapter, it was determined that the power bandwidth of an op amp is established by its slew rate and output amplitude. Here we find that another bandwidth is determined by the Bode plot of an op amp. To avoid confusion, this is called the *small-signal bandwidth*. The small-signal bandwidth can be determined from the Bode plot or from the gain-bandwidth product, which is called f_{unity}. In Fig. 9-27, f_{unity} is 1 MHz. This is the frequency at which the gain of the amplifier is unity. A gain of unity means that the gain is 1, which corresponds to 0 dB. If you know f_{unity} for an op amp, you can determine the small-signal bandwidth without resorting to a Bode plot. The break frequency can be found by dividing f_{unity} by the ratio gain:

$$f_b = \frac{f_{unity}}{A_V}$$

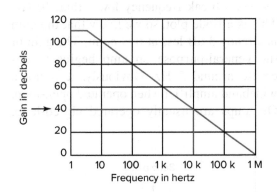

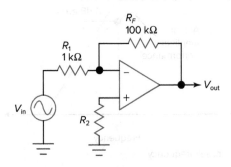

Fig. 9-27 Finding the bandwidth for a closed-loop amplifier.

EXAMPLE 9-7

Find the small-signal bandwidth for an op amp with a gain-bandwidth product of 1 MHz if the closed-loop voltage gain is 60 dB. The first step is to convert 60 dB to the ratio gain (A_V):

$$60 \text{ dB} = 20 \times \log A_V$$

Divide both sides of the equation by 20:

$$3 = \log A_V$$

Take the inverse log of both sides:

$$A_V = 1,000$$

Find the break frequency:

$$f_b = \frac{1 \text{ MHz}}{1,000} = 1 \text{ kHz}$$

The small-signal bandwidth of the amplifier is 1 kHz. Please refer to Fig. 9-27 and verify that this agrees with the Bode plot for a gain of 60 dB.

Solve the following problems.

17. Refer to Fig. 9-20. Is the amplifier operating open-loop or closed-loop?
18. Refer to Fig. 9-23. Assume that $R_1 = 470\ \Omega$ and $R_F = 47\ k\Omega$. What is the low-frequency voltage gain of the amplifier? What is the input impedance of the amplifier?
19. Refer to Fig. 9-23. You want to use this circuit for an amplifier that has an input impedance of $3,300\ \Omega$ and a voltage gain of -10. What value would you choose for R_1? For R_F?
20. Refer to Fig. 9-24. $R_1 = 47\ k\Omega$, $R_2 = 22\ k\Omega$, $R_F = 47\ k\Omega$. Calculate the voltage gain of the amplifier. What is the input impedance of the amplifier?

Questions 21 to 27 use Fig. 9-27 as a guide.

21. The desired amplifier characteristics are a voltage gain of 80 dB and an input impedance of $100\ \Omega$. Select a value for R_1.
22. Select a value for R_F.
23. Select a value for R_2 that will minimize dc offset error.
24. What is the small-signal bandwidth of the amplifier?
25. What is the gain of the amplifier at f_b?
26. What is the gain of the amplifier at 10 Hz?
27. What is the gain of the amplifier at 1 kHz?

9-5 Frequency Effects in Op Amps

We have learned that the open-loop gain of general-purpose operational amplifiers starts decreasing at a rate of 20 dB per decade at some relatively low frequency. This is caused by an internal *RC lag network* in the op amp. If you refer back to Fig. 9-15, you will find a single capacitor in the diagram. This capacitor forms one part of the lag network that determines the break frequency f_b. This capacitor is also one of the major factors that limit the slew rate of the amplifier.

Figure 9-28 summarizes RC lag networks. The *RC* circuit is shown in Fig. 9-28(*a*). It consists of a series resistor and a capacitor connected to ground. A lag network does two things: (1) it causes the output voltage to drop with increasing frequency, and (2) it causes the output voltage to *lag behind* the input voltage. Figure 9-28(*b*) shows the vector diagram for an *RC* lag network that is operating at its break frequency f_b. The resistance R and the capacitive reactance X_C are equal in this case, and the phase angle of the circuit is -45 degrees. Figure 9-28(*c*) shows two Bode plots for the *RC* lag network. The one at the top is about the same as those shown in the last section. The change in amplitude from the break frequency to a frequency 10 times higher ($10 f_b$) is -20 dB.

RC lag network

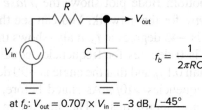

at f_b: $V_{out} = 0.707 \times V_{in} = -3\ \text{dB},\ \underline{/-45°}$

(*a*) RC lag network

$$Z = \sqrt{R^2 + X_C^2}$$

$$\theta = \tan^{-1} \frac{-X_C}{R}$$

$$= -45°$$

(*b*) Vector diagram for lag network at f_b

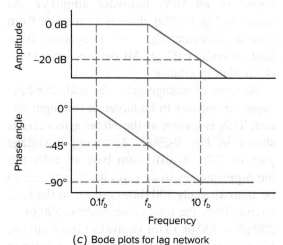

(*c*) Bode plots for lag network

Fig. 9-28 *RC* lag network.

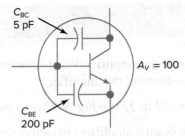

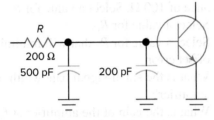

(a) Transistor interelectrode capacitances

(b) Miller equivalent input circuit

Fig. 9-29 Capacitive loading in a transistor amplifier.

The bottom Bode plot shows the *phase angle response* for the network. You can see that the angle is −45 degrees at f_b. It also shows that the angle is 0 degrees for frequencies equal to or less than $0.1 f_b$ and that the angle is −90 degrees for frequencies $>10 f_b$. As stated before, Bode plots are approximate. The maximum error points are at $0.1 f_b$ where the angle is actually −6 degrees and at $10 f_b$ where the angle is actually −84 degrees.

RC lag networks are *inherent in all amplifiers.* Transistors have *interelectrode capacitances* that form lag networks with certain resistances within the amplifier. Figure 9-29 shows how capacitive loading affects the input circuit of an NPN transistor amplifier. As shown in Fig. 9-29(a), there is a capacitor from base to collector (C_{BC}) and a capacitor from base to emitter (C_{BE}). All devices have interelectrode capacitance.

Because of voltage gain, the collector-base capacitor appears to be larger in the input circuit. This is known as the *Miller effect* and is shown in Fig. 9-29(b). Assuming a voltage gain of 100 (40 dB) from base to collector, the 5-pF interelectrode capacitance appears to be approximately 100 times greater in the base circuit. Thus, the total capacitance is 500 pF + 200 pF = 700 pF in the equivalent input circuit. If we assume the equivalent input resistance in Fig. 9-29(b) is 200 Ω, we can evaluate the

circuit as a lag network and find its break frequency:

$$f_b = \frac{1}{2\pi RC} = \frac{1}{6.28 \times 200 \ \Omega \times 700 \ \text{pF}}$$

$$= 1.14 \ \text{MHz}$$

Knowing f_b allows us to predict the frequency response of the amplifier. If we use the last example, we know that the gain will be 100 (40 dB) for frequencies below 1 MHz. We also know that it will be 37 dB at 1.14 MHz, 20 dB at 11.4 MHz, and 0 dB at 111 MHz. However, we have considered only the amplifier input circuit. The actual break frequency could be lower, depending on the output circuit.

Since it may have been some time since you worked with ac circuits, let's check the numbers another way. We will use the data from the last example: 700 pF, 200 Ω, and 1.14 MHz. Find the capacitive reactance:

$$X_C = \frac{1}{2\pi fC} = \frac{1}{6.28 \times 1.14 \ \text{MHz} \times 700 \ \text{pF}}$$

$$= 200 \ \Omega$$

Find the impedance:

$$Z = \sqrt{R^2 + X^2} = \sqrt{200^2 + 200^2} = 283 \ \Omega$$

Now, if you refer back to Fig. 9-28(a), you can see that the capacitor and resistor form a voltage divider. We can use the voltage divider equation along with the impedance and capacitive reactance:

$$V_{\text{out}} = \frac{X_C}{Z} \times V_{\text{in}} = \frac{200 \ \Omega}{283 \ \Omega} \times V_{\text{in}}$$

$$= 0.707 \times V_{\text{in}}$$

This demonstrates that the output voltage is 0.707, or −3 dB, at f_b. The phase angle can be determined with

$$\Phi = \tan^{-1} \frac{-X_C}{R} = \tan^{-1} \frac{-200 \ \Omega}{200 \ \Omega} = -45°$$

The vector diagram of Fig. 9-28(b) shows that X_C is negative and that the phase angle is also negative (it lags).

As the schematic of the op amp shows (Fig. 9-15), there are quite a few transistors. Each of them has interelectrode capacitances. *There are many lag networks in any operational amplifier.* With many lag networks, there are going to be several break points. Figure 9-30 shows a Bode plot with several break frequencies.

Interelectrode capacitance

Miller effect

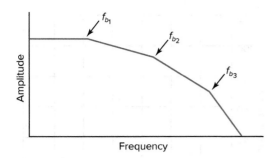

Fig. 9-30 Bode plot with several break points.

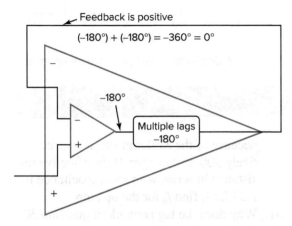

Fig. 9-31 How negative feedback can become positive.

The gain will drop at 20 dB per decade between f_{b_1} and f_{b_2}. It will drop at 40 dB per decade between f_{b_2} and f_{b_3}. It will drop at 60 dB per decade for frequencies beyond f_{b_3}. The effect of multiple lag networks accumulates.

The phase angle with multiple lag networks also accumulates. It can be −100, −150, or −180 degrees. This is a problem when an amplifier uses negative feedback. If the inherent lags add up to −180 degrees, the amplifier can become *unstable*. Figure 9-31 shows this situation. The op amp uses a connection from its output to its inverting input. This normally provides negative feedback. However, if internal lags accumulate to −180 degrees, the overall feedback goes to 0 degrees. A phase angle of 0 degrees is in-phase or *positive feedback*.

This is what can happen with positive feedback: An input signal drives the amplifier. If the amplifier has gain, a larger signal appears at its output. This larger signal is returned to the − (minus) input with a phase that *reinforces* the input signal. The input and the feedback phase add for an even greater effective input. The output responds by increasing even more. This increases the input even more. The amplifier is no longer controlled by the input signal

but by its own output. It is unstable and useless as an amplifier.

Instability is not acceptable in any amplifier. One solution is that most op amps are *internally compensated*. They have a *dominant lag network* that begins rolling off the gain at a low frequency. By the time the other lag networks (due to transistor capacitances) start to take effect, the gain has dropped below 0 dB. With the gain less than 0 dB, the amplifier cannot become unstable regardless of the actual feedback phase. Now you know why the open-loop Bode plot for the general-purpose op amp has such a low value of f_b.

Unfortunately, the internal frequency compensation limits high-frequency gain and slew rate. For this reason, some op amps are available with *external frequency compensation*. The designer must compensate the amplifier in such a way that it is always stable. This is more involved and requires more components. Figure 9-32 shows an example of an externally compensated op amp.

Another possibility is to use a high-performance op amp. This type is more costly, but it has a better slew rate and wider open-loop bandwidth than the general-purpose device. Figure 9-33 shows the Bode plot for one of these amplifiers. Notice that the open-loop gain does not reach 0 dB until a frequency of 10 MHz is reached. The small-signal bandwidth of this device is 10 times that of a general-purpose operational amplifier.

EXAMPLE 9-8

The Bode plot for a high-performance op amp illustrated in Fig. 9-33 indicates that f_{unity} is 10 MHz. Find the small-signal bandwidth for this amplifier when it operates with a closed-loop gain of 40 dB. Using Fig. 9-33 for a graphical solution is straightforward. Project across from 40 dB and then down to verify that the bandwidth is 100 kHz. The other method is to find the ratio gain and divide into f_{unity}:

$$40\ \text{dB} = 20 \times \log A_V$$
$$2 = \log A_V$$
$$A_V = 100$$
$$f_b = \frac{10\ \text{MHz}}{100} = 100\ \text{kHz}$$

Instability

Internally compensated

External frequency compensation

Positive feedback

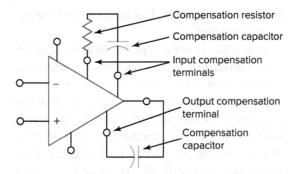

Compensation resistor
Compensation capacitor
Input compensation terminals
Output compensation terminal
Compensation capacitor

Fig. 9-32 An externally compensated op amp.

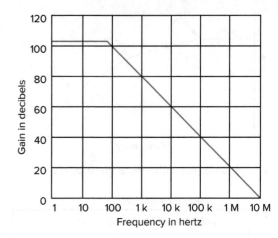

Fig. 9-33 Bode plot for a high-performance op amp.

Self-Test

Solve the following problems.

28. What does a lag network do to the amplitude of an ac signal as it increases in frequency?
29. What does a lag network do to the phase of an ac signal as it increases in frequency?
30. The compensation capacitor in a general-purpose op amp is 30 pF. However,

because of the Miller effect, it is effectively 300 times larger. If the effective resistance in series with this capacitance is 2.53 MΩ, find f_b for the op amp.
31. Why does the lag network of question 30 dominate the op amp?
32. If an op amp is designed for external compensation, why can it become unstable if the circuit is not designed correctly?

9-6 Op-Amp Applications

Operational amplifiers are widely applied. This section presents some of the most popular uses for op amps.

Summing Amplifiers

Figure 9-34 shows an operational amplifier used in the *summing mode*. Two input signals V_1 and V_2 are applied to the inverting input. The output will be the *inverted sum* of the two input signals. Summing amplifiers can be used to add

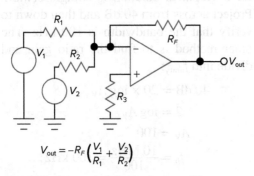

$$V_{out} = -R_F\left(\frac{V_1}{R_1} + \frac{V_2}{R_2}\right)$$

Fig. 9-34 An operational summing amplifier.

ac or dc signals. The output signal is given by the equation shown in Fig. 9-34.

Suppose, in Fig. 9-34, that all the resistors are 10 kΩ. Assume also that V_1 is 2 V and V_2 is 4 V. The output will be

$$V_{out} = -10 \text{ k}\Omega\left(\frac{2 \text{ V}}{10 \text{ k}\Omega} + \frac{4 \text{ V}}{10 \text{ k}\Omega}\right)$$

$$= -\left(\frac{2 \text{ V} \times 10 \text{ k}\Omega}{10 \text{ k}\Omega} + \frac{4 \text{ V} \times 10 \text{ k}\Omega}{10 \text{ k}\Omega}\right)$$

$$= -(2 \text{ V} + 4 \text{ V}) = -6 \text{ V}$$

The output voltage is negative because the two inputs are summed at the inverting input.

The circuit of Fig. 9-34 can be changed to scale the inputs. For example, R_1 could be changed to 5 kΩ. The output voltage will now be

$$V_{out} = -10 \text{ k}\Omega\left(\frac{2 \text{ V}}{5 \text{ k}\Omega} + \frac{4 \text{ V}}{10 \text{ k}\Omega}\right)$$

$$= -(4 \text{ V} + 4 \text{ V}) = -8 \text{ V}$$

The amplifier has scaled V_1 to 2 times its value and then added it to V_2.

Figure 9-34 could be expanded for more than two inputs. A third, fourth, and even a tenth input can be summed at the inverting input. Scaling of some or all the inputs is possible by selecting the input resistors and the feedback resistor.

Op-amp *summing amplifiers* are also called *mixers*. An audio mixer could be used to add the outputs of four microphones during a recording session. One of the advantages of inverting op-amp mixers is that there is no interaction between inputs. The inverting input is a *virtual ground*. This prevents one input signal from appearing at the other inputs. Figure 9-35 shows that the virtual ground isolates the inputs.

Subtracting Amplifiers

Op amps can be used in a *subtracting mode*. Figure 9-36 shows a circuit that can provide the difference between two inputs. With all resistors equal, the output is the nonscaled difference of the two inputs. If $V_1 = 2$ V and $V_2 = 5$ V, then

$$V_{out} = V_2 - V_1 = 5\text{ V} - 2\text{ V} = 3\text{ V}$$

It is possible to have a negative output if the voltage to the inverting input is greater than the

voltage to the noninverting input. If $V_1 = 6$ V and $V_2 = 5$ V,

$$V_{out} = 5\text{ V} - 6\text{ V} = -1\text{ V}$$

Figure 9-36 can be modified to scale the inputs. Changing R_1 or R_2 would accomplish this.

Active Filters

A filter is a circuit or device that allows some frequencies to pass through and stops (attenuates) other frequencies. Filters that use only resistors, capacitors, and inductors are called *passive filters*. Filter performance can often be improved by adding active devices such as transistors or op amps. Filters that use active devices are called *active filters*. Integrated circuit op amps are inexpensive and have made active filters very popular, especially at frequencies below 1 MHz. Active filters eliminate the need for expensive inductors in this frequency range.

Figure 9-37 shows graphs that describe the frequency response of filters. In Fig. 9-37(a) an ideal *low-pass filter* is shown. An ideal filter is often called a "brick-wall" filter. The passband includes all those frequencies that go through the filter with no attenuation (the amplitude is maximum). The stopband includes all those frequencies that don't get through the filter (the amplitude is zero and the attenuation is infinite). The transition from the passband to the stopband is immediate. Or, to say it another way, the transition bandwidth is zero. Figure 9-37(b) shows an ideal bandpass filter. It is not possible to build ideal filters. It is possible to approach the brick-wall response with elaborate filters and also with digital signal processing.

Figure 9-37(c) shows the frequency response of a real low-pass filter. Real filters differ from the ideal (brick-wall) in some or all of these ways:

- There could be ripple in the passband.
- There could be ripple in the stopband.
- There could be loss in the passband (especially in passive filters).
- The transition bandwidth is greater than zero (this is always true).
- The stopband attenuation is not infinite (this is always true).

Summing amplifier

Mixers

Passive filter

Active filter

Subtracting amplifier

Low-pass filter

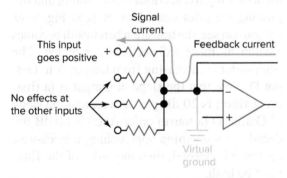

Fig. 9-35 The virtual ground isolates the inputs from one another.

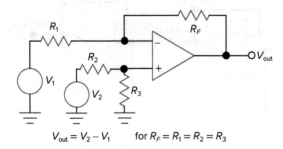

$$V_{out} = V_2 - V_1 \qquad \text{for } R_F = R_1 = R_2 = R_3$$

Fig. 9-36 An operational subtracting amplifier.

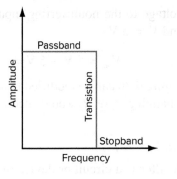

(a) Ideal low-pass frequency response

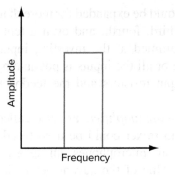

(b) Ideal band-pass frequency response

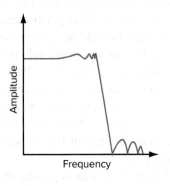

(c) Real low-pass frequency response

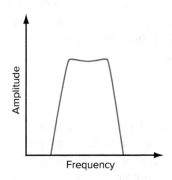

(d) Real band-pass frequency response

Fig. 9-37 Filter frequency response curves.

Usually, when people say that a filter is *sharp*, they mean the transition bandwidth is small and approaches the ideal. Sharp filters are more elaborate and therefore more costly. As to the stopband attenuation, it is made as large as necessary for each application. Sharpness and stopband attenuation are both improved by increasing the *order* of a filter, as discussed next.

The *RC* lag network that was presented earlier in this chapter is a basic low-pass filter. It's not a very sharp filter, but a *cascade* arrangement can be used to increase the filter order and improve sharpness. Look at Fig. 9-38. It shows

a cascade *RC* filter using op amps. What do the op amps do? They serve as buffers to prevent the following *RC* sections from loading and degrading the prior ones. Now look at Fig. 9-39. As you can see, the transition bandwidth becomes smaller as more *RC* sections are added. The filter order is increasing from Output A to Output D. Note that the slope of Output A (a first-order filter) is 20 dB per decade, and the slope of Output D (a fourth order filter) is 80 dB per decade. If something approaching a brick-wall response is needed, then the order of the filter must be high.

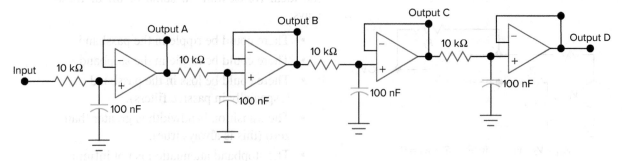

Fig. 9-38 Cascade *RC* low-pass filter.

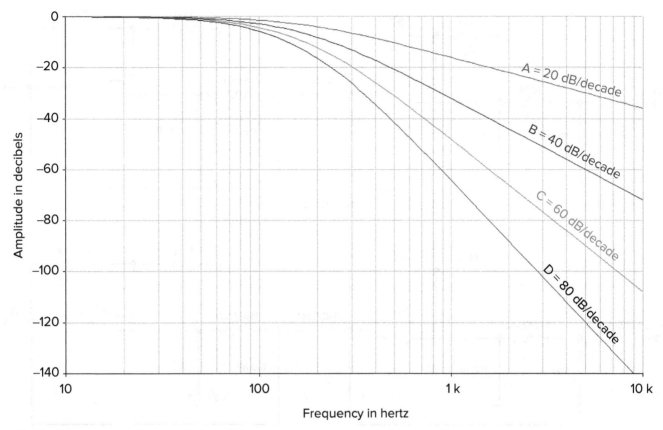

Fig. 9-39 Frequency response curves for *RC* filters.

Find the break frequency for the filter in Fig. 9-38 at Output A and then determine the amplitude at Output D at that frequency. The break frequency, as given earlier in this chapter, is

$$f_b = \frac{1}{2\pi RC} = \frac{1}{6.28 \times 10 \text{ k}\Omega \times 100 \text{ nF}}$$

$$= 159 \text{ Hz}$$

When applied to a filter, the break frequency, or −3-dB frequency, is also called the cutoff frequency. Thus, the cutoff frequency for Output A in Fig. 9-38 is 159 Hz. At output D, we find the cumulative effect of 4 *RC* sections having the same cutoff frequency. So, the amplitude at Output D at 159 Hz is 4×-3 dB $= -12$ dB. This means that the cutoff frequency for Output D is lower in frequency than the break frequency for one of the *RC* sections. It is 70 Hz.

Cascade *RC* filters are not popular because, for about the same cost, it is possible to obtain a better *knee*. Figure 9-40 shows two examples. Both of these filters use *feedback* to sharpen the knee. To understand how this works, consider C_1. It will not affect the signal going through the filter when the output of the op amp is at the same amplitude and phase as the input signal. So, if C_2 is ignored and the gain of the op amp is close to 1 (and it is), then there is little current flow in C_1 at any frequency since there is little voltage difference across it. With the op amp gain close to 1 and ignoring C_2, there is no filter action. When C_2 is considered, the picture changes. The signal at the noninverting input of the op amp will start to drop at higher frequencies due to C_2. So will the output of the op amp. There is now a signal voltage difference across C_1, and it too loads the input. The feedback sharpens the knee. When C_2 "kicks in," it causes C_1 to become active also because of the feedback.

Filter designers can choose any of several types of filter response by adjusting the feedback and the break frequency for each filter

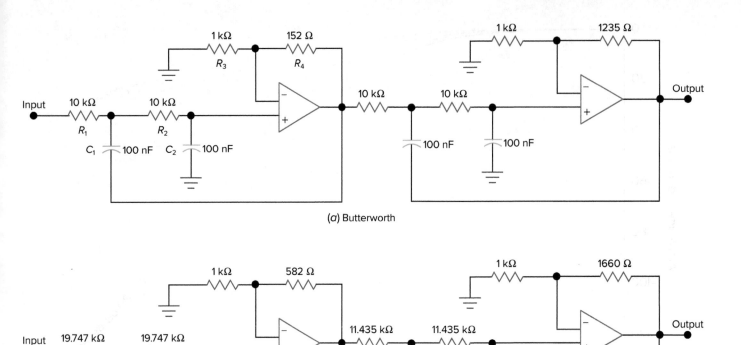

Fig. 9-40 Fourth-order low-pass filters.

section. There are tradeoffs. The Chebyshev filter shown in Fig. 9-40(*b*) has a sharp knee, but it also has 0.5 dB of ripple in the passband. The ripple doesn't show in the frequency response graph in Fig. 9-41 because 0.5 dB is a tiny portion of the large range covered by the vertical axis. The Butterworth filter has no ripple in the passband. These are known as "maximally flat filters" and are used when that type of response is important.

Figure 9-41 compares the frequency responses of the three active filters presented so far. Notice how *soft* the knee of the cascade *RC* filter is when compared with the feedback type filters. Also note that the Chebyshev filter has an even sharper knee than the Butterworth and has better attenuation in the stopband.

Table 9-2 compares some popular filter designs. The cascade *RC* filter is not included because it is seldom used. The table also rates the designs in terms of phase response and pulse response. When phase is important, a linear response is usually the best. When digital signals (pulses) are filtered, the pulse response is

usually more important than the frequency response.

People who design filters use tables of filter component values, computer-aided design, and computer simulation. Look again at Fig. 9-40. The values for the gain-setting resistors ($R_3 - R_4$) and the *RC* break frequencies ($R_1 - C_1$ and $R_2 - C_2$) can be found by consulting tables. So if a designer chooses a Chebyshev response with 1 dB of ripple and determines that an eighth-order filter will be adequate, consulting the tables will provide the necessary information. Using tables and a calculator works, but the software approach is easier because the software also gives graphs of frequency response, phase response, and pulse response during the design process. Software programs also make it easy to juggle component values to avoid costly, nonstandard parts.

Passive filters can also provide a sharp knee. Figure 9-42 shows an *LC* filter that compares favorably with the Chebyshev filter in Fig. 9-40(*b*). The bad news about the *LC* filter is that inductors in the henry range are very

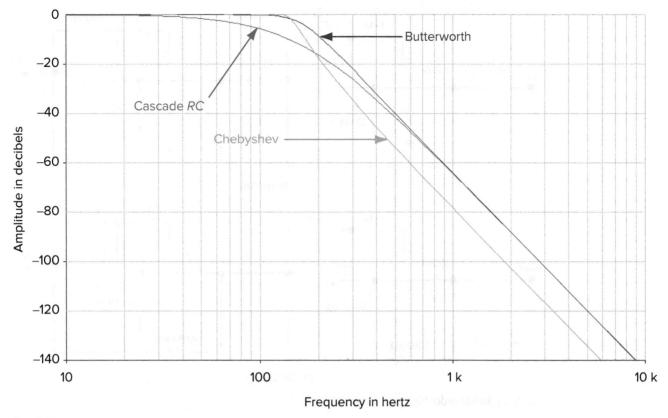

Fig. 9-41 Frequency response.

Table 9-2	Filter Characteristics				
Type	Knee	Passband Ripple	Stopband Ripple	Phase Response	Pulse Response
Butterworth	Good	No	No	Good	Good
Chebyshev	Sharp	Yes	No	Poor	Poor
Elliptic	Sharp	Yes	Yes	Poor	Poor
Bessel	Soft	No	No	Best	Best

large, very heavy, and very expensive. Today, active filters (and other technologies) have almost completely eliminated *LC* filters in low-frequency applications. They are still used when large load currents are required, such as in speaker crossover networks. They are also used above 1 MHz or so, since the inductor values are in the microhenry range at those frequencies. Small-value inductors are not large, heavy, and expensive, and are therefore practical.

High-pass filters can be realized by interchanging the resistors and capacitors as shown in Fig. 9-43. Compare this filter with the one in Fig. 9-40 to verify that the frequency determining resistors and capacitors have been

interchanged. Note that in the case of the Butterworth filters [Figs. 9-40(a) and 9-43(a)], all the component values are the same. However, this won't work for Chebyshev filters as

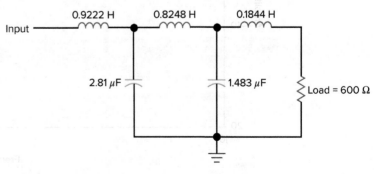

Fig. 9-42 *LC* low-pass filter.

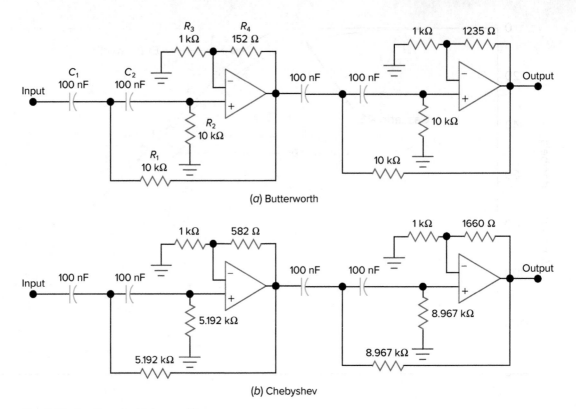

Fig. 9-43 Fourth-order high-pass filters.

Band-pass filter

you can see by comparing Fig. 9-40(b) with Fig. 9-43(b). Figure 9-44 shows the frequency response curves for the high-pass filters. The graph has been expanded to show the 0.5 dB-ripple for the Chebyshev response.

A *band-pass filter* can be realized by combining low-pass and high-pass filters. As Fig. 9-45 shows, this can be accomplished by using both capacitive and resistive feedback in each stage. The feedback capacitors provide attenuation of

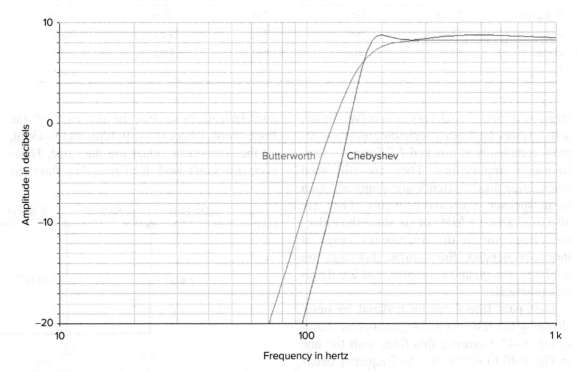

Fig. 9-44 High-pass response curves.

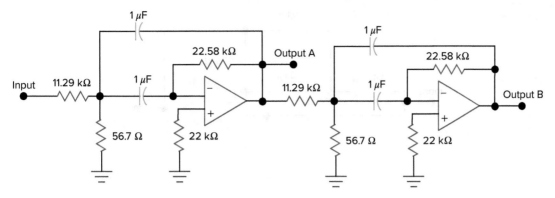

Fig. 9-45 Band-pass filter.

the lower frequencies, and the feedback resistors attenuate the higher frequencies. This arrangement is called a multiple-feedback circuit. Figure 9-46 shows the resulting band-pass frequency response. A sharper response is achieved by cascading two filters.

Figure 9-47 shows a 60-Hz *band-stop filter*. It may also be called a notch filter or a trap. It produces maximum attenuation (or minimum gain) at a single frequency (in this case, 60 Hz). Frequencies significantly above or below 60 Hz will pass through the notch filter with no attenuation. Band-stop filters are useful when a signal at one particular frequency is causing problems. For example, the 60-Hz notch filter could be used to eliminate power-line hum.

How does it work? Notice that the input signal is applied to both op-amp inputs. At that frequency where these inputs see the same signal, the output will be very small due to the common mode rejection of the amplifier. In the filter in Fig. 9-47, the resistive and capacitive feedback is arranged so that this occurs at 60 Hz. At frequencies much above or below 60 Hz, there is a differential input, and the gain is close to 1.

Figure 9-48 shows the frequency response for the 60-Hz notch filter. The optimum response curve shows a deep and sharp notch at 60 Hz. Unfortunately, the filter is not practical. The other curves shown in Fig. 9-48 show the expected performance variations of production

Band-stop filter

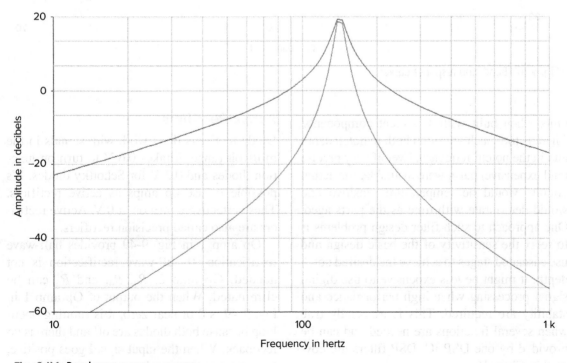

Fig. 9-46 Band-pass response curves.

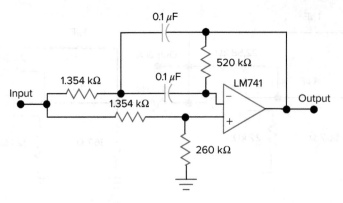

Fig. 9-47 Band-stop filter.

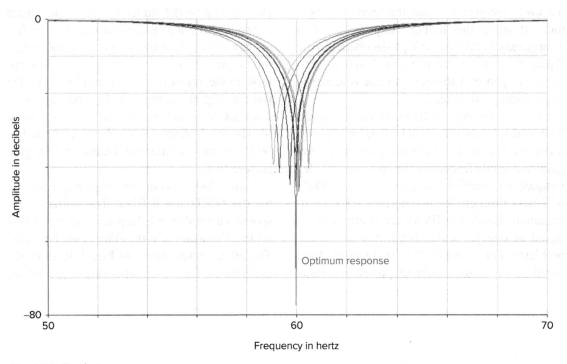

Fig. 9-48 Band-stop response curves.

units when built with 1 percent components. There is too much variation both in notch depth and in the notch frequency. Even if very precise (and expensive) parts were available, the actual circuits would be temperature sensitive and would deteriorate with time as the parts aged. One approach to such filter design problems is to relax the sensitivity of the basic design and use cascaded stages to achieve the desired notch depth. It might be less expensive to use digital signal processing when high performance and stability are required. This is especially true when several functions are needed and can be provided by one DSP IC. DSP filters are covered in Chap. 16.

Active Rectifiers

Diode rectifiers don't work with signals in the millivolt range. It takes 0.6 V to turn on junction diodes and 0.2 V for Schottky diodes. It's possible to use op amps as active rectifiers. These effectively turn on at 0 V. Active rectifiers are also called precision rectifiers.

Op amp 1 in Fig. 9-49 provides half-wave rectification. If full-wave rectification is not needed, Op amp 2, R_3, R_4, and R_5 can be eliminated. When the output of Op amp 1 in Fig. 9-49 is 0 or near zero, it is running open-loop because both diodes are off and there is no feedback. When the input signal goes positive, the output of Op amp 1 goes negative and turns

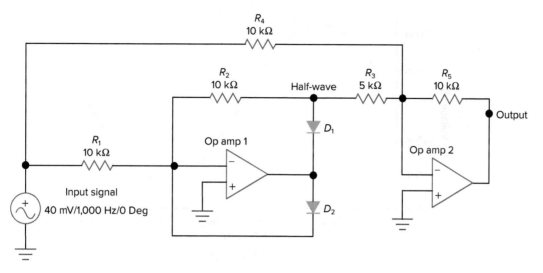

Fig. 9-49 Active rectifier.

Comparators

on D_1. This completes the feedback loop through R_2, and the gain of the circuit drops to -1. The forward-biased resistance of D_1 is small, and the gain is determined by the ratio of R_2/R_1. Thus, we expect a negative alternation at the half-wave point, which can be seen in Fig. 9-50(b).

When the input signal in Fig. 9-49 goes negative, the output of Op amp 1 goes positive and D_2 turns on. The output will remain small for the entire negative alternation because the forward-biased resistance of D_2 is much smaller than R_1. D_1 stays off during the negative alternation, and the voltage at the half-wave point is zero during this time as shown in Fig. 9-50(b).

Op amp 2 in Fig. 9-49 is an inverting adder (summing circuit). It sums the half-wave signal with the input signal. The half-wave signal is scaled by a factor of $-2(R_5/R_3 = 2)$, and the input signal is scaled by a factor of $-1(R_5/R_4 = 1)$. The result of the scaling and the summation is the full-wave signal shown in Fig. 9-50(c).

Comparators

Operational amplifiers are sometimes used as *comparators*. Also, special comparator ICs are available. These are introduced in the last section of this chapter. A comparator operates open-loop. This makes the gain very high, and the output is normally saturated in either a high or a low state. The output of a comparator is therefore a *digital signal* (has only two states). Comparators are nonlinear circuits.

Comparators are used to provide an indication of the relative state of two inputs. If the positive (noninverting) input is more positive than the negative (inverting) input, the comparator output will be at positive saturation. If the positive input is less positive than the negative input, the output will be at negative saturation. Generally, a fixed reference voltage is applied to one input. The output will then be an indication of the relative magnitude of any signal applied to the other input. Comparators are often used to determine whether a signal is above or below the reference level. Several comparator applications in this category are presented later in this section.

When the reference voltage is zero, a comparator may be called a *zero-crossing detector*. A zero-crossing detector can be used to convert a sine wave into a square wave. Two comparators can be used in a "window" circuit that is used to determine whether a signal is between two prescribed limits, as shown in the last section of this chapter.

It is often desired that the output of a comparator change states as quickly as possible. Another requirement is that the comparator output be compatible with logic inputs. A special *strobe* input may be needed in some applications so that the comparator output is active only at selected times. Special comparator ICs offer enhanced performance and additional features over op amps and are used in favor of op amps

Comparator

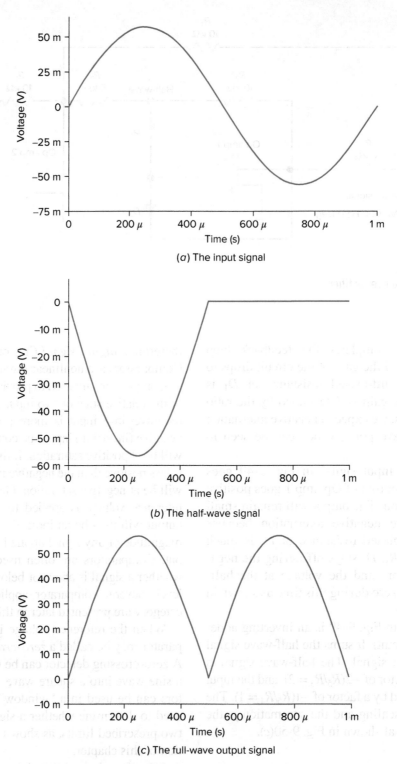

(a) The input signal

(b) The half-wave signal

(c) The full-wave output signal

Fig. 9-50 Active rectifier waveforms.

Integrator circuits

Op-amp integrator

in some applications. These are most commonly operated from a single supply voltage.

Integrator

Another way the operational amplifier may be used is in *integrator circuits*. Integration is a mathematical operation. It is a process of continuous addition. Integrators were used in analog computers. As we will see, there are other uses for integrators.

An *op-amp integrator* is shown in Fig. 9-51. Notice the capacitor in the feedback circuit.

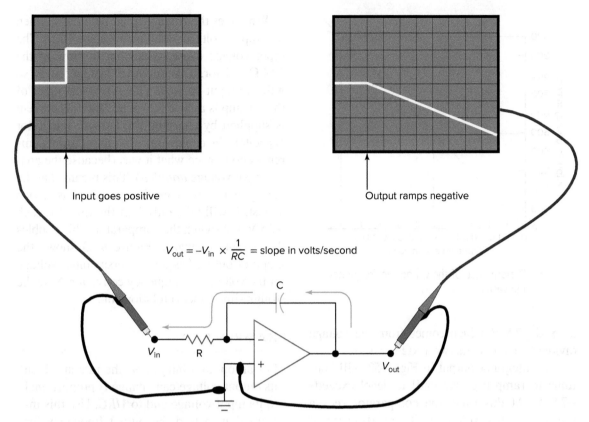

$$V_{out} = -V_{in} \times \frac{1}{RC} = \text{slope in volts/second}$$

Input goes positive

Output ramps negative

Fig. 9-51 An op-amp integrator.

Suppose a positive-going signal is applied to the input. The output must go negative because the inverting input is used. The feedback keeps the inverting input at virtual ground. The current through resistor R is supplied by charging the feedback capacitor as shown.

If the input signal in Fig. 9-51 is at some constant positive value, the feedback current will also be constant. We can assume that the capacitor is being charged by a constant current. When a capacitor is charged by a constant current, the voltage across the capacitor increases in a linear fashion. Figure 9-51 shows that the output of the integrator is ramping negative and that the ramp is linear. Notice that the slope in volts/second can be determined by V_{in} and the integrator circuit values.

Now look at Fig. 9-52. This circuit is a *voltage-to-frequency converter*. It is a very useful circuit. It uses an op-amp integrator to convert positive voltages to a frequency. If the frequency is sent to a digital counter, a *digital voltmeter* is the result. If the voltage V_{in} represents a temperature, a *digital thermometer* is the result. Voltage-to-frequency converters form the basis for many of the digital instruments now in use.

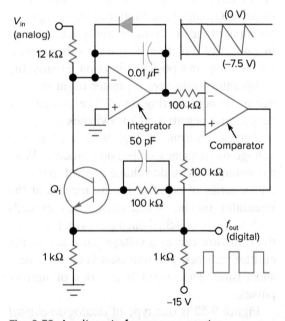

Fig. 9-52 A voltage-to-frequency converter.

What happens when a dc voltage is applied in the circuit in Fig. 9-52? If the voltage is positive, we know that the integrator will ramp in a negative direction. Note that the output of the integrator goes to a second op amp used as a comparator. It compares two inputs. One comparator input is

Voltage-to-frequency converter

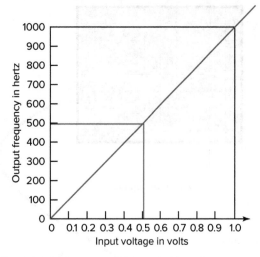

Fig. 9-53 Performance of the voltage-to-frequency converter.

Why does the frequency output double when the input voltage doubles in Fig. 9-52? The input voltage causes a current to flow in the 12-kΩ resistor. If the input voltage increases, so will the input current. The − (minus) input of the op amp is a virtual ground, and this current is supplied by charging the 0.01-μF integrator capacitor. We can assume that the charging current is now twice what it was (because the analog input voltage *doubled*). This means that the voltage across the capacitor will increase twice as fast. It will take only half the time to reach −7.5 V and switch the comparator. This doubles the output frequency. Figure 9-53 shows the graph of output frequency versus input voltage for the voltage-to-frequency converter. Note the straight-line (linear) relationship.

a fixed −7.5 V, which comes from the voltage divider formed by the two 1-kΩ resistors.

The integrator output in Fig. 9-53 will continue to ramp negative until its level exceeds −7.5 V. At this time, the comparator sees a greater negative voltage at its inverting input. This will cause the output of the comparator to go positive. This positive-going output then turns on Q_1. Since the emitter of Q_1 is negative, the input of the integrator is now quickly driven in a negative direction. This makes the integrator output go in a positive direction. Finally, the comparator again sees a greater negative voltage at its noninverting input. The comparator output goes negative, which switches off Q_1.

The waveforms in Fig. 9-52 explain the voltage-to-frequency conversion process. With a constant positive dc voltage applied to the input, a series of negative ramps appears at the integrator output. When each ramp exceeds −7.5 V, Q_1 is switched on. The current through the transistor causes a voltage pulse across the emitter resistors. The transistor is on for a very short time. The output is a series of narrow pulses.

Analog-to-digital converter

Figure 9-52 is one type of *analog-to-digital converter*. It converts a positive analog dc input voltage to a rectangular (digital) output. Ideally, circuits like this should show a linear relationship between the analog input and the digital output. For example, if the dc input voltage is exactly doubled, the output frequency should double. This means the output frequency is a linear function of the input voltage.

Light integrator circuit

EXAMPLE 9-10

For an op-amp integrator, the rate at which its output voltage can change is proportional to its input voltage and to $1/RC$. Use this information to find the output frequency for the circuit in Fig. 9-52 when the input is +1 V. The output slope is negative-going since this is an inverting integrator:

$$\text{V/s} = \text{slope} = -V_{\text{in}} \times \frac{1}{RC} =$$

$$-1 \text{ V} \times \frac{1}{12 \text{ k}\Omega \times 0.01 \text{ } \mu\text{F}} = -8{,}330 \text{ V/s}$$

Since the integrator output ramps from about 0 to −7.5 V, the output frequency can be found with

$$f_{\text{out}} = \frac{-8{,}300 \text{ V/s}}{-7.5 \text{ V}} = 1.11 \text{ kHz}$$

This agrees reasonably well with the graph in Fig. 9-53.

Figure 9-54 shows another application for an op-amp integrator followed by an op-amp comparator. The circuit is called a *light integrator* because it is used to *sum* the amount of light received by a sensor in order to achieve some desired total exposure. Light integrators have applications in areas such as photography, where exposures are critical. A simple timer could be used to control exposure, but there are problems with this approach if the light intensity varies.

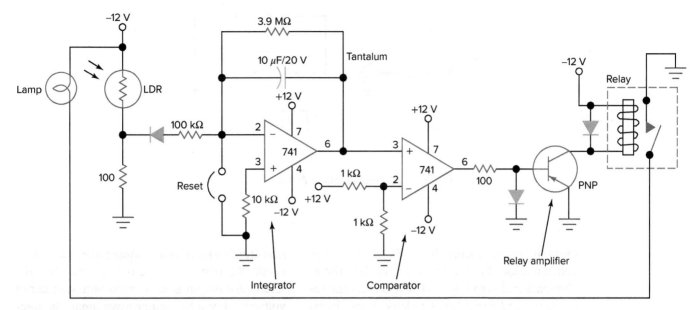

Fig. 9-54 Light-integrator circuit.

For example, the intensity of a light source may change with the power-supply voltage and the temperature and age of the lamp. Another problem is that a filter may be used for some exposures. A filter decreases light intensity. With all of these possible variations in light, a simple timer may not offer adequate control.

The circuit in Fig. 9-54 uses a *light-dependent resistor* (LDR) to measure light intensity. Its resistance drops as brightness increases. The LDR and the 100-Ω resistor form a voltage divider for the −12 V power supply. The divided negative voltage is applied to the input of the integrator. The output of the integrator ramps positive at a rate that is directly proportional to light intensity. A second op amp is used as a comparator in Fig. 9-54. Its inverting input is biased at +6 V by the voltage divider formed by the two 1-kΩ resistors. As the integrator is ramping positive, the output of the comparator will be negative until the +6-V reference threshold is crossed. Notice in Fig. 9-54 that the comparator output is applied to the base of the PNP relay amplifier. As long as the comparator output is negative, the transistor is on, and the relay contacts are closed. The closed relay contacts keep the lamp energized and the exposure continues. However, when the integrator output crosses the +6-V reference threshold, the comparator will suddenly switch to a positive output (its inverting input is now negative with respect to its noninverting input), and the relay will open. The light will now remain off until the

reset button is pressed, which discharges the integrator capacitor and begins another exposure cycle.

The circuit in Fig. 9-54 can produce very accurate exposures. Changes in light intensity are compensated for by the amount of time the relay remains closed. For example, if the light source were momentarily interrupted, an accurate exposure would still result. The integrator would stop ramping at the time of the interruption. It would hold its output voltage level until light once again reached the LDR.

The diode in the input circuit of the integrator in Fig. 9-54 prevents the integrator from being discharged if the light source is fluctuating. The diode in the base circuit of the transistor protects the transistor when the comparator output is positive. The diode will come on and prevent the base voltage from exceeding approximately +0.7 V. The diode across the relay coil prevents the inductive "kick" from damaging the transistor when it turns off.

Differentiator

A *differentiator* is the opposite of an integrator. The output of an integrator is a ramp with a slope or a rate of change that is proportional to the input amplitude. The output of a differentiator is proportional to the rate of change of the input. If a signal is integrated and then differentiated, the original signal will result. If a signal is differentiated and then integrated, the same

Light-dependent resistor (LDR)

Differentiator

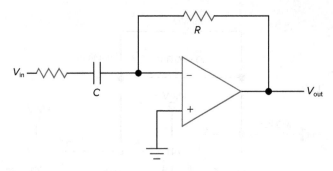

Fig. 9-55 An op-amp differentiator.

thing happens. Figure 9-55 shows the basic configuration for an inverting differentiator. Comparing this to Fig. 9-51 reveals that the resistor (R) and capacitor (C) elements are interchanged. The capacitor is now in the input circuit, and the resistor is the feedback element. An additional series input resistor is often needed to limit the high-frequency response for improved stability. It is usually small enough in value to be ignored for differentiation calculations. The basic equation is

$$V_{out} = -\left(\frac{V_{in}}{t}\right)RC$$

EXAMPLE 9-11

Use the data from Example 9-10 to demonstrate that differentiation is the opposite process of integration.

$$V_{out} = -\left(\frac{V_{in}}{t}\right)RC = -(-8{,}330 \text{ V/s}) \times 12 \text{ k}\Omega$$
$$\times 0.01 \ \mu\text{F} = 1 \text{ V}$$

Note: The unit s (seconds) cancels here because the RC product (time constant) is in seconds.

$$\text{Seconds} = \text{ohms} \times \text{farads} = RC$$

Table 9-3 compares integrator and differentiator waveforms. For a square-wave input, an integrator will produce a constant slope (triangle wave) for both input levels: one ramping positive and one ramping negative. The differentiator produces narrow high-amplitude pulses that coincide with the rise and fall of the square wave, where the rate of change is very high and is infinite for an ideal square wave. The

transition times of actual square waves are often small, the rate of change is large, and the differentiator output goes to maximum (to the rail voltages). For a triangular-wave input, the integrator will produce a parabolic wave and the differentiator a square wave. In the case of sine waves, integrators provide a phase lag and differentiators provide phase leads. Be advised that the circuits used to generate the waveforms for Table 9-3 were *inverting* integrators and differentiators. So if you look closely at the integrator output for a sine wave input, you can mentally invert the red waveform to determine that it does provide a phase lag. Using the same technique, you can visualize that the differentiator provides a phase lead.

Schmitt Trigger

There are a few op-amp circuits that use *positive feedback*. For example, Fig. 9-56 shows a signal-conditioning circuit known as a *Schmitt trigger*. This circuit is similar to a comparator, but the positive feedback gives it *two threshold points*. Assume that the op amp is powered by ±20 V and can swing about ±18 V at its output. Resistors R_1 and R_2 divide the output and establish the voltage that is applied to the noninverting input of the op amp.

When the output of the circuit in Fig. 9-56 is maximum positive, the voltage divider will produce the upper threshold point (UTP):

$$\text{UTP} = V_{max}\left(\frac{R_1}{R_1 + R_2}\right)$$
$$= +18 \text{ V}\left(\frac{2.2 \text{ k}\Omega}{2.2 \text{ k}\Omega + 10 \text{ k}\Omega}\right)$$
$$= +3.25 \text{ V}$$

Positive feedback

Schmitt trigger

Threshold points

Table 9-3 Integrator and Differentiator Input and Output Waveforms

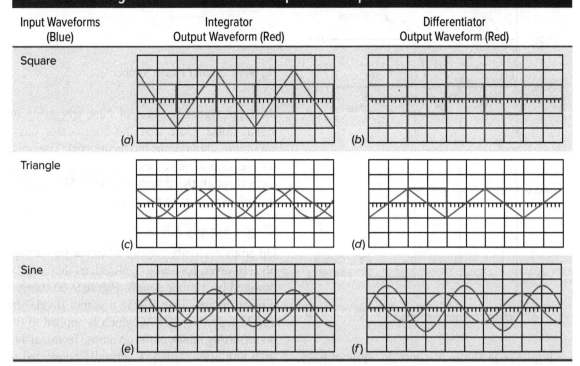

Input Waveforms (Blue)	Integrator Output Waveform (Red)	Differentiator Output Waveform (Red)
Square	(a)	(b)
Triangle	(c)	(d)
Sine	(e)	(f)

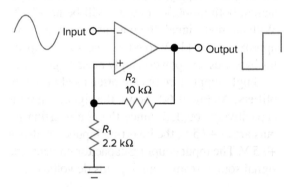

Fig. 9-56 Using an op amp as a Schmitt trigger.

When the output of this circuit is maximum negative (V_{min}), the voltage divider will produce the lower threshold point (LTP):

$$LTP = V_{min}\left(\frac{R_1}{R_1 + R_2}\right)$$

$$= -18 \text{ V}\left(\frac{2.2 \text{ k}\Omega}{2.2 \text{ k}\Omega + 10 \text{ k}\Omega}\right)$$

$$= -3.25 \text{ V}$$

Figure 9-57 shows the Schmitt trigger in operation with an input signal that exceeds the upper and lower threshold points. As the input signal is going positive, it eventually crosses the upper threshold point of +3.25 V. The inverting input of the op amp is now more positive than

the noninverting input; therefore, the output rapidly switches to −18 V. Later, the input signal starts going negative and eventually crosses the lower threshold point of −3.25 V. At this time the Schmitt trigger output goes positive to +18 V, which reestablishes the UTP. The difference between the two threshold points is called *hysteresis* and in our example is

Hysteresis

$$\text{Hysteresis} = \text{UTP} - \text{LTP}$$

$$= +3.25 - (-3.25)$$

$$= 6.5 \text{ V}$$

EXAMPLE 9-12

Calculate the hysteresis voltage for the circuit in Fig. 9-56 if the op amp is powered by a bipolar 9-V supply. We will assume that the output will swing ±8 V. The trip points are

$$\text{UTP} = +8 \text{ V} \times \frac{2.2 \text{ k}\Omega}{2.2 \text{ k}\Omega + 10 \text{ k}\Omega} = 1.44 \text{ V}$$

$$\text{LTP} = -8 \text{ V} \times \frac{2.2 \text{ k}\Omega}{2.2 \text{ k}\Omega + 10 \text{ k}\Omega} = -1.44 \text{ V}$$

The hysteresis is the difference between the two trip points:

$$\text{Hysteresis} = 1.44 \text{ V} - (-1.44 \text{ V}) = 2.88 \text{ V}$$

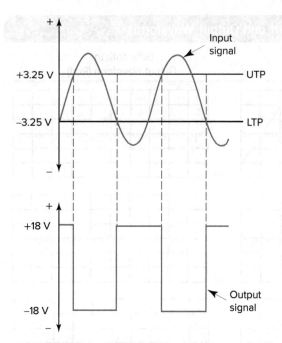

Fig. 9-57 Schmitt trigger operation.

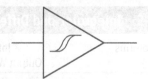

Fig. 9-58 Schmitt trigger symbol.

the input signal because of false triggering on signal noise. Note that the comparator has a *single threshold point* (no hysteresis). The noise on the signal causes extra crossings back and forth through the threshold point (TP), and extra pulses appear in the output.

Single Supply Circuits

Op amps normally require a bipolar power supply. However, for some applications they can be powered by a *single supply*. Figure 9-60 shows a typical circuit. Two 10-kΩ resistors divide the +15-V supply to +7.5 V, which is applied to the noninverting inputs of the op amps. Terminal 4 of each amplifier, which is normally connected to the negative supply, is grounded. With no input signal, both amplifier outputs will be at +7.5 V. With an input signal, the outputs can swing from approximately +14 V to +1 V. The 4.7-μF capacitor bypasses any power-supply noise to ground.

Single supply circuits are often used in ac amplifiers. As Fig. 9-60 shows, the signal source is capacitively coupled. Since the noninverting inputs are at +7.5 V, the inverting inputs are also at +7.5 V. The input coupling capacitor prevents the signal source from changing this dc voltage.

Figure 9-58 shows a schematic symbol for a Schmitt trigger. You might recognize the *hysteresis loop* inside the general amplifier symbol.

Hysteresis is valuable when *conditioning* noisy signals for use in a digital circuit or system. Figure 9-59 shows why. It shows how a Schmitt trigger output can differ from the output of a comparator. The Schmitt trigger has hysteresis. The noise on the signal does not cause false triggering, and the output is at the *same frequency* as the input. However, the comparator output is at a higher frequency than

Single supply

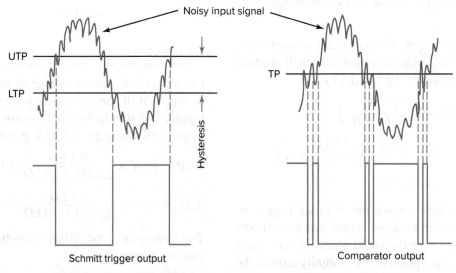

Fig. 9-59 A comparison of Schmitt trigger output with comparator output when the input is noisy.

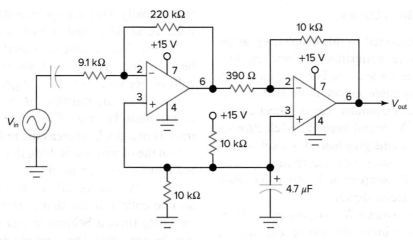

Fig. 9-60 Powering op amps from a single supply.

Solve the following problems.

33. Refer to Fig. 9-34. All resistors are the same value. If $V_1 = +1$ V and $V_2 = +2$ V, what will the output voltage be (value and polarity)?

34. Refer to Fig. 9-34. All resistors are the same value. If $V_1 = -2$ V and $V_2 = -3$ V, what will the output voltage be?

35. Refer to Fig. 9-34. All resistors are the same value. If $V_1 = +2$ V and $V_2 = -3$ V, what will the output voltage be?

36. Refer to Fig. 9-34. $R_F = 20$ kΩ, $R_1 = 10$ kΩ, and $R_2 = 5$ kΩ. If $V_1 = 2$ V and $V_2 = 1$ V, what will the output voltage be?

37. Refer to Fig. 9-35. What circuit feature prevents a signal at one of the inputs from appearing at the other inputs?

38. Refer to Fig. 9-36. All the resistors are the same value. If $V_1 = 3$ V and $V_2 = 5$ V, what will the output voltage be?

39. Refer to Fig. 9-36. All resistors are the same value. If $V_1 = 5$ V and $V_2 = 5$ V, what will the output voltage be?

40. Refer to Fig. 9-36. All resistors are the same value. If $V_1 = -2$ V and $V_2 = 1$ V, what will the output voltage be?

41. A low-pass filter is checked with a variable-frequency signal generator and an oscilloscope. The following data are collected:

$V_{out} = 10$ V peak-to-peak at 100 Hz
$V_{out} = 10$ V peak-to-peak at 1 kHz
$V_{out} = 7$ V peak-to-peak at 10 kHz
$V_{out} = 1$ V peak-to-peak at 20 kHz

What is the cutoff frequency (f_c) of the filter?

42. Refer to Fig. 9-43. What can you expect to happen to the circuit gain as the signal frequency drops below f_c?

43. Assume the gain of a band-pass filter to be maximum at 2,500 Hz. Also assume that the gain drops 3 dB at 2,800 Hz and at 2,200 Hz. What is the filter bandwidth?

44. Suppose the input to the integrator shown in Fig. 9-51 goes negative. What will the output do?

45. Refer to Fig. 9-52. Assume the converter is perfectly linear. If $f_{out} = 300$ Hz when $V_{in} = 0.3$ V, what should f_{out} be at $V_{in} = 0.6$ V?

46. Refer to Fig. 9-54. Assume linear operation. If the relay remains energized for 2 s, how long will it remain energized after the circuit is reset if the light intensity falls to one-half its original level?

47. Refer to Fig. 9-56. Assume that the output of the op amp can swing ±13 V and that R_1 is changed to a 1-kΩ resistor. What is the value of UTP, LTP, and the hysteresis?

9-7 Comparators

As already discussed in this chapter, op amps can be used as comparators. However, this doesn't work in some cases. For example, when a *digital-compatible* output is needed, an op amp used as a comparator may not meet circuit requirements. More and more *hybrid* electronics applications are emerging that are a combination of both analog and digital electronics circuits and devices. A comparator is often the place where analog meets digital.

Special *comparator ICs* are available. They are better for joining the analog and digital worlds. They provide a logic state output that indicates the relative state of two analog voltages, one of which is often a fixed reference. Comparators can signal when a voltage exceeds a reference, when a voltage is less than a reference, or when a voltage is within a specified range. Comparator ICs must change output states rapidly. They are optimized for high gain, wide bandwidth, and a fast slew rate. The *switching time* of a digital signal usually must be very fast. As Fig. 9-61 shows, the critical voltages for some (i.e. TTL) digital circuits are 0.8 and 2 V. Any transition between these two points must be fast. The white oscilloscope trace is marginal, whereas the red trace is well within the required switching time. The LM311 is a popular comparator IC.

The signal source driving a comparator can also be critical. If the signal source has a slow switching time, a Schmitt trigger configuration may be necessary. This configuration uses positive feedback and was discussed in the preceding section of this chapter. Another factor is the impedance of the signal source that drives a comparator. In the case of high-impedance sources, extra output pulses called *glitches* are possible. Again, a Schmitt trigger is a possible solution.

Hybrid

Comparator IC

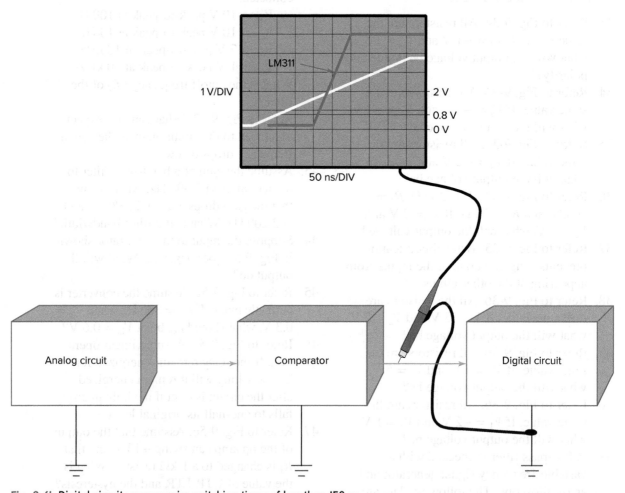

Fig. 9-61 Digital circuits may require switching times of less than 150 ns.

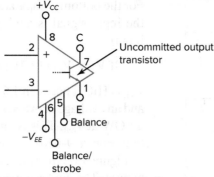

Pin numbers are for 8-pin dual in-line package

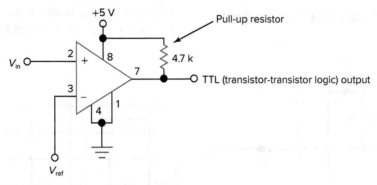

Fig. 9-62 LM3ll or TLC3ll comparator.

Some comparator ICs can operate from single or dual supplies from 5 to 30 V (or ±15 V). Others require dual supplies like op amps. Some, like the popular LM311, have an uncommitted output transistor with both the collector and the emitter terminals available. This makes the output of this device very flexible. It can be used to drive many kinds of logic circuits. Figure 9-62 shows the pinout of the LM311. The balance inputs work pretty much the same as the offset null terminals on some op amps. The strobe input is for those cases when the output is to be active only during a specified time called the *strobe interval*. Notice in Fig. 9-62 that a *pull-up resistor* may be required because of the uncommitted transistor at the output of this device.

In a *window comparator*, two comparators are used to provide two thresholds or trip points. The circuit in Fig. 9-63(a) is called an *inside* window comparator. All the resistors are 1 kΩ. It uses open drain devices. These, like open collector devices, need a pullup resistor. The output will be high when an input signal (V_{in}) falls within (inside) the window.

For the top comparator (V_{ref} is at node A and the input signal is applied to the inverting input):

If $V_{in} < V_{ref}$ (3.33V) then V_{out} = HIGH   **Strobe interval**

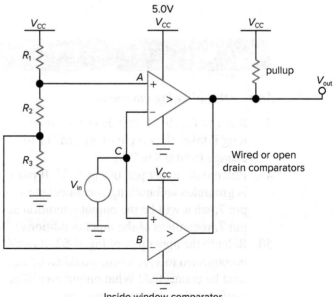

Inside window comparator
output is high when the input is within the window

Fig. 9-63(a) An Inside window comparator.

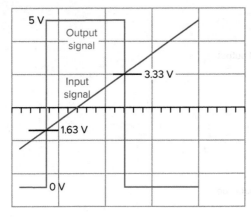

Fig. 9-63(b) Inside window comparator waveforms.

For the bottom comparator (V_{ref} is at node B and the input signal is applied to the noninverting input):

If $V_{in} > V_{ref}$ (1.67V) then V_{out} = HIGH

V_{out} is HIGH between 1.67 and 3.33 V at input C and any voltage above 3.33 *OR* below 1.67 causes a LOW at V_{out}. This is called a wired OR function. Figure 9-63(b) shows the waveforms.

Figure 9-64(a) shows an *outside* window comparator. The diodes combine the two comparator outputs. When V_{in} is below the lower threshold (V_{LL}) or when V_{in} is above the upper threshold (V_{UL}) V_{out} is high as shown in Fig. 9-64(b).

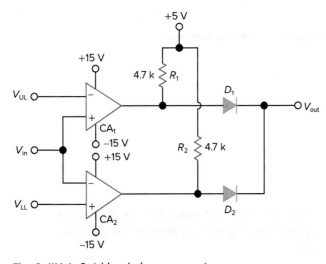

Fig. 9-64(a) Outside window comparator.

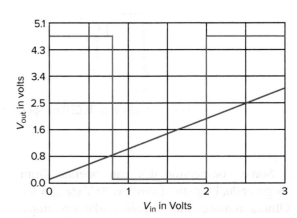

Fig. 9-64(b) Outside window comparator waveforms.

Self-Test

Answer the following questions.

48. Refer to Fig. 9-61, and determine how long it takes the output of an LM311 to change from 0.8 to 2 V.
49. Refer to the upper part of Fig. 9-62. If pin 1 is grounded and nothing is connected to pin 7, what will be the output condition at pin 7, regardless of the input conditions?
50. Refer to the upper part of Fig. 9-62. If pin 7 is connected to V_{CC}, where would an output load be connected? What output amplifier configuration does this represent?

51. Refer to Fig. 9-64(a). What are R_1 and R_2 called?
52. Refer to Fig. 9-64(a). Assume V_{UL} = 12.9 V and V_{LL} = 11.9 V. What is V_{out} when V_{in} = 12.6 V?
53. Refer to Fig. 9-64(a). Assume V_{UL} = 12.9 V and V_{LL} = 11.9 V. What is V_{out} when V_{in} = 13.0 V?
54. Refer to Fig. 9-64(a). Assume V_{UL} = 12.9 V and V_{LL} = 11.9 V. Which diode(s) is (are) forward-biased when V_{in} = 13.0 V?

Chapter 9 Summary and Review

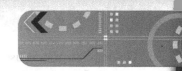

Summary

1. A differential amplifier responds to the difference between two input signals.
2. A dual (or bipolar) supply develops both positive and negative voltages with respect to ground.
3. A differential amplifier can be driven at one or both of its inputs.
4. It is possible to use a differential amplifier as an inverting or a noninverting amplifier.
5. A differential amplifier rejects common-mode signals.
6. The common-mode rejection ratio is the ratio of differential gain to common-mode gain.
7. A differential amplifier can show high CMRR for a single-ended output if the resistance of the emitter supply is very high.
8. Current sources have a very high output impedance.
9. Most op amps have a single-ended output (one output terminal).
10. Op amps have two inputs. One is the inverting input, and the other is the noninverting input. The inverting input is marked with a minus (−) sign, and the noninverting input is marked with a plus (+) sign.
11. An op amp's offset null terminals can be used to reduce dc error in the output. With no dc differential input, the output terminal is adjusted to 0 V with respect to ground.
12. Slew rate can limit the amplitude of an op-amp output and cause waveform distortion.
13. General-purpose op amps work best for dc and low ac frequencies.
14. The open-loop (no-feedback) gain of op amps is very high at 0 Hz (dc frequency). It drops off rapidly as frequency increases.
15. Op amps are operated closed-loop (with feedback).
16. Negative feedback decreases the voltage gain and increases the bandwidth of the amplifier.
17. The gain of an op-amp inverter is set by the ratio of feedback resistance to input resistance.
18. Negative feedback makes the impedance of the inverting input very low. The terminal is called a virtual ground.
19. The impedance of the noninverting input is very high.
20. The Bode plot for a standard op amp shows the gain decreasing at 20 dB per decade beyond the break frequency.
21. The actual gain at the break frequency is 3 dB lower than shown on the Bode plot.
22. The high-frequency performance of an op amp is limited by both its Bode plot and its slew rate.
23. An *RC* lag network causes amplitude to drop at 20 dB per decade beyond the break frequency.
24. An *RC* lag network causes the output to phase-lag the input by 45 degrees at the break frequency and as much as 90 degrees for higher frequencies.
25. Because of device interelectrode capacitance, *RC* lag networks are inherent in any amplifier.
26. Because of the inherent lag networks, the total phase error will be −180 degrees at some frequency. This will cause instability in an amplifier using negative feedback unless the gain is less than 1.
27. Most op amps are internally compensated to prevent instability.
28. Some op amps use external compensation to allow circuit designers to achieve better high-frequency gain and better slew rate.
29. Internally compensated op amps are easier to use and are more popular.
30. Op amps can be used as summing amplifiers.
31. By adjusting input resistors, a summing amplifier can scale some, or all, of the inputs.
32. Summing amplifiers may be called mixers. A mixer can sum several audio inputs.
33. Op amps can be used as subtracting amplifiers. The signal at the inverting input is subtracted from the signal at the noninverting input.
34. Op amps are used in active filter circuits. One of their advantages is that they eliminate the need for inductors.
35. Active filters can be cascaded (connected in series) for sharper cutoff.

36. Op-amp integrators use capacitive feedback. The output of an integrator produces a linear ramp in response to a dc input signal.
37. A comparator is a circuit that looks at two input signals and switches its output according to which of the inputs is greater.
38. An op-amp integrator and an op-amp comparator can be combined to form a voltage-to-frequency converter. This is one way to achieve analog-to-digital conversion.

39. A Schmitt trigger is a signal-conditioning circuit with two threshold points.
40. In a Schmitt trigger, the difference between the two threshold points is called hysteresis.
41. Hysteresis can prevent noise from false-triggering a circuit.
42. Op amps can be powered from a single supply voltage by using a voltage divider to bias the inputs at half the supply voltage.

Related Formulas

Common-mode rejection ratio (dB):

$$\text{CMRR} = 20 \times \log \frac{A_{V(\text{dif})}}{A_{V(\text{com})}}$$

Differential amplifier emitter current:

$$I_{E(\text{total})} = \frac{V_{EE} - 0.7}{R_E}$$

$$I_E = \frac{I_{E(\text{total})}}{2}$$

AC base resistance: $r_B = \dfrac{R_B}{\beta}$

Differential gain: $A_{V(\text{dif})} = \dfrac{R_L}{(2 \times r_E) + r_B}$

Common-mode gain: $A_{V(\text{com})} = \dfrac{R_L}{2 \times R_E}$

Op-amp power bandwidth: $\dfrac{SR}{2\pi \times V_p}$

Op-amp noninverting gain: $A_V = 1 + \dfrac{R_F}{R_1}$

Op-amp inverting gain: $A_V = -\dfrac{R_F}{R_1}$

Closed-loop gain: $A_{CL} = \dfrac{A}{AB + 1}$

Small-signal bandwidth (break frequency): $f_b = \dfrac{f_{\text{unity}}}{A_V}$

RC break frequency: $f_b = \dfrac{1}{2\pi RC}$

Integrator output slope: $V_{\text{out}} = -V_{\text{in}} \times \dfrac{1}{RC}$

Schmitt trigger trip points:

$$\text{UTP} = V_{\max}\left(\frac{R_1}{R_1 + R_2}\right)$$

$$\text{LTP} = V_{\min}\left(\frac{R_1}{R_1 + R_2}\right)$$

Schmitt trigger hysteresis: Hysteresis = UTL − LTP

Chapter Review Questions

Answer the following questions.

9-1. What name is given to an amplifier that responds to the difference between two input signals? (9-1)

9-2. A bipolar power supply provides how many polarities with respect to ground? (9-1)

9-3. Refer to Fig. 9-4. Assume that the input signal drives the base of Q_1 in a positive direction. What effect does this have on the emitter of Q_2? On the collector of Q_2? (9-1)

9-4. Refer to Fig. 9-5. Assume that both wires that connect the signal source to the amplifier pick up a hum voltage. Why can the amplifier greatly reduce this hum voltage? (9-1)

9-5. Refer to Fig. 9-5. If the output signal is taken across the two collectors, what is the output called? (9-1)

9-6. Refer to Fig. 9-5. If the single-ended output signal is 2.3 V peak-to-peak, what will be the differential output be? (9-1)

9-7. The differential input of an amplifier is 150 mV, and the output is 9 V. What is the differential gain of the amplifier? (9-1)

9-8. In using the same amplifier as in question 9-7, it is noted that a 2-V common-mode signal is reduced to 50 mV in the output. What is the CMRR? (9-1)

9-9. Refer to Fig. 9-11. Will this circuit show good common-mode rejection at either of its single-ended outputs? (9-2)

9-10. Refer to Fig. 9-13. Does this operational amplifier provide a single-ended or differential output?

9-11. What geometric shape is often used on schematic diagrams to represent an amplifier? (9-3)

9-12. What polarity sign will be used to mark the noninverting input of an operational amplifier? (9-3)

9-13. Which op-amp terminals can be used to correct for slight dc internal imbalances? (9-3)

9-14. An op amp has a slew rate of 5 V/μs. What is the power bandwidth of the op amp for a 16-V peak-to-peak output swing? (*Hint:* Don't forget to use the peak value in your calculation.) (9-3)

9-15. What is the gain of an op amp called when there is no feedback? (9-4)

9-16. What does negative feedback do to the open-loop gain of an op amp? (9-4)

9-17. What does negative feedback do to the bandwidth of an op amp? (9-4)

9-18. Refer to Fig. 9-21. To what value will R_F have to be changed in order to produce a voltage gain of 33? (9-4)

9-19. Refer to Fig. 9-23. Change R_1 to 470 Ω. What is the voltage gain? What is the input impedance? (9-4)

9-20. Refer to Fig. 9-23. What component sets the input impedance of this amplifier? (9-4)

9-21. Refer to Fig. 9-23(b). What can happen to the op amp if R_2 is very different in value compared with the parallel equivalent of R_1 and R_F? (9-4)

9-22. Refer to Fig. 9-23(b). Resistors R_1 and R_2 are 2,200 Ω, and R_F is 220 kΩ. What is the voltage gain of the amplifier? What is the input impedance? (9-4)

9-23. Refer to Fig. 9-25. Where does the maximum error occur in a Bode plot? What is the magnitude of this error? (9-4)

9-24. Refer to Fig. 9-25. The gain of the op amp is to be set at 80 dB by using negative feedback. Where will the break frequency be? (9-4)

9-25. Refer to Fig. 9-25. Is it possible to use this op amp to obtain a 30-dB gain at 100 Hz? (9-4)

9-26. Refer to Fig. 9-25. Is it possible to use this op amp to obtain a 70-dB gain at 1 kHz? (9-4)

9-27. Refer to Fig. 9-34. Assume that R_1 and R_2 are 4.7 kΩ. What impedance does source V_1 see? Source V_2? (9-6)

9-28. Refer to Fig. 9-34. Resistors R_1 and R_2 are both 10 kΩ, and R_F is 68 kΩ. If $V_1 = 0.3$ V and $V_2 = 0.5$ V, what will V_{out} be? (9-6)

9-29. Refer to Fig. 9-36. All the resistors are 1 kΩ. If $V_1 = 2$ V and $V_2 = 2$ V, what will the output voltage be? (9-6)

9-30. The cutoff frequency of a filter can be defined as the frequency at which the output drops to 70.7 percent of its maximum value. What does this represent in decibels? (9-6)

9-31. Find the break frequency for an RC lag network with 22 kΩ of resistance and 0.1 μF of capacitance. What is the phase angle of the output at f_b? (9-5)

9-32. What is the phase angle of the output from a lag network operating at 10 times its break frequency? (9-5)

9-33. What can happen in a negative-feedback amplifier if the internal lags accumulate to -180 degrees? (9-5)

9-34. Refer to Fig. 9-52. What component is used to discharge the integrator? (9-6)

9-35. Refer to Fig. 9-52. Which op amp is used to turn on Q_1? (9-6)

9-36. Refer to Fig. 9-53. Is the relationship between input voltage and output frequency linear? (9-6)

9-37. Refer to Fig. 9-56. What happens to the hysteresis as R_1 is made larger? Smaller? (9-6)

9-38. Refer to Fig. 9-59. How does the output frequency from the Schmitt trigger compare with the input frequency? (9-6)

9-39. Refer to Fig. 9-59. How does the output frequency from the comparator compare with the input frequency? (9-6)

9-40. Refer to Fig. 9-60. What is the load on the signal source, and what is the overall gain of the two stages? (9-6)

Chapter Review Problems

9-1. Find the dB CMRR for a differential amplifier with a common-mode gain of 0.5 and a differential gain of 35. (9-1)

9-2. Refer to Fig 9-8. Assume that the transistors have a current gain of 250 from base to collector and use 25 mV when estimating ac emitter resistance. Change all the resistors to 1 kΩ and find the dB CMRR. (9-2)

9-3. What is the drop across the base resistors for problem 9-2? (9-2)

9-4. What is V_{CE} for problem 9-2? (9-2)

9-5. Refer to the current source shown in Fig. 9-11. Find the total current supplied to the differential amplifier if the drop across the zener is only 4 V. (9-2)

9-6. Determine the power bandwidth for an op amp with a slew rate of 20 V/μs when it delivers a peak-to-peak output of 10 V. (9-3)

9-7. Refer to Fig. 9-21. The signal source delivers 3 mV, and the feedback resistor is changed to 470 kΩ. Find the amplitude of the output signal. (9-3)

9-8. Refer to Fig. 9-23. Find the output amplitude if the input resistor is 220 Ω and the signal source has an unloaded amplitude of 20 mV with an internal impedance of 100 Ω. (9-3)

9-9. Find the small-signal bandwidth for an op amp with a gain-bandwidth product of 20 MHz and a voltage gain of 50. (9-3)

9-10. What is the break frequency for a 560-Ω resistor and a 5-nF capacitor. (9-5)

9-11. Refer to Fig. 9-34. What is the ideal value for R_3 if the other resistors are all 10 kΩ? (9-6)

9-12. In problem 9-11, assume $V_1 = -2.5$ V and $V_2 = +2.5$ V. What is V_{out}? (9-6)

9-13. Refer to Fig. 9-51. What is the slope of V_{out} if $V_{in} = -150$ mV, the resistor is 680 kΩ, and the capacitor is 4.7 nF? (9-6)

9-14. Refer to Fig. 9-52. Find the output frequency when $V_{in} = 200$ mV. (9-6)

9-15. Refer to Fig. 9-54. Assume as an initial condition that the integrator output is 0 V (the circuit has just been reset). Calculate how long the light source will remain on if its intensity causes the LDR resistance to be 900 Ω. Use 0.6 V for the diode drop at the integrator input. (9-6)

9-16. Refer to Fig. 9-56. Change R_1 to 1,500 Ω and calculate the hysteresis voltage assuming that the output saturates at ± 12 V. (9-6)

Critical Thinking Questions

9-1. Why is CMRR a critical specification for some medical electronic equipment?

9-2. What advantage could be offered by cascading differential amplifiers?

9-3. An amplifier uses three op-amp stages in cascade. The break frequency of each individual stage is 10 kHz. Why is the small-signal bandwidth of the cascade circuit less than 10 kHz?

9-4. People who work around radiation sources may be required to wear a film badge. The purpose of the badge is to accumulate a measurement of their total dose of radiation exposure. Can you think of an electronic replacement for the film badge?

9-5. What would be the advantages of the electronic gadget described in question 9-4?

9-6. What would be the disadvantage of the electronic gadget described in question 9-4?

9-7. Would the output of a Schmitt trigger ever show any noise? Why or why not?

Answers to Self-Tests

1. dual or bipolar supply
2. nothing
3. negative; positive
4. because Q_2 acts as an emitter follower and drives Q_1
5. 4 V peak-to-peak
6. +12 V; −12 V; +24 V
7. ground
8. 100
9. 800 (58 dB)
10. $I_{R_E} = 0.933$ mA
    $I_{E(Q_1)} = 0.466$ mA
    $I_{E(Q_2)} = 0.466$ mA
    $V_{CE(Q_1)} = 5.04$ V
    $V_{CE(Q_2)} = 5.04$ V
    $V_{B(Q_1)} = -0.233$ V
    $V_{B(Q_2)} = -0.233$ V
    $A_{V(\text{dif})} = 46.6$
    $A_{V(\text{com})} = 0.562$

    CMRR = 82.9 (38.4 dB)
11. section 3
12. 180 degrees out of phase
13. single-ended
14. offset null
15. amplitude reduction and waveform distortion
16. 3.41 kHz; the small signal bandwidth is greater (0.6 MHz)
17. closed loop
18. −100; 470 Ω
19. 3,300 Ω; 33 kΩ
20. 2; 22 kΩ
21. 100 Ω
22. 1 MΩ
23. 100 Ω
24. 100 Hz
25. 77 dB

26. 80 dB
27. 60 dB
28. decreases it
29. shifts it more negative
30. 6.99 Hz
31. It occurs at a very low frequency so gain rolls off to less than 1 before the inherent lag networks can cause a problem.
32. The feedback can become positive at a frequency where the gain is greater than 1.
33. −3 V
34. +5 V
35. +1 V
36. −8 V
37. the virtual ground
38. +2 V

39. 0 V
40. +3 V
41. 10 kHz
42. It will decrease at 80 dB/decade.
43. 600 Hz
44. It will ramp positive.
45. 600 Hz
46. 4 s
47. +1.18 V; −1.18 V; 2.36 V
48. 25 ns
49. 0 V (LOW)
50. from pin 1 to ground; emitter follower
51. pull-up resistors
52. zero volts (LOW)
53. ≈ 5 V (HIGH)
54. D_1

Troubleshooting

Electronic components sometimes fail, and part of the troubleshooting process involves identifying the failed components. This is accomplished by using logic based on an understanding of circuits and by using test equipment. Today, this part of the troubleshooting process must be based on a system viewpoint and should be the next step after some very important preliminary checks have been made.

Learning Outcomes

This chapter will help you to:

10-1 *Save* troubleshooting time by performing preliminary checks. [10-1]

10-2 *Develop* and use a system view. [10-1]

10-3 *Use* procedures that prevent electrostatic discharge. [10-1]

10-4 *Troubleshoot* for the symptom of no output. [10-2]

10-5 *Troubleshoot* for reduced output. [10-3]

10-6 *Correct* distortion and noise problems. [10-4]

10-7 *Deal* with intermittent problems. [10-5]

10-8 *Troubleshoot* op-amp circuits. [10-6]

10-9 *Explain* how boundary scan can be used to troubleshoot circuits. [10-7]

10-10 *Explain* the use of thermal measurements. [10-8]

10-1 Preliminary Checks

When troubleshooting, remember the word *GOAL*. Good troubleshooting is a matter of

GOAL

1. Observing the symptoms
2. Analyzing the possible causes
3. Limiting the possibilities

The letter *L* is last in GOAL. Keep that in mind. Don't be too quick to limit the possibilities. Troubleshooting requires a *system view*. There is more than just a chance that the real cause of a problem is not where the symptoms appear. Medical doctors, who must find and deal with root cause, know this. In the human body, *referred pain* can occur. That is, a pain in one location may have its origin in an entirely different location.

System view

Look at Fig. 10-1. It shows a piece of equipment that requires troubleshooting because it is not working properly. A good technician knows to use a system view. There are all sorts of things that can go wrong with systems:

- Faulty power sources (including dead batteries)
- Bad connectors and loose connectors
- Open cables and cables connected incorrectly
- Input signals missing

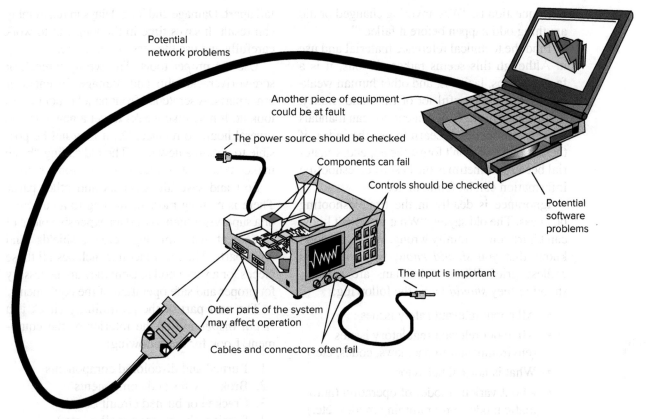

Potential
network problems

Another piece of equipment
could be at fault

The power source should be checked

Components can fail

Controls should be checked

Potential
software
problems

The input is important

Other parts of the system
may affect operation

Cables and connectors often fail

Fig. 10-1 Troubleshooting often requires a system view.

- Incorrectly set controls
- Component failures
- Network problems
- Software problems

The last two items are becoming more common than in the past. Equipment is often in "communication" with other pieces of equipment, and software runs behind the scenes and affects how things work. It is very important to take a system view and to make some preliminary checks before taking a piece of equipment apart. Overlooking software problems can cause a great loss of time and money. If there has been a recent software or firmware update, be aware that this is often a cause for malfunctions.

Not all equipment is networked to other equipment, but it still pays to use a system view. Troubleshooters who have this mind-set know how important preliminary checking is. It is amazing but true that many technicians have found that the cause of a reported "problem" was that a piece of equipment was not plugged in. Don't assume anything when troubleshooting. Don't forget to look at back panels, indicator lights, ready-access fuse holders, switch and selector

settings, plugs and cables, and so on. Check everything, and always begin with the most obvious items. This takes a little time, but it takes more time to reassemble something that should not have been taken apart in the first place.

Experience is extremely valuable. Many technical workers believe in what is called the *ten percent rule*. This rule says that 10 percent of the possibilities cause 90 percent of the problems. When you know the items that are likely to fail, you can check them early in the troubleshooting process. Obviously, beginners do not often know what belongs in the 10 percent category. Don't be reluctant to ask questions. Ask coworkers and supervisors direct questions such as, "What could be wrong with a model 360L that it won't pass diagnostic check number 4?" Ask customers indirect questions such as, "Has anything like this happened before?" Another

Ten percent rule

ABOUT ELECTRONICS

When troubleshooting, one obvious place to find information is the Internet. One can search for model numbers, user groups, symptoms, and so on.

good question is, "Was anything changed or did anything odd happen before it failed?"

Find the technical reference material and use it. Although this seems rather obvious, it is a fact that egos, laziness, and other human weaknesses waste an awful lot of time and money. Learn to use the table of contents to scan manuals for relevant sections. Learn to use the index, if there is one, and don't forget the appendix material because sometimes the best troubleshooting information can be found there.

Ignorance is deadly in the troubleshooting business. The old saying "What you don't know can't hurt you" is totally wrong. What you don't know, that you *should know*, will cause you endless grief. When technicians are troubleshooting they *should know* the following:

- All about relevant safety issues
- All about relevant regulatory issues (environmental impact laws, codes, etc.)
- What is normal behavior
- About various modes of operation (automatic modes, programming modes, etc.)
- What the various parts of a system do
- What the controls are supposed to do
- What inputs and outputs are for and how they should be connected
- Whether a device can reasonably be tested when it is removed from a system
- What role software might play in performance

A cold boot might be in order. If the equipment contains computer chips or microprocessors, always disconnect it from the power line for several minutes (merely turning it off might not force a cold boot). Also, remove any batteries for several minutes. Then restore power, and determine if the symptoms are different.

When the preliminary checks and system tests are completed and the unit is still not working, an internal inspection must be made. Do not attempt to remove the unit from its cabinet until you have disconnected it from the ac line. Be wary of charged filter capacitors. Use a voltmeter and *verify* that they are discharged.

Follow the manufacturer's procedures when taking apart equipment. Often *service literature* will show exactly how to do it. Many technicians overlook this and just start removing parts and fasteners. This may cause internal assemblies to fall apart. Damage and long delays in reassembly can result. It saves time in the long run to work carefully and use the service literature.

Use the proper tools. The wrong wrench or screwdriver can slip and damage fasteners or other parts. A scratched front panel is not nice to look at. It may take weeks to get a new one and several hours to replace. Or, it may not be possible to obtain a new one. The old saying "haste makes waste" fits perfectly in electronic repair.

Sort and save all fasteners and other parts. There is nothing more disturbing to a customer or a supervisor than to find an expensive piece of equipment with missing screws, shields, and other parts. The manufacturer includes all those pieces for a very good reason: they are necessary for proper and safe operation of the equipment.

The next part of the preliminary check is a *visual inspection* of the interior of the equipment. Look for the following:

1. Burned and discolored components
2. Broken wires and components
3. Cracked or burned circuit boards
4. Foreign objects (paper clips, etc.)
5. Bent transistor leads that may be touching (this includes other noninsulated leads as well)
6. Parts falling out of sockets or only partly seated
7. Loose or partly seated connectors
8. Leaking components (especially electrolytic capacitors and batteries)
9. Blown fuses

The last section of this chapter shows some examples.

Obvious damage can be repaired at this point. However, do not energize the unit immediately. For example, suppose a resistor is burned black. In many cases, the new one will quickly do the same. Inspect the schematic to see what the resistor does in the circuit. Try to determine what kinds of problems could have caused the overload.

Blown fuses can be replaced at this point, but once again do not energize the unit until possible causes are investigated. Figure 10-2 shows some types of fuses found on circuit boards. Some are surface-mount devices and might look like other two-lead surface-mount devices such as diodes, capacitors, resistors, or inductors. Often, printing on the circuit board will

Visual inspection

Internet Connection

Career and other information can be found at the website for the Consumer Technology Association (CTA).

Service literature

Fig. 10-2 Examples of fuses found on circuit boards.
Courtesy of Littelfuse

identify fuses with the designation F1, F2, F3, and so on. The current rating of a fuse might be marked on its body. However, the fusing characteristics (response time) and voltage ratings are often not. Consult the schematic and other service literature. *It can be dangerous* to replace a fuse with the wrong response time or with the wrong current or voltage rating!

Some technicians use an ohmmeter check *before* replacing a blown fuse. Ohmmeter testing in-circuit should *never* be attempted before one is *certain* that the power cord is disconnected, the batteries are removed, and all capacitors are discharged. A multimeter on a voltage range can verify that zero volts are across the test points before switching the meter to a resistance range. By measuring resistance between ground and the fuse terminals, one can often determine if there is a short. If there is, a new fuse will quickly blow. Another technique is to make a resistance measurement with one ohmmeter polarity and then swap the meter leads and repeat. If the resistance is the same and low with both polarities, this often is caused by a shorted or leaky component. In-circuit ohmmeter testing can be confusing because the

ohmmeter is an energy source, and several paths that are not obvious might cause a lower than expected resistance reading. However, results are usually clear-cut for opens and shorts.

The wide application of two-lead surface-mount devices has made troubleshooting more difficult. The Smart Tweezers R-C-L meter shown in Fig. 10-3 makes things easier. This meter has a built-in signal generator capable of producing four different test frequencies. The signal is applied to the device under test, and the device current and voltage and related phase angle are used to *automatically* determine and display R or C or L or both C and R as shown in the photo. The R value in the photo is the *equivalent series resistance* (ESR), which can be an important factor. A capacitor can have the correct value of capacitance and will test as "good" using an ordinary capacitance meter, yet cause a circuit to malfunction due to high series resistance. This is a common problem with electrolytics. The Smart Tweezers also have a diode mode, can be used to measure the Q (quality factor) of a coil and the D (dissipation factor) of a capacitor, and, of course, can also find a blown fuse.

Fig. 10-3 Smart Tweezers R-C-L meter.
Courtesy of Advance Devices, Inc.

Electrical check

Refer to Fig. 10-4. Suppose a visual inspection revealed that R_1 was badly burned. What kinds of other problems are likely? There are several:

1. Capacitor C_1 could be shorted.
2. Zener diode D_1 could be shorted.
3. There is a short somewhere in the regulated output circuit.
4. The unregulated input voltage is too high (this is not too likely, but it is a possibility).

Overheating

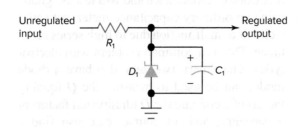

Verify voltage

Fig. 10-4 Zener shunt regulator.

When the preliminary visual inspection is complete, a preliminary *electrical check* should be made. Be sure to use an isolation transformer in those cases that require it. Remember that most variable transformers are autotransformers, and they do not provide line isolation. In some cases, it is possible to safely make *floating measurements*. However, remember that floating measurements require test equipment designed for this purpose. Line isolation and floating measurements were covered in Chap. 4 (see Figs. 4-25 and 4-26).

The first part of the preliminary electrical check involves any signs of *overheating*. Your nose may give important information. Hot electronic components often give off a distinct odor. The last section of this chapter treats thermal issues.

The next part of the electrical check is to *verify* the power-supply *voltages*. Power-supply

Table 10-1	Electrostatic Charges Generated by Technical Personnel	
	Generated Charge in Volts	
Action	High Humidity	Low Humidity
Walking across carpet	1,500	35,000
Removing item from plastic bag	1,200	20,000
Sliding off or onto plastic chair	1,500	18,000
Walking across vinyl floor	250	12,000
Sliding sleeve across laminated bench	100	6,000

Table 10-2	ESD Susceptibility for Various Device Types
Device Type	ESD Susceptibility
Microprocessor chips	As low as 10 V
EPROM devices	100 V
CMOS logic devices	250 to 3,000 V
Film resistors	300 to 3,000 V
Bipolar transistors	380 to 7,000 V
TTL logic devices	1,000 to 2,500 V

problems can produce an entire range of symptoms. This is why it can save a lot of time to check here first. Consult the manufacturer's specifications. The proper voltages are usually indicated on the schematic diagram. Some error is usually allowed. A 20 percent variation is not unusual in many circuits. Of course, if a precision voltage regulator is in use, this much error is not acceptable. Be sure to check all the supply voltages. Remember, it only takes one incorrect voltage to keep a system from working.

Damage caused by electrostatic discharge (ESD) is a serious problem, and all technicians and engineers should be aware of it. A static charge is an imbalance of electrons. Too many electrons create a negative static charge, and too few create a positive charge. Static charges can be created by friction between objects or by removing an insulated wrapper from an item. The friction causes one surface to gain electrons while the other surface loses them. When the static charge is high enough, a destructive discharge can occur. Most discharges are lower in level than workers can detect. Table 10-1 lists some typical static charges generated by ordinary activities. Table 10-2 lists the ESD susceptibility of several device types. Note that damage can occur at as low as 10 V.

Devices can be damaged in any of the following ways:

- A charged body touches the device [Fig. 10-5(a)].
- A charged device touches a grounded object or surface.

- A charged machine or tool touches the device.
- The field surrounding a charged object induces a charge into the device.

There are three types of device damage, the first one listed below being the most prevalent:

- Leakage and shorts caused by localized heating
- Oxide "punch-through"
- Fused (open) conductors

Static discharges and static induction usually cause what is known as an *ESD latent defect*. The device is damaged but continues to function within normal limits. However, it has been weakened and often fails later. There is no way to test for latent defects. So-called "instant death" occurs in only about 15 percent of electrostatic discharges where damage was actually done.

The ESD *susceptibility symbol* shown in Fig. 10-5(b) consists of a triangle, a reaching hand, and a slash through the reaching hand. The triangle means *caution*, and the slash through the reaching hand means *don't touch*. This symbol is applied directly to integrated circuits, boards, and assemblies that are static sensitive. It indicates that handling this item may result in damage from ESD if proper precautions are not taken. The ESD *protective symbol* is shown below the susceptibility symbol and also consists of a reaching hand in a triangle. An arc over the triangle replaces the slash. The arc represents an umbrella of protection. Thus, the symbol indicates ESD protection. It is applied to mats, chairs, wrist straps, garments, packaging, and other items that provide ESD protection. It also may be used on equipment such as hand tools, conveyor belts, or automated handlers that are especially designed or modified to provide ESD control.

Static discharge

ESD latent defect

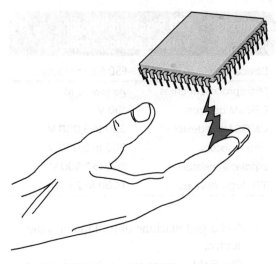

(a) ESD can damage components

ESD sensitive: Don't touch!

ESD protective material or item

(b) ESD symbols

Fig. 10-5 Electrostatic discharge (ESD).

ESD protection in work areas includes the following:

- The work surfaces must be grounded; static-dissipative materials and a dissipative floor mat might also be required [Fig. 10-6(a)].
- Technicians must use wrist straps [Fig. 10-6(b)].
- Wrist straps should be tested often.
- Technicians are sometimes required to wear special ESD footwear and smocks.
- Insulator materials should be removed from the work area or neutralized with an ionizer. Ionized gas (air) conducts to drain off static charges.
- Maintain the relative humidity around 50 percent (Table 10-1).
- Keep ICs and circuit boards in protective carriers when transferring, shipping, or storing.

What about field service where a protected work area is not available? Most technicians follow these rules:

- Assume that all components and circuit boards are susceptible to damage by ESD.
- Use as little motion as possible (Table 10-1).
- Use a wrist strap.
- Turn everything off before touching circuits or components.
- Touch a grounded case, frame, or chassis before touching any part of the circuit.
- When connecting instruments or other equipment, connect the ground leads first.
- Handle components and circuit boards as little as possible. Keep them in their protective carriers, until they are needed. Touch the carrier to a grounded case, frame, or chassis before removing the part.
- Immediately place removed parts into protective carriers.
- Use grounded soldering tools.
- Use antistatic sprays and chemicals.

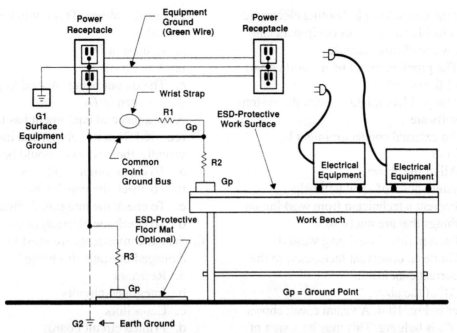

NOTES: A. G1 (surface equipment ground) or G2 (earth ground) is acceptable for ESD ground. Where both grounds are used, they are connected (bonded) together.
B. R1 is mandatory for all wrist straps.
C. R2 (for static-dissipative work surfaces) and R3 (for ESD-protective floor mats) are optional. ESD-protective flooring are connected directly to the ESD ground without R3.
D. This ESD-protected workstation complies with JEDEC Standard No. 42.

(*a*) ESD-protected workstation

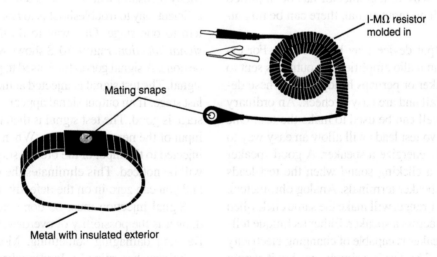

(*b*) Wrist strap

Fig. 10-6 Work area ESD prevention.

Self-Test

Choose the letter that best completes each statement.

1. The first step in troubleshooting is to take

a. Resistance readings
b. Voltage readings
c. A look at the overall system
d. The covers off all defective units

2. In some cases, troubleshooting electronic parts inside one piece of equipment can be a waste of time because
   a. The problem might be in another part of the system
   b. The problem might be with the system software
   c. An external connector might be loose
   d. All of the above
3. A system point of view will help
   a. Prevent a technician from working on things that are not broken
   b. Prevent time from being wasted
   d. Guide an observant technician to the correct conclusion
   d. All of the above
4. Refer to Fig. 10-4. A visual check shows that C_1 is bulging. This may be a sign of excessive voltage. This could have been caused by
   a. A short in D_1
   b. An open in R_1
   c. The output being shorted to ground
   d. An open in D_1
5. After a piece of equipment has been removed from its cabinet and inspected visually, the next step should be
   a. To check supply voltages
   b. To check the transistors
   c. To check the integrated circuits
   d. To check the electrolytic capacitors
6. Which components are most likely to be damaged by static discharge?
   a. Resistors
   b. Integrated circuits
   c. Capacitors
   d. Printed circuit boards

10-2 No Output

No input

There are several causes for no output from a circuit. Perhaps the most obvious is *no input*. It is worth the effort to check this early. You may find that a wire or a connector has been pulled loose. With no input signal, there can be no output signal.

Signal injection

The output device may be defective. For example, in an audio amplifier, the output is sent to a loudspeaker or perhaps headphones. These devices can fail and are easy to check. An ordinary flashlight cell can be used to make the test. One cell with two test leads will allow an easy way to temporarily energize a speaker. A good speaker will make a clicking sound when the test leads touch the speaker terminals. Analog ohmmeters, on the $R \times 1$ range, will make the same click when connected across a speaker. Either technique tells you the speaker is capable of changing electricity into sound. This test is a simple one, but it cannot be used to check the quality of a speaker.

Signal chain

If there is nothing wrong with the output device, the power supply, or the input signal, then there is a break in the *signal chain*. Figure 10-7 illustrates this. The signal must travel the chain, stage by stage, to reach the load. A break at any point in the chain will usually cause the no-output symptom.

A four-stage amplifier contains many parts. Many measurements can be taken. Therefore, the efficient way to troubleshoot is to *isolate* the problem to one stage. One way to do this is to use *signal injection*. Figure 10-8 shows what needs to be done. A signal generator is used to provide a test signal. The test signal is injected at the input to the last stage. If an output signal appears, then the last stage is good. The test signal is then moved to the input of the next-to-last stage. When the signal is injected to the input of the broken stage, no output will be noticed. This eliminates the other stages, and you can zero in on the defective circuit.

Signal injection must be done carefully. One danger is the possibility of overdriving an amplifier and damaging something. More than one technician has ruined a loudspeaker by feeding too large a signal into an audio amplifier. A high-power amplifier must be treated with respect!

Another danger in signal injection is improper connection. A schematic diagram is a

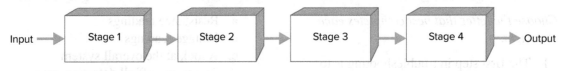

Fig. 10-7 The signal chain.

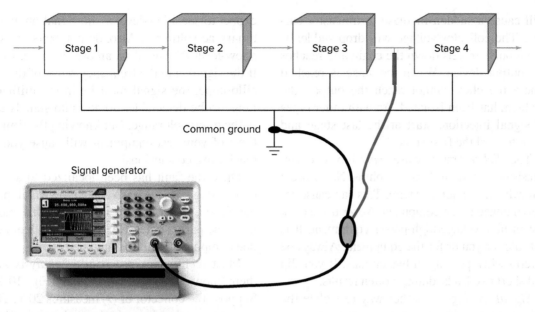

Fig. 10-8 Using signal injection.
Reproduced with Permission, Tektronix, Inc.

must. The common ground is generally used for connecting the ground lead from the generator. Assuming that the chassis is common will not always work out. If the common connection is made in error, a large hum voltage may be injected into the system. Damage can result.

Many amplifiers can be tested with *ac* signals. The signal should be capacitively *coupled* to avoid upsetting the bias on a transistor or integrated circuit. If the generator is dc-coupled, a capacitor must be used in series with the hot lead. This capacitor will block the dc component yet allow the ac signal to be injected. A 0.1-μF capacitor is usually good for audio work, and a 0.001-μF (1nF) capacitor can be used for radio frequencies (be sure that the voltage rating of the capacitor is adequate).

The test frequency varies, depending on the amplifier being tested. A frequency of 400 to 1,000 Hz is often used for audio work. A radio frequency amplifier should be tested at its design frequency. This is especially important in bandpass amplifiers. Some are so narrow that an error of a few kilohertz will cause the signal to be blocked. It may be necessary to vary the generator frequency slowly while watching for output.

Signal injection can be performed without a signal generator in many amplifiers. A resistor can be used to inject a *click* into the signal chain. The click is really a *signal pulse* caused by suddenly changing the bias on a transistor. A resistor is connected momentarily from the collector lead to the base lead as in Fig. 10-9. This

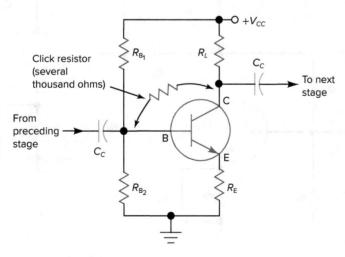

Fig. 10-9 The click test.

will cause a sudden increase in transistor current. The collector voltage will drop suddenly, and a pulse travels down the chain and reaches the output device. When the stage is reached where the click cannot reach the output, the problem has been isolated. As with other types of signal injection, start at the last stage and work toward the first stage.

Click test
The *click test* must be used *carefully*. Use only a resistor of several thousand ohms. Never use a screwdriver or a jumper wire. This can cause severe damage to the equipment. *Never* use a click test in high-voltage/high-power equipment. It is not safe for you or for the equipment. Always be careful when probing in live circuits. If you slip and short two leads, damage often results.

Voltage analysis

Signal tracing
Signal tracing is another way to isolate the defective stage. This technique may use a meter, an oscilloscope, a signal tracer, or some related instrument. Signal tracing starts at the input to the first stage of the amplifier chain. Then the tracing instrument is moved to the input of the second stage, and so on. Suppose that a signal is found at the input to the third stage but not at the input to the fourth stage. This would mean that the signal is being lost in the third stage. The third stage is probably defective.

The important thing to remember in signal tracing is the gain and frequency response of the instrument being used. For example, do not expect to see a low-level audio signal on an ordinary ac voltmeter. Also, do not expect to see a low-level RF signal on an oscilloscope. Even if the signal is in the frequency range of the oscilloscope, the signal must be in the millivolt range to be detectable. Some radio signals are in the microvolt range. Not knowing the limitations of your test equipment will cause you to reach false conclusions!

Once the fault has been localized to a particular stage, it is time to determine which part has failed. Of course, it is possible that more than one part is defective. More often than not, one component will be found defective.

Most technicians use *voltage analysis* and their knowledge of circuits. Study Fig. 10-10. Suppose the collector of Q_2 measures 20 V. The manufacturer's schematic shows that the collector of Q_2 should be 12 V with respect to ground. What could cause this large error? It is likely that Q_2 is in *cutoff*. A 20-V reading at the collector tells us that the voltage is almost the same on both ends of R_6. Ohm's law tells us that a low voltage drop means little current flow. Transistor Q_2 must be in cutoff.

Now, what are some possible causes for Q_2 to be cut off? First, the transistor could be defective. Second, R_7 could be open. Resistor R_7 supplies the base current for Q_2. If it opens, no base current will flow. This cuts off the transistor.

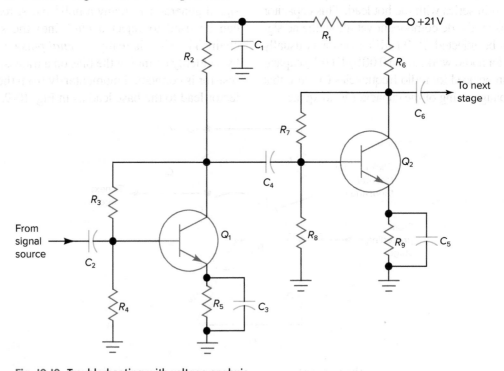

Fig. 10-10 Troubleshooting with voltage analysis.

This can be checked by measuring the base voltage of Q_2. With R_7 open, the base voltage will be zero. Third, R_9 could be open. If it opens, there will be no emitter current. This cuts off the transistor. A voltage check at the collector of Q_2 will show a little less than 21 V. The actual voltage will be determined by the divider formed by R_6, R_7, and R_8. Fourth, R_8 could be shorted. This seldom happens, but a troubleshooter soon learns that all things are possible. With R_8 shorted, no base current can flow and the transistor is cut off. The base voltage will measure zero.

Let us try another symptom. Suppose the collector voltage at Q_1 measures 0 V. A check on the manufacturer's service notes shows that it is supposed to be 11 V. What could be wrong? First, C_1 could be shorted. The combination of R_1 and C_1 acts as a low-pass filter to prevent any hum or other unwanted ac signal from reaching Q_1. If C_1 shorts, R_1 will drop the entire 21-V supply. This can be checked by measuring the voltage at the junction of R_2 and C_1. With C_1 shorted, it will be 0 V. Second, R_2 could be open. This can also be checked by measuring the voltage at the junction of R_2 and C_1. A 21-V reading here indicates R_2 must be open. Could Q_1 be shorted? The answer is no. Resistor R_5 would drop at least a small voltage, and the collector would be above 0 V.

Sometimes it helps to ask yourself what might happen to the circuit given a specific component failure. This thoughtful question-and-answer

game is used by most technicians. Again, refer to Fig. 10-10. What would happen if C_4 shorts? This short circuit would apply the dc collector potential of Q_1 to the base of Q_2. Chances are that this would greatly increase the base voltage and drive Q_2 into saturation. The collector voltage at Q_2 will drop to some low value.

What if C_2 in Fig. 10-10 shorts? Transistor Q_1 could be driven to cutoff or to saturation. If the signal source has a ground or negative dc potential, the transistor will be cut off. If the signal source has a positive dc potential, the transistor will be driven toward saturation.

The advantage of voltage analysis is that it is easy to make the measurements. Often, the expected voltages are indicated on the schematic diagram. A small error is usually not a sign of trouble. Many schematics will indicate that all voltages are to be within a ± 10 percent range.

Current analysis is not easy. Circuits must be broken to measure current. Sometimes, a technician can find a known resistance in the circuit where current is to be measured. A voltage reading can be converted to current by Ohm's law. However, if the resistance value is wrong, the calculated current will also be wrong.

Resistance analysis can also be used to isolate defective components. This can be tricky, however. Multiple paths may produce confusing readings. Refer to Fig. 10-11. An ohmmeter check is being made to verify the value of a

Current analysis

Resistance analysis

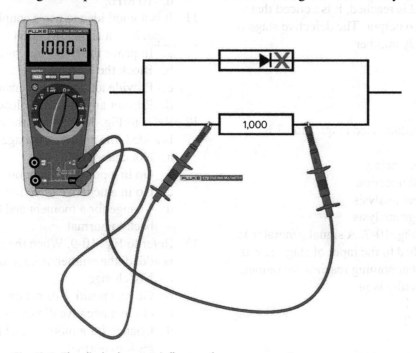

Fig. 10-11 The diode does not influence the measurement.

1,000-Ω resistor. In this case, the reading is good because the diode junction is not turned on by the meter. However, always remember that in-circuit testing usually finds other paths that make the measured resistance lower.

Even if a junction is not turned on by the ohmmeter, in many cases it is still impossible to obtain useful resistance readings. There will be other components in the circuit to draw current from the ohmmeter. Any time you are using resistance analysis, remember that a *low reading* could be caused by multiple paths.

It is usually poor practice to unsolder parts for resistance analysis unless you are reasonably sure the part is defective. Unsoldering can cause damage to circuit boards and to the parts. It is also time-consuming.

As mentioned before, most technicians use voltage analysis to locate defective parts. This is valid and effective since most circuit faults will change at least one dc voltage. However,

there is the possibility of an ac fault that breaks the signal chain without changing any of the dc readings. Some ac faults are

1. An open coupling capacitor
2. A defective coupling coil or transformer
3. A break in a printed circuit board
4. A dirty or bent connector (plug-in modules often suffer this fault)
5. An open switch or control such as a relay

To find this type of fault, signal tracing or signal injection can be used. You will find different conditions at either end of the break in the chain. Some technicians use a coupling capacitor to bypass the signal around the suspected part. The value of the capacitor can usually be 0.1 μF for audio work and 0.001 μF (1nF) for radio circuits. *Do not* use this approach in high-voltage circuits. *Never* use a jumper wire. Severe circuit damage may result from jumping the wrong two points.

Self-Test

Choose the letter that best answers each question.

7. Refer to Fig. 10-7. A signal generator is applied to the input of stage 4, then stage 3, and then stage 2. When the input of stage 2 is reached, it is noticed that there is no output. The defective stage is most likely number
   a. 1
   b. 2
   c. 3
   d. 4
8. The procedure used in question 7 is called
   a. Signal tracing
   b. Signal injection
   c. Current analysis
   d. Voltage analysis
9. Refer to Fig. 10-7. A signal generator is first applied to the input of stage 1. It is noticed that nothing reaches the output. The difficulty is in
   a. Stage 1
   b. Stage 2
   c. Stage 3
   d. Any of the stages

10. A good test frequency for audio troubleshooting is
    a. 2 Hz
    b. 1 kHz
    c. 455 kHz
    d. 10 MHz
11. It is a good idea to use a coupling capacitor in signal injection to
    a. Improve the frequency response
    b. Block the ac signal
    c. Provide an impedance match
    d. Prevent any dc shift or loading effect
12. Refer to Fig. 10-9. When the click resistor is added, the collector voltage should
    a. Not change
    b. Go in a positive direction
    c. Go in a negative direction
    d. Change for a moment and then settle back to normal
13. Refer to Fig. 10-9. When the click resistor is added, the emitter voltage should
    a. Not change
    b. Go in a positive direction
    c. Go in a negative direction
    d. Change for a moment and then settle back to normal

14. Refer to Fig. 10-10. Assume that C_4 is open. It is most likely that
    a. The collector voltage of Q_1 will read high
    b. The collector voltage of Q_1 will read low
    c. The base of Q_2 will be 0 V
    d. All the dc voltages will be correct
15. Refer to Fig. 10-10. Resistor R_7 is open. It is most likely that the
    a. Collector voltage of Q_2 will be high (near 21 V)
    b. Collector voltage of Q_2 will be low (near 1 V)
    c. Case of Q_2 will run hot
    d. Transistor will go into saturation
16. Refer to Fig. 10-10. Suppose it is necessary to know the collector current of Q_1. The easiest technique is to
    a. Break the circuit and measure it
    b. Measure the voltage drop across R_2 and use Ohm's law
    c. Measure the collector voltage
    d. Measure the emitter voltage

17. Refer to Fig. 10-10. Transistor Q_1 is defective. This will not affect the dc voltages at Q_2 because
    a. C_1 isolates the two stages from the power supply
    b. The two stages are not dc-coupled
    c. Both transistors are NPN devices
    d. All of the above
18. Refer to Fig. 10-10. You want to check the value of R_8. The power is turned off; the negative lead of the ohmmeter is applied at the top of R_8, and the positive lead is applied at the bottom. This will prevent the ohmmeter from forward-biasing the base-emitter junction. The ohmmeter reading is still going to be less than the actual value of R_8 because
    a. Capacitor C_4 is in the circuit
    b. Transistor Q_1 is in the circuit
    c. Resistors R_7 and R_6 and the power supply provide a path to ground
    d. None of the above

10-3 Reduced Output

Low output from an amplifier tells us that there is lack of gain in the system. In an audio amplifier, for example, normal volume cannot be reached at the maximum setting of the volume control. Do not troubleshoot for low output until you have made the preliminary checks described in Sec. 10-1.

Low output from an amplifier can be caused by *low input* to the amplifier. The signal source is weak for some reason. A microphone may deteriorate with time and rough treatment. The same thing can happen with some sensors. To check, try a new signal source or substitute a signal generator.

Another possible cause for low output is reduced performance in the output device. A loudspeaker defect or a poor connection may prevent normal loudness. In a speed control, slowness might be due to a defective motor. This can be checked by substituting for the output device or replacing the device with a known load and measuring the output.

In Fig. 10-12, a loudspeaker has been replaced with an 8-Ω resistor in order to measure the output power of an audio amplifier. This resistor must match the output impedance requirement of the amplifier. It must also be rated to safely dissipate the output power of the amplifier. The signal generator is usually

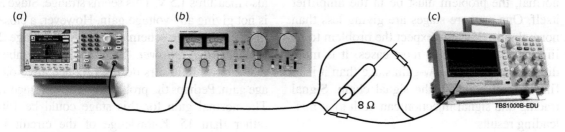

(a)   (b)   (c)

8 Ω

TBS1000B-EDU

Fig. 10-12 Replacing the speaker with a resistor.

(a) Reproduced with Permission, Tektronix, Inc. (b) donatas1205/123RF (c) Tektronix. Reprinted with permission. All Rights Reserved.

adjusted for a sinusoidal output of 1 kHz. The signal level is set carefully so as not to overdrive the amplifier being tested.

Suppose you want to check an audio amplifier for rated output power with an oscilloscope and signal generator. The specifications for the amplifier rate it at 100 W of continuous sine wave power output. How could you be sure the amplifier meets its specification and does not suffer from low output? The power formula shows the relationship between output voltage and the output resistance:

$$P = \frac{V^2}{R}$$

In this case, P is known from the specifications and R is the substitute resistor. What you must determine is the expected output voltage:

$$V^2 = PR \text{ or } V = \sqrt{PR}$$

From the known data,

$$V = \sqrt{100 \text{ W} \times 8 \text{ }\Omega} = 28.28 \text{ V}$$

The 100-W amplifier can be expected to develop 28.28 V across the 8-Ω load resistor. The oscilloscope measures peak-to-peak. Thus, it would be a good idea to convert 28.28 V to its peak-to-peak value,

$$V_{p-p} = V_{rms} \times 1.414 \times 2 = 80 \text{ V}$$

To test the 100-W amplifier, the gain control would be advanced until the oscilloscope showed an output sine wave of 80 V peak-to-peak. There should be no sign of clipping on the peaks of the waveform. If the amplifier passes this test, you know it is within specifications.

Testing some amplifiers requires other equipment. Figure 10-13 shows a two-way radio that is tested for RF output using an RF wattmeter and a 50-Ω dummy load.

If the input signal and output device are both normal, the problem must be in the amplifier itself. One or more stages are giving less than normal gain. You can expect the problem to be **Emitter follower** limited to one stage in most cases. It is more difficult to isolate a low-gain stage than it is to find a total break in the signal chain. Signal tracing and signal injection can both give misleading results.

Suppose you are troubleshooting the three-stage amplifier shown in Fig. 10-14. Your

Two-way radio

Power meter

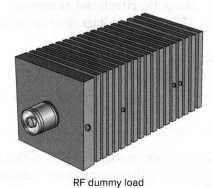
RF dummy load

Fig. 10-13 Measuring transmitter power output.
(CB Radio) Comstock Images/Alamy Stock Photo
(Power Meter) the bunwangs/Shutterstock

oscilloscope shows that the input to stage 1 is 0.1 V and the output is 1.5 V. A quick calculation gives a gain of 15:

$$A_V = \frac{1.5 \text{ V}}{0.1 \text{ V}} = 15$$

This seems acceptable, so you move the probe to the output of stage 2. The voltage here also measures 1.5 V. This seems strange. Stage 2 is not giving any voltage gain. However, a close inspection of the schematic shows that stage 2 is an *emitter follower*. You should remember that emitter followers do not produce any voltage gain. Perhaps the problem is really in stage 1. The normal gain for this stage could be 150 rather than 15. Knowledge of the circuit is required for troubleshooting the low-output symptom.

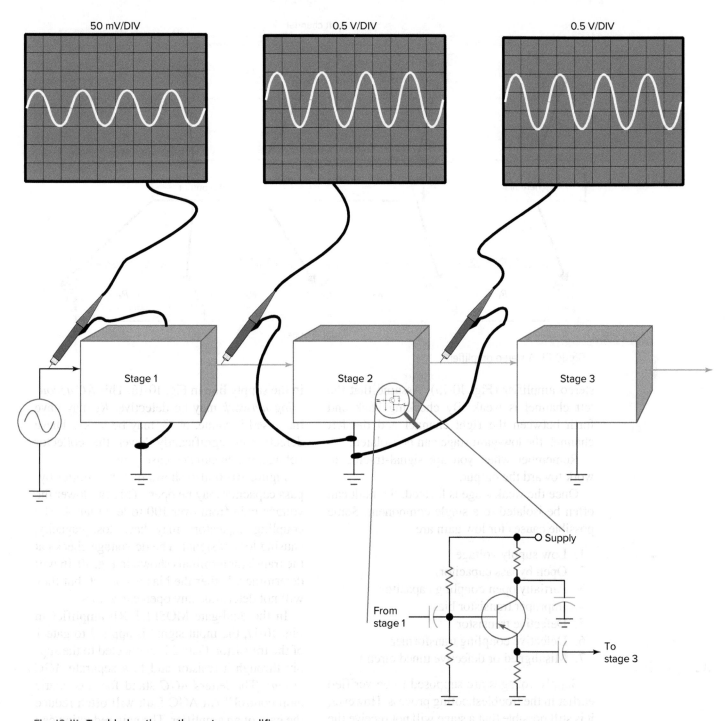

50 mV/DIV   0.5 V/DIV   0.5 V/DIV

Stage 1   Stage 2   Stage 3

Supply

From stage 1

To stage 3

Fig. 10-14 Troubleshooting a three-stage amplifier.

Some beginners believe that a dead circuit is going to be more difficult to fix than a weak one. Just the opposite is generally true. It is usually easier to find a broken link in the signal chain because the symptoms are more definite.

With experience, the low-output problem is usually not too difficult because the technician knows approximately what to expect from each stage. In addition to experience, service notes

and schematics can be a tremendous help. They might include pictures of the expected waveforms at various circuit points. If the oscilloscope shows a low output at one stage, then the input can be checked. If the input signal is normal, it is safe to assume that the low-gain stage has been found.

Sometimes the needed information is found in the equipment itself. A good example is a

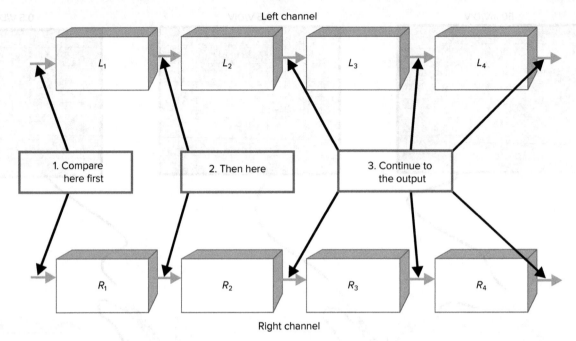

Left channel

1. Compare here first

2. Then here

3. Continue to the output

L_1 L_2 L_3 L_4

R_1 R_2 R_3 R_4

Right channel

Fig. 10-15 A stereo amplifier.

stereo amplifier (Fig. 10-15). Assume that the left channel is weak. By checking back and forth between the right channel and the left channel, the low-gain stage can be isolated.

Remember when you are signal-tracing to work toward the output.

Once the weak stage is located, the fault can often be isolated to a single component. Some possible causes for low gain are

1. Low supply voltage
2. Open bypass capacitor
3. Partially open coupling capacitor
4. Improper transistor bias
5. Defective transistor
6. Defective coupling transformer
7. Misaligned or defective tuned circuit

AGC circuit

Supply voltages are supposed to be verified earlier in the troubleshooting process. However, it is still possible that a stage will not receive the proper voltage. Note the resistor and capacitor

in the supply line in Fig. 10-16. This *RC decoupling network* may be defective. R_3 may have increased in value, or C_2 may be leaky. These defects can significantly lower the collector voltage, which can decrease gain.

Figure 10-16 also shows that the emitter bypass capacitor may be open. This can lower the voltage gain from over 100 to less than 4. The coupling capacitors may have lost capacity, causing loss of signal. The dc voltage checks at the transistor terminals shown in Fig. 10-16 will determine whether the bias is correct, but they will not determine any open capacitors.

In the dual-gate MOSFET RF amplifier in Fig. 10-17, the input signal is applied to gate 1 of the transistor. Gate 2 is connected to the supply through a resistor and to a separate *AGC circuit*. The letters *AGC* stand for "automatic gain control." An AGC fault will often reduce the gain of an amplifier. The gain reduction can be more than 20 dB. Thus, if an amplifier is controlled by AGC, this control voltage must be measured to determine if it is normal.

Figure 10-17 shows that the drain load is a tuned circuit. This circuit is adjusted for the correct resonant frequency by moving a tuning slug in the transformer. The possibility exists that someone turned the slug. This can produce a severe loss of gain at the operating frequency of the amplifier. In such cases, refer to the service notes for the proper adjustment procedure.

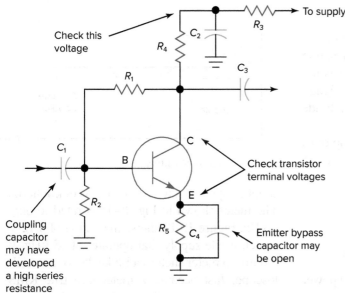

Fig. 10-16 Checking for the cause of low gain.

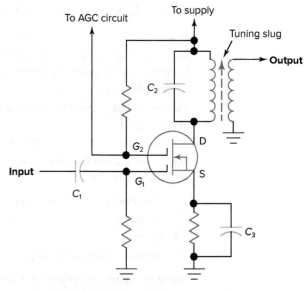

Fig. 10-17 A MOSFET RF amplifier.

Troubleshooting for loss of gain in amplifiers can be difficult. Many things can give this symptom. Voltage analysis will locate some of them. Others must be found by substitution. For example, a good capacitor can be temporarily placed in parallel with one that is suspected of being open. If gain is restored, the technician's suspicion that the original capacitor was defective is correct.

Self-Test

Choose the letter that best answers each question.

19. Refer to Fig. 10-12. The amplifier is rated at 35 W power output. Assuming a sine wave test signal, the oscilloscope should show at least
    a. 17.9 V peak-to-peak before clipping
    b. 47.2 V peak-to-peak before clipping
    c. 75.8 V peak-to-peak before clipping
    d. 99.6 V peak-to-peak before clipping
20. The normal voltage gain for an emitter-follower amplifier is
    a. 250
    b. 150
    c. 50
    d. Less than 1
21. Refer to Fig. 10-16. The power supply has been checked, and it is normal, yet the collector of the transistor is quite low in voltage. This could be caused by

    a. An open in R_1
    b. A leaky C_2
    c. An open in C_3
    d. An open in C_4
22. Refer to Fig. 10-16. The stage is supposed to have a gain of 50, but a test shows that it is much less. It is *least* likely that the cause is
    a. A defective transistor
    b. A short in C_4
    c. An open in C_4
    d. An open in C_1
23. Refer to Fig. 10-17. The stage is suffering from low gain. This could be caused by
    a. An incorrect AGC voltage
    b. A misadjusted tuning slug
    c. A short in C_2
    d. Any of the above

10-4 Distortion and Noise

Distortion and noise in an amplifier mean that the output signal contains different information from the input signal. A linear amplifier is not supposed to change the quality of the signal. The amplifier is used to increase the amplitude of the signal.

Noise can produce a variety of symptoms. Some noise problems that may be found in an audio amplifier are

1. Constant frying or hissing noise
2. Popping or scratching sound
3. Hum
4. Motorboating (a "putt-putt" sound)

Noise problems can often be traced to the power supply. In troubleshooting for this symptom, it is a very good idea to use an oscilloscope to check the various supply lines in the equipment. The checks detailed in Sec. 10-1 rely on meter readings to check the supply. An oscilloscope will show things that a meter cannot. For example, Fig. 10-18 shows a power-supply waveform with *excess* ac *ripple*. The average dc value of the waveform is correct. This means that the meter reading will be acceptable.

A meter can measure ac ripple, but there are cautions. First, the meter must block the dc component of the waveform. A meter such as the one shown in Fig. 10-11 does this when switched to AC volts (next to OFF). Today, some power

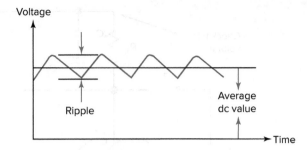

Fig. 10-18 Excessive ripple.

supplies operate at tens or hundreds of kilohertz. The meter shown in Fig. 10-11 is rated to only 1 kHz. Figure 10-19 shows the output of a 12-V switch mode supply that operates at 45 kHz. It requires a meter such as a Fluke 289 or an oscilloscope. Just because a meter is a true RMS (root mean square) device does not guarantee that it will work in all situations. Technicians must understand the limits of their equipment.

The most common noise problem is *hum*. Hum refers to the introduction of a 60-Hz signal into the amplifier. It can also refer to 120-Hz interference. If the hum is coming from the power supply, it will be 60 Hz for half-wave supplies and 120 Hz for full-wave supplies. Hum can also get into the amplifier because of a broken ground connection. High-gain amplifiers often use shielded cables in areas where the ac line frequency can induce signals into the circuitry. *Shielded cable* is used in a stereo amplifier system for connections between

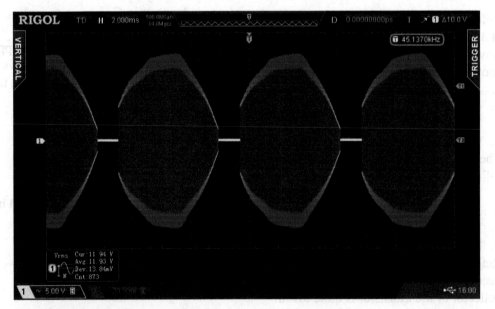

Fig. 10-19 An oscilloscope or a true RMS meter may be needed.

Charles A. Schuler/McGraw Hill

Hum

Shielded cable

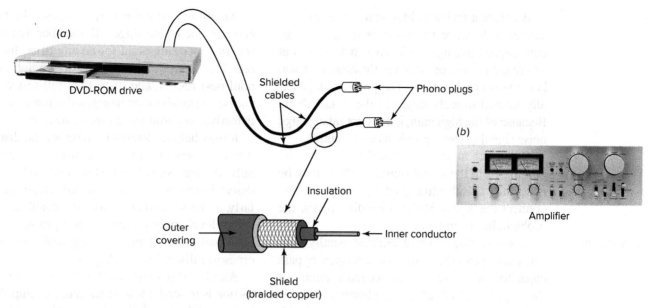

Fig. 10-20 Signal circuits often use shielded cables.
(a) vtls/123RF (b) donatas1205/123RF

components (Fig. 10-20). Be sure to check all shielded cables when hum is a problem. The braid may be open, which allows hum to get into the amplifier. Check connectors since the break in the ground shield is often found there.

Some high-gain amplifiers use metal shields around circuits to keep hum and noise out. These shields must be in the proper position and fastened securely.

Another cause for hum is poor grounding of circuit boards. In some equipment, the fasteners do double duty. They mechanically hold the board and provide an electrical contact to the chassis. Check the fasteners to make sure they are secure.

Sometimes amplifier noise can be limited to general sections of the circuit by checking the effect of the various controls. Figure 10-21 is a

block diagram of a four-stage amplifier. The gain control is located between stage 2 and stage 3. It is a good idea to operate this control to see whether it affects the noise reaching the output. If it does, then the noise is most likely originating in stage 1, stage 2, or the signal source. Of course, if the control has no effect on the noise level, then it is probably originating in stage 3 or stage 4 or in the power supply.

Another good reason for checking the controls is that they are often the source of the noise. Scratchy noises and popping sounds can often be traced to variable resistors. Special cleaner sprays are available for reducing or eliminating noise in controls. However, the noise often returns. The best approach is to replace noisy controls.

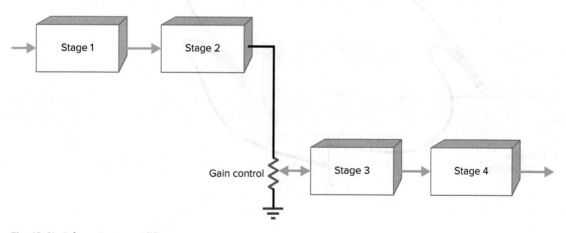

Fig. 10-21 A four-stage amplifier.

Bias error

A constant frying or hissing noise usually indicates a defective transistor or integrated circuit. Signal tracing is effective in finding out where the noise is originating. Resistors can also become somewhat noisy. The problem is generally limited to early stages in the signal chain. Because of the high gain, it does not take a large noise signal to cause problems at the output.

It is worth mentioning that the noise may be coming from the signal source itself. It may be necessary to substitute another source or disconnect the signal. If the noise disappears, the problem has been found.

Motorboating

Motorboating is a problem that usually indicates an open filter capacitor, an open bypass capacitor, or a defect in the feedback circuit of the amplifier. An amplifier can become unstable and turn into an *oscillator* (make its own signal) under certain conditions. This topic is covered in detail in Sec. 11-6 of the next chapter.

Triangular waveform

Amplifier distortion may be caused by *bias error* in one of the stages. Remember that bias sets the operating point for an amplifier. Incorrect bias can shift the operating point to a nonlinear region, and distortion will result. Of course, a transistor or integrated circuit can be defective and produce severe distortion.

It may help to determine whether the distortion is present at all times or just on some signals. A large-signal distortion may indicate a defect in the power (large-signal) stage. Similarly, a distortion that is more noticeable at low signal levels may indicate a bias problem in a push-pull power stage. You may wish to review crossover distortion in Chap. 8.

Another way to isolate the stage causing distortion is to feed a test signal into the amplifier and "walk" through the circuit with an oscilloscope. Many technicians prefer using a *triangular waveform* (Fig. 10-22) for making this test. The

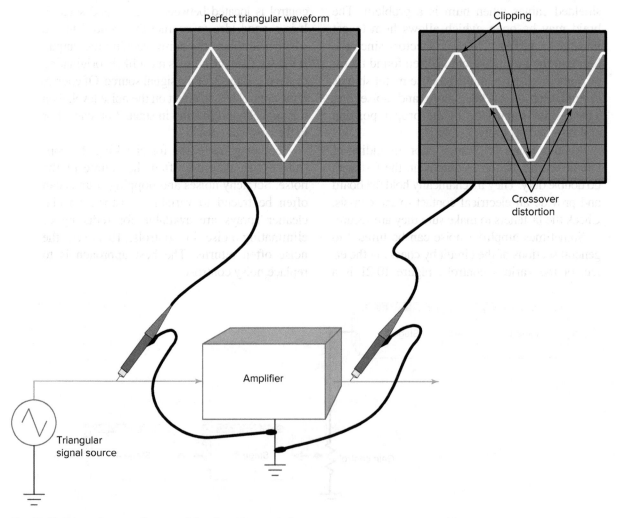

Fig. 10-22 Triangular waveform used for distortion analysis.

sharp peaks of the triangle make it very easy to spot any clipping or compression. The straight lines make it very easy to see any crossover or other types of distortion. In using such a test, it is a good idea to try different signal levels. Some problems show up at low levels and some at high levels. For example, (1) crossover distortion in a push-pull amplifier is most noticeable at low levels, and (2) operating point error in an early class A stage is most noticeable at high levels.

A spectrum analyzer or software that provides that function can be used for distortion analysis. As Fig. 10-23(a) shows, the spectrum for a pure sinewave appears as a single vertical line at one frequency (1 kHz in this illustration). All amplifiers add some harmonic distortion so energy at multiples of the test frequency is expected—the worse the distortion the larger the number of harmonics and the greater their amplitudes. Look at Fig. 10-23(b), which shows the spectrum for a triangle wave. There is energy at the odd harmonic frequencies and the energy decreases as the frequency goes higher. Spectrum analysis is more sensitive to any distortion and provides more information. For example, distortion at even harmonics compared to odd harmonics can provide clues. As another example, if an amplifier has poor high-frequency response and is tested with a triangle wave, the harmonics will decrease in amplitude more rapidly than usual (the vertical lines

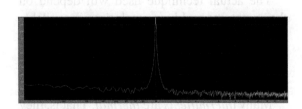

Fig. 10-23(a) Sinewave spectrum
Charles A. Schuler/McGraw Hill

Fig. 10-23(b) Triangle wave spectrum
Charles A. Schuler/McGraw Hill

at the right end will be shorter than expected). Small imperfections that cannot be seen on an oscilloscope are obvious with spectrum analysis.

Sometimes, one of the output transistors in a push-pull amplifier will fail (C-E leakage), and the output will be distorted, provided that the fuse does not blow. Looking at it on an oscilloscope will show clipping. When replacing an output transistor, common practice is to replace both of them (push-pull). NPN-PNP pairs are available with similar characteristics for good balance (complementarity). For example, the NPN MJE521 is complementary to the PNP MJE371.

Some push-pull amplifiers use a single-supply voltage and a large electrolytic output coupling capacitor. If this capacitor develops excess leakage, the current will be abnormally high and the fuse might blow, or the output could be distorted. If the capacitor develops high series resistance, the output will be reduced (less than normal volume). If the capacitor loses much of its capacitance, the output will lack low frequencies (poor or no bass).

A dual-supply push-pull amplifier might have a shutdown circuit that senses excess current or a dc voltage at an output greater than 100 mV or so. Shutdown circuits often work in conjunction with a relay that disconnects the loudspeaker in the event of a problem. A dc voltage across the speaker terminals could cause additional damage. If there is no output, check for a dc voltage before the relay. High-temperature shutdown is another possibility. Playing the system at a high volume can activate the shutdown circuit and so can using the wrong loudspeakers.

It is prudent to check for a dc offset at the output of a switch-mode power amplifier, especially if there is an unusually loud "thump" sound when the unit is turned on. Other sounds such as clicking might point to an overloaded power supply. As always, check all supply voltages early in the troubleshooting process. There might be some AM radio interference associated with a switch-mode power amplifier. A modified speaker circuit or speaker wiring can make such interference worse. Also, bypass capacitors and filter capacitors could be at fault and might cause a hissing sound in the speakers.

Choose the letter that best answers each question.

24. A stereo amplifier has severe hum in the output only when the selector is switched to TAPE. Which of the following is *least* likely to be wrong?
    a. A defective tape jack
    b. A defective shielded cable to the tape player
    c. A bad filter capacitor in the power supply
    d. An open ground in the tape player

25. An audio amplifier has severe hum all the time. Which of the following is most likely at fault?
    a. The volume control
    b. The power cord
    c. The filter in the power supply
    d. The output transistor

26. Refer to Fig. 10-21. The amplifier has a loud, hissing sound only when the volume control is turned up. Which of the following conclusions is the best?
    a. The problem is in stage 3.
    b. The problem is in stage 4.
    c. The power supply is defective.
    d. The problem is in stage 1 or 2.

27. An audio amplifier has bad distortion when played at low volume. At high volume, the distortion is only slight. Which of the following is most likely to be the cause?
    a. A bias error in the push-pull output stage
    b. A defective volume control
    c. A defective tone control
    d. Inadequate power supply voltage

28. An amplifier makes a putt-putt sound at high volume levels (motorboating). Which of the following is most likely to be the cause?
    a. Crossover distortion
    b. A defective transistor
    c. An open filter or bypass capacitor
    d. A defective speaker

29. An amplifier has bad distortion when played at high volume. Which of the following is most likely to be the cause?
    a. A cracked circuit board
    b. A bias error in one of the amplifiers
    c. A defective volume control
    d. A defective tone control

10-5 Intermittents

An electronic device is *intermittent* when it will work only some of the time. It may become defective after being on for a few minutes. It may come and go with vibration. The source of these kinds of problems can be very difficult to locate. Technicians generally agree that intermittents are the most difficult to troubleshoot.

There are two basic ways to find the cause of an intermittent problem. One way is to run the equipment until the problem appears and then use ordinary troubleshooting practice to isolate it. The second way is to use various procedures to force the problem to show up. Some of these are

1. Heat various parts of the circuit.
2. Cool various parts of the circuit.
3. Change the supply voltage.
4. Vibrate various parts of the circuit.

The actual technique used will depend on the symptoms and how much time is available to service the equipment. Some intermittents will not show up in a week of continuous operation. In such a case, it is probably best to try to make the problem occur.

Many *intermittents* are *thermal*. That is, they appear at one temperature extreme or another. If the problem shows up only at a high temperature, it may be very difficult to find with the cabinet removed. With the cabinet removed, the circuits usually run much cooler, and a thermal intermittent will not show. In such a case, it may be necessary to use a little heat to find the problem.

Figure 10-24 shows some of the ways that electronic equipment and components can safely be heated to check for thermal intermittents. The bench lamp is useful for heating many

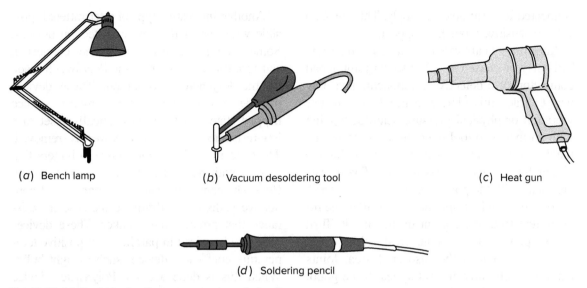

(a) Bench lamp   (b) Vacuum desoldering tool   (c) Heat gun

(d) Soldering pencil

Fig. 10-24 Methods for heating components and circuits.

components at one time. By placing a 100-W incandescent lamp near the equipment, the circuits will become quite warm after a few minutes. Be careful not to overheat the circuits. Certain plastic materials can be easily damaged. A vacuum desoldering tool makes a good heat source for small areas. Squeezing the bulb will direct a stream of hot air where needed. Be careful not to spray solder onto the circuit. A heat gun is useful for heating larger components and several parts at one time. Be careful because some heat guns can damage circuit boards and parts. Finally, an ordinary soldering pencil may be used by touching the tip to a component lead or to a metal case.

Spray coolers are available for tracing thermal intermittents. A spray tube is included to control the application closely (Fig. 10-25). This makes it easy to confine the spray to one

component at a time. Be very careful not to use just any spray coolant. Some types can generate static charges in the thousands of volts when they are used, and others may damage the environment. Sensitive devices can be damaged by static discharges, as discussed earlier.

Some intermittents are voltage-sensitive. The ac line voltage is nominally rated at 117 V. However, it can and does fluctuate. It may go below 105 V, and it may go above 130 V. Most electronic equipment is designed to work over this range. In some cases, a circuit or a component can become critical and voltage-sensitive. This type of situation may show up as an intermittent. Figure 10-26 shows one test arrangement. The equipment is

Spray tube

Circuit Freeze

Fig. 10-25 Spray cooler with tube for localizing spray.

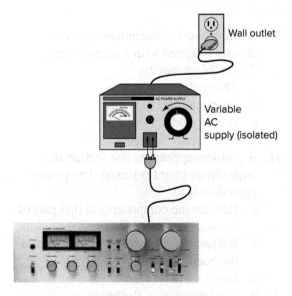

Wall outlet

Variable
AC
supply (isolated)

Fig. 10-26 Checking for a voltage-sensitive intermittent.
donatas1205/123RF

connected to a variable ac supply. This forces a voltage-sensitive problem to appear.

Intermittents are often sensitive to *vibration*. This may be caused by a bad solder joint, a bad connector, or a defective component. The only way to trace this kind of a problem is to use vibration or physical pressure. Careful tapping with an insulated tool may allow you to isolate the defect. In addition to tapping, try flexing the circuit boards and the connectors. These tests are made with the power on, *so be very careful.*

You may find it impossible to localize the intermittent to a single point in the circuit. Turn off the power and use some fresh solder to reflow every joint in the suspected area. Joints can fail electrically and yet appear to be good. Resoldering is the only way to be sure.

Do not overlook sockets. Try plugging and unplugging several times to clean the sliding contacts. The power *must be off. Never* plug or unplug connectors, devices, or circuit boards with the power on. Severe damage may result.

Circuit board connectors may require cleaning. An ordinary pencil eraser does a good job on the board contacts. Do not use an ink eraser as it is too abrasive. Use just enough pressure to brighten the contacts. Clear away any debris left by the eraser before reconnecting the board.

Another interesting type of intermittent is possible with one of the newer protection devices. Some audio speaker systems use series-connected, self-resetting fuses such as the PolySwitch made by the Raychem Corporation. These devices provide soft switching into a high-resistance tripped state and then automatically reset to a low-resistance state when power is removed. They are made from conductive polymers that expand with heat when there is excess current flow. The expansion causes a separation of conductive paths and a dramatic increase in resistance that protects the speaker. These devices may also be wired in parallel with positive temperature coefficient devices such as light bulbs. When this is done and the PolySwitch device trips, the speaker current now flows through the bulb, which heats and increases in resistance. So, the symptom in this case is a decrease in speaker volume. Without the shunt light bulb, the symptom is a total loss of volume. In either case, the volume is restored when the amplifier is turned off (or down) and the fuse is allowed to cool.

Intermittents can be tough to work on, but they are not impossible. Use every clue and test possible to localize the problem. It is far easier to check a few things than to check every joint, contact, and component in the system.

Self-Test

Choose the letter that best completes each statement.

30. A circuit works intermittently as the chassis is tapped with a screwdriver. The problem may be
    a. Thermal
    b. An open filter capacitor
    c. A cold solder joint
    d. Low supply voltage
31. A problem appears as one section of a large circuit board is flexed. The proper procedure is to
    a. Replace the components in that part of the board
    b. Reflow the solder joints in that part of the board
    c. Heat that part of the board
    d. Cool that part of the board

32. A circuit always works normally when first turned on but then fails after about 20 minutes of operation. Out of its cabinet, it works indefinitely. The problem is
    a. Thermal
    b. An open ground
    c. High line voltage
    d. The ON-OFF switch
33. The correct procedure to isolate the defect in question 32 is to
    a. Run the circuit at reduced line voltage
    b. Replace the electrolytic capacitors one at a time
    c. Remove the cabinet and heat various parts
    d. Resolder the entire circuit

10-6 Operational Amplifiers

The techniques used when troubleshooting circuits with op amps are similar to those already presented in this chapter. As always, check the obvious things first. Op amps often use a bipolar supply. You should verify *both* supply voltages early in the troubleshooting process.

In general, component failures follow a pattern. The following list is presented as a rough guide to *average failure rates*. Many devices, such as resistors, have very low failure rates and are not listed. Also, parts such as cells and batteries are not listed since they are expected to be replaced on a regular basis. The items listed first have the highest failure rates:

1. High-power devices and devices subject to transients
2. Incandescent lamps
3. Complex devices such as integrated circuits
4. Mechanical devices such as connectors, switches, and relays
5. Electrolytic capacitors

High-power devices run hot. If equipment is not powered all the time, a large number of hot-cold cycles can accumulate. The repeated expansion and contraction tend to weaken the internal connections of electronic devices. Most op amps are small-signal devices. However, a few do dissipate enough power to run hot, and some may be located where they are heated by other devices. Remember that *heat* is one of the leading causes of failures in electronic systems.

A *transient* is a brief and abnormally high voltage. Transients are tough on solid-state devices because they can break down PN junctions and degrade or destroy them. Op amps are sometimes connected to sensors through long runs of wire. These wires can act as antennas and pick up transients caused by lightning or by surges in other wires in the vicinity. Watch for repeated failures in such cases. They can indicate the need for a different wiring arrangement or the addition of transient protection devices.

Op amps are almost always integrated circuits. This makes them complex and increases their failure rate over simple devices. So if you are working on a defective circuit that contains some transistors and some integrated circuits, it is more likely that an IC is bad than a transistor (unless the transistors are high-power ones or

subject to transients). Please remember that the above list is a *rough* guideline. Experienced technicians know that anything can go wrong and expect that it will sooner or later. The idea is that their experience makes them more efficient because it tells them where to look first.

Please don't form the opinion that integrated circuits are not reliable. They are actually *more reliable* than equivalent circuitry using separate parts. The equivalent circuit for an ordinary op amp might contain 40 parts. A circuit with 40 parts is usually not as reliable as a single IC because it has so many more places for failures to occur. Just remember that a complicated device is more likely to fail than one simple device.

Average failure rates

Occasionally, an op amp can experience *latch-up*. This is not a failure but can be a troubleshooting problem. When an op amp latches, its output gets stuck at its maximum value (either positive or negative). The only way to get it back to normal is to power down and then power up again. Normal operation will be restored if the initial cause (such as an abnormally large input signal) has been removed. Op amps have a maximum common-mode range, which is the maximum signal that can be applied to both inputs without saturating or cutting off the amplifier. Latch-up usually occurs in voltage follower stages where saturation has occurred.

Latch-up

When a nonfeedback amplifier is driven into saturation, it will resume normal operation when the abnormal input signal is removed. This is always true unless the abnormal input was large enough to damage the amplifier. When feedback is used, the situation can be different. A saturated stage no longer acts as an amplifier. It acts as a resistor and passes some part of the signal through to the next stage. If the saturated stage was an inverting amplifier, *the inversion is lost*. When this happens in an amplifier with intended negative feedback, the feedback goes positive and the amplifier may then keep itself in saturation. Op amps are candidates for latch-up since they are usually operated with negative feedback.

Transient

Latch-up in operational amplifiers is not as probable as it once was. The designers of linear ICs have made changes that make it less likely to occur. Try powering down and back on again if you suspect latch-up. If the symptoms seem to indicate it, then you will have to investigate the input signal to determine if it is exceeding the

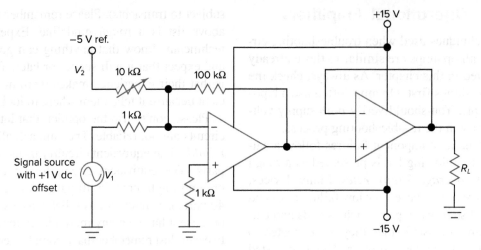

Fig. I0-27 Op-amp troubleshooting example.

common mode range of the device. Also make sure that the power supply is normal and that the positive and negative voltages are applied at the same time when the circuit is turned on.

The output of an op amp can usually go within a volt or so of the power supply values. If it is powered by ±12 V, then the output could approach +11 V or −11 V. There is usually a problem when the output of an op amp is at or near one of its extremes (unless it is serving as a comparator). It could be latch-up, but it is more likely to be a dc error at its input or in a prior stage. The dc gain in some circuitry is high. A moderate error in an early stage can drive an output stage to one of its extremes. Use a meter and check the dc voltages. When stages are dc-coupled, check earlier stages as well. Do this even though the problem may appear on an ac signal. Always remember that an amplifier near saturation or cutoff cannot provide a normal linear output swing for any signal.

Don't forget to check all of the relevant dc levels. Op amps can sum and subtract several different dc signals. All it takes is one of them to be missing or in error to throw off the dc balance of the entire circuit. Refer to Fig. 10-27. It shows a circuit with two stages. The signal source has a +1 V dc offset that must be

eliminated. This is accomplished by the first stage, where a −5-V reference is summed with the signal source. Assuming that the 10-kΩ trimmer is adjusted for 5 kΩ, we can calculate the dc output of the first stage:

$$V_{out} = -100 \text{ k}\Omega\left(\frac{+1 \text{ V}}{1 \text{ k}\Omega} + \frac{-5 \text{ V}}{5 \text{ k}\Omega}\right)$$

$$= 0 \text{ V}$$

This calculation demonstrates that the amplifier is properly designed and adjusted to eliminate the dc offset of the signal source.

What will happen in Fig. 10-27 if the 10-kΩ trimmer resistor develops an open? This will effectively remove the second term from inside the parentheses, and the output now calculates to

$$V_{out} = -100 \text{ k}\Omega\left(\frac{+1 \text{ V}}{1 \text{ k}\Omega}\right)$$

$$= -100 \text{ V}$$

Obviously, the amplifier cannot achieve this output. It will saturate at approximately −14 V. The second stage is a voltage follower, and its output will also be near −14 V.

When you are troubleshooting comparators, observe the symptoms carefully. If there is no output at all, make sure that the power supply is OK and that there is a normal signal connected to the comparator input. Other possibilities for no output include a shorted output cable, a bad comparator IC, and an open pull-up resistor.

If the symptom is noise in the output or extra output pulses, the input signal should be checked. Many comparators do not work well with high-impedance signal sources, so verify the source and verify a good connection. Finally, a comparator might need positive feedback to prevent a noisy output. Investigate whether there is a resistor that feeds back from the output to the + input and make sure that it is not open.

Self-Test

Choose the letter that best completes each statement.

34. Of the following components, the *least* likely to fail is
    a. A high-power output transistor
    b. An integrated circuit
    c. A device that runs hot
    d. A small-signal transistor
35. Latch-up in a negative feedback amplifier is caused by
    a. An open bypass capacitor
    b. An open coupling capacitor
    c. A stage driven into saturation
    d. An open in the feedback network
36. Refer to Fig. 10-27. The purpose of the −5 V reference and 10-kΩ potentiometer is
    a. To set the gain of the first stage
    b. To null the dc offset in the source
    c. To power the first op amp
    d. To adjust the frequency response of the amplifiers
37. Refer to Fig. 10-27. The connection to the signal source is defective (open). The dc across R_L will be about
    a. +14 V
    b. −14 V
    c. 0 V
    d. ±7 V

10-7 Automated Testing

Automated testing was originally developed to verify that just-manufactured units worked properly. Today, new technology has extended automated testing to all product phases. It's important to realize that troubleshooting is an important part of these distinct product phases:

- Preproduction (design phase)
- Production (manufacturing phase)
- Postproduction (customer phase)

Technicians who troubleshoot during the design phase face a wide range of issues. In the case of a new product, there can be thousands of reasons why performance is not satisfactory. In some cases, due to hardware design errors or software errors, a prototype might be working as well as it can (troubleshooting alone cannot improve it).

Production troubleshooting comes into play when products fail to pass the tests designed to verify proper operation. For high-cost items, defective units are diagnosed, repaired, retested, and then shipped. For low-cost items, failed units might still be diagnosed before disposal. The information gained often leads to improvements in design and/or manufacturing.

Postproduction troubleshooting deals with products that worked for some period of time and then failed. Technicians might diagnose and repair these products at the customer's location or at a repair facility.

Automated testing takes various forms. Sub assemblies, such as circuit boards, might be loaded into a test fixture with appropriate connectors to apply supply voltages and signals. In the past, a "bed-of-nails" fixture was often used to test circuit boards. This fixture was so named because it consisted of an array of sharp metal probes that made electrical contact with various test points on circuit boards. Figure 10-28 shows a bed-of-nails fixture, and Fig. 10-29 shows a flying probe setup, which is a similar multinode testing arrangement. With flying probes, a board will be tested in steps, with each step involving several probes in different positions. Both of these techniques are losing popularity due to more hidden connections on circuit boards. Figure 10-30 shows a ball grid array integrated circuit (BGA IC) where none of the connections are available for testing. Because this was a common problem for many manufacturers, a joint test action group (JTAG) was formed. The result of their cooperation is an automated testing technology called *boundary scan*.

Boundary scan

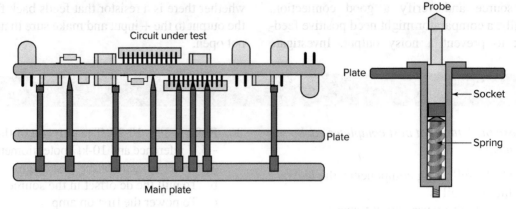

Fig. 10-28 Bed-of-nails testing.

Fig. 10-29 Flying probe testing (one position shown).

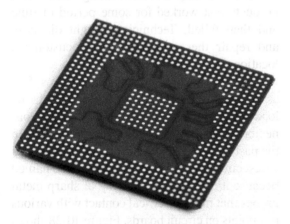

Fig. 10-30 The bottom of a BGA chip showing its connections.

KirVKV/123RF

The original group was formed in Europe, but their work has become an international standard and has been adopted by organizations such as the Institute of Electrical and Electronic Engineers (IEEE), a U.S. organization.

The IEEE 1149.1 test bus and boundary-scan architecture allows an IC, a board, or an entire product to be controlled and verified via a standard four-wire interface. Each IEEE 1149.1-compliant IC allows each functional pin of the IC to be controlled and observed via the four-wire interface. Test, debug, or initialization patterns can be loaded serially (one bit at a time) into the appropriate IC(s) via the test bus. This allows integrated circuit, board, or system functions to be observed or controlled without the physical access once provided by the bed-of-nails test.

Figure 10-31 shows how boundary scan works. A JTAG connector provides a series data path (shown in black in Fig. 10-31) through the devices. The serial data path is also called the *scan chain*. Notice the "virtual nails." There are no actual nails in boundary scan, but the information provided is the same as if the nails were probing the pins; hence the term "virtual nails." As data enters and leaves the system via the JTAG connector, two distinct kinds of information are produced:

- The circuit board traces between devices and connectors (shown in yellow in Fig. 10-31) can be verified. Both opens and shorts can be detected.
- The IC core logic functions can be verified. Thus, faulty devices can be identified.

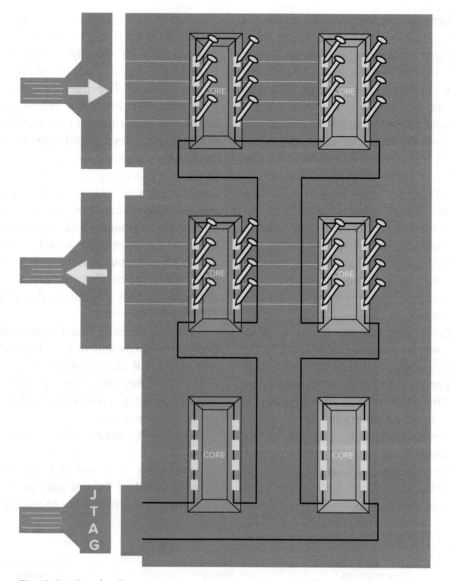

Fig. 10-31 Virtual nails.

Figure 10-32 shows the internal working of a boundary-scan chip. Each pin is connected to a cell that determines if output pins will be driven by the core logic of the chip (NO—normal output) or by the serial data coming from the JTAG connector (SO—serial output). Likewise, input pins are switchable between NI (normal input) and SI (serial input). In normal operation,

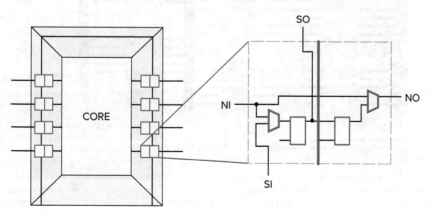

Fig. 10-32 Boundary-scan cell.

the IC performs its intended function as though the boundary-scan circuits were not present. When testing or programming, the scan logic is activated. Data can then be sent to the IC and read from it via the JTAG connector. This data can stimulate the device core, drive signals outward to the printed circuit board, sense the input pins from the board, or sense device outputs. The result is a tremendous reduction in the number of test points needed on the circuit board. The JTAG port may also be called the test access port (TAP).

Manufacturers now offer DSP chips, microprocessors, and application-specific integrated circuits (ASICs) with these same pins. Some of these also have a fifth pin for resetting the boundary-scan portion of the chip.

In addition to its use in board testing, boundary scan allows programming almost all types of complex programmable logic devices (CPLDs) and flash memories, regardless of size or package type. The programming can take place on the board, after PCB (printed circuit board) assembly. On-board programming saves money and improves throughput by reducing device handling, simplifying inventory management, and integrating the programming steps into the board production line. Figure 10-33 shows some boundary scan applications.

What about analog circuit testing? This area is addressed by IEEE 1149.4. This standard is compatible with the digital version (1149.1). Look at Fig. 10-34. Each pin of the analog ICs is controlled by five internal switches. These solid-state switches allow selective access to the analog core function of each chip and also to the external devices and networks connected to the ICs. Suppose, for example, that a measurement of device Z5 is needed. The switches can be set so that Z5 is isolated from the analog core and the source at the lower left provides a current flow through Z5 via analog bus 1. Then, other switches operate to allow the voltage across Z5 to be routed to the detector via analog bus 2. Once the current and voltage are known, the resistance of Z5 can be determined with Ohm's law.

How does analog boundary scan test circuit boards for opens and shorts? That is the function of the DR (data register) blocks shown in Fig. 10-34. They are called *digitizers* and make the analog device pins digital for performing the interconnect tests.

Whether digital or analog, boundary scan requires complex test signals. These are supplied by computers. The required software usually runs on standard PCs or notebooks. A JTAG port or adapter provides the necessary interface.

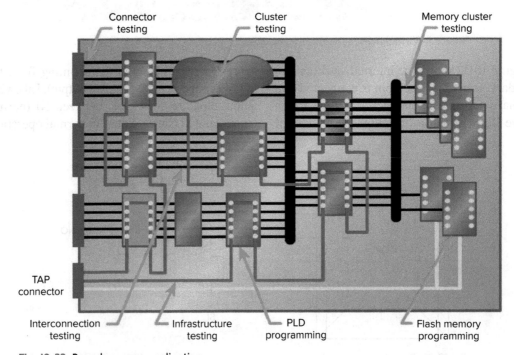

Fig. 10-33 Boundary-scan applications.

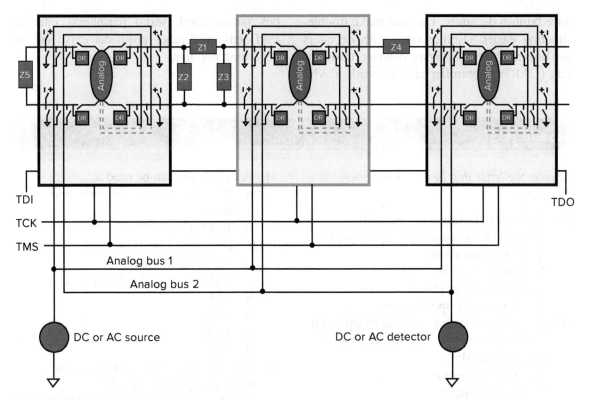

Fig. 10-34 Analog boundary scan.

Automated testing was once applied only to the manufacturing phase of products. Now, it is being applied to all phases. There is little doubt that 21st-century technicians will have to troubleshoot products and systems of ever-increasing complexity. However, they will have access to powerful tools to help them. These tools, combined with technicians' knowledge of circuits, will make their jobs interesting and rewarding.

There is little question that electronics as a field is becoming more complex. But the tools that technicians have are also advancing. Figure 10-35 shows that some oscilloscopes

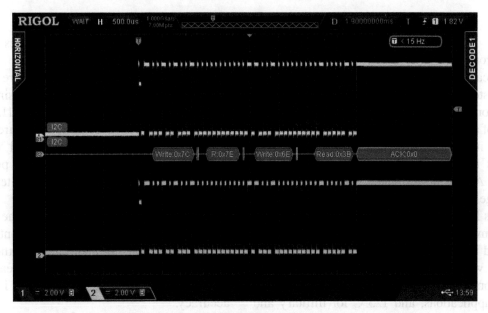

Fig. 10-35 I²C (inter-integrated circuit) decoding using an oscilloscope.
Charles A. Schuler/McGraw Hill

have built-in decoders. This can make trouble-shooting easier. Oscilloscopes with decoders for universal asynchronous receiver transmitters (UARTs), controller area network (CAN) bus (automotive), serial peripheral interface (SPI) bus, and other protocols exist and are becoming popular.

Self-Test

Choose the letter that best answers each question.

38. Automated testing was originally designed to verify product operation for
    a. The design phase
    b. The production phase
    c. The customer phase
    d. The product update phase
39. A fixture consisting of many sharp metal probes to contact test points on a circuit board is called a
    a. JTAG connector
    b. TAP connector
    c. Bed of nails
    d. Boundary-scan port
40. The minimum number of wires or connections that make up a JTAG port is
    a. 4
    b. 5
    c. 6
    d. 7

41. A JTAG port can be used to
    a. Test for circuit board opens and shorts
    b. Test for failed ICs
    c. Program FLASH memory
    d. All of the above
42. A JTAG port can also be called a
    a. TAP
    b. TDO
    c. TCK
    d. TDI
43. How can boundary scan be used to measure Z_1 in Fig. 10-34?
    a. The analog core in the middle IC will be programmed to send it a test signal.
    b. The switches apply the signal source to Z_1 and connect the detector across it.
    c. Both of the above can be used.
    d. None of the above can be used.

10-8 Thermal Issues

Electronic failures can be caused by, or lead to, overheated components such as diodes, power transistors and ICs. Traditionally, overheated components are detected by touching the surface of a component. This is *dangerous* and can result in burned fingers and/or electric shock. Shock hazards can lurk near low voltage circuits. A 5 V power supply might use ac line voltage in the near vicinity. Probing with fingers is NOT a good idea!

Conventional electronic components are designed to operate over a specified temperature range with upper limits generally set at 70°C for commercial applications, 85°C for industrial applications, and 125°C for military and automotive applications. Recent trends in the use of electronics in cars and aircraft have seen the development of reliable electronics for use at 150°C and higher.

Many multimeters have temperature probes that can be used to check for hot heat sinks and devices. Referring back to Fig. 10-11, when measuring temperature the temperature probe leads are connected to the COM and V jacks and the selector switch is set to the mV position. Figure 10-36 shows a typical DMM temperature probe.

When using a probe such as the one shown in Fig. 10-36, use the probe tip to contact the device or its heatsink and hold until the reading is stable. A dab of thermal grease can be used to make better contact and improve accuracy.

Temperature probes are not convenient to use for troubleshooting. Placing a hand near

Fig. 10-36 Temperature probe for a digital multimeter.
Courtesy of Fluke Corporation

circuits must be done with caution. Technicians can use non-contact temperature measurements. Circuit boards can be scanned for hotspots using an IR thermal imager with no need to physically touch components. Infrared thermal imaging is able to capture the temperature distribution of the whole circuit which makes it easy to see the hotspot at a glance. Figure 10-37 shows a thermal imager. If a hot spot is detected, the view can be shifted and the instrument moved closer for a more accurate measurement.

A capture from the instrument shown in Fig. 10-37 can transferred to a computer for storage and for report generation. Figure 10-38 shows one image capture and how the image has been manipulated using software. Infrared, blended, and normal views are possible.

Figure 10-39 shows examples of heat damage. In Fig. 10-39(a) and (c) there is also damage to the circuit board. In Fig. 10-39(b) and (d) the heat has caused visible damage to the device packages.

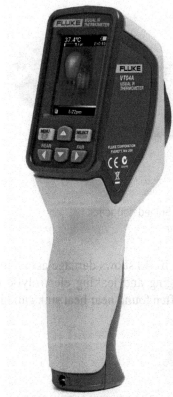

Fig. 10-37 Fluke non-contact thermometer and imager.
Courtesy of Fluke Corporation

| Ifrared | Blended | Normal |

Fig. 10-38 One capture from the Fluke IR thermometer viewed in software.
(images) Charles A. Schuler

(a) kuracisto/123RF   (b) htao/Shutterstock   (c) srapulsar38/123RF   (d) s.juchim/Shutterstock

Fig. 10-39 Damaged components.

Fig. 10-40 Bulged capacitor.
pproman/123RF

Figure 10-40 shows damage associated with heat. Bulging and leaking electrolytic capacitors are often found near heat sinks and in other hot spots.

Fig. 10-41 Blown fuses.
banchaphoto123/123RF

In Fig. 10-41 we see blown and cracked fuses. They fail due to the heat produced by excess current flow. When using thermal imaging with live circuits it might be possible to see fuses that are operating near their limits.

10-9   Software Updates

As software plays a larger role in technology, updates are becoming more important but can be the cause of equipment malfunctions. There are several classifications of software including operating systems, applications, drivers, and firmware. It is important to keep all of them up to date but there are potential issues to be aware of. It is not possible to cover all software types here and this section is limited to firmware.

Bricking is a term associated with a firmware update gone bad that has rendered a device or equipment *completely dead* and useless (except perhaps as a brick paperweight). That's why firmware updates should be:

1. Performed only when necessary.
2. Based on verified files obtained directly from the manufacturer or their website.
3. Accomplished according to the manufacturer's exact instructions. When installing from a USB drive, the drive's specifications might be an important factor.
4. Only attempted when backup power is present.

Recovery after a device is bricked can be difficult. It might require access to a JTAG port and JTAG software and firmware files. A JTAG port might be labeled as shown in Table 10-3.

Sometimes a JTAG connector cannot be located but the pin names can be found on the printed circuit board—perhaps adjacent to a microcontroller. Another possibility is the use of an AVRISP programmer (AVR microcontrollers are produced by Amtel and ISP denotes in-system programming).

It used to be that firmware was software that rarely or never needed to be changed. Now, firmware updates are common. It is not unusual for a product to have new firmware available within a year or so of the product's release. Sometimes the changes are minor and if they are not important for any given situation, it could be prudent to not install the latest firmware.

Firmware that never changes can be stored in a ROM (read only memory). If so, updating is

Table 10-3	JTAG Connector Names and Descriptions
Pin	Description
TCK	Clock signal
TDI	Test data in
TDO	Test data out
TMS	Test mode select
TRST (optional pin)	Reset

not possible. Now, firmware is more likely to be stored in EEPROM (electrically erasable programmable ROM). EEPROM is sometimes called flash memory and writing to EEPROM can be called *flashing*. Actually, EEPROM is different from flash. In an EEPROM the data can be accessed one byte (8 bits) at a time but with flash only one block (or sector) at a time. USB memory sticks are flash devices and are often called flash drives.

Firmware is found in test equipment such as oscilloscopes, and in printers, routers, and even in a room thermostat that can be accessed with a smartphone. Updating the firmware on a device with a USB port usually involves inserting a flash drive containing the firmware file into the port and turning the device on and perhaps following a few prompts. Or, first a utility menu is accessed that has firmware updating as an option. The process is specific to each device and the user manual should be consulted before attempting it. Bricking is real and should be avoided! If no USB port is available, some other connector might be used. Online updating over Ethernet or WiFi are also possibilities.

Chapter 10 Summary and Review

Summary

1. When troubleshooting, always use a system point of view. Don't ignore other equipment and/or software that could be affecting performance, and always check the obvious.
2. If the unit is ac-operated, disconnect it from the line before taking it apart.
3. Use service literature and the proper tools.
4. Sort and save all fasteners, knobs, and other small parts.
5. Make a thorough visual inspection of the interior of the equipment.
6. Try to determine why a replaced component failed before turning on the power.
7. Check for overheating.
8. Verify all power-supply voltages.
9. Lack of amplifier output may not be in the amplifier itself. There could be a defective output device or no input signal.
10. A multistage amplifier can be viewed as a signal chain.
11. Signal injection begins at the load end of the chain.
12. Signal tracing begins at the input end of the chain.
13. Voltage analysis is generally used to limit the possibilities to one defective component.
14. Some circuit defects cannot be found by dc voltage analysis. These defects are usually the result of an open device or coupling component.
15. Low output from an amplifier may be due to low input.
16. A dummy load resistor is often substituted for the output device when amplifier performance is measured.
17. Both signal tracing and signal injection can give misleading results when troubleshooting for the low-gain stage.
18. Voltage analysis will lead to some causes of low gain.
19. A capacitor suspected of being open can be checked by bridging it with a new one.
20. A linear amplifier is not supposed to change anything but the amplitude of the signal.
21. Noise may be originating in the power supply.
22. Hum refers to a 60-Hz or a 120-Hz signal in the output.
23. Hum may be caused by a defective power supply, an open shield, or a poor ground.
24. Operate all controls to see if the noise occurs before or after the control.
25. Motorboating noise (a "putt-putt" sound) means the amplifier is oscillating.
26. Distortion can be caused by bias error, defective transistors, or an input signal that is too large.
27. Thermal intermittents may show up after the equipment is turned on for some time.
28. Use heat or cold to localize thermal problems.
29. Vibration intermittents can be isolated by careful tapping with an insulated tool.
30. Failure rates are directly related to device temperature and complexity.
31. Transients can and often do damage solid-state devices.
32. An amplifier with negative feedback may have latch-up if it is driven into saturation.
33. Boundary scan is an automated testing procedure that can be applied to any phase of the life cycle of a product.

Choose the letter that best answers each question.

10-1. When troubleshooting, which of the following questions is not part of a preliminary check? (10-1)
   a. Are all cables plugged in?
   b. Are all controls set properly?
   c. Are all transistors good?
   d. Is the power supply on?

10-2. What is the quickest way to check a speaker for operation (not for quality)? (10-2)
   a. A click test using a dry cell
   b. Substitution with a good speaker
   c. Connecting an ammeter in series with the speaker
   d. Connecting an oscilloscope across the speaker

10-3. Refer to Fig. 10-2. The regulated output is zero. The unregulated input is normal. Which of the following could be the cause of the problem? (10-1)
   a. C_1 is open.
   b. D_1 is open.
   c. R_1 is open.
   d. R_1 is shorted.

10-4. Refer to Fig. 10-2. The regulated output is low. The unregulated input is normal. Which of the following could be the cause of the problem? (10-1)
   a. D_1 is open.
   b. D_1 is shorted.
   c. C_1 is open.
   d. Excessive current at the output.

10-5. Which of the following is most sensitive to ESD damage? (10-1)
   a. Rectifier diode
   b. Fuse
   c. Integrated circuit
   d. Front panel lamp

10-6. Refer to Fig. 10-7. The output is zero. Where should the signal be injected first? (10-2)
   a. At the input of stage 1
   b. At the input of stage 2
   c. At the input of stage 3
   d. At the input of stage 4

10-7. Refer to Fig. 10-7. The amplifier is dead. A known good signal has been connected to the input of stage 1. Signal tracing should begin at what point? (10-2)
   a. The output of stage 1
   b. The output of stage 2
   c. The output of stage 3
   d. The output of stage 4

10-8. Refer to Fig. 10-10. The collector of Q_1 measures almost 21 V, and it should be 12 V. Which of the following is most likely to be wrong? (10-2)
   a. Q_1 is open.
   b. C_3 is shorted.
   c. R_4 is open.
   d. Q_2 is shorted.

10-9. Refer to Fig. 10-10. The collector of Q_2 measures 2 V, and it is supposed to be 12 V. Which of the following could be the problem? (10-2)
   a. Q_1 is shorted.
   b. C_1 is open.
   c. R_2 is open.
   d. C_4 is shorted.

10-10. Refer to Fig. 10-10. Resistor R_1 is open. Which of the following statements is correct? (10-2)
   a. The collector of Q_1 will be at 0 V.
   b. Q_2 will run hot.
   c. Q_2 will go into cutoff.
   d. Q_1 will run hot.

10-11. Refer to Fig. 10-10. Resistor R_9 is open. Which of the following statements is correct? (10-2)
   a. The collector of Q_2 will be at 0 V.
   b. Q_2 will go into saturation.
   c. Q_2 will go into cutoff.
   d. Q_2 will run hot.

10-12. Refer to Fig. 10-16. Resistor R_1 is open. Which of the following is correct? (10-3)
   a. The collector voltage will be very low.
   b. The collector voltage will be very high.
   c. The transistor will be in saturation.
   d. The emitter voltage will be very high.

10-13. Refer to Fig. 10-17. Capacitor C_3 is open. Which of the following is correct? (10-3)
   a. The dc voltages will all be wrong.
   b. The transistor will run hot.
   c. Extreme distortion will result.
   d. The gain will be low.

10-14. Refer to Fig. 10-21. A scratching sound is heard as the gain control is rotated. Where is the problem likely to be? (10-4)
a. Stage 1 or 2
b. The volume control
c. Stage 3 or 4
d. The speaker

10-15. Refer to Fig. 10-21. There is severe hum in the output, but turning down the volume control makes it stop completely. Where is the problem likely to be? (10-4)
a. Third stage or fourth stage
b. Power-supply filter
c. The volume control
d. Input cable (broken ground)

10-16. An amplifier is capacitively coupled. What is the best way to find an open coupling capacitor? (10-2)
a. Look for transistors in cutoff.
b. Look for transistors in saturation.
c. Look for dc bias errors on the bases.
d. Look for a break in the signal chain.

10-17. An amplifier has a push-pull output stage. Bad distortion is noted at high volume levels only. The problem could be which of the following? (10-4)
a. Bias error in an earlier stage
b. A shorted output transistor
c. Crossover distortion
d. A defective volume control

10-18. What is probably the slowest way to find an intermittent problem? (10-5)
a. Try to make it show by using vibration.
b. Use heat.
c. Use cold.
d. Wait until it shows up by itself.

10-19. An automobile radio works normally except while traveling over a bumpy road. What is likely to be the cause of the problem? (10-5)
a. Thermal
b. An open capacitor
c. A low battery
d. A loose antenna connection

10-20. When working on electronic equipment, a grounded wrist strap may be used to prevent (10-1)
a. Ground loops
b. Electrostatic discharge

c. Thermal damage
d. Loading effect

10-21. Refer to Fig. 10-27. Suppose the −5-V reference supply fails and goes to 0 V. The voltage across R_L will be (10-6)
a. 0 V
b. +14 V
c. −14 V
d. +30 V

10-22. Refer to Fig. 10-31. The information from the test points represented by the virtual nails is extracted by reading the (10-7)
a. Signals at the top left connector
b. Signals at the center connector
c. Serial signal at TDI on the JTAG connector
d. Serial signal at TDO on the JTAG connector

10-23. Refer to Fig. 10-31. It's possible to use the JTAG port to verify the top and middle connectors and their associated circuit traces by (10-7)
a. Applying signals to the top connector and output indicators to the middle connector
b. Using a loop through cable to interconnect them
c. Both of the above
d. None of the above

10-24. Refer to Fig. 10-31. The scan path is shown as (10-7)
a. Yellow lines
b. A right-facing yellow arrow
c. A left-facing yellow arrow
d. Gray lines

10-25. Refer to Fig. 10-32. When normal operation is selected, (10-7)
a. NO, NI, and the core are active
b. NO, SI, and core are active
c. SO, NI, and the core are active
d. SO, SI, and the core are active

10-26. Analog boundary scan, compared with digital boundary scan (10-7)
a. Requires two additional busses
b. Requires five internal switches for each active device pin
c. Also requires TDI, TDO, TCK, and TMS pins
d. All of the above

Critical Thinking Questions

10-1. You are visiting a friend and notice that the sound coming from the left speaker of her stereo is distorted. You have no test equipment with you. Is there anything you can do to help her find out what is wrong?

10-2. Your automobile often won't start on Monday mornings and never fails to start at other times. Is this some sort of weird coincidence?

10-3. Can you think of any reason why stereo amplifiers sometimes fail during parties?

10-4. Technicians often put batteries in portable equipment before performing other tests, even though the customer has stated that the batteries are new. Do technicians think their customers are mistaken?

10-5. Where do technicians look when they are working on equipment that failed during a lightning storm?

10-6. Can you think of an op-amp application in which it is normal for the output to be saturated?

10-7. Component-level repair is not a widespread practice today. Is it still worthwhile to learn how electronic circuits operate?

Answers to Self-Tests

1. C	12. C	23. D	34. D
2. D	13. B	24. C	35. C
3. D	14. D	25. C	36. B
4. D	15. A	26. D	37. A
5. A	16. B	27. A	38. B
6. B	17. B	28. C	39. C
7. B	18. C	29. B	40. A
8. B	19. B	30. C	41. D
9. D	20. D	31. B	42. A
10. B	21. B	32. A	43. B
11. D	22. B	33. C	

Oscillators

Learning Outcomes

This chapter will help you to:

11-1 *Identify* oscillator circuits. [11-1]

11-2 *Apply* the concepts of gain and feedback to oscillators. [11-1]

11-3 *Predict* the frequency of operation for oscillators. [11-2, 11-3, 11-4, 11-5]

11-4 *List* causes of undesired oscillations. [11-6]

11-5 *Identify* techniques used to prevent undesired oscillation. [11-6]

11-6 *Troubleshoot* oscillators. [11-7]

11-7 *Explain* and *troubleshoot* direct digital synthesizers. [11-8, 11-9]

An amplifier needs an ac input signal to produce an ac output signal, but an oscillator doesn't. An oscillator is a circuit that creates an ac signal. Oscillators can be designed to produce many kinds of waveforms such as sine, rectangular, triangular, or sawtooth. The range of frequencies that oscillators can generate is from less than 1 Hz to well over 10 gigahertz (10 GHz = 1×10^{10} Hz). Depending on the waveform and frequency requirements, oscillators are designed in different ways. This chapter covers some of the most popular circuits, and it also discusses undesired oscillations.

II-I Oscillator Characteristics

An oscillator is a circuit that converts dc to ac as shown in Fig. 11-1(a). The only input to the oscillator is a dc power supply, and the output is ac. Most oscillators are amplifiers with *feedback* as shown in Fig. 11-1(b). If the feedback is *positive*, the amplifier may oscillate (produce alternating current).

Feedback

Many amplifiers will oscillate if the conditions are correct. For example, you probably know what happens when someone adjusts the volume control too high on a public address system. The squeals and howls that are heard are oscillations. The feedback in this case are the sound (acoustical) waves from the loudspeakers that enter the microphone (Fig. 11-2). Although acoustical feedback can produce oscillations, almost all practical oscillators use electrical feedback. The feedback circuit uses components such as resistors, capacitors, coils, or transformers to connect the input of the amplifier to the output of the amplifier.

Feedback alone will not guarantee oscillations. Look at Fig. 11-2 again. You probably know that turning down the volume control will stop the public address system from

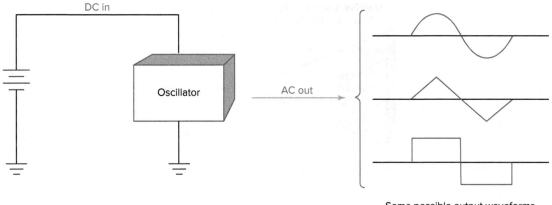

DC in

Oscillator

AC out

Some possible output waveforms

(a) Oscillators convert direct to alternating current

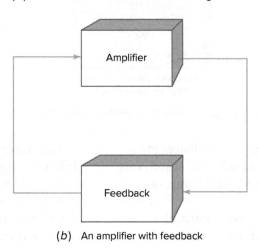

Amplifier

Feedback

(b) An amplifier with feedback

Fig. 11-1 Oscillator basics.

oscillating. The feedback is still present. But now there is not enough gain to overcome the loss in the feedback path. This is one of the two basic criteria that must be met if an amplifier is to oscillate: the amplifier *gain* must be *greater* than the loss in the feedback path. The other criterion is that the signal fed back to the input

of the amplifier must be *in phase*. In-phase feedback is also called *positive feedback*, or regenerative feedback. When the amplifier input and output are normally out of phase (such as in a common-emitter amplifier), the feedback circuit will have to produce a phase reversal. Figure 11-3 summarizes what the requirements are. The total phase shift is 180° + 180° = 360°. Note that 360° is the same phase as 0°. The feedback circuit provides the needed phase shift at the desired frequency of oscillation (f_o).

Oscillators are widely applied, for example:

In phase

Gain

1. Many digital devices such as computers, calculators, and watches use oscillators to generate rectangular waveforms that time and coordinate the various logic circuits.

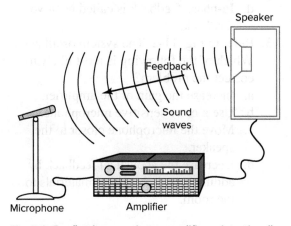

Speaker

Feedback

Sound waves

Microphone   Amplifier

Fig. 11-2 Feedback can make an amplifier unintentionally oscillate.

ABOUT ELECTRONICS

An arbitrary waveform generator can offer tens of preloaded waveforms and allow more to be programmed.

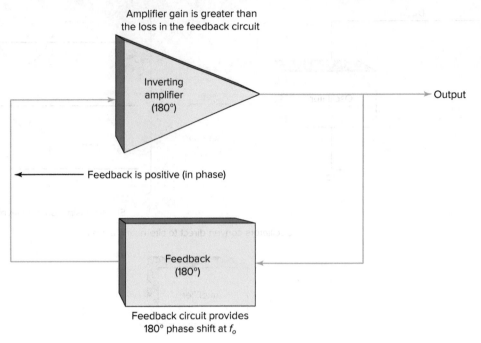

Amplifier gain is greater than
the loss in the feedback circuit

Inverting
amplifier
(180°)

Output

Feedback is positive (in phase)

Feedback
(180°)

Feedback circuit provides
180° phase shift at f_o

Fig. II-3 What's required to make an amplifier oscillate.

2. Signal generators use oscillators to produce the frequencies and waveforms required for testing, calibrating, or troubleshooting other electronic systems.
3. Touch-tone telephones, musical instruments, and remote control transmitters can use oscillators to produce the various frequencies needed.
4. Radio and television transmitters use oscillators to develop the basic signals sent to the receivers.

The various oscillator applications have different requirements. In addition to frequency and waveform, there is the question of *stability*. A stable oscillator will produce a signal of constant amplitude and frequency. Another requirement for some oscillators is the capability to produce a range of frequencies. *Variable-frequency oscillators* (VFOs) meet this need, as do *voltage-controlled oscillators* (VCOs).

Stability

Variable-frequency oscillators (VFOs)

Voltage-controlled oscillators (VCOs)

Self-Test

Choose the letter that best answers each question.

1. What conditions are required for an amplifier to oscillate?
   a. There must be feedback.
   b. The feedback must be in phase.
   c. The gain must be greater than the loss.
   d. All of the above are required.
2. Which of the following statements is *not* true?
   a. An oscillator is a circuit that converts dc to ac.
   b. An oscillator is an amplifier that supplies its own input signal.

   c. All oscillators generate sine waves.
   d. In-phase feedback is called positive feedback.
3. Refer to Fig. 11-2. The system oscillates. Which of the following is most likely to correct the problem?
   a. Increase the gain of the amplifier.
   b. Use a more sensitive microphone.
   c. Move the microphone closer to the speaker.
   d. Decrease the acoustical feedback by adding sound-absorbing materials to the room.

11-2   *RC* Circuits

It is possible to control the frequency of an oscillator by using resistive-capacitive components. One *RC* circuit that can be used for frequency control in oscillators is shown in Fig. 11-4. This circuit is called a *lead-lag network* and shows maximum output and zero phase shift at one frequency. This frequency is called the resonant frequency f_r. It can be found with this equation:

$$f_r = \frac{1}{2\pi RC}$$

The series and shunt values of R and C in Fig. 11-4 are equal.

Figure 11-5 illustrates a computer-generated amplitude and phase response for a 1.59-kHz lead-lag network. The amplitude plot rises as the frequency increases from 100 Hz until the resonant frequency is reached. The amplitude plot drops for frequencies above resonance. The phase angle plot leads for frequencies below resonance and lags for frequencies above resonance. Note that the phase response is 0 degree at resonance.

Lead-lag network

EXAMPLE 11-1

In Fig. 11-4, both resistors are 10 kΩ and both capacitors are 0.01 μF. Determine the resonant frequency of the lead-lag network. Use the equation

$$f_r = \frac{1}{2\pi RC}$$
$$= \frac{1}{6.28 \times 10 \times 10^3 \times 0.01 \times 10^{-6}}$$
$$= 1.59 \text{ kHz}$$

The lead-lag network, at resonance, shows an output voltage that is one-third the input voltage:

$$V_{dB} = 20 \times \log \frac{V_{out}}{V_{in}} = 20 \times \log \frac{1}{3}$$
$$= -9.54 \text{ dB}$$

This result agrees with the amplitude plot in Fig. 11-5. If an oscillator uses a lead-lag network in its feedback circuit, then its amplifier section will require a voltage gain greater than 3 (9.54 dB).

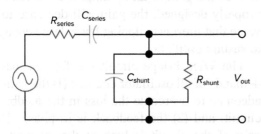

Fig. 11-4 An *RC* lead-lag network.

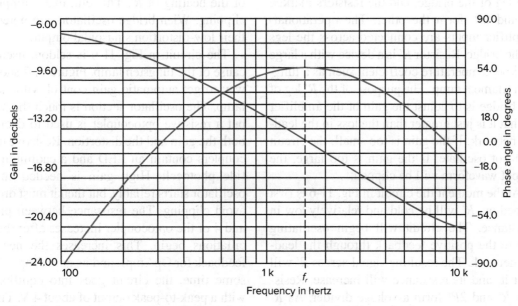

Fig. 11-5 Computer-generated amplitude and phase response of a lead-lag network.

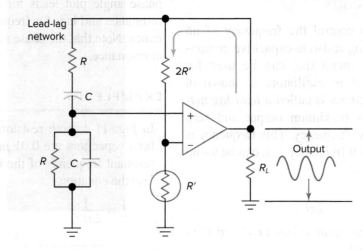

Lead-lag network

Output

Fig. 11-6 A Wien bridge oscillator.

Figure 11-6 shows how a lead-lag network can be used to control the frequency of an oscillator. Note that the feedback is applied through the lead-lag network to the noninverting input of an op amp. Feedback applied to the noninverting input is positive feedback. However, *only one frequency* will arrive at the noninverting input exactly in phase. That frequency is the *resonant frequency f_r* of the network. All other frequencies will lead or lag. This means that the oscillator will operate at a single frequency, and the output will be sinusoidal.

Wien bridge oscillator

The circuit of Fig. 11-6 is called a *Wien bridge oscillator*. The lead-lag network forms one leg of the bridge, and the resistors marked R' and $2R'$ form the other. The operational-amplifier inputs are connected across the legs of the bridge. Resistor R' is a device with a large positive temperature coefficient such as a tungsten filament lamp. The purpose of the R' leg of the bridge is to adjust the gain of the amplifier so that it is just greater than the loss in the lead-lag network. If the gain is too small, the circuit will not oscillate. If the gain is too large, the output waveform will be clipped.

At the moment the circuit in Fig. 11-6 is first turned on, R' will be cold and relatively low in resistance. The circuit will begin oscillating due to the positive feedback through the lead-lag network. The resulting signal across R' will heat it, and its resistance will increase. Resistors R' and $2R'$ form a voltage divider. As R'

increases, the voltage applied to the inverting input of the operational amplifier will increase. As we learned earlier, negative feedback decreases the gain of an op amp. If the circuit is properly designed, the gain will decrease to a value that prevents clipping but is large enough to sustain oscillation.

The Wien bridge circuit satisfies the basic demands of all oscillator circuits: (1) the gain is adequate to overcome the loss in the feedback circuit, and (2) the feedback is in phase. The gain of the circuit is high at the moment of power on. This ensures rapid starting of the oscillator. After that, the gain decreases because of the heating of R'. This eliminates amplifier clipping. Wien bridge oscillators are noted for their low-distortion sinusoidal output.

The circuit in Fig. 11-6 is seldom used because of the tungsten lamp. Figure 11-7 shows a way to get automatic gain control without the lamp. The oscillator section is much the same, but a resistive optocoupler is used to decrease both the gain and the distortion. Resistive optocouplers contain an LED and a cadmium sulfide photocell. High gain is needed so the oscillator starts reliably, but then it must drop to avoid clipping. The resistance between pins 3 and 4 of the optocoupler increases after the oscillations begin. This increases the negative feedback for OpAmp1, and its gain drops. After some time, the circuit goes into equilibrium with a peak-to-peak output of about 4 V. This is

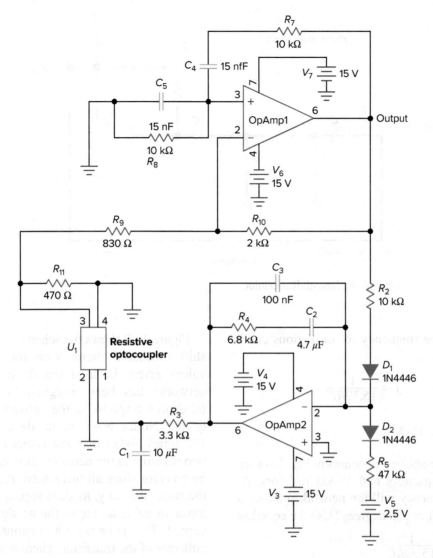

Fig. 11-7 Wien bridge oscillator with AGC.

far below the clipping level, and the output has low distortion. In fact, this circuit has a THD around 0.001 percent, which is considered excellent. This circuit is discussed further in Sec. 11-7.

EXAMPLE 11-2

Calculate the output frequency for the circuit in Fig. 11-7.

$$f = \frac{1}{6.28RC}$$

$$= \frac{1}{6.28 \times 10 \times 10^3 \times 15 \times 10^{-9}}$$

$$= 1.06 \text{ kHz}$$

There is another way to use RC networks to control the frequency of an oscillator. They can be used to produce a 180-degree phase shift at the desired frequency. This is useful when the common-emitter amplifier configuration is used. Figure 11-8 shows the circuit for a *phase-shift oscillator*. The signal at the collector is 180 degrees out of phase with the signal at the base. By including a network that gives an additional 180-degree phase shift, the base receives in-phase feedback. This is because $180° + 180° = 360°$, and $360°$ is the same as $0°$ in circular measurement.

In Fig. 11-8, the phase-shift network is divided into three separate sections. Each section is designed to produce a 60-degree phase shift, and the total phase shift will be $3 \times 60°$,

Phase-shift oscillator

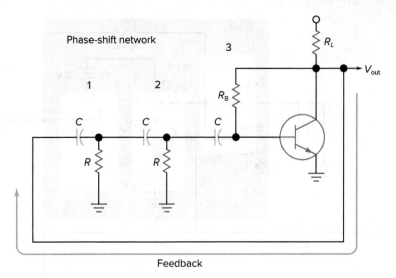

Fig. II-8 A phase-shift oscillator.

or 180°. The frequency of oscillations can be found with

$$f = \frac{1}{15.39RC}$$

EXAMPLE II-3

The phase-shift components in Fig. 11-8 are 0.1-μF capacitors and 18-kΩ resistors. At what frequency will the network produce a phase shift of 180 degrees? Use the equation

$$f = \frac{1}{15.39RC}$$

$$= \frac{1}{15.39 \times 18 \times 10^3 \times 0.1 \times 10^{-6}}$$

$$= 36.1 \text{ Hz}$$

Figure 11-9 shows the schematic of a phase-shift oscillator circuit with the component values given. Each of the three phase-shift networks has been designed to produce a 60-degree response at the desired output frequency. Notice, however, that the value of R_B is 100 times higher than the values of the other two resistors in the network. This may seem to be an error since all three networks should be the same. Actually, R_B does appear to be much lower in value as far as the ac signal is concerned. This is because it is connected to the collector of the transistor. There is an ac signal present at the collector when the oscillator is running, which is 180 degrees of phase with the base signal. This makes the voltage difference across R_B much higher than would be produced by the base signal alone. Thus, more

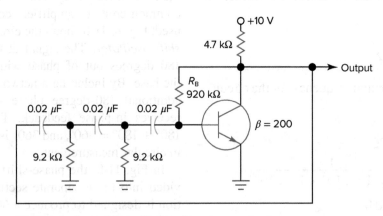

Fig. II-9 A phase-shift oscillator with component values.

signal current flows through R_B. Resistor R_B produces an ac loading effect at the base that is set by the voltage gain of the amplifier and the value of R_B:

$$r_B = \frac{R_B}{A_V}$$

This equation tells us that the actual ac loading effect r_B is equal to R_B divided by the voltage gain of the amplifier. If we assume that the gain of the amplifier is 100, then

$$r_B = \frac{920 \text{ k}\Omega}{100} = 9.2 \text{ k}\Omega$$

We may conclude that all three phase-shift networks are the same. The frequency of oscillation for Fig. 11-9 will be

$$f = \frac{1}{15.39RC}$$

$$= \frac{1}{15.39 \times 9.2 \times 10^3 \times 0.02 \times 10^{-6}}$$

$$= 353 \text{ Hz}$$

Figure 11-10 shows a computer-generated amplitude and phase plot for the RC phase-shift part of Fig. 11-9. Networks of this type produce an output voltage equal to $\frac{1}{29}$ of the input voltage at that frequency where the phase shift is

180 degrees. This represents a feedback network gain of

$$V_{dB} = 20 \times \log \frac{V_{out}}{V_{in}} = 10 \times \log \frac{1}{29}$$

$$= -29.2 \text{ dB}$$

The common-emitter amplifier in Fig. 11-9 must have a gain greater than 29.2 dB in order for the circuit to oscillate.

The circuit in Fig. 11-9 will not oscillate at exactly 353 Hz. The formula deals with only the values of the RC network. It ignores some other effects caused by the transistor. The formula is adequate for practical work.

Phase-shift oscillators often employ op amps, as shown in Fig. 11-11. The virtual ground at the inverting input allows the use of a 9.2-kΩ resistor at R_3. Since the gain of the phase-shift network is $\frac{1}{29}$, R_4 is determined by

$$R_3 \times 29 = 9.2 \text{ k}\Omega \times 29 = 267 \text{ k}\Omega$$

However, to ensure that the oscillator starts up within a reasonable time after the circuit is turned on, R_4 is made somewhat larger so the gain of the op amp is greater than 29. This is a compromise—making R_4 larger gives a faster start time but often leads to some clipping of the output. The solution to this problem is to lower the gain after the circuit starts oscillating.

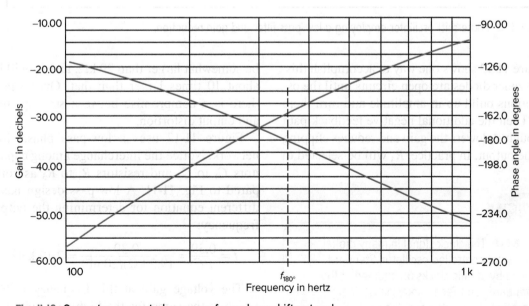

Fig. 11-10 Computer-generated response for a phase-shift network.

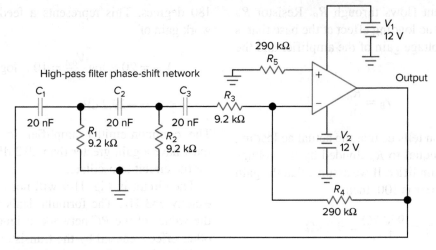

Fig. II-II Phase-shift oscillator employing an op amp.

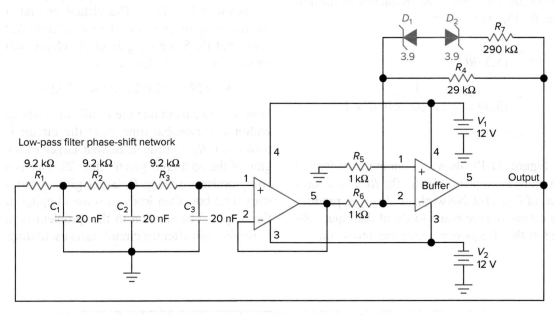

Fig. II-I2 Phase-shift oscillator employing a low-pass filter and gain reduction.

Figure 11-12 shows one way to accomplish this. The zener diodes are open circuits until the oscillations build up in amplitude and turn them on. Then the additional negative feedback path through R_7 lowers the gain and reduces clipping in the output. In practice, R_4 will be selected to be somewhat larger than 29 kΩ, and R_7 will be about 10 times larger than that. Once again, there is a compromise between start-up time and output distortion.

Figure 11-12 uses a low-pass phase-shift network. Notice the interchange among capacitors C_1 to C_3 and resistors R_1 to R_3 as compared to Fig. 11-11. A low-pass design has a different equation for determining the output frequency:

$$f = \frac{0.39}{RC} = \frac{0.39}{(9.2 \text{ k}\Omega)(20 \text{ nF})} = 2.12 \text{ kHz}$$

The voltage gain at this frequency is $\frac{1}{29}$, which is the same for the high-pass circuit at the frequency of oscillation.

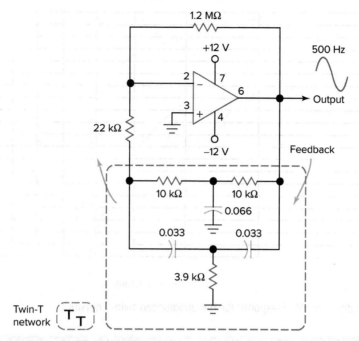

Fig. II-I3 Op-amp twin-T oscillator.

Figure 11-13 shows another type of *RC* oscillator that is based on the *twin-T network*. These networks act as notch filters and provide minimum output amplitude and a phase lag of 180 degrees at their resonant frequency. The resonant frequency of a twin-T network can be found by examining the circuit to determine which components are in series with the signal flow. Then use the values of the series components in this equation:

$$f_r = \frac{1}{2\pi RC}$$

EXAMPLE II-4

Find the resonant frequency for the twin-T network in Fig. 11-13. The series components in the network are the 10-kΩ resistors and the 0.033-μF capacitors. Use the equation

$$f_r = \frac{1}{2\pi RC}$$

$$= \frac{1}{6.28 \times 10 \times 10^3 \times 0.033 \times 10^{-6}}$$

$$= 482 \text{ Hz}$$

The twin-T network provides feedback from the output to the inverting input of the op amp. That feedback becomes positive at f_r because the twin-T network shows a 180-degree phase shift at this particular frequency. Notice that the 0.066-μF network capacitor is twice the value of each series capacitor. This is standard in a twin-T network. However, the 3.9-kΩ resistor is not standard. It is normally equal to one-half the value of each series resistor, or 5 kΩ in this case. A perfectly balanced twin-T network would provide no feedback at f_r. The intentional error allows enough positive feedback to reach pin 2 of the op amp, and a sine wave signal of approximately 500 Hz appears at the output.

Figure 11-14 shows a computer-generated response for the unbalanced twin-T network of Fig. 11-13. Because of the unbalancing, the actual resonant frequency is about 520 Hz, and the amplitude is −31 dB at that point. The op amp must provide a voltage gain greater than 35.5 (31 dB) in order for the circuit to oscillate.

Twin-T network

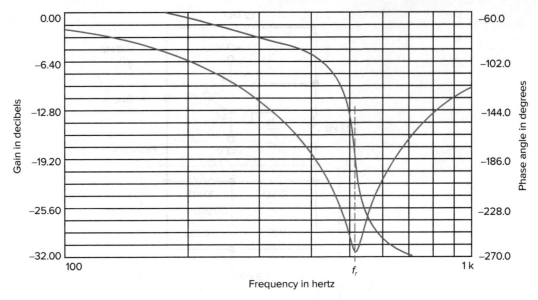

Fig. 11-14 Computer-generated response for the unbalanced twin-T network.

Self-Test

Choose the letter that best answers each question.

4. Refer to Fig. 11-4, where $R = 4,700 \Omega$ and $C = 0.02 \, \mu F$. At what frequency will V_{out} be in phase with the signal source?
   a. 486 Hz
   b. 1,693 Hz
   c. 3,386 Hz
   d. 9,834 Hz

5. Refer to Fig. 11-6, where $R = 6,800 \Omega$ and $C = 0.002 \, \mu F$. What will the frequency of the output signal be?
   a. 11.70 kHz
   b. 46.79 kHz
   c. 78.90 kHz
   d. 98.94 kHz

6. Refer to Fig. 11-6. What is the phase relationship of the output signal and the signal at the noninverting (+) input of the amplifier?
   a. 180 degrees
   b. 0 degrees
   c. 90 degrees
   d. 270 degrees

7. Refer to Fig. 11-8. What is the configuration of the transistor amplifier?
   a. Common emitter
   b. Common collector
   c. Common base
   d. Emitter follower

8. Refer to Fig. 11-9. Assume that $R_B = 820 \, k\Omega$ and the voltage gain of the circuit is 120. What is the actual loading effect of R_B as far as the phase-shift network is concerned?
   a. 1 MΩ
   b. 500 kΩ
   c. 6,833 Ω
   d. 384 Ω

9. Refer to Fig. 11-9. The capacitors are all changed to 0.05 μF. What is the frequency of oscillation?
   a. 60 Hz
   b. 141 Hz
   c. 1.84 kHz
   d. 0.95 MHz

10. Refer to Fig. 11-9. What is the phase relationship of the signal arriving at the base compared with the output signal?
    a. 0 degrees
    b. 90 degrees
    c. 180 degrees
    d. 270 degrees

11. What do phase-shift oscillators, twin-T oscillators, and Wien bridge oscillators have in common?
    a. They use *RC* frequency control.
    b. They have a sinusoidal output.
    c. They use amplifier gain to overcome feedback loss.
    d. All of the above are true.

11-3 *LC* Circuits

The *RC* oscillators are limited to frequencies below 1 MHz. Higher frequencies require a different approach to oscillator construction. Inductive-capacitive (*LC*) circuits can be used to design oscillators that operate at hundreds of megahertz. These *LC* networks are often called *tank circuits*, or flywheel circuits.

Figure 11-15 shows how a tank circuit can be used to develop sinusoidal oscillations. Figure 11-15(*a*) assumes that the capacitor is charged. As the capacitor discharges through the inductor, a field expands about the turns of the inductor. After the capacitor has been discharged, the field collapses and current continues to flow. This is shown in Fig. 11-15(*b*). Note that the capacitor is now being charged in the opposite polarity. After the field collapses, the capacitor again acts as the source. Now the

current is flowing in the opposite direction. Fig. 11-15(*c*) shows the second capacitor discharge. Finally, Fig. 11-15(*d*) shows the inductor acting as the source and charging the capacitor back to the original polarity shown in Fig. 11-15(*a*). The cycle will repeat over and over.

Inductors and capacitors are both energy storage devices. In a tank circuit, they exchange energy back and forth at a rate fixed by the values of inductance and capacitance. The frequency of oscillations is given by

$$f_r = \frac{1}{2\pi\sqrt{LC}}$$

Tank circuit

You should recognize this formula. It is the resonance equation for an inductor and a capacitor. It is based on the resonant frequency, where the inductive reactance and the capacitive reactance are equal. An energized *LC* tank circuit will oscillate at its resonant frequency.

EXAMPLE 11-5

What is the resonant frequency of the tank circuit in Fig. 11-15 if the coil is 1 μH and the capacitor is 180 pF? Apply the equation

$$f_r = \frac{1}{6.28 \times \sqrt{1 \times 10^{-6} \times 180 \times 10^{-12}}}$$

$$= 11.9 \text{ MHz}$$

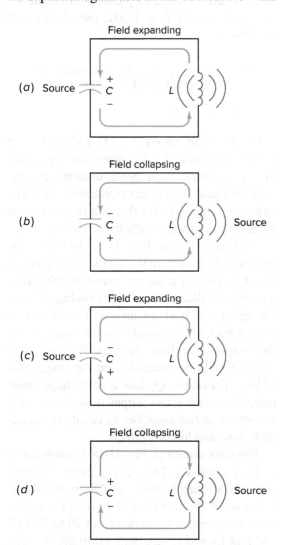

Field expanding

(*a*) Source

Field collapsing

(*b*) Source

Field expanding

(*c*) Source

Field collapsing

(*d*) Source

Fig. 11-15 Tank circuit action.

Real tank circuits have resistance in addition to inductance and capacitance. This resistance will cause the tank circuit oscillations to decay with time. To build a practical *LC* oscillator, an amplifier must be added. The gain of the amplifier will overcome resistive losses, and a sine wave of constant amplitude can be generated.

One way to combine an amplifier with an *LC* tank circuit is shown in Fig. 11-16. The circuit is called a *Hartley oscillator*. Note that the *inductor is tapped*. The tap position is important since the ratio of L_A to L_B determines the *feedback ratio* for the circuit. In practice, the feedback ratio is selected for reliable operation. This ensures that the oscillator will start every time the power is turned on. Too much feedback will cause clipping and distort the output waveform.

The transistor amplifier in Fig. 11-16 is in the common-emitter configuration. This means that a 180-degree phase shift will be required

Hartley oscillator

Feedback ratio

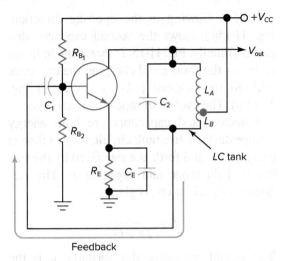

Fig. II-16 The Hartley oscillator.

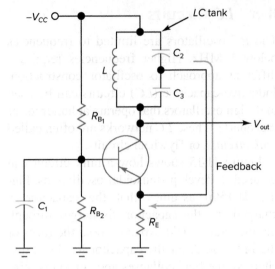

Fig. II-17 The Colpitts oscillator.

Buffer amplifier

Colpitts oscillator

somewhere in the feedback path. The tank circuit provides this phase shift. Note that the coil is tapped and that the tap connects to $+V_{CC}$. The tap is at ac ground, and there is a phase reversal across the tank. Thus, the collector signal arrives in phase at the base. Knowing the total inductance and the capacitance of the tank circuit will allow a solution for the resonant frequency. For example, if the total inductance $L_A + L_B$ is 20 μH, and the capacitance C_2 is 400 pF, then

$$f_r = \frac{1}{6.28 \times \sqrt{20 \times 10^{-6} \times 400 \times 10^{-12}}}$$

$$= 1.78 \text{ MHz}$$

Another way to control the feedback of an *LC* oscillator is to *tap* the *capacitive* leg of the tank circuit. When this is done, the circuit is called a *Colpitts oscillator* (Fig. 11-17). Capacitor C_1 grounds the base of the transistor for ac signals, and the transistor is operating as a common-base amplifier. You may recall that the input (the emitter) and the output (the collector) are in phase for this amplifier configuration. The feedback is in phase for the common-base configuration (shown in Fig. 11-17).

Capacitors C_2 and C_3 in Fig. 11-17 act in series as far as the tank circuit is concerned. Assume that $C_2 = 1,000$ pF and $C_3 = 100$ pF. Let us use the series capacitor formula to determine the effect of the series connection:

$$C_T = \frac{C_2 \times C_3}{C_2 + C_3} = \frac{1,000 \text{ pF} \times 100 \text{ pF}}{1,000 \text{ pF} + 100 \text{ pF}}$$

$$= 90.91 \text{ pF}$$

This means that 90.91 pF, along with the value of L, would be used to predict the frequency of oscillation. If $L = 1$ μH, the circuit will oscillate at

$$f_r = \frac{1}{6.28 \times \sqrt{1 \times 10^{-6} \times 90.9 \times 10^{-12}}}$$

$$= 16.7 \text{ MHz}$$

Figure 11-18 shows a VFO followed by a *buffer amplifier*. Both stages are operating in the common-drain configuration and use insulated-gate field-effect transistors. This circuit represents a design that can be used when maximum frequency stability is needed.

Transistor Q_1 in Fig. 11-18 provides the needed gain to sustain the oscillations. Transistor Q_2 serves as a buffer amplifier. This protects the oscillator circuit from loading effects. Changing the load on an oscillator tends to change both the amplitude and the frequency of the output. For best stability, the oscillator circuit should be isolated from the stages that follow. Transistor Q_2 has a very high input impedance and a low output impedance. This allows the buffer amplifier to isolate the oscillator from any loading effects.

The tank circuit in Fig. 11-18 is made up of L, C_1, C_2, and C_3. This arrangement is known as a series-tuned Colpitts, or *Clapp*, oscillator. It is one of the most stable of all *LC* oscillators. Assume that C_1 varies from 10 to 100 pF and that C_2 and C_3 are both 1,000 pF. We will use the series capacitor formula to determine

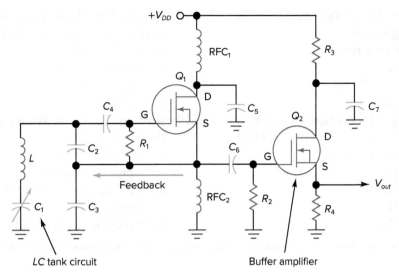

+V_{DD}

RFC$_1$

R_3

Q_1

D

C_4

G

C_5

Q_2

C_7

C_2

R_1

S

C_6

D

L

G

S

Feedback

RFC$_2$

R_2

V_{out}

C_1

C_3

R_4

LC tank circuit

Buffer amplifier

Fig. 11-18 A highly stable oscillator design.

the capacitive range of the tank circuit. When $C_1 = 10$ pF,

$$C_T = \frac{1}{1/C_1 + 1/C_2 + 1/C_3}$$

$$= \frac{1}{1/10 \text{ pF} + 1/1{,}000 \text{ pF} + 1/1{,}000 \text{ pF}}$$

$$= 9.8 \text{ pF}$$

When $C_1 = 100$ pF,

$$C_T = \frac{1}{1/100 \text{ pF} + 1/1{,}000 \text{ pF} + 1/1{,}000 \text{ pF}}$$

$$= 83.3 \text{ pF}$$

The calculations show that the effective value C_T of the capacitors is determined mainly by C_1. The stray and shunt capacities in Fig. 11-18

appear in parallel with C_2 and C_3. These stray and shunt capacities can change and cause frequency drift in *LC* oscillator circuits. The Clapp design minimizes these effects because the series-tuning capacitor has the major effect on the tank circuit.

Variable-frequency oscillators can be tuned by variable capacitors. However, variable capacitors are expensive and tend to be large. Many designs now replace the variable capacitor with a *varicap diode*. These diodes were covered in Sec. 3-4 of Chap. 3. As an example, variable capacitor C_1 in Fig. 11-18 could be replaced with a varicap diode and a bias circuit. Varying the bias voltage would tune the oscillator to various frequencies. Such a circuit would be called a voltage-controlled oscillator.

Varicap diode

Self-Test

Choose the letter that best answers each question.

12. Refer to Fig. 11-16. What is the configuration of the amplifier?
    a. Common emitter
    b. Common base
    c. Common collector
    d. Emitter follower
13. Refer to Fig. 11-16. Where is the feedback signal shifted 180 degrees?
    a. Across C_1
    b. Across R_{B_2}
    c. Across R_E
    d. Across the tank circuit

14. Refer to Fig. 11-16, where $C_2 = 120$ pF and $L_A + L_B = 1.8\ \mu$H. Calculate the frequency of the output signal.
    a. 484 kHz
    b. 1.85 MHz
    c. 5.58 MHz
    d. 10.8 MHz
15. Refer to Fig. 11-16. What is the waveform of V_{out}?
    a. Sawtooth wave
    b. Sine wave
    c. Square wave
    d. Triangle wave

16. Refer to Fig. 11-17. The amplifier is in what configuration?
    a. Common emitter
    b. Common base
    c. Common collector
    d. Emitter follower
17. Refer to Fig. 11-17, where $C_2 = 330$ pF, $C_3 = 47$ pF, and $L = 0.8\ \mu$H. What is the frequency of oscillation?
    a. 1.85 MHz
    b. 9.44 MHz
    c. 23.1 MHz
    d. 27.7 MHz
18. Refer to Fig. 11-18. Transistor Q_1 is operating as what type of oscillator?
    a. Clapp oscillator
    b. Hartley oscillator
    c. Phase-shift oscillator
    d. Buffer oscillator
19. Refer to Fig. 11-18. What is the major function of Q_2?
    a. It provides voltage gain.
    b. It provides the feedback signal.
    c. It isolates the oscillator from loading effects.
    d. It provides a phase shift.

11-4 Crystal Circuits

Piezoelectric material

Another way to control the frequency of an oscillator is to use a *quartz crystal*. Quartz is a *piezoelectric material*. Such materials can change electric energy into mechanical energy. They can also change mechanical energy into electric energy. A quartz crystal will tend to vibrate at its resonant frequency. The resonant frequency is determined by the physical characteristics of the crystal. Crystal thickness is the major determining factor for the resonant point.

Figure 11-19(*a*) shows the construction of a quartz crystal. The quartz disk is usually very thin, especially for high-frequency operation. A metal electrode is fused to each side of the disk. When an ac signal is applied across the electrodes, the crystal vibrates. The vibrations will be strongest at the resonant frequency of the crystal. When a crystal is vibrating at this frequency, a large voltage appears across the electrodes. The schematic symbol for a crystal is shown in Fig. 11-19(*b*).

Crystals can become the frequency-determining components in high-frequency oscillator circuits. They can replace *LC* tank circuits. Crystals have the advantage of producing very *stable* output frequencies. A crystal oscillator can have a stability better than 1 part in 10^6 per day. This is equal to an accuracy of 0.0001 percent. A crystal oscillator can be placed in a temperature-controlled oven to provide a stability better than 1 part in 10^8 per day.

An *LC* oscillator circuit is subject to frequency variations. Some things that can cause a change in oscillator output frequency are

1. Temperature
2. Supply voltage
3. Mechanical stress and vibration
4. Component value drift
5. Movement of metal parts near the oscillator circuit

Crystal-controlled circuits can greatly reduce all these effects.

A quartz crystal can be represented by an equivalent circuit (Fig. 11-20). The *L* and *C* values of the quartz equivalent circuit represent the resonant action of the crystal and determine what is known as the *series resonance* of the crystal. The electrode capacitance causes the crystal to also show a *parallel resonant* point. Since the capacitors act in series, the net capacitance is a little lower for parallel resonance. This makes the parallel resonant frequency slightly higher than the series resonant frequency.

The equivalent circuit of a crystal predicts that oscillations can occur in two modes: parallel and series. In practice, the parallel mode is from 2 to 15 kHz higher. Oscillator circuits may be designed to use either mode. When a

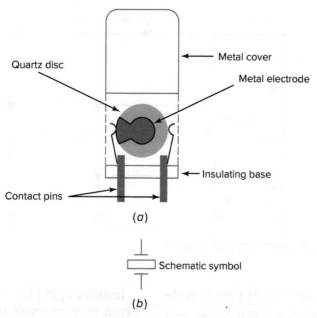

Fig. II-19 A crystal used for frequency control.

crystal is replaced, it is very important to obtain the correct type. For example, if a series-mode crystal is substituted in a parallel-mode circuit, the oscillator will run high in frequency.

Refer again to Fig. 11-20. Note that the quartz equivalent circuit also contains resistance R. This represents losses in the quartz. Most crystals have small losses. In fact, the losses are small enough to give crystals a very high Q. Circuit Q is very important in an oscillator circuit. *High Q* gives frequency stability. Crystal Qs can be in excess of 3,000. By comparison, LC tank circuit Qs seldom exceed 200. This is why a crystal oscillator is so much more stable than an LC oscillator.

Figure 11-21 shows the schematic diagram of a *crystal oscillator*. The amplifier configuration is common emitter. This means that the feedback path must provide a 180-degree phase shift for oscillations to occur. This phase shift is produced by capacitors C_1 and C_2. These capacitors form a voltage divider to control the amount of feedback. Excess feedback causes distortion and drift. Too little feedback causes unreliable operation; for example, the circuit may not start every time it is turned on. Capacitor C_3 is a *trimmer capacitor*. It is used to precisely set the frequency of oscillation. The remaining components in Fig. 11-21 are standard for the common-emitter configuration.

Crystal oscillator

High Q

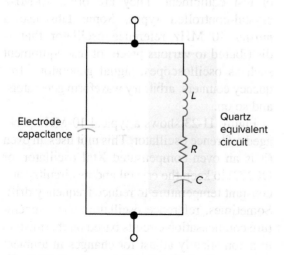

Fig. II-20 Quartz-crystal equivalent circuit.

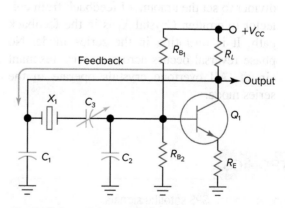

Fig. II-2I A crystal-controlled oscillator.

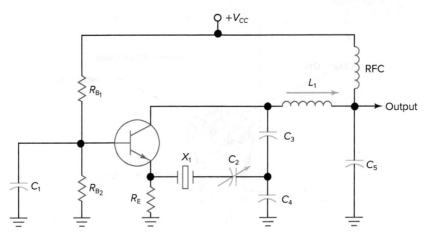

Fig. II-22 An overtone crystal oscillator.

Very high-frequency crystals present problems. The thickness of the quartz must decrease as frequency goes up. Above 15 MHz, the quartz becomes so thin that it is too fragile. Higher frequencies require the use of *overtone crystals*, which operate at *harmonics* of the fundamental frequency. Harmonics are integer multiples of a frequency. For example, the second harmonic of 10 MHz is 20 MHz, the third harmonic is 30 MHz, and so on. The use of harmonics can extend the range of crystal oscillators to around 150 MHz.

Oscillator circuits designed to use overtone crystals must include an *LC* tuned circuit. This circuit must be tuned to the correct harmonic. This ensures that the crystal will vibrate in the proper mode. Otherwise, it would tend to oscillate at a lower frequency.

Figure 11-22 is an *overtone oscillator* circuit. Capacitor C_1 grounds the base for ac signals. The transistor is in the common-base configuration, so no phase reversal is required in the feedback circuit. Capacitors C_3 and C_4 form a divider to set the amount of feedback from collector to emitter. Crystal X_1 is in the feedback path. It is operating in the series mode. No phase reversal occurs across a series resonant circuit. All overtone crystals operate in the series mode.

Inductor L_1 of Fig. 11-22 is part of the tuned circuit used to select the proper overtone. It resonates with C_3, C_4, and C_5 to form a tank circuit. Coil L_1 is adjusted to the correct overtone frequency. Capacitor C_2 is a trimmer capacitor used to set the crystal frequency. In practice, L_1 is adjusted first until the oscillator starts and works reliably. Then C_2 is adjusted for the exact frequency required.

Crystals increase the cost of oscillator circuits. This can become quite a problem in equipment such as a multichannel transmitter. A separate crystal will be required for every channel. The cost soon reaches the point where another solution must be found. This solution is a *frequency synthesizer*. There are combination digital and analog circuits that can synthesize many frequencies from one or more crystals.

Reference oscillators are part of many pieces of test equipment. They are often 10-MHz crystal-controlled types. Some labs use a *master* 10 MHz reference oscillator that is distributed to various pieces of test equipment such as oscilloscopes, signal generators, frequency counters, arbitrary waveform generators, and so on.

Figure 11-23 shows a typical 10-MHz packaged reference oscillator. This unit uses an oven (it is an oven-compensated Xtal oscillator, or OCXO) to keep the crystal and its circuitry at a constant temperature to reduce frequency drift. Sometimes, reference oscillators use temperature compensation circuits based on thermistors to automatically adjust for changes in ambient temperature, while others use microcontrollers

Overtone crystal

Harmonics

Frequency synthesizer

Overtone oscillator

ABOUT ELECTRONICS

Frequency standards can use GPS satellite signals.

Fig. II-23 Ten-megahertz reference oscillator.

Courtesy of Wenzel Associates, Inc.

for more accurate control. The oven types work well but need more power.

The best stability ratings go to the atomic type oscillators that use a gas such as rubidium or cesium. The gas is heated by radio frequency, and atomic absorption takes place when the frequency is just right. A photocell detector responds to the change in light caused by atomic absorption. Table 11-1 lists the types of reference oscillators and their major characteristics. Note the enormous range of accuracy: from 10^{-4} to 10^{-12}. 10^{-4} represents a frequency error of 100 Hz at 1 MHz or 100 parts per million (PPM). 10^{-12} is an error of only 1 Hz at 1 THz (a THz or 10^{12} Hz is a thousand billion cycles per second).

Table II-I   Reference Oscillator Characteristics

Oscillator Type	Accuracy	Aging (10 years)	Power in Watts	Weight in Grams
Crystal	10^{-5} to 10^{-4}	10–20 PPM	20×10^{-6}	20
TCXO	10^{-6}	2–5 PPM	100×10^{-6}	50
MCXO	10^{-8} to 10^{-7}	1–3 PPM	200×10^{-6}	100
OCXO	10^{-8}	2×10^{-8} to 2×10^{-7}	1–3	200–500
Rubidium	10^{-9}	5×10^{-10} to 5×10^{-9}	6–12	1,500–2,500
Cesium	10^{-12} to 10^{-11}	10^{-12} to 10^{-11}	25–40	10,000–20,000

TCXO = temperature-compensated crystal oscillator; MCXO = microprocessor-compensated crystal oscillator; and OCXO = oven-compensated crystal oscillator.

Self-Test

Choose the letter that best completes each statement.

20. The quartz crystals used in oscillators show a
    a. Piezoelectric effect
    b. Semiconductor effect
    c. Diode action
    d. Transistor action
21. An oscillator that uses crystal control should be
    a. Frequency-stable
    b. Useful only at low frequencies
    c. A VFO
    d. None of the above
22. A 6-MHz crystal oscillator has a stability of 1 part in 10^6. The largest frequency error expected of this circuit is
    a. 0.06 Hz
    b. 0.6 Hz

    c. 6 Hz
    d. 60 Hz
23. A series-mode crystal is marked 10.000 MHz. It is used in a circuit that operates the crystal in its parallel mode. The circuit can be expected to
    a. Run below 10 MHz
    b. Run at 10 MHz
    c. Run above 10 MHz
    d. Not oscillate
24. Refer to Fig. 11-21. The phase relationship of the signal at the collector of Q_1 compared with the base signal is
    a. 0 degrees
    b. 90 degrees
    c. 180 degrees
    d. 360 degrees

25. Refer to Fig. 11-21. The required phase shift is produced by
    a. R_L
    b. C_1 and C_2
    c. R_{B_1}
    d. C_3
26. Refer to Fig. 11-22. The configuration of the amplifier is
    a. Common emitter
    b. Common collector

c. Common base
d. Emitter follower
27. Refer to Fig. 11-22. The function of C_2 is to
    a. Act as an emitter bypass
    b. Adjust the tank circuit to the crystal harmonic
    c. Produce the required phase shift
    d. Set the exact frequency of oscillation

11-5 Relaxation Oscillators

Relaxation oscillator

All the oscillator circuits discussed so far produce a sinusoidal output. There is another major class of oscillators that do not produce sine waves. They are known as *relaxation oscillators*.

The outputs for these circuits are sawtooth or rectangular waveforms.

Figure 11-24 shows the circuit and waveforms for a relaxation oscillator based on a programmable unijunction transistor (PUT). Capacitor C_1 charges through resistor R_4. When

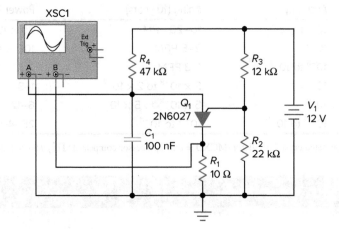

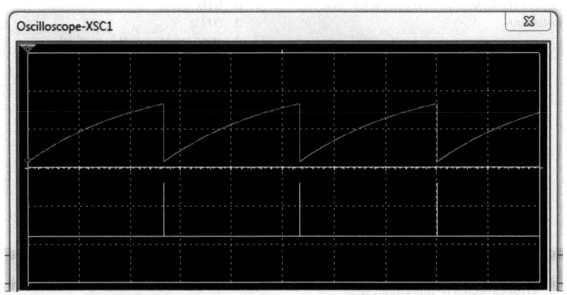

Fig. 11-24 Programmable unijunction transistor oscillator.

Source: Multisim

the voltage reaches a bit more than 8 V, the transistor fires (turns on) and rapidly discharges the capacitor. Discharge resistor R_1 is small so the discharge is rapid, as can be seen in the yellow waveform. After discharge, the cycle repeats as shown in both waveforms.

EXAMPLE 11-6

How does one make a rough guess about the size of the capacitor needed along with 10 kΩ resistor to make a 100 Hz relaxation oscillator? Begin with the equation

$$f = \frac{1}{RC}$$

Rearrange the equation to solve for C

$$C = \frac{1}{Rf} = \frac{1}{10 \times 10^3 \times 100} = 1 \ \mu F$$

The programming resistors R_2 and R_3, along with the supply voltage, set the firing voltage.

$$V_P = \frac{R_2}{R_1 + R_2} \times 12 \ V + 0.7 \ V = 8.46 \ V$$

The period of oscillation is based on the supply voltage, V_P, R_4, and the capacitor:

$$T = R \times C \times \log_e \frac{V_1}{V_1 - V_P}$$

$$= 47 \times 10^3 \times 100 \times 10^{-9} \times \log_e \frac{12}{12 - 8.46}$$

$$= 5.74 \ ms$$

$\log_e$ is the *natural log*, and it is the *ln* key on most calculators.

Figure 11-25 is the volt-ampere characteristic curve for the PUT. When the firing voltage V_P is reached, the resistance from anode to cathode drops rapidly. This is often called the *negative resistance* region (from V_P to V_V, where V_V is the valley voltage).

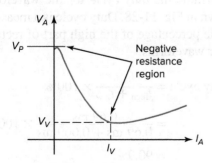

Fig. 11-25 PUT characteristic curve.

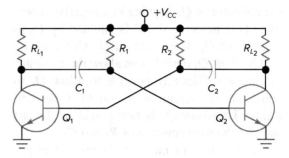

Fig. 11-26 The astable multivibrator.

Figure 11-26 shows another type of relaxation oscillator, the *astable multivibrator*. The circuit has no stable state. The circuit voltages switch constantly as it oscillates. This is in contrast to the *monostable* version, which has one stable state, and the *bistable* circuit, with two stable states. The monostable and bistable circuits will not be discussed since this chapter is limited to oscillators.

Astable multivibrators are also called *free-running flip-flops*. This name is more descriptive of how the circuit behaves. Notice in Fig. 11-26 that two transistors are used. If Q_1 is on (conducting), Q_2 will be off. After a period of time, the circuit flips and Q_1 goes off while Q_2 comes on. After a second period, the circuit flops, turning on Q_1 again and turning off Q_2. The flip-flop action continues as long as the power is applied.

Study the waveforms shown in Fig. 11-27. They are for transistor Q_1 in Fig. 11-26. Transistor Q_2's waveforms will look the same, but they will be inverted. Suppose that Q_2 has just turned on, making its collector less positive. This means

Astable multivibrator

Free-running flip-flop

Negative resistance

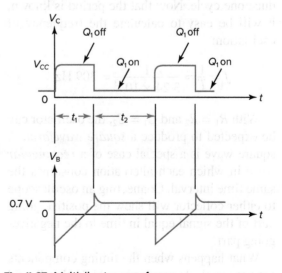

Fig. 11-27 Multivibrator waveforms.

the collector of Q_2 is going in a negative direction. This negative signal is coupled by C_2 to the base of Q_1. This turns off Q_1. Capacitor C_2 will hold off Q_1 until R_2 can allow the capacitor to charge sufficiently positive to allow Q_1 to come on. The circuit works on RC time constants. Transistor Q_1 is being held in the off state by the time constant of R_2 and C_2.

As Q_1 is turning on, its collector will be going less positive. This negative-going signal is coupled by C_1 to the base of Q_2, and Q_2 is turned off. It will stay off for a period determined by the *RC time constant* of R_1 and C_1.

RC time constant

Again, refer to Fig. 11-27. One rectangular cycle will be produced during one period. The period has two parts; thus it is equal to

$$T = t_1 + t_2$$

It takes 0.69 time constants for the *RC* network to reach the base turn-on voltage. This gives us a way to estimate the time that each transistor will be held in the off state:

$$t = 0.69RC$$

Assume that R_1 and R_2 are both 47-kΩ resistors, and C_1 and C_2 are both 0.05-μF capacitors. Each transistor should be held off for

$$t = 0.69 \times 47 \times 10^3 \times 0.05 \times 10^{-6}$$
$$= 1.62 \times 10^{-3} \text{ s}$$

The period will be twice this value:

$$T = 2 \times 1.62 \text{ ms} = 3.24 \text{ ms}$$

It will take 3.24 ms for the oscillator to produce one cycle. Now that the period is known, it will be easy to calculate the frequency of oscillation:

$$f = \frac{1}{T} = \frac{1}{3.24 \times 10^{-3}} = 309 \text{ Hz}$$

Square waveform

Rectangular wave

With $R_1 = R_2$ and $C_1 = C_2$, the oscillator can be expected to produce a *square waveform*. A square wave is a special case of a *rectangular wave* in which each alternation consumes the same time interval. Connecting an oscilloscope to either collector will show the positive-going part of the signal equal in time to the negative-going part.

What happens when the timing components are not equal? Assume in Fig. 11-26 that R_1 and

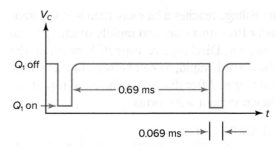

Fig. 11-28 Waveform for a nonsymmetrical multivibrator.

R_2 are 10 kΩ, $C_1 = 0.01 \ \mu$F, and $C_2 = 0.1 \ \mu$F. What waveform can be expected at the collector of Q_2? Computing both time constants will answer this question:

$$t_1 = 0.69 \times 10 \times 10^3 \times 0.1 \times 10^{-6}$$
$$= 0.69 \times 10^{-3} \text{ s}$$
$$t_2 = 0.69 \times 10 \times 10^3 \times 0.01 \times 10^{-6}$$
$$= 0.069 \times 10^{-3} \text{ s}$$

Transistor Q_1 will be held in the off mode 10 times longer than Q_2. Figure 11-28 shows the expected collector waveform for Q_1. Such a circuit is *nonsymmetrical*, and the output waveform is considered rectangular but is not a square wave.

What is the frequency of the rectangular waveform in Fig. 11-28? First, the period must be determined:

$$T = 0.69 \times 10^{-3} \text{ s} + 0.069 \times 10^{-3} \text{ s}$$
$$= 0.759 \times 10^{-3} \text{ s}$$

The frequency will be given by

$$f = \frac{1}{0.759 \times 10^{-3}} = 1318 \text{ Hz}$$

EXAMPLE 11-7

Determine the *duty cycle* for the waveform shown in Fig. 11-28. Duty cycle is a measure of the percentage of the high part of rectangular waveforms:

$$\text{Duty cycle} = \frac{t_{\text{high}}}{t_{\text{high}} + t_{\text{low}}} \times 100\%$$
$$= \frac{0.69 \text{ ms}}{0.69 \text{ ms} + 0.069 \text{ ms}} \times 100\%$$
$$= 90.9\%$$

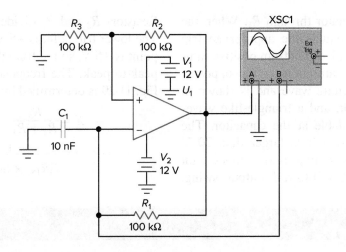

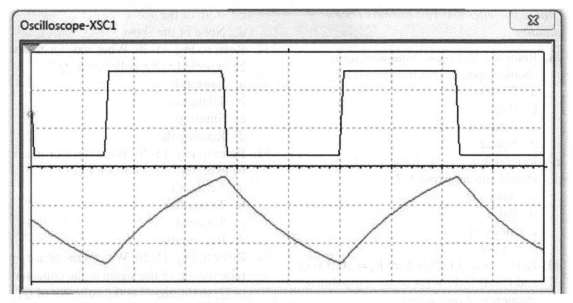

Fig. II-29 Schmitt trigger oscillator circuit and waveforms.

EXAMPLE II-8

Determine the duty cycle for a square wave. A square wave spends the same amount of time high as it does low. Assuming a unity value of 1 for the time,

$$\text{Duty cycle} = \frac{1}{1+1} \, 100\% = 50\%$$

Figure 11-29 shows another relaxation oscillator based on a Schmitt trigger. Schmitt triggers are comparators with positive feedback and were covered in Chap. 9. Resistors R_2 and R_3 form a voltage divider, and a portion of the output is applied to the noninverting input. This is positive feedback and determines the upper and lower trip points. The output is also applied

EXAMPLE II-9

Determine the output frequency for Fig. 11-29. First, find X:

$$X = \frac{R_3}{R_2 + R_3} = \frac{100 \text{ k}\Omega}{100 \text{ k}\Omega + 100 \text{ k}\Omega} = 0.5$$

Now, find the frequency (note that ln is the *natural logarithm*; use the ln key on your calculator):

$$f = \frac{1}{2R_1 C \times \ln\frac{1+X}{1-X}}$$

$$= \frac{1}{2 \times 100 \text{ k}\Omega \times 10 \text{ nF} \times \ln\frac{1.5}{0.5}} = 455 \text{ Hz}$$

to the timing capacitor through R_1. When the capacitor voltage equals the appropriate trip point, the output will switch from positive saturation to negative saturation, or the opposite. The output is a square wave and is shown in Fig. 11-29 in black, and a triangle-like waveform (red) is available at the capacitor. The output waveform is a little larger than 20 V peak-to-peak. The op amp used in this circuit is not capable of rail-to-rail output swing.

Resistors R_2 and R_3 divide the output by two. The lower trip point is −5 V, and the upper trip point is +5 V, so the triangle-like wave is 10 V peak-to-peak. The frequency of the output for Fig. 11-29 is determined by

$$X = \frac{R_3}{R_2 + R_3}$$

$$f = \frac{1}{2R_1C \times \ln\frac{1 + X}{1 - X}}$$

Self-Test

Choose the letter that best answers each question.

28. Refer to Fig. 11-24. What waveform should appear across the capacitor?
    a. Sawtooth
    b. Pulse
    c. Sinusoid
    d. Square

29. Refer to Fig. 11-24. What waveform should appear across R_3?
    a. Sawtooth
    b. Pulse
    c. Sinusoid
    d. Square

30. Refer to Fig. 11-24, where $R_4 = 10,000\ \Omega$ and $C = 0.5\ \mu F$. What is the approximate frequency of operation?
    a. 200 Hz
    b. 1,000 Hz
    c. 2,000 Hz
    d. 20 kHz

31. Refer to Fig. 11-24. What is the purpose of resistors R_2 and R_3?
    a. To set the desired intrinsic standoff ratio
    b. To set the exact frequency of oscillation

    c. Both of the above
    d. None of the above

32. Refer to Fig. 11-26. What waveform can be expected at the collector of Q_1?
    a. Sawtooth
    b. Triangular
    c. Sinusoid
    d. Rectangular

33. Refer to Fig. 11-26. What waveform can be expected at the collector of Q_2?
    a. Sawtooth
    b. Triangular
    c. Sinusoid
    d. Rectangular

34. Refer to Fig. 11-26. What is the phase relationship of the signal at the collector of Q_2 to the signal at the collector of Q_1?
    a. 0 degrees
    b. 90 degrees
    c. 180 degrees
    d. 360 degrees

35. Refer to Fig. 11-26. Assume that $C_1 = C_2 = 0.5\ \mu F$ and $R_1 = R_2 = 22\ k\Omega$. What is the frequency of oscillation?
    a. 16 Hz
    b. 33 Hz
    c. 66 Hz
    d. 99 Hz

II-6 Undesired Oscillations

It was mentioned earlier that a public address system can oscillate if the gain is too high. Such oscillations are undesired. Now that you have studied oscillators, it will be easier to understand how amplifiers can oscillate and what can be done to prevent it.

Negative feedback is often used in amplifiers to decrease distortion and improve frequency response. A three-stage amplifier is shown in simplified form in Fig. 11-30. Each stage uses the common-emitter configuration, and each will produce a 180-degree phase shift. This makes the feedback from stage 3 to stage 1

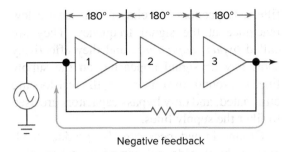

Fig. II-30 A three-stage amplifier with negative feedback.

negative. Positive feedback is required for oscillation; therefore, the amplifier should be stable. But, at very high or very low frequencies, the feedback can become positive. Transistor interelectrode capacitances form *lag* networks that can cause a phase error at high frequencies. Coupling capacitors form *lead* networks that cause phase errors at low frequencies. These effects accumulate in multistage amplifiers. The overall phase error will reach −180 degrees at some high frequency and +180 degrees at some low frequency if the amplifier uses capacitive coupling.

A system such as the one shown in Fig. 11-30 can become an oscillator at a frequency where the internal phase errors sum to ±180 degrees. If amplifier gain is high enough at that frequency, the amplifier will oscillate. Such an amplifier is unstable and useless. *Frequency compensation* can be used to make such an amplifier stable. A compensated amplifier has one or more networks added that decrease gain at the frequency extremes. Thus, by the time the frequency is reached where the phase errors total ±180 degrees, the gain is too low for oscillations to occur. A good example of this technique is modern operational amplifiers. They are internally compensated for gain reductions of 20 dB per decade. At the higher frequencies where the phase errors total −180 degrees, the gain is too low for oscillation to occur. This has already been discussed in Sec. 9-5 in Chap. 9.

Figure 11-31 shows an op-amp output waveform with conditional stability. It is close to becoming an oscillator as evidenced by the severe ringing on the bottom edge of the square wave. This can happen with op amps that are operated out of their recommended range. It has already been mentioned that some are not stable at unity gain. Also, capacitive loading at the output terminal can cause ringing and unwanted oscillation.

Another way that amplifiers can become unstable is when feedback paths occur that do not show on the schematic diagram. For example, a

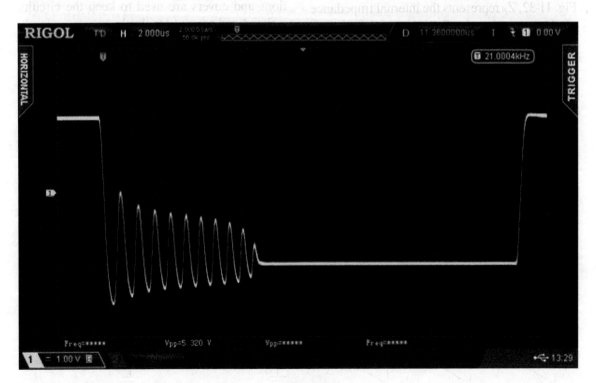

Fig. II-3I Op-amp output ringing with a square wave input signal.

Charles A. Schuler/McGraw Hill

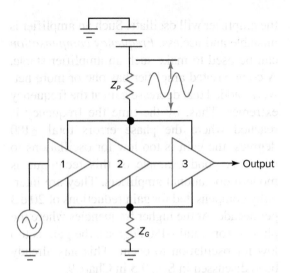

Fig. II-32 The effect of supply and ground impedances.

good power supply is expected to have a very low internal impedance. This will make it very difficult for ac signals to appear across it. However, a power supply might have a high impedance. This can be caused by a defective filter capacitor. An old battery power supply may develop a high internal impedance because it is drying out. The impedance of the power supply can provide a common load where signals are developed.

Shielding

In the simplified three-stage amplifier in Fig. 11-32, Z_P represents the internal impedance of the power supply. Suppose that stage 3 is drawing varying amounts of current because it is amplifying an ac signal. The varying current will produce a signal across Z_P. This signal will obviously affect stage 1 and stage 2. It is a form of unwanted feedback, and it may cause the circuit to oscillate.

Figure 11-33 shows a solution for the *unwanted feedback* problem. An *RC* network has been added in the power-supply lines to each amplifier. These networks act as low-pass

Unwanted
feedback

filters. Capacitors are chosen that have a low reactance at the signal frequency. They are called *bypass capacitors*, and they effectively short any ac signal appearing on the supply lines to ground. In some cases, the resistors are eliminated, and only bypass capacitors are used to filter the supply lines.

Ground impedances can also produce feedback paths that do not appear on schematics. Heavy currents flowing through printed circuit foils or the metal chassis can cause voltage drops. The voltage drop from one amplifier may be fed back to another amplifier. Refer again to Fig. 11-32. The impedance of the ground path is Z_G. As before, signal currents from stage 3 could produce a voltage across Z_G that will be fed back to the other stages. Ground currents cannot be eliminated, but proper layout can prevent them from producing feedback. The idea is to prevent later stages from sharing ground paths with earlier stages.

High-frequency amplifiers such as those used in radio receivers and transmitters are often prone to oscillation. These circuits can be coupled by stray capacitive and magnetic paths. When such circuits can "see" each other in the electrical sense, oscillations are likely to occur. These circuits must be *shielded*. Metal partitions and covers are used to keep the circuits isolated and prevent feedback.

Another feedback path often found in high-frequency amplifiers lies within the transistor itself. This path can also produce oscillations and make the amplifier useless. In Fig. 11-34, C_{bc} represents the capacitance from the collector to the base of the transistor in a tuned high-frequency amplifier. This capacitance will feed some signal back. The feedback can become

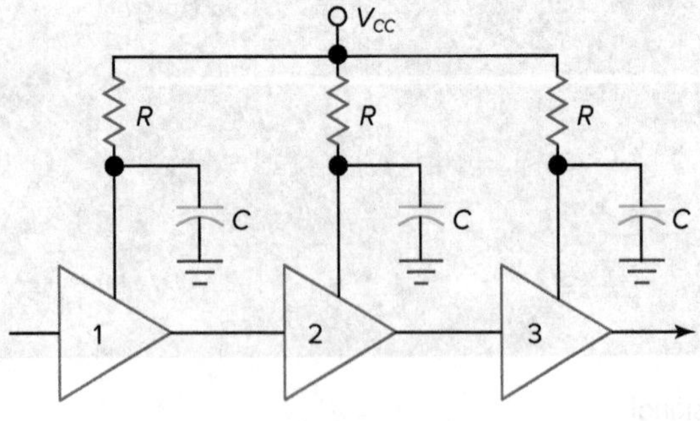

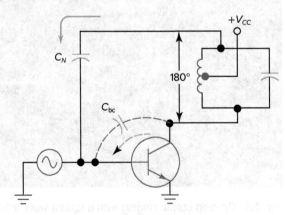

Fig. II-33 Preventing supply feedback.

Fig. II-34 Feedback inside and outside the transistor.

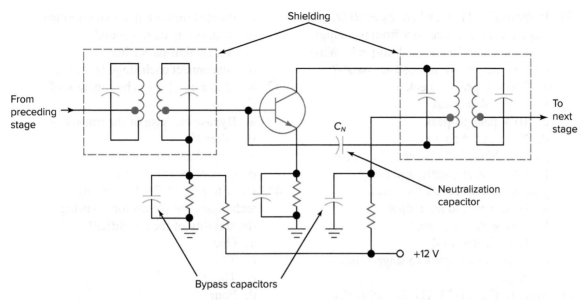

Fig. 11-35 A stabilized RF amplifier.

positive at a frequency where enough internal phase shift is produced.

Nothing can be done to eliminate the feedback inside a transistor. However, it is possible to create a second path external to the transistor. If the phase of the external feedback is correct, it can cancel the internal feedback. This is called *neutralization*. Figure 11-34 shows how a capacitor can be used to cancel the feedback of C_{bc}. Capacitor C_N feeds back from the collector circuit to the base of the transistor. The phase of the signal fed back by C_N is opposite to the phase fed back by C_{bc}. This stabilizes the amplifier. Notice that the required phase reversal is produced across the tuned circuit. Another possibility is to use a separate neutralization winding that is coupled to the tuned circuit.

Figure 11-35 is an actual radio-frequency amplifier used in a frequency modulation (FM) tuner. You will note that several of the techniques discussed in this section have been employed to stabilize the amplifier.

Neutralization

Self-Test

Choose the letter that best answers each question.

36. Examine Fig. 11-30. Assume that at some frequency extreme, the actual phase shift in each stage is 240 degrees. What happens to the feedback at that frequency?
    a. It does not exist.
    b. It becomes positive.
    c. It decreases the gain for that frequency.
    d. None of the above occurs.

37. Refer to question 36. Assume that the amplifier has more gain at that frequency than it has loss in the feedback path. What happens to the amplifier?

    a. It burns out.
    b. It short-circuits the signal source.
    c. It becomes unstable (oscillates).
    d. It can no longer deliver an output signal.

38. Why are most operational amplifiers internally compensated for gain reductions of 20 dB per decade?
    a. To prevent them from becoming unstable
    b. To prevent any phase error at any frequency
    c. To prevent signal distortion
    d. To increase their gain at high frequencies

39. Refer to Fig. 11-32, where $Z_P = 10 \, \Omega$, and stage 3 is taking a current from the supply that fluctuates 50 mA peak-to-peak. What signal voltage is developed across Z_P?
   a. 100 mV peak-to-peak
   b. 1 V peak-to-peak
   c. 10 V peak-to-peak
   d. None of the above
40. Refer to Fig. 11-32. Stage 3 draws current from the power supply, and a signal is produced across Z_P. This signal
   a. Is delivered to the output
   b. Is canceled in stage 1
   c. Is dissipated in Z_G
   d. Becomes feedback to stage 1 and stage 2
41. Refer to Fig. 11-33. The *RC* networks shown are often called decoupling networks. This is because they

   a. Prevent unwanted signal coupling
   b. Bypass any dc to ground
   c. Act as high-pass filters
   d. Disconnect each stage from V_{CC}
42. Refer to Fig. 11-34. The function of C_N is to
   a. Bypass the base of the transistor
   b. Filter V_{CC}
   c. Tune the tank circuit
   d. Cancel the effect of C_{bc}
43. Refer to Fig. 11-35. How many techniques are shown for ensuring the stability of the amplifier?
   a. One
   b. Two
   c. Three
   d. Four

11-7 Oscillator Troubleshooting

Oscillator troubleshooting uses the same skills needed for amplifier troubleshooting. Since most oscillators are amplifiers with positive feedback added, many of the faults are the same. When troubleshooting an electronic circuit, remember the word "GOAL." Good troubleshooting involves

1. Observing the symptoms
2. Analyzing the possible causes
3. Limiting the possibilities

It is possible to observe the following symptoms when troubleshooting oscillators:

1. No output
2. Reduced amplitude
3. Unstable frequency
4. Frequency error

It is also possible that two symptoms may be observed at the same time. For example, an oscillator circuit may show reduced amplitude and frequency error.

Certain instruments are very useful for proper symptom identification. A digital frequency counter is valuable when troubleshooting for

frequency error. An oscilloscope is also a good instrument for oscillator troubleshooting. As always, a voltmeter is needed for power-supply and bias-voltage checks. When using instruments in and around oscillator circuits, always remember this: oscillators can be subject to *loading effects*. More than one technician has been misled because connecting test equipment pulled the oscillator off frequency or reduced the amplitude. In some cases, an instrument may load an oscillator to the point where it will stop working altogether.

Loading effects can be reduced by using high-impedance instruments. It is also possible to reduce loading effects by taking readings at the proper point. If an oscillator is followed by a buffer stage, frequency and waveform readings should be taken at the output of the buffer. The buffer will minimize the loading effect of the test equipment.

Do not forget to check the effect of any and all controls when troubleshooting. If the circuit is a VFO, it is a good idea to tune it over its entire range. You may find that the trouble appears and disappears as the oscillator is tuned. Variable capacitors can short over a portion of their range. If the circuit is a VCO, it may be necessary to override the tuning voltage with an external power supply to verify proper operation and frequency range. Use a current-limiting resistor of around 100 kΩ to avoid loading effects and circuit damage when running this type of test.

The power supply can have several effects on oscillator performance. Frequency and amplitude are both sensitive to the power-supply voltage. It is worth knowing if the power supply is correct and stable. Power-supply checks should be made early in the troubleshooting process. They are easy to make and can save a lot of time.

It is important to review the theory of the circuit when troubleshooting. This will help you analyze possible causes. Determine what controls the operating frequency. Is it a lead-lag network, an *RC* network, a tank circuit, or a crystal? Is there a varicap diode in the frequency-determining network? Remember that loading effects can pull an oscillator off frequency. The problem could be in the next stage that is fed by the oscillator circuit.

The circuit shown earlier in Fig. 11-7 is easier to troubleshoot when you keep a few things in mind. Suppose the symptom is no output and all three supply voltages are good. This is a "fussy circuit," and the 2.5-V supply must be very close in value for correct operation.

The oscillator loop gain must be high enough for oscillations to start, so try shorting R11 if there is no output. This will increase the gain; if the oscillator starts, then the trouble is in the

AGC circuit or the optocoupler. You can connect pin 1 to the +15-V supply through a 2.2-kΩ resistor. This will turn its internal LED on and should lower the resistance and allow the oscillator to start. This will verify that the optocoupler is working as it should.

If the output is clipped, the gain is too high. You can check the optocoupler by grounding pin 1. This will turn off its internal LED, and the output resistance will go high, which should kill the oscillations. When the circuit is operating normally, the dc voltage at the output of the AGC amplifier (OpAmp2) is about +4 V. The negative input pin of the AGC amplifier is biased in a negative direction by the 2.5-V supply and in a positive direction by the rectified signal from the oscillator. The relative balance of these two signals is what sets the loop gain and allows it to drop after the oscillator starts. So the output of OpAmp2 falls from about +7 V to about +4 V as equilibrium is reached.

Unstable oscillators can be quite a challenge. Technicians often resort to tapping components and circuit boards with an insulated tool to localize the difficulty. If this fails, they may use heat or cold to isolate a sensitive component. Desoldering pencils make excellent heat sources. A squeeze on the bulb will direct a stream of hot air just where it is needed. Chemical "cool sprays" are available for selective cooling of components.

Table 11-2 is a summary of causes and effects to help you troubleshoot oscillators.

Table 11-2 Troubleshooting Oscillators

Problem	Possible Cause
No output	Power-supply voltage. Defective transistor or op-amp. Shorted component (check tuning capacitor in VFO). Open component. Severe load (check buffer amplifier). Defective crystal. Defective joint (check printed circuit board).
Reduced amplitude	Power-supply voltage low. Transistor bias (check resistors). Circuit loaded down (check buffer amplifier). Defective transistor or op-amp.
Frequency unstable	Power-supply voltage changes. Defective connection (vibration test). Temperature sensitive (check with heat and/or cold spray). Tank circuit fault. Defect in *RC* network. Defective crystal. Load change (check buffer amplifier). Defective transistor or op-amp.
Frequency error	Wrong power-supply voltage. Loading error (check buffer amplifier). Tank circuit fault (check trimmers and/or variable inductors). Defect in *RC* network. Defective crystal. Transistor bias (check resistors).

Choose the letter that best answers each question.

44. What can loading do to an oscillator?
    a. Cause a frequency error
    b. Reduce the amplitude of the output
    c. Kill the oscillations completely
    d. All of the above

45. An astable multivibrator is a little off frequency. Which of the following is least likely to be the cause?
    a. The power-supply voltage is wrong.
    b. A resistor has changed value.
    c. A capacitor has changed value.
    d. The transistors are defective.

46. A technician notes that a tool or a finger brought near a high-frequency oscillator tank circuit causes the output frequency to change.
    a. This is to be expected.
    b. It is a sign that the power supply is unstable.
    c. The tank circuit is defective.
    d. There is a bad transistor in the circuit.

47. A technician replaces the UJT in a relaxation oscillator. The circuit works, but the frequency is off a little. What is wrong?
    a. The new transistor is defective.
    b. The intrinsic standoff ratio is different.
    c. The resistors are burned out.
    d. The circuit is wired incorrectly.

11-8 Direct Digital Synthesis

There was a time when crystal-controlled oscillators were the best choice when accurate and stable high-frequency signals were needed. They are still a good choice when a small number of frequencies are required. When a large number of frequencies are needed, a phase-locked loop (PLL) frequency synthesizer or a direct digital synthesizer (*DDS*) is probably a better choice. Phase-locked loop synthesis is covered in Chap. 13.

A direct digital synthesizer can be used in place of a crystal-controlled oscillator. The basic advantage of direct digital synthesis is frequency agility. A DDS can be programmed to produce a large number of high-resolution frequencies. Sometimes, DDS systems are called *numerically controlled oscillators*. A DDS generates an output waveform that is a function of a clock signal and a tuning word that is in the form of a binary number. Figure 11-36 shows the major parts of a DDS. A frequency-tuning word sets the value for the phase increment. On each clock pulse, the phase accumulator jumps to a new location in the sine lookup table. Each sine value is then sent from the lookup table to the digital-to-analog (D/A) converter, which produces a voltage that corresponds to a sine wave at some particular phase value. Note that the D/A output shown in Fig. 11-36 is an approximation of a sine wave. After the high-frequency components are filtered out, the output is close to being sinusoidal.

Figure 11-37 shows how the output frequency of a DDS is controlled. The clock frequency is fixed. What changes is the phase increment value. In the case of the top waveform, the phase increment is 30 degrees. The phase increment is 45 degrees for the bottom waveform. Notice that the smaller phase increment produces a lower output frequency. In Fig. 11-37, for the same number of clock pulses, the smaller phase increment produces exactly $1\frac{1}{2}$ cycles of output, and the larger phase increment produces $2\frac{1}{4}$ cycles of output.

The output frequency for a DDS is given by

$$f_{\text{out}} = \frac{f_{\text{clock}} \times \Delta\phi}{2^N}$$

where

f_{out} = the output frequency
f_{clock} = the clock frequency
$\Delta\phi$ = the phase increment value
N = the bit size of the phase accumulator

The word *bit* is a contraction for *b*inary dig*it*. Binary numbers have only two characters: 0 and 1. Some commercial DDS ICs have 32-bit phase accumulators and can operate

DDS

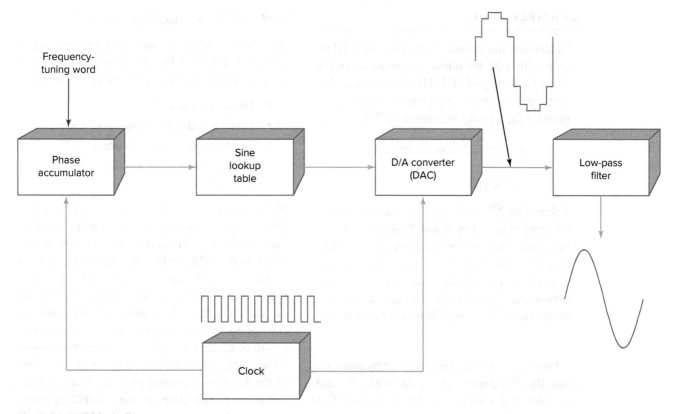

Fig. II-36 DDS block diagram.

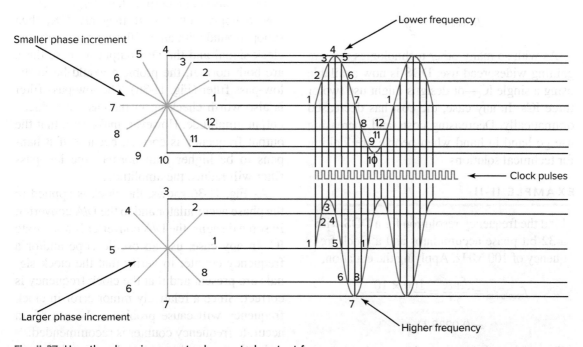

Fig. II-37 How the phase increment value controls output frequency.

at clock frequencies as high as 100 MHz or more. The largest phase increment value would be equal to the size of the phase accumulator. In that case, the output frequency would be equal to the clock frequency. In practice, this is never done, and the phase increment value is always some integer value that is smaller than 2^N.

EXAMPLE 11-10

Determine the output frequency for a DDS chip with a 32-bit phase accumulator and a clock frequency of 30 MHz. Assume that the frequency-tuning word programs the accumulator for a phase increment of 2^{30}.

$$f_{out} = \frac{f_{clock} \times \Delta\phi}{2^N} = \frac{30 \text{ MHz} \times 2^{30}}{2^{32}}$$

$$= 7.5 \text{ MHz}$$

Values like 2^{30} can be handled with a calculator that has an x^y key. Enter 2 and press the x^y key, then enter 30 and press the = key (display shows 1073741824). Also, you may notice in this example that exponents are subtracted when dividing, so the denominator becomes 2^2 (4), and 30 divided by 4 is 7.5.

Any frequency can be produced by programming the phase increment value with any integer value that is within the bit resolution of the phase accumulator. The frequency resolution of a DDS is the smallest possible frequency change:

$$f_{resolution} = \frac{f_{clock}}{2^N}$$

As with so many other technologies that are gaining widespread use, DDS is now available using a single IC—or designs might use two or three ICs. In any case, the cost has decreased dramatically. Decreasing costs and increasing use go hand in hand when people are looking for technical solutions.

EXAMPLE 11-11

Find the frequency resolution for a DDS with a 32-bit phase accumulator and a clock frequency of 100 MHz. Applying the equation,

$$f_{resolution} = \frac{f_{clock}}{2^N} = \frac{100 \times 10^6}{2^{32}}$$

$$= 0.0233 \text{ Hz}$$

This example is important because it demonstrates one of the most important features of DDS: *the ability to produce accurate, high-frequency signals in programmable millihertz steps*. Currently, DDS is the only technology available that can accomplish this.

Bit stream

11-9 DDS Troubleshooting

As always, when troubleshooting, remember the word "GOAL." Exactly what are the symptoms that you observe? Some possibilities are

- There is no output.
- There is reduced output amplitude.
- Some frequencies are wrong.
- All frequencies are wrong by a modest amount.

Next, you observe, analyze, and then limit. Don't forget to use a system viewpoint. Look at Fig. 11-36. The frequency-tuning word is applied to the DDS chip via a parallel bus or a serial bus. In the case of a parallel bus, there might be as many as 22 frequency-control pins on the DDS IC. These pins will typically be controlled by a microprocessor. So, when some frequencies are wrong, it is possible that there is a bad connection or solder joint at one or more of the frequency control pins, the DDS chip itself could be defective, or there could be a problem with the microprocessor (and that could be a software problem).

In the case of no output, don't forget to check power-supply voltages. If they are OK, then snoop around with an oscilloscope to verify the clock signal and the D/A output. In case these are both normal, the problem would be in the low-pass filter (Fig. 11-36). The low-pass filter is also worth checking in the case of reduced output amplitude. However, make sure that the output frequency is correct, because if it happens to be higher than normal, the low-pass filter will reduce the amplitude.

As Fig. 11-36 shows, the clock is applied to the phase accumulator and to the D/A converter. In some designs, the D/A converter is a separate IC. In any case, use an oscilloscope and/or a frequency counter to verify that the clock signals are present and that the clock frequency is correct. Since a relatively minor error in clock frequency will cause problems, the use of an accurate frequency counter is recommended.

In the case of a serial bus, as shown in Fig. 11-38, the DDS chip is programmed at the *data input* (pin 25 on the AD9850 IC) with a *bit stream*. Each bit is *clocked in*, one at a time, by an external signal applied to pin 7 of the IC. At the end of the 40-bit sequence, a *load* pulse is applied to pin 8 of the IC. Thus, troubleshooting

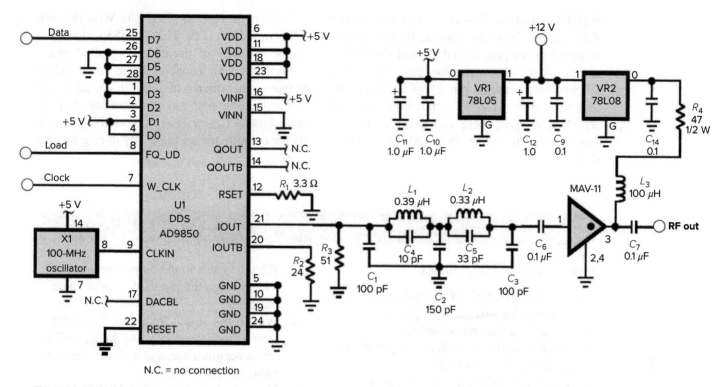

Fig. II-38 DDS circuit.

might involve verifying the data, load, and clock input signals, using an oscilloscope or a logic analyzer. These signals will often be supplied by a computer or a microprocessor, so a software problem is a possibility.

In the case of no output, *always* check the power-supply voltages as soon as possible. For Fig. 11-38, there is a 12-V input and two on-board voltage regulators. All three voltages should be verified. Table 11-3 shows that W34 is a power-down control bit. When this bit is

high, the AD9850 powers down. However, be sure to eliminate other possibilities first. As an example, the RF output at pin 21 of the IC should be checked. If an RF signal is present there, then the problem could be in the inductors or the MAV-11 output amplifier. Also, no output could be caused by the failure of the on-board 100-MHz oscillator (check for a signal on pin 9 of the integrated circuit).

In the case of low output, check R_1 in Fig. 11-38. This is the digital-to-analog converter's external

Table II-3	AD9850 40-Bit Serial-Load Word Function Assignment						
W0	Freq-b0 (LSB)	W10	Freq-b10	W20	Freq-b20	W30	Freq-b30
W1	Freq-b1	W11	Freq-b11	W21	Freq-b21	W31	Freq-b31 (MSB)
W2	Freq-b2	W12	Freq-b12	W22	Freq-b22	W32	Control
W3	Freq-b3	W13	Freq-b13	W23	Freq-b23	W33	Control
W4	Freq-b4	W14	Freq-b14	W24	Freq-b24	W34	Power-Down
W5	Freq-b5	W15	Freq-b15	W25	Freq-b25	W35	Phase-b0 (LSB)
W6	Freq-b6	W16	Freq-b16	W26	Freq-b26	W36	Phase-b1
W7	Freq-b7	W17	Freq-b17	W27	Freq-b27	W37	Phase-b2
W8	Freq-b8	W18	Freq-b18	W28	Freq-b28	W38	Phase-b3
W9	Freq-b9	W19	Freq-b19	W29	Freq-b29	W39	Phase-b4 (MSB)

LSB = least significant bit and MSB = most significant bit.

RSET connection. This resistor value sets the DAC full-scale output current. If this resistor is high in value, the output current will be reduced. Also, check the low-pass filter (the components between pin 21 of the DDS IC and pin 1 of the MAV-11 output amplifier) and R_4. If L_3 is open, there will be an extremely low output. Of course, the MAV-11 could be defective.

You might have noticed bits W34 through W39 in Table 11-3. The AD9850 also provides five bits of digitally controlled phase modulation, which enables phase shifting of its output in increments of 180, 90, 45, 22.5, and 11.25 degrees, and any combination thereof. Both bits W32 and W32 should always be 0. Either or both are set high only during factory testing.

Self-Test

Answer the following questions.

48. List three methods of producing high-stability, high-frequency signals.
49. What is another name for DDS?
50. What is used to smooth the output of the D/A converter in a DDS?
51. In any given DDS design, the clock frequency does not _____.
52. In any given DDS, the output frequency is changed by varying the _____.
53. What happens to the output frequency of a DDS when the phase increment value increases?
54. What is the output frequency for a DDS with a clock frequency of 5 MHz, a 32-bit phase accumulator, and a phase increment value of 85899346?
55. If the clock frequency in a DDS is out of tolerance, which output frequencies will suffer?

Chapter 11 Summary and Review

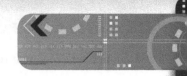

Summary

1. Oscillators convert direct current to alternating current.
2. Many oscillators are based on amplifiers with positive feedback.
3. The gain of the amplifier must be greater than the loss in the feedback circuit to produce oscillation.
4. The feedback must be in-phase (positive) to produce oscillation.
5. It is possible to control the frequency of an oscillator by using the appropriate RC network.
6. The resonant frequency of a lead-lag network produces maximum output voltage and a 0-degree phase angle.
7. The Wien bridge oscillator uses a lead-lag network for frequency control.
8. It is possible to make the lead-lag network tunable by using variable capacitors or variable resistors.
9. Phase-shift oscillators use three RC networks, each giving a 60-degree phase angle.
10. An LC tank circuit can be used in very high-frequency oscillator circuits.
11. A Hartley oscillator uses a tapped inductor in the tank circuit.
12. The Colpitts oscillator uses a tapped capacitive leg in the tank circuit.
13. A buffer amplifier will improve the frequency stability of an oscillator.
14. The series-tuned Colpitts, or Clapp, circuit is noted for good frequency stability.
15. A varicap diode can be added to an oscillator circuit to provide a voltage-controlled oscillator.
16. A quartz crystal can be used to control the frequency of an oscillator.
17. Crystal oscillators are more stable in frequency than LC oscillators.
18. Crystals can operate in a series mode or a parallel mode.
19. The parallel resonant frequency of a crystal is a little above the series resonant frequency.
20. Crystals have a very high Q.
21. Relaxation oscillators produce nonsinusoidal outputs.
22. Relaxation oscillators can be based on negative-resistance devices such as the UJT.
23. Relaxation oscillator frequency can be predicted by RC time constants.
24. The intrinsic standoff ratio of a UJT will affect the frequency of oscillation.
25. The intrinsic standoff ratio of a programmable UJT can be set by the use of external resistors.
26. The astable multivibrator produces rectangular waves.
27. A nonsymmetrical multivibrator is produced by using different RC time constants for each base circuit.
28. Feedback amplifiers can use frequency compensation to achieve stability.
29. Feedback signals can develop across the internal impedance of the power supply.
30. An RC network or a bypass capacitor is used to prevent feedback on power-supply lines.
31. High-frequency circuits often must be shielded to prevent feedback.
32. Defective oscillator symptoms include no output, reduced amplitude, instability, and frequency error.
33. Test instruments can load an oscillator circuit and cause errors.
34. Unstable circuits can be checked with vibration, heat, or cold.
35. A direct digital synthesizer produces a large number of precise frequencies.
36. Direct digital synthesizers are sometimes called numerically controlled oscillators.

Related Formulas

Lead-lag network resonant frequency:

$$f_r = \frac{1}{2\pi RC}$$

For phase-shift oscillators,

high-pass frequency:

$$f = \frac{1}{15.39RC}$$

Low-pass frequency:

$$f = \frac{0.39}{RC}$$

Twin-T network resonant frequency:

$$f = \frac{1}{2\pi RC}$$

LC tank circuit resonant frequency:

$$f_r = \frac{1}{2\pi\sqrt{LC}}$$

Equivalent series capacitance:

$$C_T = \frac{C_1 \times C_2}{C_1 + C_2} = \frac{1}{1/C_1 + 1/C_2 + 1/C_n}$$

Approximate PUT oscillator frequency:

$$f \approx \frac{1}{RC}$$

RC time constant: $T = RC$

Multivibrator time constant: $T = 0.69\,RC$

Frequency (relationship to period): $f = \frac{1}{T}$

Duty cycle (rectangular wave):

$$\text{Duty cycle} = \frac{t_{high}}{t_{high} + t_{low}} \times 100\%$$

Schmitt trigger oscillator: $X = \dfrac{R_3}{R_2 + R_3}$ and

$$f = \frac{1}{2R_1 C \times \ln\dfrac{1+X}{1-X}}$$

Direct digital synthesizer output:

$$f_{out} = \frac{f_{clock} \times \Delta\phi}{2^N}$$

Direct digital synthesizer resolution:

$$f_{resolution} = \frac{f_{clock}}{2^N}$$

Chapter Review Questions

Choose the letter that best answers each question.

11-1. When will an amplifier oscillate? (11-1)
   a. There is feedback from output to input.
   b. The feedback is in-phase (positive).
   c. The gain is greater than the loss.
   d. All of the above are true.

11-2. You want to build a common-emitter oscillator that operates at frequency f. The feedback circuit will be required to provide (11-2)
   a. A 180-degree phase shift at f
   b. A 0-degree phase shift at f
   c. A 90-degree phase shift at f
   d. Band-stop action for f

11-3. In Fig. 11-4, $R = 3,300\ \Omega$ and $C = 0.1\ \mu$F. What is f_r? (11-2)
   a. 48 Hz
   b. 120 Hz

   c. 482 Hz
   d. 914 Hz

11-4. Examine Fig. 11-4. Assume the signal source develops a frequency above f_r. What is the phase relationship of V_{out} to the source? (11-2)
   a. Positive (leading)
   b. Negative (lagging)
   c. In-phase (0 degrees)
   d. None of the above

11-5. In Fig. 11-6, $R = 8,200\ \Omega$ and $C = 0.05\ \mu$F. What is the frequency of oscillation? (11-2)
   a. 39 Hz
   b. 60 Hz
   c. 194 Hz
   d. 388 Hz

11-6. Refer to Fig. 11-6. What is the function of R'? (11-2)
  a. It provides the required phase shift.
  b. It prevents clipping and distortion.
  c. It controls the frequency of oscillation.
  d. None of the above are true.

11-7. In Fig. 11-9, assume that $R_B = 470$ kΩ and the voltage gain of the amplifier is 90. What is the actual loading effect of R_B to a signal arriving at the base? (11-2)
  a. 5,222 Ω
  b. 8,333 Ω
  c. 1 MΩ
  d. Infinite

11-8. Refer to Fig. 11-9. Assume the phase-shift capacitors are changed to 0.1 μF. What is the frequency of oscillation? (11-2)
  a. 10 Hz
  b. 40 Hz
  c. 75 Hz
  d. 71 Hz

11-9. Refer to Fig. 11-9. How many frequencies will produce exactly the phase response needed for the circuit to oscillate? (11-2)
  a. One
  b. Two
  c. Three
  d. An infinite number

11-10. Refer to Fig. 11-16. What is the major effect of C_E? (11-3)
  a. It increases the frequency of oscillation.
  b. It decreases the frequency of oscillation.
  c. It makes the transistor operate common base.
  d. It increases voltage gain.

11-11. Refer to Fig. 11-16. What would happen if C_2 were increased in capacity? (11-3)
  a. The frequency of oscillation would increase.
  b. The frequency of oscillation would decrease.
  c. The inductance of L_A and L_B would change.
  d. Not possible to determine.

11-12. In Fig. 11-17, $L = 1.8$ μH, $C_2 = 270$ pF, and $C_3 = 33$ pF. What is the frequency of oscillation? (11-3)
  a. 11 MHz
  b. 22 MHz
  c. 33 MHz
  d. 41 MHz

11-13. Refer to Fig. 11-17. What is the purpose of C_1? (11-3)
  a. It bypasses power-supply noise to ground.
  b. It determines the frequency of oscillation.
  c. It provides an ac ground for the base.
  d. It filters V_{out}.

11-14. Refer to Fig. 11-18. What is the configuration of Q_1? (11-3)
  a. Common source
  b. Common gate
  c. Common drain
  d. Drain follower

11-15. Crystal-controlled oscillators, as compared with LC-controlled oscillators, are generally (11-4)
  a. Less expensive
  b. Capable of a better output power
  c. Superior for VFO designs
  d. Superior for frequency stability

11-16. Why can the circuit in Fig. 11-21 not be used for overtone operation? (11-4)
  a. The common-emitter configuration is used.
  b. Trimmer C_3 makes it impossible.
  c. The feedback is wrong.
  d. There is no LC circuit to select the overtone.

11-17. The Q of a crystal, as compared to the Q of an LC tuned circuit, will be (11-4)
  a. Much higher
  b. About the same
  c. Lower
  d. Impossible to determine

11-18. In Fig. 11-24, $R_4 = 47$ kΩ and $C = 10$ μF. What is the approximate frequency of oscillation? (11-5)
  a. 0.21 Hz
  b. 2.13 Hz
  c. 200 Hz
  d. 382 Hz

11-19. In Fig. 11-27, $R_1 = R_2 = 10,000$ Ω, $C_1 = 0.5$ μF, and $C_2 = 0.02$ μF. What is the frequency of oscillation? (11-5)
  a. 112 Hz
  b. 279 Hz
  c. 312 Hz
  d. 989 Hz

11-20. In question 11-21, what will the rectangular output waveform show? (11-5)
  a. Symmetry
  b. Nonsymmetry
  c. Poor rise time
  d. None of the above

11-21. Refer to Fig. 11-30. How may the stability of such a circuit be ensured? (11-6)
  a. Operate each stage at maximum gain.
  b. Decrease losses in the feedback circuit.
  c. Use more stages.
  d. Compensate the circuit so the gain is low for those frequencies that give a critical phase error.

11-22. Refer to Fig. 11-32. How may signal coupling across Z_G be reduced? (11-6)
  a. By not allowing stages to share a ground path
  b. By careful circuit layout
  c. By using low-loss grounds
  d. All of the above

11-23. What is the purpose of neutralization? (11-6)
  a. To ensure oscillations
  b. To stabilize an amplifier
  c. To decrease amplifier output
  d. To prevent amplifier overload

11-24. Another name for a direct digital synthesizer is (11-8)
  a. Packaged oscillator
  b. Phase-locked loop
  c. Numerically controlled oscillator
  d. All of the above

11-25. A DDS is programmed to the desired frequency by binary information sent to its (11-8)
  a. Phase accumulator
  b. Clock circuit
  c. A/D converter
  d. D/A converter

11-26. In a DDS chip, changing the phase increment to a smaller value will (11-8)
  a. Lower the output frequency
  b. Raise the output frequency
  c. Lower the clock frequency
  d. Raise the clock frequency

11-27. What is the output frequency from a DDS if its clock frequency is 50 MHz, the size of its phase accumulator is 32 bits, and the tuning word is 56,194,128? (11-8)
  a. 1.111 MHz
  b. 894.5 kHz
  c. 756.1 kHz
  d. 654.2 kHz

11-28. What is the step size for the DDS described in question 11-27?
  a. 11.64 mHz
  b. 21.21 mHz
  c. 56.29 mHz
  d. 1.005 Hz

11-29. A DDS has a small frequency output error at any programmed frequency. What could the problem be? (11-9)
  a. The low-pass filter or D/A converter is damaged.
  b. The clock frequency is out of tolerance.
  c. The controlling microprocessor has a blown data bus.
  d. Any of the above might be the problem.

Critical Thinking Questions

11-1. Are there any other ways to keep a PA system from oscillating aside from turning down the volume?

11-2. Can digital computer technology replace oscillators?

11-3. How could an oscillator be used as a metal detector?

11-4. Almost all timekeeping instruments use some form of an oscillator. Can you think of any that do not?

11-5. Quartz is not the only piezoelectric material. Does this fact suggest anything to you?

11-6. Can you name any electronic products that are oscillators but are called something else?

11-7. What is the most powerful electronic oscillator commonly found in homes and apartments?

1. D	16. B	31. C	46. A
2. C	17. D	32. D	47. B
3. D	18. A	33. D	48. crystal-controlled oscillator, PLL, DDS
4. B	19. C	34. C	
5. A	20. A	35. C	
6. B	21. A	36. B	49. numerically controlled oscillator
7. A	22. C	37. C	
8. C	23. C	38. A	50. low-pass filter
9. B	24. C	39. D	51. change
10. C	25. B	40. D	52. frequency-tuning word (or phase increment value)
11. D	26. C	41. A	
12. A	27. D	42. D	
13. D	28. A	43. C	53. it increases
14. D	29. B	44. D	54. 0.1 MHz
15. B	30. D	45. D	55. all of them

CHAPTER 12

Communications

Learning Outcomes

This chapter will help you to:

12-1 *Define* modulation and demodulation. [12-1]

12-2 *List* the characteristics of AM, SSB, and FM. [12-1, 12-4]

12-3 *Explain* the operation of basic radio receivers. [12-2]

12-4 *Predict* the bandwidth of AM signals. [12-1]

12-5 *Calculate* the oscillator frequency for superheterodyne receivers. [12-3]

12-6 *Calculate* the image frequency for superheterodyne receivers. [12-3]

12-7 *Describe* wireless data systems. [12-5]

12-8 *Troubleshoot* receivers. [12-6]

12-9 *Troubleshoot* wireless data systems. [12-6]

The field of communications represents a large part of the electronics industry. This chapter introduces the basic ideas used in electronic communications. Once these basics are learned, it is easier to understand other applications such as television, two-way radio, telemetry, and digital data transmission. Modulation is the fundamental process of electronic communication. It allows voice, pictures, and other information to be transferred from one point to another. The modulation process is reversed at the receiver to recover the information. This chapter covers the basic theory and some of the circuits used in radio receivers and transmitters.

12-1 Modulation and Demodulation

Any high-frequency oscillator can be used to produce a *radio wave*. Figure 12-1 shows an oscillator that feeds its output energy to an antenna. The antenna converts the high-frequency alternating current to a radio wave.

A radio wave travels through the atmosphere or space at the speed of light (3×10^8 m/s). If a radio wave strikes another antenna, a high-frequency current will be induced that is a

Radio wave

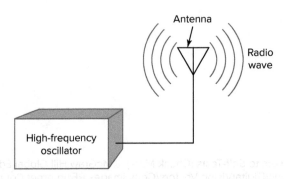

Fig. 12-1 A basic radio transmitter.

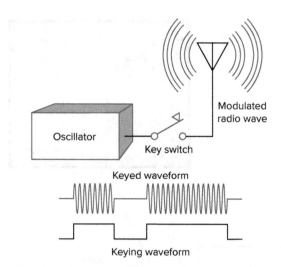

Fig. I2-2 A CW transmitter.

replica of the current flowing in the transmitting antenna. Thus, it is possible to transfer high-frequency electrical energy from one point to another without using wires. The energy in the receiving antenna is typically only a tiny fraction of the energy delivered to the transmitting antenna.

A radio wave can be used to carry information by a process called *modulation*. Figure 12-2 shows a very simple type of modulation. A key switch is used to turn the antenna current (and thus the radio wave) on and off. This is the basic scheme of *radio telegraphy*. The key switch is opened and closed according to some pattern or *code*. For example, Morse code can be used to represent numbers, letters, and punctuation. This basic modulation form is known as *interrupted continuous wave*, or CW. ICW is a form of binary signaling. The keying waveform is either high or low (1 or 0), and it can also be called amplitude shift keying (ASK). The Morse letter *A* is represented by the key waveforms in Fig. 12-2.

Continuous-wave modulation is very simple, but it has disadvantages. A code such as Morse code is difficult to learn; transmission is slower than in voice communications; and CW cannot be used for music, pictures, and other kinds of information. Today, CW is used only by some amateur radio operators.

Figure 12-3 shows *amplitude modulation*, or *AM*. In this modulation system, the intelligence or information is used to *control the amplitude* of the RF signal. Amplitude modulation overcomes the disadvantages of CW modulation.

It can be used to transmit voice, music, data, or even picture information (video). The oscilloscope display in Fig. 12-3 shows that the RF signal amplitude varies in accordance with the audio frequency (AF) signal. The RF signal could just as well be amplitude-modulated by a video signal or digital (on-off) data.

Figure 12-4 shows a typical circuit for an amplitude modulator. The audio information is coupled by T_1 to the collector circuit of the transistor. The audio voltage induced across the secondary of T_1 can either aid or oppose V_{CC}, depending on its phase at any given moment. This means that the collector supply for the transistor is not constant. It varies with the audio input. This is how the amplitude control is achieved.

As an example, suppose that V_{CC} in Fig. 12-4 is 12 V and that the induced audio signal across the secondary of T_1 is 24 V peak-to-peak. When the audio peaks negative at the top of the secondary winding, 12 V will be added to V_{CC} and the transistor will see 24 V. When the audio peaks positive at the top of the secondary winding, 12 V will be subtracted from V_{CC} and the transistor will see 0 V.

Transformer T_2 and capacitor C_2 in Fig. 12-4 form a resonant *tank circuit*. The resonant frequency will match the RF input. Capacitor C_1 and resistor R_1 form the input circuit for the transistor. Reverse bias is developed by the base-emitter junction, and the amplifier operates in class C. Signal bias was covered in Chap. 8.

An amplitude-modulated signal consists of several frequencies. Suppose that the signal from a 500-kHz oscillator is modulated by a 3-kHz audio tone. Three frequencies will be present at the output of the modulator. The original RF oscillator signal, called the *carrier*, is shown at 500 kHz on the frequency axis in Fig. 12-5. Also note that an *upper sideband* (USB) appears at 503 kHz, and a *lower sideband* (LSB) appears at 497 kHz. An AM signal consists of a carrier plus two sidebands.

Figure 12-5 is the type of display shown on a *spectrum analyzer*. A spectrum analyzer uses a cathode-ray-tube or a liquid crystal display similar to an oscilloscope. The difference is that the spectrum analyzer draws a graph of amplitude versus *frequency*. An oscilloscope draws a graph of amplitude versus *time*. Spectrum analyzers display the *frequency domain*, while

Modulation

Tank circuit

Radio
telegraphy

Carrier

Spectrum
analyzer

Amplitude
modulation
(AM)

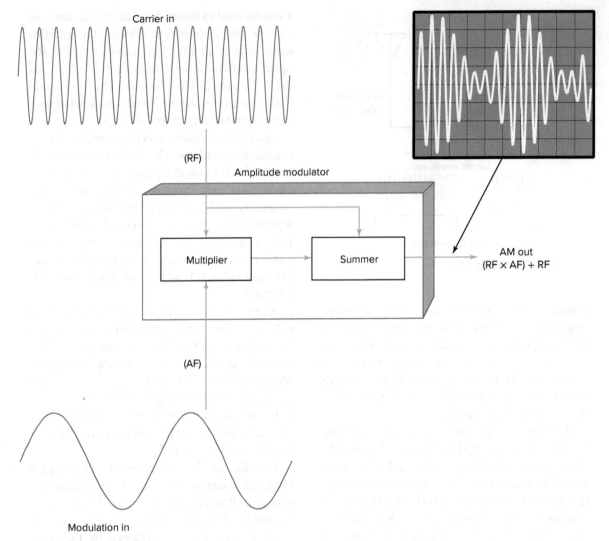

Carrier in

(RF)

Amplitude modulator

Multiplier

Summer

AM out
(RF × AF) + RF

(AF)

Modulation in

Fig. 12-3 Amplitude modulation.

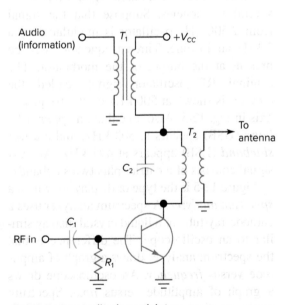

Audio
(information)

T_1

$+V_{CC}$

T_2

To
antenna

C_2

C_1

RF in

R_1

Fig. 12-4 An amplitude modulator.

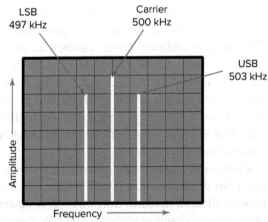

LSB
497 kHz

Carrier
500 kHz

USB
503 kHz

Amplitude

Frequency

Fig. 12-5 AM on a spectrum analyzer.

oscilloscopes display the *time domain*. Figure 12-6 shows how an AM signal looks on an oscilloscope. In Fig. 12-6(a) the carrier frequency is relatively low, so the individual cycles can be seen. In practice, the carrier frequency is relatively high and the individual cycles cannot

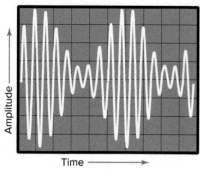

(a) When the carrier frequency is relatively low

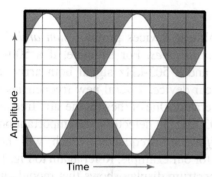

(b) When the carrier frequency is relatively high

Fig. 12-6 AM waveforms on an oscilloscope.

be seen [Fig. 12-6(b)]. Spectrum analyzers are generally more costly than oscilloscopes. They are very useful instruments for evaluating the frequency content of signals.

Since AM signals have sidebands, they must also have *bandwidth*. An amplitude-modulated signal will occupy a given portion of the available spectrum of frequencies. The sidebands appear above and below the carrier, according to the frequency of the modulating information. If someone whistles at 1 kHz into the microphone of an AM transmitter, an upper sideband will appear 1 kHz above the carrier frequency, and a lower sideband will appear 1 kHz below the carrier frequency. The bandwidth of an AM signal is *twice* the modulating frequency. For example, frequencies up to about 3.5 kHz are required for speech. An AM voice transmitter will have a minimum bandwidth of 7 kHz (2 × 3.5 kHz).

EXAMPLE 12-1

What is the bandwidth of the AM signal shown in Fig. 12-5? It can be found by subtracting:

$$BW = f_{high} - f_{low}$$
$$= 503 \text{ kHz} - 497 \text{ kHz} = 6 \text{ kHz}$$

Bandwidth is important because it limits the number of stations that can use a range of frequencies without interference. For example, the standard AM broadcast band has its lowest channel assigned to a carrier frequency of 540 kHz and its highest channel assigned to a carrier frequency of 1,600 kHz. The channels are spaced 10 kHz apart, and this allows

$$\text{No. of channels} = \frac{1,600 \text{ kHz} - 540 \text{ kHz}}{10 \text{ kHz}} + 1$$
$$= 107 \text{ channels}$$

However, each station may modulate with audio frequencies up to 15 kHz, so the total bandwidth required for one station is twice this frequency, or 30 kHz. With 107 stations on the air, the total bandwidth required would be 107 × 30 kHz = 3,210 kHz. This far exceeds the width of the AM broadcast band.

One solution would be to limit the maximum audio frequency to 5 kHz. This would allow the 107 stations to fit the AM band. This is not an acceptable solution since 5 kHz is not adequate for reproduction of quality music. A better solution is to assign channels on the basis of geographical area. The Federal Communications Commission (*FCC*) assigns carrier frequencies that are spaced at least three channels apart in any given geographical region. The three-channel spacing separates the carriers by 30 kHz and prevents the upper sideband from a lower channel from spilling into the lower sideband of the channel above it.

An AM radio receiver must recover the information from the modulated signal. This process reverses what happened in the modulator section of the transmitter. It is called *demodulation* or *detection*.

The most common AM detector is a diode (Fig. 12-7). The modulated signal is applied across the primary of T_1. Transformer T_1 is tuned by capacitor C_1 to the carrier frequency. The passband of the tuned circuit is wide enough to pass the carrier and both sidebands. The diode *detects* the signal and recovers the original information used to modulate the carrier at the transmitter. Capacitor C_2 is a low-pass filter. It removes the carrier and sideband frequencies since they are no longer needed. Resistor R_L serves as the load for the information signal.

A diode makes a good detector because it is a *nonlinear* device. All nonlinear devices can

Bandwidth

Demodulation

Detection

Nonlinear

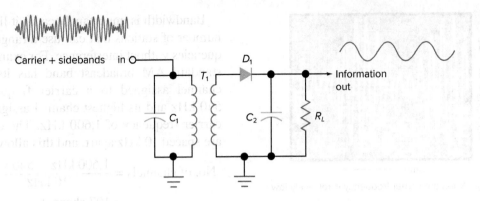

Fig. 12-7 An AM detector.

be used to detect AM. Figure 12-8 is a volt-ampere characteristic curve of a solid-state diode. It shows that a diode will make a good detector and a resistor will not.

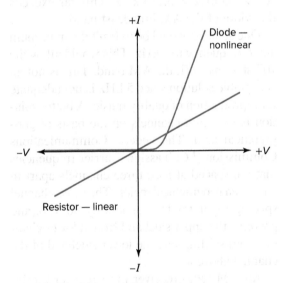

Fig. 12-8 Diodes are nonlinear devices.

Nonlinear devices produce *sum and difference frequencies*. For example, if a 500-kHz signal and a 503-kHz signal both arrive at a nonlinear device, several new frequencies will be generated. One of these is the sum frequency at 1,003 kHz. In detectors, the important one is the *difference* frequency, which will be at 3 kHz for our example. Refer again to Fig. 12-5. The spectrum display shows that modulating a 500-kHz signal with a 3-kHz signal produces an upper sideband at 503 kHz. When this signal is detected, the modulation process is reversed, and the original 3-kHz signal is recovered.

The lower sideband will also interact with the carrier. It, too, will produce a difference frequency of 3 kHz (500 kHz − 497 kHz = 3 kHz). The two 3-kHz difference signals add in phase in the detector. Thus, in an AM detector, both sidebands interact with the carrier and reproduce the original information frequencies.

A transistor can also serve as an AM detector (Fig. 12-9). The circuit shown is a

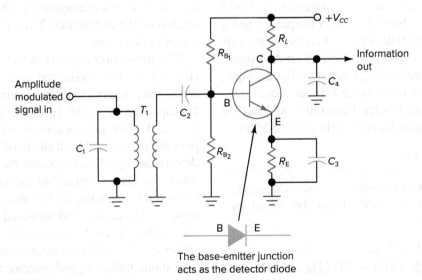

The base-emitter junction
acts as the detector diode

Fig. 12-9 A transistor detector.

Communications Technology
Digital communication systems might use a combination of phase and amplitude modulation. The global positioning satellite (GPS) system requires highly accurate time signals from an atomic clock.

Andrey Armyagov/Shutterstock

common-emitter amplifier. Transformer T_1 and capacitor C_1 form a resonant circuit to pass the modulated signal (carrier plus sidebands). Capacitor C_4 is added to give a low-pass filter action, since the high-frequency carrier and the sidebands are no longer needed after detection.

BJTs can demodulate signals because they are also nonlinear devices. The base-emitter junction is a diode. The transistor detector has the advantage of producing gain. This means that the circuit in Fig. 12-9 will produce more information amplitude than the simple diode detector in Fig. 12-7. Both circuits are useful for detecting AM signals.

The modulation-demodulation process is the basis of all electronic communication. It allows high-frequency carriers to be placed at different frequencies in the RF spectrum. By spacing the carriers, interference can be controlled. The use of different frequencies also allows different communication distances to be covered. Some frequencies lend themselves to short-range work, and others are better for long-range communications.

Self-Test

Choose the letter that best completes each statement.

1. A circuit used to place information on a radio signal is called
   a. An oscillator   c. An antenna
   b. A detector   d. A modulator

2. A CW transmitter sends information by
   a. Varying the frequency of the audio signal
   b. Interrupting the radio signal
   c. Using a microphone
   d. Using a camera

3. Refer to Fig. 12-4. The voltage at the top of C_2 will
   a. Always be equal to V_{CC}
   b. Vary with the information signal
   c. Be controlled by the transistor
   d. Be a constant 0 V with respect to ground
4. Refer to Fig. 12-4. Capacitor C_2 will resonate the primary of T_2
   a. At the radio frequency
   b. At the audio frequency
   c. At all frequencies
   d. None of the above
5. A 2-MHz radio signal is amplitude-modulated by an 8-kHz sine wave. The frequency of the lower sideband is
   a. 2.004 MHz   c. 1.996 MHz
   b. 2.000 MHz   d. 1.992 MHz
6. A 1.2-MHz radio transmitter is to be amplitude modulated by audio frequencies up to 9 kHz. The bandwidth required for the signal is
   a. 9 kHz   c. 27 kHz
   b. 18 kHz   d. 1.2 MHz

7. The electronic instrument used to show both the carrier and the sidebands of a modulated signal in the frequency domain is the
   a. Spectrum analyzer
   b. Oscilloscope
   c. Digital counter
   d. Frequency meter
8. Refer to Fig. 12-7. The carrier input is 1.5000 MHz, the USB input is 1.5025 MHz, and the LSB is 1.4975 MHz. The frequency of the detected output is
   a. 1.5 MHz
   b. 5.0 kHz
   c. 2.5 kHz
   d. 0.5 kHz
9. Diodes make good AM detectors because
   a. They rectify the carrier
   b. They rectify the upper sideband
   c. They rectify the lower sideband
   d. They are nonlinear and produce difference frequencies

12-2 Simple Receivers

Figure 12-10 shows the most basic form of an AM radio receiver. An antenna is necessary to intercept the radio signal and change it back into an electric signal. The diode detector mixes the sidebands with the carrier and produces the audio information. The headphones convert the audio signal into sound. The ground completes the circuit and allows the currents to flow.

Obviously, a receiver as simple as the one shown in Fig. 12-10 must have shortcomings. Such receivers do work but are not practical. They cannot receive weak signals (they have poor *sensitivity*). They cannot separate one carrier frequency from another (they have no *selectivity*). They are inconvenient because they require a long antenna, an earth ground, and headphones.

Sensitivity

Selectivity

Before we leave the simple circuit in Fig. 12-10, one thing should be mentioned. You have, no doubt, become used to the idea that electronic circuits require some sort of power supply. This is still the case. A radio

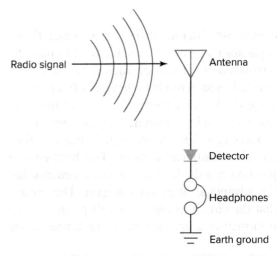

Fig. 12-10 A simple radio receiver.

signal is a wave of pure energy. Thus, the signal is the source of energy for this simple circuit.

The problem of poor sensitivity can be overcome with gain. We can add some amplifiers to the receiver to make weak signals detectable. Of course, the amplifiers will have to be

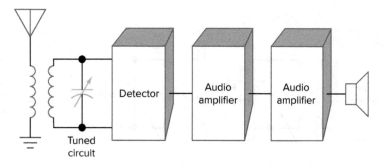

Fig. 12-11 An improved radio receiver.

energized. A power supply, other than the weak signal itself, will be required. As the gain is increased, the need for a long antenna is decreased. A small antenna is not as efficient, but gain can overcome this deficiency. Gain can also do away with the need for the headphones. Audio amplification after the detector can make it possible to drive a loudspeaker. This makes the receiver much more convenient to use.

What about the lack of selectivity? Radio stations operate at different frequencies in any given location. This makes it possible to use band-pass filters to select one out of the many that are transmitting. The resonant point of the filter may be adjusted to agree with the desired station frequency.

Figure 12-11 shows a receiver that overcomes some of the problems of the simple receiver. A two-stage audio amplifier has been added to allow loudspeaker operation. A *tuned circuit*

has been added to allow selection of one station at a time. This receiver will perform better.

The circuit in Fig. 12-11 is an improvement, but it is still not practical for most applications. One tuned circuit will not give enough selectivity. For example, if there is a very strong station in the area, it will not be possible to reject it. The strong station will be heard at all settings of the variable capacitor.

Selectivity can be improved by using more tuned circuits. Figure 12-12 compares the selectivity curves for one, two, and three tuned circuits. Note that more tuned circuits give a sharper curve (less bandwidth). This improves the ability to reject unwanted frequencies. Figure 12-12 also shows that bandwidth is measured 3 dB down from the point of maximum gain. An AM receiver should have a bandwidth just wide enough to pass the carrier and both sidebands. A bandwidth of about

Tuned circuit

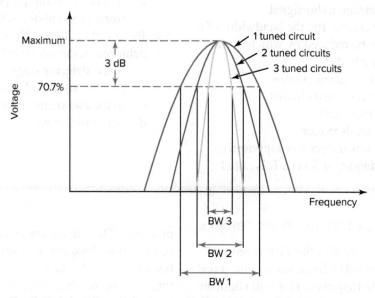

Fig. 12-12 Selectivity can be improved with more tuned circuits.

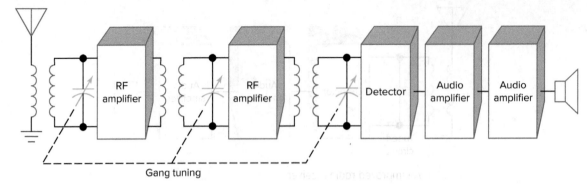

Fig. 12-13 A TRF receiver.

20 kHz is typical in an ordinary AM broadcast receiver. Too much bandwidth means poor selectivity and possible interference. Too little bandwidth means loss of transmitted information (with high-frequency audio affected the most).

A tuned radio-frequency *(TRF)* receiver can provide reasonably good selectivity and sensitivity (Fig. 12-13). Four amplifiers—two at radio frequencies and two at audio frequencies—give the required gain.

The TRF receiver has some disadvantages. Note in Fig. 12-13 that all three tuned circuits are gang-tuned. In practice, it is difficult to achieve perfect tracking. *Tracking* refers to how closely the resonant points will be matched for all settings of the tuning control. A second problem is in bandwidth. The tuned circuits will not have the same bandwidth for all frequencies. Both of these disadvantages have been eliminated in the superheterodyne receiver design that is discussed in the next section.

Tracking

Self-Test

Choose the letter that best answers each question.

10. What energizes the radio receiver shown in Fig. 12-10?
    a. The earth ground
    b. The diode
    c. The headphones
    d. The incoming radio signal
11. To improve selectivity, the bandwidth of a receiver can be reduced by which of the following methods?
    a. Using more tuned circuits
    b. Using fewer tuned circuits
    c. Adding more gain
    d. Using a loudspeaker
12. A 250-kHz tuned circuit is supposed to have a bandwidth of 5 kHz. It is noted

that the gain of the circuit is maximum at 250 kHz and drops about 30 percent (3 dB) at 252.5 kHz and at 247.5 kHz. What can be concluded about the tuned circuit?
    a. It is not as selective as it should be.
    b. It is more selective than it should be.
    c. It is not working properly.
    d. None of the above is true.
13. Refer to Fig. 12-13. How is selectivity achieved in this receiver?
    a. In the detector stage
    b. In the gang-tuned circuits
    c. In the audio amplifier
    d. All of the above

12-3 Superheterodyne Receivers

Intermediate frequency

The major difficulties with the TRF receiver design can be eliminated by fixing some of the tuned circuits to a single frequency. This will eliminate the tracking problem and the changing-bandwidth

problem. This fixed frequency is called the *Intermediate Frequency*, or simply IF. It must lie outside the band to be received. Then, any signal that is to be received must be converted to the intermediate frequency. The conversion process

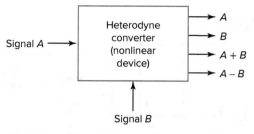

Fig. 12-14 Operation of a heterodyne converter.

is called *mixing* or *heterodyning*. A *superheterodyne receiver* converts the received frequency to the intermediate frequency.

Figure 12-14 shows the basic operation of a heterodyne converter. Here, the letters *A* and *B* represent frequencies. When signals at two different frequencies are applied, new frequencies are produced. The output of the converter contains the *sum and difference frequencies* in addition to the original frequencies. Any nonlinear device (such as a diode) can be used to heterodyne or mix two signals. The process is the same as AM detection. However, the *purpose* is different. Detection is the proper term to use when information is being recovered from a signal. The terms *heterodyning* or *mixing* are used when a signal is being converted to another frequency, such as an intermediate frequency.

Most superheterodyne circuits use a transistor mixer rather than a diode. This is because the transistor provides gain. In some cases, it can also supply one of the two signals needed for mixing.

Figure 12-15 shows a block diagram of a superheterodyne receiver. An *oscillator* provides a signal to mix with signals coming from the antenna. The mixer output contains sum and difference frequencies. If any of the signals present at the mixer output is at or very near the intermediate frequency, then that signal will reach the detector. All other frequencies will be rejected because of the selectivity of the IF amplifiers. The standard intermediate frequencies are

1. *Amplitude modulation broadcast band:* 455 kHz (or 262 kHz for some automotive receivers)
2. *Frequency modulation broadcast band:* 10.7 MHz
3. *Television broadcast band:* 44 MHz (analog TV receivers)

Shortwave and communication receivers may use various intermediate frequencies, for example, 455 kHz, 1.6 MHz, 3.35 MHz, 9 MHz, 10.7 MHz, 40 MHz, and others.

The oscillator in a superheterodyne receiver is usually set to run above the received frequency by an amount equal to the IF. To receive a station at 1,020 kHz on a standard AM broadcast receiver, for example,

$$\text{Oscillator frequency} = 1{,}020 + 455 \text{ kHz}$$
$$= 1{,}475 \text{ kHz}$$

The oscillator signal at 1,475 kHz and the station signal at 1,020 kHz will mix to produce sum and difference frequencies. The difference signal will be in the IF passband (those frequencies that the IF will allow to pass through) and will reach the detector. Another station

Mixing or heterodyning

Sum and difference frequencies

Oscillator

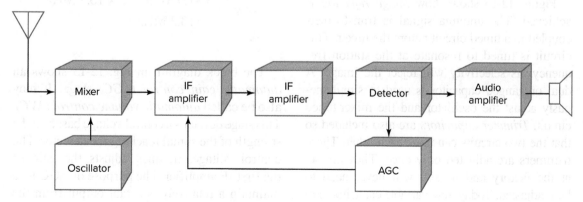

Fig. 12-15 Block diagram of a superheterodyne receiver.

operating at 970 kHz can be rejected by this process. Its difference frequency will be

$$1{,}475 \text{ kHz} - 970 \text{ kHz} = 505 \text{ kHz}$$

Since 505 kHz is not in the passband of the 455-kHz IF stages, the station transmitting at 970 kHz is rejected.

EXAMPLE 12-2

A communication receiver has an IF of 9 MHz. What is the frequency of its oscillator when it is tuned to 15 MHz? Since the oscillator usually runs above the IF,

$$f_{osc} = f_{receive} + \text{IF} = 15 \text{ MHz} + 9 \text{ MHz}$$
$$= 24 \text{ MHz}$$

It should be clear that adjacent channels are rejected by the selectivity in the IF stages. However, there is a possibility of interference from a signal not even in the broadcast band. To receive 1,020 kHz, the oscillator in the receiver must be adjusted 455 kHz higher. What will happen if a shortwave signal at a frequency of 1,930 kHz reaches the antenna? Remember, the oscillator is at 1,475 kHz. Subtraction shows that

$$1{,}930 \text{ kHz} - 1{,}475 \text{ kHz} = 455 \text{ kHz}$$

This means that the shortwave signal at 1,930 kHz will mix with the oscillator signal and reach the detector. This is called *image interference*.

Image interference

The only way to reject image interference is to use selective circuits *before* the mixer. In any superheterodyne receiver, there are always two frequencies that can mix with the oscillator frequency and produce the intermediate frequency. One is the desired frequency, and the other is the image frequency. The image must not be allowed to reach the mixer.

Image rejection

Figure 12-16 shows how *image rejection* is achieved. The antenna signal is transformer-coupled to a tuned circuit before the mixer. This circuit is tuned to resonate at the station frequency. Its selectivity will reject the image. A dual or ganged capacitor is used to simultaneously adjust the oscillator and the mixer-tuned circuit. *Trimmer capacitors* are also included so that the two circuits can track each other. These trimmers are adjusted only once. They are set at the factory and usually will never need to be readjusted. Today, few variable capacitors are used for tuning due to their size and cost.

Automatic gain control (AGC)

Automatic volume control (AVC)

Trimmer capacitor

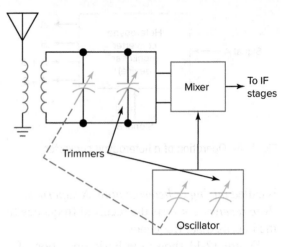

Fig. 12-16 A tuned circuit before the mixer rejects the image.

The mixer-tuned circuit is not highly selective. Its purpose has nothing to do with the adjacent channel selectivity of the receiver. This selectivity is provided in the IF stages. The purpose of the mixer-tuned circuit is to reject the image frequency that is twice the intermediate frequency above the desired station frequency. Thus, the desired station and the image are separated by 910 kHz (2 × 455 kHz) in a standard AM broadcast receiver. A signal this far away from the desired one is easier to reject, so one tuned circuit before the mixer is often all that is needed.

EXAMPLE 12-3

An FM receiver is tuned to receive a station at 91.9 MHz. Find the image frequency. We can assume an IF of 10.7 MHz and that the local oscillator runs above the desired frequency. The image frequency is found by

$$f_{image} = f_{station} + 2 \times \text{IF}$$
$$= 91.9 \text{ MHz} + 2 \times 10.7 \text{ MHz}$$
$$= 113.3 \text{ MHz}$$

The block diagram in Fig. 12-15 shows an *automatic gain control (AGC)* stage. It may also be called *automatic volume control (AVC)*. This stage develops a control voltage based on the strength of the signal reaching the detector. The control voltage, in turn, adjusts the gain of the first IF amplifier. The purpose of AGC is to maintain a relatively constant output from the receiver. Signal strengths can vary quite a bit as the receiver is tuned across the band. The AGC action keeps the volume from the speaker reasonably constant.

Automatic gain control can be applied to more than one IF amplifier. It can also be applied to an RF amplifier before the mixer, if a receiver has one. The control voltage is used to vary the gain of the amplifying device. If the device is a BJT, two options exist. The graph in Fig. 12-17 shows that maximum gain occurs at one value of collector current. If the bias is increased and current increases, the gain tends to drop. This is called *forward* AGC. The bias can be reduced, the current decreases, and so does the gain. This is called reverse AGC. Both types of AGC are used with bipolar transistors.

Different transistors vary quite a bit in their AGC bias characteristics. This is an important consideration when replacing an RF or IF transistor in a receiver. If AGC is applied to that stage, an exact replacement is highly desirable. A substitute transistor may cause poor AGC performance, and the receiver performance can be seriously degraded.

Dual-gate MOSFETs are often used when AGC is desired. These transistors have excellent AGC characteristics. The control voltage is usually applied to the second gate. You may wish to refer to Sec. 7-3 in Chap. 7 on field-effect transistor amplifiers.

Integrated circuits with excellent AGC characteristics are also available. These are widely applied in receiver design, especially as IF amplifiers.

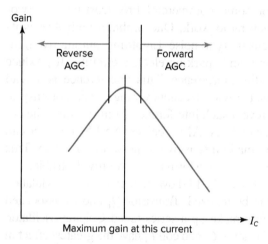

Fig. I2-I7 The AGC characteristics of a transistor.

Self-Test

Choose the letter that best answers each question.

14. You want to receive a station at 1,160 kHz on a standard AM receiver. What must the frequency of the local oscillator be?
    a. 455 kHz   c. 1,615 kHz
    b. 590 kHz   d. 2,000 kHz
15. A standard AM receiver is tuned to 1,420 kHz. Interference is heard from a shortwave transmitter operating at 2,330 kHz. What is the problem?
    a. Poor image rejection
    b. Poor AGC action
    c. Inadequate IF selectivity
    d. Poor sensitivity

16. Which of the following statements about the oscillator in a standard AM receiver is true?
    a. It is fixed at 455 kHz.
    b. It oscillates 455 kHz above the dial setting.
    c. It is controlled by the AGC circuit.
    d. It oscillates at the dial frequency.
17. Refer to Fig. 12-15. The receiver is properly tuned to a station at 1,020 kHz that is modulated by a 1-kHz audio test signal. What frequency or frequencies are present at the input of the detector stage?
    a. 1 kHz
    b. 454, 455, and 456 kHz
    c. 1,020 kHz
    d. 1,020 and 1,475 kHz

18. In question 17 what frequency or frequencies are present at the input of the audio amplifier?
    a. 1 kHz
    b. 454, 455, and 456 kHz
    c. 1,020 kHz
    d. 1,020 and 1,475 kHz

19. It is noted that a receiver uses an NPN transistor in the first IF stage. Tuning a strong station causes the base voltage to become more positive. The stage
    a. Is defective
    b. Uses reverse AGC
    c. Uses forward AGC
    d. Is not AGC-controlled

12-4 Other Modulation Types

Frequency
modulation
(FM)

Frequency modulation, or *FM*, is an alternative to amplitude modulation. Frequency modulation has some advantages that make it attractive for some commercial broadcasting and two-way radio work. One problem with AM is its sensitivity to noise. Lightning, automotive ignition, and sparking electric circuits all produce radio interference. This interference is spread over a wide frequency range. It is not easy to prevent such interference from reaching the detector in an AM receiver. An FM receiver can be made insensitive to noise interference. This noise-free performance is highly desirable.

Figure 12-18 shows how frequency modulation can be realized. Transistor Q_1 and its associated parts make up a series-tuned Colpitts oscillator. Capacitor C_3 and coil L_1 have the greatest effect in determining the frequency of oscillation. Diode D_1 is a varicap diode. It is connected in parallel with C_3. This means that as the capacitance of D_1 changes, so will the resonant frequency of the tank circuit. Resistors R_1 and R_2 form a voltage

divider to bias the varicap diode. Some positive voltage (a portion of V_{DD}) is applied to the cathode of D_1. Thus, D_1 is in reverse bias.

A varicap diode uses its depletion region as the dielectric. More reverse bias means a wider depletion region and less capacitance. Therefore, as an audio signal goes positive, D_1 will reduce in capacitance. This will shift the frequency of the oscillator up. A negative-going audio input will reduce the reverse bias across the diode. This will increase its capacitance and shift the oscillator to some lower frequency. The audio signal is *modulating the frequency* of the oscillator.

The relationship between the modulating waveform and the RF oscillator signal can be seen in Fig. 12-19(a). Note that the amplitude of the modulated RF waveform is constant. Compare this with the AM waveform in Fig. 12-6. Figure 12-19 also shows two ways to send digital or binary information. In Fig. 12-19(b) the frequency is shifted higher or lower according to the modulating waveform (FSK . . . frequency shift keying). In Fig. 12-19(c), the phase is

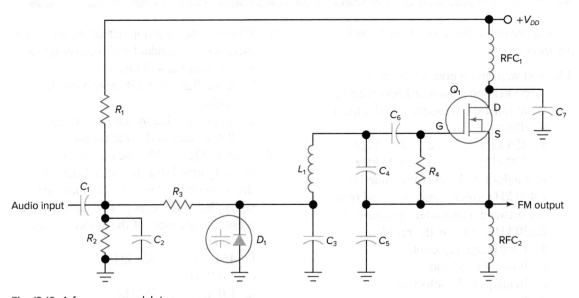

Fig. 12-18 A frequency modulator.

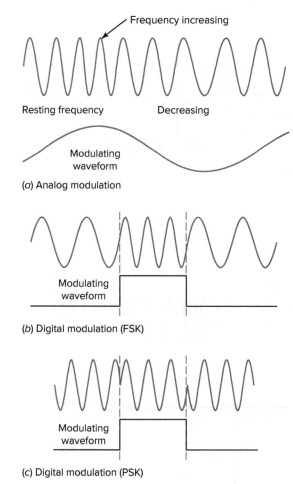

(a) Analog modulation

(b) Digital modulation (FSK)

(c) Digital modulation (PSK)

Fig. 12-19 FM and FSK waveforms.

controlled by the modulating waveform (PSK . . . phase-shift keying). Earlier, we saw that ASK can send binary information. Today, binary modulation methods are used more often than analog methods. Even the older mainstays of electronic communication, such as radio and television, are increasingly using binary modulation methods.

Amplitude modulation produces sidebands, as does FM (Fig. 12-20). Suppose an FM

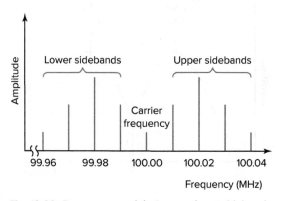

Fig. 12-20 Frequency modulation produces sidebands.

transmitter is being modulated with a steady 10-kHz (0.01-MHz) tone. This transmitter has an operating (carrier) frequency of 100 MHz. The frequency domain graph shows that *several* sidebands appear. These sidebands are spaced 10 kHz apart. They appear above and below the carrier frequency. This is one of the major differences between AM and FM. An FM signal generally requires more bandwidth than an AM signal.

The block diagram for an FM superheterodyne receiver (Fig. 12-21) is quite similar to that for the AM receiver. However, you will notice that a *limiter* stage appears after the IF stage and before the detector stage. This is one way that an FM receiver can reject noise. Figure 12-22 shows what happens in a limiter stage. The input signal is very noisy. The output signal is noise-free. By limiting or by amplitude clipping, the noise spikes have been eliminated. Some FM receivers use two stages of limiting to eliminate most noise interference.

Limiting cannot be used in an AM receiver. The amplitude variations carry the information to the detector. In an FM receiver, the frequency variations contain the information. Amplitude clipping in an FM receiver will not remove the information, just the noise.

Detection for FM is more complicated than for AM. Since FM contains several sidebands above and below the carrier, a simple nonlinear detector will not demodulate the signal. A double-tuned *discriminator* circuit is shown in Fig. 12-23. It serves as an FM detector. The discriminator works by having two resonant points. One is above the carrier frequency, and one is below the carrier frequency.

In the frequency response curves for the discriminator circuit (Fig. 12-24), f_o represents the correct point on the curves for the carrier. In a superheterodyne receiver, the station's carrier frequency will be converted to f_o. This represents a frequency of 10.7 MHz for broadcast FM receivers. The heterodyning process allows one discriminator circuit to demodulate any signal over the entire FM band.

Refer to Figs. 12-23 and 12-24. When the carrier is unmodulated, D_1 and D_2 will conduct an equal amount. This is because the circuit is operating where the frequency response curves cross. The amplitude is equal for both tuned circuits at this point. The current through R_1

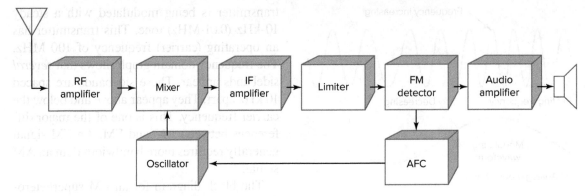

Fig. 12-21 Block diagram of an FM receiver.

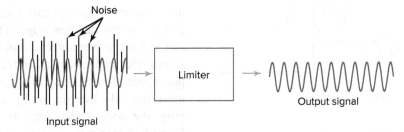

Fig. 12-22 Operation of a limiter.

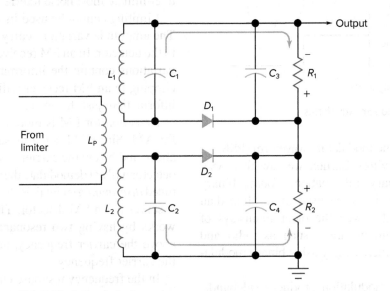

Fig. 12-23 A discriminator.

will equal the current through R_2. If R_1 and R_2 are equal in resistance, the voltage drops will also be equal. Since the two voltages are series-opposing, the output voltage will be zero. When the carrier is unmodulated, the discriminator output is zero.

Suppose the carrier shifts higher in frequency because of modulation. This will increase the amplitude of the signal in L_2C_2 and decrease the amplitude in L_1C_1. Now there will be more voltage across R_2 and less across R_1. The output goes positive.

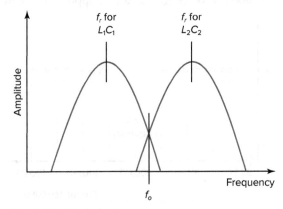

Fig. 12-24 Discriminator response curves.

What happens when the carrier shifts below f_o? This moves the signal closer to the resonant point of L_1C_1. More voltage will drop across R_1, and less will drop across R_2. The output goes negative. The output from the discriminator circuit is zero when the carrier is at rest, positive when the carrier moves higher, and negative when the carrier moves lower in frequency. The output is a function of the carrier frequency.

The output from the discriminator can also be used to correct for drift in the receiver oscillator. Note in Fig. 12-21 that the FM detector feeds a signal to the audio amplifier and to a stage marked AFC. The letters *AFC* stand for *automatic frequency control*. If the oscillator drifts, f_o will not be exactly 10.7 MHz. There will be a steady dc output voltage from the discriminator. This dc voltage can be used as a control voltage to correct the oscillator frequency and eliminate the drift. Some receivers use the discriminator output to drive a tuning meter as well. A zero-center meter shows the correct tuning point. Any tuning error will cause the meter to deflect to the left or to the right of zero.

Frequency modulation discriminator circuits work well, but they are sensitive to amplitude. This is why one or two limiters are needed for noise-free reception. The *ratio detector* provides a simplified system; it is not nearly as sensitive to the amplitude of the signal. This makes it possible to build receivers without limiters and still provide good noise rejection.

Figure 12-25 shows a typical ratio detector circuit. Its design is based on the idea of dividing a signal voltage into a ratio. This ratio is equal to the ratio of the voltages on either half of L_2. With frequency modulation, the ratio shifts, and an audio output signal is available at the center tap of L_2. Since the circuit is ratio-sensitive, the input signal amplitude may vary over a wide range without causing any change in output. This makes the detector insensitive to amplitude variations such as noise.

There are several other FM detector circuits. Some of the more popular ones are the *quadrature detector*, the *phase-locked loop detector*, and the *pulse-width detector*. PLLs are covered in Chap. 13. These circuits are likely to be used in conjunction with integrated circuits. They usually have the advantage of requiring no alignment or only one adjustment. Alignment for discriminators and ratio detectors is more time-consuming.

Automatic frequency control (AFC)

Single sideband (SSB) is another alternative to amplitude modulation. Single sideband is a subclass of AM. It is based on the idea that both sidebands in an AM signal carry the same information. Therefore, one of them can be eliminated in the transmitter with no loss of information at the receiver. The carrier can also be eliminated at the transmitter. Therefore, an SSB transmitter sends only one sideband and no carrier.

Single sideband (SSB)

Energy is saved by not sending the carrier and the other sideband. Also, the signal will occupy only half the original bandwidth. Single sideband is much more efficient than AM. It has an effective gain of 9 dB. This is equivalent to increasing the transmitter power eight times!

Ratio detector

The carrier is eliminated in an SSB transmitter by using a *balanced modulator*

Balanced modulator

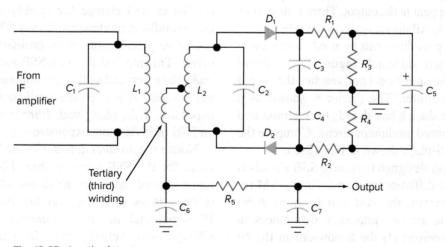

Fig. 12-25 A ratio detector.

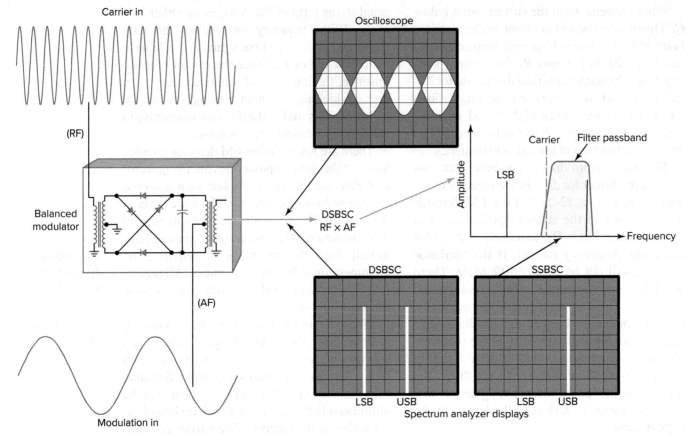

Fig. 12-26 Generation of DSBSC and SSBSC.

Double-sideband suppressed carrier (DSBSC)

Product detector

Beat-frequency oscillator (BFO)

(Fig. 12-26). The result is a *double-sideband suppressed carrier (DSBSC)* signal. Note that a balanced modulator produces only the product of the RF and AF signals. Take time to compare Fig. 12-26 to Figs. 12-3 and 12-5. Also note that the oscilloscope display in Fig. 12-26 is not the same as the one shown in Fig. 12-6(*b*).

The diodes in Fig 12-26 are connected so that no carrier can reach the output. However, when audio is applied, the circuit balance is upset and sidebands appear at the output. There is no carrier in the output. All the energy is in the sidebands.

A band-pass filter can be used to eliminate the unwanted sideband. Figure 12-26 shows that only the upper sideband reaches the output of the transmitter. The carrier is shown as a broken line since it has already been eliminated by the balanced modulator circuit. Compare the spectrum displays shown in Fig. 12-26.

A receiver designed to receive SSB signals is only a little different from an ordinary AM receiver. However, the cost can be quite a bit more. There are two important differences in the SSB receiver: (1) the bandwidth in the IF amplifier will be narrower and (2) the missing

carrier must be replaced by a second (local) oscillator so detection can occur. You should recall that the carrier is needed to mix with the sidebands (or sideband) to produce the difference (audio) frequencies.

Single sideband receivers usually achieve the narrow IF bandwidth with crystal or mechanical filters. These are more costly than inductor-capacitor filters. An SSB receiver must be very stable. Even a small drift in any of the receiver oscillators will change the quality of the received audio. A moderate drift, say 500 Hz, will not be very noticeable in an ordinary AM receiver. This much drift in an SSB receiver will make the recovered audio very unnatural sounding or unintelligible. Stable oscillators are more expensive. This, along with filter costs, makes an SSB receiver more expensive.

Notice the *product detector* in the block diagram for the SSB receiver (Fig. 12-27). This name is used since the audio output from the detector is the difference product between the IF signal and the *beat-frequency oscillator (BFO)* signal. Actually, all AM detectors are product detectors. They all use the difference

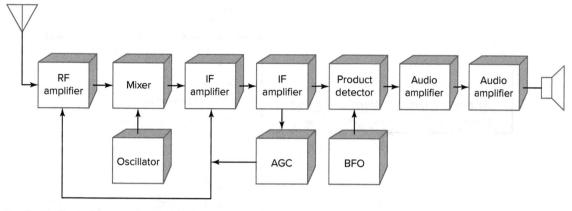

Fig. 12-27 Block diagram of an SSB receiver.

frequency product as their useful output. An ordinary diode detector can be used to demodulate an SSB signal if it is supplied with a BFO signal to replace the missing carrier. Or, the diode-balanced modulator shown earlier can be used as a balanced demodulator.

The BFO in an SSB receiver can be fixed at one frequency. In fact, it is often crystal-controlled for the best stability. A small error between the BFO frequency and the carrier frequency of the transmitted signal can be corrected by adjusting the main tuning control. The main difference between tuning an AM receiver and an SSB receiver is the need for critical tuning in the latter. Even a slight tuning

error of 50 Hz will make the received audio sound unnatural.

The critical tuning of the SSB makes it undesirable for most radio work. It is useful when maximum communication effectiveness is needed. Since it is so efficient, in terms of both power and bandwidth, it is popular in citizens band radio, in amateur radio, and for some military communications.

We will briefly examine some details about digital modulation. This is a vast subject, and only a few ideas are presented here. One possible method is multiple quadrature amplitude modulation (MQAM). Figure 12-28 shows a block diagram for 16QAM. It is so named because it can

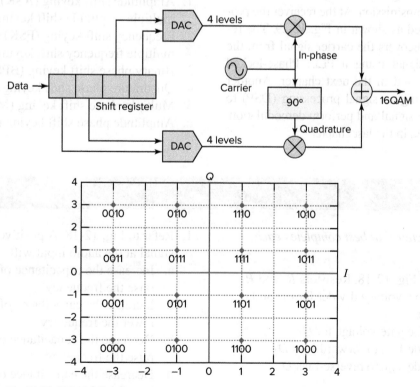

Fig. 12-28 16QAM diagrams.

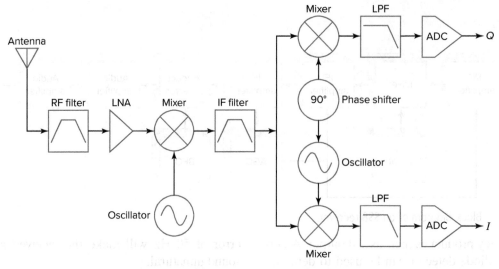

Fig. 12-29 Quadrature amplitude modulation receiver.

send 16 symbols at a time. Think of a symbol as a binary number from 0000 (decimal 0) to 1111 (decimal 15). A data shift register provides temporary storage for each symbol. Once a symbol has been sent, the next one is loaded into the shift register. The D/A converters each convert two bit inputs to four analog levels. The two resulting analog signals are sent to balanced modulators. Here, they are mixed with both an in-phase (I) and quadrature (Q) signal, which are then added together for transmission. At the receiver the process is reversed as shown in Fig. 12-29. The receiver often recovers the carrier signal from the transmitted signal using a PLL. Phase-locked loops are covered in the next chapter. Another method uses digital signal processing (DSP) to sample the IF signal and perform demodulation. DSP is covered in the last chapter.

As the bottom of Fig. 12-28 shows, the 16 possible binary combinations are mapped onto a *constellation diagram* showing all the possibilities. Each state is defined by a specific amplitude and phase. As examples, the symbol 1111 is represented as +1, 45 degrees and 0000 as –3, 225 degrees.

Some additional binary modulation methods include the following:

1. Amplitude shift keying (ASK) and multiple amplitude shift keying (MASK)
2. Frequency shift keying (FSK) and multiple frequency shift keying (MFSK)
3. Binary phase shift keying (BPSK)
4. Quadrature phase shift keying (QPSK)
5. Multiple phase shift keying (MPSK)
6. Amplitude phase shift keying (APSK)

Self-Test

Choose the letter that best completes each statement.

20. Refer to Fig. 12-18. Resistors R_1 and R_2
    a. Form a voltage divider for the audio input
    b. Set the gate voltage for Q_1
    c. Divide V_{DD} to forward-bias D_1
    d. Divide V_{DD} to reverse-bias D_1

21. Refer to Fig. 12-18. A positive-going signal at the audio input will
    a. Increase the capacitance of D_1 and raise the frequency
    b. Increase the capacitance of D_1 and lower the frequency
    c. Decrease the capacitance of D_1 and raise the frequency
    d. Decrease the capacitance of D_1 and lower the frequency

22. Frequency modulation as compared to amplitude modulation
    a. Can provide better noise rejection
    b. Requires more bandwidth
    c. Requires complex detector circuits
    d. All of the above
23. The function of the limiter stage in Fig. 12-21 is to
    a. Reduce amplitude noise
    b. Prevent overdeviation of the signal
    c. Limit the frequency response
    d. Compensate for tuning error
24. The purpose of the AFC stage in Fig. 12-21 is to
    a. Reduce noise
    b. Maintain a constant audio output (volume)
    c. Compensate for tuning error and drift
    d. Provide stereo reception
25. Refer to Figs. 12-23 and 12-24. Assume that f_o is 10.7 MHz. If the signal from the limiter is at 10.65 MHz
    a. The output voltage will be 0 V
    b. The output voltage will be negative
    c. The output voltage will be positive
    d. Resistor R_1 will conduct more current than R_2
26. Refer to question 25. If the signal from the limiter is at 10.7 MHz, then
    a. Diode D_1 will conduct the most current
    b. Diode D_2 will conduct the most current
    c. The output voltage will be zero
    d. None of the above

27. Refer to Fig. 12-25. The advantage of this FM detector as compared with a discriminator is that it
    a. Is less expensive (uses fewer parts)
    b. Can drive a tuning meter
    c. Can provide AFC
    d. Rejects amplitude variations
28. Refer to Fig. 12-26. The carrier input is 455 kHz, and the audio input is a 2-kHz sine wave. The output frequency or frequencies are
    a. 2 kHz
    b. 455 kHz
    c. 453, 455, and 457 kHz
    d. 453 and 457 kHz
29. Single sideband as compared with amplitude modulation is
    a. More efficient in terms of bandwidth
    b. More efficient in terms of power
    c. More critical to tune
    d. All of the above
30. Refer to Fig. 12-27. The purpose of the BFO circuit is
    a. To correct for tuning error
    b. To replace the missing carrier so detection can occur
    c. To provide noise rejection
    d. All of the above
31. Refer to Fig. 12-27. The IF bandwidth of this receiver, compared with an ordinary AM receiver, is
    a. Narrower
    b. The same
    c. Wider
    d. Indeterminate

12-5 Wireless Data

Wireless data includes wireless networks and radio-frequency identification systems. A wireless local area network (*WLAN*) uses radio-frequency technology to transmit and receive data over the air. A wireless identification system uses RF to read tags that contain information (data) about the objects they are affixed to. These wireless data systems are generally license-free, and they operate in the ISM bands and are subject to interference.

The industrial, scientific, and medical (ISM) bands were initially reserved for license-free

use of radio-frequency energy for purposes *other than communications*. Applications in the ISM bands include radio-frequency welding, microwave ovens, and medical diathermy machines. The strong emissions of these devices can create electromagnetic interference and disrupt radio communication for other devices sharing the same frequencies. In general, communications and data equipment operating in these bands must accept any interference generated by ISM equipment. In 1985, the FCC opened up the ISM bands for WLANs and mobile communications. In 1997, it added

WLAN

additional bands in the 5-GHz range. Some of the ISM bands are crowded. As an example, the following is a list of devices that might (and often do) use the 2.4-GHz frequency in the ISM band:

1. WiFi
2. Bluetooth
3. Radio-frequency identification (RFID)
4. Baby monitors
5. Wireless microphones
6. Wireless headphones
7. Wireless video cameras
8. Remote controllers (e.g., keyboards, trackballs, mice)
9. Garage door openers
10. Local (in-home) video/audio distribution systems

Let's start with the first item in the list, WiFi (Wireless Fidelity). The Institute of Electrical and Electronics Engineers (IEEE) has established the IEEE 802.11 standard, which is predominant for WLANs.

When radio communications and other devices share a frequency range, there is potential for interference. One way to reduce such interference is to use a wideband approach called *spread spectrum*. A second signal called a *key* (which modulates the carrier in addition to the data signal) increases the bandwidth of the transmitted signal (spreads its spectrum). The spreading signal is removed at the receiver. Interfering signals are rejected because they do not contain the key. Only the desired signal, which has the correct key, will be recognized at the receiver. This means that other spread spectrum signals not having the right key will be rejected. This allows different spread spectrum devices to be active simultaneously in the same band.

One way to increase the capacity of a communication medium is to use frequency division multiplexing (FDM). FDM uses multiple carriers that are sent simultaneously over the medium. However, FDM has an inherent problem: wireless signals can travel multiple paths from transmitter to receiver (by bouncing off metal objects, buildings, mountains, and even passing cars). *Orthogonal FDM* deals with this *multipath problem* by splitting carriers into smaller subcarriers and then broadcasting those simultaneously. This reduces multipath distortion and also reduces RF interference. The subcarriers' specific frequencies are "orthogonal,"

Spread spectrum

or noninterfering, to each other, allowing for greater throughput. *Orthogonal* means right angle (a phase relationship of 90 degrees). For example, a sine wave signal and a cosine wave signal at the same frequency are orthogonal signals.

The speed at which a WLAN performs depends on many things, from the efficiency of the wired network to the configuration of the building and the type of WLAN in use. As a general rule, data throughput decreases as the distance between the WLAN access point and the wireless client (user) increases. For example, a notebook computer with wireless access will often show a decrease in performance (i.e., more time will be required for downloads and uploads) as it is moved away from the access point.

The 802.11 standards support multiple data rates to accommodate the loss of signal strength while maintaining low error rates. The WLAN client constantly performs operations to detect and automatically set the best possible speed. The data rates are listed as a series of numbers to correspond to throughput at various ranges, as shown in Table 12-1. The frequency at which 802.11b and 802.11g are transmitted allows them to penetrate solid materials, allowing a maximum range of about 300 feet (but at reduced speed). The 802.11a protocol experiences a steeper decline in throughput as distance increases from the access point and generally has less range. The range and transmission speed are affected by the environment in which the WLAN is deployed.

Table 12-1 can be helpful when troubleshooting WLANs. It shows that distance is important and speed can suffer. It also warns of incompatibilities. An 802.11a device cannot communicate with 802.11b, 802.11g, or 802.11n devices. An 802.11b device will communicate with 802.11g only if the devices are designed for dual-mode operation. Additional troubleshooting information for WLANs can be found in the next section of this chapter. The sixth entry in Table 12-1 lists 802.11y, which is intended for much longer distances. Its uses include extending data communications (e.g., the Internet) to rural areas and mobile applications. Another IEEE standard, 802.16, has established the parameters for what is known as WiMAX, which also is intended for long distances and mobile networks. WiMAX offers an alternative to 3G (third-generation

Table 12-1 IEEE 802.11 Specifications (WiFi)

802.11 Protocol	Released	Freq. (GHz)	Bandwidth (MHz)	Data Rate per Stream (Mbit/s)	Allowable MIMO[1] Streams	Modulation	Approx. Range in Meters	
							Indoors	Outdoors
	June 1997[2]	2.4	20	1, 2	1	DSSS[3] FHSS[4]	20	100
a	Sept. 1999	5	20	6, 9, 12, 18, 24, 36, 48, 54	1	OFDM[5]	35	120
b	Sept. 1999	2.4	20	5.5, 11	1	DSSS	38	140
g	June 2003	2.4	20	6, 9, 12, 18, 24, 36, 48, 54	1	OFDM DSSS	38	140
n	Oct. 2009	2.4/5	20	7.2, 14.4, 21.7, 28.9, 43.3, 57.8, 65, 72.2	4	OFDM	70	250
			40	15, 30, 45, 60, 90, 120, 135, 150			70	250
y[6]	Sept. 2008	3.7	20	6, 9, 12, 18, 24, 36, 48, 54	1	OFDM	—	5,000
ac	Dec. 2013	5	80	1,300–1,700	8	OFDM	70	250
ax	Feb. 2021	2.4, 5, 6	20, 40, 80, 160	9600	8	QAM		

[1]Multiple input and multiple output antennas at both transmitter and receiver (smart antenna technology).

[2]The original specification was released as IEEE 802.11 in 1997 followed by 802.11a and 802.11b in 1999.

[3]Direct-sequence spread spectrum.

[4]Frequency-hopping spread spectrum.

[5]Orthogonal frequency-division multiplexing.

[6]IEEE 802.11y is licensed by the FCC in the United States.

Note: The WiFi certified logo on a device denotes that it has met interoperability testing requirements to ensure that compatible products from different vendors will work together. Be careful when mixing subtypes (e.g., 802.11a is not compatible with 802.11b).

mobile telecommunications) and 4G wireless networks.

At the opposite end of the distance charts, we find Bluetooth devices and networks (IEEE 802.15). Bluetooth technology is designed for personal area networks (PANs) and for appliances that don't require large data flows (e.g., printers, keyboards, mice, personal computers, and mobile phones). Many recent automobiles use Bluetooth to link cell phones to the sound system for truly hands-free operation while driving. One does not have to touch the phone to answer it or to access its stored database and dial out. Spoken commands such as "call home" and "call Elaine" are recognized.

Bluetooth uses the same 2.4-GHz spectrum as some versions of 802.11. Bluetooth data rates are usually much slower than 802.11 (less than 1 Mbps). However, Bluetooth 2.0 can run as fast as 3 Mbps, and 3.0 can run up to 24 Mbps. Versions 4 and 5 add low-energy specifications that are important for battery-operated devices.

Bluetooth 1.0 and 2.0 are slower, as they use a protocol where each data packet must be received and checked for error by the client before another one can be sent. This creates higher overheads and reduces the transmission speeds, but it also allows for more accurate transmission of the data. With 3.0, faster speeds are achieved by using the 802.11 protocol, which basically allows the Bluetooth protocol to piggyback onto a WiFi signal when transferring large amounts of data, like videos, music, and photos. (See Table 12-2.) The typical range of Bluetooth is about 30 feet, and the chips use much less power; thus, it is better suited for portable devices.

Bluetooth frequencies are all located within the 2.4-GHz ISM band. There are 79 Bluetooth channels, spaced 1 MHz apart. Channel 1 starts at 2.402 GHz, and channel 79 finishes at 2.480 GHz. Bluetooth limits interference with a *hopping* carrier (*frequency-hopping spread spectrum*, or FHSS). A Bluetooth transmission only remains on a given frequency for a brief time, and

Table 12-2 A Comparison of Bluetooth and WiFi

	Bluetooth	WiFi
Data rate	Low (800 kbps)	High (11 Mbps)
Hardware requirement	Bluetooth adaptor on every device on the network	Wireless adaptors on all the devices on the network, a wireless router, and/or wireless access points
Ease of use	Fairly simple. Can be used to connect up to seven devices at a time. It is easy to switch between devices or find and connect to a device	More complex and requires configuration of hardware and software
Typical devices	Mobile phones, mouse, keyboards, printers, office and industrial automation devices	Notebook computers, desktop computers, servers
Range	10 meters	100 meters
Security	More secure than WiFi as it covers shorter distances and has a 2-level password protection	Less secure. Has the risks associated with any other network. If someone accesses one part, the rest can also be accessed.
Power consumption	Low	High
Frequency	2.4 GHz	2.4/5 GHz

if interference is present, the data will be resent later when the signal has changed to a different channel that is clear of interference. The rate is 1,600 hops per second, and the system spreads over all the available channels using a predetermined sequence.

WiFi and Bluetooth both occupy a section of the 2.4-GHz ISM band that is 83 MHz wide. WiFi uses direct sequence spread spectrum (DSSS) instead of FHSS. Its carrier does not hop or change frequency and remains centered on one channel that is 22 MHz wide. While there is room for 11 overlapping channels in this 83-MHz-wide band, there is only room for three nonoverlapping channels. So there can be no more than three different WiFi networks operating in close proximity. When Bluetooth and WiFi are operating in close proximity, the single 22-MHz-wide WiFi channel occupies the same frequency space as 22 of the 79 Bluetooth channels. When a Bluetooth transmission occurs on a frequency that lies within the frequency space occupied by a simultaneous WiFi transmission, interference can occur, depending on the strength of each signal.

When a Bluetooth device encounters interference on a channel, it deals with the problem by hopping to the next channel and trying again.

Of course, this can cause degraded throughput (slow transfers). When WiFi encounters interference, it decreases the data rate from 11 Mbps to 5.5, 2, or even 1 Mbps in an effort to lower the error rate caused by interference. If a WiFi device encounters interference from a Bluetooth transmission and slows its transmission rate, it will then spend more time than before transmitting on a frequency available to Bluetooth. This can have the effect of *increasing* the likelihood of interference between the two. Data are not lost, but the data throughput may slow to an intolerable level.

The Internet of Things (IoT) is making a growing impact on how we work, play, and interact with our environment. This also includes managing our health. Smart devices are used as heart monitors and calorie counters, and the sensors often connect using wireless technology. The IoT is making increasing use of the ISM band.

Some medical sensors use ZigBee (IEEE 802.15.4), which is an array of communication standards that can be used for low-cost and low-bandwidth applications and devices. ZigBee devices and networks are intended to be simpler and less expensive than WiFi or Bluetooth. Applications include wireless light switches,

Table 12-3	ZigBee and Z-Wave				
ZIGBEE AND Z-WAVE SPECIFICATIONS AND CAPABILITIES					
Technology	Frequency	Modulation	Data Rate	Range	Applications
ZigBee	2.4 to 2.483 GHz	QPSK	250 kbits/s	10 m	Home automation, smart grid, remote control
Z-Wave	908.42 MHz	GFSK	9.6/40 kbits/s	30 m	Home automation, security

electric utility meters, traffic management systems, and other consumer and industrial equipment that requires short-range, low data rate, wireless communication.

Z-Wave is another wireless specification. It operates at a different frequency and offers a lower data rate, as shown in Table 12-3.

Radio-frequency identification is similar to bar code identification. With RFID, electromagnetic coupling in the RF portion of the electromagnetic spectrum is used to gather data. An RFID system consists of an antenna, a reader, and transponders, or tags, which are integrated circuits containing the RF circuitry and the data. RFID systems can be used just about anywhere, such as clothing tags, warehouse pallet tags, implanted pet tags, and packaged food tags. The tag can carry such information as a pet owner's name and address, a part or batch number, or the date of manufacture. Vehicle manufacturers can use RFID systems to move parts and assemblies through an automated line. Hospitals can use RFID tags to ensure patients receive the proper tests and medications.

The key difference between RFID and bar code technology is that RFID eliminates the need for line-of-sight reading. Also, RFID scanning can be done at greater distances than bar code scanning. High-frequency RFID systems (850 MHz to 950 MHz and 2.4 GHz to 2.5 GHz) offer transmission ranges of more than 90 feet. In an RFID system, the transponder that contains the data to be transmitted is called a tag. *Active tags* have an internal battery to power them and are usually *writable* in addition to being *readable* (they may be called *read-write tags*). Active tags generally can transmit data over longer distances. An active tag is larger than a passive tag and has a limited life span. *Passive tags* get their operating power from the

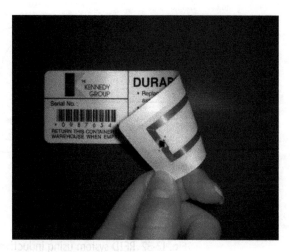

Fig. 12-30 Passive RFID tag.
Courtesy of The Kennedy Group

reader (via radio waves). They are smaller and lighter than active tags but have a shorter communication range and require more powerful readers. Passive tags are generally read-only and must be replaced to change the data. Figure 12-30 shows an example of a passive tag.

Passive tags are read in one of three ways:

1. Capacitive coupling
2. Inductive coupling
3. Backscatter coupling

Capacitive coupling is often used with "smart cards," where the card is placed in a reader and electrodes capacitively couple the reader to the integrated circuit embedded in the card. Inductive coupling is based on the mutual inductance between two circuits, and backscatter coupling uses RF energy reflected from the tag (Fig. 12-31).

With RFID inductive coupling, both the tag and the reader have "antenna" coils. When the tag is located near the reader, the field from the reader coil will inductively couple to the coil in the tag. A voltage will be induced in the tag that

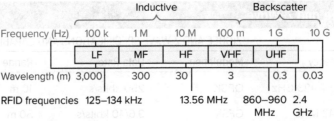

Frequency (Hz)	100 k	1 M	10 M	100 m	1 G	10 G
	LF	MF	HF	VHF	UHF	

Inductive   Backscatter

Frequency (Hz)   100 k   1 M   10 M   100 m   1 G   10 G

Wavelength (m)   3,000   300   30   3   0.3   0.03

RFID frequencies   125–134 kHz   13.56 MHz   860–960   2.4
                                                    MHz   GHz

(Other RFID frequencies include 5 to 7 MHz, 433 MHz, and 5.25 to 5.8 GHz.)

Fig. 12-31 Inductive and backscatter coupling.

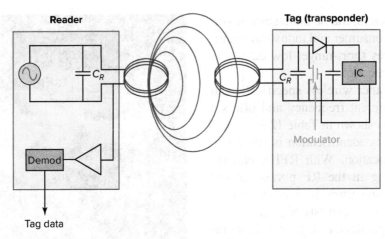

Fig. 12-32 RFID system using inductive coupling.

is rectified, filtered, and used to power the tag circuitry. Figure 12-32 shows a diode and filter capacitor in the transponder that converts the RF energy to dc to energize the integrated circuit (IC). Both the reader and the tag coils are resonated by capacitors (C_R). This affords a much more efficient transfer of energy and also provides rejection of signals at other frequencies that could cause interference. Figure 12-32 also shows a modulator FET that is switched on and off according to the binary memory data stored in the tag. Because the two tuned circuits are coupled, this modulation can be detected (demodulated), and the recovered data sent on to a computer or LAN.

RFID inductive coupling is a so-called near field effect. The distance between the coils should be less than 0.1 wavelength for reliable operation. The relationship between wavelength (λ) and frequency (f) is given by

$$\lambda = c/f = \frac{\text{speed of light}}{\text{frequency}}$$
$$= \frac{3 \times 10^8 \text{ meters per second}}{\text{Hz}} = x \text{ meters}$$

EXAMPLE 12-4

Use Fig. 12-31 to determine the maximum tag distance for inductive coupling operating in the HF (high-frequency) portion of the spectrum. The illustration shows the frequency to be 13.56 MHz; thus, the wavelength is

$$\lambda = c/f = \frac{3 \times 10^8}{13.56 \times 10^6} = 22 \text{ meters}$$

$$0.1 \lambda = 2.2 \text{ meters}$$

RFID backscatter coupling operates beyond the near field region. When the signal reaches the tag's antenna, current flows in the antenna, and some of the signal is scattered back (reflected) to the transceiver, as shown in Fig. 12-33. Since the tag is passive, some of the RF energy is rectified and filtered to provide dc for energizing the tag's circuits. A field-effect transistor modulator places the tag data onto the backscatter signal by switching on and off in accord with tag's digital data.

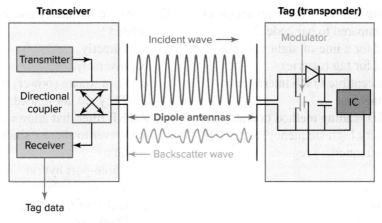

Transceiver

Transmitter

Directional coupler

Receiver

Tag data

Incident wave →

Dipole antennas

← Backscatter wave

Tag (transponder)

Modulator

IC

Fig. I2-33 RFID system using backscatter coupling.

The transceiver in Fig. 12-33 contains a *directional coupler* (also called a duplexer), which is a special type of RF filter that allows the transmitter and receiver sections to share one antenna. After passing through the directional coupler, the backscatter signal is amplified and demodulated to recover the tag's data.

Self-Test

Choose the letter that best completes each statement.

32. Devices such as microwave ovens operate in the
    a. ISM bands
    b. AM bands
    c. FM bands
    d. HF bands

33. In regard to frequency spectrum allocations, ISM refers to
    a. Internet security method
    b. Integrated safe modem
    c. Intelligent site map
    d. Industrial, scientific, and medical

34. Which of the following would be used to build a wireless local area network?
    a. Bluetooth
    b. RFID
    c. WiFi
    d. FAX

35. Spectrum spreading can be accomplished by modulating the carrier with a secondary signal called the
    a. Key
    b. Orthogonal
    c. Sync byte
    d. FDM

36. Signals arriving at a receiver might be a combination of direct travel and reflected travel, which is called
    a. Backscatter
    b. Phantom distortion
    c. Multipath
    d. Echoplexing

37. In systems using the 802.11 standards, interference or weak signals are often a cause for
    a. Security breaches
    b. Decreased bit rates
    c. Lost e-mail messages
    d. All of the above

38. The major application for Bluetooth technology is for
    a. WLANs
    b. PANs
    c. Video baby monitors
    d. None of the above

39. Which of the following uses frequency-hopping spread spectrum?
    a. 802.11b
    b. 802.11g
    c. 802.11n
    d. Bluetooth

40. Which of the following is an advantage of RFID as compared to bar codes?
   a. No need for a line-of-sight viewpoint
   b. No need for tag batteries
   c. Less susceptible to RF interference
   d. Lower tag cost

41. Which RFID reading method operates in the VHF (very high-frequency) and UHF (ultra-high-frequency) regions of the spectrum?
   a. Capacitive
   b. Inductive
   c. Backscatter
   d. Near field

42. How are frequency and wavelength related?
   a. Directly.
   b. Inversely.
   c. As a square (power) function.
   d. They are not related.

43. An RF filter that allows a transmitter and a receiver to share a single antenna is called a
   a. Three-port hybrid
   b. Two-port hybrid
   c. Either of the above
   d. Duplexer

12-6 Troubleshooting

Radio receiver troubleshooting is very similar to amplifier troubleshooting. Most circuits in a receiver are amplifiers. The material covered in Chap. 10 on amplifier troubleshooting is relevant to receiver troubleshooting. For example, Sec. 10-1 on preliminary checks should be followed in exactly the same way.

You should view a receiver as a signal chain. If the receiver is dead, the problem is to find the broken link in the chain. Signal injection should begin at the output (speaker) end of the chain. However, a receiver involves gain at different frequencies. Several signal-generator frequencies will be involved. You must use both an audio generator and an RF generator. Figure 12-34 shows the general scheme of signal injection in a superheterodyne receiver.

It is also possible to make a click test in most receivers. Use the same procedure discussed in Chap. 10 on amplifier troubleshooting. This will work in the audio and IF stages. The noise generated by the sudden shift in transistor bias should reach the speaker. It is also possible to test the mixer with the click test. The oscillator may respond to the click test, but the results would not be conclusive. It is possible that the oscillator is not oscillating, or it may be oscillating at the wrong frequency.

If we assume that the signal chain is intact from the first IF to the speaker, then the problem must be in the mixer or the oscillator. Checking the oscillator is not too difficult. An oscilloscope or frequency counter could be used. A voltmeter with an RF probe is another possibility, but there would be no way to tell whether the frequency was correct. Some technicians prefer to tune for the oscillator signal by using a second receiver. Place the second receiver very close to the receiver being tested.

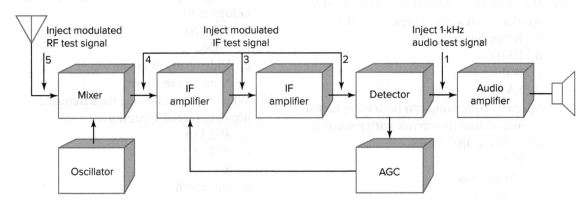

Fig. 12-34 Signal injection in a superheterodyne receiver.

Set the dial on the second receiver above the dial frequency on the receiver under test. The difference should equal the IF of the receiver under test. This is based on the fact that the oscillator is supposed to run above the dial setting by an amount equal to the IF. Now, rock one of the dials back and forth a little. You should hear a carrier (no modulation). This tells you that the oscillator is working, and it also indicates whether the frequency is nearly correct.

If the receiver sounds distorted on strong stations, the problem could be in the AGC circuit. This can be checked with a voltmeter. Monitor the control voltage as the receiver is tuned across the band. You should find a change in the control voltage from no station (clear frequency) to a strong station. The service notes for the receiver usually will indicate the normal AGC range.

If the receiver has poor sensitivity, again it is possible that the AGC circuit is defective. Since AGC can produce several symptoms, it should be checked early in the troubleshooting process.

Poor sensitivity can be difficult to troubleshoot. A dead stage is usually easier to find than a weak stage. Signal injection may work. It is normal to expect less injection for a given speaker volume as the injection point moves toward the antenna. Some technicians disable the AGC circuit when making this test. This can be done by clamping the AGC control line with a fixed voltage from a power supply. A current-limiting resistor around 10 kΩ should be connected in series with the supply to avoid damaging the receiver.

Poor sensitivity can be caused by a leaky detector diode. Disconnect one end of the diode from the circuit and check its forward and reverse resistance with an ohmmeter. Diode testing was covered in Sec. 3-3 of Chap. 3.

Improper *alignment* is another possible cause of low gain and poor sensitivity. All the IF stages must be adjusted to the correct frequency. Also, the oscillator and mixer tuned circuits must track for good performance across the band. If the receiver has a tuned RF stage, then three tuned circuits must track across the band.

Alignment is usually good for the life of the receiver. However, someone may have tampered with the tuned circuits, or a part may have been replaced that upsets the alignment. Do not attempt alignment unless the service notes and the proper equipment are available.

Intermittent receivers and noisy receivers should be approached by using the same techniques described in Chap. 10 for amplifier troubleshooting. In addition, you should realize that receiver noise may be due to some problem outside the receiver itself. Some locations are very noisy, and poor receiver performance is typical. Compare performance with a known receiver to verify the source of the noise.

It should also be mentioned that receiver performance can vary considerably from one model to another. Many complaints of poor performance cannot be resolved with simple repairs. Some receivers simply do not work as well as others.

A superheterodyne receiver may have a total gain in excess of 100 dB. Unwanted feedback paths or coupling of circuits may cause oscillations. If the receiver squeals only when a station is tuned in, the problem is likely to be in the IF amplifier. If the receiver squeals or motorboats constantly, a bypass capacitor or AGC filter capacitor may be open. If it's a portable receiver, try fresh batteries. Always check to be sure that all grounds are good and that all shields are in place. In some cases, improper alignment can also cause oscillation.

Interference from nearby transmitters is becoming an increasingly complex problem. When a transmitting antenna is located close to other receiving equipment, problems are likely to occur. Some interference problems can be difficult to solve. Figure 12-35 shows a few techniques that may be successful. Solving interference problems is often a process of trying various things until progress is noted. Try the easiest and least expensive cures first.

Receiver interference can often be traced to nontransmitting equipment. Computers, computer peripherals, light dimmers, touch-controlled lamps, and even power lines are known sources of radio and television interference. The best way to verify if a device is causing a problem is to turn it off. Touch-controlled lamps should be disconnected from the wall outlet to determine if they are causing the interference.

The ISM band is an experiment by the FCC in unregulated spectrum sharing. WiFi 802.11 devices share this band with microwave ovens, cordless phones, Bluetooth, wireless video, microwave links, game controllers, motion detectors, ballasts for fluorescent lighting, ZigBee,

Interference

Alignment

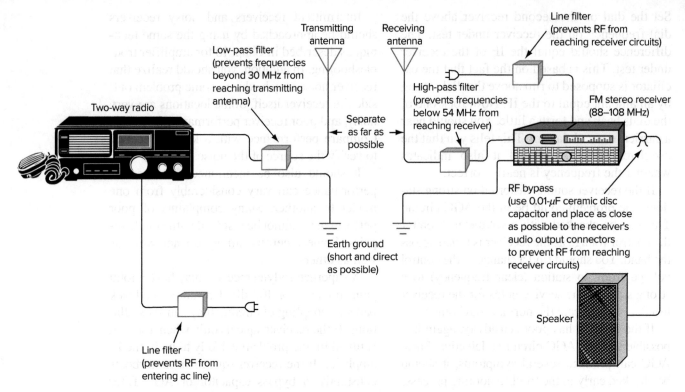

Transmitting antenna

Receiving antenna

Low-pass filter (prevents frequencies beyond 30 MHz from reaching transmitting antenna)

High-pass filter (prevents frequencies below 54 MHz from reaching receiver)

Line filter (prevents RF from reaching receiver circuits)

FM stereo receiver (88–108 MHz)

Two-way radio

Separate as far as possible

RF bypass (use 0.01-μF ceramic disc capacitor and place as close as possible to the receiver's audio output connectors to prevent RF from reaching receiver circuits)

Earth ground (short and direct as possible)

Speaker

Line filter (prevents RF from entering ac line)

Fig. 12-35 Steps to prevent radio interference.

Table 12-4	WiFi Signal Levels		
# of Bars	SNR (dB)	Error Rate	Speed
5	>40	Very low	Very fast
4	30–40	Low	Fast
3	20–30	Moderate	Moderate
2	10–20	High	Low
1	<10	Too high	Too slow
0	—	—	—

medical scanners, and other devices. Some of these are designed to operate in the ISM band (microwave ovens) and some are not (fluorescent lights). In any case, interference is an ongoing and increasingly critical issue. A WiFi device needs some minimum signal strength for proper operation. A crude but useful set of values is shown in Table 12-4. This table is based on the familiar bar graph display seen on many computer screens and smartphones. Figure 12-36 shows how the bit error rate of a WiFi signal varies with signal strength. As expected, stronger signals have a lower rate of error. This is why some situations call for repeaters or amplifiers to boost the signal.

What is not shown in the table is the signal to interference plus noise ratio (SINR). As shown in Fig. 12-37, the SNR might be one value, but when an interference source such as a microwave oven turns on (point A on the time line), the SINR drops. This illustration is important for troubleshooting. It shows how a WiFi network can slow down or stop working from time to time. Digital data is often sent frame by frame or packet by packet. After a frame is received, a checksum is used to flag any errors in that frame. With errors, frames have to be resent. This is why things slow down so much when there is a poor SNR or SINR.

Troubleshooting RF devices like WiFi can be challenging, as many IT workers have found. Back in the days when an Ethernet cable and device were installed, they normally kept working once any initial kinks were worked out. With WiFi, the physical layer can stop working or slow down at *any time*. Bad connections caused by loose wires can arc, and even those can cause interference.

Wireless local area network (WLAN) troubleshooting can be challenging. There are helpful tools available. Also, these tools can be used in the design phase of a wireless network. Wireless *site surveys* are commonly used both to establish and maintain networks. Changes, even something as minor as a notebook that a new employee connects to a wireless network, can affect WLAN performance. Other factors such as interference from nearby WLANs also play a role. This is why regular site surveys that are conducted with tools like those available from Tamograph are important. Figure 12-38 shows one such software product. Figure 12-39 shows a companion product. It is a dual band spectrum analyzer. Together, the two (software and hardware) can perform a site survey.

Another tool is a *network sniffer* such as CommView, shown in Fig. 12-40. It provides WLAN monitoring and analysis. It provides a picture of network traffic. It helps to identify network problems and to troubleshoot software and hardware. It even allows analysis, recording, and playback of voice over internet protocol (VOIP) traffic. Some of its functions include:

- Scanning for WiFi stations and access points (APs).
- Capturing WLAN traffic.
- Viewing node and channel statistics.
- Viewing IP statistics: IP addresses, ports, sessions, etc.

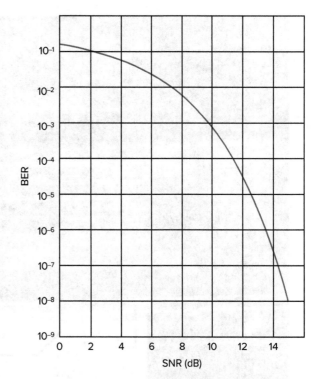

Fig. 12-36 Bit error rate (BER) as a function of signal strength.

- Viewing protocol "pie" charts.
- Monitoring bandwidth use.
- Browsing captured and decoded packets in real time.
- Searching for strings or hex data in captured packet contents.

All tools have limitations and may not give the needed results due to improper use. Network technicians (including some electronics technicians) have quite a challenge in certain installations. Rogue access points, in addition to bring your own device (BYOD) company policies, add another dimension to the many challenges at hand. Frequent site surveys are now common.

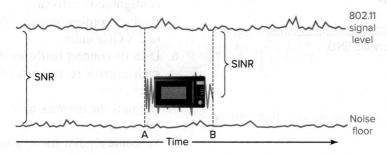

Fig. 12-37 SNR and SINR.

Charles A. Schuler/McGraw Hill

Fig. I2-38 Tamograph site survey desktop.
TomoSoft

Fig. I2-39 Wi-Spy DBx dual-band spectrum analyzer (2.4 GHz and 5 GHz).

MetaGeek, LLC

Here is a list of some actions that can be taken:

1. Choose the location of APs carefully. Routers usually work best in a central location. Gain antennas and repeaters may serve well in some cases. Signal reflections cause multipath signal distortion. Sometimes minor physical relocations make a big difference.
2. Move routers off the floor and away from walls and metal objects. Some routers have ceiling mounting brackets.
3. Upgrade to a diversity antenna router or a smart antenna router. Some of these can beam (focus) RF energy in the right directions and suppress interference.
4. Change channels using the router configuration software.
5. Replace cordless phones with 900 MHz or 5.8 GHz units.
6. Directly connect hardware devices such as printers, and turn off their WiFi electronics.
7. Upgrade the network to 802.11n or 802.11ac.
8. Schedule regular surveys and security sweeps to eliminate problems and rogue devices before they cause trouble.

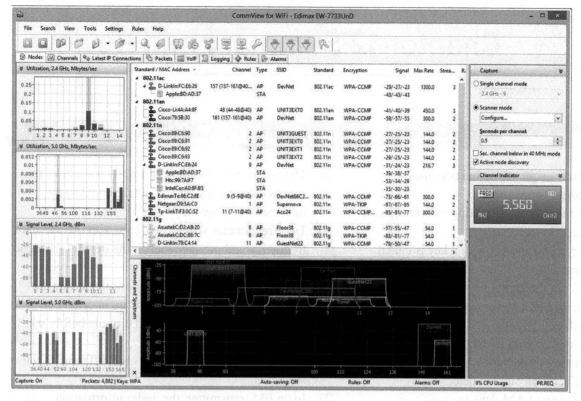

Fig. 12-40 TamoSoft CommView for WiFi.

TomoSoft

Self-Test

Choose the letter that best answers each question.

44. A 1-kHz test signal can be used for testing which stage of a superheterodyne receiver?
    a. Mixer
    b. IF
    c. Detector
    d. Audio

45. You want to check the oscillator of a superheterodyne receiver by using a second receiver. If the dial is set at 980 kHz, where should the oscillator be heard on the second receiver?
    a. 525 kHz
    b. 980 kHz
    c. 1,435 kHz
    d. 1,610 kHz

46. A receiver sounds distorted only on the strongest signals. Where would the fault likely be found?
    a. In the AGC system
    b. In the loudspeaker
    c. In the audio amplifier
    d. In the volume control

47. What would cause poor sensitivity in a receiver?
    a. A defective mixer
    b. A defective IF amplifier
    c. A weak detector
    d. All of the above

48. What results from improper alignment?
    a. Poor sensitivity
    b. Dial error
    c. Oscillation
    d. All of the above

Summary

1. A high-frequency oscillator signal becomes a radio wave at the transmitting antenna.
2. Modulation is the process of putting information on the radio signal.
3. Turning the signal on and off with a key is called CW modulation.
4. When AM is used, the signal has three frequency components: a carrier, a lower sideband, and an upper sideband.
5. The total bandwidth of an AM signal is twice the highest modulating frequency.
6. Demodulation is usually called detection.
7. A diode makes a good AM detector.
8. Other nonlinear devices, such as transistors, can also be used as AM detectors.
9. A simple AM receiver can be built from an antenna, a detector, headphones, and a ground.
10. Sensitivity is the ability to receive weak signals.
11. Selectivity is the ability to receive one range of frequencies and reject others.
12. Gain provides sensitivity.
13. Tuned circuits provide selectivity.
14. The optimum bandwidth for an ordinary AM receiver is about 15 kHz.
15. A superheterodyne receiver converts the received frequency to an intermediate frequency.
16. Tuning a radio receiver to different stations does not change the passband of the IF amplifiers.
17. The mixer output will contain several frequencies. Only those in the IF passband will reach the detector.
18. The standard IF for the AM broadcast band is 455 kHz.
19. The receiver oscillator will usually run above the received frequency by an amount equal to the intermediate frequency.
20. Two frequencies will always mix with the oscillator frequency and produce the IF: the desired frequency and the image frequency.
21. Adjacent-channel interference is rejected by the selectivity of the IF stages. Image interference is rejected by one or more tuned circuits before the mixer.
22. The AGC circuit compensates for different signal strengths.
23. In an FM transmitter, the audio information modulates the frequency of the oscillator.
24. Frequency modulation produces several sidebands above the carrier and several sidebands below the carrier.
25. Frequency-modulation detection can be achieved by a discriminator circuit.
26. Discriminators are sensitive to amplitude; thus, limiting must be used before the detector.
27. A ratio detector has the advantage of not requiring a limiter circuit for noise rejection.
28. Single sideband (SSB) is a subclass of AM.
29. Receiver troubleshooting is similar to amplifier troubleshooting.
30. The signal chain can be checked stage by stage by using signal injection.
31. A leaky detector can cause poor sensitivity.
32. Good alignment is necessary for proper receiver performance.

Choose the letter that best answers each question.

12-1. Which portion of a transmitting station converts the high-frequency signal into a radio wave? (12-1)
   a. The modulator
   b. The oscillator
   c. The antenna
   d. The power supply

12-2. What is the modulation used in radio telegraphy called? (12-1)
   a. CW
   b. AM
   c. FM
   d. SSB

12-3. An AM transmitter is fed audio as high as 3.5 kHz. What is the bandwidth required for its signal? (12-1)
   a. It is dependent on the carrier frequency.
   b. 3.5 kHz.
   c. 7.0 kHz.
   d. 455 kHz.

12-4. An AM demodulator uses the difference frequency between what two frequencies? (12-2)
   a. USB and LSB
   b. Sidebands and the carrier
   c. IF and the detector
   d. All of the above

12-5. Which of the following components is useful for AM detection? (12-2)
   a. A tank circuit
   b. A resistor
   c. A capacitor
   d. A diode

12-6. Refer to Fig. 12-4. Assume the audio input is zero. The carrier output to the antenna will (12-1)
   a. Fluctuate in frequency
   b. Fluctuate in amplitude
   c. Be zero
   d. None of the above

12-7. Refer to Fig. 12-10. What serves as the energy source for this receiver? (12-2)
   a. The radio signal.
   b. The detector.
   c. The headphones.
   d. There is none.

12-8. Refer to Fig. 12-11. How may the selectivity of this receiver be improved? (12-2)
   a. Add more audio gain
   b. Add more tuned circuits
   c. Use a bigger antenna
   d. All of the above

12-9. A tuned circuit has a center frequency of 455 kHz and a bandwidth of 20 kHz. At what frequency or frequencies will the response of the circuit drop to 70 percent? (12-2)
   a. 475 kHz
   b. 435 kHz
   c. 435 and 475 kHz
   d. 445 and 465 kHz

12-10. An AM receiver has an IF amplifier with a bandwidth that is too narrow. What will the symptom be? (12-3)
   a. Loss of high-frequency audio
   b. Poor selectivity
   c. Poor sensitivity
   d. All of the above

12-11. What is the major advantage of the superheterodyne design over the TRF receiver design? (12-3)
   a. It eliminates the image problem.
   b. It eliminates tuned circuits.
   c. It eliminates the need for an oscillator.
   d. The fixed IF eliminates tracking problems and bandwidth changes.

12-12. An AM superheterodyne receiver is tuned to 1,140 kHz. Where is the image? (12-3)
   a. 865 kHz
   b. 1,315 kHz
   c. 2,050 kHz
   d. 2,850 kHz

12-13. Refer to Fig. 12-16. The dial of the receiver is set at 1,190 kHz. Which statement is true? (12-3)
   a. The mixer tuned circuit should resonate at 1,190 kHz.
   b. The oscillator circuit should resonate at 1,645 kHz.
   c. The difference mixer output should be at 455 kHz.
   d. All of the above are true.

12-14. An FM receiver is set at 93 MHz. Interference is received from a station transmitting at 114.4 MHz. What is the problem caused by? (12-3)
  a. Poor selectivity in the RF and mixer tuned circuits
  b. Poor selectivity in the IF stages
  c. Poor limiter performance
  d. Poor ratio detector performance

12-15. What FM receiver circuit is used to correct for frequency drift in the oscillator? (12-4)
  a. AGC
  b. AVC
  c. AFC
  d. All of the above

12-16. A transistor in an FM receiver is controlled by decreasing its current as the received signal becomes stronger. What is this an example of? (12-3)
  a. Forward AGC
  b. Reverse AGC
  c. Stereo reception
  d. None of the above

12-17. How does frequency modulation compare with amplitude modulation with regard to the number of sidebands produced? (12-4)
  a. Frequency modulation produces the same number of sidebands.
  b. Frequency modulation produces fewer sidebands.
  c. Frequency modulation produces more sidebands.
  d. Frequency modulation produces no sidebands.

12-18. What is the function of a limiter stage in an FM receiver? (12-4)
  a. It rejects adjacent-channel interference.
  b. It rejects image interference.
  c. It rejects noise.
  d. It rejects drift.

12-19. The output of the device in Fig. 12-23 is connected to a zero-center tuning meter. How will the meter respond when a station is correctly tuned? (12-4)
  a. It will indicate in the center of its scale.
  b. It will deflect maximum to the right.
  c. It will deflect to the left.
  d. It depends on the station.

12-20. Which of the following circuits is not used for FM demodulation? (12-4)
  a. Diode detector
  b. Discriminator
  c. Ratio detector
  d. Quadrature detector

12-21. The output of a balanced modulator is called (12-4)
  a. SSB
  b. DSBSC
  c. FM
  d. None of the above

12-22. An SSB transmitter runs 100 W. What power will be required in an AM transmitter to achieve the same range? (12-4)
  a. 5 W
  b. 20 W
  c. 800 W
  d. 1,200 W

12-23. What is the bandwidth of an SSB signal as compared to that of an AM signal? (12-4)
  a. About two times
  b. About the same
  c. About half
  d. About 10 percent

12-24. What must be done to demodulate an SSB signal? (12-4)
  a. Replace the missing carrier
  b. Use two diodes
  c. Use a phase-locked loop detector
  d. Convert it to an FM signal

12-25. Which of the following test signals would be the least useful for troubleshooting an AM broadcast receiver? (12-5)
  a. 1-kHz audio
  b. 455-kHz modulated RF
  c. 1-MHz modulated RF
  d. 10.7-MHz frequency-modulated RF

12-26. An FM receiver works well, but the dial accuracy is poor. The problem is most likely in the (12-5)
  a. Detector
  b. Oscillator
  c. IF amplifiers
  d. Limiter

Critical Thinking Questions

12-1. Can you identify some uses for radio frequencies other than communication?

12-2. A shortwave listener tells you that some stations can be received at two different frequencies. Are these stations transmitting on two frequencies, or is there another explanation? How can you find out?

12-3. Federal Aviation Agency (FAA) rules prohibit passengers on commercial flights from using radio receivers. Why?

12-4. How can a personal computer interfere with radio and television reception?

12-5. Can you think of any significant difference between vehicular cellular telephones and vehicular CB radios?

12-6. The AM broadcast band ranges from 540 to 1,600 kHz, for a total bandwidth of a little over 1 MHz. A single television channel is allocated 6 MHz. Why is one television channel wider in bandwidth than the entire AM band?

Answers to Self-Tests

1. D	13. B	25. B	37. B
2. B	14. C	26. C	38. B
3. B	15. A	27. D	39. D
4. A	16. B	28. D	40. A
5. D	17. B	29. D	41. C
6. B	18. A	30. B	42. B
7. A	19. C	31. A	43. D
8. C	20. D	32. A	44. D
9. D	21. C	33. D	45. C
10. D	22. D	34. C	46. A
11. A	23. A	35. A	47. D
12. D	24. C	36. C	48. D

Integrated Circuits

Learning Outcomes

This chapter will help you to:

13-1 *Compare* integrated circuit technology with discrete technology. [13-1]

13-2 *Explain* the photolithographic process used to make ICs. [13-2]

13-3 *Make* calculations for 555 timer circuits. [13-3]

13-4 *Identify* analog, digital, and mixed-signal ICs. [13-4, 13-5]

13-5 *Identify* 12C and SPI communication protocols. [13-6]

13-6 *Troubleshoot* circuits with ICs. [13-7]

An integrated circuit (IC) can be the equivalent of dozens, hundreds, or thousands of separate electronic parts. Digital ICs, such as microprocessors, can equal millions of parts. An Intel Xeon microprocessor has billions of parts. Now, digital and mixed-signal ICs are finding more applications in analog systems.

13-1 Introduction

The integrated circuit was introduced in 1958. It has been called the most significant technological development of the 20th century. Integrated circuits have allowed electronics to expand at an amazing rate. Much of the growth has been in the area of digital electronics. Lately, analog ICs have received more attention, and the designation "mixed-signal" is now applied to ICs that combine digital and analog functions.

Electronics is growing rapidly. One major reason is the advance in performance while costs remain stable and sometimes decrease. Circuits have become a lot smaller, more reliable, and more energy efficient. Witness the many portable devices available today. Integrated circuit technology is the major force behind the growth in the electronics industry. Consider that many systems and devices that we now enjoy were not practical or even possible a decade ago.

Discrete circuits use individual resistors, capacitors, diodes, transistors, and other devices to achieve the circuit function. These individual or discrete parts must be interconnected. The usual approach is to use a circuit board. This method, however, increases the cost of the circuit. The board, assembly, soldering, and testing all make up a part of the cost.

Integrated circuits do not eliminate the need for circuit boards, assembly, soldering, and testing. However, with ICs the number of discrete parts can be reduced. This means that the circuit boards can be smaller, often use less power,

Discrete circuit

and cost less to produce. It may also be possible to reduce the overall size of the equipment by using integrated circuits, which can reduce costs in the chassis and cabinet.

Integrated circuits may lead to circuits that require fewer alignment steps at the factory. This is especially true with digital devices. Alignment is expensive, and fewer steps mean lower costs. Also, variable components are more expensive than fixed components, and if some components can be eliminated, savings are realized. Finally, variable components are not as reliable as fixed components.

Integrated circuits may also increase performance. Certain ICs work better than equivalent discrete circuits. A good example is a modern integrated voltage regulator. A typical unit may offer 0.03 percent regulation, excellent ripple and noise suppression, automatic current limiting, and thermal shutdown. An equivalent discrete regulator may contain dozens of parts, cost six times as much, and still not work as well!

Reliability is related indirectly to the number of parts in the equipment. As the number of parts goes up, the reliability comes down. Integrated circuits make it possible to reduce the number of discrete parts in a piece of equipment. Thus, electronic equipment can be made more reliable by the use of more ICs and fewer discrete components.

Integrated circuits are available in an increasing number of package styles. Figure 13-1

Reliability

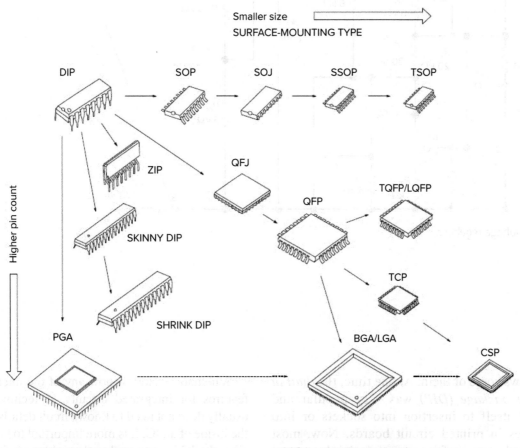

DIP (dual in line package)
SOJ (small outline J leads)
TSOP (thin small outline package)
QFJ (quad flat J leads)
TQFP (thin quad flat package)
TCP (tape carrier package)
BGA (ball grid array)
CSP (chip-size package)

SOP (small outline package)
SSOP (shrink small outline package)
ZIP (zigzag in line package)
QFP (quad flat package)
LQFP (low-profile quad flat package)
PGA (pin grid array)
LGA (land grid array)

Fig. 13-1 IC package styles.

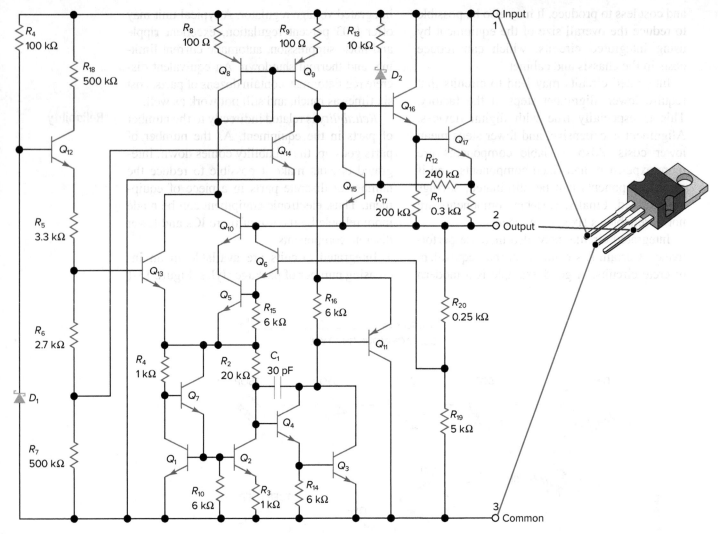

Fig. 13-2 Integrated circuit voltage regulator.

Dual in line package (DIP)

shows some of them. At one time, the *dual in line package (DIP)* was very popular and lent itself to insertion into sockets or into holes in printed circuit boards. Now, most boards use surface mount technology, and sockets are almost a thing of the past. A few integrated circuits, such as voltage regulators, use three-leaded packages such as the TO-220, which is also used for some power transistors. Thus, one cannot always properly identify a component with a casual glance. Service literature and part numbers are a must.

Schematics seldom show any of the internal features for integrated circuits. A technician usually does not need to know circuit details for the inside of an IC. It is more important to know what the IC is supposed to do and how it functions as a part of the overall circuit. Figure 13-2 shows the schematic for an IC voltage regulator and its package. The way this part appears on a schematic diagram is a rectangle with three leads. The leads are labeled, numbered, or both. A technician will check the voltages and possibly the waveforms on the three leads to verify operation or identify a defective IC.

Choose the letter that best answers each question.

1. When was the integrated circuit developed?
   a. 1920
   b. 1944
   c. 1958
   d. 1983
2. What is an electronic circuit that is constructed of individual components such as resistors, capacitors, transistors, diodes, and the like called?
   a. An integrated circuit
   b. A chassis
   c. A circuit board
   d. A discrete circuit
3. The use of ICs in a design can
   a. Decrease the number and size of parts
   b. Lower cost
   c. Increase reliability
   d. All of the above
4. What is the only sure way to identify a part as an integrated circuit?
   a. Look at the package style.
   b. See how it is connected to other parts.
   c. Check the schematic or part number.
   d. Count the pins.
5. When will a technician need the internal schematic for an integrated circuit?
   a. Very seldom
   b. When troubleshooting
   c. When making circuit adjustments
   d. When taking voltage and waveform readings

13-2 Fabrication

Placing over 1 million transistors on a piece of silicon the size of a fingertip is intricate work. The current precision is less than one micron, with one-tenth of a micron now being used. A micron is only about one one-hundredth the diameter of a human hair. X-ray lithography can extend fabrication precision into the 20 nanometer realm.

The fabrication process is applied to thin wafers of silicon. There are eight basic steps. Some of these steps are repeated many times, making the total number of steps one hundred or more. The entire process usually takes from 10 to 30 days. The steps are performed in a clean room where dust, temperature, and humidity are all very closely controlled. The eight basic steps are as follows:

1. Deposition (forming an insulating layer of silicon dioxide, or SiO_2, on the silicon wafer)
2. Photolithography (exposing a light-sensitive layer through a patterned photomask)
3. Etching (removing patterned areas using plasma gas or chemicals)
4. Doping (placing donor and acceptor impurities into the wafer by diffusion or by using ion implantation)
5. Metallization (forming interconnects and connection pads by depositing metal)
6. Passivation (applying a protective layer)
7. Testing (using probes to check each circuit for proper electrical function)
8. Packaging (separating wafers into chips, after which the chips are mounted and bonded/wired, and the packages are sealed)

Figure 13-3 shows how the wafers of silicon needed to make ICs are manufactured. (*a*) Chunks of polysilicon (multiple, small crystalline grains) are loaded into a quartz crucible and heated to around 2,500°F using RF induction coils. A seed crystal about the diameter of a pencil is lowered into the crucible until it contacts the top of the molten silicon. The seed is colder than the molten silicon and causes the silicon to start solidifying (freezing) at the contact point. The seed is rotated and slowly withdrawn from the crucible at about 1.5 mm per minute. The result (*b*) is a monocrystalline (one-crystal) silicon ingot that weighs over 181 kg and is about 400 mm in diameter. The ingot is trimmed and ground (*c* and *d*) and then sliced (*e*) into wafers that are about 1.5 mm thick. The wafer edges are

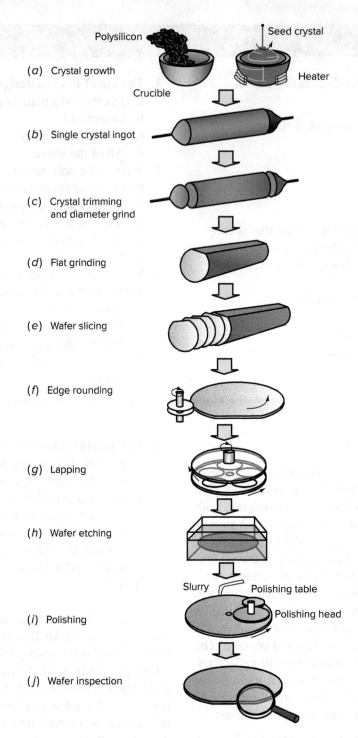

(a) Crystal growth

Polysilicon
Seed crystal
Crucible
Heater

(b) Single crystal ingot

(c) Crystal trimming
and diameter grind

(d) Flat grinding

(e) Wafer slicing

(f) Edge rounding

(g) Lapping

(h) Wafer etching

(i) Polishing

Slurry
Polishing table
Polishing head

(j) Wafer inspection

Fig. 13-3 Silicon wafer production.

Photoresist

Photomask

rounded (f), then the wafers are lapped (g) to remove the saw marks caused by the slicing. Acid etching (h) further improves surface quality, and a final polishing (i) provides a mirror finish. After inspection (j), the wafers are ready for processing.

The wafers are exposed to pure oxygen to form a layer of SiO$_2$. Next, the wafers are coated with *photoresist*, which is a material that hardens when exposed to light. The exposure is made through a *photomask*. Each mask has a pattern that will be transferred to the wafer. The unhardened areas of the photoresist, caused by the opaque areas of the photomask, wash away during the developing step. The wafer is then etched to remove the silicon dioxide and expose the patterned areas of the substrate. The exposed areas act as windows to allow penetration by impurity atoms. The remains of the photoresist are removed with chemicals or plasma gas. Figure 13-4 shows the major steps in this mostly photolithographic process. The wafer is reoxidized, and the photolithographic sequence is repeated from 8 to 20 times,

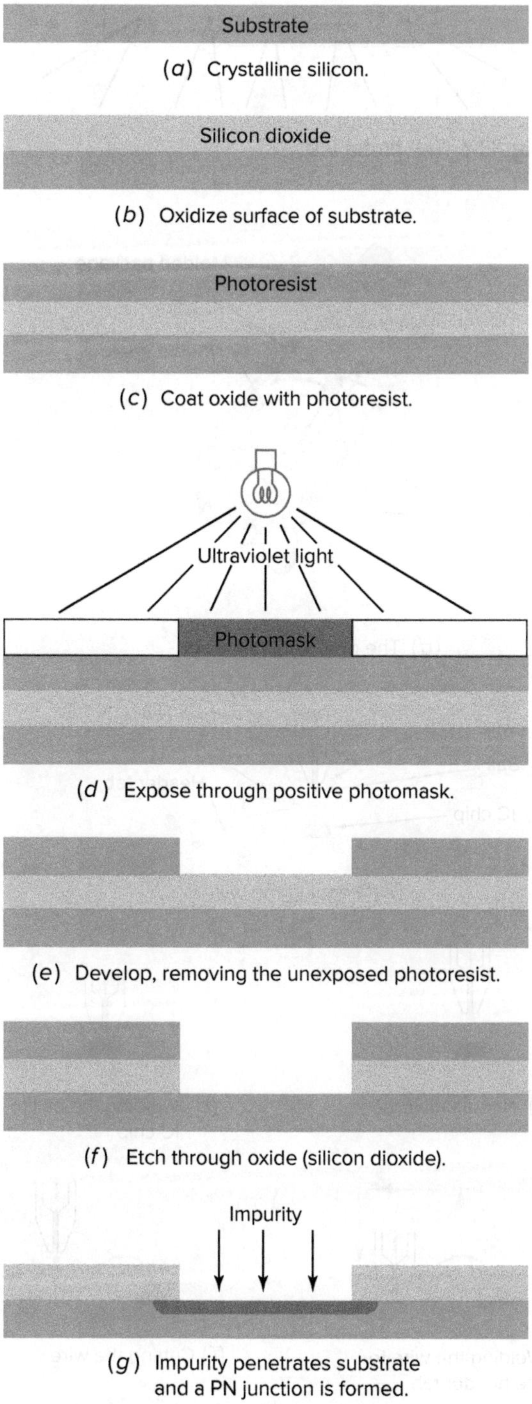

depending on the complexity of the IC being manufactured. Thus, photolithography is considered the *core process* in IC fabrication.

Core process

When the basic circuit has finally been completed, the surface is *passivated* using a silicon nitride coating. This coating acts as an insulator and also serves to protect the surface from damage and contamination.

Passivated

The wafer size back in 1971 was about 2 inches in diameter. Now, wafers larger than 12 inches in diameter are being processed. This means that ICs are being manufactured in ever-increasing batch sizes, and that's one of the reasons costs are decreasing. A large wafer will yield hundreds or thousands of individual chips (Fig. 13-5). Some of the individual chips might be defective. Figure 13-6 shows that needle-sharp probes are used to electrically test each chip. The defective ones are marked with a dot of ink for later disposal. The wafer is cut apart with a diamond saw, and the good circuits, now called chips, are mounted onto metal headers as shown in Fig. 13-7. The chip pads and header tabs are connected with very fine wire. Ball bonding, or more likely ultrasonic bonding, is used to make the connections. The package is then sealed. Plastic packages are most common and ceramic or metal packages are used for military or other critical applications.

The same fabrication techniques used for an IC are used to make a microelectromechanical system (MEMS). Such systems have at least some elements with mechanical functions, and these elements may or may not move. Figure 13-8 shows a partial view of a micro machine with gears that do move. One gear tooth in this machine is about 1 micron (1×10^{-6} m) wide, and one gear is about 13 microns in diameter. It is amazing that these little machines are batch-processed on silicon wafers like ICs. They vary from basic devices with no moving parts to extremely complex electromechanical systems with multiple moving elements under the control of integrated microelectronics.

MEMS can be used as

- Sensors such as accelerometers
- Gyroscopes
- Printheads
- Digital light projectors
- Biosensors and drug delivery devices

Substue

(*a*) Crystalline silicon.

Silicon dioxide

(*b*) Oxidize surface of substrate.

Photoresist

(*c*) Coat oxide with photoresist.

Ultraviolet light

Photomask

(*d*) Expose through positive photomask.

(*e*) Develop, removing the unexposed photoresist.

(*f*) Etch through oxide (silicon dioxide).

Impurity

(*g*) Impurity penetrates substrate and a PN junction is formed.

Fig. 13-4 The photolithographic process.

(a) Design the circuit.

(b) Design the layout.

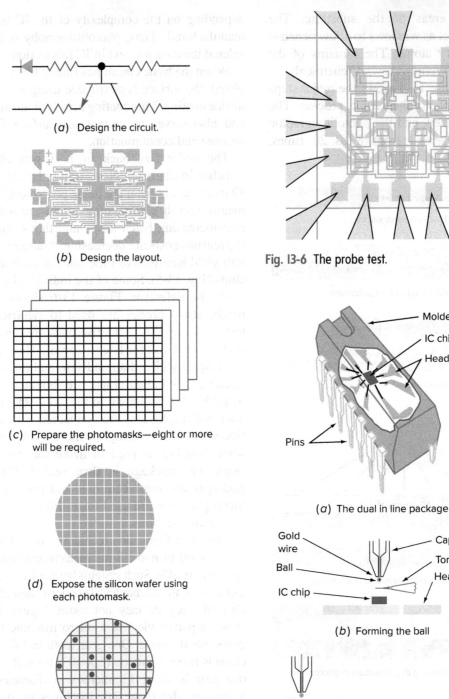

Fig. 13-6 The probe test.

(c) Prepare the photomasks—eight or more will be required.

(d) Expose the silicon wafer using each photomask.

(e) Run probe test and scribe the wafer.

(f) Saw into individual chips.

(g) Mount chip into package—bond and seal.

Fig. 13-5 The major steps in making an IC.

(a) The dual in line package

Molded package
IC chip
Header tabs
Pins

(b) Forming the ball

Gold wire
Ball
IC chip
Capillary needle
Torch
Header tab

(c) Needle lowered

(d) Welding the ball to IC chip

(e) Welding the wire to the header tab

(f) Cutting the wire

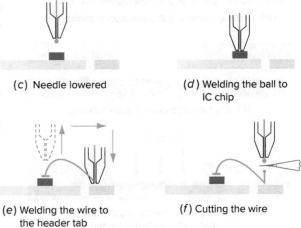

Fig. 13-7 The ball-bonding process.

Fig. I3-8 Part of an MEMS (Sandia National Laboratories).
Courtesy of Sandia National Laboratories

- Microwave filters and RF switches
- Material and gas sensors

This list gives only a small sample. New devices and novel applications appear on a regular basis. As an example, MEMS accelerometers can be found in smart phones, game controllers, fall detectors, tablet computers, and even some TV remote controls.

A general overview of IC fabrication has been presented so far, and more detail about transistor, diode, resistor, and capacitor circuit functions will now be offered. Figure 13-9 shows one way to fabricate an NPN junction transistor. A P-type substrate is shown. An N^+ layer is diffused into the substrate to form the collector of the transistor. N^+ means that more than the average number of impurity atoms enter the crystal. This is called heavy doping and serves to lower the resistance of the collector. An N layer is then formed over the substrate using an *epitaxial* process. Epitaxy is the controlled growth on a crystalline substrate of a crystalline layer, called an epilayer. The epilayer exactly duplicates the properties and crystal structure of the substrate. The epilayer is oxidized and exposed through a photomask. After developing, a P-type impurity such as boron is diffused into the windows until the substrate is reached. This electrically isolates an entire region on the N-type epilayer. This is called the *isolation diffusion* and allows separate electrical functions to exist in a single layer.

Refer again to Fig. 13-9. Photolithography opens a window, and a P-type impurity can be diffused in to form the base of the transistor. Later, an N-type diffusion will form the emitter. Polarity reversals by repeated diffusions will eventually saturate the crystal so their number is usually limited to three. Since emitters are normally heavily doped in any case, the process is designed so that the emitter diffusion is the last one.

The transistor has now been electrically isolated, and its three regions have been formed. To be useful, it must be connected. Once again, the wafer is oxidized, and photolithography is used to open up windows as shown in Fig. 13-10. These expose the connection points for the emitter, base, and collector. Aluminum is evaporated and then deposited onto the surface of the wafer to make contact through the windows. Photolithography is used to pattern the metal layer. Etching removes the unwanted aluminum, and Fig. 13-10(*c*) and (*d*) show what remains. Complex ICs can have two or even three separate aluminum layers separated by dielectric layers.

While the transistors are being formed, diodes are also being formed. Figure 13-11 shows a PN-junction diode in an IC. Notice that it

Isolation diffusion

Epitaxial

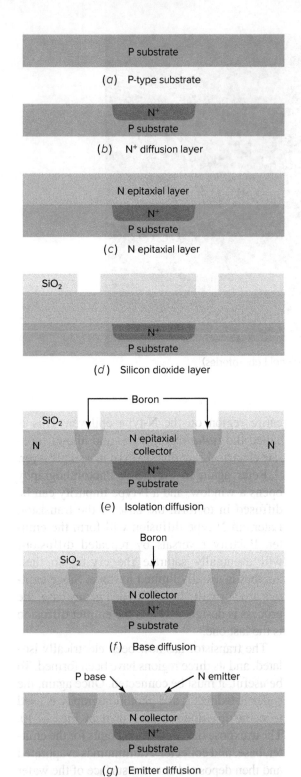

(a) P-type substrate

(b) N$^+$ diffusion layer

(c) N epitaxial layer

(d) Silicon dioxide layer

(e) Isolation diffusion

(f) Base diffusion

(g) Emitter diffusion

Fig. I3-9 Forming an NPN-junction transistor.

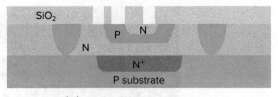

(a) Oxide layer with openings

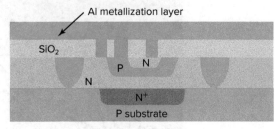

(b) Aluminum evaporated onto the wafer

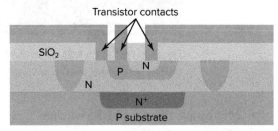

(c) The unwanted aluminum etched away

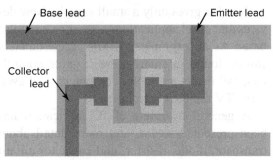

(d) Top view showing the remaining aluminum

Fig. I3-I0 Connecting the transistor.

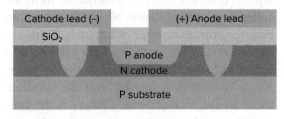

Fig. I3-II Forming a junction diode.

looks a lot like the transistor in Fig. 13-9. The collector-base junction is used as a diode, so no emitter diffusion is needed.

Figure 13-12 shows how a capacitor might be formed. The N-type region acts as one plate, with an aluminum layer as the other, and silicon

dioxide serves as the insulator. Another approach is to use a reverse-biased PN junction as a capacitor. Both methods are used.

Figure 13-13 illustrates resistor formation. Different values of resistance are realized by controlling the size of the N channel and the

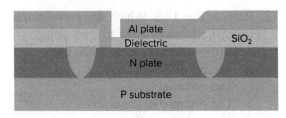

Fig. 13-12 Forming an MOS capacitor.

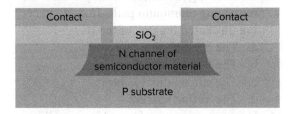

Fig. 13-13 Forming a resistor.

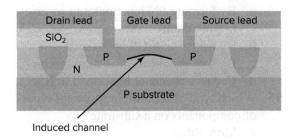

Induced channel

Fig. 13-14 Forming an MOS transistor.

level of doping. Once again, heavy doping produces less resistance.

An MOS transistor is shown in Fig. 13-14. Notice the insulating (SiO₂) layer between the gate and the channel. MOS transistors take up less space than BJTs and are often preferred for that reason.

IC components have certain limitations when compared with discrete components:

- Resistor accuracy is limited. However, resistors in hybrid ICs can be laser trimmed to overcome this.
- Very low and very high resistor values are not practical.
- Inductors are usually not practical.

- Only small values of capacitance are practical.
- PNP transistors tend not to perform as well as discrete types.
- High-voltage components are not practical.
- Power dissipation is usually limited to modest levels.

On the other hand, there are a couple of advantages for integrated components:

- Since all components are formed together, matched characteristics are easy to obtain.
- Since all components exist in the same structure, thermal tracking is inherent.

Of course, the biggest advantage is the huge savings in cost. Often, a single IC that costs less than a dollar can replace hundreds or thousands of discrete components that would cost hundreds of dollars.

So far, the discussion has been limited to monolithic (single-stone) ICs. Hybrid ICs combine several technologies. For example, a hybrid IC might contain monolithic ICs, film resistors, chip capacitors, and discrete transistors on a ceramic substrate. Hybrid ICs are generally more expensive than monolithic ICs. The advantages of hybrid ICs are as follows:

- Power levels can reach into the kilowatt region.
- Precision components can be used.
- Laser trimming can be used.

Self-Test

Choose the letter that best answers each question.

6. How are monolithic ICs made?
   a. On ceramic wafers
   b. By batch-processing on silicon wafers

   c. As miniature assemblies of discrete parts
   d. None of the above
7. What is the core process used in making monolithic ICs called?
   a. Photolithography
   b. Wave soldering

c. Electron-beam fusion
d. Acid etching

8. Refer to Fig. 13-4(*e*). How was the window produced?
   a. By boron diffusion
   b. With electron-beam milling
   c. By washing the unexposed area away
   d. By stencil cutting

9. What is the purpose of the probe test?
   a. To check the photoresist coatings
   b. To verify the ball-bonding process
   c. To count how many ICs have been processed
   d. To eliminate bad IC chips before packaging

10. What process is used to wire the chip pads to the header tabs?
    a. Ball or ultrasonic bonding
    b. Soldering
    c. Epoxy
    d. Aluminum evaporation

11. Refer to Fig. 13-9. Why is this called a monolithic IC?
    a. A hybrid structure is used.
    b. Everything is formed in a single slab of silicon.
    c. The base of the transistor is in the collector.
    d. All of the above are correct.

12. Refer to Fig. 13-9. Which step prevents the transistor from shorting to other components being formed at the same time?
    a. The N^+ diffusion layer
    b. The epitaxial layer
    c. The silicon dioxide layer
    d. The isolation diffusion

13. Refer to Fig. 13-10(*d*). What prevents the remaining aluminum paths from contacting unwanted regions of the IC?
    a. The N^+ diffusion layer
    b. The P-type substrate
    c. The silicon dioxide layer
    d. The Teflon spacers

14. How are capacitors formed in monolithic integrated circuits?
    a. By forming PN junctions and reverse-biasing them
    b. By using the MOS approach
    c. Both of the above
    d. Capacitors cannot be formed in ICs

15. Which type of IC combines several types of components on a substrate?
    a. Monolithic
    b. Silicon
    c. Digital
    d. Hybrid

13-3 The 555 Timer

The NE555 IC timer offers low cost and versatility. It is available in the 8-pin mini-DIP and in the miniature molded small outline package (MSOP).

The 555 provides stable time delays or free-running oscillation. The time-delay mode is *RC*-controlled by two external components. Timing from microseconds to hours is possible. The oscillator mode requires three or more external components, depending on the desired output waveform. Frequencies from less than 1 Hz to 500 kHz with duty cycles from 1 to 99 percent can be attained.

Figure 13-15 shows the major sections of the 555 timer IC. It contains two voltage comparators, a bistable flip-flop, a discharge transistor, a resistor divider network, and an output amplifier with up to 200-mA current capability. There are three divider resistors, and each is 5 kΩ. This divider network sets the *threshold* comparator trip point at two-thirds of V_{CC} and the *trigger* comparator at one-third of V_{CC}. V_{CC} may range from 4.5 to 16 V.

Suppose that $V_{CC} = 9$ V in Fig. 13-15. In this case the trigger point will be 3 V ($\frac{1}{3} \times 9$ V) and the threshold point 6 V ($\frac{2}{3} \times 9$ V). When pin 2 goes below 3 V, the trigger comparator output switches states and sets the flip-flop to the high state, and output pin 3 goes high. If pin 2 returns to some value greater than 3 V, the output stays high because the flip-flop "remembers" that it was set. Now, if pin 6 goes above 6 V, the threshold comparator switches states and resets the flip-flop to its low state. This does two things: the output (pin 3) goes low, and the discharge transistor is turned on. Note that the

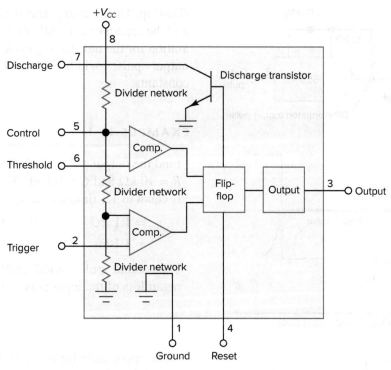

Fig. 13-15 Functional block diagram of the NE555 IC timer.

output of the 555 timer is *digital*; it is either high or low. When it is high, it is close to V_{CC}, and when it is low, it is near ground potential.

Pin 6 in Fig. 13-15 is normally connected to a capacitor that is part of an external *RC* timing network. When the capacitor voltage exceeds $\frac{2}{3}V_{CC}$, the threshold comparator resets the flip-flop to the low state. This turns on the discharge transistor, which can be used to discharge the external capacitor in preparation for another timing cycle. Pin 4, the reset, gives direct access to the flip-flop. This pin overrides the other timer functions and pins. It is a digital input, and when it is taken low (to ground potential), it resets the flip-flop, turns on the discharge transistor, and drives output pin 3 low. Reset may be used to halt a timing cycle. The reset function is ordinarily not needed, so pin 4 is typically tied to V_{CC}. Once the 555 is triggered and the timing capacitor is charging, additional triggering (pin 2) will not begin a new timing cycle.

Figure 13-16 shows the IC timer connected for the *one-shot mode* or *monostable mode*. This mode produces an *RC*-controlled output pulse that goes high when the device is triggered. The timer is *negative-edge* triggered.

The timing cycle begins at t_1 when the trigger input falls below $\frac{1}{3}V_{CC}$. The trigger input must return to some voltage greater than $\frac{1}{3}V_{CC}$ before the time-out period. In other words, the trigger pulse cannot be wider than the output pulse. In those cases where it is, the trigger input must be *ac-coupled* as

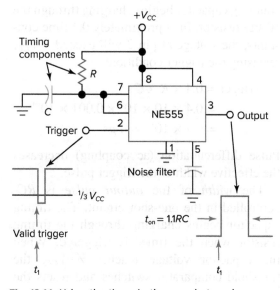

One-shot (monostable) mode

Fig. 13-16 Using the timer in the one-shot mode.

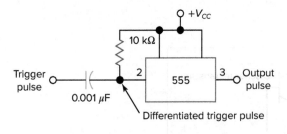

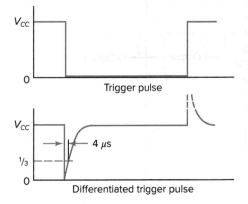

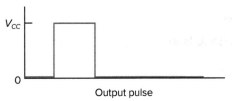

Fig. 13-17 An ac-coupled trigger pulse.

Pulse stretcher

Free-running (astable) mode

Differentiate

shown in Fig. 13-17. The 0.001-μF coupling capacitor and 10-kΩ resistor *differentiate* the input trigger pulse. The waveforms show that the negative edge of the trigger pulse causes pin 2 to drop to 0 V. This triggers the timer, and the coupling capacitor begins charging through the 10-kΩ resistor. In approximately 0.4 time constants, the voltage at pin 2 will exceed $\frac{1}{3}$ V_{CC}, releasing the trigger condition:

$$\begin{aligned} \text{Trigger} &= 0.4 \times R \times C \\ &= 0.4 \times 10 \times 10^3 \times 0.001 \times 10^{-6} \\ &= 0.4 \times 10^{-5} = 4 \ \mu\text{s} \end{aligned}$$

Pulse differentiation (ac coupling) decreases the effective width of the trigger pulse.

The *width* of the *output pulse* is RC-controlled in the one-shot circuit. The timing capacitor begins charging through the timing resistor when the timer is triggered. When the capacitor voltage reaches $\frac{2}{3}$ V_{CC}, the threshold comparator switches and resets the

flip-flop. The discharge transistor is turned on, and the capacitor is rapidly discharged in preparation for the next timing cycle. The resulting output pulse width is equal to 1.1 time constants.

EXAMPLE 13-1

Find the output pulse width for Fig. 13-16 if $R = 10$ kΩ and $C = 0.1$ μF. The pulse width is equal to 1.1 time constants:

$$\begin{aligned} t_{\text{on}} &= 1.1 \, RC = 1.1 \times 10^3 \times 0.1 \times 10^{-6} \\ &= 1.1 \text{ ms} \end{aligned}$$

The output pulse width will be 1.1 ms regardless of the input pulse width.

One application for the one-shot mode is to use it as a *pulse stretcher*. A pulse stretcher is often handy for troubleshooting digital logic circuits. A very narrow pulse can be stretched to give a visible flash of light from an LED indicator.

Figure 13-18 shows the timer configured for the *free-running* or *astable mode*. The trigger (pin 2) is tied to the threshold (pin 6). When the circuit is turned on, timing capacitor C is discharged. It begins charging through the series combination of R_A and R_B. When the capacitor voltage reaches $\frac{2}{3}$ V_{CC}, the output drops low, and the discharge transistor comes on. The capacitor now discharges through R_B. When the capacitor reaches $\frac{1}{3}$ V_{CC}, the output

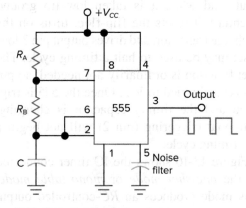

Fig. 13-18 The astable mode.

switches high, and the discharge transistor is turned off. The capacitor now begins charging through R_A and R_B again. The cycle will repeat continuously with the capacitor charging and discharging and the output switching high and low.

The charge path for the astable circuit is through two resistors, and the time that the output will be held high is given by

$$t_{high} = 0.69 (R_A + R_B)C$$

Assume that both timing resistors in Fig. 13-18 are 10 kΩ and that the timing capacitor is 0.1 μF. The output will remain high for

$$t_{high} = 0.69 (10 \times 10^3 + 10 \times 10^3) 0.1 \times 10^{-6}$$
$$= 1.38 \text{ ms}$$

The discharge path is through only one resistor (R_B), so the time that the output is held low is shorter:

$$t_{low} = 0.69 (R_B)C$$
$$= 0.69(10 \times 10^3) 0.1 \times 10^{-6}$$
$$= 0.69 \text{ ms}$$

The output waveform is nonsymmetrical. The total period can be found by adding t_{high} to t_{low}. The output frequency will be equal to the reciprocal of the total period. Or the output frequency can be found with

$$f_o = \frac{1.45}{(R_A + 2R_B) C}$$
$$= \frac{1.45}{(10 \times 10^3 + 20 \times 10^3) 0.1 \times 10^{-6}}$$
$$= 483 \text{ Hz}$$

The *duty cycle D* of a rectangular waveform is the percentage of time that the output is high. It can be found by dividing the total period of the waveform into the time that the output is high. For the astable circuit in Fig. 13-18, it can be found from

$$D = \frac{R_A + R_B}{R_A + 2R_B} \times 100\%$$

Assuming two 10-kΩ timing resistors gives

$$D = \frac{10 \times 10^3 + 10 \times 10^3}{10 \times 10^3 + 20 \times 10^3} \times 100\%$$
$$= 66.7\%$$

EXAMPLE 13-2

Calculate the output frequency and duty cycle for the circuit in Fig. 13-18 if $R_A = 1$ kΩ, $R_B = 47$ kΩ, and $C = 1$ μF. Is the output a square wave? The output frequency is given by

$$f_o = \frac{1.45}{(R_A + 2R_B)C}$$
$$= \frac{1.45}{(1 \text{ k}\Omega + 2 \times 47 \text{ k}\Omega)1 \text{ } \mu\text{F}} = 15.3 \text{ Hz}$$

The duty cycle:

$$D = \frac{R_A + R_B}{R_A + 2R_B} \times 100\%$$
$$= \frac{1 \text{ k}\Omega + 47 \text{ k}\Omega}{1 \text{ k}\Omega + 2 \times 47 \text{ k}\Omega} \times 100\%$$
$$= 50.5\%$$

When R_A is relatively small in value, the output approaches being a square wave.

The circuit shown in Fig. 13-18 cannot be used to produce a *square wave*. A square wave is a rectangular wave with a 50 percent duty cycle. The circuit also cannot provide waveforms with duty cycles smaller than 50 percent. The problem is that the timing capacitor charges through both resistors but discharges only through R_B. The duty-cycle equation shows that making R_A equal to 0 Ω will provide a 50 percent duty cycle. However, this can damage the IC, since there would be no current limiting for the internal discharge transistor.

Figure 13-19 shows a modification that permits *duty cycles* of 50 percent or less. A diode has been added in parallel with R_B. This diode bypasses R_B in the charging circuit. Now, the timing capacitor charges through R_A only and discharges through R_B as before. The following equations are appropriate for the modified circuit:

Square wave

Duty cycle

$$t_{high} = 0.69(R_A)C$$
$$t_{low} = 0.69(R_B)C$$
$$\text{Period} = T = t_{high} + t_{low}$$
$$f_o = \frac{1}{T} = \frac{1.45}{(R_A + R_B)C}$$
$$D = \frac{R_A}{R_A + R_B} \times 100\%$$

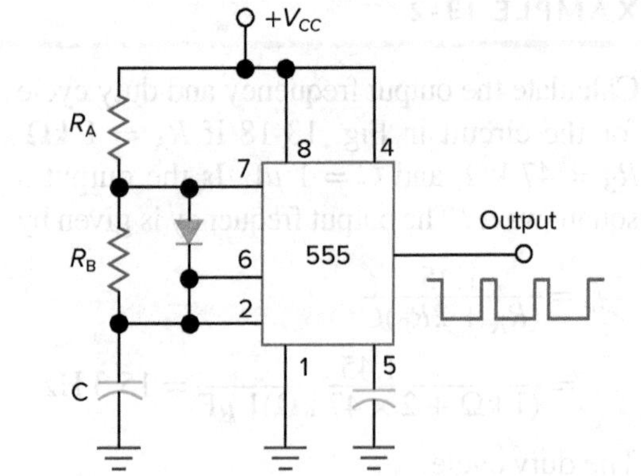

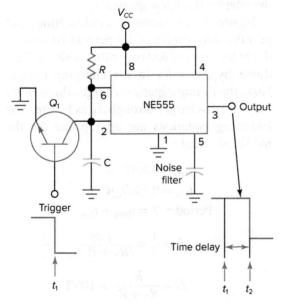

Fig. 13-19 Achieving duty cycles of 50 percent or less.

Time-delay mode

Figure 13-20 shows the NE555 operating in the *time-delay mode*. This mode calls for the output to change state at some determined time *after* the trigger is received. The time-delay circuit does not use the internal discharge transistor. Operation begins with Q_1 on, which keeps the timing capacitor discharged. Timing begins when the trigger signal goes low, turning Q_1 off. This allows timing capacitor C to begin charging through resistor R. When the capacitor reaches the threshold, the output switches to a low state. If $R = 47$ kΩ and $C = 0.5$ μF, the time delay can be found by

$$t_{\text{delay}} = 1.1 \times R \times C$$
$$= 1.1 \times 47 \times 10^3 \times 0.5 \times 10^{-6}$$
$$= 2.59 \times 10^{-2}\ \text{s} = 2.59\ \text{ms}$$

Fig. 13-20 Using the timer in the time-delay mode.

EXAMPLE 13-3

Select resistor values for the circuit in Fig. 13-19 that will produce a 1-kHz square wave when the timing capacitor is 0.01 μF. Beginning with the frequency equation,

$$f_o = \frac{1.45}{(R_A + R_B)C}$$

Rearranging gives

$$R_A + R_B = \frac{1.45}{f_o \times C}$$
$$= \frac{1.45}{1 \times 10^3\ \text{Hz} \times 0.01 \times 10^{-6}\ \text{F}}$$
$$= 145\ \text{k}\Omega$$

A square wave has a 50 percent duty cycle, so the resistors should be equal in value. Each resistor must be half of 145 kΩ:

$$R_A = R_B = \frac{145\ \text{k}\Omega}{2} = 72.5\ \text{k}\Omega$$

If the trigger signal goes high again before the IC times out, the output will not go low. This feature is useful in circuits such as security alarms where some time must be provided to exit an area before arming the alarm circuit.

For the 555 timer applications discussed so far, the control input (pin 5) has not been used. This input has been bypassed to ground with a noise capacitor (typically 0.01 μF) to prevent erratic operation. By applying a voltage at this pin, it is possible to vary the threshold comparator's trip point above or below the $\frac{2}{3} V_{CC}$ value. This feature opens other possibilities and allows the timer IC to function as a voltage-controlled oscillator or as a pulse-width modulator. Figure 13-21 shows the waveforms when a control signal is applied to an astable circuit.

EXAMPLE 13-4

Determine how much time is available to leave a protected area after arming an alarm that uses the circuit in Fig. 13-20, assuming $R = 470$ kΩ and $C = 50$ μF. The time delay is equal to 1.1 time constants:

$$t_{\text{delay}} = 1.1 \times 470\ \text{k}\Omega \times 50\ \mu\text{F} = 25.9\ \text{s}$$

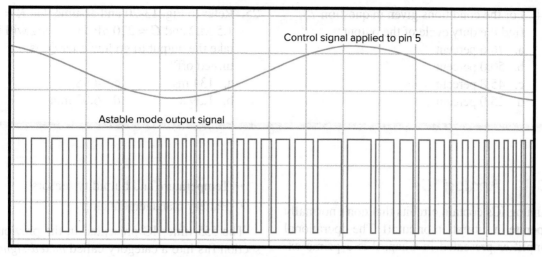

Fig. 13-21 Modulating the output of an astable circuit.

Self-Test

Choose the letter that best answers each question.

16. Refer to Fig. 13-16 and assume that V_{CC} is equal to 12 V. The IC will trigger when pin 2 drops below
    a. 2 V
    b. 4 V
    c. 6 V
    d. 8 V

17. Refer to Fig. 13-16 and again assume that V_{CC} is 12 V. The timing capacitor will begin to discharge when it reaches
    a. 2 V
    b. 4 V
    c. 6 V
    d. 8 V

18. Refer to Fig. 13-16 and assume that $C = 0.5\ \mu F$ and $R = 100\ k\Omega$. For a valid trigger, the output pulse width will be
    a. 220 μs
    b. 1.58 ms
    c. 55 ms
    d. 1.5 s

19. If the trigger input pulse width to a 555 operating in the one-shot mode is greater than the desired output pulse width, the trigger must be
    a. AC-coupled (differentiated)
    b. Inverted

    c. Amplified
    d. All of the above

20. Refer to Fig. 13-18 and assume that $R_A = 4.7\ k\Omega$, $R_B = 10\ k\Omega$, and $C = 0.01\ \mu F$. The output signal will be a
    a. Square wave
    b. Single rectangular pulse
    c. Rectangular wave
    d. Ramp wave

21. For the conditions given in question 20, find the output frequency.
    a. 898 Hz
    b. 5.87 kHz
    c. 18.9 kHz
    d. 155 kHz

22. For the conditions given in question 20, find the duty cycle.
    a. 59.5 percent
    b. 45.3 percent
    c. 33.7 percent
    d. 21.1 percent

23. Refer to Fig. 13-19 and assume that $R_A = R_B = 22\ k\Omega$, and $C = 0.005\ \mu F$. The output frequency will be
    a. 567 Hz
    b. 1.06 kHz
    c. 2.22 kHz
    d. 6.59 kHz

24. For the conditions given in question 23, find the duty cycle of the output.
   a. 76.6 percent
   b. 50.0 percent
   c. 45.8 percent
   d. 25.0 percent

25. Refer to Fig. 13-20 and assume that $R = 1.5$ MΩ and $C = 220$ μF. How long will it take the output to go low after Q_1 is turned off?
   a. 134 ms
   b. 1.39 s
   c. 4.98 s
   d. 6.05 min

13-4 Analog ICs

Analog ICs contain circuits that don't normally operate in saturation or cutoff. The operational amplifier presented in Chap. 9 is a prime example. The voltage regulator shown earlier in this chapter is another. Analog ICs include the following:

- Audio amplifiers
- RF and IF amplifiers (radio frequency and intermediate frequency)
- Modulators and mixers (used in communications)
- Operational amplifiers
- Instrumentation amplifiers (precision op amps)
- Voltage regulators
- Voltage references

- Temperature and humidity sensors
- Battery management chips

The 555 timer IC presented in the previous section fits into a category called *mixed-signal* ICs. These contain or use both digital and analog functions. Additional mixed-signal ICs are presented in the next section of this chapter.

Figure 13-22 shows the National Semiconductor LM1875 audio power amplifier. It is housed in a plastic package and can provide up to 25 W of power to an 8-Ω speaker. Its total harmonic distortion is less than one-tenth of 1 percent and its signal to noise ratio is 95 dB or better, so it qualifies for use in component stereo and home theater applications. It is both overvoltage and short-circuit protected. As the schematic shows, its use requires only a few external parts. Considering the low cost of this IC,

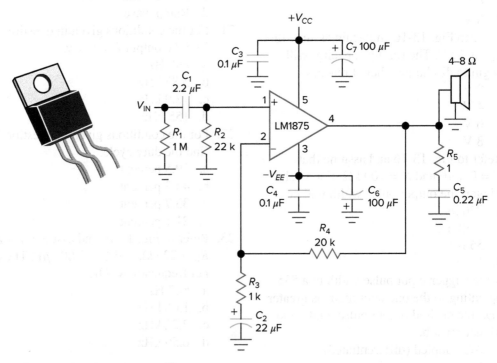

Fig. 13-22 IC audio power amplifier.

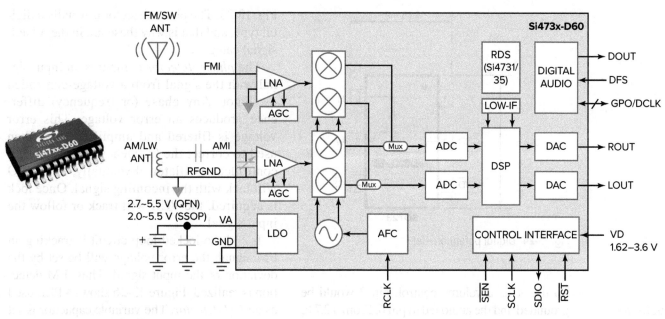

Fig. 13-23 IC radio receiver.
Courtesy of Silicon Labs

few designers would consider a discrete design if this device would serve instead.

Figure 13-23 shows an integrated circuit radio receiver. It works on four bands: AM, FM, SW (2 to 26 MHz), and LW (150 to 280 kHz). It features digital tuning, includes digital audio out, and uses digital signal processing for IF filtering.

Self-Test

Choose the letter that best answers each question.

26. Analog ICs contain
    a. Circuits that are normally driven to cutoff
    b. Circuits that are normally driven to saturation
    c. Both of the above
    d. None of the above
27. Mixed-signal ICs contain
    a. Both analog and digital circuits or functions
    b. Only analog circuits or functions
    c. Only digital circuits or functions
    d. Both BJTs and FETs

28. The 555 timer is an example of what type of IC?
    a. Analog   c. Mixed signal
    b. Digital   d. None of the above
29. Refer to Fig. 13-22. What is the purpose of C3, C4, C6, and C7?
    a. Muting   c. Equalization
    b. Bypassing   d. All of the above
30. Refer to Fig. 13-23. The IC package is an example of
    a. Surface mount technology
    b. A socketed chip for easy field replacement
    c. Both of the above
    d. None of the above

13-5 Mixed-Signal ICs

Mixed-signal ICs combine analog and digital circuit functions to provide improved performance and features not possible or practical using analog functions alone. The digital potentiometer shown in Fig. 13-24 provides 32 output levels at pin 5 (W represents a *wiper* contact, equivalent to a potentiometer center terminal).

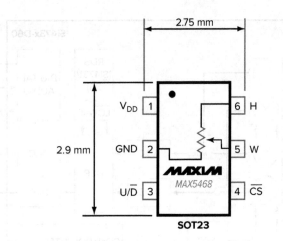

Fig. 13-24 Digital potentiometer.

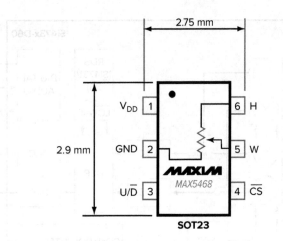

SOT23

V_DD 1   6 H
GND 2   5 W
U/D̄ 3   4 C̄S̄

MAXIM
MAX5468

2.75 mm
2.9 mm

FM detector

Up count mode

Down count mode

Tone decoding

Frequency synthesizer

To use it as a volume control, pin 2 would be grounded and the audio fed to pin 6. From +2.7 to +5.5 V dc is needed at pin 1. When pin 4 is low, the serial interface is active. The mode is set by the logic state of pin 3 at the moment pin 4 goes from high to low. At this moment, the chip will go into the *up count mode* if pin 3 is high. If pin 3 is low, the *down count mode* is selected. After the mode has been selected, the volume can be adjusted by sending digital pulses to pin 3.

Why is the Maxim IC in Fig. 13-24 considered a mixed-signal device? It easily qualifies for the digital function since both its mode and its output level are set by digital signals. However, it provides an analog circuit function, equivalent to what is accomplished by using a potentiometer. When you consider that early television receivers with remote control used motors, gear boxes, and rotating shaft potentiometers to turn the sound up and down, it's easy to appreciate the elegant simplicity of devices like this.

Phase-locked loops are interesting mixed-signal ICs. The major sections are shown in

Fig. 13-25. The phase detector is usually a digital type, and that is why these are in the mixed-signal category.

The *phase detector* compares an input signal with the signal from a voltage-controlled oscillator. Any phase (or frequency) difference produces an error voltage. This error voltage is filtered and amplified. It is then used to correct the frequency of the voltage-controlled oscillator. Eventually, the VCO will lock with the incoming signal. Once lock is acquired, the VCO will track or follow the input signal.

If a phase-locked-loop circuit is tracking an FM signal, the error voltage will be set by the deviation of the input signal. Thus, FM detection is realized. Figure 13-26 shows a PLL used as an *FM detector*. The variable capacitor is set so that the voltage-controlled oscillator operates at the center frequency of the FM signal. As modulation shifts the signal frequency, an error voltage is produced. This error voltage is the detected audio output. Phase-locked loops make very good FM detectors.

Phase-locked loops are also used as *tone decoders*. These are useful circuits that can be used for remote control or signaling by selecting different tones. In Fig. 13-27, two phase-locked loop ICs are used to build a dual-tone decoder. The output will go high *only* when *both* tones are present at the input. This type of approach is less likely to be accidentally tripped by false signals. Telephone touch-tone dialing systems use dual tones for this reason.

Frequency synthesizers have replaced older tuning methods in many electronic communications systems. Some use phase-locked loops combined with digital dividers to provide a range of precisely controlled output

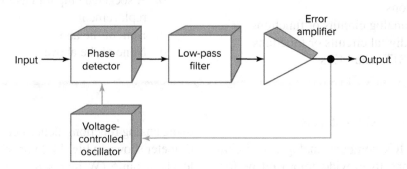

Fig. 13-25 Block diagram for a phase-locked loop (PLL).

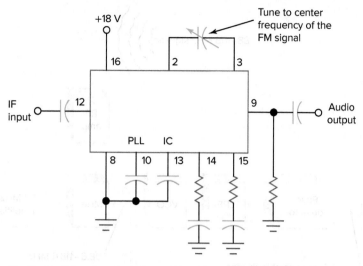

Fig. 13-26 Using the phase-locked loop for FM detection.

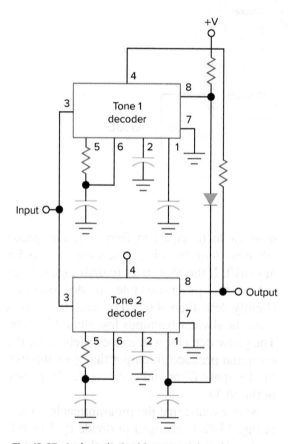

Fig. 13-27 A phase-locked-loop tone decoder.

frequencies. Figure 13-28 shows a partial block diagram for a synthesized FM receiver. Such a receiver is desirable because it is easy to tune and makes locating a given station easy. Analog dials on FM receivers are often several hundred kilohertz in error, and it may take some time to find a station even when the frequency is

known. A synthesized receiver is also very stable—so stable that no automatic frequency control circuit is needed.

The FM broadcast band extends from 88 to 108 MHz. The channels are spaced 0.2 MHz apart. The number of channels is found by

$$\text{No. of channels} = \frac{\text{frequency range}}{\text{channel spacing}}$$

$$= \frac{108 \text{ MHz} - 88 \text{ MHz}}{0.2 \text{ MHz}}$$

$$= 100$$

Figure 13-28 shows that a phase-locked loop, two crystal oscillators, and a programmable divider can be used to synthesize 100 FM channels. The stability of the output is determined by the stability of the crystals. Since crystal oscillators are among the most stable available, drift is not a problem. One input to the phase detector is derived by dividing a 10-MHz signal by 50. This produces a signal of 0.2 MHz and is called the *reference signal*. Note that the frequency of the reference signal is equal to the channel spacing. In a PLL synthesizer, the reference frequency is usually equal to the smallest frequency change that must be programmed.

Reference signal

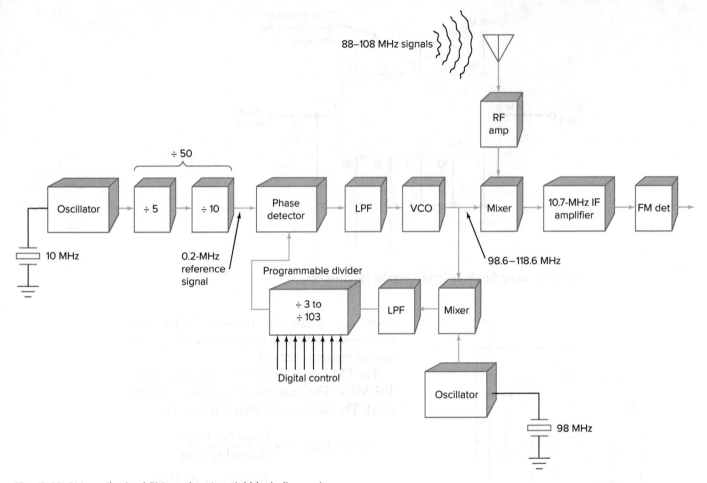

88–108 MHz signals

RF amp

÷ 50

Oscillator → ÷ 5 → ÷ 10 → Phase detector → LPF → VCO → Mixer → 10.7-MHz IF amplifier → FM det

10 MHz

0.2-MHz reference signal

Programmable divider

÷ 3 to ÷ 103

Digital control

LPF → Mixer

98.6–118.6 MHz

Oscillator

98 MHz

Fig. 13-28 PLL synthesized FM receiver (partial block diagram).

VCO

Figure 13-28 also shows a *voltage-controlled oscillator* (*VCO*). It feeds the receiver mixer (to the right) and the synthesizer mixer (below). Suppose you wanted to tune in a station that broadcasts at 91.9 MHz. To do so, the VCO would have to produce a signal higher than the station frequency by an amount equal to the IF frequency. The VCO frequency should be 91.9 MHz + 10.7 MHz = 102.6 MHz. The synthesizer mixer would subtract the second crystal-oscillator frequency of 98 MHz from the VCO frequency of 102.6 MHz to produce a difference of 4.6 MHz (102.6 MHz − 98 MHz = 4.6 MHz). This signal would be sent through a low-pass filter to a *programmable divider*. Assume that the divider is currently programmed to divide by 23. Therefore, the second input to the phase detector in Fig. 13-28 is 0.2 MHz (4.6 MHz ÷ 23 = 0.2 MHz), which is the same as the first input. All frequency synthesizers show both inputs to the phase

detector to be equal in frequency and phase *when the loop is locked*. The loop corrects for any drift. If the VCO tries to drift low, the signal to the programmable divider becomes slightly less than 4.6 MHz, and the output from the divider becomes less than 0.2 MHz. The phase detector will immediately sense the error and produce an output that goes through the low-pass filter and corrects the frequency of the VCO.

Now, assume that the programmable divider in Fig. 13-28 is changed to divide by 103. Immediately, the bottom input to the phase detector becomes much less than 0.2 MHz since we are dividing by a much larger number. The phase detector responds to this error and develops a control signal that drives the VCO higher and higher in frequency. As the VCO reaches 118.6 MHz, the system starts to stabilize. This is because 118.6 MHz − 98 MHz = 20.6 MHz, and 20.6 MHz ÷ 103 = 0.2 MHz, which is equal to the reference frequency. Any time the divider is programmed to a new number, the phase detector will develop a correction signal that will drive the VCO in the direction that will eliminate the error, and once again both inputs to the phase detector will become equal. Refer to Fig. 13-28 and verify that the entire FM band is covered by the synthesizer and that the channel spacing is equal to the reference frequency of 0.2 MHz.

The digital control signals to the programmable divider in Fig. 13-28 come from a front-panel keypad, a remote control, or perhaps a scanning circuit controlled by up and down push buttons. The user of the receiver programs the desired station frequency into the receiver. The frequency information is converted to the correct digital code and is sent to the programmable divider. This blend of digital and analog circuits is very common today. Very large scale integration (VLSI) chips that contain most of the synthesizer circuitry in one package are available. It is also worth mentioning that frequency synthesizers open up new areas of performance, such as channel memory, band scanning, and automatic channel change at a prescribed time.

Analog-to-Digital Conversion

The analog signals that come from sources such as sensors and microphones are *continuous*. Their voltage value changes smoothly with time. Another type of signal, called *discrete*, can change in voltage value only at specified points in time. A continuous signal must be changed into a discrete signal to make it manageable for digital processing. The process of changing a continuous signal involves *sampling* it at points in time. Usually, a sample-and-hold circuit is employed for the first part of this process. In Fig. 13-29(a), a continuous signal is shown, and in Fig. 13-29(b), its discrete counterpart is shown. Sampling begins at time t_0.

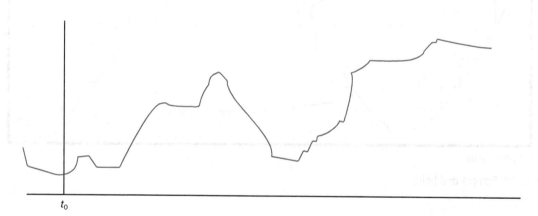

(*a*) Input to a sample-and-hold amplifier

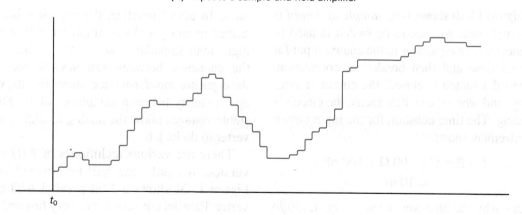

(*b*) Output of the sample-and-hold amplifier

Fig. 13-29 Changing a continuous signal into a discrete signal.

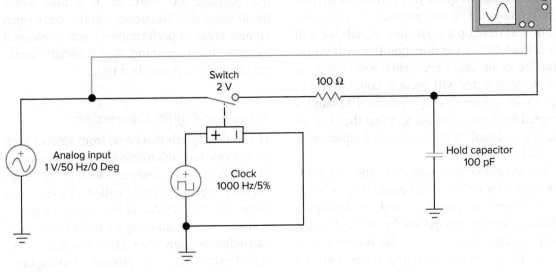

(a) Sample-and-hold circuit

Switch
2 V

Analog input
1 V/50 Hz/0 Deg

+ |

Clock
1000 Hz/5%

100 Ω

Hold capacitor
100 pF

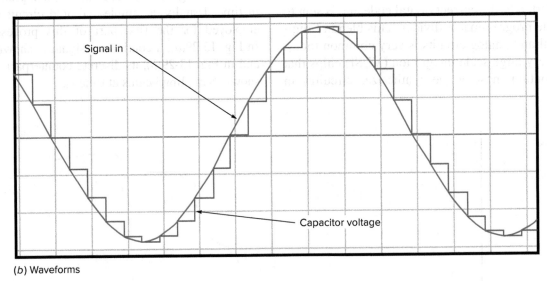

Signal in

Capacitor voltage

(b) Waveforms

Fig. 13-30 Sample and hold.

Sample and hold

Figure 13-30 shows how *sample and hold* is accomplished. An electronic switch is used to connect a *hold capacitor* to the analog input for a brief time and then break that connection. When the switch is closed, the circuit is sampling, and when the switch opens, the circuit is holding. The time constant for the input circuit is extremely short:

$$T = R \times C = 100 \text{ } \Omega \times 100 \text{ pF}$$
$$= 10 \text{ ns}$$

Flash converter

That's why the blue waveform in Fig. 13-30(*b*) shows instantaneous changes in voltage (note the vertical steps). After the switch opens, the capacitor voltage holds because there is no load

on it. In actual practice, the capacitor is connected to an operational amplifier with a very high input impedance so as not to discharge the capacitor between samples. Notice that the capacitor waveform (blue) shows that the voltage is steady between sampling points. These stable voltages allow the analog-to-digital converter to do its job.

There are various techniques of A/D conversion, but only one will be covered here. Figure 13-31 shows a 3-bit parallel A/D converter. Parallel converters are very fast and are often called *flash converters*. The problem with flash converters is that they are elaborate when a lot of bits are required. The word *bit* is

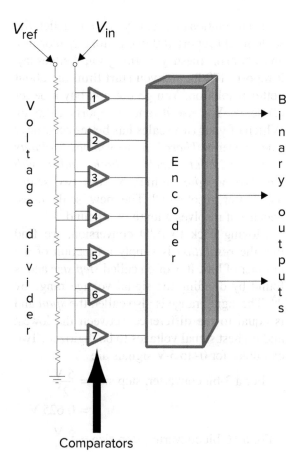

Fig. 13-31 A 3-bit parallel (flash) A/D converter.

a contraction of two words: *binary* and *digit*. There are only two binary digits: 0 and 1. Thus, an A/D converter changes each signal sample into some number of bits (0's and 1's). Here are some examples of the output of a 3-bit A/D:

- 000 (this binary number is equal to decimal 0)
- 011 (this binary number is equal to decimal 3)
- 101 (this binary number is equal to decimal 5)
- 111 (this binary number is equal to decimal 7)

Binary 0 is often called *LOW*, and binary 1 is often called *HIGH*. If there is any chance of confusion between binary and decimal, binary numbers may be written with a *subscript* 2, and more familiar decimal numbers with a *subscript* 10:

$$111_2 = 7_{10}$$

The 3-bit converter in Fig. 13-31 uses eight resistors in the voltage divider and seven comparators:

Number of resistors required $= 2^N = 2^3 = 8$

Number of comparators required $= 2^N - 1$

$$= 2^3 - 1 = 7$$

where $N =$ the number of bits.

Flash converters are practical when eight or fewer bits are enough. A flash converter for CD-quality audio would not be practical because 16 bits are required for each sound sample. Even with IC technology, the following calculations show that the circuit would be elaborate:

Number of resistors required $= 2^{16} = 65,536$

Number of comparators required $= 2^{16} - 1$

$$= 65,535$$

Returning to Fig. 13-31, we can see that each comparator has two inputs and that the input signal V_{in} is applied to every comparator. We also see that a reference voltage V_{ref} is applied to a voltage divider. We will assume a 5-V reference. The divider provides a different voltage to each comparator. The input signal will be compared to seven different voltages, so each comparator will trip at a different voltage value, as shown in Table 13-1. Each resistor in the voltage divider in Fig. 13-31 is the same value. You can solve for the voltages shown in Table 13-1 by using the voltage divider equation with $V_{ref} = 5$ V.

Note in Table 13-1 that none of the comparator outputs is high when the input signal is 0. This is because the voltage divider places all comparator inputs at some voltage greater

Table 13-1	Comparator Outputs	
Analog Input, V	Comparator Outputs	Data (Binary) Outputs
0.000	0000000	000
0.625	0000001	001
1.250	0000011	010
1.875	0000111	011
2.500	0001111	100
3.125	0011111	101
3.750	0111111	110
4.375	1111111	111

than 0. When the input signal reaches 0.625 V, only the bottom comparator trips since it is located at the lowest point on the voltage divider. At 1.25 V, the bottom two comparators are tripped. As the input signal voltage increases, additional comparators go high. Finally, at inputs of 4.375 V and above, all the comparator outputs are high. This data format is sometimes called the *thermometer code*. The thermometer code is not directly useful in most cases, so the encoder shown in Fig. 13-31 converts it to the binary data output form listed in Table 13-1.

Suppose you are troubleshooting the A/D converter shown in Fig. 13-31. What would you typically expect? If the circuit is normal, $V_{ref} =$ 5 V, and when V_{in} is fixed at 2.5 V, data 0 pin will be LOW (close to 0 V), data 1 pin will be the same, and data 2 pin will be HIGH (close to 5 V). Compare this with Table 13-1. Most often, all of Fig. 13-31 is in one IC so the comparator outputs will not be available for your measurements. If V_{in} varies with time, the data output pins will switch back and forth between about 0 and 5 V, and an oscilloscope will show a rectangular waveform at these pins.

Look again at Table 13-1. What if the input signal changes from 1.25 to 1.35 V and then to 1.75 V? What happens? The answer is: *nothing*. This particular A/D converter cannot *resolve* these changes. You should commit the following definitions to memory:

- *Resolution:* the ability to distinguish values, or the fineness of a measurement
- *Accuracy:* the conformity of a measured value with an accepted standard

Now you are ready for some interesting examples. This is necessary since the two words just defined are among those most abused in our language. *High resolution does not guarantee accuracy*—although it does imply it. If you visit a machine shop, you will see devices called *micrometers* that can measure in increments of one ten-thousandth of an inch. In other words, their resolution is 0.0001 in. However, if the micrometer has been dropped and bent, its *accuracy* could be far less. The bent micrometer still resolves 0.0001 in., but maybe it's off by 0.1 in.! The bent micrometer reports 0.5521 in. when measuring a true value of 0.4521 in.

Here is another example of resolution and accuracy: Digital bathroom scales typically

have a resolution of 1 lb. If you go on a diet, the scale might report 180 lb on Monday morning and 179 lb on Tuesday morning. On Wednesday, it reports 179 lb, and you start thinking about better resolution. You go out and buy a better (?) scale. You use it, and it reports 181.5 lb. Which of your two scales has better accuracy? The answer is, *There is no way to be absolutely certain without checking both scales with standard weights*. Which of your two scales has better resolution? The new scale does, because it resolves a tenth of a pound.

Moving back to A/D conversion, we find that the *resolution* is simply a function of the number of bits. It is often called *step size* and is found by dividing the signal voltage range by 2^N. The signal range is also called the *span* and is equal to the difference between the lowest and highest signal voltages to be digitized. Two examples for 0- to 5-V signals are

$$\text{For a 3-bit converter, step size} = \frac{5\text{ V}}{2^3}$$
$$= 0.625\text{ V}$$
$$\text{For a 16-bit converter, step size} = \frac{5\text{ V}}{2^{16}}$$
$$= 76.3\ \mu\text{V}!$$

We can conclude that a 3-bit A/D converter would not provide the needed resolution for high-quality audio, but a 16-bit converter would.

People appreciate compact disk audio because 16 bits are used for every sound sample. The high resolution makes the reproduction sound as good as the original. But remember that accuracy is also required in many applications. A scale might resolve a tenth of a pound but be off by 5 lb. Another scale that resolves only 1 lb but is never off by more than 1 lb is a better scale. Unfortunately, most people believe that the scale that resolves a tenth of a pound *has* to be more accurate, and this is not always the case. The accuracy of A/D conversion is a function of the devices used and the reference voltage. Other things, such as temperature, can also affect accuracy. So, a 7-bit converter will *always* have less resolution than an 8-bit converter, but it could have better accuracy.

Digital-to-Analog Conversion

As is true with A/D converters, D/A converters are most often ICs, and various types are used. Only one type will be presented here.

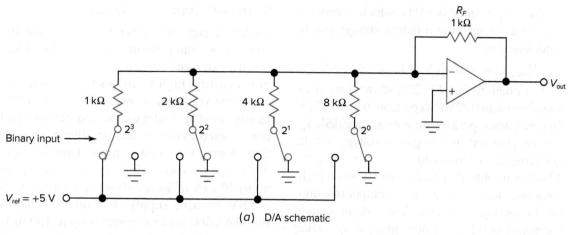

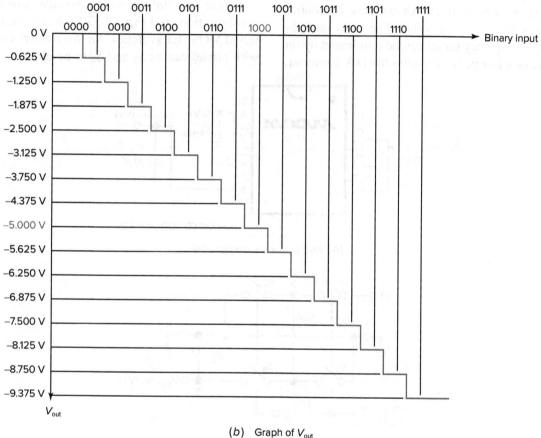

(b) Graph of V_{out}

Fig. 13-32 A 4-bit D/A converter.

Figure 13-32(a) shows the schematic for a 4-bit D/A converter. Each single-pole double-throw (SPDT) switch represents a binary input. When a switch is set to ground, the binary input is LOW or 0; the other switch position is HIGH or 1. As shown in Fig. 13-32(a), the 4-bit input happens to be at 1000_2. What will V_{out} be? Using standard op-amp theory,

$$V_{out} = -V_{ref} \times \frac{R_F}{R_{in}} = -5 \text{ V} \times \frac{1 \text{ k}\Omega}{1 \text{ k}\Omega} = -5 \text{ V}$$

The resolution (smallest step size) of the D/A converter circuit shown in Fig. 13-32(a) can be found by solving for the output voltage produced when the least significant bit (2^0) is high:

$$V_{out} = -V_{ref} \times \frac{R_F}{R_{in}} = -5 \text{ V} \times \frac{1 \text{ k}\Omega}{1 \text{ k}\Omega}$$

$$= -0.625 \text{ V}$$

The largest input is 1111_2, which is equal to 15_{10}, and the maximum output voltage can be determined by

$$V_{\text{out(max)}} = -0.625 \text{ V} \times 15 = -9.375 \text{ V}$$

The graph in Fig. 13-32(b) shows a staircase waveform with 0.625-V steps, from 0 to −9.375 V. In cases when positive outputs are needed, V_{ref} can be changed to a negative voltage, or the converter can be followed by an inverter. Figure 13-32(b) implies that there will be some high-frequency noise inherent in the output of a digital to analog converter. The output cannot change smoothly, as it does in a purely analog system. The staircase shows how the output steps from value to value. The steps (and the high-frequency noise) can be eliminated by the use of a low-pass filter after the D/A converter.

Switched Capacitor Devices

Switched capacitor circuits provide voltage conversion, integration, and filtering. They simplify designs and provide a better degree of versatility. Figure 13-33 shows a switched capacitor voltage converter. These are commonly used in battery-powered devices and can be used to invert, double, divide, or multiply a positive input voltage. One of these devices, the Maxim MAX1044, can supply up to 10 mA of load current from a 1.5-V to a 10-V positive supply. Maxim also makes the MAX660 for load currents up to 100 mA.

Figure 13-33(a) shows the voltage inverter configuration, and Fig. 13-33(b) shows the internal MOSFET switches. An inverter (triangle with circle) causes S_2 and S_4 to be opposite to

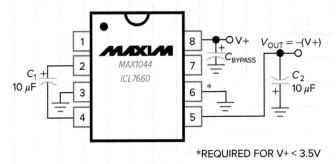

(a) Voltage inverter configuration

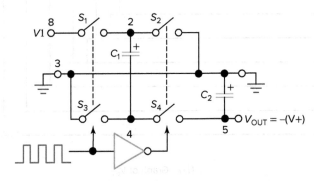

(b) The internal switches

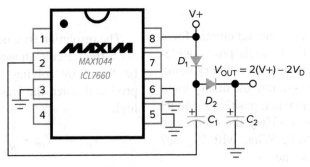

(c) Voltage doubler configuration

Fig. 13-33 Switched capacitor voltage converter.

the condition of S_1 and S_3. When 2 and 4 are open, 1 and 3 are closed. An internal oscillator toggles the switches at a 10 kHz rate. With 1 and 3 closed, C_1 quickly charges to V+. Then 1 and 3 open, and 2 and 4 close. This connects C_1 and C_2 in parallel, charging C_2. Note that the top plates of both capacitors are positive, making V_{out} negative with respect to ground.

Figure 13-33(c) shows the voltage doubler configuration. Two external diodes are required in addition to the two capacitors. Schottky diodes are recommended to limit the diode voltage loss. How does this circuit work? We can simplify the analysis by realizing that switches 3 and 4 in Fig. 13-33(b) can be ignored because

pin 4 of the IC is not used in the doubler circuit. Pin 2 is the key. It is switched between V+ and ground by switches 1 and 2. With pin 2 grounded, C_1 charges to V+ through D_1. Then pin 2 switches to V+, which acts in series with the charge on C_1, and a doubled voltage is applied to C_2 via D_2.

Switched capacitor integrators provide a circuit function that is equivalent to the op-amp integrator discussed in Chap. 9. Figure 13-34(a) shows an example. You should recall that the slope of V_{out} is given by

$$\text{Slope} = -V_{in} \times \frac{1}{RC}$$

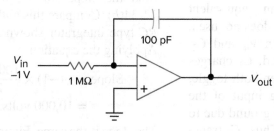

(a) Conventional integrator

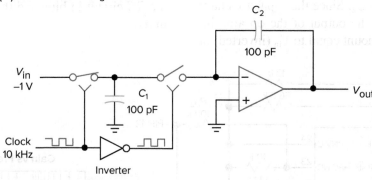

(b) Switched capacitor integrator

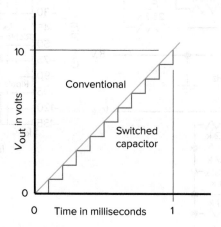

(c) Integrator time response

Fig. 13-34 Switched capacitor integrator.

EXAMPLE 13-5

Find the slope of V_{out} for Fig. 13-34(a) and determine V_{out} after 1 ms. Applying the equation,

$$Slope = -V_{in} \times \frac{1}{RC}$$

$$= -(-1) \times \frac{1}{1M \times 100p}$$

$$= 10,000 \text{ V/s}$$

Assuming an initial V_{out} of 0 V, the instantaneous output is found by multiplying the slope times the period:

$$V_{out(inst.)} = Slope \times period$$

$$= 10,000 \text{ V/s} \times 1 \text{ ms} = 10 \text{ V}$$

Figure 13-34(b) shows an equivalent switched capacitor integrator. It does not use a resistor. It switches C_1 between V_{in} and C_2. With the left-hand switch closed, C_1 charges quickly to V_{in}. The switches then both toggle, connecting C_1 to the inverting input of the op amp, which acts as a virtual ground due to the negative feedback via C_2. Thus, C_1 transfers its charge to C_2. Since the capacitors have the same value, the output of the op amp increases by an amount equal to V_{in} (inverted) or

+1 V in this case. Every time C_1 discharges into the virtual ground, the output will step more positive by another 1. The output slope of switched capacitor integrators such as the one shown in Fig. 13-34(b) is given by

$$Slope = -V_{in} \times \frac{C_1 \cdot f_{clock}}{C_2}$$

Why bother with switched capacitor integrators? The answer is, because they are controlled by clock frequency. This makes it possible to build timing circuits and filters that can be adjusted on the fly. This is one of the major reasons why mixed-signal ICs are gaining in popularity.

EXAMPLE 13-6

Find the slope for the integrator in Fig. 13-34(b). Compare this with the slope of the RC type integrator shown in Fig. 13-34(a). Applying the equation,

$$Slope = -(-1) \times \frac{100p \cdot 10k}{100p}$$

$$= 10,000 \text{ volts per second}$$

The slope is the same. Figure 13-34(c) shows the RC circuit in blue and the digital circuit in red.

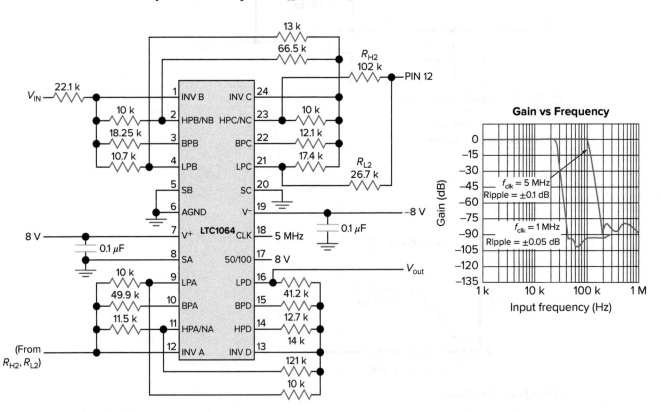

Fig. 13-35 Switched capacitor low-pass filter.

Figure 13-35 shows a switched capacitor low-pass filter. The graph shows that the filter's cutoff is about 25 kHz with a 1-MHz clock and about 100 kHz with a 5-MHz clock.

Self-Test

Choose the letter that best answers each question.

31. Refer to Fig. 13-24. When used as a volume control, the output signal at pin 5 is controlled by
    a. Digital pulses applied to pin 6
    b. The supply voltage at pin 1
    c. Digital pulses applied to pin 3
    d. None of the above

32. Refer to Fig. 13-25. The error signal is proportional to
    a. The amplitude of the input signal
    b. The amplitude of the VCO
    c. The phase error between the input and VCO
    d. None of the above

33. Refer to Fig. 13-27. The output will go high when the input is presented with
    a. Tone 1
    b. Tone 2
    c. Tone 1 and tone 2
    d. All of the above

34. Refer to Fig. 13-28. The loop is locked. What is the input frequency to the bottom of the phase detector?
    a. 100 MHz
    b. 10 MHz
    c. 1 MHz
    d. 0.2 MHz

35. Analog signals are
    a. Discrete
    b. Continuous
    c. Digital
    d. Binary

36. An A/D converter needs a steady analog voltage during the conversion process. This is accomplished by
    a. A shift register
    b. An encoder
    c. An array of comparators
    d. A sample-and-hold circuit

37. Which of the following defines the resolution for an N-bit A/D converter?
    a. $2N-1$
    b. $2N$
    c. 2^N
    d. $2N^2$

38. What kind of signal is expected at the output of a D/A converter?
    a. Discrete
    b. Continuous
    c. Sinusoidal
    d. Analog

39. What type of filter typically follows a D/A converter?
    a. High-pass
    b. Low-pass
    c. Band-pass
    d. Band-stop

40. Refer to Fig. 13-33(b). If the supply is 9 V, what is V_{out}?
    a. +9 V
    b. +18 V
    c. −18 V
    d. −9 V

41. Refer to Fig. 13-33(c). If the supply is 9 V and the diodes are Schottky types, what is V_{out}?
    a. −18 V
    b. +16.6 V
    c. +17.6 V
    d. None of the above

42. Refer to Fig. 13-34. If the clock frequency in Fig. 13-34(b) is changed to 5 kHz, the slope of V_{out} shown in Fig. 13-34(c) will
    a. Not change
    b. Be twice as steep
    c. Be half as steep
    d. None of the above

43. Refer to Fig. 13-35. If the chip is clocked at pin 1 from an external source, the cutoff frequency of the filter will then be controlled by
    a. The supply voltage
    b. The frequency of the external clock
    c. The phase of the external clock
    d. The amplitude of the external clock

13-6 IC Communication Protocols

Integrated circuits can communicate with each other using a bus protocol such as **I2C** or **SPI**. An MEMS accelerometer chip might have both I2C and SPI interfaces but, generally, one or the other will be available for circuit board chip to chip communication.

What Is I2C?

I2C denotes *I*nter *I*ntegrated *C*ircuit. It's a two-wire synchronous (clocked) serial communication protocol used for short distances; one wire is used for the data (SDA) and another wire is used for the clock (SCL) as shown in Fig. 13-36(*a*). It can support a large number (128 or more) of devices on a single bus and work at various data rates from 100 kbits per second (standard mode) up to 5 Mbits per second (ultra-fast mode).

The data bus is bidirectional, which means the controller is able to both send and receive agent data. Each agent on the bus is addressable with a unique 7-bit or 10-bit address.

Data are transferred in packets, which can be the address of the device, a read/write command, or the data being transferred. The process is in six parts:

1. A start signal will be generated by the controller which signals the agents to listen to the bus and prepare to receive data.
2. The controller will send a 7-bit agent address plus one bit denoting direction. If the direction bit is a 0, the controller writes data to the agent. If it is a 1, the controller reads the agent data.
3. Each agent compares the address sent by the controller with its own address. The agent that matches the address returns an ACK (acknowledgment) bit by pulling SDA low.
4. When the controller receives an ACK from the agent, it will start transmitting or receiving data.
5. After transmitting each data frame, the agent returns another ACK bit to the controller to confirm that the frame is successfully received.
6. When the data transfer is completed, the controller will send a stop signal which signals the release of the bus to other devices and the bus will enter an idle state.

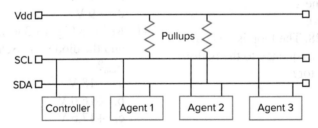

Fig. 13-36(*a*) I2C block diagram.

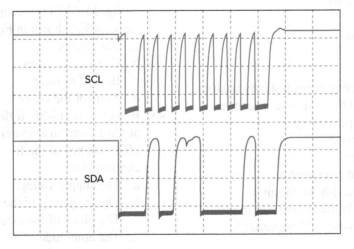

Fig. 13-36(*b*) I2C waveforms.

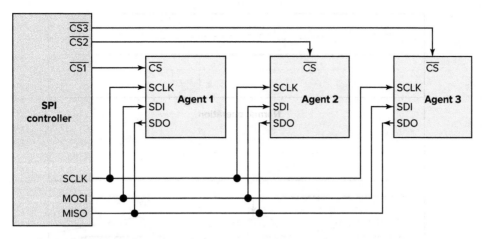

Fig. 13-37(a) SPI block diagram.

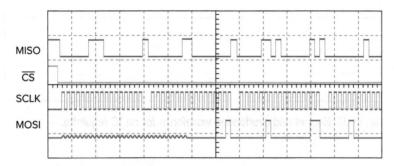

Fig. 13-37(b) SPI waveforms.

What Is SPI?

The *S*erial *P*eripheral *I*nterface is a four-wire communication protocol. The wires are MOSI (controller data to agent), MISO (agent data to controller), SCLK (a serial clock generated by the controller), and $\overline{CS}$ (agent chip select—active when at logic 0). The clock signal is in the MHz range.

Figure 13-37(a) shows an arrangement where an individual chip select for each agent is required from the controller. There is also a daisy chain arrangement (not shown) where the agents are configured such that the chip select signals are tied together and data propagate from one agent to the next. Communication is always started by the controller. Like I2C, it is a synchronous (clocked) communication protocol.

An SPI sequence follows:

1. The controller pulls CS low to activate the agent.
2. The controller sends the clock signal.
3. The controller sends data over MOSI to the agent.
4. The agent returns data using MISO.

What Is CAN?

Another serial communication protocol, CAN (*C*ontroller *A*rea *N*etwork), exists. It uses two wires and was developed for in-vehicle networks. It is a message-oriented transmission protocol and is widely used in trucks and automobiles. It is only mentioned here for completeness.

13-7 Troubleshooting

Troubleshooting procedures for equipment using integrated circuits are about the same as those covered in Chap. 10. The preliminary checks, signal tracing, and signal injection can all be used to locate the general area of the problem.

The real key to good troubleshooting of complex equipment is a sound knowledge of the overall block diagram. This diagram gives the symptoms meaning. It is usually possible to quickly limit the difficulty to one area when the function of each stage is known. It is really not important if the stage uses ICs or discrete circuits. The function of the stage is what helps to determine if it could be causing the symptom or symptoms.

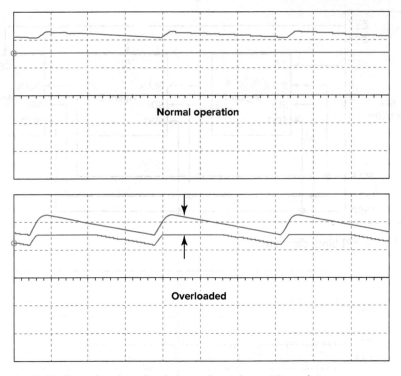

Fig. 13-38 Normal and overloaded waveforms for an IC regulator.

Consider an IC voltage regulator. What is it supposed to do? Look at the normal waveforms shown in Fig. 13-38. The red waveform is the output of the voltage regulator, and the blue waveform is the input voltage to the regulator. With normal operation the input voltage is higher, and there may be some ac ripple, as shown. The difference between the two is called margin. This is an important idea when troubleshooting; the regulator needs some voltage margin to work.

Look again at Fig. 13-38, and consider the overloaded waveforms. Now the output is not a straight line, it has quite a bit of ac ripple. If a technician is using a voltmeter, a lower-than-normal dc output voltage will be measured. The black arrows show that the regulator is actually working some of the time, but only when there is enough margin.

What about the situation where the waveforms are normal but the output is wrong? IC regulators have tolerance. As shown in Table 13-2, some error is normal. Examples of actual part numbers include LM7805, LM7812, and LM7815A. The table values are for a constant temperature, so actual in-circuit values may be different.

Table 13-2	Voltage Regulator IC Tolerance	
Worst-Case Low Voltage	LMXXXX Parts (Nominal V)	Worst-Case High Voltage
4.8	5.0	5.2
8.65	9.0	9.35
11.5	12.0	12.5
14.4	15.0	15.6
	LMXXXXA Parts (Nominal V)	
4.9	5.0	5.1
8.82	9.0	9.16
11.75	12.0	12.25
14.75	15.0	15.3

If the output of an IC voltage regulator is going off and on, be aware that many ICs have internal temperature sensing and internal shutdown circuitry. In some cases, the output can be normal for a time after power on and then drop as the IC heats up. Also, some have a feature that allows an initial surge of current. Others will immediately go into current limiting if too much current flows. Current limiting is covered in Chap. 15.

The 555 IC circuits rely on resistor and capacitor values for timing. Errors in frequency, pulse width, or duty cycle may be due to discrete component tolerance issues. Look back at Fig. 13-19; the diode affects duty cycle. If it is open, the duty cycle will be larger than normal and the frequency will be lower. If Q1 in Fig. 13-20 is open, the output will be stuck low. As always, verify supply voltages early in the process.

Next, we will look at a PLL FM detector. Figure 13-39 shows three waveforms. The modulating signal is a 100-Hz sine wave. The PLL frequency and carrier frequency are both 25 kHz in the top waveform, and the demodulated output (the white signal) is correct. In the second case, the carrier frequency is shifted up to 25.6 kHz, and the demodulated white signal is distorted. If the carrier is shifted down to 24.4 kHz, again the white signal is distorted. A PLL FM detector will not produce proper demodulation if the carrier frequency is shifted too far. This can be caused by tuning error in a superheterodyne receiver (covered in Chap. 12). Also, since PLLs have some limited capture range, a detector can malfunction if the FM deviation is too large.

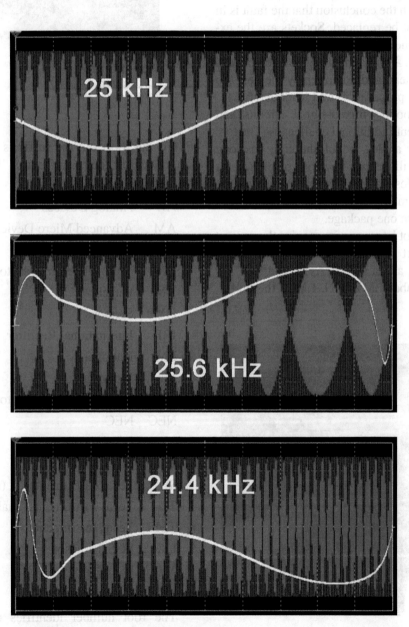

Fig. 13-39 PLL FM demodulator waveforms.
Source: Multisim

The lock range of a PLL is more than its capture range. Once a loop has acquired a signal, it can stay locked over a broader range of frequencies that can acquire lock.

When troubleshooting mixed-signal ICs, look for clock signals. If any are missing, circuit operation will not be normal. When troubleshooting I2C or SPI, the data busses can be monitored for activity. Also, the chip's function can be checked. For example, if the chip is a digital potentiometer, the analog input and output signals can be verified. As for the bus signals, modern oscilloscopes often provide decoding for I2C and SPI.

If you reach the conclusion that the fault is in an IC, it must be replaced. Sockets are the exception, not the rule. Thus, a tricky desoldering job is in store. Avoid damaging the circuit board with excess heat, and do not apply the heat too long. Use the proper tools and work carefully.

Troubleshooting mixed-signal circuits and devices presents additional challenges. The tool of choice for analog troubleshooting has long been the oscilloscope. The tool of choice for digital troubleshooting is often the logic analyzer. Many manufacturers now combine both instruments in one package.

Last but not least, repairs require the correct parts. IC part numbers often have a prefix, a root number, and a suffix. As an example, for the part number LM741CN, LM is the prefix,

Wafer turntable.

Jeff Maloney/Media Bakery

741 is the root, and the suffix is CN. Prefixes often identify the manufacturer. The following prefixes are only a partial list:

AD	Analog Devices
AM	Advanced Micro Devices
DG	Siliconix
DM	National Semiconductor (digital)
HM	Hitachi
HYB	Siemens
IRF	International Rectifier
LM	National
MC	Motorola
NDS	National Semiconductor
NEC	NEC
SD	SGS Thomson
SI	Siliconix
SN	Texas Instruments, TI (standard)
TL	Texas Instruments (analog, linear)
TMS	Texas Instruments
UA	SGS Thomson
XR	Exar Corp.
Z	Zilog

The root number identifies the part and will often be the same among several manufacturers. An LM741 and a UA741 are both

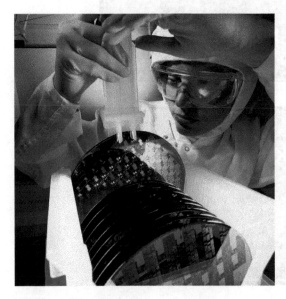

Working in the clean room.

Mark Joseph/Getty Images

general-purpose op amps with similar ratings and characteristics. The suffix can identify the package type, the operating temperature, the supply voltage, and so on. An LM741CN is an 8-pin DIP with a temperature range of 0 to 70°C, an LM741IN is an 8-pin DIP with a temperature range of 0 to 85°C, and an LM741H is packaged in a metal can.

Self-Test

Choose the letter that best answers each question.

44. What can cause error in the output voltage of an IC voltage regulator?
   a. Low input voltage
   b. Excess power supply ripple
   c. Excess load current
   d. All of the above
45. The circuit in 13-18 works, but the frequency at pin 3 is wrong. What could be at fault?
   a. R_A is out of tolerance.
   b. R_B is out of tolerance.
   c. C is out of tolerance.
   d. Any of the above may be at fault.
46. A PLL FM detector can correctly demodulate
   a. A signal in its capture range
   b. A signal above its capture range
   c. A signal below its capture range
   d. None of the above

Chapter 13 Summary and Review

Summary

1. Discrete circuits use individual components to achieve a function.
2. Integrated circuits decrease the number of discrete components and reduce cost.
3. Integrated circuits can reduce the size of equipment and the power required, and eliminate some factory alignment procedures.
4. Integrated circuits often outperform their discrete equivalents.
5. It is possible to increase the reliability of electronic equipment by using more ICs and fewer discrete components.
6. ICs are available in a variety of package styles.
7. Monolithic integrated circuits are batch-processed into 10-mil-thick silicon wafers.
8. The core process in making monolithic ICs is photolithography.
9. Photoresist is the light-sensitive material used to coat the wafer.
10. Aluminum is evaporated onto the wafer to interconnect the various components.
11. A monolithic IC uses a single-stone type of structure.
12. A hybrid IC combines several types of components on a common substrate.
13. The 555 timer can be used in the monostable mode, the astable mode, and the time-delay mode.
14. The output of a 555 timer IC is a digital signal.
15. The 555 timer uses three identical internal resistors in its voltage divider.
16. The internal divider sets trip points at one-third and two-thirds of the supply voltage.
17. The pulse width of a timer IC is controlled by external parts.
18. Applying a voltage to the control pin of the 555 timer allows it to be used as a VCO or as a variable-pulse-width modulator.
19. Analog ICs contain circuits that are not normally in saturation or cutoff.
20. Mixed-signal ICs combine analog and digital circuit functions.
21. A phase-locked loop compares an incoming signal with a reference signal and produces an error voltage proportional to any phase (or frequency) difference.
22. Phase-locked loops are used as FM detectors, as tone decoders, and as part of frequency synthesizers.
23. Switched capacitor ICs provide voltage conversion, integration, and filtering.
24. Check the power-supply voltages first when troubleshooting IC stages.
25. When troubleshooting ICs, check the dc voltages at all of the pins.
26. Always remove and insert socketed ICs with the power turned off.
27. An oscilloscope can be used to troubleshoot I2C and SPI signaling.

Related Formulas

555 one-shot mode: $t_{on} = 1.1\,RC$

555 astable mode: $t_{high} = 0.69(R_A + R_B)C$

$$t_{low} = 0.69\,R_B C$$

$$f_{out} = \frac{1.45}{(R_A + 2R_B)C}$$

$$\text{Duty cycle} = \frac{R_A + R_B}{R_A + 2R_B} \times 100\%$$

555 astable with diode in parallel with R_B:

$$t_{high} = 0.69\,R_A C$$

$$t_{low} = 0.69\,R_B C$$

$$f_{out} = \frac{1.45}{(R_A + R_B)C}$$

$$\text{Duty cycle} = \frac{R_A}{R_A + R_B} \times 100\%$$

555 time delay mode: $t_{\text{delay}} = 1.1\ RC$

A/D and D/A: Resolution $= 2^N$

$$\text{Step size} = \frac{\text{span}}{2^N}$$

Op amp RC integrator (inverting):

$$\text{Slope} = -V_{\text{in}} \times \frac{1}{RC}$$

$$V_{\text{out (inst.)}} = \text{slope} \times \text{period}$$

Switched capacitor integrator:

$$\text{Slope} = -V_{\text{in}} \times \frac{C_1 \times f_{\text{clock}}}{C_2}$$

TLC04 switched capacitor filter:

$$f_{\text{clock}} = \frac{1}{0.69\ R_{\text{CLK}} C_{\text{CLK}}}$$

$$f_{\text{cutoff}} = \frac{f_{\text{clock}}}{50}$$

$$f_{\text{max}} = \frac{f_{\text{clock}}}{2}$$

Chapter Review Questions

Choose the letter that best completes each statement.

13-1. A monolithic integrated circuit contains all of its components (13-2)
   a. On a ceramic substrate
   b. In a single chip of silicon
   c. On a miniature printed circuit board
   d. On an epitaxial substrate

13-2. A discrete circuit uses (13-2)
   a. Hybrid technology
   b. Integrated technology
   c. Individual electronic components
   d. None of the above

13-3. Refer to Fig. 13-1. When troubleshooting ICs, one may find a pin by (13-1)
   a. Counting counterclockwise from pin 1 (top view)
   b. Counting clockwise from pin 1 (bottom view)
   c. Both of the above
   d. None of the above

13-4. Refer to Fig. 13-1. One may find pin 1 on an IC by (13-1)
   a. Looking for the long pin
   b. Looking for the short pin
   c. Looking for the wide pin
   d. Looking for package markings and/or using data sheets

13-5. When electronic equipment is inspected, a positive identification of ICs can be made by (13-1)
   a. Using service literature and part numbers
   b. Counting the package pins

   c. Finding all TO-3 packages
   d. All of the above

13-6. Refer to Fig. 13-2. A technician needs the schematic (13-1)
   a. Seldom
   b. For choosing a replacement
   c. For troubleshooting
   d. To determine how to insert the replacement IC

13-7. The major semiconductor material used in making ICs is (13-2)
   a. Silicon
   b. Plastic
   c. Aluminum
   d. Gold

13-8. When monolithic ICs are made, the following is exposed to ultraviolet light: (13-2)
   a. Silicon dioxide
   b. Aluminum
   c. Photomask
   d. Photoresist

13-9. Which type of IC is capable of operating at the highest power level? (13-2)
   a. Discrete
   b. Hybrid
   c. Monolithic
   d. MOS

13-10. The pads on the IC chip are wired to the header tabs (13-2)
   a. By plastic conductors
   b. With photoresist

c. By ultrasonic bonding or ball-bonding
d. In a diffusion furnace

13-11. Refer to Fig. 13-9. Assume that the last boron diffusion [step (*f*)] was not performed. The component available is (13-2)
   a. An inductor
   b. A diode
   c. A resistor
   d. An MOS transistor

13-12. The function of the isolation diffusion is (13-2)
   a. To insulate the transistors from the substrate
   b. To insulate the various components from one another
   c. To improve the collector characteristics
   d. To form PNP transistors

13-13. The various components in a monolithic IC are interconnected to form a complete circuit by (13-2)
   a. The aluminum layer
   b. Ball bonding
   c. Printed wiring
   d. Tiny gold wires

13-14. Refer to Fig. 13-11. If this structure is to be used as a capacitor, the dielectric will be (13-2)
   a. The isolation diffusion
   b. The silicon dioxide
   c. The substrate
   d. The depletion region

13-15. Refer to Fig. 13-15. Many applications do not use pin (13-3)
   a. 8
   b. 7
   c. 4
   d. 2

13-16. Refer to Fig. 13-15. The signal at pin 3 is (13-3)
   a. Analog
   b. Continuous
   c. Digital
   d. All of the above

13-17. Refer to Fig. 13-16. A check with an accurate oscilloscope shows that the output pulse is only half as long as it should be. The problem is in (13-3)
   a. The timing resistor
   b. The timing capacitor
   c. The IC
   d. Any of the above

13-18. Refer to Fig. 13-16. You want to make the output pulse 1 s long. A 1-μF capacitor is already in the circuit. The value of the timing resistor should be (13-3)
   a. 1 kΩ
   b. 90 kΩ
   c. 220 kΩ
   d. 0.909 MΩ

13-19. Refer to Fig. 13-19. You want to build a square-wave oscillator with an output frequency of 38 kHz. Assume that a 0.01-μF capacitor is already in the circuit. The values for R_A and R_B are (13-3)
   a. $R_A = R_B = 189\ \Omega$
   b. $R_A = R_B = 3,798\ \Omega$
   c. $R_A = 1,899\ \Omega$ and $R_B = 3,798\ \Omega$
   d. None of the above

13-20. In Fig. 13-20, $R = 18$ kΩ and $C = 4.7\ \mu$F. The output will switch low, after the trigger, in (13-3)
   a. 18.2 ms
   b. 93.1 ms
   c. 188 ms
   d. 0.82 s

13-21. A phase-locked-loop IC makes an excellent tone decoder or (13-5)
   a. Voltage regulator
   b. FM demodulator
   c. Television IF amplifier
   d. Power amplifier

13-22. The reference frequency in a synthesizer is usually equal to (13-5)
   a. The VCO frequency
   b. The output frequency
   c. The crystal frequency
   d. The channel spacing

13-23. A digital potentiometer is an example of (13-5)
   a. An analog IC
   b. A digital IC
   c. A mixed signal IC
   d. None of the above

13-24. How many output levels can be achieved by a 12-bit D/A converter? (13-5)
   a. 256
   b. 1,024
   c. 4,096
   d. 8,192

13-25. A sample-and-hold circuit works by (13-5)
   a. Storing a voltage across a capacitor
   b. Storing a voltage in an inductor
   c. Latching a voltage into a flip-flop
   d. All of the above

13-26. A 13-bit A/D converter, when compared with a 12-bit A/D converter, always has more (13-5)
   a. Accuracy
   b. Resolution
   c. Speed
   d. Temperature stability

13-27. Switched capacitor voltage converters can provide (13-5)
   a. Increased output voltage
   b. Inverted output voltage
   c. Reduced output voltage
   d. All of the above

13-28. The slope of V_{out} from a switched capacitor integrator is controlled by (13-5)
   a. V_{in}
   b. The clock frequency

   c. The capacitor values
   d. All of the above

13-29. The cutoff frequency of a switched capacitor filter is controlled by (13-5)
   a. The amplitude of the input signal
   b. The clock frequency
   c. The clock phase
   d. The clock amplitude

13-30. Which of the following ICs is most likely to require a clock signal? (13-6)
   a. Mixed signal
   b. Analog
   c. Op amp
   d. Audio power amplifier

13-31. When troubleshooting a product that uses ICs, the first step is to (13-6)
   a. Replace the ICs one by one until it starts working normally
   b. Apply signal tracing
   c. Apply signal injection
   d. Check for power

Critical Thinking Questions

13-1. The photolithographic process used to make ICs is based on ultraviolet light. There is also a related process called *x-ray lithography*. Can you think of any reason for using x-rays to make ICs?

13-2. Several companies are experimenting with *fault-tolerant* ICs that are capable of repairing themselves. What kinds of applications might they be used for?

13-3. Mixed-signal ICs combine linear and digital functions. What are some examples?

13-4. IC manufacturers often license their designs to other manufacturers. This gives other corporations the right to make and sell their designs. Why would the original manufacturer do this?

13-5. Some electronic equipment contains ICs with part numbers that cannot be referenced in catalogs, data manuals, substitution guides, or reference books. Why would this be?

1. C	13. C	25. D	37. C
2. D	14. C	26. D	38. A
3. D	15. D	27. A	39. D
4. C	16. B	28. C	40. D
5. A	17. D	29. B	41. C
6. B	18. C	30. A	42. C
7. A	19. A	31. C	43. B
8. C	20. C	32. C	44. D
9. D	21. B	33. C	45. D
10. A	22. A	34. D	46. A
11. B	23. D	35. B	
12. D	24. B	36. D	

Electronic Control Devices and Circuits

Learning Outcomes

This chapter will help you to:

14-1 *Calculate* efficiency in control circuits. [14-1]

14-2 *Identify* the schematic symbols for thyristors. [14-2, 14-3]

14-3 *Explain* the operation of thyristors. [14-2, 14-3]

14-4 *Define* conduction angle in thyristor circuits. [14-2, 14-3]

14-5 *Explain* commutation in thyristor circuits. [14-2, 14-3]

14-6 *Discuss* servomechanisms. [14-4]

14-7 *Explain* the operation of photovoltaic and LED controllers. [14-5]

14-8 *Troubleshoot* control circuits. [14-6]

Control of loads is an important application area. For example, a control circuit may be used to accurately set and maintain the speed of a motor. Lights and heating elements can also be regulated with control circuits. The adjustable resistor, or rheostat, can be used to control loads. This chapter describes solid-state control devices and circuits that work much more efficiently than rheostats. It also shows how feedback can be used in control circuits.

14-1 Introduction

Figure 14-1 shows the use of a rheostat to control the brightness of an incandescent lamp. It is obvious that as the rheostat is adjusted for more *resistance*, the circuit current will decrease and the lamp will dim. The rheostat gets the job done but wastes energy. Solving a typical circuit will show why. In order to dim the lamp in Fig. 14-2, the rheostat has been set for a resistance of 120 Ω. This makes the total circuit resistance

Resistance control

$$R_T = 120\ \Omega + 120\ \Omega = 240\ \Omega$$

The circuit current can now be found by using Ohm's law:

$$I = \frac{V}{R} = \frac{120\ \text{V}}{240\ \Omega} = 0.5\ \text{A}$$

This, of course, is less current than when the rheostat is set for no resistance:

$$I = \frac{120\ \text{V}}{120\ \Omega} = 1\ \text{A}$$

Ohm's law has shown that setting the resistance of the rheostat equal to the load resistance halves the current flow.

Now let's investigate the power dissipated in the load. The current flow is 0.5 A, and the load resistance is 120 Ω:

$$P = I^2R = (0.5\ \text{A})^2 \times 120\ \Omega = 30\ \text{W}$$

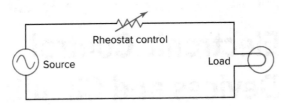

Fig. 14-1 A simple rheostat control circuit.

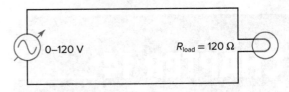

Fig. 14-3 Voltage control.

Efficiency

Switch control

When the rheostat is set at no resistance, the power is

$$P = (1 \text{ A})^2 \times 120 \ \Omega = 120 \text{ W}$$

The rheostat controls the power dissipated in the load. It has been shown that the power dissipated in the load drops to one-fourth when the current is halved. This is to be expected since power varies as the square of the current.

It is time to look at the *efficiency* of the rheostat control circuit. At full power, the rheostat is set for no resistance. Therefore, no power will be dissipated in the rheostat:

$$P = (1 \text{ A})^2 \times 0 \ \Omega = 0 \text{ W}$$

At one-fourth power, the rheostat dissipation is

$$P = (0.5 \text{ A})^2 \times 120 \ \Omega = 30 \text{ W}$$

This is not an efficient circuit. Half of the total power is dissipated in the control device when the current is halved for an efficiency of only 50 percent. As the resistance of the rheostat is increased, the circuit efficiency decreases. In a high-power circuit, the poor efficiency will produce a high cost of operation. The rheostat will have to be physically large to dissipate the heat safely.

The previous analysis was simplified. It assumed that the resistance of the incandescent lamp remains constant. It does not. However, the conclusions are correct. Rheostat control is inefficient.

What are the alternatives? One is *voltage control*. Figure 14-3 shows such a circuit. As the voltage of the source is adjusted from 0 to

120 V, the power dissipated in the load will vary from 0 to 120 W. This method is much more efficient than the rheostat control circuit. Since there is only one resistance in the circuit in Fig. 14-3, there is only one place to dissipate power. The efficiency of the circuit will always be 100 percent.

Unfortunately, voltage control is not easy to obtain. There is no simple and inexpensive way to control line voltage. A variable transformer is a possibility, but it would be a large and expensive item for a high-power circuit.

To be efficient, a control device should have very low resistance. A *switch* is an example of an efficient control device. When the switch in Fig. 14-4 is closed, 1 A of current flows. The power dissipated in the load is 120 W. If the switch has very low resistance, then very little power is dissipated in the switch. When the switch is open, no current flows. With no current, there cannot be any dissipation in the switch. Thus, there is never any significant power dissipation in a switch.

You may be wondering what the circuit of Fig. 14-4 has to do with dimming a lamp or controlling the speed of a motor. It seems that only on-off control is available. This is usually the case with ordinary mechanical switches. However, think for a moment about a very fast switch. Suppose this fast switch can open and close 60 times per second and is closed only half the time. What do you think the condition of the lamp will be? Since the lamp will be connected to the source only half the time, it will

Voltage control

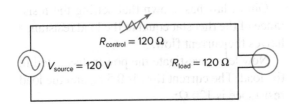

Fig. 14-2 Analyzing a rheostat control circuit.

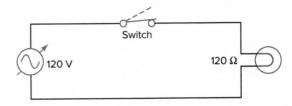

Fig. 14-4 Switch control.

operate at reduced intensity, and the control device (the fast switch) will run cool.

Mechanical switches cannot serve in this capacity. Even if they could be made to operate quickly, they would wear out in a short time.

An *electronic* (solid-state) *switch* is needed. Fast operation will allow the lamp to dim without any noticeable flicker, and the electronic switch will run cool. The next section covers such a control device.

Electronic switch

Self-Test

Choose the letter that best answers each question.

1. A load has a constant resistance of 60 Ω. A rheostat is connected in series with the load and set for 0 Ω. How much power is dissipated in the load if the line voltage is 120 V?
   a. 0 W
   b. 60 W
   c. 120 W
   d. 240 W
2. What is the circuit efficiency in question 1?
   a. 0 percent
   b. 25 percent
   c. 50 percent
   d. 100 percent
3. Refer to question 1. Everything is the same except the rheostat is set for a resistance of 30 Ω. How much power is dissipated in the load?
   a. 18 W
   b. 36 W
   c. 107 W
   d. 120 W
4. What is the circuit efficiency in question 3?
   a. 11 percent
   b. 67 percent
   c. 72 percent
   d. 100 percent

5. The resistance of a certain load is constant. The current through the load is doubled. The load power will increase
   a. 1.25 times
   b. 2.00 times
   c. 4.00 times
   d. Not enough information is given
6. Why is it not efficient to use a control resistor or a rheostat to vary load dissipation?
   a. Much of the total power is dissipated in the control.
   b. Power is set by voltage, not by circuit resistance.
   c. Power is set by current, not by circuit resistance.
   d. Loads do not show constant resistance.
7. Why is there no power dissipation in a perfect switch?
   a. When the switch is closed, its resistance is zero.
   b. When the switch is open, the current is zero.
   c. Both of the above are true.
   d. None of the above are true.

14-2 The Silicon-Controlled Rectifier

One of the most popular electronic switches is the *silicon-controlled rectifier (SCR)*. This device is easier to understand if we first examine the two-transistor equivalent circuit shown in Fig. 14-5. The circuit shows two directly connected transistors, one an NPN and the other a PNP. The key to understanding this circuit is to recall that BJTs do not conduct until base current is applied. It can be seen in Fig. 14-5 that each transistor must be on to supply the other with base current.

Silicon-controlled rectifier (SCR)

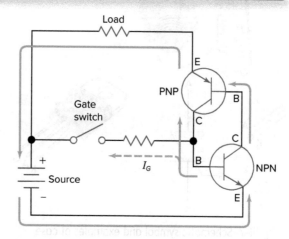

Fig. 14-5 A two-transistor switch.

How does the circuit in Fig. 14-5 turn on? Notice that a *gate switch* has been included. When the source is first connected, no current will flow through the load because both transistors are off. When the gate switch is closed, the positive side of the supply is applied to the base of the NPN transistor. This forward-biases the base-emitter junction, and the NPN transistor turns on. This applies base current to the PNP transistor and it turns on. With both transistors on, current flows through the load.

What happens in Fig. 14-5 when the gate switch opens? Will the transistors shut off and stop the load current? No, because once the transistors are on, they supply each other with base current. Once triggered by the gate circuit, the transistors in Fig. 14-5 continue to conduct until the source is removed or the load circuit is

opened. The two-transistor switch can be turned on by a gating current, but removing the gating current will not turn the switch back off. Such a circuit is often called a *latch*. Once triggered, a latch stays on.

The two-transistor switch in Fig. 14-5 is efficient. When the transistors are off, they show very high resistance, and the current and power dissipation approach zero. When the transistors are on, they are in saturation (turned on hard) and show low resistance. This means low power dissipation in the switch.

Figure 14-6 shows a way to simplify a two-transistor switch. A single four-layer device will do the same job. Study Fig. 14-6 and verify that it is the equivalent of the two transistors shown in Fig. 14-5. The *four-layer diode*, as shown in Fig. 14-6, is an important electronic control device. It is called a diode because it conducts in one direction and blocks in the other. It is usually called a silicon-controlled rectifier (SCR). The device is turned on by applying forward bias across the gate-cathode junction.

Figure 14-7 shows the SCR symbol and a few case styles. The electron flow is the same as for an ordinary diode, from cathode (K) to anode (A). The letter *K* is sometimes used to represent a cathode terminal in electronics. The small package is a TO-92, the larger one is a TO-220, and the largest is a stud mount type. These SCRs can have threads sizes of 10-32, ¼-28, ½-20, or ¾-16.

Figure 14-8 is a volt-ampere characteristic curve for an SCR. It shows device behavior for both forward bias (+V) and reverse bias (−V). As in ordinary diodes, very little current flows when the device is reverse-biased until the reverse breakover voltage is reached. Reverse breakover is avoided by using SCRs with

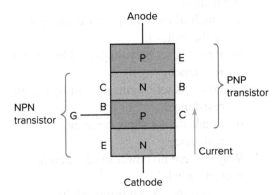

Fig. 14-6 A four-layer diode, or silicon-controlled, rectifier.

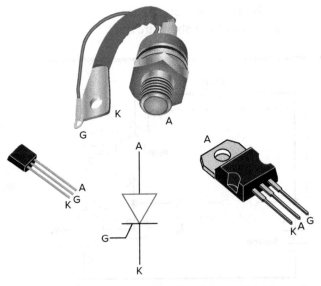

Fig. 14-7 Schematic symbol and examples of case style for SCRs.

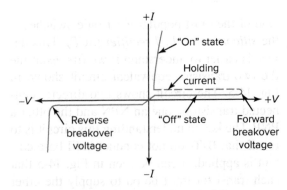

Fig. 14-8 An SCR volt-ampere characteristic curve.

ratings greater than the circuit voltages. The forward-bias portion of the volt-ampere curve is very different when compared to that of an ordinary diode. The SCR stays in the off state until the *forward breakover* voltage is reached. Then the diode switches to the on state. The drop across the diode decreases rapidly, and the current increases. The *holding current* is the minimum flow that will keep the SCR latched on.

Figure 14-8 is only part of the story because it does not show how gate current affects the characteristics of the SCR. Refer to Fig. 14-9. Gate current IG_1 represents the smallest of the three values of gate current. You can see that when gate current is low, a high forward-bias voltage is required to turn on the SCR. Gate current IG_2 is greater than IG_1. Note that less forward voltage is needed to turn on the SCR when the gate current is increased. Finally, IG_3 is the highest of the three gate currents shown. It requires the least forward bias to turn on the SCR.

In ordinary operation, SCRs are not subjected to voltages high enough to reach forward breakover. They are switched to the on state with a *gate pulse* large enough to guarantee turn-on even with relatively low values of forward-bias voltage. Once triggered on by gate current, the device remains on until the current flow is reduced to a value lower than the holding current.

Now that we know something about SCR characteristics, we can better understand some applications. Figure 14-10 shows the basic use of an SCR to control power in an ac circuit. The load could be a lamp, a heating element, or a

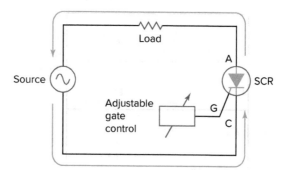

Fig. 14-10 Using an SCR to control ac power.

motor. The SCR will conduct in only the direction shown, so this is a *half-wave* circuit. The adjustable gate control determines when the SCR is turned on. Turnoff is automatic and occurs when the ac source changes polarity and reverse-biases the SCR.

Figure 14-11 shows the waveforms for the circuit in Fig. 14-10. The red waveforms are the load current (or load voltage, as they are shaped the same). The blue waveforms are the gate voltage. Note that the load current is zero until the gate pulses turn on the SCR. The circuit remains on (latched) until the source waveform (not shown in Fig. 14-11) reverses polarity. SCRs act as diodes and will not conduct when reverse-biased.

Gate pulse

The bottom pair of waveforms in Fig. 14-11 show low power. The gate pulses arrive late in the ac cycle, thus the SCR is on for only a brief period of time. At half-power, the gate pulses arrive at the moment when the source is at its peak value. For high power, the gate pulses arrive early in the cycle, and the SCR is on for most of the positive alternations. However, the negative alternations are not used and the circuit is considered a half-wave controller. It is possible to use two SCRs for full-wave control, and there are other methods that will be covered later.

The waveforms in Fig. 14-11 illustrate *conduction angle control.* The larger the conduction angle, the greater the load power. Also, circuits of this type are said to use *phase control.* As the phase angle of the gate waveform advances, the load power increases. Thus, the adjustable gate control block shown in Fig. 14-10 varies the phase of the gate pulses, with the source voltage serving as the phase reference.

Conduction angle control

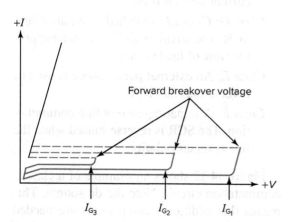

Fig. 14-9 The effect of gate current on breakover voltage.

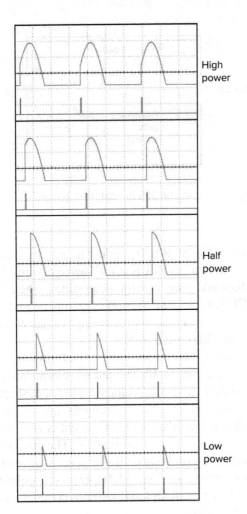

High
power

Half
power

Low
power

Commutation

Fig. 14-11 SCR waveforms.

Silicon-controlled rectifiers serve well in power-switching and power-control applications. They are available with voltage ratings ranging from 6 to around 5,000 V and with current ratings ranging from 0.25 to around 2,000 A. Even higher voltage and current ratings are possible by using series and parallel combinations of SCRs. Silicon-controlled rectifiers with moderate ratings can switch a load of hundreds of watts with a gate pulse of a few microwatts that lasts for a few microseconds. This performance represents a power gain of over 10 million and makes the SCR one of the most sensitive control devices available today.

Turning off an SCR requires that the anode-cathode circuit be zero-biased or reverse-biased. Reverse bias achieves the fastest possible turnoff. In either case, the turnoff will not be complete until all of the current carriers in the center junction of the device are able to

recombine. Recombination is a process of free electrons filling holes to eliminate both types of carriers. Recombination takes time. The time that elapses after current flow stops and before forward bias can be applied without turning the device on is the "turnoff" time. It can range from several microseconds to several hundred microseconds, depending on the construction of the SCR.

An SCR can be shut down by interrupting current flow with a series switch. Another possibility is to close a parallel switch, which would reduce the forward bias across the SCR to zero. In ac circuits, turnoff is usually automatic because the source periodically changes polarity. Whatever method is used, the process of shutting down an SCR is called *commutation*. Mechanical switches are seldom suitable for commutation of SCRs. A third approach is called forced commutation and includes six classes or categories of operation:

Class A: Self-commutated by resonating the load. A coil and capacitor effectively form a series resonant circuit with the load. Induced oscillations reverse-bias the SCR.

Class B: Self-commutated by an *LC* circuit. A coil and capacitor form a resonant circuit across the SCR. Induced oscillations reverse-bias the SCR.

Class C: C- or *LC*-switched by a second load-carrying SCR. A second SCR turns on and provides a discharge path for a capacitor or inductor-capacitor combination that reverse-biases the first SCR. The second SCR also provides load current when it is on.

Class D: C- or *LC*-switched by an auxiliary SCR. The auxiliary SCR does not support the flow of load current.

Class E: An external pulse source is used to reverse-bias the SCR.

Class F: Alternating-current line commutation. The SCR is reverse-biased when the line reverses polarity.

Figure 14-12 shows an example of a class D commutation circuit. Note the dc source. This means that additional components are needed for commutation. No load current can flow until SCR_1 is gated on. The load current flows

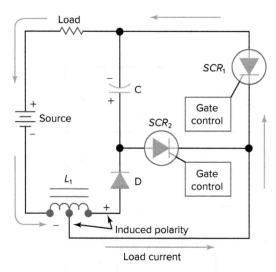

Fig. 14-12 Class D SCR commutation circuit.

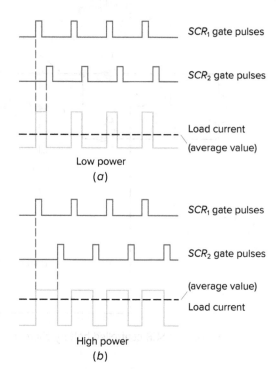

Fig. 14-13 Silicon-controlled (a) and (b) rectifier commutation waveforms.

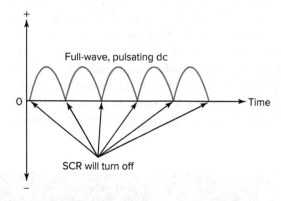

Fig. 14-14 Commutation with full-wave, pulsating direct current.

as shown, and the left-hand portion of L_1 is in the load circuit. As the load current increases through L_1, a magnetic field expands and induces a positive voltage at the right-hand terminal of L_1. This positive voltage charges capacitor C, as shown in Fig. 14-12. Diode D prevents the capacitor from discharging through the load, the source, and the inductor. When SCR_2 is gated on, the capacitor is effectively connected across SCR_1. Note that the positive plate of the capacitor is applied through SCR_2 to the cathode of SCR_1. Also note that the negative plate of the capacitor is applied to the anode of SCR_1. The capacitor voltage reverse-biases SCR_1 and it turns off.

In Fig. 14-12, SCR_1 supports the flow of load current. When it is gated on, load current begins to flow. SCR_2 is used to turn SCR_1 off. When it is gated on, load current stops. Load power can be controlled by the relationship between the gate timing pulses to the two SCRs. Look at Fig. 14-13(a). It shows that SCR_2 is gated on soon after SCR_1. The load current pulses are short in duration since the turnoff comes so soon after the turn-on. Now look at Fig. 14-13(b). It shows more delay for the gate pulses to SCR_2. The load current pulses are longer in time and more power dissipates in the load.

It is possible to achieve *full-wave control* with an SCR by combining it with a full-wave rectifier circuit. Figure 14-14 shows full-wave pulsating direct current. If an SCR is used in this type of circuit, it will no longer be

forward-biased at those times when the waveform drops to 0 V. The current in the SCR will drop to some value less than the holding current, and the SCR will turn off.

Figure 14-15 shows a battery charger that uses full-wave, pulsating direct current. Diodes D_1 and D_2, along with the center-tapped transformer, provide full-wave rectification. SCR_1 is in series with the battery under charge. It will be turned on early in each alternation by gate current applied through D_4 and R_4. Commutation is automatic in this circuit, as shown earlier in Fig. 14-14.

Full-wave control

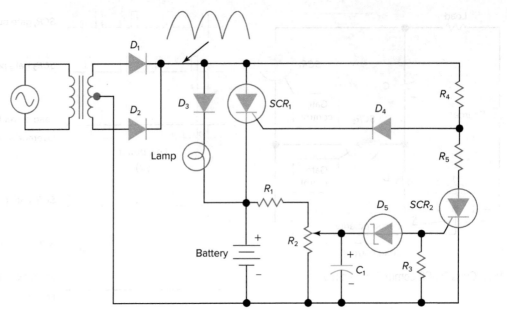

Fig. 14-15 An SCR-controlled battery charger.

The battery charger in Fig. 14-15 also features automatic shutdown when full charge is reached. As the battery voltage increases with charge, the voltage across R_2 also increases. Eventually, at the peak of the line, D_5 starts breaking down, and SCR_2 is gated on. As the battery voltage climbs even higher, the angle of SCR_2 keeps advancing (it is now coming on before the line peaks) until SCR_2 is eventually triggering before the input alternation is large enough to trigger SCR_1. With SCR_2 on, the voltage-divider action of R_4 and R_5 cannot supply enough voltage to forward-bias D_4 and gate SCR_1 on. The heavy charging has ceased. The battery is now trickle-charged through D_3 and the lamp (which lights to signal that the battery has reached full charge). The cutout voltage can be adjusted by R_2. Diode D_3 prevents battery discharge through SCR_2 in the event of a power failure.

Self-Test

Choose the letter that best answers each question.

8. Refer to Fig. 14-5. Assume that the source voltage has just been applied and the gate switch has not been closed. What can you conclude about the load current?
   a. The load current will equal zero.
   b. The load current will gradually increase.
   c. It will be mainly determined by V_{source} and the load resistance.
   d. The load current will flow until the gate switch is closed.

9. Refer to Fig. 14-5. Assume the gate switch has been closed and then opened again. What can you conclude about the load current?
   a. It will go off and then on.
   b. It will go on and then off.
   c. It will come on and stay on.
   d. None of the above is true.

10. How are SCRs normally turned on?
   a. By applying a reverse breakover voltage
   b. By applying a forward breakover voltage
   c. By a separate commutation circuit
   d. By applying gate current

11. How can an SCR be turned off in the shortest possible time?
    a. By zero-biasing it
    b. By reverse-biasing it
    c. By reverse-biasing its gate lead
    d. None of the above
12. What happens to the value of forward breakover voltage required to turn on an SCR as more gate current is applied?
    a. It is not changed.
    b. It increases.
    c. It decreases.
    d. None of the above is true.
13. Refer to Fig. 14-10. Assume that the adjustable gate control is set for maximum power dissipation in the load. What should the load waveform look like?
    a. Half-wave, pulsating direct current
    b. Full-wave, pulsating direct current
    c. Pure direct current
    d. Sinusoidal alternating current (same as the source)
14. How does an SCR control load dissipation in a circuit such as that shown in Fig. 14-10?
    a. The resistance of the SCR is adjustable.
    b. The source voltage is adjustable.
    c. The load resistance is adjustable.
    d. The conduction angle is adjustable.
15. Refer to Fig. 14-15. What is the function of SCR_2?
    a. It provides class D commutation for SCR_1.
    b. It limits the charging current to some safe value.
    c. It prevents SCR_1 from coming on when full charge is reached.
    d. It controls the conduction angle of SCR_1 when charging is started.

14-3 Full-Wave Devices

The SCR is a *unidirectional device*. It conducts in one direction only. It is possible to combine the function of two SCRs in a single structure to obtain *bidirectional conduction*. The device in Fig. 14-16 is called a *triac* (triode ac semiconductor switch). The triac may be considered as two SCRs connected in inverse parallel. When one of the SCRs is in its reverse-blocking mode, the other will support the flow of load current. Triacs are full-wave devices. They have limited ratings as compared with SCRs. They are available with current ratings up to about 40 A and voltage ratings to about 600 V. SCRs are capable of handling much more power, but triacs are more convenient for many low- and medium-power ac applications.

Figure 14-16 shows that the three triac connections are called main terminal 1, main terminal 2, and gate. The gate polarity usually is measured from gate to main terminal 1. A triac may be triggered by a gate pulse that is either positive or negative with respect to main terminal 1. Also, main terminal 2 can be either positive or negative with respect to main terminal 1

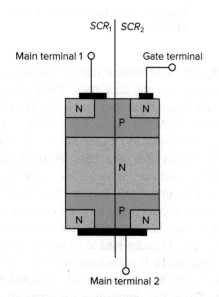

Triac

Fig. 14-16 The structure of a triac.

when triggering occurs. There are a total of four possible combinations or triggering modes for a triac. Table 14-1 summarizes the four modes for triac triggering. Note that mode 1 is the most sensitive. Mode 1 compares with ordinary SCR triggering. The other three modes require more gate current.

Table 14-1 Triac Triggering Mode Summary

Mode	Gate to Terminal 1	Terminal 1 to Terminal 2	Gate Sensitivity
1	Positive	Positive	High
2	Negative	Positive	Moderate
3	Positive	Negative	Moderate
4	Negative	Negative	Moderate

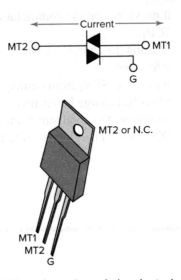

Fig. 14-17 Triac schematic symbol and a typical device.

Fig. 14-18 Another triac package.

Static switch

Figure 14-17 shows the schematic symbol for a triac and one case style. The load current flows between main terminal 1 (MT1) and main terminal 2 (MT2). Sometimes MT1 is called anode 1 (A1) and MT2 is anode 2 (A2). The red arrow shows that the triac is bidirectional, meaning load current can flow in both directions. The metal tab *may* be insulated (N.C. means no connection). It is not connected in a BTA41 triac, but in a BTB41, it is connected to MT2 (A2). This can lead to an unfortunate mistake! Figure 14-18 shows another triac package. This device has an electrically isolated metal case. Triacs are convenient for controlling (or switching) ac power. Silicon-controlled rectifiers are used when high-power levels are encountered. Both devices are in the thyristor family. **Thyristor** The term *thyristor* can refer to either an SCR or a triac. Thyristors may be used to perform static switching of ac loads. A static switch is one with no moving parts. Switches with moving parts are subject to wear, corrosion, contact bounce, arcing, and the generation of interference. Static

switching eliminates these problems. Most triacs are designed for 50 to 400 Hz and make good static switches over this frequency range. SCRs can operate to approximately 30 kHz.

Figure 14-19 shows a schematic diagram for a simple three-position *static switch*. In position 1, there is no gate signal and the triac remains off. In position 2, the triac is gated at every other alternation of the source, and the load receives half power. In position 3, the triac is gated at every alternation, and the load receives full power. The three-position switch is

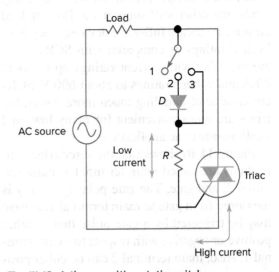

Fig. 14-19 A three-position static switch.

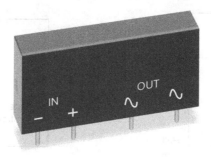

(a) Typical package style

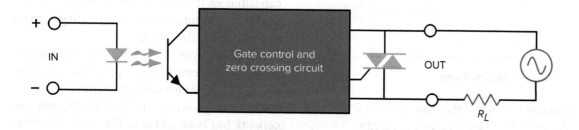

(b) What's inside the package

Fig. 14-20 Solid-state relay.

mechanical, but it operates in a low-current part of the circuit where arcing is not a problem.

Solid-state relays (SSRs) use static switching and optical isolation to make it safe and easy to control line-operated loads from logic level circuits. Figure 14-20 shows a typical unit. There is no electrical connection between the IN and OUT terminals. The breakdown voltage is the maximum safe potential difference from input to output. It is typically 4,000 V. The input is a digital signal, usually 0 V for off and 4 V for on. Most SSRs require about 2 mA of input current for turn on. The load side is rated from 24 to 600 V_{rms} at currents up to 8 A (depending on the particular part). The output holding current is 30 mA.

Figure 14-20(b) shows that the SSR includes a *zero crossing circuit*. Its purpose is to limit load surge when the relay turns on. If the triac happened to gate on during a line voltage peak, a large current surge could result. Current surges can cause damage. Interference is another reason why turn-on near the source peaks is undesirable. The sudden increase in current can cause radio-frequency interference. The zero crossing circuit allows the triac to be gated on only when the ac line is at or near a zero crossing. This limits both surge current and interference.

SSRs provide basic on-off load control. A different arrangement is used when smooth load control is needed. Light dimmers are an example of smooth load control. Figure 14-21 shows an adjustable gate control driving a triac.

Figure 14-22 shows the waveforms for the circuit in Fig. 14-21. The red waveforms are the load current, and the blue waveforms are the gate voltage. The load current is zero until the gate pulses turn on the triac. The triac

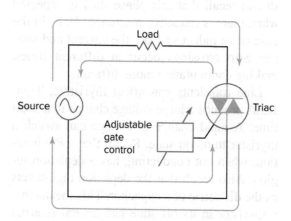

Fig. 14-21 Using a triac to control ac power.

Zero crossing circuit

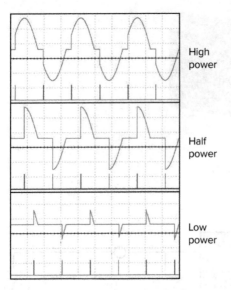

Fig. 14-22 Triac waveforms.

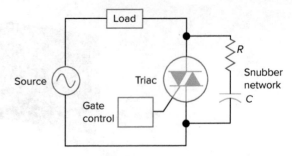

Fig. 14-23 A snubber network.

RC snubber network

remains on until the source waveform (not shown in Fig. 14-22) reverses polarity.

The bottom waveforms in Fig. 14-22 show low power. The gate pulses arrive late in each alternation. At half power, the gate pulses arrive at the moment when each alternation is at its peak value. For high power, the gate pulses arrive early, and most of each alternation flows in the load. This is conduction angle, or phase control, as was discussed for the SCR. The adjustable gate control block shown in Fig. 14-21 varies the phase of the gate pulses, with the source voltage serving as the phase reference. You should now compare Figs. 14-22 and 14-11.

Commutation can be more complicated in triac circuits. With ac, the triac should commutate (turn off) at each zero-voltage point (the zero crossings). Commutation is not a problem with resistive loads. With inductive loads (such as motors), the current lags the voltage. You should recall that this phase shift is expected when there is inductive reactance. Thus, in the case of an inductive load, the current and voltage zero crossings occur at different times, making commutation more difficult.

Line transients can affect thyristors. Transients produce a large voltage change in a short time. A rapid change in voltage can switch a thyristor to its on state. Recall that a PN junction, when not conducting, has a depletion region. Also recall that the depletion region acts as the dielectric of a capacitor. This means that a thyristor in its off state has several internal capacitances. A sudden voltage change across

Trigger device

Diac

the thyristor terminals will cause the internal capacitances to draw charging currents. These charging currents can act as a gating current and switch the device on.

Inductive loads and transients are problem areas in triac control. These problems can be reduced by special networks that limit the rate of voltage change across the triac. An *RC snubber network* has been added in Fig. 14-23. Snubber networks divert the charging current from the thyristor and help prevent unwanted turn-on.

Triac gating circuits vary from application to application. A triac may simply be switched on or off. Or, it may be phase controlled for various conduction angles. There are many gating circuits in use, and they range from simple to complex. Figure 14-24 shows two simple triac gating circuits. Figure 14-24(a) uses a variable-resistor in series with the gate lead. As *R* is set for less resistance, the triac will gate on sooner, and the conduction angle will increase. This will result in an increase in load power. This approach does not provide control over the entire 360-degree range and has poor symmetry. The positive alternations will have a different conduction angle than the negative alternations. This is due to different gating modes (Table 14-1). The circuit is also temperature-sensitive. Figure 14-24(b) shows improved operation. This circuit has the advantage of providing a broader range of control. The setting of R_1 will control how rapidly C_1 and C_2 charge. Decreasing R_1 will advance the firing point and increase the load power.

The better gate-trigger circuits use a negative-resistance device to turn on the triac. These devices show a rapid decrease in resistance after some critical turn-on voltage is reached. *Triggering devices* with this negative-resistance quality include neon lamps, unijunction transistors, two-transistor switches, and *diacs*.

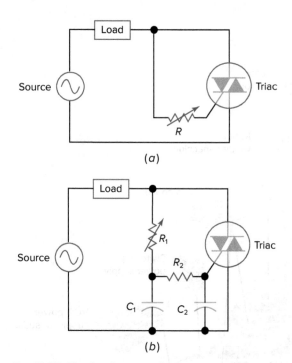

Fig. 14-24 Simple triac gate circuits.

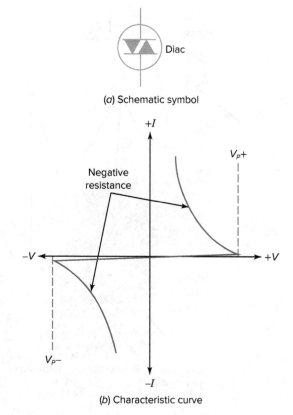

(a) Schematic symbol

(b) Characteristic curve

Fig. 14-25 Diac.

The schematic symbol for a diac is shown in Fig. 14-25(a). The diac is a bidirectional device and is well suited for gating triacs. The characteristic curve for a diac is shown in Fig. 14-25(b). The device shows two breakover points V_p+ and V_p-. If either a positive or a negative voltage reaches the breakover value, the diac rapidly switches from a high-resistance state to a low-resistance state.

Figure 14-26 shows a popular circuit that combines a diac and a triac to give smooth power control. Resistors R_1 and R_2 determine how rapidly C_3 will charge. When the voltage across C_3 reaches the diac breakover point, the diac fires. This provides a complete path for C_3 to discharge into the gate circuit of the triac. The discharge of C_3 on the triac turns.

Figure 14-26 also includes two components to suppress *radio-frequency interference (RFI)*. Triacs switch from the off state to the on state in 1 or 2 μs. This produces an extremely rapid increase in load current. Such a current step contains many harmonics. A harmonic is an integer multiple of some frequency. For example, the third harmonic of 1 kHz is 3 kHz. The harmonic energy in triac control circuits extends to several megahertz and can produce severe interference to AM radio reception. The energy level of the harmonics falls off as the frequency

increases. Interference from thyristors is more of a problem at the lower radio frequencies. Capacitor C_1 and inductor L_1 in Fig. 14-26 form a low-pass filter to prevent the harmonic energy from reaching the load wiring and radiating. This will reduce the interference to a nearby AM radio receiver.

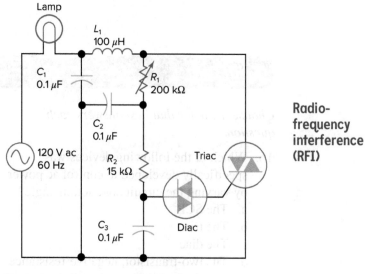

Radio-frequency interference (RFI)

Fig. 14-26 A diac-triac control circuit.

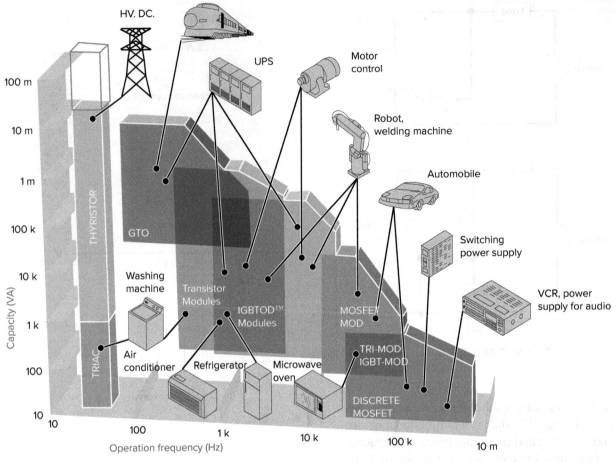

Fig. 14-27 Power capacities and operating frequencies of semiconductor devices.

Figure 14-27 shows the power range and frequency range for various solid-state devices. Thyristors, in the form of SCRs, are the power champions with ratings well into the megawatts. Gate turnoff devices (GTOs) are similar to SCRs but use a pulse of negative gate current to force device turnoff. Unlike SCRs, GTOs can be turned off at their gate terminals. Power MOSFETs and IGBTs were covered in Chap. 5.

Self-Test

Choose the letter that best answers each question.

16. Which of the following devices was specifically developed to control ac power by varying the circuit conduction angle?
    a. The SCR
    b. The triac
    c. The diac
    d. The two-transistor, negative-resistance switch

17. In a triac circuit, how may the load dissipation be maximized?
    a. Hold the conduction angle to 0 degrees.
    b. Hold the conduction angle to 180 degrees.
    c. Hold the conduction angle to 270 degrees.
    d. None of the above

18. Suppose a triac is used to control the speed of a motor. Also assume that the motor is highly inductive and causes loss of commutation. What is the likely result?
    a. The triac will short and be ruined.
    b. The triac will open and be ruined.
    c. Power control will be lost.
    d. The motor will stop.
19. Which of the following events will turn on a triac?
    a. A rapid increase in voltage across the main terminals
    b. A positive gate pulse (with respect to terminal 1)
    c. A negative gate pulse (with respect to terminal 1)
    d. Any of the above
20. Why may a voltage transient cause unwanted turn-on in thyristor circuitry?
    a. Because of internal capacitances in the thyristor
    b. Because of arc-over
    c. Because of a surge current in the snubber network
    d. All of the above

21. Which of the following is a solid-state, bidirectional, negative-resistance device?
    a. Neon lamp
    b. Diac
    c. UJT
    d. Two-transistor switch
22. How many breakover points does the volt-ampere characteristic curve of a diac show?
    a. One
    b. Two
    c. Three
    d. Four
23. What is an advantage of a zero crossing switch?
    a. Commutation circuits are never needed.
    b. Static switching is eliminated.
    c. Snubber networks can be used for greater conduction angle.
    d. Surges and RFI are reduced.

14-4 Feedback in Control Circuitry

Electronic control circuits can be made more effective by using *feedback* to automatically adjust operation should some change be sensed. For example, suppose a thyristor is used to control the speed of a motor. After the motor has been set for speed, assume the load on the motor increases. This will tend to slow down the motor. It is possible, by using feedback, to make the speed of the motor constant even though the mechanical load is changing.

Figure 14-28 shows the diagram for a motor-speed control that uses feedback to improve performance. R_1, R_2, D_1, and C_1 form an adjustable

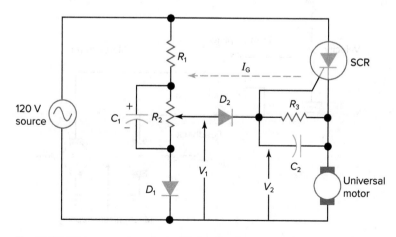

Fig. 14-28 Motor-speed control with feedback.

dc power supply. Diode D_1 rectifies the ac line. The SCR will fire earlier during a positive alternation if the wiper arm of R_2 is moved up toward R_1. This is because V_1 will be more positive and D_2 will be forward-biased sooner. This increases motor speed. The circuit cannot achieve full speed with a 120-V motor, however, because the SCR will conduct only on positive alternations. Sometimes a circuit of this type will be used with special universal motors rated at 80 V to allow full-speed operation on 120 V. Universal motors are so named because they can be energized on alternating current or direct current. They may be identified by their construction, which uses a segmented brass commutator at one end of the armature. Brushes are used to make electrical contact with the rotating commutator.

Servomechanism

Positive alternations in Fig. 14-28 will gate the SCR on when D_2 is forward-biased. This occurs when V_1 is more positive than V_2 by about 0.6 V. V_1 is determined by the setting of R_2 and the instantaneous line voltage; V_2 is determined by the counterelectromotive force (*CEMF*) of the *motor*. The residual magnetism in a universal motor gives it some of the characteristics of a generator. Therefore, the CEMF is determined by the motor's magnetic structure, its iron characteristics, and its *speed*. If the mechanical load on the motor is increased, the motor tends to slow down. The drop in speed will decrease the CEMF V_2. This means that V_1 will now exceed V_2 at some earlier point in the alternation. Thus, the SCR is gated on sooner, and

Motor CEMF

Velocity servo

Tachometer

Error amplifier

the motor speed is stabilized. On the other hand, if the mechanical load is decreased, the motor tries to speed up. This increases V_2, and now V_1 will exceed V_2 later in the alternation. The SCR is on for a shorter period of time, and again the motor speed is stabilized.

The performance of the motor-speed control circuit shown in Fig. 14-28 is adequate for some applications. However, many motors do not develop a CEMF signal that can be used to stabilize speed. It may be necessary to arrange for other types of feedback to make motor speed independent of mechanical load. In some systems, the feedback may relate to the angular position of a shaft rather than its speed. Feedback systems that sense and control position are called *servomechanisms*. Feedback systems that control speed are called *servos*. However, today the distinction is not as important as it once was, and you may find systems that control quantities other than position being classified as servomechanisms. In general terms, a servomechanism is a controller that involves some mechanical action and provides automatic error correction. A servomechanism or servo in its most elementary form consists of an amplifier, a motor, and a feedback element.

Figure 14-29 shows the basic arrangement for a *velocity servo*. The motor is mechanically coupled to a *tachometer*. A tachometer is a small generator, and its output voltage is proportional to its shaft speed. The faster the motor in Fig. 14-29 runs, the greater the output voltage from the tachometer. The *error amplifier*

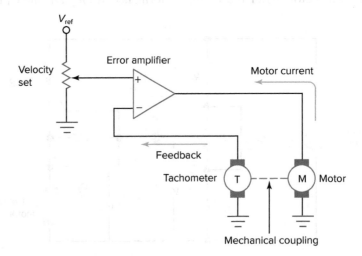

Fig. 14-29 Velocity servo.

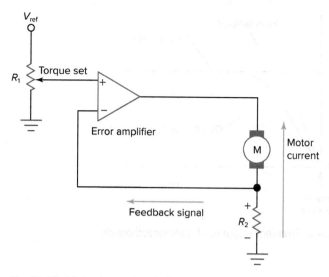

Fig. 14-30 Motor-torque control system.

compares the voltage from the velocity-set potentiometer with the feedback voltage from the tachometer. If the load on the motor increases, the motor tends to slow down. This causes the output from the tachometer to decrease. Now the error amplifier sees less voltage at its inverting input. The amplifier responds by increasing the positive output voltage to the motor. The motor torque (twisting force) increases, and the speed error is greatly reduced. Changing the position of the velocity-set potentiometer will make the motor operate at a different speed. Therefore, the velocity servo provides both speed regulation and speed control.

Figure 14-30 shows a motor-*torque control* system. The torque of the motor is controlled by the current that flows through it. Resistor R_2 provides a feedback voltage that is proportional

to motor current. This feedback voltage is compared to the reference voltage that is divided by R_1. Suppose that the load on the motor increases its torque output. The motor will draw more current, and this increased current will increase the voltage drop across R_2. The inverting input of the error amplifier is going in a positive direction. This will make the output of the amplifier go less positive. The motor current will decrease, and the torque output will be held constant.

A positioning servomechanism is shown in Fig. 14-31. The motor drives a potentiometer through a mechanical reduction system (gear train). Many turns of the motor will result in one turn of the potentiometer shaft. The angle of the potentiometer shaft determines the voltage at the wiper arm. This voltage is fed back to

Torque control

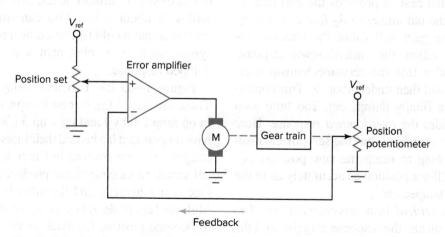

Fig. 14-31 Positioning servomechanism.

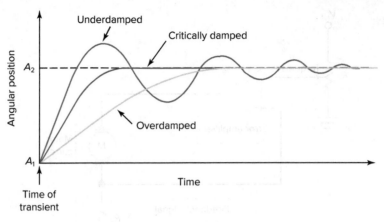

Fig. 14-32 Transient response of a servomechanism.

the error amplifier. The motor is a dc type that reverses rotation when its supply voltage reverses. Any error between the two potentiometer settings in Fig. 14-31 will cause the amplifier output to drive the motor in the direction that will reduce the error. Therefore, the *position* of the gear train can be *controlled* by adjusting the position-set potentiometer.

The response and accuracy of a servomechanism are functions of *gain*. The more gain the error amplifier has, the greater the positioning accuracy. This is often referred to as the *stiffness* of a servomechanism. Stiffness is usually desirable for fast response to commands and for high positioning accuracy. However, too much gain causes problems. For example, suppose the position-set potentiometer in Fig. 14-31 is suddenly changed. This introduces an abrupt error or transient into the system. Figure 14-32 shows three ways a servomechanism can respond to a transient. The *critically damped* response is the best. It provides the best change from A_1 (the old angle) to A_2 (the new angle). Raising the gain will cause the transient response to follow the *underdamped* response curve. Notice that the servomechanism overshoots A_2 and then undershoots it. This continues until it finally damps out. Too little loop gain provides the *overdamped* response. Here there is no overshoot, but the servomechanism takes too long to reach the new position, A_2. Also, it will not position as accurately as in the critically damped case.

Gain is *critical* in a servomechanism. Too little gain makes the response sluggish and the accuracy poor. Too much gain causes damped

oscillations when a transient is introduced into the system. In fact, a servomechanism may oscillate violently and continuously if the gain is too high. The gain of most servomechanisms is adjustable for best stiffness and transient response.

Oscillations will occur in any feedback system when the gain is greater than the loss and the feedback is positive. It is usually possible to increase gain and still avoid oscillations by controlling the phase angle of the feedback loop. *Phase-compensation* networks are used in most servo systems to improve performance.

Figure 14-33 shows a computer simulation of a servo system with and without phase compensation. A 0.01-μF compensation capacitor greatly improves the response. With no phase compensation (switch open), the response is underdamped. The simulation shows that the circuit settles in about 10 ms. With compensation, the response is almost ideal, and the circuit settles in about 2 ms. You can imagine the problems that would be caused by a positioning system such as a robot arm with an underdamped response.

Figure 14-33 has two *time delays* or *lags*. These are due to the feedback capacitors found in op amp 2 (OA_2) and op amp 3 (OA_3), which have outputs that lag behind their inputs. Op-amp integrators were covered in Chap. 9. In a typical servo, an electric motor produces two lags. One is mechanical, and the other is electrical. Multiple lags or delays cause negative feedback to become positive feedback as the frequency goes up. In Fig. 14-33, the feedback is positive

Position control

Gain

Stiffness

Critically damped

Underdamped

Overdamped

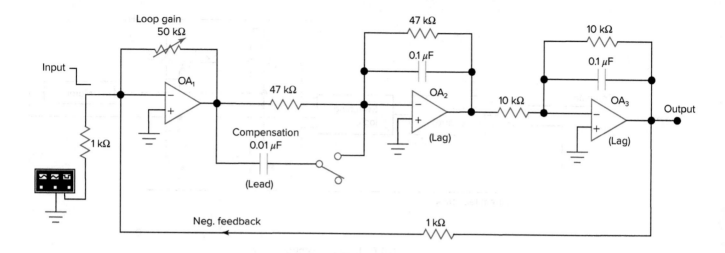

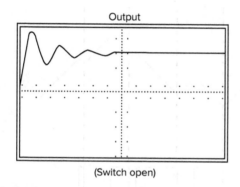

Fig. 14-33 Simulation of a servo.

at a frequency of about 500 Hz. This circuit will oscillate at 500 Hz if the gain is increased. Likewise, some servos will physically oscillate if the gain is increased.

Lag circuits are also called integrators, time delays, or low-pass filters. Which name is used has a lot to do with the particular application. The important idea here is that multiple lags, such as those found in any motor, cause negative feedback to become positive, and positive feedback can cause an underdamped response or continuous oscillation.

Lead circuits are also called differentiators, anticipators, or high-pass filters. They are used in servos to compensate for the unavoidable lags found in motors and other mechanisms. A lead circuit has a phase angle opposite to that of a lag circuit; therefore, a lead circuit can cancel a lag. Lead circuits are used to compensate for delays in applications such as temperature control systems.

Figure 14-34 shows a general block diagram of a typical servo system. There is more than one feedback path because systems are

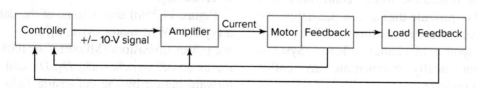

Fig. 14-34 Servo general block diagram.

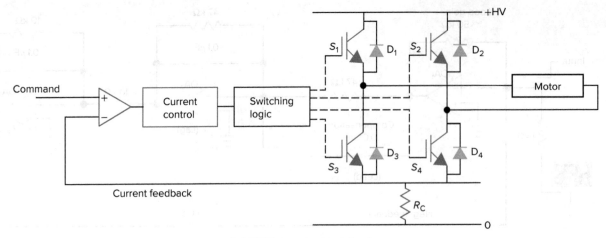

(a) PWM current control circuit

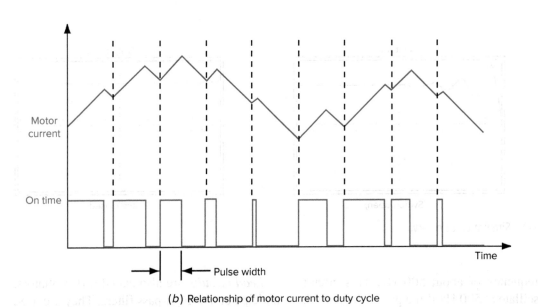

(b) Relationship of motor current to duty cycle

Fig. 14-35 Pulse-width modulation.

often designed to control position, velocity, and/or acceleration. The controller contains the algorithms (software routines) to close the desired servo loop(s), handles machine interfacing (inputs/outputs, programming terminals, etc.), and contains the necessary compensation functions. Many controllers now use DSP to provide digital compensation and to achieve advanced features such as automatic response to changing loads. Systems that automatically compensate are called *adaptive systems*.

The signal from the controller can be digital or analog. If it's analog, it usually varies between ±10 V as shown in Fig. 14-34. The control signal drives an amplifier, which in turn supplies the motor current. Pulse-width modulation (PWM) is usually preferred because of its efficiency.

Figure 14-35(a) shows some of the details of a typical PWM motor drive. S_1, S_2, S_3, and S_4 are power transistors (MOSFET or IGBT) acting as on-off switches. D_1, D_2, D_3, and D_4 are forward-biased by the collapsing field of the

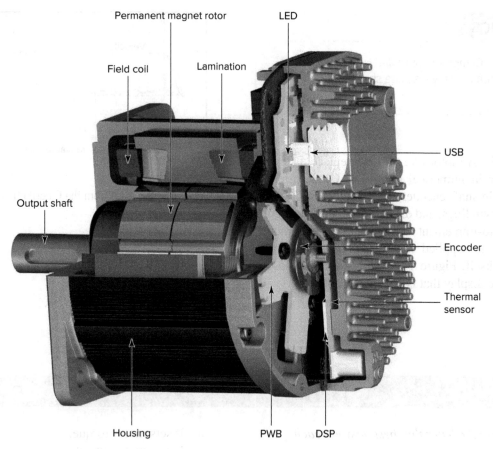

Permanent magnet rotor

LED

Field coil

Lamination

USB

Output shaft

Encoder

Thermal sensor

Housing

PWB

DSP

Fig. 14-36 ClearPath servo motor.
Courtesy of Teknic

motor when their associated transistors turn off. These are sometimes called *freewheeling diodes*. The motor is connected as a bridge tied load. This load connection was presented in Chap. 8. Current can be sent through the motor in either direction by activating the appropriate switches.

The bus voltage in Fig. 14-35(a) is depicted as +HV. R_C is used to measure the motor current. The switch on-time is determined by the difference between the current called for by the controller and the actual motor current. A current control circuit compares both signals at each time interval (typically 50 ms or less) and activates the switches accordingly (this is done by the switching logic circuit, which also performs basic protection functions). Figure 14-35(b) shows the relationship between the pulse width (on-time) and the motor current. Electric motors

are an inductive load. The current rise time depends on the bus voltage and the load inductance. The slope of the current rise is proportional to V/L. Therefore, certain minimum load inductance requirements are necessary depending on the bus voltage. With small values of L (inductance) and a large voltage, the slope of the current increase would be steep, and the current could exceed a safe value.

As mentioned several times before, operating the control devices [$S_1 - S_4$ in Fig. 14-35(a)] in a digital mode provides the best efficiency. Years ago, servos used analog amplifiers as motor drivers. These amplifiers were larger and ran a lot hotter. Today, analog servo drives are generally used only in a limited number of low-power applications.

Figure 14-36 shows the internal construction of a permanent magnet servo motor offered by

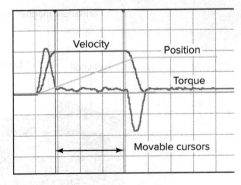

Fig. 14-37 A screen shot from the ClearPath control software.

Teknic. A motor like this can replace a stepper motor in numerical control (NC). It offers a built-in shaft encoder, solid-state control of the field windings, and digital signal processing for easy-to-implement motion control. This motor, with the supplied software, can automatically tune itself. Figure 14-37 shows the oscilloscopelike display that is part of the software.

Self-Test

Choose the letter that best answers each question.

24. Refer to Fig. 14-28. What would cause V_2 to increase?
    a. A shorted SCR
    b. The motor slowing down
    c. The motor speeding up
    d. A decrease in source voltage
25. Refer to Fig. 14-28. What should happen when V_2 increases?
    a. The SCR should gate earlier in the alternation.
    b. The SCR should gate later in the alternation.
    c. There will be no change in conduction angle.
    d. D_2 will be forward-biased all the time.
26. Refer to Fig. 14-28. What produces the feedback signal?
    a. D_2
    b. C_1
    c. R_4
    d. The motor
27. Refer to Fig. 14-28. What is the function of R_2?

a. It sets motor torque.
b. It sets motor speed.
c. It sets motor position.
d. All of the above are true.
28. Suppose a chemical plant operator uses a remote pressure sensor to monitor gas flow. When the pressure goes too high, the operator closes a circuit that runs a motor and controls a valve. Why would this *not* qualify as a servomechanism?
    a. The system is not automatic.
    b. No mechanical action is involved.
    c. Gas pressure has nothing to do with servomechanisms.
    d. All of the above are true.
29. Refer to Fig. 14-29. Assume that you have measured the output of the error amplifier for 1 minute and noted no change. If the servomechanism is working properly, what can you conclude?
    a. The error detector has only one input signal.
    b. The motor is gradually slowing down.
    c. The motor speed is stable.
    d. The motor is gradually speeding up.

30. Refer to Fig. 14-29. Assume that you are monitoring the output of the error amplifier. You note that as the mechanical load increases, the output goes more positive. As the mechanical load decreases, the output goes less positive.

What can you conclude about the servo circuit?
a. It is working properly.
b. It is missing its reference signal.
c. It is not working at all.
d. It is connected to a defective power supply.

14-5 Managing Energy

As the depletion of finite resources increases and issues such as pollution and climate change loom ever larger, it is clear that more efficient use of energy is a high priority. Electronic control circuits can assist in making better use of energy. This section explores only a small segment of this broad topic: controllers for photovoltaic arrays and controllers for light-emitting diodes.

Smaller (i.e., less than 300 peak watts or so) photovoltaic (PV) systems are often straightforward. The PV arrays are simply connected directly to loads, inverters, or storage batteries. Larger photovoltaic arrays are often connected through controllers. Photovoltaic cells were introduced in Chap. 3. You may recall that their output varies widely with solar intensity. As an example, Fig. 14-38 shows three different maximum power points (MPPs), with the red curve (MPP_1) representing the highest and the blue curve (MPP_3) the lowest. The red curve represents the array's electrical characteristics for the most intense light.

In Fig. 14-38, the maximum power point tracking (MPPT) controller presents a much more optimum load on the PV array (compared to a fixed load). When illumination changes, so does the load presented to the panel. The general method is to use a dc-to-dc converter with adjustable pulse-width modulation to charge batteries or power inverters. In this way, the

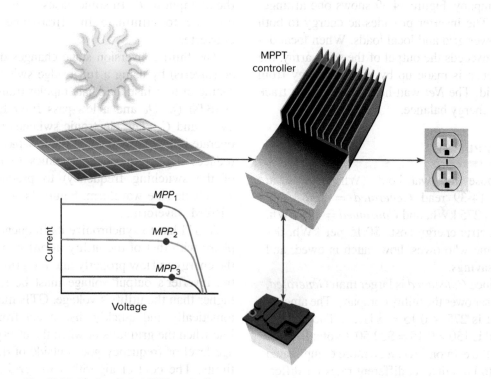

Fig. 14-38 Photovoltaic MPPT controller.

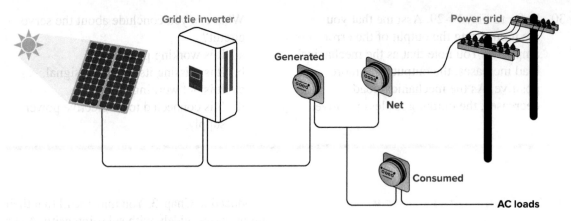

Fig. 14-39 MPPT grid-tie inverter system.

PWM duty cycle can be varied to present a wide range of optimum loads for the PV array for various light levels and sun angles. The overall efficiency can increase from less than 40 to more than 90 percent by employing an MPPT controller. Power point tracking should not be confused with *panel tracking*, by which the physical mount of a PV panel allows the plane of the panel surface to follow (track) the angle of the sun throughout the day.

MPPT controllers include *grid-tie systems*, by which excess energy can be sold to the utility company. Figure 14-39 shows one arrangement. The inverter provides ac energy to both the power grid and local loads. When local demand exceeds the output of the solar array, the difference is made up by buying energy from the grid. The *Net* watt-hour meter keeps track of the energy balance.

EXAMPLE 14-1

Suppose the watt-hour (Wh) meters in Fig. 14-39 read *Generated* = 150 kWh, *Net* = 275 kWh, and *Consumed* = 425 kWh. If electric energy costs $0.15 per kWh, determine who owes, how much is owed, and any savings.

Since *Consumed* is larger than *Generated*, the user owes the utility company. The amount owed is $275 \times 0.15 = \$41.25$. The amount saved is $150 \times 0.15 = \$22.50$. (Note: In practice, the economics can be more complicated due to laws that set different rates for different situations.)

Figure 14-40 shows a simplified schematic diagram for a grid-tie inverter (GTI). The dc voltage from the solar array is stepped up by a *boost converter* formed with inductor L_1, enhancement-mode MOSFET Q_1, diode D_1, and capacitor C_2. Chapter 15 covers dc boost circuits. One of the input dc busses (usually negative) has to be grounded, and the ac output is connected to the power grid; thus the inverter must provide isolation between the input and output. This function is provided by transformer T_1. Transistors Q_2–Q_5 provide the ac input to T_1. In some cases, T_1 is a step-up type to eliminate the first-stage boost converter.

The third conversion stage changes dc into ac (inverts) by using a full bridge switch consisting of four insulated gate bipolar transistors (IGBTs) Q_6–Q_9 and a low-pass filter formed by L_3 and C_4. The electronic switches Q_6–Q_9 operate in PWM mode, and the low-pass filter reduces high-frequency harmonics (multiples of the switching frequency) to produce an acceptable sine waveform. Figure 14-41 shows a PWM waveform.

A GTI has to synchronize its frequency and phase with that of the utility in order to allow the energy to flow properly into the grid. Also, the inverter's output voltage must be slightly higher than the utility's voltage. GTIs must automatically and quickly disconnect from the line when the grid fails or when the utility voltage level or frequency goes outside of defined limits. The control algorithm for grid-tie inverters is complicated and is normally done with microcontrollers.

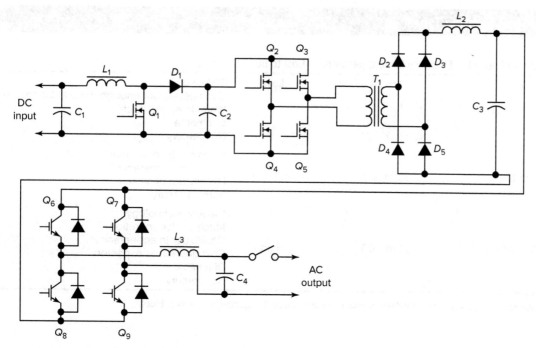

Fig. 14-40 Partial schematic diagram for a grid-tie inverter.

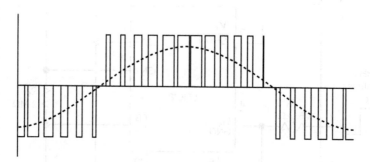

Fig. 14-41 PWM signal for inverting dc to sinusoidal ac.

Residential and commercial lighting will make more use of LEDs in the future. After a slow start, there are now 12-W LED lamps available that produce as much light as 100-W incandescent lamps. The efficiency gains will mean not only lower utility costs but reduced global carbon emissions. The U.S. Department of Energy estimates that replacing regular light bulbs with LEDs could save as much as 200 terawatt-hours annually (the equivalent of lighting over 100 million homes). Over a 10-year period, the savings would exceed $100 billion. Table 14-2 compares three light sources.

Current LED lamps use several LED in series. Each LED consumes about 1 W. Figure 14-42 shows a circuit that would be suitable for a battery-operated device. The 2.49-Ω resistor acts as a current sensor, and the LT1618 IC regulates the average current delivered to the series string using PWM. Circuits for fixed lighting rectify the ac line voltage to dc and also control the series LED string with PWM.

Table 14-2   Comparison of Lighting Types

	Life (in thousands of hours)	Efficiency (lumens per watt)	Cost to Buy	Notes
Incandescent	1	7 to 24	Low	Oldest technology Can operate hot enough to be a fire hazard Warm in appearance* Dimmable No mercury
CFL†	3 to 10	30 to 100	Medium	Much cooler operation Cooler in appearance* Usually not dimmable Some mercury
LED	30 to 80	30 to 200	High	Newest technology Much cooler operation Cool/bluer in appearance* Cost decreases expected Dimmable No mercury

* The appearance is called the *color temperature* with warmer lighting having more reds and cooler lighting having more blues.
† Compact fluorescent light.

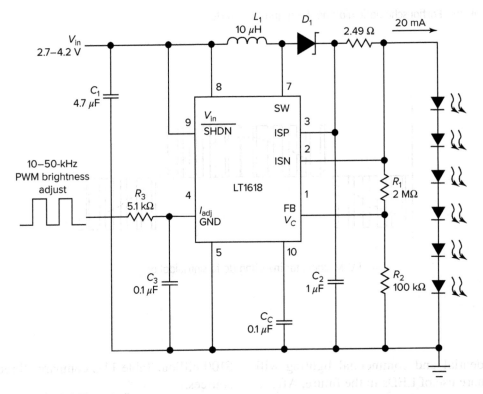

Fig. 14-42 LED controller.

Self-Test

Choose the letter that best answers each question or completes the statement.

31. MPPT represents
    a. Maximum power point tracking
    b. Medium photon penetration transitions
    c. Metal panel power tracking
    d. More power per track

32. Which curve in Fig. 14-38 would demand the lowest load resistance for best efficiency (best power transfer)?
    a. Red
    b. Green
    c. Blue
    d. Black (the axis)
33. Which type of solar power system provides the utility grid with surplus energy?
    a. Hydroponic
    b. Grid tie
    c. MPPT
    d. All of the above
34. What is the function of L_3 and C_4 in Fig. 14-40?
    a. Change dc to sinusoidal ac
    b. Change ac to dc
    c. Smooth the output waveform
    d. Provide protection when the grid goes out of normal range
35. The sine signal shown in Fig. 14-41 is
    a. Equal to the peak value of the rectangular signal
    b. Equal to the frequency of the rectangular signal
    c. Equal to the phase of the rectangular signal
    d. Equal to the average value of the rectangular signal
36. CFL stands for
    a. Capacitor-filtered inductor
    b. Compact fluorescent light
    c. Charge flowing lighting
    d. None of the above
37. What controls the brightness in an LED controller?
    a. AM
    b. FM
    c. PM
    d. PWM

14-6 Troubleshooting Electronic Control Circuits

Technicians who troubleshoot thyristor control circuits must be aware of their limitations and know safe procedures. The thyristor control circuits covered in this chapter can be used with only certain kinds of loads. Severe damage to the load and the control circuit may result if improper connections are attempted. The general rule for SCR and triac control circuits is this: *Never* attempt to use them with *ac-only* equipment. This type of equipment includes the following:

1. Fluorescent lamps (unless specially designed for thyristor control)
2. Radios
3. Television receivers or sound systems
4. Induction motors (including those on fans, record players, tape players, washing machines, large equipment such as air compressors, and so on)
5. Transformer-operated devices (such as soldering guns, model-train power supplies, battery chargers, and so on)

In general, it is safe to use thyristor control circuits with *resistive loads*. These include incandescent lamps, soldering pencils, heating elements, and so on. It is also safe to use thyristor control circuits with universal (ac/dc) motors. These motors usually are found in portable power tools such as drills, saber saws, and sanders. When in doubt, check the manufacturer's specifications. Also be sure that the wattage rating of the load does not exceed the wattage rating of the control circuit.

The general safety rules for analyzing and troubleshooting electronic control circuits are the same as those for any line-operated circuit. It is dangerous to connect line-operated test equipment to thyristor control circuits. A ground loop is likely to cause damage and perhaps severe electric shock. Even if the test equipment is battery-operated, danger still exists. A battery-operated oscilloscope may seem safe, but remember that the cabinet and the probe grounds may reach a dangerous potential when directly connected into power circuits.

If a control circuit is for light duty, it may be possible to use an isolation transformer. Then it is safe to use test instruments for analyzing the circuit. Be sure the wattage rating on the isolation transformer is adequate before attempting this approach.

AC-only load

Resistive load

Suppose you are troubleshooting a thyristor motor-speed control. You notice that the motor always runs at top speed. The speed control has no effect. Assume that you have already completed the usual preliminary checks and have found nothing wrong. What is the next step? Ask yourself what kinds of problems could cause the motor to always run at top speed. Could the thyristor be open? No, because that should stop the motor. Could the thyristor be shorted? Yes, that is a definite possibility. Is it time to change the thyristor? No, the analysis is not over yet. Are there any other causes for the observed symptoms? What if the gating circuit is defective? Could this make the motor run at top speed? Yes, it could.

The last part of the troubleshooting process is to limit the possibilities. How can this be done? One way is to shut off the power. Then disconnect the gate lead of the thyristor. Turn the power back on. Does the motor run at top speed again? If it does, the thyristor is, no doubt, shorted. It should be replaced. What if the motor will not run at all? This means the thyristor is good. With its gate lead open, it will not come on. The problem is in the gate circuit.

There are many types of gate circuits. It will be necessary to study the circuit and determine its principle of operation. If the circuit uses a unijunction transistor, it will be necessary to determine whether the UJT pulse generator is working as it should. If the circuit uses a diac, it will be necessary to determine whether the diac is operating properly. It may be possible to use resistance analysis (with the power off) to find a defect. A resistor may be open. A capacitor may be shorted. Some solid-state device may have failed.

Some modern equipment has a lot of trouble-shooting assistance designed in. The LED shown in Fig. 14-36 displays three colors and over 30 blink codes. These codes can be deciphered to identify servo problems such as

- Overspeed
- Physical limit exceeded
- Overtorque
- Overtemperature
- Failed hard stop
- Direction error

A red blink code means the motor has failed.

At this point, you should begin to realize that the answers are usually not in the manuals or the textbooks. A good troubleshooter understands the basic principles of electronic devices and circuits. This knowledge will allow a logical and analytic process to flow. It is not always easy. Highly skilled technicians "get stuck" from time to time. However, usually they do not keep retracing the same steps over and over. Once a particular fact is confirmed, it is noted on paper or mentally. Using paper is best because another job or quitting time can interfere. It is too easy to forget what has and has not been checked.

Different technicians use somewhat different approaches to troubleshooting. All good technicians have these things in common, however:

1. They work safely and use a system view.
2. They follow the manufacturer's recommendations.
3. They find and use the proper service literature.
4. They use a logical and orderly process.
5. They observe, analyze, and limit the possibilities.
6. They keep abreast of technology.
7. They understand how devices and circuits work in general terms. They understand what each major stage is supposed to do.
8. They are skilled in the use of test equipment and tools.
9. They are neat; use the proper replacement parts; and put all the shields, covers, and fasteners back where they belong.
10. They check their work carefully to make sure nothing was overlooked.
11. They never consider modifying a piece of equipment or defeating a safety feature just because it is convenient at the time.

Technicians who have developed these skills and habits are in demand and always will be.

Choose the letter that best answers each question.

38. Which of the following loads should never be connected to a thyristor control circuit?
    a. Incandescent lamps
    b. Soldering irons
    c. Soldering guns
    d. Soldering pencils

39. Why is it not safe to connect a line-operated oscilloscope across a triac in a light dimmer?
    a. A ground loop could cause damage.
    b. The cabinet and controls of the scope could assume line potential.
    c. Both of the above are possible.
    d. None of the above are possible.

40. Refer to Fig. 14-28. Assume that the SCR is open. What is the most likely symptom?
    a. The motor will run at top speed (no control).
    b. The motor will speed up and slow down.
    c. Diode D_1 will be burned out by the overload.
    d. The motor will not run.

41. In Fig. 14-28, assume D_2 is open. What is the most likely symptom?
    a. The motor will run at top speed (no control).
    b. The motor will be damaged.
    c. The motor will not run at all.
    d. Resistor R_4 will burn up.

42. Refer to Fig. 14-33. With the switch open, the response is
    a. Overdamped
    b. Underdamped
    c. Critically damped
    d. None of the above

43. Refer to Fig. 14-35. An oscilloscope connected to the gate lead of S_1 should show:
    a. A sine wave
    b. A sawtooth wave
    c. A triangular wave
    d. A rectangular wave

44. Refer to Fig. 14-35. If any of the diodes fail by opening, an associated transistor could fail because
    a. The motor CEMF will exceed its breakdown rating
    b. The slope of current rise and drop will decrease
    c. The slope of current rise and drop will increase
    d. None of the above

45. Refer to Fig. 14-35. An oscilloscope connected to the minus input of the op amp shows no signal. This implies that
    a. The motor current is at its maximum value
    b. The motor current is at its 50 percent value
    c. The motor current is zero
    d. None of the above

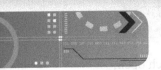

Chapter 14 Summary and Review

Summary

1. A rheostat can be used to control circuit current.
2. Rheostat control is not efficient since much of the total circuit power is dissipated in the rheostat.
3. Voltage control is much more efficient than resistance control.
4. Switches dissipate little power when open or closed.
5. A fast switch can control power in a circuit without producing undesired effects such as flicker.
6. Switch control is much more efficient than resistance control.
7. A latch circuit can be formed from two transistors: one an NPN and the other a PNP.
8. A latch circuit is normally off. It can be turned on with a gating current.
9. Once the latch is on, it cannot be turned off by removing the gate current.
10. A latch can be turned off by interrupting the load circuit or by applying reverse bias.
11. A four-layer diode or silicon-controlled rectifier is equivalent to the NPN-PNP latch.
12. An SCR, like an ordinary diode, conducts from cathode to anode.
13. An SCR, unlike an ordinary diode, does not conduct until turned on by a breakover voltage or by gate current.
14. In ordinary operation, SCRs are gated on and not operated by breakover voltage.
15. The SCR is a half-wave device.
16. Commutation refers to turning off an SCR.
17. The SCR is a unidirectional device since it conducts in only one direction.
18. The triac is a bidirectional device since it conducts in both directions.
19. Triacs are capable of full-wave ac power control.
20. Triacs are useful as static switches in low- and medium-power ac circuits.
21. The term *thyristor* is general and can be used in referring to SCRs or triacs.
22. A snubber network may be needed when triacs are used with inductive loads or when line transients are expected.
23. Negative-resistance devices are often used to trigger thyristors.
24. A diac is a bidirectional, negative-resistance device.
25. Diacs are often used to gate triacs.
26. Feedback can be used in control circuits to provide automatic correction for any error.
27. A load such as a motor may provide its own feedback signal.
28. A separate sensor such as a tachometer may be required to provide the necessary feedback signal.
29. A servomechanism is any control system using feedback that represents mechanical action.
30. Servomechanisms provide automatic control.
31. Servomechanism loop gain determines positional accuracy (stiffness) and transient response.
32. Too much loop gain may cause oscillations in a servomechanism.
33. Thyristor control circuits may be safely used with universal (ac/dc) motors.
34. The wattage rating of a thyristor control circuit must be greater than its load dissipation.
35. Some problems in a thyristor control circuit may be isolated by opening the gate lead.

Chapter Review Questions

Choose the letter that best answers each question.

14-1. Refer to Fig. 14-1. Suppose the load resistance is constant at 80 Ω, and the source voltage is 240 V. What will the load dissipation be if the rheostat is set for 160 Ω? (14-1)
 a. 80 W
 b. 168 W
 c. 235 W
 d. 411 W

14-2. What is the dissipation in the rheostat in question 14-1? (14-1)
 a. 62 W
 b. 160 W
 c. 345 W
 d. 590 W

14-3. What is the efficiency of the circuit in question 14-1? (14-1)
 a. 33 percent
 b. 68 percent
 c. 72 percent
 d. 83 percent

14-4. Suppose the resistance of a load is constant. What will happen to the power dissipation in the load if the current is increased to three times its original value? (14-1)
 a. The power will drop to one-third its original value.
 b. The power will remain constant.
 c. The power will increase 3 times.
 d. The power will increase 9 times.

14-5. Why is resistance control so inefficient? (14-1)
 a. Resistors are very expensive.
 b. The control range is too restricted.
 c. Much of the circuit power dissipates in the control device.
 d. None of the above are true.

14-6. Refer to Fig. 14-5. What is the purpose of the gate switch? (14-2)
 a. To turn the transistor switch on and off
 b. To commutate the NPN transistor
 c. To provide an emergency shutdown feature (safety)
 d. To turn on the transistor switch

14-7. Refer to Fig. 14-5. The transistors are on. How can they be shut off? (14-2)
 a. By opening the gate switch
 b. By closing the gate switch
 c. By opening the load circuit
 d. By increasing the source voltage

14-8. Refer to Fig. 14-5. Which of the following terms best describes the way the circuit works? (14-2)
 a. Latch
 b. Resistance controller
 c. Rheostat controller
 d. Linear amplifier

14-9. How is a silicon-controlled rectifier similar to a diode rectifier? (14-2)
 a. Both can be classed as thyristors.
 b. Both support only one direction of current flow.
 c. Both are used to change alternating current to pulsating direct current (rectify).
 d. Both have one PN junction.

14-10. What is the effect of increasing the gate current in an SCR? (14-2)
 a. The reverse breakover voltage is improved.
 b. The forward breakover voltage is increased.
 c. The forward breakover voltage is decreased.
 d. The internal resistance of the SCR increases.

14-11. Refer to Fig. 14-10. What is the maximum conduction angle of this circuit? (14-2)
 a. 45 degrees
 b. 90 degrees
 c. 180 degrees
 d. 360 degrees

14-12. Refer to Fig. 14-10. If the load is a motor, what should the motor do if the conduction angle is increased? (14-2)
 a. Slow down
 b. Stop
 c. Gradually slow down
 d. Speed up

14-13. Why is thyristor control more efficient than resistance control? (14-2)
 a. Thyristors are less expensive.
 b. Thyristors are easier to mount on a heat sink.
 c. Thyristors vary their resistance automatically.
 d. Thyristors are solid-state switches.

14-14. Refer to Fig. 14-12. What happens when SCR_1 is gated on? (14-2)
 a. The load comes on.
 b. The load goes off.
 c. The capacitor turns off SCR_2.
 d. SCR_2 comes on.

14-15. Refer to Fig. 14-12. What happens when SCR_2 is gated on? (14-2)
 a. The load comes on.
 b. The load goes off.
 c. The capacitor turns off SCR_2.
 d. SCR_1 comes on.

14-16. Turning off a thyristor is known as (14-2)
 a. Gating
 b. Commutating
 c. Forward-biasing
 d. Interrupting

14-17. Which of the following devices was developed specifically for the control of ac power? (14-3)
 a. The SCR
 b. The UJT
 c. The snubber
 d. The triac

14-18. Refer to Fig. 14-21. Suppose the load is an incandescent lamp and the conduction angle of the circuit is decreased. What will happen to the lamp? (14-3)
 a. Nothing will happen.
 b. It will dim.
 c. It will produce more light.
 d. It will flicker violently.

14-19. Refer to Fig. 14-21. The load is operating at full power. What is the conduction angle of the circuit? (14-3)
 a. 45 degrees
 b. 90 degrees
 c. 180 degrees
 d. None of the above

14-20. What is the chief advantage of a triac as compared with a silicon-controlled rectifier? (14-3)
 a. It costs less to buy.
 b. It runs much cooler.
 c. The triac is bidirectional.
 d. All of the above are advantages.

14-21. Refer to Fig. 14-23. What is the function of the snubber network? (14-3)
 a. It prevents false commutation.
 b. It reduces television interference.
 c. It helps reduce unwanted turn-on.
 d. It helps the gate control circuit work sooner.

14-22. Refer to Fig. 14-26. Which component turns on and then gates the triac? (14-3)

 a. Capacitor C_3
 b. Capacitor C_1
 c. Resistor R_2
 d. The diac

14-23. Refer to Fig. 14-26. Which component or components have been added to reduce radio interference? (14-3)
 a. Inductor L_1 and capacitor C_1
 b. Capacitor C_3
 c. The diac
 d. Capacitor C_2 and resistor R_2

14-24. Some devices exhibit a rapid decrease in resistance after some turn-on voltage is reached. What are they called? (14-3)
 a. Negative-resistance devices
 b. FETs
 c. Linear resistive elements
 d. Voltage-dependent resistors

14-25. Refer to Fig. 14-28. What will happen if the SCR shorts from anode to cathode? (14-4)
 a. The motor will stall.
 b. The motor will run above its top normal speed.
 c. V_2 will fall to zero.
 d. None of the above will occur.

14-26. Refer to Fig. 14-28. What will happen if D_2 burns out (opens)? (14-4)
 a. The motor will burn out.
 b. The motor will run at above half speed.
 c. The motor will run at below half speed.
 d. The motor will not run.

14-27. Refer to Fig. 14-29. What symptom would appear if the tachometer coupling is loose and is slipping on its shaft? (14-4)
 a. None, because the speed is regulated.
 b. The motor will slow down and stop.
 c. The motor will run fast.
 d. The reference signal will become unstable.

14-28. Which of the following devices should never be operated from a thyristor power-control device? (14-6)
 a. A washing machine motor
 b. A heater
 c. Christmas tree lights
 d. A soldering iron

14-1. Which of the power-control circuits presented in this chapter qualify as linear circuits? Why?

14-2. Could a BJT be used as a linear dc power controller? Would there be any disadvantage to such an application?

14-3. Is there a way to connect two SCRs so they will provide full-wave control?

14-4. Some companies manufacture optically coupled triac drivers. They consist of infrared LEDs optically coupled to photodetectors with triac outputs. Can you think of any application for these components?

14-5. What technical term can be used to describe the "cruise control" feature that is found on some vehicles?

Answers to Self-Tests

1. D	13. A	25. B	37. D
2. D	14. D	26. D	38. C
3. C	15. C	27. B	39. C
4. B	16. B	28. A	40. D
5. C	17. D	29. C	41. C
6. A	18. C	30. A	42. C
7. C	19. D	31. A	43. D
8. A	20. A	32. A	44. A
9. C	21. B	33. B	45. C
10. D	22. B	34. C	
11. B	23. D	35. D	
12. C	24. C	36. B	

Design Elements: Answers to Self-Tests (Check Mark): ©McGraw Hill Global Education Holdings, LLC; Horizontal Banner (Futuristic Banner): ©touc/DigitalVision Vectors/Getty Images RF; Internet Connection (Globe): ©Shutterstock/Sarunyu_foto; Vertical Banner (Hazard Stripes): ©Ingram Publishing

CHAPTER 15

Regulated Power Supplies

Learning Outcomes

This chapter will help you to:

15-1 *Perform* basic calculations for power-supply regulator circuits. [15-1]

15-2 *Explain* the use of feedback in voltage regulator circuits. [15-2]

15-3 *Identify* the types of current regulation. [15-3]

15-4 *Identify* crowbar circuits. [15-3]

15-5 *Identify* switch-mode regulators and their characteristics. [15-4]

15-6 *Explain* boost power factor correction and its safety implications. [15-5]

15-7 *Troubleshoot* regulated power supplies. [15-6]

Chapter 4 covered rectification, filtering, and zener diode shunt regulation. This chapter builds on those concepts and shows how basic power-supply performance is enhanced to meet the needs of modern electronic systems.

15-1 Open-Loop Voltage Regulation

Voltage regulation is one of the most important power-supply characteristics. It is the measure of a supply's ability to maintain a constant output voltage. *Open loop* means that feedback is not used to hold the output constant. The next section of this chapter examines the use of feedback (closed-loop) regulator circuits.

Consider Fig. 15-1. It is a graph of the performance of a typical nonregulated power supply. You can see that the output drops 6 V (ΔV) as the load on the power supply is increased from 0 to 5 A. Also note that the power supply delivers its rated 12 V only when it is fully loaded. When less of a load is taken, the output is greater than 12 V.

Now examine Fig. 15-2. It illustrates the *line regulation* curve for the same power supply. It shows that the output voltage drops as the line voltage falls below its nominal 120 V value. It also shows that high line voltage will increase the output above normal. Line voltage does change. In fact, the word *brownout* refers to a condition of low line voltage caused by heavy use of electrical power. Brownouts are common in cities during very hot weather. Power companies are often forced to reduce line voltage under severe load conditions to prevent equipment failure.

When the conditions of low line voltage and high load current are combined, a rather low output voltage will result in a nonregulated supply. Conversely, if high line voltage occurs when the

Line regulation

Brownout

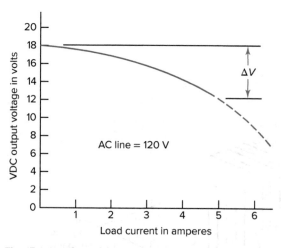

Fig. I5-I Load regulation curve for a I2-V, 5-A power supply.

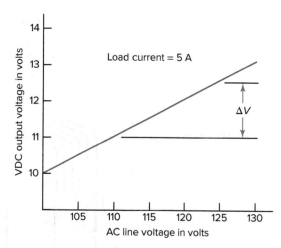

Fig. I5-2 Line regulation curve for a I2-V, 5-A power supply.

load current is low, a rather high output voltage will occur. Thus, it can be seen that load changes and line changes have significant effects on non-regulated power-supply output voltages.

One answer to this problem is to use a special power transformer. Figure 15-3 shows the construction of an ordinary (linear) power transformer. There are two major flux (magnetic flow) paths, and the primary and secondary coils are both wound around the center of the laminated core. The core in such a transformer is designed to be linear. That is, the core will not saturate. Now look at the *ferroresonant transformer* in

Fig. 15-4. It differs in several important ways. There are separate windows for the primary and secondary windings. There are air gaps in the shunt flux path. Finally, there is a *resonating capacitor* across the secondary.

The transformer shown in Fig. 15-4 can be used to build a power supply with a much more stable output voltage than the typical unregulated supply. As line voltage is applied to the primary, the main magnetic path excites the secondary. Part, or all, of the secondary winding is tuned by a resonating capacitor (usually several microfarads). As the secondary goes

Ferroresonant transformer

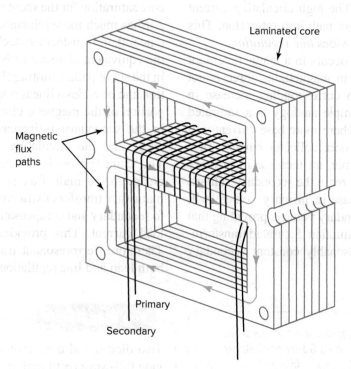

Fig. I5-3 Linear power transformer construction.

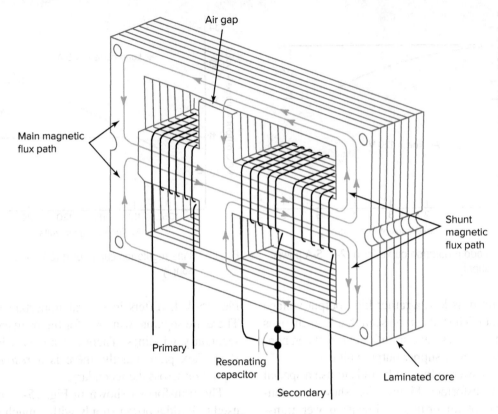

Fig. 15-4 Ferroresonant transformer construction.

into resonance, large currents flow in the capacitor and the resonant part of the secondary. The circulating current in a parallel resonant circuit is much greater than the line current when the circuit Q is high. The high circulating current drives the main flux path into saturation. This *core saturation* provides *line regulation*.

Core saturation

Core saturation occurs in a magnetic circuit when an increase in magnetizing force is not accompanied by a corresponding increase in flux density. A simple analogy is a saturated transistor circuit where more base current will not produce any more collector current. In a saturated transformer, an increase in primary voltage will not increase the secondary voltage. Similarly, a decrease in primary voltage will not affect the secondary voltage, providing that the core stays in saturation. Saturated transformers produce a reasonably constant secondary

Load regulation

output over some range of primary voltage (typically 90 to 140 V).

Another feature of the ferroresonant transformer in Fig. 15-4 is that the air gaps prevent core saturation for the shunt magnetic flux path. Air has much more reluctance (magnetic resistance) than transformer steel. The air gaps are the equivalent of series resistors and limit flux in the shunt path. Limiting flux prevents saturation and provides a linear response for the shunt portion of the magnetic circuit. If load current in the secondary is increased, the circuit Q drops and the circulating current decreases. The shunt flux will decrease, allowing an increase in the main flux path. An increase in main flux transfers extra energy from primary to secondary and compensates for the increased load current. This provides *load regulation*. Therefore, ferroresonant transformers provide both load and line regulation.

ABOUT ELECTRONICS

No More Power Outages The typical uninterruptable power supply (UPS) uses lead-acid batteries and a 60-Hz oscillator to replace the line voltage in the case of a power failure.

You May Recall

Two diodes and a center-tapped winding provide full-wave rectification.

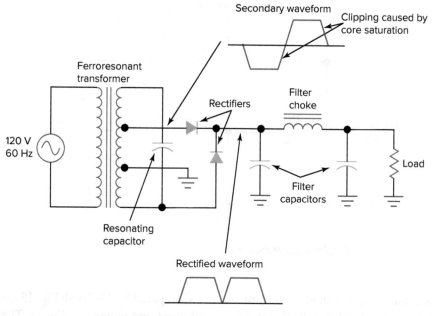

Fig. 15-5 Ferroresonant supply schematic.

Figure 15-5 shows the use of a ferroresonant transformer in a dc power supply. Notice that in this case the resonating capacitor is across the entire secondary winding. Also notice that the secondary waveform is clipped. This is caused by core saturation. The clipped sine wave has several advantages. It is easier to filter because the resulting rectified waveform has less ripple content than ordinary full-wave, pulsating direct current. A second advantage is that the clipped waveform has a lower peak voltage, which is easier for the rectifiers to handle. These types of power supplies are *very* reliable and provide good voltage regulation for both changing line voltage and changing load current. They are also noted for their efficiency. Unfortunately, the ferroresonant transformer is a large, heavy, and expensive component.

Figure 15-6 shows another answer to the regulation problem. This circuit was presented in Chap. 4. It uses a zener *diode* connected in parallel with the load. A zener diode will drop a relatively constant voltage when operating in reverse breakover. Therefore, the load will also see a relatively constant voltage. You can see from Fig. 15-6 that the zener current and load current add in resistor R_Z.

A problem with zener shunt regulators is that the diode dissipation is too large in some applications. For example, if the regulator shown in Fig. 15-6 is used to supply 12 V at 1 A, a high-wattage zener will be required. Assume the unregulated dc input to be 18 V. Resistor R_Z will have to drop 6 V (18 V − 12 V = 6 V). If the desired zener current is 0.5 A, R_Z can be found using Ohm's law:

$$R_Z = \frac{V}{I}$$

$$= \frac{6\text{ V}}{1\text{ A} + 0.5\text{ A}} = 4\ \Omega$$

Next, the power dissipation in the diode is

$$P_D = V \times I$$

$$= 12\text{ V} \times 0.5\text{ A} = 6\text{ W}$$

However, if the load is removed from the regulator, all of the current will flow through the zener diode, and its dissipation increases to

$$P_D = V \times I$$

$$= 12\text{ V} \times 1.5\text{ A} = 18\text{ W}$$

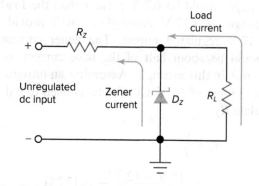

Fig. 15-6 Zener shunt regulator.

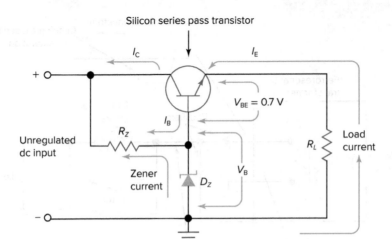

Fig. 15-7 Amplified zener regulator.

For good reliability, a zener diode rated for a power dissipation of at least 2×18 W (36 W) would be required. High-power zener diodes are too expensive for most applications.

Figure 15-7 shows a way to reduce the zener dissipation. The circuit is often called an *amplified zener regulator*. The zener diode is used to regulate the base voltage of a power transistor called a *series pass transistor*. If there is a reasonably constant 0.7-V drop across the base-emitter junction of the pass transistor, the emitter voltage and the load voltage will also be reasonably constant.

The current through R_Z in Fig. 15-7 is the sum of the base current (I_B) and the zener current. Assuming a load current of 1 A and a transistor β of 49, the base current is

$$I_B = \frac{I_E}{\beta + 1}$$

$$= \frac{1 \text{ A}}{50} = 0.02 \text{ A}$$

Because of the base-emitter drop, the zener voltage should be 0.7 V greater than the load voltage. A 12.7-V zener diode will provide a 12-V regulated output. The zener current should be about half of the base current, or 10 mA in this example. Assuming an unregulated input of 18 V, Ohm's law is used to calculate R_Z:

$$R_Z = \frac{V}{I}$$

$$= \frac{18 \text{ V} - 12.7 \text{ V}}{0.02 \text{ A} + 0.01 \text{ A}} = 177 \ \Omega$$

Compare Fig. 15-7 with Fig. 15-6 using identical input and output conditions. The worst-case zener dissipation occurs at zero load current. In Fig. 15-7, 30 mA will flow in the zener diode if the load is removed from the regulator. The zener current increases because with no load current, the base current drops to zero. Therefore, all 30 mA must flow in the zener. The zener dissipation will therefore increase:

$$P_D = V \times I$$

$$= 12.7 \text{ V} \times 0.03 \text{ A} = 0.381 \text{ W}$$

A 1-W zener will be safe under all operating conditions in Fig. 15-7. It should now be obvious why the amplified zener regulator is preferred for high-current applications. The circuit does require a series pass transistor, but this is a less expensive component than a high-wattage zener diode.

Figure 15-8 shows a *negative* amplified *regulator*. The pass transistor is PNP, and the circuit regulates a negative voltage referenced to the ground terminal. Note that the zener diode cathode is grounded. Compare this connection with that shown in Fig. 15-7.

Figure 15-8 shows another component that is often found in amplified zener regulators. An electrolytic capacitor bypasses the base of the transistor to ground. This capacitor is typically around 50 μF and, in conjunction with R_Z, forms a low-pass filter. This filter helps to remove noise and ripple present at the unregulated dc input. Also, zener diodes generate noise and the capacitor is useful for eliminating it from the output of the regulator. Most amplified zener-regulated power supplies use this capacitor.

Amplified zener regulator

Series pass transistor

Negative regulator

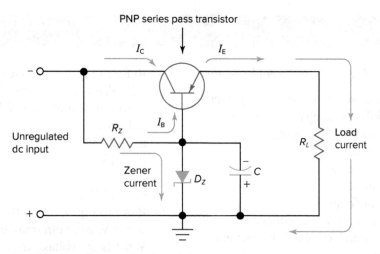

Fig. 15-8 Negative amplified regulator.

EXAMPLE 15-1

Select a value for R_Z in Fig. 15-8 if D_Z is a 5.7-V zener, the load current is 2 A, $\beta = 25$, the unregulated input is 9 V, and the desired zener current is 10 mA. Also, determine the worst-case zener dissipation. Begin by finding the base current:

$$I_B = \frac{I_E}{\beta + 1} = \frac{2\text{ A}}{25 + 1} = 76.9\text{ mA}$$

The total current in R_Z is the sum of the base current and the zener current:

$$I_{RZ} = I_B + I_{ZD}$$

$$= 76.9\text{ mA} + 10\text{ mA} = 86.9\text{ mA}$$

The drop across R_Z is the unregulated input voltage minus the zener voltage:

$$V_{RZ} = 9\text{ V} - 5.7\text{ V} = 3.3\text{ V}$$

Ohm's law can now be used to find R_Z:

$$R_Z = \frac{V_{RZ}}{I_{RZ}} = \frac{3.3\text{ V}}{86.9\text{ mA}} = 38.0\ \Omega$$

The worst-case zener dissipation is

$$P_D = V \times I$$

$$= 5.7\text{ V} \times 86.9\text{ mA} = 0.495\text{ W}$$

Figure 15-9 shows a dual-polarity *(bipolar) power supply*. This circuit provides both a positive and a negative regulated voltage with respect to the ground terminal. Notice that transformer T_1 has two secondary windings. Each secondary is center-tapped and supplies a full-wave rectifier circuit. Capacitors C_1 and C_2 filter the rectifier outputs. Q_1 and Q_2 are series pass transistors.

Bipolar power supply

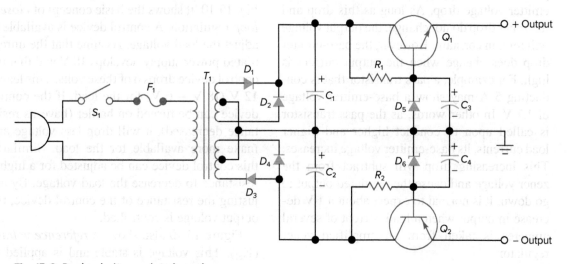

Fig. 15-9 Dual-polarity regulated supply.

Choose the letter that best answers each question.

1. Load regulation is a measure of a power supply's ability to keep a constant output under conditions of changing
   a. Line voltage
   b. Current demand
   c. Temperature
   d. Oscillator frequency

2. Why must a capacitor be connected across the secondary of a ferroresonant transformer?
   a. To filter out ac ripple
   b. To change dc to ac
   c. To cause core saturation
   d. To eliminate radio-frequency interference

3. Refer to Fig. 15-6. If $R_Z = 15\ \Omega$, the unregulated input is 12 V, and the zener operates at 6 V, what is the diode dissipation when the load current is 0 A?
   a. 4.8 W
   b. 2.4 W
   c. 1.2 W
   d. 0 W

4. Refer to Fig. 15-8. Assume a dc input of −20 V, a zener voltage of 14.4 V, and a silicon pass transistor. What is the load voltage?
   a. −20 V
   b. −15.1 V
   c. −14.4 V
   d. −13.7 V

5. Refer to Fig. 15-9. Both zeners are rated at 6.8 V, and both transistors are silicon. What is the voltage from the + output terminal to the − output terminal?
   a. 12.2 V
   b. 10.6 V
   c. 6.8 V
   d. 6.1 V

6. Refer to Fig. 15-9. Resistor R_1 is open (infinite resistance). What symptom can be expected?
   a. The − output will be zero.
   b. Both outputs will be zero.
   c. Both outputs will be high.
   d. The + output will be zero.

15-2 Closed-Loop Voltage Regulation

Feedback

The amplified zener regulators discussed in the previous section depend on a constant base-emitter voltage drop. As long as this drop and the zener drop do not change, the output voltage will remain constant. However, the base-emitter drop does change when the output current is high. For example, a pass transistor that is conducting 5 A may show a base-emitter voltage of 1.7 V. In other words, as the pass transistor is called upon to conduct higher and higher load currents, its base-emitter voltage increases. This increasing drop will subtract from the zener voltage and cause the regulated output to go down. It is normal to expect about a 1-V decrease in output when a load current of several amperes is taken from an amplified zener regulator.

Reference voltage

Open-loop regulators cannot provide highly stable output voltages, especially when large changes in load current are expected. *Feedback* can be used to improve regulation. Examine Fig. 15-10. It shows the basic concept of *closed-loop* regulation. A control device is available to adjust the load voltage. Assume that the unregulated power supply develops 18 V and that the control device drops 6 of these volts. This leaves 12 V (18 V − 6 V) for the load. If the control device can be turned on harder (have its resistance decreased), it will drop less voltage and make more available for the load. Similarly, this control device can be adjusted for a higher resistance to decrease the load voltage. By adjusting the resistance of the control device, the output voltage is controlled.

Figure 15-10 also shows a *reference voltage* (V_{ref}). This voltage is stable and is applied to

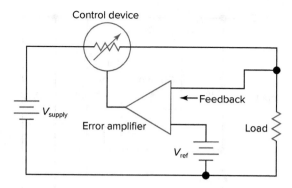

Fig. 15-10 Closed-loop regulation.

one of the inputs of an *error amplifier*. The other input to the error amplifier is feedback from the load. This feedback allows the amplifier to compare the load voltage with the reference voltage. Any change in load voltage will create a differential signal at the input of the error amplifier. This difference represents an error, and the amplifier adjusts the drive to the control device to decrease the error. If the output voltage tends to drop because of an increased load, the error is sensed and the control device is turned on harder to eliminate the drop in output. The feedback and error amplifier stabilize the output voltage.

Figure 15-11 shows a schematic diagram for a feedback (closed-loop) regulator. Transistor Q_1 is a series pass transistor and serves as the control device. The zener diode produces the reference voltage. Transistor Q_2 is the error amplifier. R_1 is the load resistor for Q_2. R_2 and R_3

form a voltage divider for the output voltage and provide feedback to Q_2. The emitter voltage of Q_2 is zener-regulated, and its base voltage is proportional to the output voltage. This allows Q_2 to amplify any error between the reference and the output.

Assume, in Fig. 15-11, that the load demands more current, causing a decrease in output voltage. The divider now sends less voltage to the base of Q_2. Transistor Q_2 responds by conducting less current, and less voltage will drop across R_1. The base voltage of Q_1 goes up, and Q_1 is turned on harder, which increases the output voltage. If you trace all of the changes, you will see that output voltage change is reduced by the feedback and the error amplifier.

Error amplifier

EXAMPLE 15-2

Calculate the zener diode current in Fig. 15-11 when the unregulated input is 16 V, the zener is 5.1 V, $\beta_{Q_1} = 35$, $R_1 = 47\ \Omega$, $R_2 = 1\ k\Omega$, $R_3 = 1\ k\Omega$, and $R_L = 5\ \Omega$. Also, find the zener current when the load is disconnected. This problem takes several steps to solve. The voltage at the base of the error amplifier is found first:

$$V_{B(Q_2)} = V_{DZ} + 0.7\ V = 5.1\ V + 0.7\ V$$
$$= 5.8\ V$$

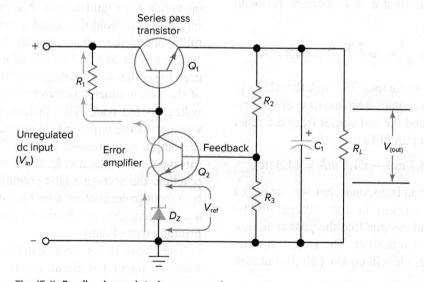

Fig. 15-11 Feedback-regulated power supply.

V_{out} is determined by the voltage divider and $V_{B(Q_2)}$:

$$V_{B(Q_2)} = V_{out} \times \frac{R_3}{R_3 + R_2}$$

$$5.8\text{ V} = V_{out} \times \frac{1\text{ k}\Omega}{2\text{ k}\Omega}$$

Solving for V_{out},

$$V_{out} = 2 \times 5.8\text{ V} = 11.6\text{ V}$$

With V_{out} known, the base voltage of the series pass transistor is readily determined by adding 0.7 V:

$$V_{B(Q_1)} = V_{out} + 0.7\text{ V}$$

$$= 11.6\text{ V} + 0.7\text{ V} = 12.3\text{ V}$$

The drop across R_1 is calculated with:

$$V_{R_1} = V_{in} - V_{B(Q_1)}$$

$$= 16\text{ V} - 12.3\text{ V}$$

$$= 3.7\text{ V}$$

Ohm's law will give us the current in R_1:

$$I_{R_1} = \frac{V_{R_1}}{R_1} = \frac{3.7\text{ V}}{47\ \Omega} = 78.7\text{ mA}$$

We now have to determine how much of this current flows through the error amplifier and the zener diode. The load current is calculated using Ohm's law:

$$I_L = \frac{11.6\text{ V}}{5\ \Omega} = 2.32\text{ A}$$

The pass transistor base current is found next:

$$I_{B(Q_1)} = \frac{I_E}{\beta + 1} = \frac{2.32\text{ A}}{36} = 64.4\text{ mA}$$

Therefore, of the total 78.7 mA that flows in R_1, 64.4 mA comes from the base of the pass transistor, and the rest comes from the zener and the error amplifier:

$$I_{D_Z} = 78.7\text{ mA} - 64.4\text{ mA} = 14.3\text{ mA}$$

When the load is disconnected, we can ignore the small current in the voltage divider network and assume that the pass transistor needs no base current. The error amplifier and zener diode will conduct all the current:

$$I_{D_Z\text{(no load)}} = 78.7\text{ mA}$$

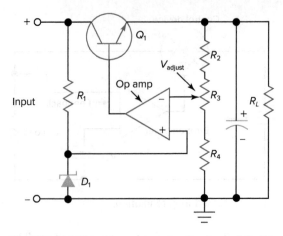

Fig. 15-12 Using an op amp in a feedback-regulated power supply.

The ability of a feedback power supply to stabilize output voltage is related to the gain of the error amplifier. A high-gain amplifier will respond to very small changes in output voltage and will provide excellent voltage regulation. Examine Fig. 15-12. An op amp is used as the error amplifier. Op amps are capable of very high gain. Resistor R_1 and the zener diode form a reference voltage for the noninverting input of the op amp. Resistors R_2, R_3, and R_4 form a voltage divider. If the output voltage goes down, there will be a decrease in voltage at the inverting input of the op amp. This decreasing voltage is negative-going and will cause the output of the op amp to go in a positive direction. The positive-going output is applied to the pass transistor and turns it on harder. This tends to increase the output and eliminate the change. The op amp's high gain means that the circuit in Fig. 15-12 can hold the output to within several millivolts, so the voltage regulation is excellent.

The circuit in Fig. 15-12 is adjustable. R_3 is used to set the output voltage. As the wiper arm of the potentiometer is moved toward R_4, less voltage is fed back. This increases the output voltage. As the wiper arm is moved toward R_2, the output is decreased. Adjustable outputs of this type are common in feedback regulators. In practice, the voltage-adjust potentiometer may be a front-panel control, a rear-panel control, or a small trimmer potentiometer mounted on a printed circuit board.

The circuit in Fig. 15-12 can be improved by using an integrated circuit in place of D_1, the zener diode. These integrated circuits are called *adjustable voltage regulators* or programmable

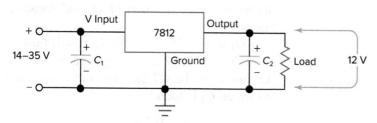
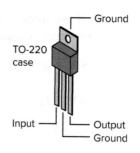

Fig. 15-13 Integrated-circuit voltage regulator.

voltage references. They have three leads and can be adjusted over a range of reference voltages with two resistors. A TL431 programmable voltage reference is shown later in this chapter (Fig. 15-41). Such circuits are more accurate and stable than zener diodes and have a low ac impedance. This will reduce the ripple voltage at the + input of the error amplifier. Thus, replacing the zener diode with a TL431 will make the load voltage more stable, and there will be less ripple across R_L.

The trend in electronics is to integrate as many circuit functions as is practical onto a single chip of silicon. Regulators have not escaped this trend. Refer to Fig. 15-13. It shows an *integrated-circuit voltage regulator*. It is supplied in the TO-220 case and has three leads. The pass transistor, error amplifier, reference circuit, and protection circuitry are all on one chip. The 7812 IC provides 12 V at load currents up to 1.5 A. Typically, it will hold the output within 12 mV over the full range of load currents.

Capacitor C_1 in Fig. 15-13 is required if the IC regulator is located more than several inches from the main power-supply filter capacitor. These ICs are often used as "on-card" regulators. In this configuration, each circuit board in a system has its own voltage regulator so they can be some distance from the main power-supply filter. Capacitor C_2 is optional and can be used to improve the way the regulator responds to rapidly changing load currents.

Some IC regulators, such as the one shown in Fig. 15-13, operate at a fixed output voltage. The 78*XX* series of regulators is typical of this type. The 7805 provides 5 V, the 7812 provides 12 V, and the 7815 provides 15 V. This series is also available in the larger TO-3 case for higher current applications. They all provide a simple and inexpensive alternative to discrete regulators and are widely applied.

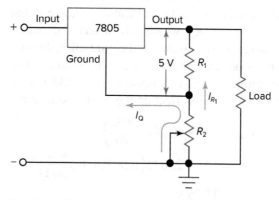

Fig. 15-14 Adjustable output from a fixed regulator.

Figure 15-14 shows a way of obtaining an *adjustable output* from a 5-V IC regulator. R_1 and R_2 form a voltage divider. Notice that the ground lead of the regulator is connected to the center of the divider rather than to the circuit ground. By adjusting R_2, the output voltage can be varied from 5 V to some higher voltage.

Two currents flow through R_2 in Fig. 15-14. One is the divider current through R_1. Since the voltage across R_1 must always be 5 V, Ohm's law may be used to find the divider current. The second current through R_2 is I_Q, the quiescent current of the 7805 IC. It is typically 6 mA. The load voltage can be found by adding the regulator voltage (5 V) to the drop across R_2. For

IC voltage regulator

Adjustable output

ABOUT ELECTRONICS

Power and Measurement

- Some modern and portable test equipment can perform the function of a strip-chart recorder. This function is invaluable for investigating power source fluctuations.
- Technicians who troubleshoot power supplies often use dummy loads.

example, if we assume R_1 and R_2 to each be 250 Ω, the output voltage is

$$V_{out} = 5\text{ V} + R_2\left(I_Q + \frac{5\text{ V}}{R_1}\right)$$

$$= 5\text{ V} + (250\ \Omega)\left(0.006\text{ A} + \frac{5\text{ V}}{250\ \Omega}\right)$$

$$= 11.5\text{ V}$$

The output has been adjusted to 11.5 V even though a fixed 5-V regulator is used.

Any increase or decrease in the quiescent current in Fig. 15-14 will cause a change in the drop across R_2, which will affect the output voltage. I_Q is sensitive to the unregulated input, the load current, and the temperature. For example, a 1-mA increase in I_Q would not be unusual, and its effect on the output voltage is

$$V_{out} = 5\text{ V} + R_2\left(I_Q + \frac{5\text{ V}}{R_1}\right)$$

$$= 5\text{ V} + 250\ \Omega\left(0.007\text{ A} + \frac{5\text{ V}}{250\ \Omega}\right)$$

Tracking

$$= 11.75\text{ V}$$

So, the output shift due to the increase in I_Q is 11.75 V − 11.5 V, or 0.25 V (250 mV). This shows that adjusting a fixed regulator with a divider degrades its regulation. Since the output is normally held within 12 mV, a 250-mV change is relatively large. Resistor R_2 in Fig. 15-14 should not be too large or the regulation will suffer even more. Values around 100 Ω are practical.

Current-boost circuit

Fixed regulators, such as the 7805, can supply 1.5 A. If more current is required, the *current-boost circuit* in Fig. 15-15 can be used. Transistor Q_1 is used to supply the extra load current. Resistor R_1 determines when Q_1 will turn on and begin sharing the load current. As the IC regulator current increases, the voltage drop

across R_1 will increase. This drop is applied to the base-emitter junction of Q_1 and forward-biases it.

If Q_1 in Fig. 15-15 is silicon, it will turn on when its base-emitter voltage reaches 0.7 V. Assume R_1 to be 4.7 Ω. The current required to turn on Q_1 is found by

$$I = \frac{V}{R} = \frac{0.7\text{ V}}{4.7\ \Omega} = 0.149\text{ A}$$

The IC regulator will conduct all of the load current up to 149 mA. As the load demand exceeds this value, the drop across R_1 turns on Q_1, and it will assist the IC to supply the load. A current-boost circuit, such as the one shown in Fig. 15-15, can provide as much as 10 A by using a high-current transistor to share the load current.

Operational-amplifier circuits often require bipolar power supplies. These power supplies provide both a positive and a negative output voltage with respect to ground. Sometimes these power supplies are adjustable and must *track* one another. A tracking power supply is one in which one or more outputs are correlated to a controller. If the controller output changes, so must the output of the correlated supplies change.

Figure 15-16 shows a dual-tracking regulator. The 7805 is a fixed 5-V regulator. The 7905 is also a fixed 5-V device, but it regulates a negative voltage with respect to ground. Neither regulator is directly grounded in Fig. 15-16. The ground leads are driven by operational amplifiers OA_1 and OA_2. This provides adjustable output voltage in a fashion related to the circuit studied in Fig. 15-14. However, the very low output impedance of the op amps ensures that any quiescent current change in the IC regulators will have only a small effect on output voltage.

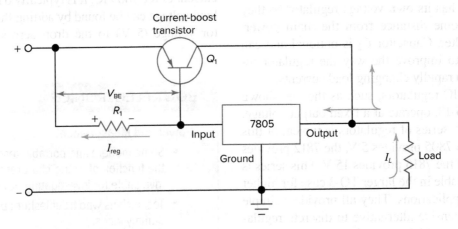

Fig. 15-15 Current-boost circuit.

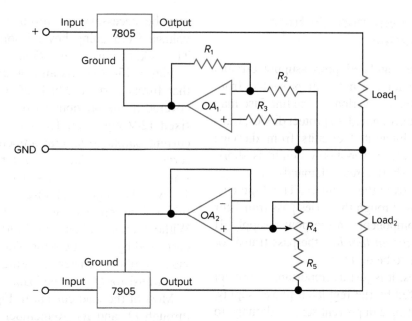

Fig. 15-16 Dual-tracking regulator.

Resistors R_4 and R_5 in Fig. 15-16 divide the negative output and apply the result to both op amps. The op amp OA_2 is connected in the non-inverting mode. As the wiper of R_4 is adjusted toward R_5, the output of OA_2 drives the ground lead of the 7905 in a negative direction. This increases the negative output across load 2. At the same time, OA_1 acts as an inverting amplifier.

The negative-going signal at the wiper of R_4 becomes a positive-going signal for the ground lead of the 7805. This increases the positive output voltage across load 1. Resistor R_4, therefore, controls both outputs in this tracking regulator. The positive output is correlated to the negative output, and any change in negative output will be tracked by the positive output.

Self-Test

Choose the letter that best answers each question.

7. Refer to Fig. 15-11. If the unregulated dc input is 18 V and the collector-to-emitter voltage of Q_1 is 12 V, then the voltage across the load resistor will be
   a. 18 V
   b. 12 V
   c. 6 V
   d. 0 V

8. Refer to Fig. 15-11. If the zener diode drops 4.7 V and the collector-to-emitter voltage of Q_2 is 4 V, what is the voltage across the load resistor? (Assume a silicon pass transistor and a light load current.)
   a. 4 V
   b. 8 V
   c. 12 V
   d. 16 V

9. Refer to Fig. 15-12. Which component provides the reference voltage?
   a. D_1
   b. Q_1
   c. R_3
   d. The op amp

10. Refer to Fig. 15-14. The quiescent current is 5 mA. Resistor R_1 is 100 Ω and R_2 is 200 Ω. What is the output voltage?
    a. 2 V
    b. 5 V
    c. 12 V
    d. 16 V

11. Refer to Fig. 15-15. R_1 is 3.3 Ω and the load current is 150 mA. Transistor Q_1 is silicon. Transistor Q_1 will conduct
    a. No load current
    b. About half the load current
    c. All of the load current
    d. All of the load current in excess of 100 mA

12. Refer to Fig. 15-15. The load current is 4 A. The regulator current is 0.5 A. What is the collector current in Q_1?
    a. 0.5 A
    b. 3.5 A
    c. 4.0 A
    d. 4.5 A

15-3 Current and Voltage Limiting

Some of the regulated power-supply circuits discussed so far are not well protected from damage caused by overloads. The line fuse may not blow quickly enough to protect diodes, transistors, and integrated circuits from damage in the event the power-supply output is short-circuited. A short circuit demands very high current flow from the regulator. This high current will flow through the series pass transistor and other components in the power supply. If

there is no *current limiting*, the pass transistor is very likely to be destroyed.

Sometimes, it is just as important to protect the circuits fed by the regulated power supply. Current limiting can prevent serious damage to other circuits. For example, a component in a direct-coupled amplifier may short. The shorted component may overbias an expensive transistor or integrated circuit. The overbiased device can take enough current from the power supply to destroy itself. If the power supply is current-limited, the expensive component may be protected from burnout.

Current limiting is helpful, but circuits can also be damaged by too much voltage. A fault in the regulator can cause the output voltage to go up to the nonregulated value. For example, the input to a 5-V regulator may be 10 V. If the pass transistor shorts, the output will go up to 10 V instead of the normal 5 V. This abnormally high voltage will be applied to all devices in the system that are connected to the 5-V supply. Many of them could be destroyed. Therefore, it

may be necessary to prevent a power-supply voltage from going beyond some safe value. This is known as *voltage limiting*.

Figure 15-17 shows an example of a circuit that limits current. Much of this circuit was covered in the previous section. The 7812 is a fixed 12-V regulator. Transistor Q_1 boosts the current output to several amperes. The 78*XX* series of IC regulators is *internally* current-limited. A 7812 IC will supply no more than 1.5 A if its output is short-circuited. However, this will not protect transistor Q_1 in Fig. 15-17. Without additional current limiting, it could be destroyed by a short circuit. Transistor Q_2 provides additional current limiting to protect the pass transistor Q_1 and the load.

Most of the load current in Fig. 15-17 flows through Q_1 and R_2. Remember that the 7812 will conduct enough current to produce base-emitter bias for the pass transistor. This bias is produced by the drop across R_1. Now, suppose that the load demands too much current. This current will cause enough voltage to drop across R_2 to turn on Q_2. In this application, R_2 serves as a *current-sensing* resistor. With Q_2 on, there is now a second path for the regulator current. It will flow from the collector of Q_2 to the emitter of Q_2 and on to the + terminal of the input. This second path will reduce the current through R_1. If the current through R_1 is reduced, the voltage drop across R_1 must also be reduced. This reduced voltage means less forward bias for the pass transistor Q_1. With less bias, the pass transistor will not conduct as much current, and the circuit goes into current limiting. Even if the

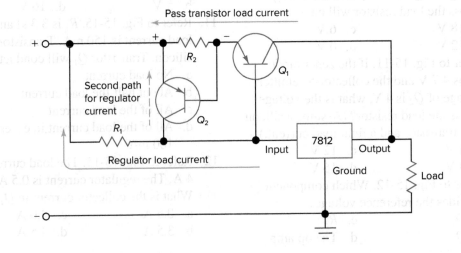

Fig. 15-17 Current-limit circuit.

load is a short circuit, the current will be limited to some predetermined value.

The maximum current permitted by the limiting action of the circuit shown in Fig. 15-17 is determined by R_2. Q_2 requires about 0.7 V of base-emitter bias to turn on and begin the current limiting action. If R_2 is a 0.1-Ω resistor, it will drop 0.7 V when it conducts 7 A ($V = I \times R$). Some of the load current flows through the 7812 IC (perhaps about 0.5 A). Thus, when the load demands more than 7.5 A, Q_2 will come on and the current will be limited from increasing much beyond this value. Making R_2 larger limits the current to less than 7.5 A. Remember, without the current-limiting action, a short circuit will often destroy the pass transistor. Also, the IC regulator must have internal current limiting or it may be damaged by an overload.

EXAMPLE 15-3

For Fig. 15-17, $R_1 = 4.7\ \Omega$ and $R_2 = 0.22\ \Omega$. Determine the values of load current required to turn on Q_1 and Q_2. Q_1 will turn on when R_1 drops 0.7 V:

$$I_{\text{load}} = \frac{0.7\ \text{V}}{R_1} = \frac{0.7\ \text{V}}{4.7\ \Omega} = 149\ \text{mA}$$

Q_2 comes on when R_2 drops 0.7 V:

$$I_{\text{load}} = \frac{0.7\ \text{V}}{R_2} = \frac{0.7\ \text{V}}{0.22\ \Omega} = 3.18\ \text{A}$$

The design employed in Fig. 15-17 is known as *conventional current limiting*. Figure 15-18 shows a graph of circuit performance when conventional current limiting is used. The graph shows that the output voltage remains constant at 12 V as the load current increases from 0 to 5 A. As the load increases beyond 5 A, the output voltage begins to drop rapidly. A short circuit will be limited to a little more than 5 A. As the output voltage drops from 12 to 0 V, the curve is in the constant-current region. This is where the circuit operates when it is in current limiting.

Conventional current limiting may not completely protect the pass transistor. Even if the transistor is rated to conduct the amount of current in the constant-current region, it may over-dissipate and be damaged or destroyed if the short persists. For example, a type 2N3055

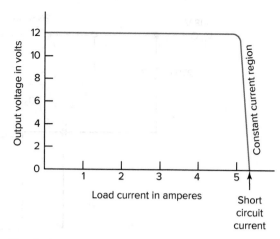

Fig. 15-18 Conventional current-limiting performance.

transistor is rated at 15 A and 117 W. Therefore, it may appear to be safe if operated in the constant-current region, as shown in Fig. 15-18. However, this may not be true. Even though 5 A is only one-third the rating of a 2N3055, the transistor can still be destroyed by too much collector dissipation. Suppose the output is shorted. Zero volts will appear across the load, and all the unregulated power supply must drop across the pass transistor. For a 12-V power supply, the unregulated input will probably be around 18 V. The transistor dissipation will be

$$P_C = V_{\text{CE}} \times I_C$$
$$= 18\ \text{V} \times 5\ \text{A} = 90\ \text{W}$$

Since 90 W is less than 117 W, the transistor is operating within its limits. But the 117-W rating is based on a junction temperature of 25°C (77°F). When a transistor is dissipating 90 W, it is going to get *very* hot. A large heat sink will help, but it is likely that the junction temperature will exceed 65°C. At this temperature, the maximum collector dissipation is less than 90 W. Power transistors must be *derated* for temperatures over 25°C. Thus, the 2N3055 will be damaged or destroyed if the short circuit lasts long enough for the transistor temperature to exceed 65°C. Conventional current limiting may provide protection only when short circuits are momentary.

Figure 15-19(a) shows a voltage regulator that uses *foldback current limiting*. An analysis of this circuit will help you understand how this improved type of current limiting works. There are actually two important current limits that can be calculated, and one of them is partly determined by the output voltage. V_{out} is

Conventional current limiting

Foldback current limiting

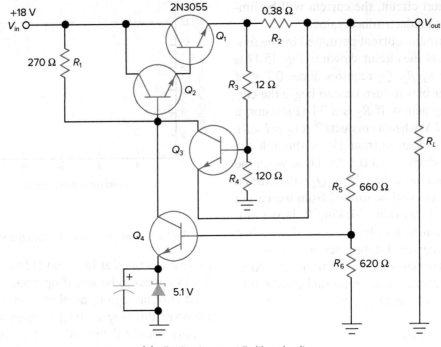

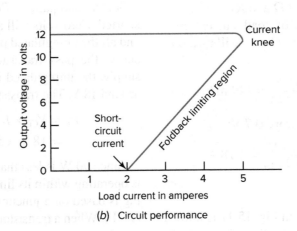

(b) Circuit performance

Fig. 15-19 Foldback current limiting.

established by the zener reference voltage and the R_5–R_6 voltage divider. We can assume that V_{out} will turn on the error amplifier (Q_4). Allowing for a 0.7-V drop from base to emitter, the base voltage of Q_4 will be

$$V_{B(Q_4)} = 5.1 \text{ V} + 0.7 \text{ V} = 5.8 \text{ V}$$

This voltage must be equal to some fraction of V_{out}, as determined by the voltage divider:

$$V_{B(Q_4)} = V_{out} \times \frac{R_6}{R_6 + R_5}$$

$$5.8 \text{ V} = V_{out} \times \frac{620 \ \Omega}{620 \ \Omega + 660 \ \Omega}$$

$$V_{out} = 12.0 \text{ V}$$

The short-circuit current in Fig. 15-19(a) is set by R_2, R_3, and R_4. With the output shorted, V_{out} is zero and the drop across R_2 will be high enough to turn on Q_3, the current limit transistor. With Q_3 on, drive current is diverted from Q_2, and the pass transistor starts to shut down. Since R_3 and R_4 form a voltage divider, Q_3 does not see the full drop across R_2:

$$V_{BE(Q_3)} = 0.7 \text{ V}$$

$$= V_{R_2} \times \frac{120 \ \Omega}{120 \ \Omega + 12 \ \Omega}$$

$$0.7 \text{ V} = V_{R_2} \times 0.909$$

$$V_{R_2} = 0.770 \text{ V}$$

Now Ohm's law is used to find the current in R_2. This is the current flow when the output is shorted:

$$I_{SC} = \frac{V_{R_2}}{R_2} \times \frac{0.770 \text{ V}}{0.38 \text{ }\Omega} = 2.03 \text{ A}$$

The maximum load current in Fig. 15-19(a) is greater than the short-circuit current. Its value is calculated by assuming that the output voltage is normal. V_{out} establishes V_E for Q_3, and V_B for Q_3 is 0.7 V higher, or 12.7 V:

$$V_{B(Q_3)} = (V_{R_2} + V_{out}) \times \frac{R_4}{R_4 + R_3}$$

$$12.7 \text{ V} = (V_{R_2} + 12 \text{ V}) \times \frac{120 \text{ }\Omega}{120 \text{ }\Omega + 12 \text{ }\Omega}$$

$$12.7 \text{ V} = 0.909 \text{ } V_{R_2} + 10.9 \text{ V}$$

$$V_{R_2} = 1.97 \text{ V}$$

The current in R_2 is found with Ohm's law:

$$I_{R_2} = I_{max} = \frac{1.97 \text{ V}}{0.38 \text{ }\Omega} = 5.18 \text{ A}$$

This demonstrates that the maximum load current is significantly higher than the short-circuit current in regulators that use foldback current limiting.

Figure 15-19(b) shows the performance graph for Fig. 15-19(a). This type of protection folds back the current flow once some preset limit is reached. Note that 5 A is the limiting point. However, in this case, the current begins to decrease instead of remaining constant near 5 A. If the overload is a short circuit, the current folds back to a value near 2 A. This greatly

limits the dissipation in the pass transistor for short circuits. If we again assume an unregulated input of 18 V, the collector dissipation will be

$$P_C = V_{CE} \times I_C$$
$$= 18 \text{ V} \times 2 \text{ A} = 36 \text{ W}$$

A dissipation of 36 W is much more reasonable for a 2N3055 transistor. The transistor will now be safe up to a junction temperature of 150°C. A good heat sink will be able to maintain the junction below this temperature. With foldback current limiting and a good heat sink, the pass transistor will be able to withstand a short circuit for an indefinite period of time.

We have already learned that the 78XX series of IC regulators features internal current limiting. This current limiting is of the conventional type. Another popular IC regulator, the 723, is capable of both types of current limiting. This IC is available in the dual in-line package. Figure 15-20 shows it connected in a circuit to provide conventional current limiting. In this circuit, R_1 and R_2 divide an internal reference voltage to set the output voltage between 2 and 7 V. R_3 is the current-sensing resistor. When the drop across this resistor reaches about 0.7 V, the regulator goes into conventional current limiting.

Figure 15-21 shows the 723 regulator configured for an output greater than 7 V and for foldback current limiting. R_4 and R_5 determine the output voltage. R_1, R_2, and R_3 determine the current knee and the short-circuit current [refer to Fig. 15-19(b)].

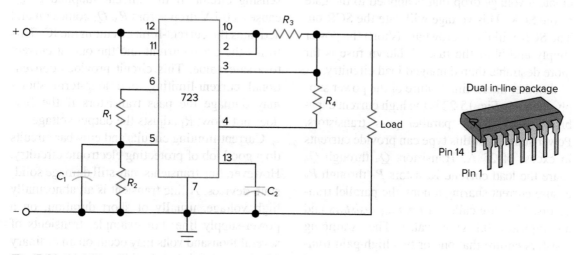

Fig. 15-20 An IC regulator configured for conventional current limiting.

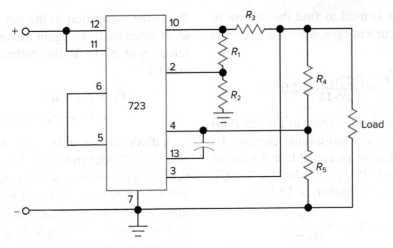

Fig. 15-21 An IC regulator configured for foldback current limiting.

Overcurrent circuits protect systems from damage. Sometimes, though, a power supply will fail and destroy other circuits even though current-limiting circuitry is included. We have learned that the series pass transistor is used to drop the unregulated voltage to the desired value. If the pass transistor shorts from emitter to collector, the entire unregulated voltage will be applied to all loads connected to the power supply. When this happens, many circuits may be damaged. Some form of *overvoltage* protection may be needed to prevent this from happening.

Figure 15-22 shows the schematic diagram for a high-current power supply with *crowbar protection*. A crowbar is a circuit that shorts the power supply when some voltage limit is exceeded. Zener diode D_1 is part of the crowbar circuit. Normally it will not conduct. However, if the output voltage goes too high, D_1 will turn on and the resulting current through R_9 will create a voltage drop that is applied to the gate of the SCR. This voltage will gate the SCR on. The SCR will then "crowbar" (short) the power supply and blow the fuse. A blown fuse is far more desirable than damaged load circuitry.

Another interesting feature of the power supply shown in Fig. 15-22 is the high current capability provided by parallel pass transistors. Power supplies of this type can provide currents in excess of 25 A. Transistors Q_3 through Q_6 share the load current. Resistors R_5 through R_8 ensure current sharing among the parallel transistors. They are called *swamping resistors* and are typically 0.1 Ω in value. The swamping resistors ensure that one or two high-gain transistors will not "hog" more than their share of

the load current. Suppose, for example, that Q_5 has a higher β than the other three pass transistors. This would tend to make it conduct more than its share of the load current, and that would make it run hotter than the other transistors. Since β increases with temperature, it would then conduct more of the load current. It would again increase in temperature and so on. This condition is called *thermal runaway* and could destroy Q_5. The swamping resistor decreases the chance of thermal runaway because it drops more voltage if the current in Q_5 increases. This drop subtracts from the transistor's forward bias and reduces the current in Q_5. Therefore, the swamping resistors in Fig. 15-22 help ensure current sharing among the four pass transistors.

In Fig. 15-22, Q_2 is called a *driver transistor*. The IC regulator cannot supply enough current for four transistors, and Q_2 boosts the drive from the IC regulator. R_3 and Q_1 form a current-sensing circuit. If the current supplied to Q_2 causes a 0.7-V drop across R_3, Q_1 comes on and activates the current-limit circuit in the IC. This limits the drive current and the output current to a safe value. This circuit provides conventional current-limiting, and long-term shorts may damage the pass transistors if the fuse does not blow; R_1 adjusts the output voltage.

Current-limiting circuits and crowbar circuits do a good job of protecting electronic circuitry. However, line transients may still damage solid-state devices. A line transient is an abnormally high voltage, usually of short duration, on a power-supply line. For example, transients of several thousand volts may occur on an ordinary 120-V ac circuit in a building. Such transients

Thermal runaway

Crowbar protection

Driver transistor

Swamping resistors

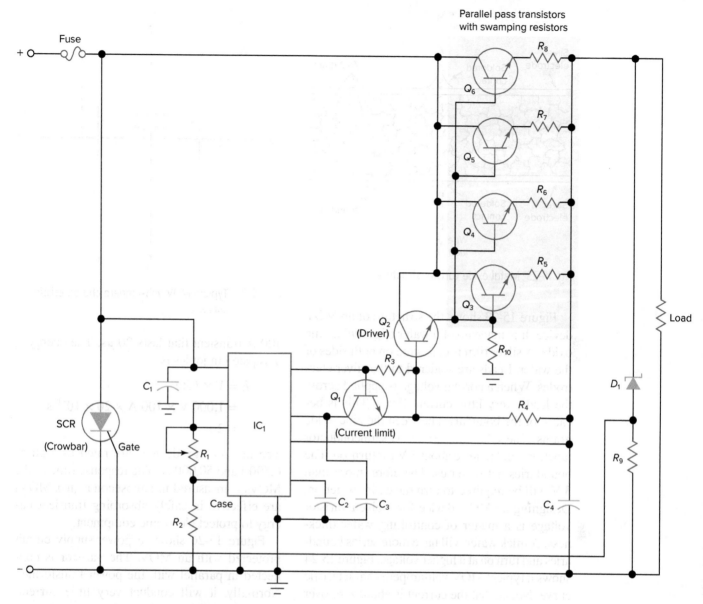

Fig. 15-22 High-current supply with crowbar protection.

are caused by lightning, equipment failures, and the switching of inductive loads such as motors and transformers. Studies predict that one 5,000-V transient can be expected each year on every 120-V service circuit in this country. More occurrences of lower voltage transients can be expected. Many electronic equipment failures are caused by *line transients*.

Transients lasting several microseconds are capable of damaging circuits, contacts, and insulation. Protection devices such as crowbars and spark gaps (ionizing breakdown protection devices) are too slow-acting. Also, these devices may not be self-clearing. That is, they may remain in conduction after the transient has

passed, and this characteristic can cause further problems. Voltage-clipping devices are considered better choices for protecting electronic circuitry from transients. These devices include selenium cells, zener diodes, and *varistors*.

Varistors are voltage-dependent resistors. Their resistance is not constant as it is with ordinary resistors. As the voltage across a varistor increases, its resistance decreases. This feature makes them valuable for clipping transients. Varistors are made from silicon carbide or, more recently, zinc oxide. Zinc oxide varistors are usually called metal oxide varistors (*MOVs*) and are widely applied for protecting electronic equipment from line transients.

Varistor

Line transient

MOV

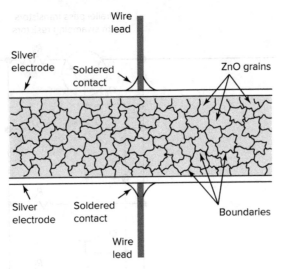

Fig. 15-23 Metal oxide varistor structure.

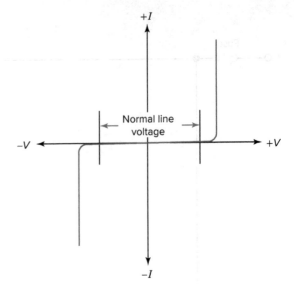

Fig. 15-24 Typical MOV volt-ampere characteristic curve.

Figure 15-23 shows the structure of an MOV device. It is made up of a wafer of granular zinc oxide. A silver film is deposited on both sides of the wafer. Leads are soldered to the silver electrodes. When a normal voltage is applied across the leads, very little current flows. This is because of the boundaries between the zinc oxide grains. These boundaries act as semiconductor junctions and require about 3 V for turn-on. The boundaries act in series. Therefore, more than 3 V will be required to turn the entire wafer on. Designing an MOV device for a given varistor voltage is a matter of controlling wafer thickness. A thick wafer will have more series boundaries and turn on at a higher voltage. Figure 15-24 shows a typical MOV volt-ampere characteristic curve. Notice that the current is about zero over the normal line-voltage range. Also note that if the line voltage exceeds normal, a very sharp increase in current can be expected.

Figure 15-25 shows four package styles for MOV devices made by General Electric. The small axial devices can safely absorb 2 joules (J) of energy and conduct 100 A during a transient. A joule is equal to a watt-second (1 W × 1 s). Suppose that an MOV device absorbs a 1,000-V,

100-A transient that lasts 20 μs. The energy E dissipated in joules is

$$E = V \times I \times t$$
$$= 1{,}000 \text{ V} \times 100 \text{ A} \times 20 \times 10^{-6} \text{ s}$$
$$= 2 \text{ J}$$

The high-energy devices are rated as high as 6,500 J and 50,000 A. The response time of the MOVs is measured in nanoseconds (ns). MOVs are effective in safely absorbing transient energy to protect electronic equipment.

Figure 15-26 shows a power-supply circuit protected with an MOV. The varistor is connected in parallel with the power transformer. Normally, it will conduct very little current. A transient will turn the varistor on and much of the transient energy will be absorbed. After the transient passes, the MOV will return to its high-resistance state, and the circuit will resume normal operation. A long-term transient will cause the fuse to blow, and it will have to be replaced. Note the schematic symbol for the varistor in Fig. 15-26. The line drawn through it shows a nonlinear resistance characteristic.

Self-Test

Choose the letter that best answers each question.

13. Current-limited power supplies can prevent damage to

    a. Rectifier diodes and power transformers
    b. Pass transistors
    c. Other circuits in the system
    d. All of the above

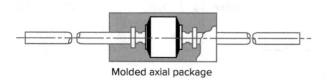
Molded axial package

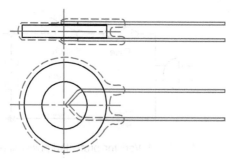
Radial lead package

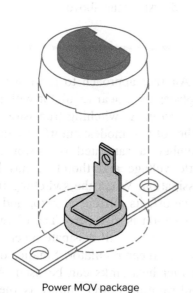
Power MOV package

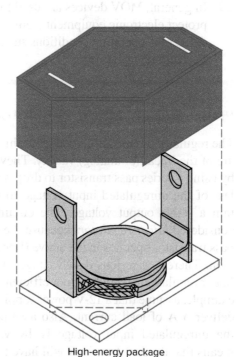
High-energy package

Fig. 15-25 General Electric MOV package styles.

14. Refer to Fig. 15-17. Both transistors are silicon; R_1 is 10 Ω, and R_2 is 0.2 Ω. At what load will Q_1 begin to provide current?
   a. 0 A
   b. 0.07 A
   c. 3.57 A
   d. 7.07 A
15. Refer to Fig. 15-17. Both transistors are silicon; R_1 is 10 Ω, and R_2 is 0.2 Ω. At what load will current limiting begin?
   a. 0 A
   b. 0.07 A
   c. 3.57 A
   d. 7.07 A
16. Foldback current limiting has the advantage of
   a. Better pass transistor protection for long-term overloads

   b. A defined turn-on point
   c. Circuit simplicity
   d. All of the above
17. A crowbar is a power-supply circuit that provides
   a. Conventional current limiting
   b. Foldback current limiting
   c. Temperature control
   d. Voltage limiting
18. Refer to Fig. 15-22. What is the function of resistors R_5 through R_8?
   a. To ensure current sharing among Q_3 through Q_6
   b. To adjust the crowbar trip point
   c. To provide current sensing to shut down Q_2
   d. To improve the voltage regulation

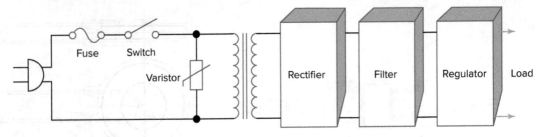

Fig. 15-26 Varistor-protected power supply.

19. In general, MOV devices are used to protect electronic equipment from dangerous operating conditions such as
    a. High temperatures
    b. Line transients
    c. Overcurrent
    d. All of the above

15-4 Switch-Mode Regulators

The regulator circuits discussed up to this point are of the linear (or analog) variety. They work by using a series pass transistor to drop more or less of the unregulated input voltage to maintain a stable output voltage. The circuits are considered *linear regulators* because the series pass transistor operates in the active (linear) region. There is a serious disadvantage to using linear regulation and that is poor efficiency. For example, assume that a 12-V power supply must deliver 5 A of load current. Also assume that the unregulated input voltage is 18 V. This means that the pass transistor will have to drop the extra 6 V. The power dissipated in the pass transistor will be 30 W (6 V × 5 A). This dissipation is wasteful and requires a large heat sink.

The efficiency of the linear regulator can be calculated by comparing the useful output power to the input power. The useful output power is 60 W (12 V × 5 A). The input power is 90 W (18 V × 5 A). Efficiency is given by

$$\eta = \frac{P_{out}}{P_{in}} \times 100\%$$

$$= \frac{60 \text{ W}}{90 \text{ W}} \times 100\% = 66.7\%$$

The overall efficiency of the power supply will be less than 66.7 percent. This is because of additional losses in the transformer, the rectifiers, and other parts of the circuit. Linear power supplies usually have overall efficiencies of less than 50 percent. This means that much of the electrical energy will be wasted in the form of heat.

Another approach to power-supply design replaces the linear regulator with a *switching transistor*. A switching transistor operates in either of two modes: cutoff or saturation. Remember, a saturated transistor drops very little voltage and therefore has low power dissipation. When the switching transistor is in cutoff, its current is zero, and the power dissipation is also zero. Therefore, a switching regulator will dissipate much less energy than a linear regulator. Smaller devices and smaller heat sinks can be used. A compact, cool-running power supply is the result. In fact, a switching power supply can be less than one-third the weight and volume of an equivalent linear power supply, and it will cost less to operate.

Switching regulators store electric energy in capacitors, inductors, or transformers. Table 15-1 shows a summary of power-supply regulators.

Charge pumps use capacitors to store electric energy and to generate an output voltage that is higher or lower than the input voltage. Regulated charge pumps can also invert the input voltage (change polarity). Generally, the load current that can be drawn from a charge pump is limited to tens of milliamperes. Some charge pumps are capable of handling up to 125 mA, such as the MAX1595 shown in Fig. 15-27. This IC generates either 3.3 V or 5 V from a 1.8 to 5.5-V input. The regulator will step up or step down the input voltage to maintain a constant output voltage. It uses a 1-MHz switching frequency to allow the use of ceramic capacitors as small as 1 μF for 125 mA of output current.

Linear regulation

Table 15-1   Summary of Power-Supply Regulators

Topology/Type	Arrangement	Strong Points	Limitations
Linear (step-down)		• Inexpensive • Can be very small • Low no-load current • Low noise/low EMI • Usually the best solution for smaller loads	• V_{out} must be less than V_{in}. • Inefficient at high-input voltages and/or large loads. • Can require a large heat sink.
Charge pump (boost or invert polarity)		• Inexpensive • Small • Can boost or invert	• Limited output power. • Limited range of input/output voltage ratio.
Buck (step-down)		• Lowest peak current • Efficient • Modest cost • Low-ripple current in output-filter capacitor • Simple inductor • Low switch-stress voltage	• V_{out} must be less than V_{in}. • High-side switch.
Boost (step-up)		• Low peak current • Low-side switch • Simple inductor • Low switch-stress voltage	• V_{out} must be greater than V_{in}. • Output cannot be completely turned off by removing drive. • No short-circuit protection.
Buck-boost (invert polarity)		• Simple inductor	• Negative output only. • High-side switch. • High peak currents.
Flyback (step-down, step-up, or invert polarity)		• Isolated output • Can offer multiple outputs • Steps up/down, inverts • Low-side switch	• Transformer instead of inductor. • High peak currents. • High switch-stress voltage.

Figure 15-28 shows the basic configuration for a *step-down* (buck) switching *regulator*. When S_1 is closed, load current flows through L_1 and through the switch into the unregulated input. The current through L_1 creates a magnetic field, and energy is stored there. When S_1 opens, the magnetic field in L_1 begins to collapse. This generates a voltage across L_1, which forward-biases D_1. Load current is now supplied by energy that was stored in inductor L_1. After a short period of time, S_1 is closed again and the

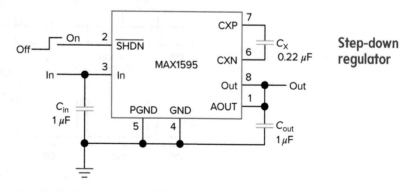

Fig. 15-27 Charge-pump switching regulator.

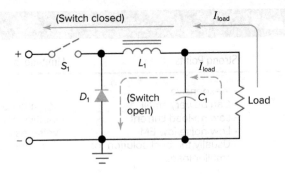

Fig. I5-28 Step-down configuration.

inductor is recharged. L_1 acts as a smoothing filter to maintain load current during those periods of time when S_1 is open. C_1 helps to filter the load voltage. The overall result is reasonably pure dc at the load even though the switch is opening and closing. The step-down configuration of Fig. 15-28 supplies less load voltage than the unregulated input voltage. As you will see later, it is also possible to have a step-up configuration in switch-mode power supplies.

You May Recall

Modulation means that one signal controls some feature of another signal.

Pulse-width modulation

Duty cycle

Switching power supplies regulate output voltage by using *pulse-width modulation*. Examine Fig. 15-29. The waveform in Fig. 15-29(*a*) shows a rectangular wave with a *duty cycle* of 50 percent. Notice that the average value of the waveform is half of the peak value. Now look at Fig. 15-29(*b*). This rectangular wave has a duty cycle of much less than 50 percent, and the average value is much less than half of the peak value. Rectangular waves are used to

drive the switch-mode regulators. By modulating (controlling) the duty cycle of the rectangular wave, the average load voltage can be controlled. The load voltage is smoothed by the filter action of inductors and capacitors to provide a low-ripple direct current.

Figure 15-30 shows a more complete circuit for a step-down switch-mode regulator. Q_1 is the switch, and it is driven by a rectangular wave with a varying duty cycle. An error amplifier compares a portion of the output voltage with a reference voltage. If the load on the power supply increases, the output tends to drop. This error is amplified, and a control signal is applied to the pulse generator, which increases the duty cycle of its output. The switching transistor is now turned on for longer periods of time. This increased duty cycle produces a higher average dc voltage, and the output voltage goes back toward normal. L_1 and C_1 eliminate the ripple; D_2 turns on when the switching transistor cuts off and allows the inductor to discharge through the load.

Switch-mode voltage regulators tend to be more complicated than linear voltage regulators. However, ICs can help to simplify designs. Look at Fig. 15-31. It shows a 78S40 IC that contains much of the circuitry needed for switch-mode operation. The oscillator is built in (integrated) and can be set to the desired frequency of operation by C_1, an external component. The typical switch-mode regulator operates at 20 kHz or above. Higher frequencies mean smaller magnetic cores in transformers and inductors. Smaller filter capacitors can also be used. Remember that capacitive reactance goes down as frequency goes up. This means that far fewer microfarads are required to filter 20-kHz ripple than 60-Hz ripple. Therefore,

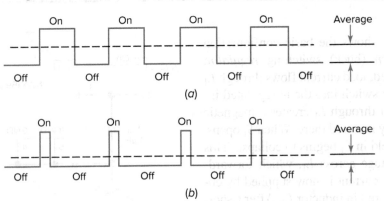

Fig. I5-29 Using pulse-width modulation to control average voltage.

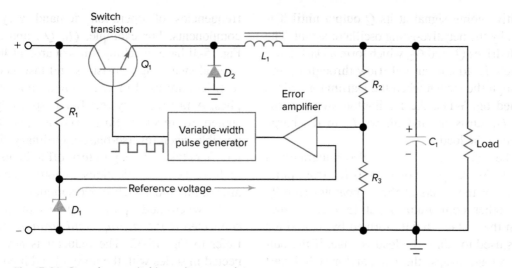

Fig. 15-30 Step-down switching voltage regulator.

many of the components can be much smaller and lighter than they would be in a 60-Hz power supply.

Figure 15-31 shows that pin 14 of the IC provides another input to the oscillator. R_1 is connected to pin 14 and acts as a current-sensing resistor. If too much load current flows, the voltage drop across R_1 will reach 0.3 V, and the oscillator duty cycle will be reduced. This will protect the IC and other components from damage. The oscillator output is combined with the output of a comparator (error amplifier) in a logical AND gate. An AND gate will allow the oscillator signal to go positive for the period of time that the comparator output is also high. Thus, this gate controls the pulse width supplied to the latch. A latch is a digital storage circuit. In this application, it will produce a

AND gate

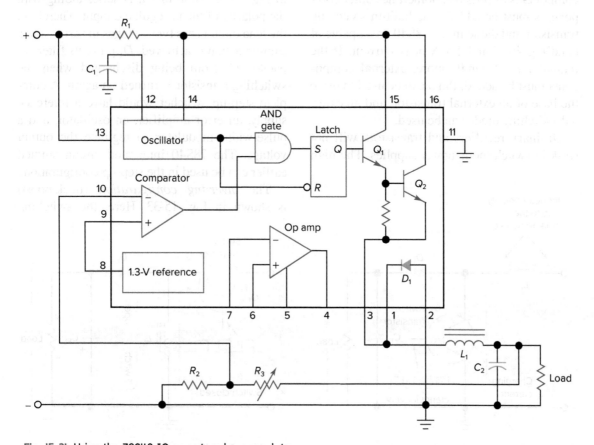

Fig. 15-31 Using the 78S40 IC as a step-down regulator.

positive-going signal at its Q output until it is reset by the negative-going oscillator signal. The latch drives Q_1 and Q_2, which form a Darlington switch. Load current will flow through L_1 and through the switch when the Darlington pair is turned on. When the Darlington switch turns off, D_1 turns on and allows L_1 to discharge through the load.

Step-up configuration

The reference voltage is also integrated in Fig. 15-31. A voltage of 1.3 V is fed from pin 8 to one of the inputs of the comparator (pin 9). The other comparator input (pin 10) comes from the voltage divider formed by R_2 and R_3; R_3 is used to adjust the load voltage. If the output voltage drops, the comparator will invert this drop and send a positive-going signal to the gate. The gate will then allow more of the positive-going oscillator signal to reach the latch. The latch will provide a higher duty-cycle drive to Q_1. This raises the average output voltage and eliminates much of the error.

There is also an op amp on the chip in Fig. 15-31. It is not used in this application. The IC manufacturer includes it to make other designs easier by eliminating as many external components as possible. Sometimes other components must be added. The built-in switching transistor and diode in the 78S40 are capable of handling 40 V and 1.5 A peak current. If the regulator must handle more, external components must be added. Pin 3 can be used to drive the base of an external transistor, and an external switching diode can be used.

Ordinary rectifiers and transistors will not work in switch-mode power supplies. The high

frequencies of operation demand very fast components. For example, Q_1, Q_2, and D_1 in Fig. 15-31 have switching times of around 400 ns. Special switching transistors and fast-recovery rectifiers are used in switch-mode power supplies. A fast-recovery rectifier is specially designed to recover (turn off) as quickly as possible when reverse-biased. Ordinary silicon rectifiers take too long to turn off to be used in high-frequency applications. Schottky rectifiers are common in switch-mode supplies.

A switch-mode power supply can also be connected in the *step-up configuration* (boost). Refer to Fig. 15-32. The inductor is now connected in series with the unregulated input, and the switching transistor is connected to ground. When the transistor is turned on by the positive-going part of the rectangular wave, a charging current flows through the transistor and through the inductor. This charging current stores energy in the inductor's magnetic field. When the rectangular wave goes negative, the transistor turns off. The field in the inductor begins to collapse. This induces a voltage across the inductor. The polarity of the induced voltage is shown in Fig. 15-32. Note that it is series-aiding with the polarity of the unregulated input. Therefore, the load circuit sees two voltages in series, and a step-up action is achieved. D_1 prevents filter capacitor C_1 from being discharged when the switching transistor is turned on again. A complete step-up switcher would have a reference supply, an error amplifier, an oscillator, and a pulse-width modulator to regulate the output voltage. The 78S40 integrated circuit studied earlier can be used in the step-up configuration.

The *inverting configuration* (buck-boost) is shown in Fig. 15-33. Here, the switching

Inverting configuration

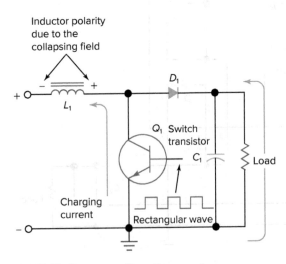

Fig. 15-32 Step-up configuration.

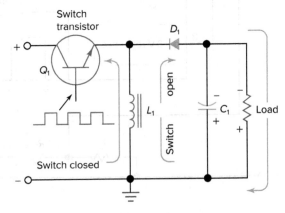

Fig. 15-33 Inverting configuration.

transistor is in series and the inductor is connected to ground. When the transistor is turned on, current flows through L_1, as shown, and charges it. When the transistor is turned off, the field collapses, and the induced voltage at the top of the inductor is negative with respect to ground. D_1 is forward-biased by this induced voltage, and the current flows through L_1, through D_1, and down through the load. The top of the load resistor is negative with respect to ground. Inverting regulators are useful in systems where a positive power supply energizes most of the circuits and one negative voltage is needed. The 78S40 can be used in the inverting configuration.

Figure 15-34 illustrates a *converter*. A converter is a circuit that changes direct current to alternating current and then changes the alternating current back to direct current again. Converters can be considered dc transformers and are used for step-up and step-down action and for isolation. Q_1 and Q_2 are driven by out-of-phase rectangular waves. They will never be on at the same time. The collector current of each

transistor flows through the primary of T_1. Alternating voltage is induced across the secondary of T_1. D_1 and D_2 form the familiar full-wave rectifier arrangement. D_3 serves the same purpose as it did in the step-down configuration (Fig. 15-28). There are periods of time when both transistors are off and L_1 will discharge to maintain load current. D_3 is forward-biased by the discharge of L_1 and completes the circuit. The circuit will work without D_3, but then the discharge current will flow through rectifiers D_1 and D_2 and through the secondary of T_1. This discharge path is not desirable since it will increase the dissipation in the rectifiers and the transformer.

Regulation is provided by pulse-width modulation in Fig. 15-34. Resistors R_1 and R_2 provide a sample of the output voltage for the inverting input of the op amp. The other input to the op amp is a reference voltage. Any error is amplified and controls the pulse width of the rectangular wave supplied to the two switching transistors.

A converter circuit such as the one shown in Fig. 15-34 will often work off the ac line. The line voltage must first be rectified, filtered, and

Converter

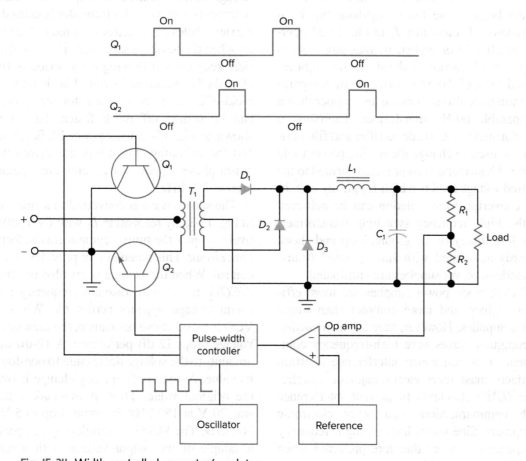

Fig. 15-34 Width-controlled converter/regulator.

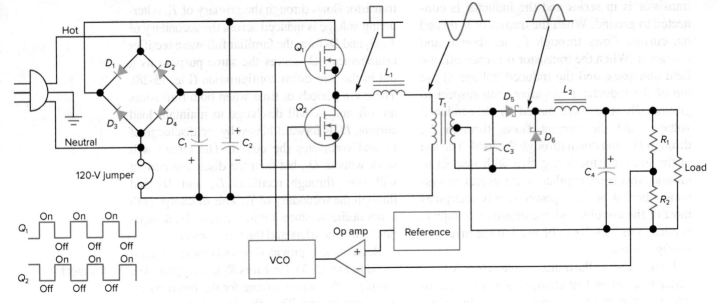

Fig. l5-35 Frequency-controlled sine wave converter.

Sine wave converter

then applied to the converter. This may seem too complicated, but in practice it still produces a more efficient and compact power supply. It is more compact because the 60-Hz power transformer has been eliminated, and it is more efficient because the linear regulator has been eliminated. Transformer T_1 in Fig. 15-34 operates at 20 kHz or higher. Its magnetic core is only a small fraction of the size and weight required for a 60-Hz transformer with comparable ratings. It also uses much less copper than a comparable 60-Hz transformer. Therefore, a transformerless 60-Hz dc rectifier and filter circuit are used to change the ac line power to dc power. Then, that dc power is transformed to the desired voltage level in a high-frequency switching converter. Line isolation can be achieved in the high-frequency switching transformer; therefore, many of the ground loop and shock hazards associated with line-operated (transformerless) power supplies are eliminated.

Tank circuit

VCO

Noise

Switch-mode power supplies are more efficient, lighter, and more compact than linear power supplies. However, they are also *noisier*. Rectangular waves have high-frequency components that can cause interference. Certain products must meet electromagnetic interference (*EMI*) standards to prevent interference with communications and other electronic equipment. Sine waves have no high-frequency components and are therefore preferred when interference is a problem.

EMI

A *sine wave converter* design is shown in Fig. 15-35. It uses power field-effect transistors and frequency control of the dc output voltage. The power FETs do not have the problems of storage time associated with bipolar transistors. Storage time in a bipolar transistor is caused by carriers (holes and electrons) stored in the crystal when the device is saturated. The stored carriers keep current flowing for a period of time after the base-emitter forward bias is removed. Field-effect transistors do not store carriers and can be turned off much faster. The circuit shown in Fig. 15-35 uses power FETs (Q_1 and Q_2) for switching. The FETs are driven with out-of-phase square waves and are operated around 200 kHz.

The square wave is converted to a sine wave in Fig. 15-35 by resonating L_1 with C_3. T_1 effectively couples the tuning components to form a *tank circuit*. This tuned circuit provides voltage control. When the voltage-controlled oscillator (*VCO*) is tuned to the resonant frequency, maximum voltage appears across C_3. When the VCO is tuned above resonance, the tank circuit voltage drops 12 dB per octave. A 12-dB drop amounts to the voltage decreasing to one-fourth its value. An octave frequency change is twice the original value. Thus, if the tank voltage was 20 V at 150 kHz, it would drop to 5 V at 300 kHz. The VCO is controlled by comparing a sample of the output voltage with a reference voltage. Any error produces a frequency change, and the output voltage is adjusted up or down to reduce the error.

In Fig. 15-35, D_5 and D_6 are *Schottky rectifiers*. These diodes can be turned off very quickly and therefore make good high-frequency rectifiers. The diodes used in the 200-kHz power supply have a turn-off time of about 50 ns.

In Fig. 15-35, D_1 through D_4 and C_1 and C_2 form a *bridge doubler circuit*. With the 120-V

jumper installed, the 120-V ac line is doubled to about 240 V dc. In order to operate the power supply from the 240-V ac line, the jumper is removed, and the circuit acts as a bridge rectifier and again provides about 240 V dc. Line isolation and voltage transformation take place in T_1. Since it operates around 200 kHz, it is tiny compared with a 60-Hz power transformer.

Schottky rectifier

Bridge doubler circuit

Self-Test

Choose the letter that best answers each question.

20. What is the function of D_1 in Fig. 15-31?
    a. It provides overvoltage protection.
    b. It prevents C_2 from discharging when Q_2 switches on.
    c. It rectifies the square wave into smooth direct current.
    d. It allows L_1 to discharge when Q_2 is switched off.
21. How is voltage regulation achieved in Fig. 15-31?
    a. Q_1 and Q_2 act as a variable resistor to drop excess voltage.
    b. The 1.3-V reference is pulse-modulated.
    c. Pulse-width modulation takes place at the base of Q_1 and Q_2.
    d. The oscillator is frequency-modulated.
22. Refer to Fig. 15-32. Assume that the input voltage is 5 V and that the average voltage induced across L_1 is 7 V. What is the load voltage?
    a. −2 V   c. 7 V
    b. 5 V   d. 12 V

23. Why is the inverting configuration used in switch-mode power supplies?
    a. To produce an opposite-polarity power-supply voltage
    b. To isolate circuits from the ac line
    c. To step up direct current
    d. To step down direct current
24. Which of the following is not considered an advantage of pulse-width-modulated power supplies?
    a. Small size
    b. Low EMI
    c. Cool running
    d. High efficiency
25. How does the circuit in Fig. 15-41 achieve voltage regulation?
    a. By pulse-width modulation
    b. By zener clamping
    c. By frequency modulation
    d. All of the above
26. D_5 and D_6 in Fig. 15-41 are
    a. Zener regulators
    b. Metal oxide varistors
    c. Schottky rectifiers
    d. Fast-recovery field-effect diodes

15-5 Power Factor Correction

In section 5 of Chap. 4, the effect of capacitor input filters on the ac line was introduced. The power factor for a 1 ampere supply with a 1,000 μF filter is around 0.35 and the total harmonic distortion is more than 30 percent. These values are not anywhere close to meeting the specifications of Energy Star 80 Plus (U.S. EPA). Years ago, filter chokes were used to help with the filtering. Adding a series choke (inductor)

raises the power factor and lowers the THD. However, that measure will not meet standards and 60 Hz chokes are large, heavy, and expensive. Harmonic distortion and poor power factor can both be reduced with a switch mode technology known as *boost PFC correction*.

Figure 15-36 shows that the basic circuit is simple and straightforward. Transistor Q is switched OFF and ON rapidly to make the input current from the 120-volt source sinusoidal

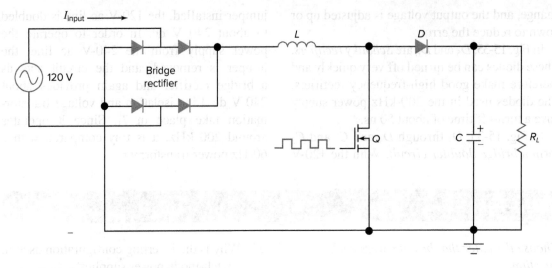

Fig. 15-36 Boost Power Factor Correction.

and in phase with the source voltage. This improves the power factor and reduces the line THD. The gate drive to transistor Q is usually supplied by an IC such as an LT8312 PFC boost controller.

Figure 15-37 shows the current in inductor L graphed by a circuit simulator. The switching frequency is relatively high and it is difficult to clearly identify the individual pulses. However, it is clear that the shape of the envelope is exactly what we expect for full-wave rectified line voltage. Figure 15-38 shows that the line current and voltage are in phase providing much less harmonic distortion. Other than minor dead times, the current and voltage are in phase (good power factor) and the harmonic distortion is minimal.

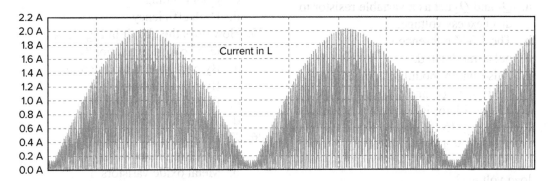

Fig. 15-37 The current in inductor L.

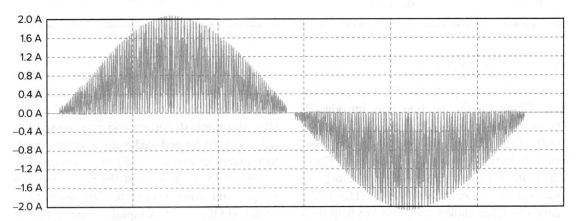

Fig. 15-38 One cycle of ac line current with PFC.

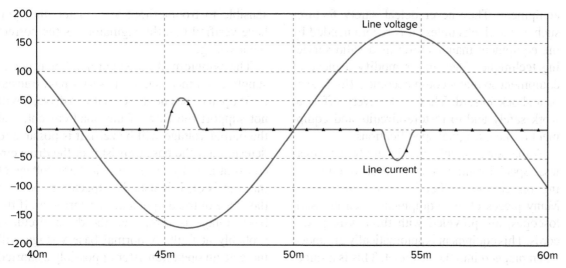

Fig. 15-39 One cycle of ac line current without PFC.

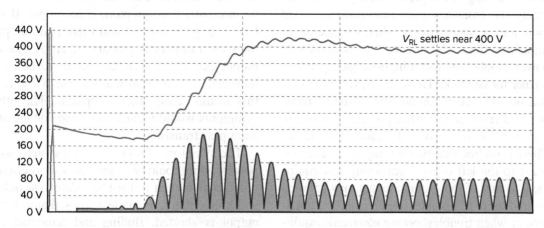

Fig. 15-40 The voltage across R_L is near 400 volts.

Comparing the power factor corrected circuit with a typical capacitor input filter circuit as shown in Fig. 15-39 shows a significant difference. Without PFC, the ac line current is in the form of narrow pulses. These pulses cause low power factor and high THD. You might want to refer back to Fig. 4-19(d) in Chap. 4 where the narrow current pulses are explained.

The voltage across the capacitor and R_L in a PFC boost circuit typically reaches around 400 volts as shown in Fig. 15-40. That is a rather high voltage and is far too much for most applications. A practical power supply will have another stage, perhaps like the step-down circuit shown in Fig. 15-28.

Safety

Never be lulled by the words *low voltage*. It should now be very clear that a so-called low voltage power supply can develop intermediate voltages that are very dangerous. Also, don't forget that charged capacitors might be lethal long after a power-down.

15-6 Troubleshooting Regulated Power Supplies

The first and foremost consideration when you are troubleshooting any power supply is *safety*. In general, high-voltage power supplies are the most dangerous. However, it must be emphasized that all electronic circuits must be treated with care and respect. A switch-mode power supply designed to deliver 5 V may develop several hundred volts in an earlier stage (this was presented in the prior section). Safe workers always use good procedures. They know circuit principles, they have and use all relevant literature, and they use the correct tools and test

Safety

equipment. They never defeat safety features such as interlocks unless it is recommended by the equipment manufacturer as a valid servicing technique. They never modify a piece of equipment so that it could become a fire hazard or a shock hazard. They make every effort to work safely and to restore circuits and equipment to meet original specifications.

Some power-supply circuits are transformerless. Special caution must be used when working on circuits without power transformers. Many pieces of test equipment, such as oscilloscopes, are provided with three-wire power cords. This equipment is automatically grounded when plugged into the ac outlet. This is a safety feature and prevents the case and ground leads from reaching a dangerous voltage. Unfortunately, the grounded leads can create a *ground loop* and a short circuit when connected to transformerless equipment. A switch-mode power supply may use a high-frequency transformer for isolation. However, a ground loop is still possible when you are analyzing the rectifier and filter circuits that precede the switch and transformer section. Examine Fig. 15-35. There is no line isolation for any of the components to the left of T_1. Connecting test equipment to any of those components may create a ground loop. If possible, use an isolation transformer when troubleshooting electronic equipment. The isolation transformer will prevent ground loops. Ground loops are covered in more detail in Chap. 4.

Ground loop

Some of the symptoms that can be observed in a regulated power supply are

1. No output
2. Low output
3. High output
4. Poor regulation and/or instability
5. Excessive ripple or noise
6. High temperature and possibly a burning odor
7. A clicking and/or squealing sound

Many of the symptoms have already been discussed in Chap. 4. It is recommended that you review the earlier section on power-supply troubleshooting. The difficulty could be in a rectifier or filter circuit that precedes the regulator. The troubleshooting information presented here assumes normal operation of the circuits preceding the regulator. It is not

sensible to troubleshoot a regulator until you have verified that the regulator has the correct input voltage.

The symptom of no output in a linear power supply often means that the series pass transistor is open or has no drive. With no drive, it will not support the flow of any load current, and the output voltage will be zero. It is possible to determine if the pass transistor or the drive circuit is at fault by measuring the base voltage of the transistor. It is usually expected to be near the value of the output (emitter) terminal. If the base voltage is zero, then the drive circuit is probably at fault. A normal base voltage will indicate an open transistor or possibly a shorted output. Refer to Fig. 15-7. If R_Z opens, the base voltage will be zero and the pass transistor will be in cutoff. The symptom is no output. If the base voltage is normal and there is no output, the pass transistor is probably open. A shorted output can also cause the voltage to be zero; however, in a circuit such as the one in Fig. 15-7, this would cause other symptoms. The other symptoms would include low base voltage and a hot pass transistor.

The symptom of no output in a switch-mode power supply could be caused by a short circuit, a defective switch transistor, a defective pulse-width modulator, or a defective oscillator. If the output is shorted, finding and removing the short should restore normal power-supply operation. An oscilloscope can be used to determine if the base or bases of the switching transistors are driven with the correct signal. With no drive, the transistors will not come on, and the output will be zero. Signal tracing with an oscilloscope will allow limiting the defect to the oscillator or the modulator. If the drive is normal, the switching transistors may be defective. The fault may also be in the high-frequency transformer, inductor, rectifiers, or filter capacitors.

If a power supply is current-limited, a short circuit may not cause extreme symptoms. This is especially true if the circuit uses foldback current limiting. It may be necessary to disconnect the power supply from its loads to determine if the fault is in the power supply or somewhere else in the system. Another technique is to measure the output current. This involves breaking the circuit and measuring the current drain on the power supply. Remember to start with a high range on the ammeter since the current could be above normal. Since it is usually not convenient to break into circuits for current measurements, try to use Ohm's law instead. Many power supplies use a resistor for current sensing. If you know that the power-supply current flows through the resistor and you know the value of the resistor, it is possible to measure the voltage drop across the resistor and use Ohm's law to calculate the current. If the resistor is in tolerance, this method will provide enough accuracy for troubleshooting purposes.

Troubleshooting for low output is similar to troubleshooting for no output. Again, the base voltage at the series pass transistor should be investigated. If it is low, then the power supply may be overloaded. Check for an overload by measuring current or disconnecting loads as discussed before. If the power supply is not overloaded, investigate the reference circuit and error amplifier to determine why the base voltage is low. For example, refer to Fig. 15-8. Three things could go wrong to make the reference voltage low: R_Z might be high in value, D_Z could be defective, or the capacitor could be leaky. Refer to Fig. 15-12. The problems here could include R_1, D_1, the op amp, or the divider network. Finally, if the power supply you are troubleshooting has more than one output, check to see if it is a tracking power supply. If it is an overloaded or defective controller power supply will cause errors in the controlled supply or supplies.

High output in a regulated power supply is often caused by a shorted pass transistor. Pass transistors are hardworking parts and are therefore prone to fail. When they fail, they often short from emitter to collector. When shorted, they drop no voltage, and the output goes up to the value of the unregulated input. An ohmmeter test with the transistor out of the circuit will provide conclusive evidence. You may also test with an ohmmeter while the transistor is in the circuit. Be sure that the power supply is unplugged and all the capacitors have discharged. Check from emitter to collector. Reverse the ohmmeter leads and check again. Zero ohms in both directions usually indicates that the pass transistor is shorted.

A shorted switching transistor can cause various symptoms, depending on the circuit configuration. Refer to Fig. 15-30. If Q_1 shorts, the output voltage will be too high. Now refer to Fig. 15-32. Here, if Q_1 shorts, the output will be

zero, and the unregulated power supply will be overloaded. A line fuse may blow in this case.

The output voltage in most regulated power supplies should be quite stable. Changes indicate that something is wrong. If the power-supply voltage varies from normal to some voltage less than normal, there may be an intermittent overload on the power supply. As before, the load current must be checked to determine if it is too high. If the power-supply voltage varies above normal, the power supply itself is unstable. Check the reference voltage. It must be stable. Any change in reference voltage will cause the output to change. Check the base of the pass transistor. With a steady load on the power supply, the base voltage should be constant. An intermittent may be found in the pass transistor itself, the error amplifier, or the voltage divider. If the power supply is located near a source of radio-frequency energy such as a transmitter, the source could be causing instability. This is usually easy to diagnose, since turning off the source would make the power supply return to normal operation. Extra shielding and bypassing may be required if a power supply must operate in a strong RF field.

Regulated linear power supplies may go into oscillation. It is normal for switchers to oscillate, but not for linear regulators. A capacitor can open, and the power supply may oscillate under some load conditions. If the power-supply voltage seems unstable, use an oscilloscope and view the output waveform. It should look like pure direct current (a straight line on the scope). Any ac content may be a result of oscillation in the regulator. Check the output capacitors and especially the bypass capacitors on any ICs in the power supply. Check for bad solder joints. Refer to Fig. 15-22. Capacitors C_1 through C_3 are very important for stability. A defect in one of these or an associated solder joint could cause oscillations.

Excessive ripple or noise on the output of a regulated power supply is usually due to the failure of a filter or a bypass capacitor. Electrolytic capacitors are widely applied in power-supply circuits. These capacitors may have a shorter life than most other electronic components. They can slowly dry out, and their effective series resistance may increase. Their capacitance may also drop. They will not be nearly as effective for filtering and bypassing. Integrated circuits and transistors can also develop noise problems. If the capacitors are all good, then the IC voltage regulator could be defective. An oscilloscope can be used to probe for the source of the noise.

Power transistors and transformers can safely run hot in some equipment. A device can be too hot to touch yet be operating normally. Probes are available for measuring the temperature of heat sinks, solid-state devices, and transformers. If a power supply seems too hot, check for an overload first. If the current and voltage are normal, the power supply may be safe. Check the manufacturer's specifications. If there is an odor of burning parts, the power supply is probably *not* safe. Troubleshoot the power supply using voltage readings. Sometimes it is necessary to turn the power supply off between readings to allow the part or parts to cool. Minimize the damage as much as possible. As always, make sure the power supply is not overloaded, since this is the most frequent cause of hot and burning components.

A clicking or squealing sound may be heard in switch-mode power supplies. If they are defective or overloaded, they can make sounds. The sounds are caused by the oscillator circuits operating at the wrong frequency. One of the reasons for running a switcher above 20 kHz is to keep it above the range of human hearing. If you can hear a switcher, then it may be running abnormally low in frequency because of an overload or a defect in the power supply. A clicking sound may mean that the power supply is overloaded and is shutting down. Every time it tries to start up again, it makes a click. The first step will be to reduce the load on the power supply. If readings are normal and the sounds stop, the circuits fed by the switcher are probably overloading it. (Completely unloading a switcher is not a good idea since many of them do not produce normal outputs under this condition.)

Troubleshooting switch-mode power supplies demands safe work habits and proper test equipment. The first section of a switcher is a line rectifier and filter. Voltage doublers may be used. Therefore, lethal dc voltages should be expected even in 5-V switch-mode power supplies. The frequencies and waveforms found in switchers are beyond the capabilities of many meters. You have learned that pulse-width modulation is used to control the output voltage in many switchers. Since the duty cycle is changing and the peak voltage is not, an ordinary ac meter may not properly indicate circuit action. A true root-mean-square meter with a frequency rating at least as high as the power-supply operating frequency will be required for accurate testing. True rms meters indicate the correct rms (effective) value for all ac waveforms. Most meters indicate the correct rms value for sinusoidal ac only. Since waveforms are so important in switchers, most technicians prefer the oscilloscope for troubleshooting. If the power supply uses frequency control, such as the one in Fig. 15-35, a frequency meter may be useful in testing. The VCO must operate near or above the resonant frequency of the tank circuit. As the load on the power supply is increased, the VCO frequency should drop to come closer to resonance. This can be seen on an oscilloscope as an increase in the period of the waveform. Period and frequency are reciprocals.

Figure 15-41 shows a *flyback* switching power supply that operates in the *critical conduction mode*. This mode is defined by the current flow in the primary of transformer T_1 ramping up to some peak value, ramping down to zero current, and then *immediately* ramping up again. Another possibility is *continuous conduction mode* where the primary current starts to ramp up before it decays to zero. A third possibility is *discontinuous conduction mode* where the current remains at zero for some time before starting up again. The advantage of the *critical* conduction mode is that the peak current is lower for a given amount of load demand. The lower peak current leads to lower power dissipation (losses in the switching circuit) and therefore improved efficiency and reliability. A critical conduction mode power supply is also self-protecting in the case of a shorted output. This is a popular circuit, and the one shown in Fig. 15-41 works over a range of ac line voltages from 85 to 270 V and line frequencies from 50 to 60 Hz.

The waveforms in Fig. 15-42 show the important relationships for a critical conduction mode flyback supply:

- The transformer current reaches zero before increasing again.
- An increase in load current increases the peak transformer current.

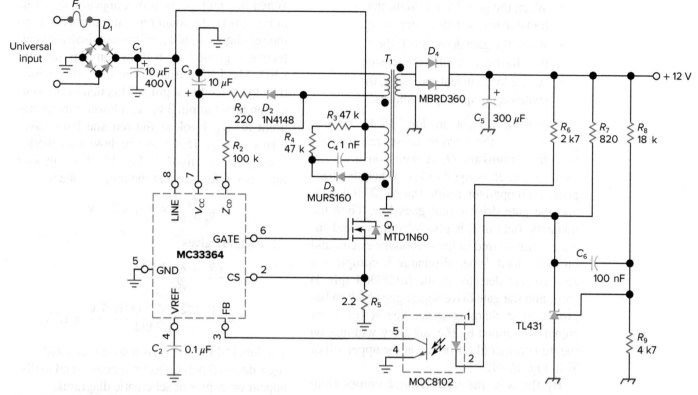

Fig. l5-4l Flyback supply.

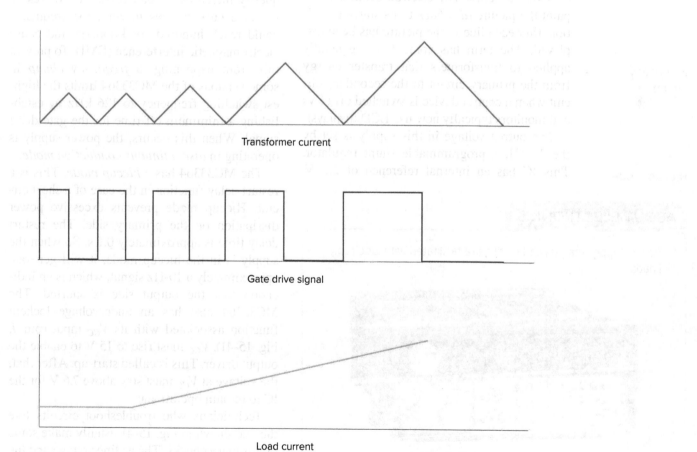

Transformer current

Gate drive signal

Load current

Fig. l5-42 Flyback supply waveforms.

- When the gate drive is high, the transformer current is increasing.
- When the gate drive is off, the transformer current is decreasing.
- As the load current increases, the switching frequency decreases.

The peak current in Fig. 15-41 is programmed by the current sense resistor, R_5. Switching transistor Q_1 remains on until the signal across R_5 is equal to V_{FB}. When that happens, a comparator inside the MC33364 trips and the gate drive signal goes low. Then the magnetic field in T_1 begins to collapse, and energy is transferred to the secondary circuit and on to the load. When discharge is complete, a zero current detector in the MC33364 (pin 1) trips, and the gate drive signal goes high to begin the next charging cycle. The zero current signal is supplied by the auxiliary winding on the transformer (the winding at the upper left of T_1 in Fig. 15-41).

By the way, the term *flyback* comes from televisions and computer monitors that use cathode-ray tubes. The electron beam used to paint the picture *flies back* to its starting position after each line in the picture has been displayed. The term has come to be generally applied to transformers that transfer energy from the primary circuit to the secondary circuit when a control device is switched off (TVs and monitors typically now use LCD displays).

The output voltage in this supply is set by the TL431, a programmable shunt regulator. This IC has an internal reference of 2.5 V.

When this voltage, or more, appears across R_9 in Fig. 15-41, the shunt regulator turns on. With the regulator on, pin 2 of the optoisolator is effectively grounded, turning on the internal LED. The LED turns on the transistor, which then loads feedback pin 3. This makes the voltage go down at pin 3 to set a lower current trip point for Q_1. Look at the red and blue waveforms in Fig. 15-42 to see how this works. Voltage divider R_8-R_9 in Fig. 15-41, along with the 2.5-V reference, set the output voltage:

$$\frac{R_9}{R_8 + R_9} \times V_{out} = 2.5 \text{ V}$$

Rearranging gives

$$V_{out} = \frac{(R_8 + R_9)2.5 \text{ V}}{R_9}$$

$$= \frac{(18 \text{ k}\Omega + 4.7 \text{ k}\Omega)2.5 \text{ V}}{4.7 \text{ k}\Omega} = 12 \text{ V}$$

The label "4 k7" for R_9 is a style that avoids using a decimal point. Decimal points tend to disappear on copies of schematic diagrams.

As Fig. 15-42 shows, the switching frequency increases as load current decreases. If the load current goes to zero, the frequency could reach hundreds of kilohertz and cause electromagnetic interference (EMI). To prevent this from happening, a *frequency clamp* in some versions of the MC33364 limits the highest switching frequency to 126 kHz by establishing a minimum off time for the gate drive signal. When this occurs, the power supply is operating in *discontinuous conduction mode*.

The MC33364 has a *hiccup mode*. This is a restart delay function in the case of a short circuit. Hiccup mode prevents excessive power dissipation on the primary side. The restart delay time is approximately 0.1 s. So when the supply is in the hiccup mode, it will generate approximately a 10-Hz signal, which is an indication that the output side is shorted. The MC33364 also has an undervoltage lockout function associated with its V_{CC} input (pin 7, Fig. 15-41). V_{CC} must rise to 15 V to enable the output driver. This is called start-up. After that, the voltage at V_{CC} must stay above 7.6 V for the IC to remain operational.

Technicians who troubleshoot circuits like the one shown in Fig. 15-41 usually make some preliminary checks. The ac line voltage and the fuse are a good place to start. As Table 15-2

Frequency clamp

Hiccup mode

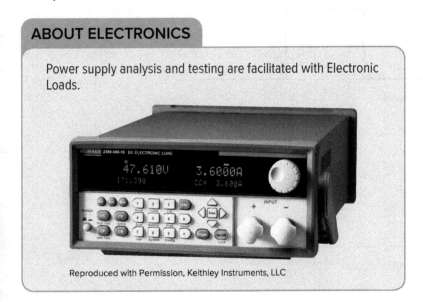

Table 15-2 Symptoms and Possible Causes

Symptom	Possible causes
F_1 blown	D_1 shorted, C_1 shorted, Q_1 shorted, MC33364
No output or very low output (fuse OK)	D_1 open, Q_1 open, T_1 open, D_4, C_5 shorted, TL431 shorted, optoisolator, short in load circuit, R_5 open, Q_1 open, MC33364, other faults
Unstable output voltage	TL431 defective, optoisolator, MC33364, intermittent overload in output circuit
High output voltage	R_8, R_9, TL431, optoisolator, R_6 open, MC33364
Low output voltage	Overload, R_5, R_8, R_9, TL431, C_1, C_5, optoisolator, Q_1, MC33364
Noisy output	C_6, C_5, C_2, C_1

shows, a blown fuse is usually caused by a shorted component. Disconnect the supply from the ac line, and *make sure that C_1 is discharged* before using an ohmmeter to find the short.

Sometimes, a symptom will scream out an answer to those who understand the circuit. A perfect example is an output voltage of 2.5 V rather than the expected 12 V. Now, what could cause that? If R_9 in Fig. 15-41 is open, there will be no voltage divider action, and the optoisolator will turn on when the output reaches the 2.5-V reference level.

A voltage check at the drain terminal of Q_1 in Fig. 15-41 should show about 160 V dc. This assumes a 120-V ac line. If this voltage is only 100 V or so, C_1 could be open. If this voltage is normal, check the dc voltage at pin 3 of the MC33364. If it's very low or zero, the MC33364, the optoisolator, or the TL431 might have failed. Many technicians use a divide-and-conquer strategy with circuits like this. Since the output side and the input side are linked by the optoisolator, it's possible to break that link to help determine which side has a problem. Opening the connection from pin 5 of the optoisolator to pin 3 of the MC33364 allows the use of a resistor from pin 3 to ground. A 1-kΩ resistor is a good start. If this allows the supply to develop some output voltage, then it is starting to look like the input circuit is OK. Making the resistor smaller should lower the output voltage, and making it larger should increase the output voltage. Failing that, the waveform across R_5 should react to the value of the test resistor (as shown in red in Fig. 15-42). If this checks out, the problem is probably on the output side.

Is the voltage at pin 7 in Fig. 15-41 OK? Remember, it has to reach 15 V to start up the IC

and then remain at 7.7 V or so. Perhaps C_3 is shorted. Is the supply operating in the hiccup mode? An oscilloscope connected to pin 6 should show a 10-Hz waveform of low-duty cycle. Knowledge of the normal waveforms shown in Fig. 15-42 will be very helpful when troubleshooting.

Is the output low and does something smell hot? If the transformer is quite hot, it might have a shorted turn. If Q_1 has shorted, verify the snubber network across the primary of T_1. It's made up of D_3, R_3, R_4, and C_4. Its job is to suppress the voltage transient associated with Q_1 turning off. Without the snubber, Q_1 can be damaged.

Troubleshooting a linear supply like the one shown in Fig. 15-43 can be a challenge for a technician. Today, schematics usually *do not* have voltages listed, so it might be necessary to use basic calculations and your knowledge of how things work to come up with some on your own. Three important voltages in this circuit can be used when troubleshooting. The first is +38 V shown at the top. This is determined by multiplying the ac input voltage to the bridge rectifier by 1.414: 2.8 × 1.414 = 39.6. Subtracting 1.4 V for two diode drops in the bridge gives about 38 V, which is expected with light loads.

Another important voltage in Fig. 15-43 is +11.2 V, shown at the output of OA_2. The zener is a 5.6-V device, so it is reasonable to assume that the voltage across the amplifier output and its inverting input is 5.6 V. Knowing that both op-amp inputs are normally at the same voltage leads to the conclusion that the voltage across the 10-kΩ resistor to ground is also 5.6 V. Thus, the voltage at the output of OA_2 is expected to be 2 × 5.6 V = 11.2 V. Lastly, the second zener

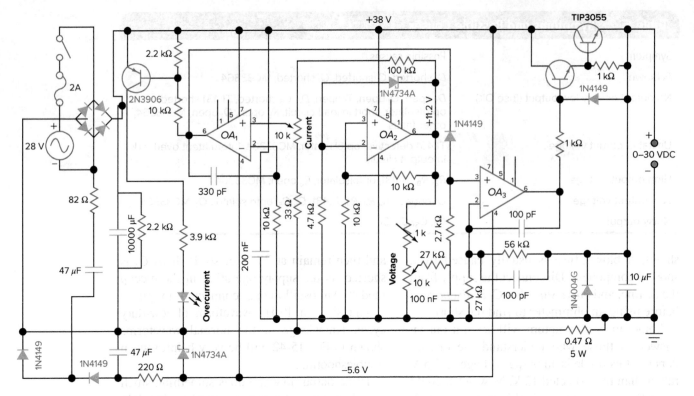

Fig. 15-43 Power-supply troubleshooting example.

diode at the bottom of the schematic regulates the negative supply terminals for OA_1 and OA_3 to −5.6 V. These three voltages are important for understanding how this supply works, and they should be verified early in the troubleshooting process.

How does the current limiting in Fig. 15-43 operate? Let's make another calculation by assuming the current potentiometer is set to its midpoint. The voltage from the wiper to ground will be

$$V_{wiper} = 11.2 \text{ V} \times \frac{5 \text{ k}\Omega}{110 \text{ k}\Omega} = 0.509 \text{ V}$$

The 33-Ω resistor has been ignored. Notice that this voltage is applied to the noninverting input of OA_1, along with the drop across the 0.47-Ω resistor. If the load current happens to be 1.083 A, then the drops are equal and OA_1 is on the verge of crossing into negative saturation. OA_1 serves as a comparator in this circuit. When the power supply is not in current limiting, the output of OA_1 is at positive saturation and has no effect on OA_3, which serves as the error amplifier for the power supply. When the drop across the 0.47-Ω resistor is more than the voltage across the wiper and the bottom of the current set potentiometer, OA_1 switches to

negative saturation, while OA_3 and the pass transistor turn off. Also, the overcurrent LED comes on at this time.

Finally, OA_3 in Fig. 15-43 is a noninverting amplifier with a gain of 3. This gain is set by the 56-kΩ and 27-kΩ resistors:

$$AV = 1 + \frac{R_F}{R_G} = 1 + 2.07 \approx 3$$

With a gain of 3, the range of output voltage for the supply is 0 to 3 × 11.2 V, or 0 to 33.6 V, ignoring the drop across the output transistors. This supply also has a 10-turn, 1-kΩ variable resistor that serves for fine adjustment of the output voltage. Generally speaking, this is a 0- to 30-V dc supply with a current capability of about 2 A. If you analyze the comparator trip point when maximum current is flowing through the 0.47-Ω resistor, it is about 2.17 A. This same circuit can be used for higher current by decreasing the value of the 100-kΩ resistor in series with the current limit potentiometer. Other components will have to be modified as well.

A knowledge of basic circuit theory and device behavior can make a difficult task manageable. When you are confronted with a circuit that is not working and you have no specified voltages to check, it may be possible to come up with some. Once these are at hand, the process for finding the defect is more efficient.

The final step in the repair process is replacing the defective part or parts. An exact replacement is usually the best. One exception is an upgraded part that is recommended by the manufacturer. Substitutions may affect the performance, reliability, and safety of a system. Some *components* are *special*. The Schottky rectifier (D_4) in Fig. 15-41 is a good example. There is no way that ordinary rectifiers will work in this circuit. The high frequency would cause tremendous dissipation in ordinary rectifiers. They would probably burn up in a short period of time and could cause damage to other components in the power supply. Capacitor C_5 in Fig. 15-41 is also somewhat critical. It must handle high peak currents without overheating. In this kind of application, parasitic resistance and inductance must be minimal. Some designs use several capacitors in parallel to decrease the parasitic resistance and inductance. You should recall that inductances and resistances in parallel have a lower effective value. Some implementations of Fig. 15-41 use three 100-μF capacitors in parallel in place of C_5. Other implementations use a special high-voltage, high-current capacitor designed for minimum loss. This is another example of how important it is to choose replacement parts carefully. An unwary technician might select a single capacitor or a standard capacitor as a replacement and have that capacitor fail after a few hours or days of operation. Also, there could be extra ripple in the output. When possible, use exact replacements.

Special components

Lead dress is important when replacing components. Lead dress refers to the length and position of the leads on a part. Leads that are too long can make some circuits unstable. It was mentioned before that linear IC voltage regulators can become unstable and oscillate. It is absolutely necessary that some bypass capacitors have very short leads. Always install replacement parts with the same lead dress as the originals.

Lead dress

Self-Test

Choose the letter that best answers each question.

27. An open-series pass transistor produces the symptom of
    a. Low output voltage
    b. High output voltage
    c. Unstable output voltage
    d. No output voltage
28. A series pass transistor with a collector-to-emitter short produces the symptom of
    a. Low output voltage
    b. High output voltage
    c. Unstable output voltage
    d. No output voltage
29. Refer to Fig. 15-17. The output voltage is zero. The input is on the high end of normal. There is no short or overload in the load circuit; in fact, the load current is zero. The defect is in
    a. The 7812 IC   c. Q_2
    b. Q_1 d. R_2
30. Refer to Fig. 15-22. D_1 is shorted. What is the symptom?
    a. High output
    b. Excessive ripple voltage

    c. Low output
    d. A blown fuse
31. Why should an isolation transformer be used when troubleshooting?
    a. To prevent shock
    b. To prevent circuit damage
    c. To prevent a ground loop
    d. All of the above
32. Refer to Fig. 15-31. Q_2 is shorted from collector to emitter. What is the symptom?
    a. There is no output.
    b. There is high output.
    c. The reference voltage on pin 8 will be over 1.3 V.
    d. None of the above are symptoms.
33. Refer to Fig. 15-26. The varistor is open. What is the symptom?
    a. No output
    b. Low output
    c. High output
    d. No symptom, but lost transient protection

Chapter 15 Summary and Review

Summary

1. In a nonregulated power supply, output voltage varies with the line voltage and the load current.
2. The output voltage tends to drop as the load on a power supply is increased.
3. Open-loop voltage regulators do not use feedback to control the output voltage.
4. A ferroresonant transformer with a saturated core can be used to regulate voltage.
5. Ferroresonant transformers use a resonating capacitor as part of their secondary circuit.
6. A zener diode shunt regulator is not practical in high-current applications because a high-power zener is required.
7. A series pass transistor can be used in conjunction with a zener diode to form a practical high-current power supply.
8. Negative regulators often use PNP pass transistors, while positive regulators use NPN pass transistors.
9. Dual-polarity (bipolar) power supplies provide both negative and positive voltages with respect to ground.
10. Better voltage regulation is obtained with feedback (closed loop) power-supply operation.
11. Feedback power supplies use an error amplifier to compare the output voltage to a reference voltage.
12. Zener diodes are often used to provide a reference voltage in feedback-operated power supplies.
13. Op amps can be used as error amplifiers in regulated power supplies.
14. Integrated circuit voltage regulators provide fixed or variable output voltages in an easy-to-use package.
15. Adjusting a fixed IC voltage regulator with a resistive divider somewhat degrades its voltage regulation.
16. A current-boost transistor can be used with IC voltage regulators to provide more load current.
17. Tracking power supplies have a controller output and one or more controlled outputs. Any change in the controller will be tracked by the controlled outputs.
18. Shorting the output of a regulated power supply may damage the series pass transistor and other components in the power supply.
19. Current-limited power supplies protect themselves and the load circuits connected to them.
20. Foldback current limiting is superior to conventional current limiting for preventing damage caused by long-term overloads.
21. Some IC voltage regulators can be configured for either type of current limiting.
22. A crowbar circuit provides voltage limiting by shorting the supply.
23. Swamping resistors can be used to ensure current sharing among parallel pass transistors.
24. Line transients can be clipped by varistors.
25. Metal oxide varistors turn on in nanoseconds and can safely handle hundreds or thousands of amperes.
26. Switch-mode regulators are more efficient than linear regulators, and result in smaller and lighter power supplies.
27. Switch-mode power supplies operate at very high frequencies, allowing smaller transformers and filter components to be used.
28. Pulse-width modulation can be used to control the output voltage in switch-mode supplies.
29. Increasing the duty cycle of a waveform increases its average voltage.
30. Switch-mode power supplies use high-speed transistors, fast-recovery rectifiers, or Schottky rectifiers.
31. A converter is a circuit that changes direct current to alternating current and then back to direct current again.
32. Switch-mode supplies are noisier than linear types and can cause electromagnetic interference.
33. Sine wave converters solve the noise and EMI problems associated with switchers.
34. An isolation transformer should be used when servicing or troubleshooting electronic equipment to avoid ground loops.

35. An open pass transistor (or no drive to the transistor) will cause the symptom of no output in a linear regulator.
36. A shorted pass transistor will cause the output to be abnormally high.
37. No output, low output, overheating, or a blown fuse are indications of an overloaded power supply.

38. An error in the reference voltage will cause an error in output voltage.
39. Switchers generate waveforms and frequencies beyond the capabilities of many meters.
40. When replacing parts, use exact replacements when possible and pay attention to lead dress.

Chapter Review Questions

Choose the letter that best answers each question.

15-1. Electrical brownouts are (15-1)
   a. Caused by lightning and accidents
   b. Periods of low line voltage
   c. Periods of high line voltage
   d. Line transients
15-2. Refer to Fig. 15-5. What is the function of the resonating capacitor? (15-1)
   a. To prevent damage from line transients
   b. To change pulsating direct current to pure direct current
   c. To cause high circulating currents in the secondary
   d. None of the above
15-3. Refer to Fig. 15-6. What happens to the zener diode dissipation if the load is disconnected? (15-1)
   a. It stays the same.
   b. It decreases.
   c. It increases.
   d. It goes to zero.
15-4. Refer to Fig. 15-8. C is open. What is the most likely symptom? (15-1)
   a. Excessive noise and ripple across the load
   b. Low output voltage
   c. No output voltage
   d. High output voltage
15-5. Refer to Fig. 15-9. What is the function of Q_1 and Q_2? (15-1)
   a. They are error amplifiers.
   b. They establish the reference voltage.
   c. They provide overcurrent protection.
   d. They are series pass transistors.

15-6. What happens to the base-emitter voltage in a transistor as that transistor is called upon to support more current flow? (15-2)
   a. It drops.
   b. It increases.
   c. It remains constant.
   d. It approaches 0 V at high current.
15-7. Refer to Fig. 15-11. Assuming that the circuit is working normally, what will happen to the series pass transistor when the load demands more current? (15-2)
   a. It is driven toward cutoff.
   b. It dissipates less power.
   c. It is turned on harder.
   d. None of the above will happen.
15-8. Refer to Fig. 15-12. How will the output of the op amp be affected if the load suddenly demands less current? (15-2)
   a. It will go more positive.
   b. It will go less positive.
   c. It will not change.
   d. It will shut down.
15-9. Refer to Fig. 15-12. What is the purpose of R_3? (15-2)
   a. To adjust the output voltage
   b. To adjust the voltage gain of the op amp
   c. To adjust the reference voltage
   d. To adjust the voltage-limiting point
15-10. Linear IC voltage regulators, such as the 78*XX* series, are useful (15-2)
   a. For decreasing costs in power-supply designs
   b. As on-card regulators
   c. In decreasing the number of discrete parts in supplies
   d. All of the above

15-11. Refer to Fig. 15-14. What is the disadvantage of this circuit? (15-2)
   a. The voltage regulation is somewhat degraded.
   b. It is too costly.
   c. It is difficult to troubleshoot.
   d. All of the above are true.

15-12. Refer to Fig. 15-14. Assume that the quiescent IC current is 6 mA, R_1 is 220 Ω, and R_2 is 100 Ω. What is the load voltage? (15-2)
   a. 4.35 V
   b. 5.00 V
   c. 7.87 V
   d. 9.00 V

15-13. Refer to Fig. 15-15. Q_1 is silicon and R_1 is 12 Ω. At what value of current will the external pass transistor turn on and help to supply the load? (15-2)
   a. 0.006 A
   b. 0.022 A
   c. 0.058 A
   d. 1.25 A

15-14. Refer to Fig. 15-16. What is the function of R_4? (15-2)
   a. To adjust the negative output voltage
   b. To adjust the positive output voltage
   c. To adjust both outputs
   d. None of the above

15-15. Refer to Fig. 15-17. Q_2 is open. What is the symptom? (15-3)
   a. No output
   b. Low output
   c. High output
   d. No current limiting

15-16. Refer to Fig. 15-22. Q_2 is open. What is the symptom? (15-3)
   a. No output
   b. Low output
   c. High output
   d. No current limiting

15-17. Refer to Fig. 15-22. What could happen if R_1 is adjusted for too much output voltage? (15-3)
   a. The IC may overheat.
   b. The crowbar may blow the fuse.

   c. The current limiting may change to foldback.
   d. All of the above are true.

15-18. Refer to Fig. 15-26. What is the function of the varistor? (15-3)
   a. It prevents brownouts from spoiling regulation.
   b. It resonates the transformer.
   c. It provides overcurrent protection.
   d. It suppresses line transients.

15-19. A 5-A-rated power supply is normal but supplies only 2 A when short-circuited. This supply is protected by (15-3)
   a. Foldback current limiting
   b. Conventional current limiting
   c. An MOV device
   d. A slow-blow fuse

15-20. Compared to switchers, linear power supplies with the same ratings are (15-4)
   a. Heavier
   b. Larger
   c. Less efficient
   d. All of the above

15-21. Refer to Fig. 15-30. What is the function of L_1? (15-4)
   a. It takes on a charge when the transistor is on.
   b. It dissipates its charge when the transistor is turned off.
   c. It helps smooth the load voltage.
   d. All of the above are true.

15-22. Refer to Fig. 15-30. What is the function of D_2? (15-4)
   a. It regulates the output voltage to the error amplifier.
   b. It turns on when Q_1 is off to keep load current flowing.
   c. It provides overcurrent protection.
   d. All of the above are true.

15-23. Refer to Fig. 15-31. Suppose the load suddenly demands less current. What happens to the signal supplied to the base of Q_1? (15-4)
   a. The peak-to-peak amplitude goes down.
   b. The duty cycle increases.
   c. The duty cycle decreases.
   d. The square wave changes to a sine wave.

15-24. Refer to Fig. 15-31. R_1 is damaged and has increased in value. What is the symptom? (15-4)
   a. Excessive output ripple
   b. High output voltage
   c. Output dropping as the supply is loaded
   d. The IC running hot

15-25. Refer to Fig. 15-32. Diode D_1 is open. What is the symptom? (15-4)
   a. No output voltage
   b. High output voltage
   c. Reverse output polarity
   d. C_1 burning up

15-26. Why are switch-mode power supplies operated at frequencies so much above 60 Hz? (15-4)
   a. To limit dissipation in transistors and diodes
   b. To allow smaller transformers and filters
   c. So that pulse-width modulators can be used
   d. All of the above

15-27. Refer to Fig. 15-34. How could the output voltage be increased? (15-4)
   a. By increasing the oscillator frequency
   b. By decreasing the oscillator frequency
   c. By removing D_3
   d. By keeping Q_1 and Q_2 on longer

15-28. Refer to Fig. 15-41. What is the purpose of C_3? (15-4)
   a. It resonates L_1 and changes the square waves to sine.
   b. It changes the frequency of the VCO.
   c. It provides transient protection.
   d. All of the above are true.

15-29. Refer to Fig. 15-41. What do Q_1 and Q_2 accomplish? (15-4)
   a. They control voltage by linear resistance change.
   b. They change the ac line power to pulsating dc power.
   c. They change direct current to alternating current.
   d. They provide conventional current limiting.

15-30. Refer to Fig. 15-41. Where is isolation from the ac line accomplished? (15-4)
   a. In the bridge rectifier
   b. By C_1 and C_2
   c. In T_1
   d. In the Schottky diodes

Chapter Review Problems

15-1. You are troubleshooting a power supply with three outputs: one controller and two controlled. Which section of the power supply should be verified first? Why?

15-2. It is desired to use a crowbar circuit to protect equipment that is remotely located. How could the basic crowbar design be modified so that the equipment would automatically come back on line after the fault cleared?

15-3. Is there any situation when the modified design of question 15-2 could perform in an undesirable way?

15-4. Can you think of any physical (nonelectrical) problems that could cause intermittent operation in power supplies?

15-5. Why does some battery-operated equipment contain voltage regulators?

15-6. What type of power-supply circuit would you expect to find in a photographer's battery-operated electronic flash unit? Why?

Answers to Self-Tests

1. B	10. D	19. B	28. B
2. C	11. A	20. D	29. A
3. B	12. B	21. C	30. D
4. D	13. D	22. D	31. D
5. A	14. B	23. A	32. B
6. D	15. C	24. B	33. D
7. C	16. A	25. C	
8. B	17. D	26. C	
9. A	18. A	27. D	

Digital Signal Processing

Learning Outcomes

This chapter will help you to:

16-1 *Explain* the popularity of DSP. [16-1]

16-2 *Discuss* the conversion of continuous signals to discrete form. [16-1]

16-3 *Sketch* the block diagram for a typical DSP system. [16-1]

16-4 *List* some advantages of DSP. [16-1]

16-5 *Explain* how signals are represented in time and frequency. [16-3]

16-6 *Explain* the operation and design of digital filters. [16-2, 16-4]

16-7 *Discuss* other applications. [16-5]

16-8 *List* some limitations of DSP. [16-6]

16-9 *Troubleshoot* DSP systems. [16-7]

The theories that digital signal processing (DSP) are based on were first proposed by two 19th-century scientists, Fourier and Laplace (pronounced four-ee-aay and la-ploss). Little did they know the many ways that their contributions would apply to 21st-century technologies. Fourier was working on heat flow, and Laplace was working on planetary motion. They developed mathematical techniques that were useful for their own efforts. Today, their techniques are used to design digital filters (and to do many other things as well). Digital computers arrived in the 1940s, and by the 1950s, a few engineers and scientists were using computers to simulate analog circuits. Digital signal processing began to emerge as a separate discipline. Then, in the early 1980s, DSP integrated circuits arrived. This changed everything because, for the first time, DSP became a practical solution for a wide range of problems. Today, DSP is the fastest growing segment of the semiconductor market. Many technical workers now need a working knowledge of DSP.

16-1 Overview of DSP Systems

It's common practice to divide systems into one of two worlds: analog or digital. An analog signal has an infinite variety of values as time goes on. For example, if the ac line voltage is viewed on a conventional oscilloscope, the display is a smooth sine curve. On that curve, an instantaneous value might be 100 V, 99.8 V, or 99.885 V. Given unlimited resolution, there are an infinite number of possible values. The ac signal continuously (smoothly) changes over time. Such a signal is often called analog, but a better term is *continuous signal*. The term *analog signal* dates back to early (now obsolete) analog computers where circuits were *analogous*

Continuous signal

Analog signal

to physical systems. Today, when people say "analog," they probably mean *continuous*. However, the term *analog* is commonly used and will be used in this chapter.

Transducers

Digital signals are noncontinuous. They jump from one allowed value to another as time moves on. There is a limited number of values because binary numbers represent the signals. The number of values, or voltages, in a digital system is determined by the number of bits in each binary number:

$$\text{Number of voltages or values} = 2^n$$

where n = the number of bits. Most DSP systems operate over a range of 8 to 24 bits. An 8-bit system has only 256 allowed values, while a 24-bit system has over 16 million. It's obvious that high-resolution systems use a lot of bits.

Discrete signal

Digital signals are also called *discrete signals*. In DSP writings and discussions, signals are often called continuous or discrete (rather than analog or digital). Now a third term can be defined: *quantization*, which is the process of converting from continuous to discrete.

Quantization

Samples

You can and should continue to use the terms *analog* and *digital*. Few people would call an analog-to-digital converter (ADC) a *continuous-to-discrete converter*. It's ironic, but language that is technically correct might not serve as well as common usage.

Let's apply the terms *continuous* and *discrete* to the block diagram shown in Fig. 16-1, which represents a DSP system. The input signal is almost always continuous. Input signals often come from *transducers*, a fancy name for a device that converts some physical value into an electrical value. A microphone is a transducer that converts sound into a voltage. Microphones change sounds waves into a continuous electrical signal.

The first stage in Fig. 16-1 is an amplifier. As you already know, amplifiers are used to boost signal levels to some useful level. An antialiasing filter follows the amplifier. This is a low-pass filter that keeps higher frequencies, such as noise, out of the rest of the system. The need for the antialiasing filter is illustrated in Fig. 16-2, where the conversion process from continuous to discrete is viewed as a series of snapshots or *samples*. Look at Fig. 16-2. Two continuous signals are snapped or sampled every 0.25 ms. The sampling process starts at 0 s and ends at 1.25 ms. Notice that the samples (dots) are identical for both the 1-kHz and the

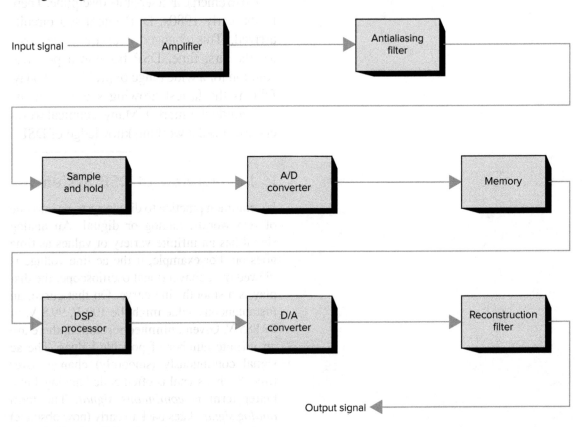

Fig. 16-1 A typical DSP system.

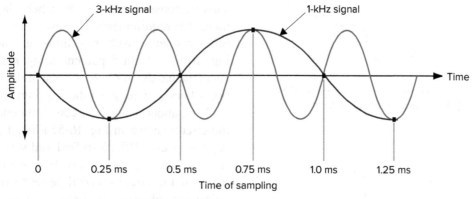

Fig. 16-2 Two signals being sampled at 0.25-ms intervals.

3-kHz signals. This is a problem, since the discrete signal would be the same for both frequencies. To say it another way, the 3-kHz signal has been aliased to 1 kHz. The fix is to use a low-pass filter that passes 1 kHz and attenuates 3 kHz, as shown in Fig. 16-3.

Returning to Fig. 16-1, we see that a sample-and-hold circuit follows the antialiasing filter. Figure 16-4 shows a typical input and output for a sample-and-hold circuit. This circuit and the A/D converter are usually combined into one IC. It's important to realize that the signal coming out of the A/D converter is a quantized version of the original input signal. A quantized signal is a series of binary words. In the case of an 8-bit converter, they could be

- 01110101 (first sample)
- 00011011 (second sample)
- 00011000 (third sample)
- 00001111 (fourth sample)

Some DSP ICs contain the sample-and-hold circuit plus the analog to digital converter.

Next in Fig. 16-1, we find the memory. This is a storage area for the binary numbers. Then comes the DSP processor. Here is where the numbers get crunched. As you will learn later in this chapter, the most important thing that happens here is called MAC (*m*ultiply-and-*ac*cumulate). The discrete samples, which are now in the form of binary numbers, are multiplied several or many times by fixed values called *coefficients*, the multiplied values are summed together, and the output is sent to the D/A converter (or DAC). The output of the D/A converter in Fig. 16-1 would look something like the waveform in Fig. 16-4(*b*). The reconstruction filter, a low-pass type, smoothes the signal to look something like the waveform shown in Fig. 16-4(*a*). Reconstruction filters are also called anti-imaging filters. Generally, the input to and the output from DSP systems are both continuous signals. Inside DSP systems,

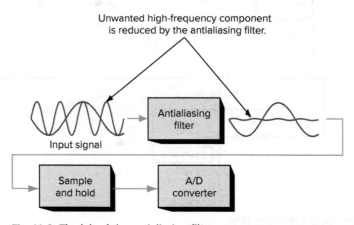

Fig. 16-3 The job of the antialiasing filter.

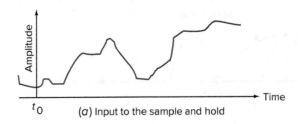

(a) Input to the sample and hold

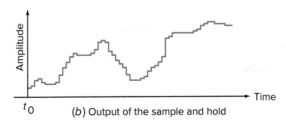

(b) Output of the sample and hold

Fig. 16-4 Sample-and-hold waveforms.

the signals are in discrete form since they are represented by binary numbers.

This all seems like a lot of bother. Why is DSP becoming so popular? A case study is in order. Suppose we need a low-pass filter with a very sharp cutoff at 1 kHz, no more than 1 dB of passband ripple, and 80 dB of attenuation at 2 kHz. Active filters were covered in Chap. 9, and as stated there, Chebyshev filters are sharp and often a good choice when some passband ripple is acceptable. A Chebyshev filter that meets the requirements is shown in Fig. 16-5. It uses four op amps and is an eighth-order design. Figure 16-6(a) shows the frequency response.

This response shows that the Chebyshev filter meets the requirements. However, there will be problems with production units, as shown in Fig. 16-6(b). With 5 percent component tolerance, the passband ripple of most of the production filters will *not* meet the requirements.

What about using 1 percent components for the circuit shown in Fig. 16-5? First, 1 percent capacitors are difficult to find and very expensive. Second, components drift with time, and many of the circuits would cease to meet requirements after months or years of use. Third, temperature change would affect filter performance. Chapter 9, in the section on active filters, describes a notch filter with similar problems.

A DSP filter can meet the design requirements and would not require precision components. Every production unit would work exactly the same (even years later), and normal temperature variations would have no effect on performance. DSP can also provide functions that would be impossible or very difficult with other approaches.

Many of the circuit functions that formerly were achieved with analog circuits are being replaced with digital technology. As always, the major force driving this change is economics. Digital and mixed-signal integrated circuits can often replace analog circuits at a lower cost. At the same time, the newer designs are often smaller and offer features that would not be possible using a strictly analog approach.

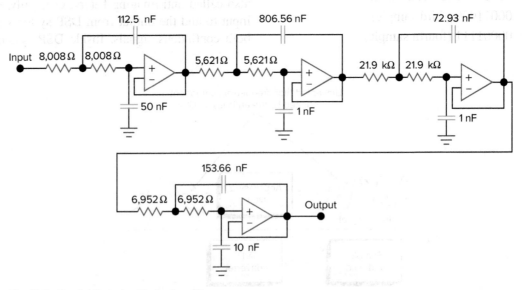

Fig. 16-5 An eighth-order Chebyshev filter.

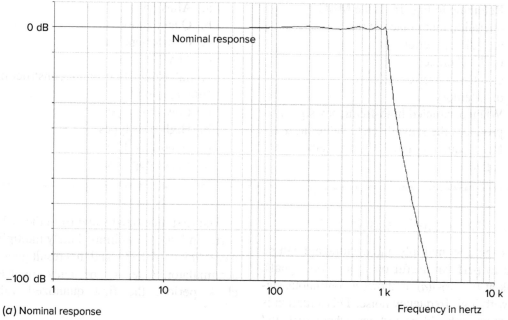

(a) Nominal response

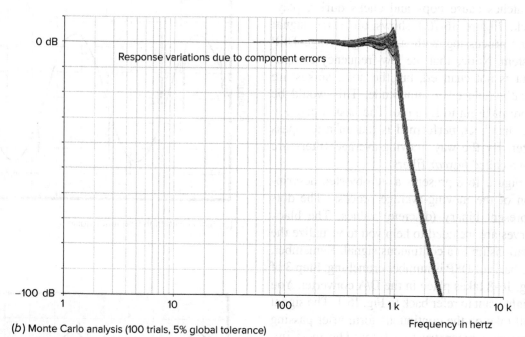

(b) Monte Carlo analysis (100 trials, 5% global tolerance)

Fig. 16-6 Frequency response curves for the Chebyshev filter.

Self-Test

Choose the letter that best answers each question.

1. The electrical signal from a microphone is best described as a(n)
   a. Quantized signal
   b. Discrete signal
   c. Aliased signal
   d. Continuous signal

2. Referring to Fig. 16-1, quantization takes place in the
   a. Antialiasing filter
   b. A/D converter
   c. Memory
   d. DSP processor

3. Referring to Fig. 16-1, the MAC operation takes place in the
   a. A/D converter
   b. DSP processor
   c. D/A converter
   d. Memory
4. When a sampled high-frequency signal produces the same discrete values as a low-frequency signal, the problem is called
   a. Aliasing
   b. Quantization
   c. Component error
   d. MAC
5. Antialiasing filters and reconstruction filters are both
   a. Low-pass types
   b. High-pass types
   c. Band-pass types
   d. Band-stop types

16-2 Moving-Average Filters

Let's see how multiply-and-accumulate can be used to perform useful operations on signals. Look at Fig. 16-7(a). It shows a continuous signal with high-frequency noise. This situation is common with old recordings where dirt and scratches cause pops and clicks during playback. Figure 16-7(b) shows the same signal after processing with a moving-average DSP system. Notice that the low-frequency component is not changed, but the noise spikes are greatly reduced in amplitude. This would be more pleasant to listen to. The basic operation that has been performed is that of a low-pass filter. By the way, moving-average filters are **Boxcar** often called *boxcar* filters.

Figure 16-8 presents a step-by-step description of the moving-average process. The dots represent binary (discrete) values. The black curves are included to help you to visualize the relationship to a continuous signal. Remember, the basis of DSP is number crunching. Step 3 of Fig. 16-8 takes place in the D/A converter. You might want to refer back to Fig. 16-1. The signal will take on the continuous form after passing through a reconstruction filter. The most important idea in Fig. 16-8 is that the noise spike is reduced in amplitude by the moving average process.

If you were to take a quantized signal in the form of a series of numbers and manually perform the process described in Fig. 16-8, you would probably use a calculator to divide by 3 after adding 3 sequential values. DSP chips are optimized for MAC operations. Look at Fig. 16-9. No divisions are used. In this drawing, an × in a circle represents multiply, and the + in the circle represents add or accumulate.

The first quantized value from the A/D converter in Fig. 16-9 is immediately multiplied by a *coefficient* of 0.333, and this result goes to the accumulator. After a time delay equal to one clock period, the first quantized value is

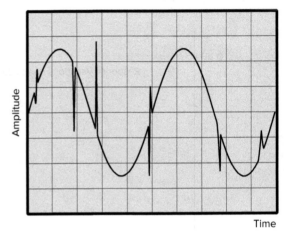

(a) Signal with high-frequency noise

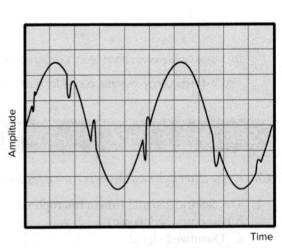

(b) Signal after processing with moving-average filter

Fig. 16-7 Moving-average low-pass filter time response.

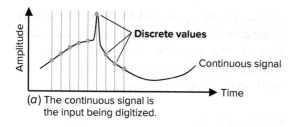

(a) The continuous signal is the input being digitized.

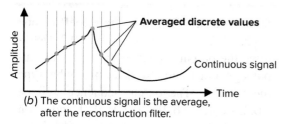

(b) The continuous signal is the average, after the reconstruction filter.

1. Add the first 3 discrete values
2. Calculate their average
3. Convert the average to analog
4. Add 2nd, 3rd, and 4th values
5. Calculate the average
6. Convert to analog
7. Add 3rd, 4th, and 5th values
8. Calculate the average
9. And so on . . .

Fig. 16-8 How the moving-average low-pass filter works.

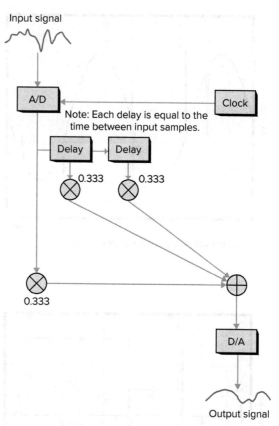

Note: Each delay is equal to the time between input samples.

Fig. 16-9 Realization of a moving-average filter using multiply-and-accumulate.

available to be multiplied by the second *coefficient* (also 0.333) and summed with the second quantized signal value after it's been multiplied by the first coefficient. After two delay periods, the first quantized value is multiplied by the third *coefficient* (again 0.333) and summed with the second quantized and multiplied value and the third quantized and multiplied value.

Suppose the output of the A/D converter in Fig. 16-9 is constant at a value of 1. The sequence of outputs from the accumulator will be 0.333, 0.666, 0.999, 0.999, 0.999, and so on. Ignoring rounding, the output settles to the average value of the input after a few clock cycles. A signal with a steady value of 1 is a dc signal with a frequency of 0 Hz. This moving-average filter is a low-pass type and will pass such a signal with no attenuation.

The MAC process is formally called *convolution*. The signal to be processed is *convolved* with the coefficients. The coefficients, taken together, are called the coefficient set. Remember, the signal must be in discrete form. By changing the number of coefficients and their values, all types of filters can be realized. Look at Fig. 16-10. This time, we want to keep the high-frequency information. Figure 16-10(b) shows the signal after processing with a high-pass

moving-average filter. The low-frequency hum has been attenuated.

Convolution is written as

$$y_{(n)} = x_{(n)} * h_{(n)}$$

where $y_{(n)}$ represents the output sequence (the discrete output signal)

$x_{(n)}$ represents the input sequence (the discrete input signal)

$h_{(n)}$ represents the coefficients

* is the convolution symbol

n is the sample number

Unfortunately, * is also the symbol for multiplication in several computer languages. Be wary of this, as it can be confusing. Multiplication and convolution are not the same. Convolution is shift–multiply–accumulate–shift–multiply–accumulate . . . and so on. In this chapter, we use · or × to indicate multiplication.

Figure 16-11 shows the details of the moving-average high-pass filter. Effectively, this filter calculates the signal average and subtracts that average from the signal. This removes the low-frequency content. However, once again,

Coefficient

Convolution

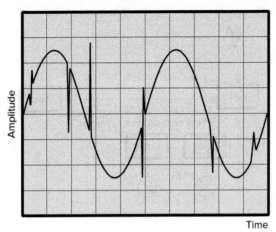

(a) Signal with low-frequency noise (hum)

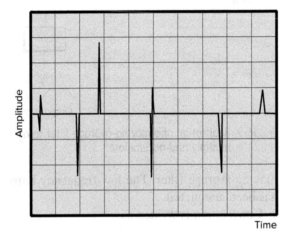

(b) Signal after processing with moving-average high-pass filter

Fig. 16-10 Moving-average high-pass filter time response.

Adaptive systems

DSP chips are optimized for MAC operations. Subtraction is realized by using negative coefficients and summing the negative products with the signal. Notice that the third coefficient is +1 and that all the others are negative.

EXAMPLE 16-1

Suppose a DSP filter uses these coefficients: −0.2, −0.2, −0.2, 1.0, −0.2, and −0.2. Determine the output sequence for a constant input signal of 1 and also determine what type the filter is. The output sequence will be −0.2, −0.4, −0.6, 0.4, 0.2, 0, 0, 0, and so on (the output then stays at zero). The input signal has constant amplitude, so it is a dc signal with a frequency of 0 Hz. The filter eliminates the dc component, so it's a high-pass type.

Now comes the really interesting part. If you compare Figs. 16-9 and 16-11, you will see that the basic structure is the same. This means that a simple software change can alter the function of a DSP filter from low pass to high pass. This is one of the best features of this technology. Because software is so easy to change compared to hardware, DSP systems can be updated at very low cost. Also, DSP systems can automatically adjust to changing conditions. These are called *adaptive systems*, and they can provide functions that are not possible using op amps or some similar approach.

Self-Test

Choose the letter that best answers each question.

6. Suppose a moving-average low-pass filter uses the coefficients 0.25, 0.25, 0.25, and 0.25. What will the output sequence be with an input signal at a constant value of 2?
   a. 0.1, 0.1, 0.1, 0.1, 0.1, 0.1, and so on
   b. 0.5, 1.0, 1.5, 2.0, 2.0, 2.0, and so on
   c. 1.0, 2.0, 3.0, 4.0, 4.0, 4.0, and so on
   d. None of the above

7. A moving-average high-pass filter uses the coefficients −0.25, −0.25, 1.0, −0.25, and −0.25. What will the output sequence be with an input signal at a constant value of 2?
   a. −0.5, −1.0, 1.0, 0.5, 0, 0, 0, and so on
   b. −0.25, −1.25, 1.25, 1.75, 2.0, 2.0, 2.0, and so on
   c. 0.5, 1.5, 1.0, 0, −1.5, −1.5, −1.5, and so on
   d. None of the above

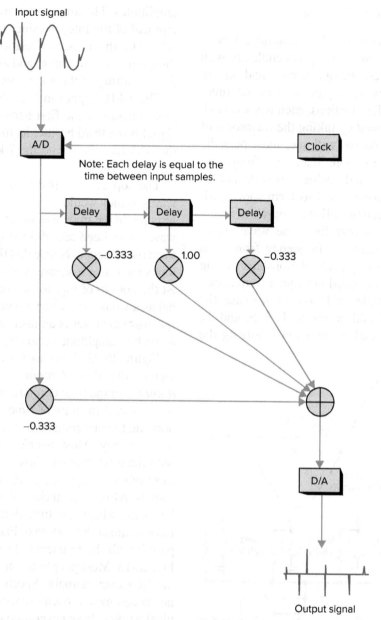

Input signal

Note: Each delay is equal to the time between input samples.

A/D

Clock

Delay

Delay

Delay

−0.333

1.00

−0.333

−0.333

D/A

Output signal

Fig. 16-11 High-pass realization using MAC.

8. What is the formal name given to the process of shift, multiply, add . . . shift, multiply, add . . . that goes on inside all DSP processors?
   a. Quantization
   b. Low-pass filtering
   c. High-pass filtering
   d. Convolution

9. Refer to Fig. 16-11 and assume a clock frequency of 1 MHz. What is the value of each delay?

   a. 0.1 μs
   b. 1.0 μs
   c. 1.5 μs
   d. 2.0 μs

10. What kinds of systems can change their characteristics on the fly as environmental factors change?
    a. Chebyshev systems
    b. LCR systems
    c. Differential op-amp systems
    d. Adaptive systems

16-3 Fourier Theory

Periodic function

Fundamental

Fourier series

Time domain

Frequency domain

We begin this section with a definition: A *periodic function* is one that repeats endlessly with time. Sine waves, triangle waves, and square waves are prime examples of periodic functions. As periodic functions, each has a period, which can be found by taking the reciprocal of the frequency. As we will see, most periodic functions contain more than one frequency. Only sine waves and cosine waves do not. A series of frequencies used to form some periodic functions can be called a *Fourier series*.

Figure 16-12 shows four sine wave signal sources. They range in frequency from 1 to 7 kHz. There are special relationships among the frequencies here. All the higher frequencies are integer multiples of 1 kHz. In this case, the integers are all odd numbers (1, 3, 5, and 7). The second special relationship is among the

amplitudes. The amplitudes are equal to the reciprocal of the integer values (1/1, 1/3, 1/5, and 1/7). The third special relationship is that all the sources are in phase, and all produce 0 V at the beginning of the 1-kHz period.

The 1-kHz signal in Fig. 16-12 is called the *fundamental* or the first harmonic. The 3-kHz signal is the third harmonic, the 5-kHz signal is the fifth harmonic, and the 7-kHz signal is the seventh harmonic.

The top of Fig. 16-12 shows the four sine waves as they would appear on a four-trace oscilloscope. In the middle of the figure, the four sine waves have been added for display on a single-trace oscilloscope. Notice that the sum waveform looks more like a square wave than a sine wave. At the bottom of Fig. 16-12, we see the sum signal as it would appear on a spectrum analyzer. A spectrum analyzer is an instrument that displays a graph of amplitude versus frequency.

Figure 16-13 shows two ways to look at any signal: the *time domain* and the *frequency domain*. In the time domain, the horizontal axis is calibrated in units of time. In the frequency domain, the horizontal axis is calibrated in units of frequency. Most people are more familiar with the time domain. This is because it is used most often to explain signals and operations on signals. Also, many technical workers use oscilloscopes, which are time-domain instruments. Look again at the bottom of Fig. 16-12, and compare it with the frequency-domain viewpoint of Fig. 16-13. Most people are not as familiar with the frequency domain. Spectrum analyzers are not as common as oscilloscopes, and many technical workers have never used one.

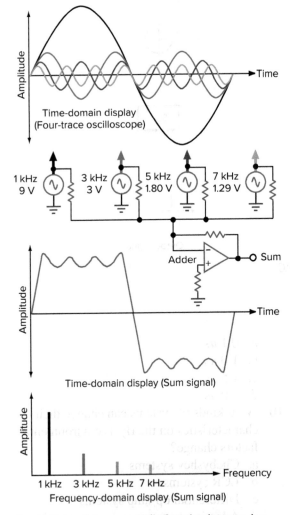

Fig. 16-12 Realizing a periodic function by summing sine waves.

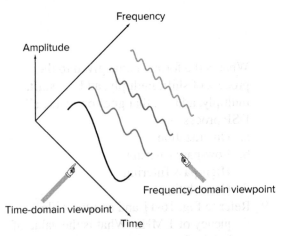

Fig. 16-13 Two views of a signal: the time domain and the frequency domain.

EXAMPLE 16-2

Determine the frequency of the 11th harmonic of a 100-Hz square wave. This is easy to find:

$$11 \times 100 \text{ Hz} = 1.1 \text{ kHz}$$

EXAMPLE 16-3

Determine the amplitude of the 10th harmonic of a 100-Hz square wave. It is zero because square waves have zero amplitude at the even harmonic frequencies.

Fourier theory states that any periodic function can be synthesized using sine waves. Generally, using more sine waves produces a better result. The square wave shown in Fig. 16-12 is synthesized using four harmonics. Figure 16-14 shows what happens with a large number of harmonics. The sum more closely approaches an ideal square wave. The rise and fall times are approaching zero, and the tops and bottoms are beginning to flatten out. However, there are spikes due to something called *Gibbs phenomenon*. These never go away, even if an extremely large number of harmonics is used. This is the major

limitation of Fourier's theory. It is not possible to synthesize ideal periodic functions that have *discontinuities*. Discontinuities are events that occur in zero time. An ideal square wave changes from a maximum positive value to a maximum negative value in zero time. There is no such thing in the real world.

Discontinuities

There is a very important principle at work here. If an electronic system must process and transfer pulses or square waves with very small rise and fall times, then that system will require a very large bandwidth. This is why there is no such thing as an ideal square wave in any physical or electrical system. Such a system would require infinite bandwidth, and that is not possible.

We now know that any periodic function can be viewed in either the time domain or the frequency domain. It's very important to realize that both viewpoints are valid, and either can be used for any signal. Can one form be derived from the other? Yes, and this is the job of the *Fourier transform*. A transform is a mathematical tool to change one representation into another to make calculations easier. For example, it is possible to make multiplication easier by transforming the numbers to be multiplied into logarithms. The logarithms are added, and the antilogarithm of the sum yields the product of the original numbers. Addition is a lot easier (and less error prone) than multiplication, and this was a popular technique before the days of calculators and computers.

Fourier transform

Gibbs phenomenon

EXAMPLE 16-4

Assuming that the pseudo-square waves shown in Figs. 16-12 and 16-14 are both 1 MHz, what are their bandwidths? Since square wave harmonics are odd, bandwidth is found by

$$BW = \text{fundamental} \times (2N - 1)$$

where N is the number of odd harmonics. For Fig. 16-12,

$$BW = 1 \text{ MHz} \times (7)$$
$$= 7 \text{ MHz}$$

For Fig. 16-14,

$$BW = \text{fundamental} \times (2N - 1)$$
$$= 1 \text{ MHz} \times (39) = 39 \text{ MHz}$$

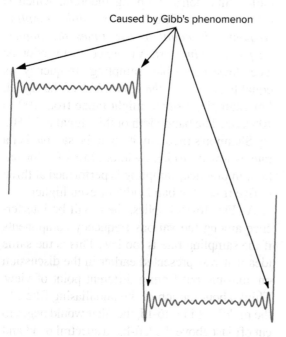

Caused by Gibb's phenomenon

Fourier square wave (20 odd harmonics)

Fig. 16-14 Gibb's phenomenon.

EXAMPLE 16-5

Find the bandwidth of a 1-MHz sine wave. Sine waves exist only at one frequency (they have no harmonics). The bandwidth of a single sine wave is zero. It produces a single vertical line on a spectrum analyzer.

EXAMPLE 16-6

Find the bandwidth of a 1-MHz cosine wave. Cosine waves also exist at only one frequency (they also contain no harmonics). The bandwidth of a single cosine wave is zero. It produces a single vertical line on a spectrum analyzer.

Note: The concept of zero bandwidth assumes a perfect sine or cosine wave: no distortion of any kind (including no noise) and absolute frequency stability. These requirements can't be met in the real world, so many people say that the bandwidth approaches zero.

The Fourier transform is used to convert from the time domain to the frequency domain, and the inverse Fourier transform is used to convert from the frequency domain to the time domain. You will see an example of this in the next section of this chapter.

Sampling frequency

A common application of the Fourier transform is a so-called real-time spectrum analyzer. These use computer chips or DSP chips to do the math on the discrete version of a time-domain signal, perhaps coming from a microphone. The output is used to drive a graphics display. The block diagram that is shown in Fig. 16-1 could just as well represent a spectrum analyzer as a moving-average filter. Again, the flexibility of DSP should be clear. Software determines function.

Discrete Fourier transform

To make a spectrum analyzer, the DSP processor would be programmed to perform a *discrete Fourier transform*, often referred to as the DFT. The general idea of the DFT is that a quantized time-domain signal is multiplied by sine coefficients of various frequencies and the products accumulated to produce a result. Each of the various results is called a bin. What accumulates in the bins represents the spectrum of the input signal. So, again, it's basically a MAC process. A special version of the DFT called the fast Fourier transform (FFT) provides better calculation efficiency. This is important because a high-resolution spectrum requires lots of bins and a rather large number of calculations.

Just as the DFT is used to transfer from the discrete time domain to the discrete frequency domain, the inverse discrete Fourier transform (IDFT) will transfer the discrete frequency domain to the discrete time domain. As such, the IDFT can be used to convert the frequency specifications for filters into the time information needed to implement those filters. In other words, the IDFT can be used to find a filter's coefficients, as will be seen later.

It's time to take a second look at sampling. Refer to Fig. 16-15. In this figure and for the rest of this chapter, f_s is the symbol for the *sampling frequency*. At the top of Fig. 16-15, you can see a continuous signal and its spectrum. If the signal is sampled at four times its highest frequency, the spectrum is repeated indefinitely and there are gaps in between (see the middle of the figure). Moving to the bottom of Fig. 16-15, if the signal is sampled at two times its highest frequency, there are no gaps. Clearly, this is some form of a limit since any lower sampling rate would cause the spectra to overlap and information to be lost. This concept is called Shannon's sampling theorem, which is often stated as: *The lowest-possible sampling frequency is equal to two times the highest frequency of interest.* A more correct version is: *The lowest-possible sampling frequency is equal to two times the bandwidth* of the signal. For example, a signal might range from 100 to 105 kHz. The bandwidth of this signal is 5 kHz. By Shannon's theorem, it could be sampled at a rate as low as 10 kHz without loss of information. In practice, sampling is performed at three to five times the bandwidth or even higher.

As Fig. 16-15 implies, there will be interference among the various frequency components if the sampling rate is too low. This is the same idea that was presented earlier in the discussion on aliasing, but from a different point of view. Think about the job of the antialiasing filter. In the middle of Fig. 16-15, the filter would begin to cut off just above the left-hand spectral band and then reach some reasonable attenuation before the second spectral band, centered at f_s, begins.

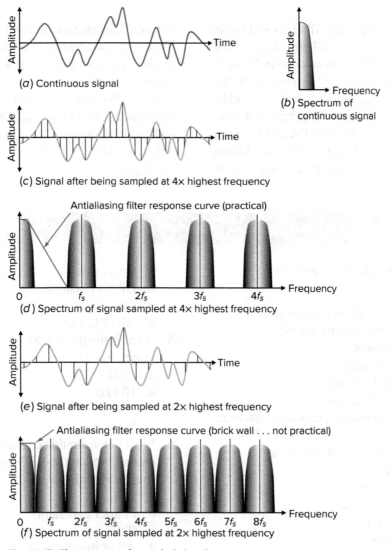

(a) Continuous signal

(b) Spectrum of continuous signal

(c) Signal after being sampled at 4× highest frequency

Antialiasing filter response curve (practical)

(d) Spectrum of signal sampled at 4× highest frequency

(e) Signal after being sampled at 2× highest frequency

Antialiasing filter response curve (brick wall . . . not practical)

(f) Spectrum of signal sampled at 2× highest frequency

Fig. 16-15 The spectra of sampled signals.

At the bottom of Fig. 16-15, the antialiasing filter would need to be unrealistically sharp. DSP systems can use higher sampling rates to ease the requirements of the antialiasing filter. In fact, it is often possible to get by with a simple *RC* filter by using a high sampling frequency.

Figure 16-15 is directly related to the information about amplitude modulation and sidebands presented in Chap. 12. There, it was shown that multiplying an information signal, such as sound, times a carrier signal produces upper and lower sidebands. In Fig. 16-15, the sampling frequency acts as the carrier, and the continuous signal is the information, which could be audio. The difference here is that the samples are snapshots in time that act as very narrow pulses. The Fourier series for a rectangular waveform with a very small duty cycle is a series of both even and odd harmonics that do

not decrease in amplitude at the higher harmonic frequencies. In Chap. 12, the carrier was a sine wave that has only one frequency, so only one set of sidebands was produced. Figure 16-15 shows that the process of quantization, or sampling, produces a spectrum that approaches an infinite bandwidth.

EXAMPLE 16-7

Investigate the possible use of a simple *RC* antialiasing filter for a DSP speech application with a 50-kHz sampling frequency. The highest critical frequency for speech is 3 kHz. Using the spectral display shown in the center of Fig. 16-15 as an aid, the left-hand band will extend from 0 Hz to 3 kHz, and the next band will start at 47 kHz and extend to

53 kHz. The antialiasing filter should begin its cutoff at 3 kHz or so and provide adequate attenuation at 47 kHz. As covered in Chap. 9, the slope of a single *RC* network is 6 dB/octave. 3 kHz × 2 × 2 × 2 × 2 = 48 kHz. You should recall that one octave is a doubling of frequency. So, the distance between the highest speech frequency and the lowest frequency that will cause an alias is four octaves. The attenuation will be about 24 dB. This would be adequate for a communication-quality speech system. It's important to understand why the antialiasing filter is significant. Without it, signal components in the vicinity of 47 kHz would be converted to audible frequencies. These audible artifacts would interfere with the audio and impair intelligibility.

Self-Test

Choose the letter that best answers each question.

11. What name is given to the lowest frequency in a Fourier series?
    a. Fundamental
    b. First harmonic
    c. Both of the above
    d. None of the above
12. A periodic signal displays on a spectrum analyzer as a single vertical line. What would the signal look like on an oscilloscope?
    a. Square wave
    b. Sawtooth wave
    c. Pulse wave
    d. Sine wave
13. A sawtooth waveform drops instantaneously from some positive value to zero. That falling edge represents a
    a. Discontinuity
    b. Fourier series
    c. Missing harmonic
    d. Limited bandwidth
14. The sum of a long Fourier series for a sawtooth function will show distortion called
    a. Crossover
    b. Gibb's phenomenon
    c. Clipping
    d. All of the above
15. What is the bandwidth of an ideal 1-kHz square wave?
    a. 1 kHz
    b. 10 kHz
    c. 1 MHz
    d. Infinite
16. What name would be given to the software routine used in the DSP chip of a real-time spectrum analyzer?
    a. FFT
    b. Fourier generator
    c. IDFT
    d. All of the above
17. According to Shannon's sampling theorem, the lowest possible sampling frequency for an information signal with a 100-kHz bandwidth is
    a. 50 kHz
    b. 100 kHz
    c. 200 kHz
    d. 400 kHz

16-4 Digital Filter Design

Moving-average filters work, but there are better designs. It is possible to use Fourier theory to select the coefficients that will be used in the convolution process. Figure 16-16 shows a low-pass filter that was designed using the inverse discrete Fourier transform for a rectangular window filter. There are other design methods as well.

The general architecture shown in Fig. 16-16 is the same as that presented earlier. Here, the IDFT was used to select the values for the coefficients, and there are more of them. Digital filter terminology assigns the name *tap* to each coefficient. The *order* of this type of filter is equal to the number of taps. Figure 16-16 is a 9-tap (ninth-order) filter. Generally, filter

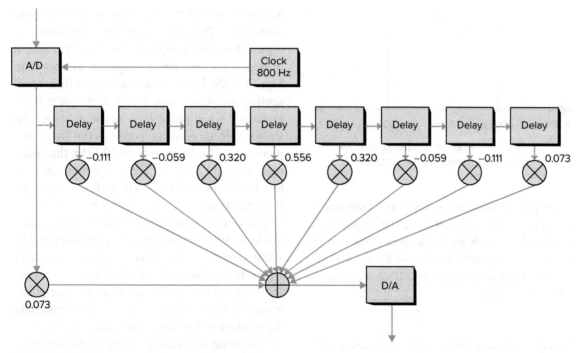

Fig. 16-16 Block diagram of a 9-tap FIR low-pass filter.

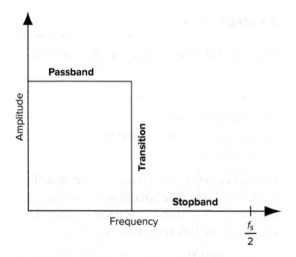

Fig. 16-17 The frequency response curve of an ideal low-pass filter.

sharpness improves as the order increases. This filter is sharper than the third-order filter shown in Fig. 16-9.

Look at Fig. 16-17, a representation of an ideal low-pass filter. Notice that the horizontal axis stops at $f_s/2$. This is true for all digital filters because of the limit imposed by the Shannon sampling theorem. The frequency $f_s/2$ is sometimes called the Nyquist frequency or the Nyquist limit. No digital system can properly handle any frequency beyond its Nyquist limit. The filter in Fig. 16-16 was designed to have the following characteristics:

- Filter type = low pass
- Filter order = 9
- Cutoff frequency = 200 Hz
- f_s = 800 Hz

The IDFT equations (for a rectangular window filter) are

$$h_{(0)} = \frac{K}{N}$$

$$h_{(n)} = \frac{1}{N} \cdot \frac{\sin(\pi n K/N)}{\sin(\pi n/N)}$$

where $h_{(0)}$ is the zeroth coefficient

      $h_{(n)}$ is the nth coefficient

      N is the filter order

      K is the number of discrete frequency samples in the passband

      π is the mathematical constant

      n is the coefficient number

The frequency spacing between samples equals the sampling frequency divided by $N - 1$. For our example,

$$f_{\text{spacing}} = \frac{f_s}{N-1} = \frac{800 \text{ Hz}}{8} = 100 \text{ Hz}$$

$$K = \frac{\text{bandwidth}}{f_{\text{spacing}}} + 1 = \frac{400 \text{ Hz}}{100 \text{ Hz}} + 1 = 5$$

As shown in Fig. 16-18, our example has 5 samples ($K = 5$) within its bandwidth.

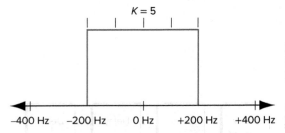

Fig. 16-18 Determining *K*: the number of discrete frequency samples in the passband.

For people just learning DSP, one of the strangest ideas is often that of *negative frequencies*. Look at Fig. 16-18 again. It shows that our example filter is graphed from −400 to +400 Hz. What is a negative frequency? It's a mathematical concept akin to

$$j = \sqrt{-1}$$

Perhaps you have been exposed to the square root of −1 as an aid for the solution of complex quantities such as impedance. Square roots of negative numbers don't actually exist in the real world, but their use in equations makes life easier when working with vector quantities.

Negative frequencies make life easier when dealing with certain kinds of signal operations. A prime example is amplitude modulation, which produces a pair of sidebands. The lower sideband is the mirror image of the upper sideband and can be attributed to negative frequencies that interact with the carrier frequency.

For more support of this idea, take another look at Fig. 16-15. Notice that the sampling process produces copies of the original spectrum shown at the top plus mirror images of it at all multiples of f_s. This is shown both in the middle of the figure and at the bottom. This makes it reasonable to declare that the original signal contains both positive and negative frequencies.

Don't worry too much about the IDFT equations. People who design digital filters mostly use software programs to find the coefficients. However, two examples will show that finding the coefficients for a filter like the one in Fig. 16-16 can be done with a calculator. Set your calculator for radian mode if you want to try Example 16-9.

Checking Fig. 16-16, we find that the zeroth coefficient has been placed as the center coefficient of the filter and that the third coefficient

has been placed three positions away in both directions. This *symmetry of coefficients* is typical for filter designs of this type because it provides a linear phase response as shown in Fig. 16-19(b). Look closely at the red line representing the phase response. It starts out at 0 degrees at 0 Hz. It phase lags (negative angle) and shows a straight-line response as the frequency increases. At a frequency of 100 Hz, the phase reaches −180 degrees. Here, it jumps abruptly to +180 degrees. This is called *phase wrapping*. It occurs because the graph has been restricted to phase angles between ±180 degrees. On a circle, +180 degrees is exactly the same point as −180 degrees. The phase response is linear when the jumps occur at ±180 degrees. Figure 16-19(b) shows that the filter has a linear phase response throughout both the passband and the transition regions. The phase response in the stopband is *nonlinear* because the jumps do not occur at ±180 degrees.

EXAMPLE 16-8

Find the 0th coefficient for the following filter:

- Filter type = low pass
- Filter order = 9
- Cutoff frequency = 200 Hz
- f_s = 800 Hz

Using Fig. 16-18 to assist, we see that the number of discrete frequency samples in the passband is equal to 5 since the samples are spaced at 100-Hz intervals:

$$\frac{800 \text{ Hz}}{N - 1} = 100 \text{ Hz}$$

$$h_{(0)} = \frac{K}{N} = \frac{5}{9} = 0.556$$

EXAMPLE 16-9

Find the third coefficient for the same filter.

$$h_{(n)} = \frac{1}{N} \cdot \frac{\sin(\pi n K/N)}{\sin(\pi n/N)}$$

Using radian mode gives

$$h_{(3)} = \frac{1}{9} \cdot \frac{\sin(\pi 3 \cdot 5/9)}{\sin(\pi 3/9)} = -0.111$$

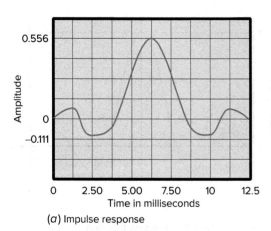

(a) Impulse response

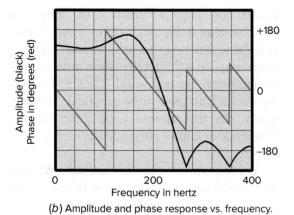

(b) Amplitude and phase response vs. frequency.

Fig. 16-19 Impulse, amplitude, and phase responses for the FIR low-pass filter.

EXAMPLE 16-10

Find the phase response for Fig. 16-19(b) at a frequency of 120 Hz. By inspection, the angle is 150 degrees. It is just as correct to say it is −210 degrees, which is 30 degrees of negative rotation beyond −180 degrees:

$$-210 = -180 - 30$$

Figure 16-19(b) shows the amplitude versus frequency response of the filter as a black plot. Notice that there is ripple in both the passband and stopband and that the transition region is not particularly sharp. We are going to see that both can be improved.

Figure 16-19(a) shows the *impulse response* of the filter. This is what happens at the filter output when a very narrow pulse is fed into the filter. Generally, it's assumed that the amplitude of the input pulse is unity or 1. As the pulse moves through the filter, the coefficients multiply it.

Because the pulse is narrow, the output is actually a graph of the coefficients. Thus, Fig. 16-19(a) is a graph of the coefficient values shown in Fig. 16-16 after smoothing by a reconstruction filter. Figure 16-19(a) is important because it demonstrates that the output of the filter will always return to 0 after the impulse passes through. This leads to the name of this filter. It is called a *finite impulse response*, or FIR, filter. The moving-average filters presented earlier are also FIR types.

Finite impulse response

EXAMPLE 16-11

Determine the impulse response of Fig. 16-9. The output would increase from 0 to 0.333, remain there for a time equal to two clock periods, and then fall back to zero. The reconstruction filter will make it appear as a hump, rather than as a rectangular pulse.

The filter shown in Fig. 16-16 can be modified to provide a high-pass response by inverting the sign of every other coefficient starting with the one at the left. Figure 16-20(a) shows the impulse response of the high-pass filter. It returns to 0 so the filter type is again FIR. Figure 16-20(b) shows the high-pass frequency response in black. The phase response is shown in red, and once again, it is linear in the passband and transition regions and nonlinear in the stopband. This phase response is always true for FIR filters with symmetrical coefficients.

How does one approach the ideal or "brick-wall" response with FIR filters? With active filters, as covered in Chap. 9, it was shown that increasing the filter order made the transition region sharper. So it is also with FIR filters. Figure 16-21 shows the frequency response for a 51-tap FIR low-pass filter. The transition is sharp, but the ripple is excessive for many applications. The ripple is caused by Gibb's phenomenon.

The IDFT equation presented earlier in this section is known as a *sinc* function. Sinc functions take the general form

Impulse response

$$t = \frac{\sin(x)}{x}$$

Figure 16-22 shows a plot of the sinc function from $x = -10$ to $x = +10$. Compare this plot with Fig. 16-19(a) to see that the filter

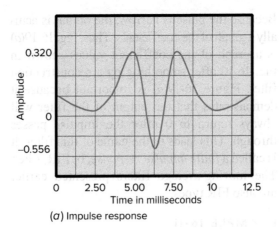

(a) Impulse response

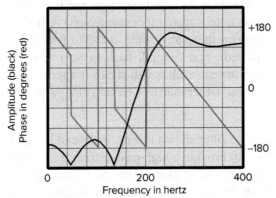

(b) Amplitude and phase response vs. frequency.

Fig. 16-20 Impulse, amplitude, and phase responses for the FIR high-pass filter.

coefficients are indeed sinc values. In fact, the filter design method used in this section is known formally as the *windowed sinc method*.

Figure 16-23 shows why the filter's response is subject to Gibb's phenomenon. Figure 16-23(a) implies that the sinc function is infinite—no

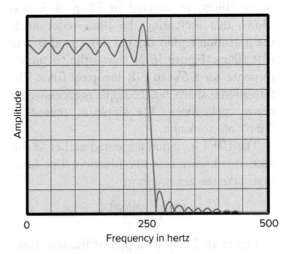

Fig. 16-21 Frequency response of a 51-tap, FIR low-pass filter.

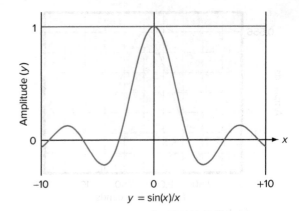

Fig. 16-22 A graph of the sinc function.

matter how large x becomes, the amplitude never reaches 0. Figure 16-23(b) shows a *truncated* sinc function. An infinite coefficient set would require infinite calculation time and that won't work in the real world. The sinc function has to be truncated to be practical. Unfortunately, the truncations create discontinuities that produce Gibb's phenomenon as described before.

A compromise can be reached. We can trade off some filter sharpness to reduce the ripple. Passing the coefficients through a nonrectangular window does this. A *rectangular* window has no effect on the filter coefficients, as shown in Fig. 16-24(a). Said another way, a rectangular window is like having no window at all. A *triangular* window will suppress the ripple since it will gradually reduce the amplitudes of the filter coefficients as shown in Fig. 16-24(b). A *Blackman* window does an even better job of reducing the ripple. The equation for a Blackman window is

$$w_{(n)} = 0.45 + 0.5 \cos(2\pi n/N) + 0.08 \cos(4\pi n/N)$$

where $w_{(n)}$ is the nth Blackman window value

    N is the filter order

    n is the number of the window value (n ranges from $-N/2$ to $N/2$)

    π is the mathematical constant

Figure 16-24(c) shows the shape of the Blackman window. The important idea is that the purpose of a window is to make a function smoothly approach 0 at both ends.

As before, don't worry about this equation. Filter design software has this window, plus other window options, programmed in. Computers mostly perform tedious calculations.

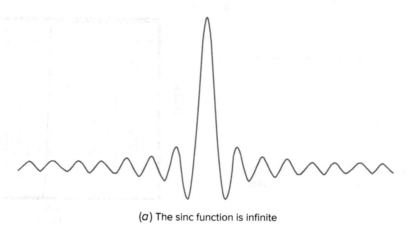

(a) The sinc function is infinite

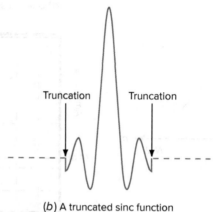

(b) A truncated sinc function

Fig. 16-23 A practical sinc function is truncated.

However, it is possible to design FIR filters with a calculator. Don't forget to use radian mode with the Blackman equation. After the $h_{(n)}$ sequence is found, using the IDFT equation presented earlier, the $w_{(n)}$ values are found using the Blackman equation. Finally, the actual filter coefficients are found by multiplying:

$$\text{Coefficient}_{(n)} = h_{(n)} \cdot w_{(n)}$$

We now have the tools to design practical FIR filters. Figure 16-25 shows both the linear and the log response for a 200-tap filter with a Blackman window. The passband ripple is gone, and the stopband ripple is at −74 dB and can be seen on only the log plot. Notice that this is a very sharp filter. The number of taps (filter order) needed for a given response can be estimated with

$$N \approx \frac{4}{\dfrac{\text{TBW}}{f_s}}$$

where N = the filter order

     TBW = the transition bandwidth

      f_s = the sampling frequency

EXAMPLE 16-12

Determine the required filter order when the sampling frequency is 1 kHz, the cutoff frequency is 250 Hz, and the transition bandwidth is 20 Hz.

$$N \approx \frac{4}{\dfrac{\text{TBW}}{f_s}} \approx \frac{4}{\dfrac{20}{1,000}} \approx 200$$

This agrees well with the response shown in Fig. 16-25.

Other window functions allow different trade-offs in filter performance. The Blackman window does a good job of reducing the passband and stopband ripple, but it increases the transition bandwidth. Filter design software allows various choices, and an optimum design for a given situation amounts to experimenting with different filter orders and different window functions.

The filter response shown in Fig. 16-25 is quite good, but a 200-tap filter is rather elaborate.

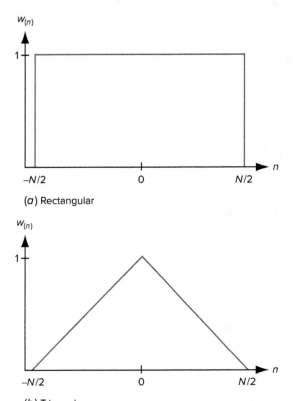

(a) Rectangular

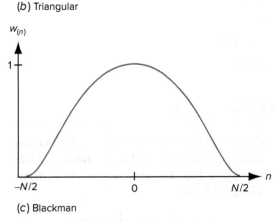

(b) Triangular

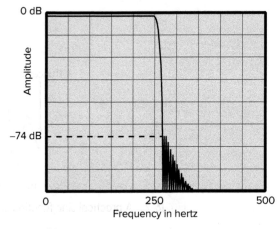

(c) Blackman

Fig. 16-24 Three window functions.

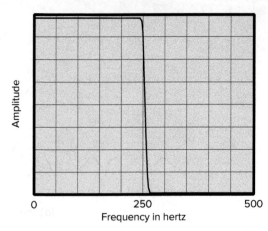

(a) Linear response

(b) Log response

Fig. 16-25 Frequency response of a 200-tap, FIR low-pass filter with a Blackman window.

Depending on the application, it might not be possible to process the incoming data fast enough. Real-time systems must process data on the fly. Imagine a communication system where there is a 1-minute delay between the time when one talker stops and the reply starts coming back. This would be confusing and annoying.

Designers have another option called infinite impulse response (IIR) filters. These filters use feedback to sharpen the filter response without having to resort to using a large number of taps. Look at Fig. 16-26. There are two sets of filter coefficients. The **a** coefficients are called feed-forward coefficients, and they work exactly as described before. The **b** coefficients are feedback coefficients. Delayed copies of the accumulator output multiply the **b** coefficients, and these results are fed back into the accumulator. You were exposed to a similar idea in Chap. 9 in the section on active filters, where feedback was used to sharpen the knee of the filter response.

IIR filters can also be called *recursive* filters. They have a different impulse response than FIR filters. Recall that the impulse response of all FIR filters decays to 0 after the impulse passes through the system. In theory, the impulse response of IIR filters never quite reaches 0 because the feedback makes the decay *exponential*. You were probably exposed to this general idea when you learned about capacitors charging and discharging. They charge and discharge exponentially so, in

Recursive

Exponential

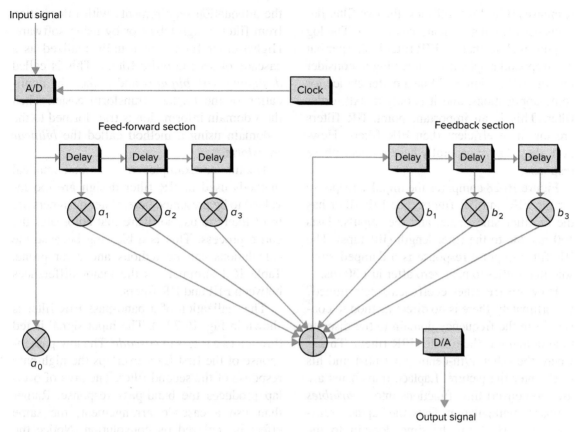

Fig. 16-26 Block diagram of an IIR filter.

theory, they never reach full charge or full discharge. However, you also learned that after five *RC* time constants, they could be considered to be at full charge or full discharge. The same is true with IIR filters. After some period of time after an impulse, the output does settle to zero. A practical filter has to eventually settle to zero or it would be useless, unless one needs an oscillator. IIR filters can oscillate if they are improperly designed. All feedback systems have the potential to become unstable and oscillate.

Figure 16-27 compares an FIR filter with an IIR filter. Both were designed using software. The specifications entered into the computer were

$$f_s = 1 \text{ kHz}$$
$$f_{pass} = 250 \text{ Hz}$$
$$f_{stop} = 283 \text{ Hz}$$

The software put the FIR at 101 taps with a Hamming window. The IIR filter ended up as a Butterworth type with six feed-forward coefficients and six feedback coefficients. The linear

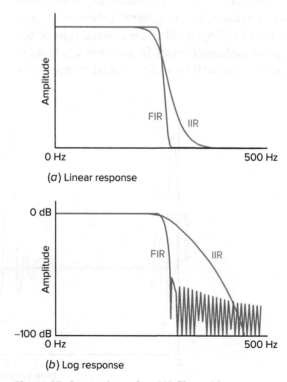

(*a*) Linear response

(*b*) Log response

Fig. 16-27 Comparison of an FIR filter with an IIR filter.

responses [Fig. 16-27(a)] show the two filter designs perform in a comparable manner. The log responses show that the FIR filter is sharper but has stopband ripple. It is interesting to consider that the IIR design could be a better choice for some applications, and it is only a sixth-order filter. This is an important point. IIR filters are far more efficient than FIR filters. However, they do *not* generally have a linear phase response.

Figure 16-28 compares the impulse response for the FIR and IIR filters. The FIR filter has the familiar sinc shape, and the impulse lasts 100 ms due to the filter length (101 taps). The IIR filter impulse response is a damped sinusoid that settles to near zero after just 30 ms.

How are IIR filter coefficients determined? Unfortunately, there is no direct method to convert from the frequency domain to the discrete time domain as there is for FIR filters. This is where the other 19th-century scientist and his work enter the picture. Laplace transforms are used to convert time functions into *s variables* for easier manipulation. Thus, the Laplace transform converts from the time domain to the *s domain*. The discrete time domain is usually called the *z domain*. In fact, in many DSP books and articles, each delay element is labeled as z^{-1}.

In general terms, IIR filters can be designed by starting with the general information presented in Chap. 9. After the desired type of response is chosen (Bessel, Butterworth, Chebyshev, or elliptic), the filter order is found by matching the attenuation requirements with information from filter design tables or by using software. Higher-order filters are usually realized as a cascade of second-order filters. This is called *designing with biquads* and makes the application of the Laplace transform easier. Then the *s* domain information is transformed to the *z* domain using a method called the *bilinear transform*.

Details or examples of the mathematical methods used in IIR filter design are too involved to be presented here. Filter designers are most likely to use software that automates the entire process. This is a blessing because the calculations can be tedious and error prone. Table 16-1 summarizes the major differences between FIR and IIR filters.

The realization of a band-pass FIR filter is shown in Fig. 16-29(a). The input signal is fed through two filters in *cascade*. The low-pass response of the first filter overlaps the high-pass response of the second filter. The area of overlap produces the band-pass response. Rather than use a cascade arrangement, the same effect is achieved by convolution. Notice the convolution symbol in Fig. 16-29(a). The filter coefficients are convolved, which produces the band-pass response. Figure 16-30 shows that this is equivalent to cascading two separate filters. An arbitrary signal input sequence of 5, 4, and 3 is used to demonstrate the concept using numbers. The first signal value to enter the first filter coefficient set is s_0 and produces an

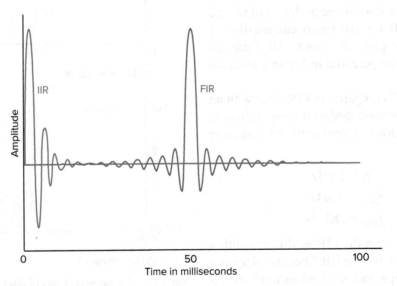

Fig. 16-28 Comparison of the impulse responses.

Table 16-1   Comparison of FIR and IIR Filters

Characteristic	FIR	IIR
Efficiency	Low	High
Speed	Slow	Fast
Overflow	Not likely	Likely
Stability	Guaranteed	Design issue
Phase response	Generally linear	Generally nonlinear
Analog modeling	Not directly	Yes
Design/noise analysis	Straightforward	Complex
Arbitrary filters	Straightforward	Complex

output at $t_0 = 0.5$. After one clock delay, the output is

$$t_{1(\text{out})} = 5 \cdot 0.5 + 4 \cdot 0.1 = 2.9$$

After convolution with the first set of coefficients, the signal is then sent to the second filter in the middle of Fig. 16-30. The bottom of the same illustration shows the convolution of the two filters. Notice that the output sequence is identical for the cascade output in the center and for the convolved set at the bottom.

Convolution of FIR filter coefficients can also be used to increase filter performance. If you refer back to Fig. 16-25(b), you will see that the passband ripple is down 74 dB. This is the best that can be done with a Blackman window. However, the filter could be convolved with itself to further reduce the ripple. Of course, this makes the filter almost twice as long.

EXAMPLE 16-13

A 51-tap FIR filter is convolved with itself. What is the resulting filter order or length?

$$\text{Order} = 2N - 1 = 101 \text{ taps}$$

Finally, as Fig. 16-29(b) shows, a band-stop filter can be achieved by summing two filter outputs. Here, the passbands do not overlap. The signal could be sent to two filters in parallel with their outputs connected to a summing amplifier. An easier way to do it is to just sum the coefficients of the two filters. Referring to

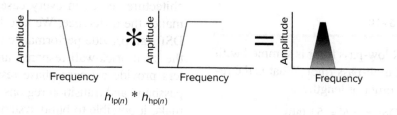

$h_{\text{lp}(n)} * h_{\text{hp}(n)}$

(a) The convolution of a low-pass filter and a high-pass filter equals a band-pass filter

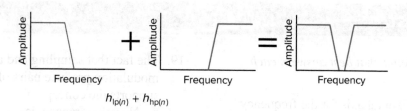

$h_{\text{lp}(n)} + h_{\text{hp}(n)}$

(b) The addition of a low-pass filter and a high-pass filter equals a band-stop filter

Fig. 16-29 Realization of band-pass and band-stop filters.

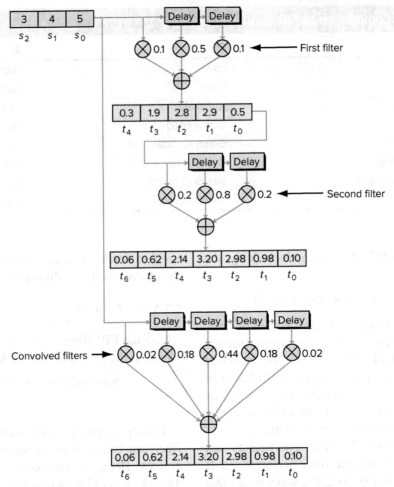

Fig. 16-30 Equivalence of cascaded filters with convolved filters.

Fig. 16-30, the coefficients for the summed filter would be 0.3, 1.3, and 0.3.

EXAMPLE 16-14

A 51-tap FIR low-pass filter is summed with a 51-tap FIR high-pass filter. What is the resulting filter order or length?

Order $= N = 51$ taps

We now see that it is possible to realize any filter response. Amazingly, one basic MAC architecture serves in every case. The software makes the difference. We also have found that DSP can provide performance that approaches the ideal brick-wall response and that FIR filters provide a linear phase response over the passband and transition regions. These features make it possible to build systems that are very difficult using analog techniques.

Self-Test

Choose the letter that best answers each question.

18. The horizontal axis for the frequency response graph of a digital filter ends at
    a. $2f_s$
    b. f_s
    c. $f_s/2$
    d. $f_s/4$

19. The fact that sampling and amplitude modulation produce pairs of sidebands supports the concept of
    a. Negative frequencies
    b. Fourier square waves
    c. Gibb's distortions
    d. All of the above

20. Refer to Fig. 16-17. When the transition region is a vertical line, the filter is called
    a. Low-pass
    b. High-pass
    c. Brick-wall
    d. Butterworth
21. Refer to Fig. 16-18. If $N = 21$ and $f_s = 800$ Hz, $K =$
    a. 3
    b. 9
    c. 11
    d. 18
22. Determine $h_{(0)}$ for question 21 above.
    a. 0.333
    b. 0.409
    c. 0.487
    d. 0.524
23. The coefficients for FIR filters are almost always arranged to be symmetrical to obtain
    a. A sharper transition
    b. Linear phase response
    c. Elimination of Gibb's distortion
    d. All of the above
24. Refer to Fig. 16-19(b). The jump in the white line at 100 Hz is called
    a. Phase wrapping
    b. Gibb's phenomenon
    c. A nonlinear response
    d. None of the above
25. Refer to Fig. 16-19(b). The phase response of the filter above 265 Hz is
    a. Linear
    b. Nonlinear
    c. Imaginary
    d. None of the above
26. Refer to Fig. 16-19(a). The impulse response is known as a

a. Sine function
b. Cosine function
c. Laplace transform
d. Sinc function

27. Refer to Fig. 16-21. The pass-band and stopband ripple are caused by
    a. The triangular window
    b. The Blackman window
    c. Gibb's phenomenon
    d. $h_{(0)}$
28. Windows are used in FIR filter designs to
    a. Reduce Gibb's phenomenon
    b. Convert low pass to high pass
    c. Convert low pass to band pass
    d. All of the above
29. Which of the following filters has an impulse response that is a sine wave that decays exponentially?
    a. 11-tap FIR
    b. Sixth-order IIR
    c. 49-tap FIR band-pass
    d. Windowed sinc type
30. When a signal passes through two FIR filters in cascade, the same effect can be obtained by
    a. Adding the filter's coefficients
    b. Multiplying the filter's coefficients
    c. Dividing the filter's coefficients
    d. Convolving the filter's coefficients
31. Refer to Fig. 16-26. The b coefficients are sometimes called the
    a. FIR coefficients
    b. Feedback coefficients
    c. Feed-forward coefficients
    d. Blackman coefficients

16-5 Other DSP Applications

The number of DSP applications grows constantly. The following represents a partial list:

- Filtering
- Modulation and demodulation
- Image enhancement and compression
- Motion control and positioning
- Seismography
- Radar
- Sonar
- Noise reduction and echo cancellation
- Speech recognition
- Interference rejection

Figure 16-31(a) shows a partial block diagram for an audio CD player. This is an example of a *multirate system*. Multirate systems are those DSP systems in which the sampling rate is changed. Sampling rates are changed for various reasons:

- To improve performance
- To marry various system components that operate at different rates

Multirate system

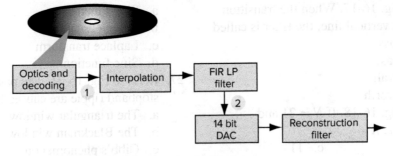

(a) Partial block diagram of audio CD player

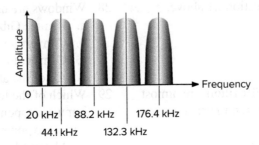

(b) Spectrum at point 1

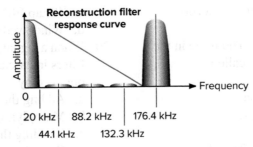

(c) Spectrum at point 2

Fig. 16-31 Partial block diagram of a CD player.

- To allow the use of more simple analog filters
- To make processing faster

Figure 16-31(*b*) shows part of the spectrum for the discrete signal coming from the decoding circuit. Notice that the sampling frequency is 44.1 kHz and that the bandwidth of the sampled audio is 20 kHz. This leaves only a small guard band between the base signal and the first alias band that is centered at 44.1 kHz, the sampling frequency. Notice that the second stage in Fig. 16-31(*a*) is labeled *interpolation* (also called *up-sampling*). This is a process used to increase the sample rate of a discrete signal. In this case, the sampling frequency is increased four times from 44.1 kHz to 176.4 kHz by stuffing three zeros between every discrete value coming from the decoding section. This is known as *zero stuffing* or *zero insertion*.

The number of zeros to be stuffed is one less than the integer multiplication factor desired for f_s.

EXAMPLE 16-15

How would a discrete signal be interpolated to a new sampling frequency eight times higher than the original? Eight minus one is seven. Therefore, seven zeros would be stuffed between each discrete signal value.

An FIR low-pass filter follows the interpolation block in Fig. 16-31(*a*). Its function is to remove the aliases associated with multiples of the *old* sampling frequency. It has a cutoff just above 20 kHz. Figure 16-31(*c*) shows attenuation of three images above 20 kHz. The spectral groups centered at 44.1 kHz, 88.2 kHz, and

Zero stuffing

Zero insertion

132.3 kHz have been attenuated by the FIR filter. There is still a spectral group centered at 176.4 kHz since that is the new sampling frequency produced by interpolation. The noticeably relaxed requirements for the reconstruction filter are also shown in Fig. 16-31(c). Compare this with what would be required without interpolation by examining Fig. 16-31(b). Without interpolation, the reconstruction filter would need a very small transition bandwidth. It would have to cut off around 20 kHz and fall sharply to offer significant attenuation at 24.1 kHz (44.1 kHz − 20 kHz). That would require an elaborate analog filter circuit.

Figure 16-31 is a good example of why digital is replacing analog. The interpolation and FIR stages are both digital. As such, they don't suffer from component errors, aging, or temperature drift. Also, practical digital filters can approach ideal performance. Thanks to interpolation, there is no need for a complicated analog reconstruction filter. A simple type will work, as shown by the response curve in Fig. 16-31(c).

Although we have shown zero stuffing (interpolation) and the following FIR low-pass filter as separate stages in Fig. 16-31, they can be combined for greater efficiency. Why bother multiplying three out of four filter coefficients times zero? To gain efficiency, one can figure out which coefficients will be multiplied by nonzero data and perform only those operations. DSP programmers use tables and counters to alter the coefficients for each new input sample to accomplish the same thing.

A time-domain viewpoint of what interpolation can do for a system is shown in Fig. 16-32. Part (a) shows an ideal sine wave in blue and the DAC output in red. Notice the large amount of error. Part (b) shows a significant improvement achieved by using interpolation to increase the sampling rate by a factor of four. By the way, both parts of Fig. 16-32 were prepared by simulating a 10-bit DAC. It is interesting to note that the audio data encoded on CDs is 16-bit, but some CD players provide adequate performance with 14-bit DACs, thanks to interpolation.

DSP is replacing other methods in communications system. As discussed in Chap. 12, single sideband (SSB) transmitters are more efficient than AM transmitters. SSB conserves both power and spectrum. Because AM sidebands are mirror images of one another, it is possible

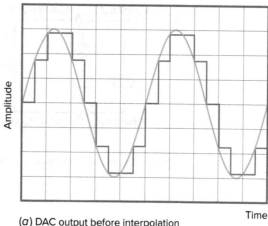

(a) DAC output before interpolation

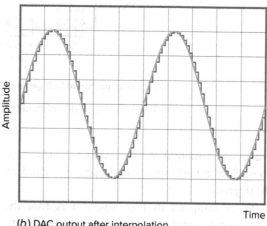

(b) DAC output after interpolation

Fig. 16-32 An advantage of interpolation.

to send only one and convey the same information. In Chap. 12, the filter method of SSB generation was presented. Here, we will look at another method.

Figure 16-33(a) shows the phasing method of SSB generation. Phase-shift networks are used to cancel one of the sidebands. The symbols are defined as follows:

V_m is the modulating voltage (perhaps 1 V)

f_m is the modulating frequency (300 Hz to 3 kHz for voice)

V_c is the carrier voltage (perhaps 4 V)

f_c is the carrier frequency (perhaps 400 kHz)

t is the instantaneous time

π is the mathematical constant

Don't let the equations in Fig. 16-33(a) scare you off. Graphing them for fixed voltages and frequencies just produces sine waves or cosine

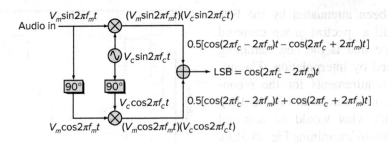

(a) Phasing method of SSB generation using continuous signal processing

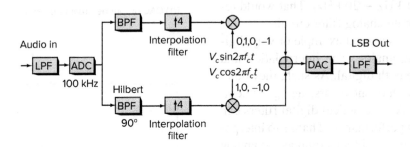

(b) Phasing method of SSB generation using digital signal processing

Fig. 16-33 Two methods of generating a single sideband signal.

waves as a function of time. Also, a cosine wave looks like a sine wave that is phase shifted by 90 degrees.

What happens in Fig. 16-33(a) is that the voice input signal gets converted to a sine component and a cosine component. These components are mostly called the *in-phase* component and the *quadrature* component, respectively. The quadrature component is obtained by a 90-degree phase-shift network. Each component is multiplied by a carrier frequency, which also is in two parts. The products are added, and phase cancellation eliminates one of the sidebands. Note that the output contains only the lower sideband. The lower sideband is a spectrally inverted copy of the original signal. The carrier has translated it up in frequency. A trig identity was used in Fig. 16-33(a) to place the two components to be added in the form that shows the cancellation of the upper sideband.

Figure 16-33(a) works with analog circuits. However, it does not necessarily work very well. The math works perfectly, but the real world dictates that it is impossible to produce an accurate 90-degree phase shift over any practical bandwidth. This results in incomplete cancellation of

one of the sidebands and that means interference for other users of the spectrum.

With DSP, it's easy to shift a band of frequencies by 90 degrees. This is usually accomplished with a special FIR function called the *Hilbert transform*. When a signal is applied to a Hilbert filter, the output result is a quadrature signal. Figure 16-33(b) shows a DSP SSB generator. The input audio is applied to an antialiasing filter. The relatively high sampling frequency of the ADC relaxes the requirements for this filter. For voice, it would have a cutoff frequency of about 3 kHz and provide enough attenuation at 50 kHz ($\frac{1}{2}$ of f_s). Next in Fig. 16-33(b), the discrete signal splits into two paths, and one is phase shifted by a Hilbert filter. Interpolation filters then up-sample both signals to 400 kHz. *Up-sample* is another way to describe an increase in the sampling frequency. Also, interpolation can be symbolized with an up-arrow, as shown in Fig. 16-33(b).

Now, it's time to multiply these signals by the carrier. Here is where a clever trick can be employed. Four samples of a sine wave can be represented by 0, 1, 0, and −1, and four samples of a cosine wave by 1, 0, −1, and 0. What this means is that *no multiplications are needed*!

Each sample will simply take three forms: (1) not changed, (2) zeroed, and (3) inverted. Each output sample from the interpolation filters is *effectively* multiplied by four samples of the carrier signal. The two signals are then added, which cancels the USB, and this result is sent on to the DAC. All of the stages in Fig. 16-33(*b*) are digital except the two low-pass filters.

EXAMPLE 16-16

Find the instantaneous values of a 1-V peak sine wave at 0, 90, 180, and 270 degrees.

$$(1 \text{ V}) \cdot \sin(0°) = 0 \text{ V}$$
$$(1 \text{ V}) \cdot \sin(90°) = 1 \text{ V}$$
$$(1 \text{ V}) \cdot \sin(180°) = 0 \text{ V}$$
$$(1 \text{ V}) \cdot \sin(270°) = -1 \text{ V}$$

EXAMPLE 16-17

Find the instantaneous values of a 1-V peak cosine wave at 0, 90, 180, and 270 degrees.

$$(1 \text{ V}) \cdot \cos(0°) = 1 \text{ V}$$
$$(1 \text{ V}) \cdot \cos(90°) = 0 \text{ V}$$
$$(1 \text{ V}) \cdot \cos(180°) = -1 \text{ V}$$
$$(1 \text{ V}) \cdot \cos(270°) = 0 \text{ V}$$

Figure 16-34 shows how SSB detection can be accomplished using DSP. The intermediate frequency (IF) spectrum is limited with a band-pass filter and then changed to a discrete time-domain signal by the analog-to-digital converter.

This discrete signal is then split into two paths and multiplied by the carrier frequency and by the carrier frequency phase shifted by 90 degrees. As before, if four samples of the carrier frequency are used for each IF sample, then no actual multiplications need take place. Next, *decimation* filters decrease the sampling rate by a factor of four. Decimation is also called *down-sampling*. Decimation is the opposite of interpolation. Here, three out of four discrete time samples are discarded. This effectively reduces the sampling frequency to one-fourth of what it was. The bandwidth is likewise reduced.

Decimation is useful because it makes more time available for the detection process. Remember, most DSP systems are real-time systems and must process the information on the fly without incurring noticeable delays. Finally, the resampled signals in Fig. 16-34 are sent through FIR band-pass filters, one of which provides a 90-degree phase shift. When the two signals are added, the lower sideband is selected, and the DAC and the reconstruction filter provide the audio output.

Figure 16-35 shows a DSP AM detector. Notice that a lot more arithmetic is required for AM detection. Since this takes more time, decimation is often a must to lower the sampling frequency. The detection process for AM is based on the fact that the envelope or shape of an AM signal, as viewed in the time domain on an oscilloscope, is the same as the audio waveform. So, to recover the audio signal using DSP, it is necessary to find the vector sum of the in-phase and quadrature parts. Again, quadrature

Decimation

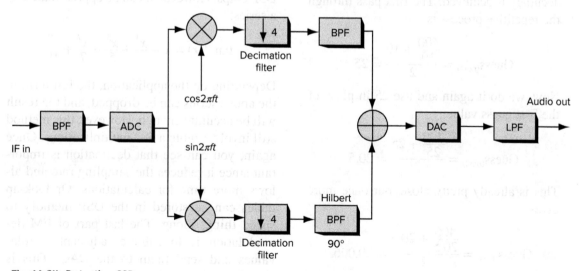

Fig. 16-34 Detecting SSB.

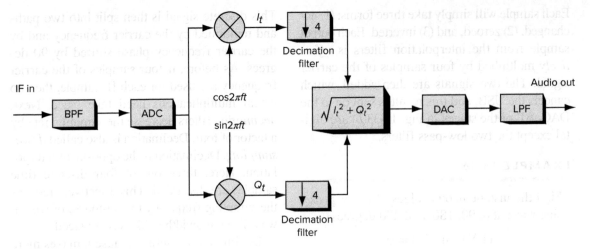

Fig. 16-35 Detecting AM.

means phase shifted by 90 degrees. You may re-call that the Pythagorean theorem can be used:

$$\text{Vector sum} = \sqrt{I_t^2 + Q_t^2}$$

where I_t = the in-phase signal

Q_t = the quadrature signal

DSP chips multiply, so the squaring function is not a problem. DSP chips do not directly extract square roots, however. This function can be achieved using a method first described by Isaac Newton:

$$\text{Guess}_{(next)} = \frac{\dfrac{\text{Number}}{\text{Guess}} + \text{Guess}}{2}$$

EXAMPLE 16-18

Use a first guess of 10 as the square root of 400, and repeat the process until reasonable accuracy is achieved. The first pass through the repetitive process is

$$\text{Guess}_{(next)} = \frac{\dfrac{400}{10} + 10}{2} = 25$$

Now, we do it again and use 25 in place of the first guess value:

$$\text{Guess}_{(next)} = \frac{\dfrac{400}{25} + 25}{2} = 20.5$$

This is already pretty close, but once more gives

$$\text{Guess}_{(next)} = \frac{\dfrac{400}{20.5} + 20.5}{2} = 20.006$$

DSP can be used to achieve all known forms of modulation and demodulation. By changing the software, an AM detector can be converted to an FM detector. There are no coils or capaci-tors to adjust and no problems with component accuracy or components drifting with time and temperature.

As one last example, consider FM detection. When an audio signal modulates the frequency of a carrier, it is called frequency modulation (FM). Doing so also produces instantaneous changes in the phase of the carrier. Thus, Fig. 16-35 can be converted to an FM detector by computing the phase angle for each discrete sample of I_t and Q_t using

$$\phi_{(carrier)} = \tan^{-1}\left(\frac{Q_t}{I_t}\right)$$

The arc tan or $\tan^{-1}$ function is not built into DSP chips. However, it can be approximated by a series:

$$\tan^{-1}(x) = x - \frac{x^3}{3} + \frac{x^5}{5} - \frac{x^7}{7} + \frac{x^9}{9}$$

Depending on the application, the last term in the above series can be dropped, and the result will be accurate enough. However, this method still involves quite a few multiplications. Once again, you can see that decimation is impor-tant since it reduces the sampling rate and al-lows more time for calculations. Or look-up tables can be stored in the DSP memory to speed things along. The last part of FM de-modulation is to subtract adjacent angular values and send them to the DAC. This is done because angular change from sample to sample is proportional to the original audio modulating signal:

$$\text{Audio} = \phi_{(\text{prior sample})} - \phi_{(\text{current sample})}$$

where ϕ is the phase angle of the discrete signal sample.

EXAMPLE 16-19

Use the above series to find the arc tan of 0.5. Applying the equation,

$$\tan^{-1}(0.5) = 0.5 - \frac{0.5^3}{3} + \frac{0.5^5}{5} - \frac{0.5^7}{7} + \frac{0.5^9}{9}$$

$$= 0.463684275 \text{ radians } (26.567°)$$

Figure 16-36 shows a radio transceiver that makes significant use of DSP. The Flex 6700 uses both field programmable gate array (FPGA) and DSP integrated circuits. An FPGA can be viewed as a "sea of gates." A hardware description language such as Verilog can be used to implement FIR filters, FFTs, and other DSP applications. The block diagram in Fig. 16-36 shows that the FPGA comes before the DSP for signals entering the system from the antennas. An FPGA takes advantage of a parallel or "pipelined" architecture to speed up signal processing. Incoming signals are converted to digital (by the ADC blocks) and are processed by the FPGA before moving on to the DaVinci™ DSP processor. The Flex 6700 offers up to eight separate but simultaneous communication bands, as shown in the screen display in Fig. 16-37.

DSP also serves well in motion control. Phase compensation of a servo system was presented in Chap. 14. It was shown there that motors act as lag networks and that lead networks can be used to compensate for this. Control-loop compensation allows more accurate positioning and speed control and limits the amount of overshoot and undershoot in a motion system. A loop can be compensated using DSP instead of RC networks and op amps. There are several performance advantages with the digital method, and it is being used more in new designs.

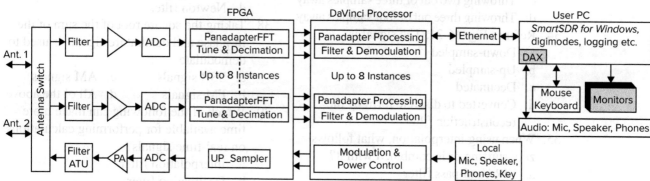

Fig. 16-36 Flex Radio 6700 transceiver.

Courtesy of FlexRadio Systems

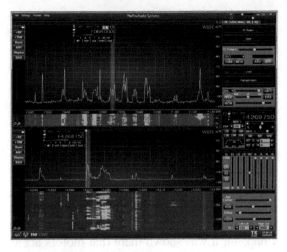

Fig. 16-37 Screen display for Flex 6700.

Courtesy of FlexRadio Systems

Adaptive software

A common problem with motion-control systems is that loop compensation is optimum for only one set of load conditions. This can be a serious problem because the physical or mechanical load varies widely in many practical applications. Consider elevators. They should accelerate and decelerate smoothly and position accurately whether they are empty or full. Have you ever noticed that a crowded elevator might stop with a bump or even overshoot a little? DSP systems can use *adaptive software* that adjusts filter coefficients (to provide varying phase compensation) on the fly.

DSP chip manufacturers often package a DSP core onto a chip along with peripheral circuits for a given application. Look at Fig. 16-38, which shows the functional block diagram for Analog Devices ADMC401 single-chip DSP-based high-performance motor controller. It provides the pulse-width modulation generators needed to control a motor. It also contains an encoder interface (for sensing position or velocity) and an eight-channel analog-to-digital converter. This single chip provides almost all the electronics needed for many motion-control applications. It's easy to see why DSP is becoming so popular.

Self-Test

Choose the letter that best answers each question.

32. Multirate systems are those that use
   a. Decimation
   b. Interpolation
   c. More than one sampling frequency
   d. All of the above

33. Increasing the sampling frequency from 50 to 150 kHz would involve
   a. Inserting two zeros between every sample
   b. Inserting three zeros between every sample
   c. Throwing two out of three samples away
   d. Throwing three out of four samples away

34. Refer to Fig. 16-31. The signal is
   a. Down-sampled
   b. Up-sampled
   c. Decimated
   d. Converted to discrete form by the reconstruction filter

35. When using interpolation, what follows zero stuffing or is combined with it?
   a. FIR band-pass filter
   b. FIR band-stop filter
   c. FIR low-pass filter
   d. FIR high-pass filter

36. When signals have two components and one of them is phase shifted by 90 degrees, they are called
   a. Sine and sinc
   b. Cosine and sinc
   c. Upper and lower
   d. In-phase and quadrature

37. Which of the following produces a 90-degree phase shift for all the frequencies within its operating range?
   a. Hilbert filter
   b. Interpolation filter
   c. Decimation filter
   d. Newton filter

38. Taking the square root of the sum of the in-phase and quadrature signals is used to demodulate
   a. SSB signals   c. AM signals
   b. FM signals   d. All of the above

39. Which of the following can make more time available for performing calculations on real-time signals?
   a. Interpolation filter
   b. Decimation filter
   c. Hilbert filter
   d. All of the above

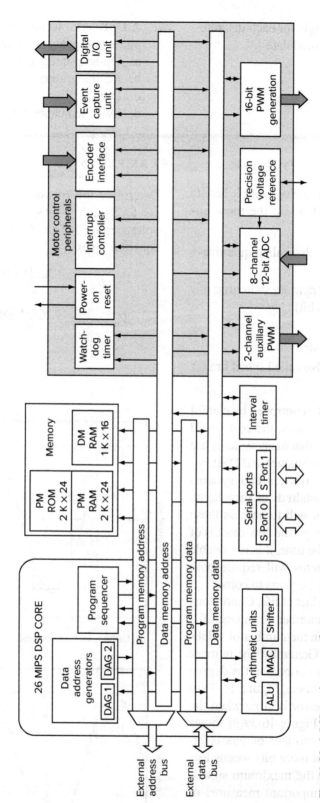

FIGURE 16-38 A DSP motion-control processor.

40. Finding the phase angle for each discrete sample is used to demodulate
    a. SSB signals
    b. FM signals
    c. AM signals
    d. All of the above

41. A DSP servo would most likely provide phase compensation using
    a. Op-amp integrators
    b. *RC* integrators
    c. Digital filters
    d. Switched capacitor differentiators

16-6 Limitations of DSP

DSP is a very capable technology, but it has its own set of limitations. Here are the most important ones:

- Quantization error (also called quantization noise)
- Idle channel noise (random switching of the least significant bit)
- Aliasing
- Limited dynamic range
- Frequency limits (also called speed limits)
- Clock feedthrough

Most of the above can be controlled by careful design. However, in some cases, the cost would be prohibitive. This means that many systems use a combination of analog techniques and DSP. For example, radio-frequency devices and systems often operate at tens or hundreds of megahertz. DSP chips are not fast enough to process these signals in real time. Although the new models of DSP chips get faster all the time, it is reasonable to expect that many systems will require both types of signal processing for years to come.

Quantization error is a fact of life. Continuous signals have an infinite number of values. Discrete signals are limited in the number of values that can be represented. Generally, the number of bits determines the resolution and the degree of quantization error or noise. Figure 16-39(*a*) shows that quantization error is large when the number of bits is small. Figure 16-39(*b*) shows what happens when the number of bits is increased. It is obvious that more bits means less error. Noise voltage and the maximum signal-to-noise ratio are both important measures of quantization error. These are calculated with

$$V_{noise(rms)} = V_{full\text{-}scale} \cdot 0.28 \frac{9}{2^n}$$

Maximum signal-to-noise ratio$_{(dB)}$ = $6.02n + 1.76$

where n = the number of bits in both of the above equations.

EXAMPLE 16-20

Determine the rms quantization noise voltage for 8- and 12-bit systems when the signal voltage range is from 0 to 5 V. Applying the equation gives

$$V_{noise(rms)} = \frac{5\text{ V} \cdot 0.289}{2^8} = 5.64\text{ mV}$$

$$V_{noise(rms)} = \frac{5\text{ V} \cdot 0.289}{2^{12}} = 353\ \mu\text{V}$$

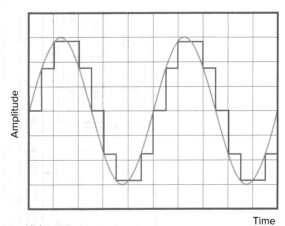

(*a*) Large quantization error

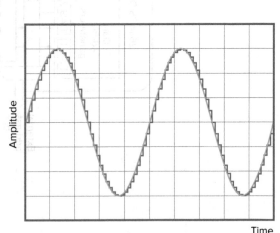

(*b*) Less quantization error

Fig. 16-39 Quantization error.

It is possible to reduce the noise level by increasing the number of bits. However, a large number of bits dictates higher cost. Also, high-resolution converters are slower. So, in the case of radio-frequency signals in the microvolt region, A/D conversion is not practical or even possible. One of the highest resolution applications is in the field of seismology. In this field, sensitive transducers convert the earth's vibrations into electrical signals. Often, 24-bit A/D converters are used. This is practical because the vibrations of interest occur at relatively low frequencies. The speed requirements of the A/D converters are much less than what is needed for audio or radio applications. Also, the processing of seismographic data often does not have to be done in real time. DSP is widely used in fields such as oil exploration and earthquake research.

EXAMPLE 16-21

Find the maximum signal-to-noise ratio for a 12-bit DSP system. Applying the equation,

$$\text{Signal-to-noise ratio} = 6.02 \cdot 12 + 1.76$$
$$= 74 \text{ dB}$$

The number of bits that a DSP chip can manipulate at one time can limit its performance for some applications. Today, there are two main types of DSP chips: 16- or 24-bit *fixed-point* processors and 32-bit *floating-point* processors. A fixed-point processor stores all numbers (including signal values and filter coefficients) as either 16- or 24-bit two's-complement integers. In two's complement, the positive values are stored as simple binary numbers, and negative numbers are stored in two's-complement form. All negative numbers have a sign bit of 1, and all positive numbers have a sign bit of 0. The sign bit is the one at the far left and is also called the most significant bit. Two's complement is formed by subtracting the value from zero. Several examples are shown in Table 16-2. Notice that the sign bit is 1 for the negative numbers. The two's-complement system is commonplace because only one hardware adder is required.

Table 16-2 Some Samples of 8-bit, Two's-Complement Numbers

Decimal Number	Two's-Complement Number
+127	01111111
+15	00001111
0	00000000
−1	11111111
−6	11111010
−128	10000000

Internally, intermediate values resulting from arithmetic operations are kept at 32- or 48-bit precision in fixed-point processors. Fixed-point devices are less expensive, faster, and usually have fewer external pins. It can be more difficult to develop software for them. Since the cost of software development is spread over the number of products, fixed-point devices are less costly when volume is high.

The typical 32-bit floating-point DSP chip stores numbers using a 24-bit mantissa and an 8-bit exponent. You might recall that in a number such as 3.56×10^6, 3.56 is the mantissa and 6 is the exponent. So, the resolution is only 24 bits, and quantization error can still be a limitation in areas like professional-quality audio. Table 16-3 shows some general application areas for the two main types of DSP chips. An "X" means that chip type is often a better choice. Remember, this table is general and exceptions exist.

Overflow errors, truncation, and rounding are all possible and can limit the performance of DSP systems. Software simulators can be used to verify that such errors do not occur or that they are not serious enough to cause faulty operation. In the case of IIR filters, such errors can even cause instability.

You might be wondering about high-resolution DSP systems. Seismography has been mentioned before. Other examples are medical imaging and astronomy. Luckily, these areas often don't require real-time computing. The data are processed in large computers where 64-bit resolution is commonplace. Even if a single image takes several minutes of processing time, the results can more than justify the wait.

Fixed point

Floating point

Table 16-3 Fixed versus Floating Point

Application Area	Fixed-Point DSP	Floating-Point DSP
High-volume products	×	
Adaptive systems		×
Cost-sensitive products	×	
Large dynamic range of signals		×
Image processing		×
Short design time a must		×
Simple and straightforward designs	×	
High-speed applications	×	

Self-Test

Choose the letter that best answers each question.

42. What is the rms noise voltage for a 10-bit system with a signal range of 0 to 3 V?
    a. 0.847 mV
    b. 1.19 mV
    c. 3.76 mV
    d. None of the above

43. What is the maximum signal-to-noise ratio for the system in question 42?
    a. 38 dB
    b. 44 dB
    c. 53 dB
    d. 62 dB

44. Which of the following would represent a negative number in a DSP system that uses two's-complement representation?
    a. 0101000101000111
    b. 1001111100010101
    c. 0001100111111011
    d. 0010010010011111

45. DSP chips that use the type of numbers shown in question 44 are called
    a. Fixed-point processors
    b. Floating-point processors
    c. Adaptive processors
    d. None of the above

46. DSP chips that store numbers as a 24-bit mantissa and an 8-bit exponent are called
    a. Fixed-point processors
    b. Floating-point processors
    c. Adaptive processors
    d. None of the above

47. Which of the following DSP application areas would not have to operate in real time?
    a. Audio filtering in home entertainment systems
    b. Cellular phone systems
    c. Motion control
    d. Medical imaging

16-7 DSP Troubleshooting

Embedded systems

Many DSP applications fit into the category called *embedded systems*. Embedded systems are those where hardware and software are integrated into one chip or product. The operation of an embedded system is controlled by a program that is stored in read-only memory (ROM). The software program is often called **Firmware** *firmware*. The ROM is sometimes found inside the processor chip and sometimes in a separate

chip. Most embedded systems are specialized computers and do not need operating systems. However, some can be upgraded in the field by loading new software into programmable read-only memory. Much of the information presented in this section can be applied to troubleshooting embedded systems in general.

Working on embedded systems requires a delicate touch. A good technician tries to *do no harm*. The clearance between pins on many surface-mount chips is only a few thousandths of an inch. Even when the probe being used is needle sharp, a slip will contact adjacent pins and short them together. The chip could be damaged or blown when this happens. Technicians look for other points on circuit boards to do their probing: perhaps a trace, a pad, a via, a resistor lead, a connector pin, and so on. Special tools and probes are available for safely working on high-density boards and are a must. Sometimes the boards are designed with field service in mind, and test points and connectors are provided.

Some DSP chips operate at speeds approaching 1 GHz. Even 50 MHz is a radio frequency. What this means is that a test lead is only a dc ground. A test lead looks like an impedance at high frequencies; everything is there—resistance, capacitance, and inductance. For example, if a 6-inch test lead is used to ground some point in a high-speed circuit, an oscilloscope might show only a shift in the dc level at that point. The ac signal might not be all that different! Another feature of high-speed circuits is that the waveforms viewed on an oscilloscope look rather strange when compared with the nice rectangular waveforms shown in books. Also, a waveform can appear at a point even when there is no device directly driving that point. Wires and traces act like antennas. As frequency goes up, they work better as antennas. Just bringing an oscilloscope probe near a high-frequency circuit will usually cause signals to show on the screen. One last point: An oscilloscope probe is never quite grounded. With higher frequencies and longer ground leads, a probe is even less grounded. So, the waveform viewed is often the sum of several signals: the one you want plus several others picked up by the ground lead. When things don't look quite right, try a shorter ground lead or a different ground point. You will often be amazed at how much effect this can have.

There is more than one kind of troubleshooting. If a design technician is working on a prototype, then the list of possible problems is long. The circuit board design might be flawed, the wrong chips could be installed, parts can be inserted backward, the software could be the wrong version, the software could be buggy, and so on. Design technicians must now be wary of radio-frequency interference (RFI). As clock speeds increase, RFI problems abound. Printed circuit layout techniques that were adequate in slower designs become unworkable. For example, a circuit layout that worked well with a 50-kHz, 12-bit A/D converter could reduce the performance of a 500-kHz, 16-bit converter to the equivalent of 10 bits.

If a field repair technician is working on a system that worked yesterday, then the list of possible problems is shorter. However, if someone "updated" the system yesterday, then the list is longer. Perhaps the software was updated to the wrong version! There is little sense in troubleshooting a system without first learning its recent history. Why spend hours or days tracking down a problem that could have been identified by asking a few basic questions? If a system was serviced recently, there is a possibility that something did not get put back together properly. Because of RFI, it is important that every screw, shield, and connector be in place and securely fastened.

Some DSP systems are dependent on numerical values stored in nonvolatile memory. This kind of memory retains its contents even if the power is removed. However, the word *nonvolatile* does not guarantee that the contents cannot become corrupted. When this happens, a system might not even start. Technicians need to be clear on what tools are at their disposal. Sometimes, software or firmware can be crossloaded to units to achieve an update or fix a bug.

This section assumes that you have already read Chap. 10. Much of what was presented there also applies to troubleshooting embedded systems. The material on electrostatic discharge (ESD) is a must, as is the safety material. When servicing a high-energy system, like a large machine or motor controller, please remember that a possibility exists for sudden and unexpected motion. Personnel could be harmed or expensive equipment could be damaged. Troubleshooting such a system requires experience and special knowledge, and in some cases, the *high-energy sections should be powered off!*

A lab notebook is a necessity for design technicians working on prototypes. Technicians sketch test configurations and waveforms, record version numbers (including the software), and make notes about ideas and where the process stopped at quitting time. Long weekends can erase one's memory. Also, teamwork might be required, and a team member could be off sick. Once something has been verified, it should be noted. Going around in circles is a waste of time. A sketch of the block diagram is often just as useful as the schematic. Signal analysis from section to section will verify which parts are working, which seem not to be working, and which seem to be doing something but not what is expected. Recording the signals on the block diagram will often point a technician in the right direction.

The test equipment to be used varies according to the situation. Design technicians sometimes use logic analyzers. These allow viewing many signals at the same time. Embedded systems have busses: address, data, and control. Connecting a logic analyzer to the busses is usually difficult and time-consuming, since 30 or more signals might be involved. Logic analyzers are seldom used in field service work for that reason. If they are, usually a special *test pod* allows rapid and easy connection to the busses. However, this feature has to be designed into the system in the form of a special connector and is not the norm. Boundary scan troubleshooting is becoming more prevalent. This concept was explained in Chap. 10. These products have a special connector, sometimes called the JTAG port, which can be used for troubleshooting in the field.

Test pod

Emulators

More and more of the signals transmitted in a DSP system are found only in digital form, represented as sequences of binary samples rather than the analog signals experienced technicians are used to probing and analyzing. "Old hands" at analog troubleshooting may become frustrated when trying to make sense of digital signals in newer DSP systems even if they are able to access the signals' binary samples with a logic analyzer. Except perhaps for simple sine waves, digital signals just don't make sense when viewed as tables of numbers or timing diagrams on a logic analyzer.

An interesting new technique proposed for reconstructing signals inside a DSP may soon provide a solution to this problem. Edwin Suominen, a former technician and DSP design engineer, invented this system, described in U.S. Patent 6,052,748. Using the system, technicians will be able to convert digital signals floating around inside a DSP into analog signals that can be viewed with familiar analog test equipment like oscilloscopes and spectrum analyzers.

Technicians having access to the new reconstruction technique will be able to select particular binary samples and reconstruct them into an equivalent analog signal without needing to know anything about the sample rate. A buffering system automatically controls the reconstruction to ensure that the analog signal will be faithfully represented even if the samples appear in bursts or at irregular intervals. If an external bus is available, the samples can be selected from it without any special DSP code by "sniffing" for a particular bus address with a logic analyzer and triggering the reconstruction device with the logic analyzer's trigger output. Alternatively, the technician can select samples from a properly configured boundary scan port.

If the DSP system is designed for test (and engineers are starting to think more about this), I/O pins of the DSP may be dedicated to a probe port (serial or parallel) from which samples of selected internal signals can be extracted and reconstructed. The system may be designed so that the DSP continuously spits out multiplexed samples of important internal signals, or it may allow the technician to select samples appearing at particular registers or addresses for reconstruction.

Emulators might be used for troubleshooting and debugging during a product's initial design phase. An emulator is a separate computer that runs the DSP software that is under development. A special cable connects the emulator to the system under development (called the *target system*). This allows efficient software debugging while exercising the software in an environment that is the same or similar to the final product. Emulators are not normally used in field servicing.

Technicians should store setup configurations for complex instruments, such as logic analyzers and oscilloscopes. Some of these instruments have floppy disks for this purpose. Log the name and number of each program in your notebook, along with a brief description of its purpose. Assign a unique name or code and record it on the diskette and in the notebook. Some oscilloscopes can store waveforms and transfer them to a computer. Doing so often saves a lot of time. As systems become more complicated, human memory is less likely to serve well. Documenting progress and procedures, even in field servicing, is becoming a necessity for many technicians.

Assume nothing when troubleshooting. Every power connection should be verified with an oscilloscope (meters don't show ripple unless it's really bad). Every ground connection should be verified the same way. It's not unusual to find 10 or more pins grounded on a DSP chip. Verify every one of them. The trace on the scope should go to 0 V dc with no noise or an acceptable level of noise. If there is too much noise, make sure the ground lead of the oscilloscope probe is as short as possible and connected properly. Some DSP chips use 3.3-V and 1.6-V supply voltages. The 3.3-V supply powers the parts of the chip that communicate with external circuits. The 1.6-V supply powers the internal logic core. Make sure that both supply voltages are clean and correct.

Surface mount technology and fine-pitch circuit boards have increased the ratio of assembly faults to component faults. Some industry practitioners report 10 assembly faults for each component fault. Figure 16-40 shows a chip on a board. The solder joints are under the chip's pins and are thus hidden from view. Even with magnification and careful inspection, it's often not possible to spot a defective solder joint. One method is to apply pressure directly to the pins using an insulator to force electrical contact between any unsoldered pin and its trace.

Is there a clock signal? Is it clean and is the frequency correct? If the DSP chip creates its own signal using an external crystal, be advised that an oscilloscope probe can load the circuit and kill the oscillations. In those cases, look for a buffered clock output or perhaps a strobe signal or control signal that is derived from the main clock. The address and data bus lines will normally display rectangular waveforms when the clock is running.

Know the limits of your equipment. An oscilloscope with a 100-MHz bandwidth will display a 50-MHz rectangular wave as a sine wave. You should understand why because of what was presented earlier in this chapter. Another point: Digital scopes are subject to aliasing. If they sample below the Nyquist limit, a 50-MHz clock signal could look like a 1-kHz signal. Never forget that test equipment is valuable because it provides information. However, if the information is false, then *it is worse than no information at all to the unsuspecting.*

Embedded systems are digital systems. There are only two states to worry about, right? Wrong. There are three: logic high, logic low, and high impedance (also called tri-state). When a circuit point is not supposed to be high impedance, it is said to be *floating.* Other inputs connected to a floating output will also be floating. Bad solder joints, dirty or loose connectors, defective sockets, blown outputs, and so on cause floating outputs and lines. When technicians discover erratic operation, sometimes they can affect system behavior just by passing their fingers over parts of the circuit board. Floating lines act like antennas!

There is a legitimate reason for some pins and lines to be tri-stated. This is so that more than one device can control a bus. For example, suppose three devices (A, B, and C) have outputs connected to the same bus line. When devices A and B are idle (tri-stated), then device C can pull the line high or low as needed. This is normal and is not the same as a floating line caused by a bad solder joint or a blown IC. Shorts can also cause floating levels because two outputs can fight for control of a bus line.

Unfortunately, oscilloscopes won't always identify a tri-state or floating condition. One possibility is to use a 1-kΩ resistor and the oscilloscope at the same time. Connect the scope probe to the pin or line, and then use the resistor to force the line low (resistor to ground) and then high (resistor to the supply). If the

Floating

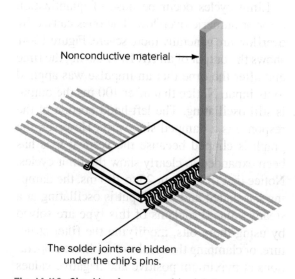

Nonconductive material →

The solder joints are hidden under the chip's pins.

Fig. 16-40 Checking for open solder joints.

scope shows any significant change in level, then that line is high impedance (floating or tri-stated). If it's not supposed to be, the problem (or at least one of them) has been found.

For most DSP chips, the *minimum logic high* voltage is 1.8 to 2.2 V. The *maximum logic low* input is 0.8 V. So, any dc logic level between 0.8 and 1.8 V is bad and indicates a fault. Generally, logic low voltages are 0.2 V or lower, and logic high voltages are 3 V or greater.

To save time, troubleshooters sometimes have their oscilloscopes set to 2 V/division. This is a compromise for viewing two waveforms. Unfortunately, they might miss an invalid logic level with that resolution. Look closely and switch to 1 V/division if a logic low looks a little too high or if a logic high looks a little too low. Remember, levels between 0.8 and 1.8 V are bad and almost always indicate a fault.

What about logic probes and pulsers? They can provide a lot of information and are easy to use. In field service work, sometimes the correct conclusion can be reached simply by verifying where there is activity (a pulsing probe light) and where there is not.

Don't forget that the embedded software has to begin operation at the correct address (the beginning of the program) when power is applied. Without a valid *reset signal*, the system might not function at all or might behave in a strange or erratic manner. The reset signal makes the embedded system start up from the beginning of the program. Sometimes the power-on reset pulse is missing due to a bad capacitor or transistor. In some embedded systems, the reset pulse is supposed to appear when the system is plugged in and not when it is turned on. In these systems, the processor is always powered when the system is plugged in (connected to the ac line). A technician can force a system reset or a chip reset by pulsing the reset bus or a reset pin. In the case of a control bus where the reset signal is common to several chips, it's a good idea to use a logic probe in conjunction with the pulser to verify that the reset pulse appears on all of the chips in the system.

In the case of a simple DSP filter, it might be possible to get a "look" at the filter's impulse response by pulsing the input and looking at the output with an oscilloscope. A logic pulser will create an input impulse that will be convolved with the FIR filter coefficients. This should

show up on the oscilloscope as a smooth curve with the scope probe connected after the reconstruction filter. It might be possible to recognize the shape of the impulse response and identify the type of filter. One thing to remember is that any input filter, such as the antialias circuit, will stretch the pulse so the output will not be exactly proportional to the coefficients stored in the processor. In the case of an IIR filter, the response will be a damped sine wave, as presented earlier in this chapter.

The impulse response test is a simple technique that might not be appropriate for more complex DSP systems. More sophisticated analysis of internal DSP signals and filter impulse responses will be practical with availability of the digital signal reconstruction technique discussed earlier.

In the case of noise, remember that some noise is present in all electronic systems. In DSP systems you can expect

- Quantization noise
- Idle channel noise
- Clock feedthrough

In the case of a new design (prototype), there is a possibility of an unwanted output due to another factor. When the input to a DSP becomes constant at zero, the output is expected to eventually settle to zero. In some cases, this does not happen and the output oscillates. These undesired oscillations are known as *limit cycles*. Limit cycles can appear in IIR filters because they use feedback. They cannot happen in FIR filters.

Limit cycles occur because of quantization error or numeric overflow. The ones caused by overflow are generally more severe. Figure 16-41 shows the output of an IIR filter during the time and after the time that an impulse was applied to its input. Notice that after 100 ms the output is still oscillating. The left-hand portion of the response is a damped sine wave. However, the graph is clipped because the vertical axis has been expanded to clearly show the limit cycles. Notice that from about 25 to 100 ms, the damping has ended and the output is oscillating at a steady value. Problems of this type are solved by using more bits, modifying the filter structure, or clamping the results of arithmetic operations at maximum positive and negative values (to control overflow).

Reset signal

Limit cycles

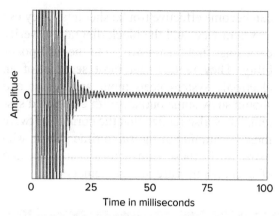

Fig. 16-41 Limit cycles.

When the noise is excessive in a system that was previously working well, there is a possibility that a bypass capacitor is open, or a shield is missing or not secure. Don't forget to check input cables and connectors. Perhaps a ground connection has failed. Also, don't forget that test leads can act as antennas and introduce noise into a system. The noise might not be originating in the digital section. Check the input signal and its connections. Also, a preamplifier might be at fault. Some important questions to ask include the following:

- Is the noise noticeable only when there is no input signal?
- Does the noise stop when the input signal is removed?
- Does the noise stop when the preamplifier is powered off?

- Is the noise random and evenly distributed across the frequency range?
- Is the noise mostly at one or two frequencies?

A technician will have to know or learn enough about the system to proceed with the verification process. For example, a front-panel switch on an instrument or a limit switch on a motor control might be connected to an *interrupt input* on the processor. Interrupt inputs force the processor to suspend what it was doing and jump to a special place in the program memory where the interrupt service routine is stored. So, the technician will need to know how the inputs are connected and what is supposed to happen when an input is activated. The technician might also need to know how the software works, at least in general terms. For example, perhaps an interrupt service routine is supposed to read a position by accessing an A/D converter. In this case, a strobe signal should appear at the converter a millisecond or two following the interrupt signal. Depending on the system fault, a logic probe might provide enough information. However, it's also possible that the time delay between the interrupt pulse and the strobe could be critical. In this case, the oscilloscope can be set up to trigger on the interrupt so that the delay time until the strobe appears can be measured. Does the embedded processor have an interrupt acknowledge output? This too is a good place to look to

Interrupt input

find out what is happening—or what is not happening when it should. Some problems can be diagnosed with a counter. For example, counting interrupts and interrupt acknowledges can provide important information about a system.

Embedded systems can be difficult to troubleshoot, but it is possible. Technicians who know basic theory, know how to use test equipment, know how the system is supposed to work, and have the necessary mechanical skills can become effective troubleshooters. They use a system view and the block diagram to verify each stage and then move on to the next possibility. They realize that most sections of an electronic system need input and power and should show some output. With embedded systems, they also realize that the correct software must be running at the same time. Those who learn to do these things in a consistent and efficient manner are in high demand.

Self-Test

Choose the letter that best answers each question.

48. The control program for an embedded system is most likely stored in/on a
    a. USB drive
    b. Hard drive
    c. CD drive
    d. Read-only memory

49. When probing signals in a high-frequency DSP system, the ground lead on an oscilloscope should be
    a. Disconnected
    b. As short as possible
    c. As long as possible
    d. Replaced with a resistor

50. What is the name of a design and development tool that runs the software being tested while connected to the target system?
    a. Emulator
    b. Logic analyzer
    c. Oscilloscope
    d. JTAG

51. What will the display look like when an oscilloscope with a 100-MHz bandwidth is used to view a 75-MHz square wave?
    a. Square but with Gibb's effect
    b. Low-duty cycle rectangular
    c. High-duty cycle rectangular
    d. Sine

52. A logic level between 0.8 and 1.8 volts in a 5 volt system is:
    a. Low
    b. High
    c. Floating
    d. None of the above

53. Floating levels can be caused by
    a. Opens
    b. Shorts
    c. Blown chips
    d. All of the above

54. What causes an embedded processor to jump to the beginning of program memory?
    a. Reset signal
    b. Ramp up of V_{CC}
    c. Ramp down of V_{CC}
    d. All of the above

55. What type of signal causes a processor to stop what it is doing and jump to an event-handling routine that is stored in memory?
    a. Address signal
    b. Data signal
    c. JTAG signal
    d. Interrupt signal

Chapter 16 Summary and Review

Summary

1. In DSP terms, analog signals are often called continuous, and digital signals are called discrete.
2. The number of bits determines the resolution of a digital signal.
3. Quantization is the process of converting from continuous to discrete.
4. The rate of quantization is called the sampling frequency.
5. Signals higher than half the sampling frequency will be aliased into the frequency range of interest. These must be attenuated with an antialiasing filter before quantization.
6. The core DSP operation is multiplying the discrete time samples by coefficients and accumulating the products (MAC). The formal name for this process is convolution.
7. A reconstruction filter or anti-imaging filter follows the output of the DAC in DSP systems.
8. The symbol * represents convolution.
9. One of DSP's strongest points is that software controls what happens. Software is very easy to change compared with hardware.
10. Adaptive DSP systems change their operation on the fly.
11. Sine waves, square waves, and triangle waves are examples of periodic functions.
12. The lowest frequency in a Fourier series is called the fundamental, or the first harmonic.
13. Any signal can be viewed from two viewpoints: the time domain and the frequency domain.
14. Fourier synthesis of periodic waveforms with discontinuities (such as a square wave) produces distortion due to Gibb's phenomenon.
15. Ideal rectangular waves (zero rise and fall times) don't exist since they would require an infinite bandwidth.
16. The Fourier transform can be used to convert from the time domain to the frequency domain, and the inverse Fourier transform converts from the frequency domain to the time domain.
17. Shannon's sampling theorem states that the lowest sampling frequency that can be used to represent any signal is two times the bandwidth of the signal. In practice, signals are sampled at three times their bandwidth or higher.
18. Using a higher sampling frequency eases the requirements for the antialiasing filter.
19. The process of quantization produces a signal approaching infinite bandwidth.
20. The order of an FIR filter is equal to the number of taps, or coefficients.
21. Sharper filters (small transition bandwidth) are realized by increasing filter order (more taps).
22. One-half the sampling frequency is sometimes called the Nyquist limit.
23. FIR means finite impulse response.
24. The impulse response of an FIR filter is the same as its coefficient set.
25. FIR filters mostly use symmetrical coefficients to obtain a linear phase response.
26. Filter ripple is due to Gibb's phenomenon caused by truncating the sinc function.
27. Filter ripple can be reduced by smoothing the coefficients using a window function.
28. IIR means infinite impulse response.
29. IIR filters use feedback and are sometimes called recursive filters.
30. IIR filters can be unstable (they can oscillate).
31. Band-pass filters can be realized by cascading low-pass and high-pass filters or by convolving their coefficient sets.
32. Band-stop filters can be realized by adding the outputs of low-pass and high-pass filters or by adding their coefficient sets.
33. Multirate DSP systems use more than one sampling rate.
34. The process of increasing the sampling rate of a discrete signal is called interpolation.
35. The process of decreasing the sampling rate of a discrete signal is called decimation.
36. Interpolation and decimation can relax the requirements for the analog portions of a system.

37. When a signal has two components: in-phase and quadrature, the phase relationship between the two is 90 degrees.
38. A Hilbert filter provides a 90-degree phase shift for all frequencies within its passband.
39. Modulation and demodulation can be accomplished with DSP.
40. Quantization noise can be reduced by increasing the number of bits.
41. High-resolution digital systems require lots of bits. This costs more, and they tend to be slower.
42. High-resolution digital systems may not have to operate in real time.
43. DSP chips are available in two forms: fixed-point and floating-point.
44. In fixed-point chips, numbers are represented in two's-complement form.
45. In floating-point chips, numbers are represented as a mantissa and an exponent.
46. Embedded systems integrate hardware and software into one chip or product.
47. In an embedded system, the software is often called firmware and is stored in read-only memory (ROM).

48. The ground lead of an oscilloscope probe should be as short as possible when working on high-speed digital systems.
49. DSP systems might not work due to software problems and memory contents that have become corrupted.
50. Logic analyzers are instruments that allow many waveforms to be viewed simultaneously.
51. Emulators are used mostly during the design phase of a product and allow the software to be exercised on the target system.
52. Some DSP systems have boundary scan ports.
53. Both power and ground circuit points should be verified.
54. Fine-pitch circuit boards and surface-mount technology have increased the ratio of assembly faults to component faults.
55. Digital signals in the range of 0.8 to 1.8 V are not valid and usually indicate a fault.
56. DSP systems need a reset signal so that operation will begin at the correct point in the firmware.
57. Verifying interrupt signals can be a good troubleshooting technique.

Chapter Review Problems

16-1. Which of the following terms would be used to describe a signal that is restricted to 256 voltage levels? (16-1)
 a. Analog
 b. Linear
 c. Discrete
 d. None of the above
16-2. The resolution of a digital signal can be increased by (16-1)
 a. Increasing the number of bits
 b. Decreasing the number of bits
 c. Decreasing the sampling frequency
 d. Decimating the signal
16-3. The process of converting from the continuous time domain to the discrete time domain is called (16-1)
 a. Sampling
 b. Quantization
 c. A/D conversion
 d. All of the above

16-4. The filter that prevents signals above half the sampling frequency from showing up as signals in the desired passband is (16-1)
 a. The reconstruction filter
 b. The anti-imaging filter
 c. The antialiasing filter
 d. A high-pass type
16-5. MAC stands for (16-1)
 a. Multiplex and commutate
 b. Modulate and communicate
 c. Multiply and accumulate
 d. Mark all capacitors
16-6. DSP is favored over analog in many cases because it (16-1)
 a. Eliminates the need for precision components
 b. Eliminates the effects of temperature and component aging
 c. Is controlled by software
 d. All of the above

16-7. The formal name for MAC is (16-2)
  a. Convolution
  b. Filtering
  c. Averaging
  d. None of the above

16-8. The symbol for convolution is (16-2)
  a. ·
  b. X
  c. *
  d. ⊕

16-9. What's the easiest way to change a DSP low-pass filter to a high-pass filter? (16-2)
  a. Reverse the position of the capacitors and resistors
  b. Change the software
  c. Select a new DSP chip
  d. Replace convolution with multiplication

16-10. During filtering, the incoming discrete signal samples are convolved with (16-2)
  a. The coefficients
  b. The sum of all samples
  c. The difference of all samples
  d. The phase difference

16-11. DSP systems that automatically adjust to changing conditions are called (16-2)
  a. Moving-average filters
  b. Boxcar filters
  c. Negative coefficients
  d. Adaptive

16-12. Which of the following waveforms would not be considered periodic? (16-3)
  a. Human speech
  b. Sine
  c. Cosine
  d. Square

16-13. What is the amplitude of the eighth harmonic of a 1-kHz square wave? (16-3)
  a. Zero
  b. One-eighth of the amplitude of the 1-kHz component
  c. One 16th of 1 kHz
  d. None of the above

16-14. Which of the following instruments displays signals in the frequency domain? (16-3)
  a. Oscilloscope
  b. Spectrum analyzer

  c. DMM
  d. Logic analyzer

16-15. In Fourier synthesis, Gibb's phenomenon is caused by (16-3)
  a. The amplitude being too high
  b. The fundamental frequency being too high
  c. Discontinuities
  d. Using cosine waves instead of sine waves

16-16. Converting from the time domain to the frequency domain is accomplished with the (16-3)
  a. Laplace transform
  b. Fourier transform
  c. Use of logarithms
  d. Use of inverse logarithms

16-17. The lowest-possible sampling frequency that can be used to represent a signal is equal to twice the bandwidth of that signal. This rule is called (16-3)
  a. Fourier's theorem
  b. Newton's theorem
  c. Relativity
  d. Shannon's theorem

16-18. The process of sampling or quantization produces a signal having a bandwidth that approaches (16-3)
  a. Half of the sampling frequency
  b. The sampling frequency
  c. Two times the sampling frequency
  d. Infinity

16-19. What is another name for a tap in a digital filter? (16-4)
  a. Adder
  b. Subtractor
  c. Coefficient
  d. IDFT

16-20. The sampling frequency divided by two can be called the (16-4)
  a. Fourier limit
  b. Laplace limit
  c. Newton limit
  d. Nyquist limit

16-21. The fact that the processes of both AM and quantization produce upper and lower sidebands that are mirror images of each other demonstrates the concept of (16-4)
  a. Negative frequencies
  b. Gibb's phenomenon

c. Nyquist's theorem
d. All of the above
16-22. FIR filters are usually designed with symmetrical coefficients to achieve (16-4)
a. Less passband ripple
b. Less stopband ripple
c. Linear phase response in the passband and transition band
d. Smaller transition bandwidth
16-23. The impulse response of an FIR filter is a picture of its (16-4)
a. Coefficients
b. Adder gain
c. MAC gain
d. Sin(x)/cos(x) function
16-24. In general, to make the transition bandwidth smaller, one must (16-4)
a. Use only IIR filters
b. Increase filter order
c. Give up a linear phase response
d. Accept a lot of ripple in the passband
16-25. Ripple is reduced by smoothing the ends of the truncated sinc function by using a(n) (16-4)
a. Analog prefilter
b. Analog postfilter
c. Window function
d. All of the above
16-26. IIR filters are also called (16-4)
a. Recursive filters
b. Hilbert filters
c. Newton filters
d. High-order filters
16-27. Convolving the coefficient set of two digital filters is equivalent to (16-4)
a. Adding the outputs of the two filters
b. Subtracting the outputs of the two filters
c. Dividing the outputs of the two filters
d. Cascading the two filters
16-28. Up-sampling is (16-5)
a. Another name for interpolation
b. Used to increase the sampling rate
c. Used to relax analog filter requirements in a DSP system
d. All of the above

16-29. Down-sampling is (16-5)
a. Another name for decimation
b. Used to decrease the sampling rate
c. Used to make more time available for processing
d. All of the above
16-30. Up-sampling by a factor of five would be achieved by (16-5)
a. Stuffing five zeros between each discrete sample
b. Stuffing four zeros between each discrete sample
c. Stuffing three zeros between each discrete sample
d. None of the above
16-31. Down-sampling by a factor of five would be achieved by (16-5)
a. Discarding every fifth sample
b. Keeping every fifth sample
c. Keeping every fourth sample
d. None of the above
16-32. If a signal in a system is considered in-phase, what would it be called after passing through a Hilbert filter? (16-5)
a. Quadrature
b. Inverted
c. Interpolated
d. Decimated
16-33. AM detection can be achieved with DSP by (16-5)
a. Summing I and Q
b. Squaring I and Q
c. Taking the square root of $I^2 + Q^2$
d. Finding the inverse tangent of $I^2 + Q^2$
16-34. Which of the following is not considered a limitation of DSP? (16-6)
a. Quantization noise
b. Limited dynamic range
c. Aliasing
d. Stability as the product ages
16-35. The signal-to-noise ratio of a DSP system can be improved by (16-6)
a. Increasing the number of bits
b. Decreasing the number of bits
c. Eliminating the antialiasing filter
d. Eliminating the anti-imaging filter

16-36. Fixed-point DSP chips are often better choices for (16-6)
  a. Adaptive systems
  b. Off-line (non-real-time) applications
  c. One-of-a-kind designs
  d. High-volume designs

16-37. Floating-point DSP chips store numbers (16-6)
  a. In straight binary format
  b. In two's-complement format
  c. In sign and magnitude format
  d. None of the above

16-38. Which of the following is not caused by having a limited number of bits to represent values? (16-6)
  a. Clock noise feedthrough
  b. Truncation errors
  c. Rounding errors
  d. Overflow errors

16-39. The term *firmware* is used to describe (16-7)
  a. Operating systems
  b. Programs stored on CDs
  c. Programs stored in ROM
  d. None of the above

16-40. Which of the following employees would normally face a larger range of possible troubleshooting problems and causes? (16-7)
  a. A technician who works on design prototypes
  b. A field service technician
  c. A production technician
  d. An installation technician

16-41. Which of the following is a good first step in troubleshooting? (16-7)
  a. Push on all of the ICs
  b. Remove all shields and covers
  c. Find out the recent history of the device or system
  d. Tighten all shields and covers

16-42. Which of the following explains why bringing a hand near a circuit can sometimes affect operation? (16-7)
  a. The heat from the hand makes a component drift.
  b. The shadow of the hand affects an optoisolator.
  c. The hand is emitting infrared radiation.
  d. A floating line is acting as an antenna.

16-43. If three devices (A, B, and C) share control of a bus signal, what is the required state of A and C when B takes control? (16-7)
  a. High impedance
  b. Low impedance
  c. Logic high
  d. Logic low

16-44. Which of the following static voltages at a digital test point spells trouble? (16-7)
  a. 0.1 V
  b. 1.2 V
  c. 3.0 V
  d. 3.5 V

16-45. Which signal commands an embedded processor to start operation at the beginning of the control program? (16-7)
  a. Interrupt
  b. Address-ready strobe
  c. Data-ready strobe
  d. Reset

16-46. A technician notes that there is excess noise and that it stops when the input signal is disconnected. This problem is due to (16-7)
  a. Idle channel noise
  b. Clock feedthrough
  c. Quantization noise
  d. None of the above

Critical Thinking Questions

16-1. This book has covered a variety of analog circuit functions such as gain, attenuation, clipping, and others. Can these be accomplished using DSP? How?

16-2. Suppose a multirate DSP system must change the sampling frequency by a factor of 1.5. How could this be done? (*Hint:* Combine interpolation and decimation.)

Critical Thinking Questions...continued

16-3. DSP is often used for echo cancellation in telephone systems. Why?

16-4. Op-amp integrators accumulate an input voltage over a period of time. Is there a way to do this using DSP?

16-5. Why might DSP offer the best solution to the problem of acoustics in enclosed spaces, where the sound changes according to the number of people in the enclosure?

16-6. When graphic artists and photographers use a computer to sharpen soft-looking images, are they using DSP? If so, can you think of how the process might work?

16-7. Can you think of a way to reverse the process posed in question 16-6?

16-8. Why can't a 100 percent digital, two-way radio be built for human speech?

Answers to Self-Tests

1. D	15. D	29. B	43. D
2. B	16. A	30. D	44. B
3. B	17. C	31. B	45. A
4. A	18. C	32. D	46. B
5. A	19. A	33. A	47. D
6. B	20. C	34. B	48. D
7. A	21. C	35. C	49. B
8. D	22. D	36. D	50. A
9. B	23. B	37. A	51. D
10. D	24. A	38. C	52. C
11. C	25. B	39. B	53. D
12. D	26. D	40. B	54. A
13. A	27. C	41. C	55. D
14. B	28. A	42. A	

Appendix A
Solder and the Soldering Process

From a Simple Task to an Art

Soldering is the process of joining two metals together by the use of a low-temperature melting alloy. Electronic soldering has become more demanding as surface mount technology and lead-free solders have become prevalent. More care is needed to achieve good results and high reliability. The importance of having high standards of workmanship cannot be overemphasized. Faulty solder joints remain a cause of equipment failure, and because of that, soldering is a critical skill.

The material contained in this appendix is designed to provide the student with both the fundamental knowledge and the practical skills needed to perform many of the high-reliability soldering operations encountered in today's electronics. Covered are the fundamentals of the soldering process, the proper selection of irons/tips/materials, and the use of the soldering station. Wave soldering and reflow soldering techniques are used in the manufacture of electronic equipment. This appendix focuses on rework soldering, which is usually a part of the repair process.

The key concept in this appendix is high-reliability soldering. Much of our present technology is vitally dependent on the reliability of countless, individual soldered connections. High-reliability soldering was developed in response to early failures with space equipment. Since then the concept and practice have spread into military and medical equipment. We have now come to expect it in everyday electronics as well.

The Advantages of Soldering

Soldering is the process of connecting two pieces of metal to form a reliable electrical path. Why solder them in the first place? The two pieces of metal could be put together with nuts and bolts, or some other kind of mechanical fastening. The disadvantages of these methods are threefold. First, the reliability of the connection cannot be ensured because of vibration and shock. Second, because oxidation and corrosion are continually occurring on the metal surfaces, electrical conductivity between the two surfaces would progressively decrease. A soldered connection does away with both these problems. There is no movement in the joint and no interfacing surfaces to oxidize. A continuous conductive path is formed, made possible by

the characteristics of the solder itself. Third, during manufacturing, hundreds or thousands of joints can be realized at the same time.

The Nature of Solder

Solder used in electronics is a low-temperature melting alloy made by combining various metals in different proportions. The most common types of solder were made from tin and lead. When the proportions are equal, it is known as 50/50 solder—50 percent tin and 50 percent lead. Similarly, 60/40 solder consists of 60 percent tin and 40 percent lead. The percentages are usually marked on the various types of solder available; sometimes only the tin percentage is shown. The chemical symbol for tin is Sn; thus Sn 63 indicates a solder that contains 63 percent tin.

Pure lead (Pb) has a melting point of 327°C (621°F), and pure tin has a melting point of 232°C (450°F). When they are combined into a 60/40 solder, the melting point drops to 190°C (374°F)—lower than either of the two metals alone. Today, lead-free solders are mandated for many manufacturing and repair procedures. Table A-1 shows both lead and lead-free alloys. All the alloys listed are available in wire form for repair work, and two are available in paste form. Paste solders are used in the reflow process for manufacturing printed circuits with surface-mount devices. Paste solder is also sometimes used in rework.

As listed in Table A-1, 60/40 solder begins to melt at 361°F (183°C) but is not fully melted until the temperature

Table A-1	Some Common Solders and Typical Temperatures		
Alloy	Melting Temperature	Work Temperature	Available in Paste Form
63% tin, 37% lead	361°F/183°C	680–770°F/ 360–410°C*	Yes
60% tin, 40% lead	361–374°F/ 183–190°C**	680–770°F/ 360–410°C*	No
96.5% tin, 3% silver, 0.5% copper	422–428°F/ 217–220°C**	698–788°F/ 370–420°C	Yes
96.5% tin, 3.5% silver	430°F/221°C	698–788°F/ 370–420°C	No

*As high as 842°F/450°C when soldering a copper ground plane.
**This alloy has a plastic range.

reaches 374°F (190°C). Between these two temperatures, the solder exists in a plastic (semiliquid) state—some, but not all, of the solder has melted. The same is true for the tin/silver/copper alloy shown in Table A-1. Two of the solder alloys have no plastic range. They transition directly from solid to liquid (or from liquid to solid) as they are heated (or cooled). These are called *eutectic* alloys or *eutectic* solders.

When using solders with a plastic range, it is especially important to avoid vibration or movement of the joint during the cool-down period. When this happens, the joint tends to have a dull, grainy appearance. Such joints are unreliable and are rejected by careful workers and quality inspectors. However, lead-free solder joints *are inherently dull and grainy*, as shown in Fig. A-1, where the two joints to the left are lead-free and the two to the right are leaded. Thus, workers and inspectors must learn new visual inspection guidelines for lead-free soldering.

In some situations, it is difficult to maintain a stable joint during cooling, for example, when wave soldering is used with a moving conveyor line of circuit boards during the manufacturing process. In other cases it may be necessary to use minimal heat to avoid damage to heat-sensitive components. In both these situations, eutectic solder is the preferred choice, since it changes from a liquid to a solid during cooling with no plastic range.

The Wetting Action

To someone watching the soldering process for the first time, it looks as though the solder simply sticks the metals together like a hot-melt glue, but what actually happens is

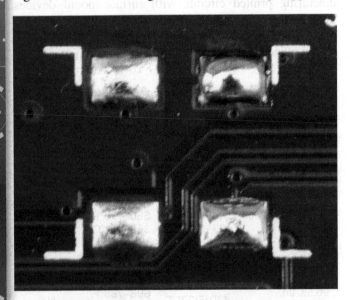

Fig. A-I Appearance of (*a*) a lead alloy solder joint and (*b*) a lead-free joint.

Charles A. Schuler

far different. A chemical reaction takes place when the hot solder comes into contact with the copper surface. The solder dissolves and penetrates the surface. The molecules of solder and copper blend together to form a new metal alloy, one that is part copper and part solder and has characteristics all its own. This reaction is called wetting and forms the intermetallic bond between the solder and copper.

Proper wetting can occur only if the surface of the copper is free of contamination and oxide films that form when the metal is exposed to air. Also, the solder and copper surfaces need to have reached the proper temperature. Even though the surface may look clean before soldering, there may still be a thin film of oxide covering it.

When solder is applied, it acts like a drop of water on an oily surface because the oxide coating prevents the solder from coming into contact with the copper. No reaction takes place, and the solder can easily be scraped off. For a good solder bond, surface oxides must be removed during the soldering process.

The Role of Flux

Reliable solder connections can be accomplished only on clean surfaces. Some sort of cleaning process is essential in achieving successful soldered connections, but in most cases it is insufficient. This is due to the extremely rapid rate at which oxides form on the surfaces of heated metals, thus creating oxide films that prevent proper soldering. To overcome these oxide films, it is necessary to utilize materials, called *fluxes*, which consist of natural or synthetic rosins and sometimes additives called activators.

It is the function of flux to remove surface oxides and keep them removed during the soldering operation. This is accomplished because the flux action is very corrosive at or near solder melt temperatures and accounts for the flux's ability to rapidly remove metal oxides. It is the fluxing action of removing oxides and carrying them away, as well as preventing the formation of new oxides, that allows the solder to form the desired intermetallic bond.

Flux must activate at a temperature lower than solder so that it can do its job prior to the solder flowing. It volatilizes very rapidly; thus it is mandatory that the flux be activated to flow onto the work surface and not simply be volatilized by the hot iron tip if it is to provide the full benefit of the fluxing action.

There are varieties of fluxes available for many applications. For example, in soldering sheet metal, acid fluxes are used; silver brazing (which requires a much higher temperature for melting than that required by tin/lead alloys) uses a borax paste. Each of these fluxes removes oxides and, in many cases, serves additional purposes. The fluxes used in electronic hand soldering are the pure

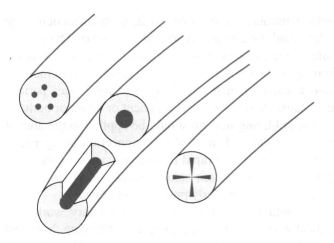

Fig. A-2 Flux cored wire solders.

rosins, rosins combined with mild activators to accelerate the rosin's fluxing capability, low-residue/no-clean fluxes, or water-soluble fluxes. Acid fluxes or highly activated fluxes should *never* be used in electronic work. Various types of flux-cored solder are in common use. They provide a convenient way to apply and control the amount of flux used at the joint (Fig. A-2).

Soldering Irons

In any kind of soldering, the primary requirement, beyond the solder itself, is heat. Heat can be applied in a number of ways—conductive (e.g., soldering iron, wave, vapor phase), convective (hot air), or radiant (IR). Here, we are mainly concerned with the conductive method, which uses a soldering iron.

Soldering stations come in a variety of sizes and shapes but consist basically of three main elements: a resistance heating unit; a heater block, which acts as a heat reservoir; and the tip, or bit, for transferring heat to the work. The standard production station is a variable-temperature, closed-loop system with interchangeable tips and is made with ESD-safe plastics.

Controlling Heat at the Joint

Controlling tip temperature is not the real challenge in soldering; the real challenge is to control the heat cycle of the work—how fast the work gets hot, how hot it gets, and how long it stays that way. This is affected by so many factors that, in reality, tip temperature is not the critical factor.

The first factor that needs to be considered is the relative thermal mass of the area to be soldered. This mass may vary over a wide range. Consider a single land on a single-sided circuit board. There is relatively little mass, so the land heats up quickly. But on a double-sided board with plated-through holes, the mass is more than doubled. Multilayered boards may have an even greater mass, and

that's before the mass of the component lead is taken into consideration. Lead mass may vary greatly, since some leads are much larger than others. Further, there may be terminals (e.g., turret or bifurcated) mounted on the board. Again, the thermal mass is increased and will further increase as connecting wires are added.

Each connection to be soldered, then, has its particular thermal mass. How this combined mass compares with the mass of the iron tip, the "relative" thermal mass, determines the time and temperature rise of the work. With a large work mass and a small iron tip, the temperature rise will be slow. With the situation reversed, using a large iron tip on a small work mass, the temperature rise of the work will be much more rapid—even though the temperature of the tip is the same.

Now consider the capacity of the iron itself and its ability to sustain a given flow of heat. Essentially, irons are instruments for generating and storing heat, and the reservoir is made up of both the heater block and the tip. The tip comes in various sizes and shapes; it's the pipeline for heat flowing into the work. For small work, a conical (pointed) tip is used, so that only a small flow of heat occurs. For large work, a large chisel tip is used, providing greater flow. Table A-2 shows some various tip styles and sizes.

Table A-2 Examples of Soldering Iron Tips		
Drawing	Size in Inches	Description
	0.031	30° chisel
	0.047	30° bent chisel extended
	0.063	30° bent chisel
	0.063	60° chisel
	0.078	60° chisel
	0.094	30° chisel
	0.125	90° chisel extended
	0.203	Chisel
	0.250	Single-sided chisel

The heat reservoir is replenished by the heating element, but when an iron with a large tip is used to heat massive work, the reservoir may lose heat faster than it can be replenished. Thus the size of the reservoir becomes important: a large heating block can sustain a larger outflow longer than a small one. An iron's capacity can be increased by using a larger heating element, thereby increasing the wattage of the iron. These two factors, block size and wattage, are what determine the iron's recovery rate.

If a great deal of heat is needed at a particular connection, the correct temperature with the right size tip is required, as is an iron with a large enough capacity and an ability to recover fast enough. Relative thermal mass, then, is a major consideration for controlling the heat cycle of the work.

A second factor of importance is the surface condition of the area to be soldered. If any oxides or other contaminants cover the lands or leads, there will be a barrier to the flow of heat. Then, even though the iron tip is the right size and has the correct temperature, it may not supply enough heat to the connection to melt the solder. In soldering, a cardinal rule is that a good solder connection cannot be created on a dirty surface. Before attempting to solder, the work should always be cleaned with an approved solvent to remove any grease or oil film from the surface. In some cases pretinning may be required to enhance solderability and remove heavy oxidation of the surfaces prior to soldering.

A third factor to consider is thermal linkage—the area of contact between the soldering iron tip and the work. Figure A-3 shows a tip touching a round lead. The contact occurs only at the point indicated by the "+," so the linkage area is very small. The contact area can be greatly increased by applying wire solder to the point of contact between the tip and workpiece. This solder heat bridge drastically improves the thermal linkage and ensures rapid heat transfer into the work.

It should now be apparent that there are many more factors than just the temperature of the iron tip that affect how quickly any particular connection is going to heat up. In reality, soldering is a very complex control problem, with a number of variables to it, each influencing the other. And what makes it so critical is time. The general rule for high-reliability soldering on printed circuit boards is to apply heat for no more than 2 seconds from the time solder starts to melt. Applying heat for longer periods may cause damage to the component or board or both.

The soldering iron tip should be applied to the area of maximum thermal mass of the connection being made. This will permit the rapid thermal elevation of the parts being soldered. Molten solder always flows toward the heat source of a properly prepared connection.

For soldering and desoldering, a primary workpiece indicator is heat rate recognition—observing how fast heat flows into the connection. In practice, this means observing the rate at which solder melts, which should be within 1 to 2 seconds. This indicator encompasses all the variables involved in making a satisfactory solder connection with minimum heating effects, including the capacity of the iron and its tip temperature, the surface conditions, the thermal linkage between the tip and the workpiece, and the relative thermal masses involved.

If the iron tip is too large for the work, the heating rate may be too fast to be controlled. If the tip is too small, it may produce a "mush" kind of melt; the heating rate will be too slow, even though the temperature at the tip is the same. A general rule for preventing overheating is, *Get in and get out as fast as you can.* That means using a heated iron you can react to—one giving a 1- to 2-second dwell time on the particular connection being soldered.

Selecting the Soldering Iron and Tip

A good all-around soldering station for electronic soldering is a variable-temperature, ESD-safe station with a pencil-type iron and tips that are easily interchangeable, even when hot (Fig. A-4). The soldering iron tip should always be fully inserted into the heating element and tightened. This will allow for maximum heat transfer from the heater to the tip.

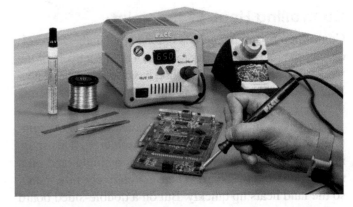

Fig. A-4 Temperature controlled soldering station.
Courtesy of PACE Woldwide

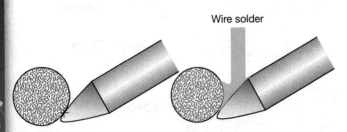

Wire solder

Fig. A-3 Increasing contact area for improved heat flow.

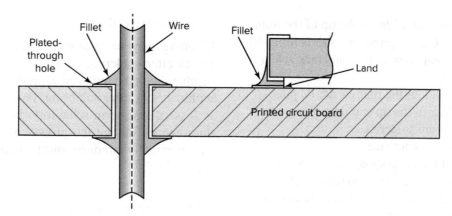

Fig. A-5 Solder fillets.

The tip should be removed daily to prevent an oxidation scale from accumulating between the heating element and the tip. A bright, thin-tinned surface must be maintained on the tip's working surface to ensure proper heat transfer and to avoid contaminating the solder connection.

The plated tip is initially prepared by holding a piece of flux-cored solder to the face to tin the surface when it reaches the lowest temperature at which solder will melt. Once the tip is up to operating temperature, it will usually be too hot for good tinning, because of the rapid oxidation at elevated temperatures. The hot tinned tip is maintained by wiping it lightly on a damp sponge to shock off the oxides. When the iron is not being used, the tip should be coated with a layer of solder.

Making the Solder Connection

The soldering iron tip should be applied to the area of maximum thermal mass of the connection being made. This will permit the rapid thermal elevation of the parts being soldered. Molten solder always flows toward the heat of a properly prepared connection.

When the solder connection is heated, a small amount of solder is applied to the tip to increase the thermal linkage to the area being heated. The solder is then applied to the opposite side of the connection so that the work surfaces, not the iron, melt the solder. Never melt the solder against the iron tip and allow it to flow onto a surface cooler than the solder-melting temperature.

Solder, with flux, applied to a cleaned and properly heated surface will melt and flow without direct contact with the heat source and provide a smooth, even surface, feathering out to a thin edge (Fig. A-5). The resulting shape is called a *fillet*. Improper soldering will exhibit a built-up, irregular appearance and poor filleting. The parts being soldered must be held rigidly in place until the temperature decreases to solidify the solder. This will prevent a disturbed or fractured solder joint.

Selecting cored solder of the proper diameter will aid in controlling the amount of solder being applied to the connection (e.g., a small-gauge solder for a small connection; a large-gage solder for a large connection).

Mounting an SMT IC: An Example

The integrated circuit used for this example is a 16 pin TSSOP with 0.65 mm lead spacing. Figure A-6 shows that the IC is to be mounted to an adapter board that will allow common electronic breadboards to be employed later on. No, the SMT revolution has not killed off breadboarding. Note: a lighted magnifier is a huge help for this kind of work.

The soldering iron tip size in Fig. A-6 is okay for this job. Wetting action and the adapter board's solder mask allow a result that is free of short circuits. Here is a brief description of the process:

1. Turn the IC to orient pin 1.
2. Flood the area with liquid flux. Repeat as necessary.
3. Clean the tip of the iron with a damp sponge.

Fig. A-6 Mounting an IC on an adapter board.
Charles A. Schuler/McGraw Hill

4. Apply a small blob of solder to the tip of the iron.
5. Tack solder one IC pin (ignore any shorts for now).
6. Move to the opposite row of pins and tack solder another pin or two.
7. Now solder or resolder the four corner IC pins.
8. Apply more solder to the tip and drag along *one side* until all pins are soldered.
9. Do the same on the other side.
10. Clean the tip with a damp sponge and repeatedly draw it away from the IC body to remove shorts.
11. Repeat sponge cleaning and pin/trace clearing until all the shorts are removed.
12. Reapply liquid flux as many times as needed during the short removal process.

Figure A-7 shows the final result. It can take some practice but hand soldering SMT devices can be accomplished. Note: *It does require practice.* It is a good idea to learn on jobs like the one shown here before tackling the repair of an important piece of equipment.

The art of soldering requires knowledge of how the process works, the proper tools and materials, lots of practice, and careful inspection. Most solder joints involve fillets, and these take on a characteristic appearance. Figure A-5 shows examples of fillets. Experience dictates how these should appear for high-reliability joints. Generally, a properly shaped fillet indicates clean conditions (good wetting action), proper soldering temperature and duration, and the correct amount of solder.

Flux Removal

Cleaning may be required to remove certain types of fluxes after soldering. If cleaning is required, the flux residue should be removed as soon as possible, preferably within 1 hour after soldering. Failure to clean can result in loss of long-term reliability. For example, flux residues can encourage the growth of metal dendrites that can eventually produce short circuits between closely spaced lands.

The Convection Process

Because of the increased use of surface mount devices, convection (hot-air) soldering and desoldering are now preferred in many cases. The Pace ST 325, shown in Fig. A-8, offers controlled hot air for both soldering and desoldering. It also provides a vacuum for the removal of devices once they are desoldered.

With convection, both the air flow and the air temperature are controlled to achieve the desired results. Table A-3 shows a chart of recommended starting points for both soldering and desoldering.

What the Law Requires

There is not sufficient space here to list the laws that apply to electronic soldering, since they vary by country. In many European countries, restriction of hazardous substances (RoHS) and waste from electrical and electronic equipment (WEEE) standards are enforced.

Fig. A-7 The completed adapter.

Charles A. Schuler/McGraw Hill

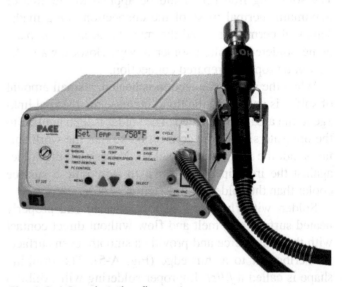

Fig. A-8 A Pace hot-air reflow system.

Courtesy of PACE Woldwide

Component		Nozzle	Process	Parameter	Substrate (PCB Type)			Reflow Cycle
Outline	Type	Recommended Type	Remove or Install	(Temperature and Blower Speed)	Low Mass	Medium Mass	High Mass	Time (sec.)
	PBGA	Appropriate Size V-A-N Nozzle	Remove	Temperature (°C)	371	371	371	77
				Blower Speed	5	5	5	
			Install	Temperature (°C)	371	371	371	90
				Blower Speed	3	3	4	
	PLCC (J Lead)	Appropriate Size Box Nozzle	Remove	Temperature (°C)	371	371	371	30
				Blower Speed	7	8	8	
			Install	Temperature (°C)	371	371	371	30
				Blower Speed	7	8	8	
	PQFP	Appropriate Size Box Nozzle	Remove	Temperature (°C)	316	371	371	18
				Blower Speed	6	7	7	
			Install	Temperature (°C)	316	371	371	18
				Blower Speed	6	7	7	
	SOIC	Appropriate Size Pattern Nozzle	Remove	Temperature (°C)	316	316	371	15
				Blower Speed	7	7	7	
			Install	Temperature (°C)	316	316	371	15
				Blower Speed	7	7	7	
	Chip Component	Appropriate Size Single Jet Nozzle	Remove	Temperature (°C)	371	371	371	11
				Blower Speed	6	6	8	
			Install	Temperature (°C)	371	371	371	12
				Blower Speed	5	6	7	

Appendix B
Thermionic Devices

Thermionic devices (vacuum tubes) dominated electronics until the early 1950s. Since that time, solid-state devices have all but completely taken over. Today, vacuum tubes are used only in special applications in communications and by some audiophiles and amateur radio enthusiasts.

Thermionic emission involves the use of heat to liberate electrons from an element called a *cathode*. The heat is produced by energizing a filament or heater circuit within the tube. A second element, called the *anode*, can be used to attract the liberated electrons. Since unlike charges attract, the anode is made positive with respect to the cathode.

A third electrode can be placed between the cathode and the anode. This third electrode can exert control over the movement of electrons from cathode to anode. Thus, it is called the *control grid*. The control grid is often negative with respect to the cathode. This negative charge repels the cathode electrons and prevents them all from reaching the anode. In fact, the tube can be cut off by a high negative grid potential. Figure B-1 shows the schematic symbol for a three-electrode vacuum tube and the polarities involved.

The vacuum tube shown in Fig. B-1 is an amplifier. The signal to be amplified can be applied to the control grid. As the signal goes in a positive direction, more plate current will flow. As the signal goes in a negative direction, less plate current will flow. Thus, the plate current is a function of the signal applied to the grid. The signal power in the grid circuit is much less than the signal power in the plate circuit. The vacuum tube is capable of good power gain.

Vacuum tubes may use extra grids located between the control grid and the plate to provide better operation. The extra grids improve gain and high-frequency performance. The tube in Fig. B-1 is called a *triode* vacuum tube (the heater is not counted as an element). If a screen grid is added, it becomes a *tetrode* (four electrodes). If a screen grid and a suppressor grid are added, it becomes a *pentode* (five electrodes).

Vacuum tubes make excellent high-power amplifiers. It is possible to run some vacuum tubes with plate potentials measured in thousands of volts and plate currents measured in amperes. These tubes offer output powers of several thousand watts. It is even possible to develop 2,000,000-W amplitude-modulated RF output by using four special tetrodes. This is an example of the outstanding power capacity of vacuum tubes.

The cathode-ray tube was used for the display of graphs, pictures, or data. Today, the liquid crystal display (LCD) serves these functions. Figure B-2 shows the basic structure. The cathode is heated and produces thermionic emission. A positive potential is applied to the first anode, the second anode, and the aquadag coating. This positive field accelerates the electrons toward the screen. The inside of the screen is coated with a chemical phosphor that emits light when hit by a stream of electrons.

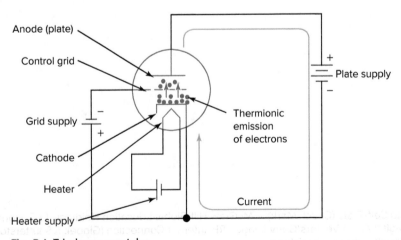

Fig. B-I Triode vacuum tube.

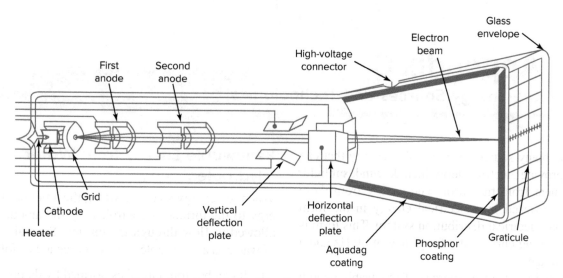

First anode

Second anode

High-voltage connector

Electron beam

Glass envelope

Grid

Cathode

Heater

Vertical deflection plate

Horizontal deflection plate

Aquadag coating

Phosphor coating

Graticule

Fig. B-2 A cathode-ray tube using electrostatic deflection.

As shown in Fig. B-2, the electrons are focused into a narrow beam. This makes it possible to produce a small dot of light on the screen. The deflecting plates can move the beam vertically and horizontally. For example, a positive voltage applied to the top vertical deflecting plate will attract the beam and move it up. The dot of light can be positioned anywhere on the screen.

The grid shown in Fig. B-2 makes it possible to control the intensity of the beam. A negative voltage applied to the grid will repel the cathode electrons and prevent them all from reaching the screen. A high negative voltage will completely stop the electrons, and the dot of light will go out.

By controlling the position and the intensity of the dot, any type of picture information can be presented on the screen. Because the phosphor will retain its brightness momentarily and because the eye will retain the image for a brief period, the effect of the moving dot is to produce what seems to be a complete picture on the screen. If this is done repeatedly, a movie effect is produced. This is how a television picture tube works. Colors can be shown by using several different chemical phosphors.

The deflection system may be different from that shown in Fig. B-2. *Magnetic deflection* uses coils around the neck of the cathode-ray tube. When a current flows through the coil, the resulting magnetic field will deflect the electron beam. Television picture tubes generally used magnetic deflection. Oscilloscopes generally used electrostatic deflection.

Appendix C
Renewable Energy Sources and Technologies

Some of the many sources of renewable energy that can be used to provide electric energy include wind, sunlight, water, and heat from the earth. The energy from these sources is converted into electrical energy in the form needed by our electrical distribution system. This form is usually alternating current at a frequency of 60 Hz and a specified voltage.

The conversion of many sources of renewable energy to 60 Hz ac is done by some type of turbine (i.e., wind, water, steam, etc.) turning the shaft of an ac generator. For a given generator with a fixed number of field poles, the output frequency is determined by the revolutions per minute (RPM) of the armature. Armature RPM is controlled by the amount of energy (fuel) provided to the turbine connected to the generator shaft. The RPM must be held constant if a constant output frequency is to be maintained. If the RPM is held constant, the output voltage can be controlled by varying the current in the coils of the field poles. Before a generator is connected to an electric power grid, its frequency and output voltage must be adjusted to match that of the grid.

Once the generator is connected to the power grid, its frequency is synchronized with all the other generators connected to the grid. Now the frequency, and thus the RPM, of the armature and the turbine are locked into that of all the other generators connected to the grid. Increasing the energy (fuel) provided to the turbine can only cause a minute change in the armature's RPM and thus a very minute change in the frequency and voltage of the grid and all of the other generators connected to the grid. However, increasing energy input to the turbine does increase the power output of the generator by increasing the current output from the armature. Of course, decreasing fuel input to the turbine also produces a very minute change in the grid voltage and frequency, but the change is in the opposite direction.

Once the input energy to a turbine has been set to accommodate the requirements of the loads connected to the grid, any significant changes in the loads on the grid will cause minute changes in the frequency and voltage of the grid. This minute change is detected by electronic detectors and circuits that constantly monitor the grid frequency and voltage and make appropriate changes in the energy input to the turbines driving the generators to counteract the minute changes detected.

Renewable Energy from Water (Hydroelectricity)

When the energy source is water, the amount of input energy to the turbine is controlled by the amount of water allowed to flow through the turbine. Several advantages of water as a renewable energy source are as follows:

1. It can be stored in lakes, formed by damming a river, and used when needed either day or night. It is a very reliable source.
2. It is a green energy source that does no damage to the environment during operation.
3. In areas where water flow tends to be inadequate at times, water can be pumped into a man-made lake when water flow is high and excess electrical energy can be used to power water pumps.
4. It is one of the cheaper sources (about 4 to 11 ¢/kWh) but limited to locations (rivers) where dams can be located.
5. It is a highly developed, proven technology.

Energy can also be obtained from the tidal movement of water in and out of a larger bay or inlet that is connected to the main body of ocean water through a narrow inlet that can be gated off at high tide and allowed to flow out through a water turbine as the tide goes out (recedes). To be economically viable, this approach to using water energy requires large, regular tidal swings and the right soil conditions for building and operating the gates needed to open and close the waterway to the bay or inlet. Locations that meet all these conditions are rather limited.

Renewable Energy from Wind

The most common wind turbine in use today is the three-blade rotor (propeller) facing into the wind. The wind energy input depends on the wind speed and the pitch of the blades. Since the rotor turns at a low speed (typically less than 30 RPM), the rotor shaft drives a gearbox that turns the ac generator at the RPM needed to produce a 60-Hz frequency. Wind speed and direction has large variations during any 24 hours, so it is continuously monitored and the blade pitch adjusted and/or the platform holding the turbine and generator rotated so that desired turbine (rotor) speed is maintained and the rotor remains facing into the wind. Once the generator's frequency and the voltage are synchronized with the grid, the generator is connected to the grid, and its frequency and voltage are essentially locked to the grid's frequency and voltage. Now, when the electronic circuitry detects any minute changes in the grid frequency and voltage, or load, the blade pitch and/or platform-to-wind orientation are automatically adjusted to correct for the detected changes.

When the rotor needs to be stopped for maintenance or because of severe wind/weather conditions, the generator is disconnected from the grid. Then the blade pitch is reduced to a minimum, the platform is rotated so that the rotor is parallel to the wind direction, and a brake system that stops the rotor shaft from turning is activated.

While wind energy systems are often said to be the cheapest form of alternative energy (4 to 6 ¢/kWh), their reliability is often low. Many times, the wind speed is too low or too high to produce the frequency and voltage required by the grid. Also, wind energy systems require a lot of maintenance that requires downtime.

Renewable Energy from Solar (Sun) Energy

Solar energy can be converted into electric energy by a PV (photovoltaic) solar panel. A single panel typically has 10 to 20 square feet of surface area to expose to sunlight and produces 170 to 350 W of dc power. A PV solar panel has many individual PV cells. A single cell produces about 0.5 V at about 1.5 A. Individual cells are connected in series to increase the panel output voltage, and series groups are connected in parallel to increase the panel's output current and power.

The output of PV solar panels is dc. The dc output can be converted to 50 or 60 Hz ac by an electronic inverter. Inverters that can handle dc inputs to 1,000 V dc and provide 400 V, three-phase ac output are available. Some inverters can provide 2.4 MW of ac power output. These inverters can be up to 98 percent efficient. As long as the frequency is correct and the voltage is stable, the output voltage can be transformed to the level needed by the grid before the PV solar panel system is connected to the grid.

A problem with PV cells and PV panels being connected in series is that when one cell or panel in a series receives reduced sunlight energy, it reduces not only the current output of that one cell or panel, but the current output of the other cells or panels in series with it. This can make a very significant reduction in the power output of the PV panels and the input into the main dc-to-ac inverter. The effects of this problem can be minimized by using a low-power inverter on the output of each panel and then combining the outputs of all of these small inverters to obtain the required voltage and current.

Although the cost of operating and maintaining a PV solar farm (system) is small, the cost of building the system has been high because building solar panels has been expensive. Thus, the cost per kWh has been high. Technical developments over the past few years have reduced these costs. Building a PV solar farm requires a sizable amount of land which, in some locations, can significantly add to construction costs.

Renewable Energy from Geothermal (Heat) Sources

The core of the earth is massive and hot. Thus, it is a source of geothermal energy that will be available for eons. A lot of geothermal energy is used to heat buildings. To produce electric energy, geothermal energy is used to drive a steam turbine/three-phase, 60-Hz electric generator. The output of the turbine/generator is connected to a grid after its output frequency and voltage are adjusted to the correct values. As mentioned before, adjustment is made by varying the amount of energy input into the turbine and the magnetic strength of the generator's field poles.

The ease and cost of obtaining geothermal energy suitable for driving a steam turbine vary tremendously from place to place. In some areas it is available at, or very close to, the surface of the earth. In other locations, holes over a mile deep must be drilled to find steam or water hot enough to produce steam when pumped to the earth's surface or core material (rocks) that are hot enough to produce steam when water is injected into the hole. Of course, the cost of drilling these deep holes will vary with the type of material that must be drilled through.

Because of the variables mentioned above, the cost of converting geothermal energy to electric energy varies widely. New advances in the use of lower temperatures to produce steam (vapor) from liquids other than water are predicted to lower the range of cost per kWh. Lower-temperature geothermal energy can be obtained from shallower holes. This will significantly lower the cost per kWh.

Glossary of Terms and Symbols

Term	Definition	Symbol or Abbreviation
AC component	The fluctuating or changing value of a waveform or signal. Pure direct current has no ac component.	
Active	An operating region between saturation and cutoff. The current in an active device is a function of its control bias.	
Active filter	An electronic filter using active gain devices (usually operational amplifiers) to separate one frequency, or a group of frequencies, from all other frequencies.	
Alias	A quantized signal improperly represented as a lower frequency signal (aliases are avoided by the use of an adequate sampling frequency and an antialias filter).	
Amplifier	A circuit or device designed to increase the level of a signal.	
Amplitude modulation	The process of using a lower frequency signal to control the instantaneous amplitude of a higher frequency signal. Often used to place intelligence (audio) on a radio signal.	AM
Analog	That branch of electronics dealing with infinitely varying quantities. Often referred to as linear electronics.	
Analog to digital	A circuit or device used to convert an analog signal or quantity to digital form (usually binary).	A/D
Anode	That element of an electronic device that receives the flow of electron current.	
Attenuator	A circuit used to decrease the amplitude of a signal.	
Automatic frequency control	A circuit designed to correct the frequency of an oscillator or the tuning of a receiver.	AFC
Automatic gain control	A circuit designed to correct the gain of an amplifier according to the level of the incoming signal.	AGC
Automatic volume control	A circuit designed to provide a constant output volume from an amplifier or radio receiver.	AVC
Avalanche	The sudden reverse conduction of an electronic component caused by excess reverse voltage across the device.	
Balanced modulator	A special amplitude modulator designed to cancel the carrier and leave only the sidebands as outputs. It is used in single sideband transmitters.	
Barrier potential	The potential difference that exists across the depletion region in a PN junction.	
Base	The center region of a bipolar junction transistor that controls the current flow from emitter to collector.	B
Beat frequency oscillator	A radio receiver circuit that supplies a carrier signal for demodulating code or single sideband transmissions.	BFO
Beta	The base-to-collector current gain in a bipolar junction transistor. Also called h_{FE}.	β
Bias	A controlling voltage or current applied to an electronic circuit or device.	
Bipolar	Having two polarities of carriers (holes and electrons).	
Bleeder	A fixed load designed to discharge (bleed off) filters.	

Term	Definition	Symbol or Abbreviation
Block diagram	A drawing using a labeled block for every major section of an electronic system.	
Blocking capacitor	A capacitor that eliminates the dc component of the signal.	
Bode plot	A graph showing the gain or phase performance of an electronic circuit at various frequencies.	
Bootstrap	A feedback circuit usually used to increase the input impedance of an amplifier. May also refer to a circuit used to start some action when the power is first applied.	
Break frequency	A frequency where the response or gain of a circuit decreases 3 dB from its best response or gain.	f_b
Brick wall	The frequency response of an ideal filter (the transition from the pass band to the reject band is immediate). Real filters that approach a brick wall response are said to be *sharp*.	
Bricked	A term used to describe a device or instrument rendered inoperable by an attempted software update.	
Bypass	A low-pass filter employed to remove high-frequency interference from a power supply line or a component such as a capacitor that provides a low-impedance path for high-frequency current.	
Capacitive coupling	A method of signal transfer that uses a series capacitor to block or eliminate the dc component of the signal.	
Capacitive input filter	A filter circuit (often in a power supply) using a capacitor as the first component in the circuit.	
Carrier	A movable charge or particle in an electronic device that supports the flow of current. Also refers to an unmodulated radio or television signal.	
Cascade	One after the other. The output of the first circuit connects to the input of the second, and so on. Circuits that are cascaded include amplifiers and filters.	
Cathode	That element of an electronic device that provides the flow of electron current.	
Characteristic curves	Graphic plots of the electrical and/or thermal behavior of electronic circuits or components.	
Choke input filter	A filter circuit (often in a power supply) using a choke or an inductor as the first component in the circuit.	
Clamp	A circuit for adding a dc component to an ac signal. Also known as a dc restorer.	
Clapp oscillator	A series-tuned Colpitts configuration noted for its good frequency stability.	
Class	One way to categorize an amplifier based on bias and conduction angle.	
Clipper	A circuit that removes some part of a signal. Clipping may be undesired in a linear amplifier or desired in a circuit such as a limiter.	
Coefficient	A fixed value used in the multiply-and-accumulate process of a DSP system (the coefficient number is often denoted by the subscript n). Digital filter coefficients are also called taps.	$h_{[n]}$
Collector	The region of a bipolar junction transistor that receives the flow of current carriers.	
Colpitts oscillator	A circuit with a capacitively tapped tank circuit.	

Term	Definition	Symbol or Abbreviation
Common base	An amplifier configuration where the input signal is fed into the emitter terminal and the output signal is taken from the collector terminal.	CB
Common collector	An amplifier configuration where the input signal is fed into the base terminal and the output signal is taken from the emitter terminal. Also called emitter follower.	CC
Common emitter	The most widely applied amplifier configuration, where the input signal is fed into the base terminal and the output signal is taken from the collector circuit.	CE
Common-mode rejection ratio	The ratio of differential gain to common-mode gain in an amplifier. It is a measure of the ability to reject a common-mode signal and is usually expressed in decibels.	CMRR
Commutation	The interruption of current flow. In thyristor circuits, it refers to the method of turning the control device off.	
Comparator	A high-gain amplifier that has an output determined by the relative magnitude of two input signals.	
Complementary metallic oxide semiconductor	An integrated circuit containing both P-channel and N-channel transistors. Most integrated circuits use this structure.	CMOS
Complementary symmetry	A circuit designed with opposite polarity devices such as NPN and PNP transistors.	
Conditional stability	A term applied to a circuit that is not perfectly stable. It can ring under certain conditions (exhibit oscillations that ebb with time).	
Conduction angle	The number of electrical degrees that a device is on.	
Continuous signal	A signal with an infinite number of amplitudes (also called an analog signal).	
Continuous wave	A type of modulation where the carrier is turned off and on following a pattern such as Morse code.	CW
Converter	A circuit that transforms dc from one voltage level to another. Also refers to a circuit that changes frequency.	
Convolution	The formal name for the multiply-and-accumulate process that is used in digital signal processing to combine signal samples and coefficients. The convolution symbol is an asterisk ($y_{[n]} = x_{[n]} * h_{[n]}$).	*
Coupling	The means of transferring electronic signals.	
Critical conduction mode	When the charging current in a transformer is turned on at the exact moment when the discharge current reaches zero. Flyback circuits can operate in this mode.	CCM
Crossover distortion	Disturbances to an analog signal that affect only that part of the signal near the zero axis or average axis.	
Crowbar	A protection circuit used to blow a fuse or otherwise turn a power supply off in the event of excess voltage.	
Crystal	A piezoelectric transducer used to control frequency, change vibrations into electricity, or filter frequencies. Also refers to the physical structure of semiconductors.	
Current gain	The feature of certain electronic components and circuits where a small current controls a large current.	A_I
Current limiter	A circuit or device that prevents current flow from exceeding some predetermined limit.	
Current mirror	A circuit that produces a stable current and is often used in integrated circuits.	
Curve tracer	An electronic device for drawing characteristic curves on a cathode-ray tube.	

Term	Definition	Symbol or Abbreviation
Cutoff	That bias condition where no current can flow.	
Darlington	A circuit using two direct-coupled bipolar transistors for very high current gain.	
DC component	The average value of a waveform or signal. Pure ac averages zero and has no dc component.	
Decibel	One-tenth of a bel. A logarithmic ratio used to measure gain and loss in electronic circuits and systems.	dB
Decimation	Decreasing the sampling frequency in a DSP system by discarding discrete samples. Also known as down-sampling.	
Demodulation	The recovery of intelligence from a modulated radio or television signal. Also called detection.	
Depletion	The condition of no available current carriers in a semiconducting crystal. Also refers to that mode of operation for a field-effect transistor where the channel carriers are reduced by gate voltage.	
Diac	A silicon bilateral device used to gate other devices.	
Differential amplifier	A gain device that responds to the difference between its two input terminals.	
Digital	That branch of electronics dealing with finite and discrete signal levels. Most digital signals are binary: they are either high or low.	
Digital filter	A system that separates signal frequencies by using digital signal processing (DSP).	
Digital multimeter	An instrument with a digital display that measures several electrical quantities such as voltage, current, and resistance.	DMM
Digital to analog	A circuit or device used to convert a digital signal into its analog equivalent.	D/A
Digital signal processing	A system using A/D and D/A converters plus a microprocessor to alter some characteristic of an analog signal.	DSP
Diode	A two-terminal electronic component. Diodes usually allow current to flow in only one direction. Different types of diodes can be used for rectification, regulation, tuning, triggering, and detection. They can also be used as indicators.	
Direct digital synthesis	A method of generating waveforms based on a lookup table and a phase accumulator.	DDS
Discontinuity	A change in the amplitude of a signal that occurs in zero time. An ideal square wave is an example since it instantaneously changes from maximum to minimum.	
Discrete circuit	An electronic circuit made up of individual components (transistors, diodes, resistors, capacitors, etc.) interconnected with wires or conducting traces on a printed circuit board.	
Discrete Fourier transform	A mathematical procedure that converts a discrete time domain signal to the discrete frequency domain.	DFT
Discrete signal	A signal with a limited number of amplitudes (also called a digital signal).	
Discriminator	A circuit used to detect frequency-modulated signals.	
Distortion	A change (usually unwanted) in some aspect of a signal.	
Doping	A process of adding impurity atoms to semiconductor crystals to change their electrical properties.	

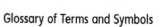

Term	Definition	Symbol or Abbreviation
Drain	That terminal of a field-effect transistor that receives the current carriers from the source.	(symbol) D
Dual power supply	A power supply that produces positive and negative outputs with reference to ground. Also called a bipolar supply.	
Effective series resistance	A parasitic resistance in a component. Often a factor in electrolytic capacitors, since they can dry out and develop a high resistance.	ESR
Efficiency	The ratio of useful output from a circuit to the input.	η
Electromagnetic interference	A form of interference to and from electronic circuits resulting from the radiation of high-frequency energy.	EMI
Electrostatic discharge	A potentially destructive flow of electrons due to the buildup of a charge imbalance caused by friction between two nonconductors.	ESD
Embedded system	Those systems where the hardware and software are combined in one or several ICs.	
Emitter	That region of a bipolar junction transistor that sends the current carriers on to the collector.	(symbol) E
Enhancement mode	That operation of a field-effect transistor where the gate voltage is used to create more carriers in the channel.	
Epitaxial	A thin, deposited crystal layer that forms a portion of the electrical structure of certain semiconductors.	
Error amplifier	A gain device or circuit that responds to the error (difference) between two signals.	(symbol)
Fast Fourier transform	A faster computing procedure for converting discrete time domain signals to the discrete frequency domain that is based on an efficient method of number crunching using powers of two.	FFT
Feedback	The application of a portion of the output signal of a circuit back to the input of the circuit. Any of a number of closed-loop systems where an output is connected to an input.	
Ferroresonant	A special type of power-supply transformer using a resonating capacitor and a saturated core to provide both load and line voltage regulation.	(symbol)
Field-effect transistor	A solid-state device that uses a terminal (gate) voltage to control the resistance of a semiconducting channel.	FET (symbol)
Filter	A circuit designed to separate one frequency, or a group of frequencies, from all other frequencies.	
Finite impulse response	The output of the system always decays to zero after the input returns to zero (a DSP system with no feedback).	FIR
Firmware	Software that never changes (or seldom changes). It is usually stored in an IC (see embedded system).	
First harmonic	The lowest frequency in a Fourier series.	
Flip-flop	An electronic circuit with two states. Also known as a multivibrator. May be free-running (as an oscillator) or exhibit one or two stable states.	
Flyback	A class of inductive circuits where energy is transferred during the collapse of the magnetic field in a coil or transformer.	
Foldback current limiting	A type of current limiting in which the current decreases beyond the threshold point as the load resistance drops.	
Fourier series	A number of sine waves that are added to construct or synthesize a periodic function.	

Term	Definition	Symbol or Abbreviation
Fourier transform	A mathematical procedure to convert time domain signals to the frequency domain.	
Frequency division multiplexing	The use of two or more carrier frequencies sent on a single medium. Its purpose is to increase the amount of information that can be sent in a given period of time.	FDM
Frequency domain	A viewpoint where the signal amplitude is plotted versus the signal frequency (a spectrum analyzer display is an example).	
Frequency modulation	The process of using a lower frequency signal to control the instantaneous frequency of a higher frequency signal. Often used to place intelligence (audio) on a radio signal.	FM
Frequency multiplier	A circuit where the output frequency is an integer multiple of the input frequency. Also known as a doubler, tripler, etc.	
Frequency synthesis	A method of generating many accurate frequencies without resorting to multiple crystal-controlled oscillators. Usually based on PLL or DDS technology.	
Gain	A ratio of output to input. May be measured in terms of voltage, current, or power. Also known as amplification.	A or G
Gain-bandwidth product	The high frequency at which the gain of an amplifier is 0 dB (unity).	f_t
Gallium arsenide	A semiconducting material used in high-frequency applications.	Ga
Gate	That terminal of a field-effect transistor that controls drain current. Also the terminal of a thyristor used to turn the device on.	G —|[symbol]
Gibbs phenomenon	Distortions in a periodic signal composed of a Fourier series that are caused by discontinuities in the periodic signal.	
Ground loop	A short (or otherwise unwanted) circuit across the ac line caused by grounded test equipment or some other ground path not normally intended to conduct current.	
Hard saturation	When a device such as a transistor has more than enough input signal to turn it on fully.	
Hartley oscillator	A circuit distinguished by its inductively tapped tank circuit.	
Heterodyne	The process of mixing two frequencies to create new (sum and difference) frequencies.	
Hilbert transform	A DSP operation that delays (or phase shifts) a discrete signal by 90 degrees.	
Holes	Positively charged carriers that move opposite in direction to electrons and can be found in semiconducting crystals.	
House numbers	Nonregistered device numbers peculiar to the manufacturer.	
Hysteresis	A dual threshold effect exhibited by certain circuits.	[symbol]
I²C	A serial communication protocol used for short-distance signaling between processors and other ICs.	
Image	The second, unwanted frequency that a heterodyne converter will interact with to produce the intermediate frequency.	

Term	Definition	Symbol or Abbreviation
Impedance match	The condition where the impedance of a signal source is equal to the impedance of the signal load. It is desired for best power transfer from source to load.	
Integrated circuit	The combination of many circuit components into a single crystalline structure (monolithic), onto a supporting substrate (thick-film), or a combination of the two (hybrid).	IC
Integrator	An electronic circuit that provides continuous summation of signals over some period of time.	
Intermediate frequency	A standard frequency in a receiver that all incoming signals are converted to before detection. Most of the gain and selectivity of a receiver are produced in the intermediate-frequency amplifier.	IF
Intermittent	A fault that only appears from time to time. It may be related to mechanical shock or temperature.	
Interpolation	Increasing the sampling frequency in a DSP system by stuffing zeros between discrete samples (also called up-sampling).	
Intrinsic standoff ratio	In a unijunction transistor, the ratio of the voltage required to fire the transistor to the total voltage applied across the transistor.	η
Inverse Fourier transform	A mathematical procedure that converts a frequency domain signal to a time domain signal.	
Inverting	An amplifier where the output signal is 180 degrees out of phase with the input.	
Latch	A device that, once triggered on, tends to stay on. Also a digital circuit for storing one of two conditions.	
Lead dress	The exact position and length of electronic components and their leads. Can affect the way circuits (especially high-frequency ones) perform.	
Lead-lag network	A circuit that provides maximum amplitude and zero phase shift for one (the resonant) frequency. It produces a leading angle for frequencies below resonance and a lagging angle for frequencies above resonance.	
Leakage	In semiconductors, a temperature-dependent current that flows under conditions of reverse bias.	
Light-emitting diode	A two-terminal device that produces visible or invisible light.	LED
Limit cycles	Undesired oscillations in a digital signal processor caused by quantization error or numeric overflow.	
Limiter	A circuit that clips off the high-amplitude portions of a signal to reduce noise or prevent another circuit from being overdriven.	
Line transient	An abnormally high voltage of short duration on the ac power line.	
Linear	A circuit or component where the output is a straight-line function of the input.	
Majority carriers	In an N-type semiconductor, the electrons. In a P-type semiconductor, the holes.	
Metal oxide semiconductor	A discrete or integrated semiconductor device that uses a metal and an oxide (silicon dioxide) as an important part of the device structure.	MOS

Term	Definition	Symbol or Abbreviation
Metal oxide varistor	A device used to protect sensitive circuitry and equipment from line transients.	MOV
Minority carriers	In an N-type semiconductor, the holes. In a P-type semiconductor, the electrons.	
Mixed-signal circuit	A circuit that contains both analog and digital functions. Many integrated circuits are mixed-signal devices.	
Modulation	The process of controlling some aspect of a periodic signal such as amplitude, frequency, or pulse width. Used to place intelligence (such as audio, video or data) on a radio or television signal.	
Multipath	Radio signals reflect off various objects, and the received signal can be compromised when the various signal components arrive at different times. Multipath distortion can cause data errors and poor performance in wireless networks.	
Multiply and accumulate	The basic process used in DSP. Signal samples and coefficients are multiplied and accumulated. The formal name is *convolution*.	MAC
Multirate system	A DSP system where more than one sampling frequency is used or where the sampling frequency is changed by interpolation or decimation or both.	
Neutralization	The application of external feedback in an amplifier to cancel the effect of internal feedback (inside the transistor).	
Noise	Any unwanted portion of, or interference to, a signal.	
Noninverting	An amplifier where the output signal is in phase with the input signal.	
Numerically controlled oscillator	Another name for a direct digital synthesizer (DDS).	NCO
Nyquist frequency	One-half of the sampling frequency in a DSP system. Also called the *Nyquist limit* since it represents the highest frequency that can be properly handled by the system.	$f_s/2$
Offset	An error in the output of an operational amplifier caused by imbalances in the input circuit.	
Open circuit	A condition of infinite resistance or infinite impedance and zero current flow.	
Operational amplifiers	High-performance amplifiers with inverting and noninverting inputs. They are usually in integrated circuit form and can be connected for a wide variety of functions and gains.	Op amp
Operating point	The average condition of a circuit as determined by some control voltage or current. Also called the quiescent point.	
Opto-isolator	An isolation device that uses light to connect the output to the input. Used where there must be an extremely high electrical resistance between input and output.	
Oscillator	An electronic circuit for generating various ac waveforms and frequencies from a dc energy source.	
Oscilloscope	An instrument that displays a graph of time versus amplitude (usually voltage).	
Pass transistor	A transistor connected in series with a load to control the load voltage or the load current.	
Periodic function	One that repeats over and over with time (sine waves, square waves, and triangular waves are examples).	

Term	Definition	Symbol or Abbreviation
Phase-locked loop	An electronic circuit that uses feedback and a phase comparator to control frequency or speed.	PLL
Phase-shift oscillator	An oscillator circuit characterized by an RC phase-shift network in its feedback path.	
Photovoltaic	A device that converts light energy into electrical energy.	PV
Pi filter	A low-pass filter using a shunt input capacitor, a series inductor, and a shunt output capacitor.	
Power amplifier	An amplifier designed to deliver a significant level of output voltage, output current, or both. Also known as a large-signal amplifier.	
Power factor correction	A circuit to reduce ac line distortion and harmonics caused by capacitor input filters.	
Power gain	The ratio of output power to input power. Often expressed in decibels.	A_P or G_P
Printed circuit	A lamination of copper foil on an insulating substrate such as fiberglass or epoxy resin. Portions of the foil are removed, leaving circuit paths to interconnect electronic components to form complete circuits.	PC
Product detector	A special detector for receiving suppressed carrier transmissions, such as single sideband.	
Programmable	A device or circuit in which the operational characteristics may be modified by changing a programming voltage, current, or some input information.	
Programmable unijunction transistor	A negative resistance device used in timing and control circuits that fires (turns on) at a predetermined voltage, which is established by two resistors. These have replaced unijunction transistors, which are not programmable.	PUT
Pulsating direct current	Direct current with an ac component (i.e., the output of a rectifier).	
Pulse-code modulation	A signal is represented by a series of binary numbers. Such signals are found at the output of analog-to-digital converters. They can be in serial form (1 bit at a time) or in parallel form (8, 16, 24, or 32 bits at a time).	PCM
Pulse-width modulation	Controlling the width of rectangular waves for the purpose of adding intelligence or controlling the average dc value.	PWM
Pure alternating current	Alternating current with no dc component. It has an average value of zero.	
Pure direct current	Direct current with no ac component. Pure direct current has no ripple or noise and is a straight line on an oscilloscope.	
Push-pull	A circuit using two devices, where each device acts on one-half of the total signal swing.	
Quadrature	A signal-to-signal phase relationship of 90 degrees.	
Quantization	The process of converting a continuous signal to a discrete signal (also known as analog-to-digital [A/D] conversion).	
Quantization error	The difference between the original continuous signal values and the quantized (discrete) values. This error decreases as the number of bits increases.	
Radio-frequency choke	A coil used to block or eliminate radio (high) frequencies.	RFC
Ratio detector	A circuit used to detect frequency-modulated signals.	

Term	Definition	Symbol or Abbreviation
Rectification	The process of changing alternating to direct current.	
Recursive filter	One that uses feedback. In a DSP system, the output will show an *infinite impulse response* (IIR). The response of an IIR system will decay exponentially after the input goes to zero.	
Regulator	A circuit or device used to hold some quantity constant.	
Relaxation oscillator	Those oscillators characterized by RC timing components to control the frequency of the output signal.	
Resettable fuse	A fuse that goes into a high resistance state when excess current flows and returns to a low resistance state when the current decreases (polymeric positive temperature coefficient, PPTC).	PPTC
Ripple	The ac component in the output of a dc power supply.	
Sample	A single value obtained during the quantization process (the sample number is often denoted by the subscript n).	$x_{[n]}$
Sampling frequency	The rate at which a continuous signal is converted to a discrete signal.	f_s
Saturation	The condition where a device, such as a transistor, is turned on hard. When a device is saturated, its current flow is limited by some external load connected in series with it.	
Schmitt trigger	An amplifier with hysteresis used for signal conditioning in digital circuits.	
Schottky diode	A rectifier with a low forward voltage drop and superior performance at high frequencies.	
Selectivity	The ability of a circuit to select, from a broad range of frequencies, only those frequencies of interest.	
Semiconductors	A category of materials having four valence electrons and electrical properties between conductors and insulators.	
Sensitivity	The ability of a circuit to respond to weak signals.	
Servomechanism	A control circuit that regulates motion or position.	
Sidebands	Frequencies above and below the carrier frequency created by modulation.	
Signal to noise	A ratio of the desired signal to either the noise or the interference.	SNR and SINR
Silicon	An element. The semiconductor material currently used to make almost all solid-state devices such as diodes, transistors, and integrated circuits.	
Silicon-controlled rectifier	A device used to control heat, light, and motor speed. It will conduct from cathode to anode when it is gated on.	SCR
Single sideband	A variation of amplitude modulation. The carrier and one of the two sidebands are suppressed.	SSB
Slew rate	The measure of the ability of a circuit to produce a large change in output in a short period of time.	
Small-signal bandwidth	The total frequency range of an amplifier in which its gain for small signals is within 3 dB of its best gain.	
Soft saturation	When a device, such as a transistor, has just enough input signal to turn it on fully.	
Source	That terminal of a field-effect transistor that sends the current carriers to the drain.	
Spectrum analyzer	An instrument that displays a graph of frequency versus amplitude.	
SPI	A four-wire communication protocol used for short-distance signaling between processors and other ICs.	

Term	Definition	Symbol or Abbreviation
Static switch	A switch with no moving parts, generally based on thyristors.	
Superheterodyne	A receiver that uses the heterodyne frequency conversion process to convert the frequency of an incoming signal to an intermediate frequency.	
Surface-mount technology	A method of printed circuit fabrication in which the component leads are soldered on the component side of the board and do not pass through holes in the boards.	SMT
Surge limiter	A circuit or component (often a resistor) used to limit turn-on surges to some safe value.	
Swamping resistor	A resistor used to swamp out (make insignificant) individual component characteristics. Can be used to ensure current sharing in parallel devices.	
Switch mode	A circuit where the control element switches on and off to achieve high efficiency.	
Tank circuit	A parallel *LC* circuit.	
Tap	A coefficient used in a digital filter.	
Temperature coefficient	The number of units change, per degree Celsius change, from a specified temperature.	
Thermal grease	A substance used to coat semiconductors to improve thermal transfer.	
Thermal imager	An infrared camera that displays the temperature of objects or circuits.	
Thermal runaway	A condition in a circuit where temperature and current are mutually interdependent and both increase out of control.	
Thermal washer	A washer used to improve thermal transfer with semiconductors that are mounted on heat sinks.	
Thyristor	The generic term referring to control devices such as silicon-controlled rectifiers and triacs.	
Time domain	A signal viewpoint where amplitude is plotted versus time (an oscilloscope display is an example).	
Total harmonic distortion	The ratio of a desired signal to unwanted frequency components (harmonics). Can be expressed as a percentage or using the decibel scale.	THD
Transducer	A device that converts a physical effect to an electrical signal (a microphone is an example). Also can refer to a device that converts an electrical signal to a physical effect (a motor is an example).	
Transimpedance amplifier	An amplifier that converts current to voltage.	
Transistor	Any of a group of solid-state amplifying or controlling devices that usually have three leads.	
Triac	A full-wave, bidirectional control device that is equivalent to two silicon-controlled rectifiers (*triode ac* switch).	
Troubleshooting	A logical and orderly process to determine the fault or faults in a circuit, a piece of equipment, or a system.	
Twin-T network	A circuit containing two branches, each arranged in the form of the letter *T,* that can be used as a notch filter or to control the frequency of an oscillator.	
Unijunction transistor	A transistor used in control and timing applications. It turns on suddenly when its emitter voltage reaches the firing voltage.	UJT

Term	Definition	Symbol or Abbreviation
Varactor diode	A two-terminal device that can be used as a voltage-controlled variable capacitor.	
Variable-frequency oscillator	An oscillator with an adjustable output frequency.	VFO
Varistor	A nonlinear resistor. Its resistance is a function of the voltage across it.	
Virtual ground	An ungrounded point in a circuit that acts as a ground as far as signals are concerned.	
Voltage-controlled oscillator	An oscillator circuit where the output frequency is a function of a dc control voltage.	VCO
Voltage gain	The ratio of amplifier output voltage to input voltage. Often expressed in decibels.	A_V or G_V
Voltage multipliers	Direct current power-supply circuits used to provide transformerless step-up of ac line voltage.	
Voltage regulator	A circuit used to stabilize voltage.	
Window	A method of smoothing DSP filter coefficients (or discrete samples) to reduce the ripple caused by Gibbs phenomenon.	
Window comparator	A circuit that responds to a certain range or window of voltage values.	
Wireless local area network	A radio-frequency communication system for two-way data transfers among digital devices and systems.	WLAN
Zener diode	A diode designed to operate in reverse breakover with a stable voltage drop. It is useful as a voltage regulator.	
Zero-crossing detector	A comparator that changes states when its input crosses the zero volt point.	

Index

D

D/A (digital-to-analog) converters, 6
Damped sine wave, 244–245
Damping, 71
  critical, 478
  under-/over-, 478
Dark current, 125
Darlington circuit, 186
Data
  input, 376
  wireless, 403–409
dBA scale, 155
dBm scale, 156
dc (direct current)
  for amplification, 106
  circuits, 10–14
  pure *vs.* pulsating, 71
  in rectification, 66–71
dc analysis, 169
dc blocking capacitors. *See* Coupling capacitors
dc component, 10
dc feedback, 206
dc restorer, 47
DDR SDRAM (double data rate, synchronous dynamic
    random-access memory), 97
DDS. *See* Direct digital synthesis (DDS)
Dead band, 237
Decibels (dB), 151
Decibel (dB) voltage gain, 152
Decimation, 565
Decoupling network, 322
Demodulation, 384–389
Demodulators, 7
Demultiplexer, 7
Depletion mode, 122–123, 203
Depletion region, 35
Derated devices, 90, 507
Designing with biquads, 558
Detection (demodulation), 387
Detectors, 7
Device temperature, 126
Diacs, 472–473
Diamonds, 23
Die, 129
Difference, 7
Difference frequency, 388, 393
Differential amplifiers, 257–266
  analysis, 261–266
  defined, 257
Differential gain, 263
Differential output, 258
Differentiate (555 timer), 432
Differentiator, 293–294
Differentiators, 7
Digital circuits, 4–6
  vs. analog circuits, 4–6
Digital communication systems, 389

Digital electronic device, 4
Digital multimeters (DMMs), 42–44
Digital output (555), 431
Digital potentiometer, 438
Digital signal processing (DSP), 3, 5–6
  applications of, 561–570
  filter design and, 550–560
  Fourier theory and, 546–550
  limitations of, 570–572
  with moving-average filters, 542–544
  overview, 537–541
  troubleshooting, 572–578
Digital-to-analog conversion and, 444–446
Digital-to-analog (D/A) converters, 5
Digital voltmeter, 291
Digitizers, 336
Diodes, 34–60
  characteristic curves of, 38–41
  four-layer, 464
  freewheeling, 40
  germanium, 37, 39–40, 44
  lead identification, 41–44
  as nonlinear devices, 388–389
  photovoltaic energy sources, 56–60
  PN-junction, 34–37
  silicon used for making, 24
  testing, 43
  types and applications, 44–56
  zener, 40, 46, 88–90, 497–498
DIP (dual in line package), 422
DIP (dual inline package), 125
DIPS (dual-inline packages), 97
Direct coupling, 185–186
Direct digital synthesis (DDS), 374–378
  circuit, 377
  overview, 374–377
  troubleshooting, 376–378
Directional coupler, RFID system, 409
Direct sequence spread spectrum (DSSS), 406
Discharges, capacitor, 76, 77
Discontinuities, 547
Discontinuous conduction mode, 526
Discrete circuits, 420
Discrete Fourier transform, 548
Discrete signal, 441, 538
Discriminator circuit, 397, 398
Dissipating power, 126, 133
Distortion, 162
  slew-rate, 269
  small-signal amplifiers, 209–210
  troubleshooting, 324–327
Distributed capacitance, 179
Diver transformer, 233
Dividers, 4, 7
DMMs (digital multimeters), 42–44
Dominant lag network, 279
Donor impurity, 26

T

Tachometer, 476, 477
Tank circuit, 243–244, 385, 520
TAP (test access port), 336
Tapped cap, 358
Tapped inductor, 357
Target system, 574
Teknic, 481
Telecommunications, 16
Telephone switching, 4
Television, 2
Temperature compensation, 241
Temperature probe, 338–339
Temperature sensitivity, 185
Ten percent rule, 307
Test access port (TAP), 336
Testing
    automated, 333–338
    diodes, 42
    gain, 120
    in-circuit, 122
    transistors, 119–122
Testing, transistors, 119–122
Test pod, 574
Texas Instruments, 3, 241, 242
THD (total harmonic distortion), 156, 196
Theoretical maximums, 232
Thermal capacitance, 135
Thermal carriers, 25, 26
Thermal equivalent circuit, 134, 135
Thermal imager, 339
Thermal intermittents, 328, 329
Thermal issues, 338–340
Thermal mass, 135
Thermal resistances, transistor, 134, 135
Thermal runaway, 510
Thermometer code, 444
Thomson, J. J., 1
Three-phase alternating current, 74–75
Three-stage amplifier, troubleshooting, 320, 321
Threshold (555), 430
Threshold points, 294
Thyristor, 470
Time constant of the circuit, 76
Time-delay mode (555 timer), 434
Time delays (time lags), 434, 478–479
Time domain, 386, 546
Timers, 7, 8
Tone decoding, 438
Torque control, 477
Total harmonic distortion (THD), 156, 196
Tracking, 392
    closed-loop voltage regulation, 502–503, 504–505
    maximum power point, 60–61
Transcendental equation, 176
Transducers, 538
Transfer characteristic curve, 116

Transformed load, 235
Transformer-coupled amplifier, 187–189, 232
Transformer coupling, 231–233
Transients, 272, 331
Transimpedance amplifier, 272
Transistor junction, checking, 120
Transistors, 105–144
    amplification and, 105–107
    applications, 110
    characteristics curves, 113–116, 113–117
    data, 117–118
    family tree, 124
    function of, 107, 124
    parameters, 117
    power, 126–137
    silicon used for making, 24
    silicon *vs.* germanium, 115–116, 116–117, 122
    structures and symbols, 107, 108
    as switches, 138–144
    switching, 514
    testing, 119–122
    types, 107–112, 122–126
TRF (tuned radio-frequency) receiver, 392
Triac, 469–472
Triangular waveform, 326–327
Triangular window, 554, 556
Trigger, 8
Trigger (555), 429
Triggering devices, 472
Trimmer capacitors, 361, 394
Troubleshooting
    automated testing, 333–338
    communications, 410–415
    component-level, 8
    control circuits, 487–489
    defined, 91
    digital signal processing, 572–578
    direct digital synthesis, 376–378
    for distortion and noise, 324–328
    integrated circuits (ICs), 451–455
    intermittents, 328–330
    for no output, 314–319
    operational amplifiers, 331–333
    oscillators, 372–373
    power supply, 66, 91–94
    preliminary checks, 306–314
    for reduced output, 319–323
    regulated power supplies, 523–531
    switching transistors, 139, 141
    thermal issues, 338–340
Truncated sinc function, 554
Tuned circuit, 391
Tuned radio-frequency (TRF) receiver, 392
Tuning diodes, 53
Turns ratio, 187
20 dB per decade, 275
Twin-T network, 355–356